W9-DIG-657

DIGITAL ELECTRONICS

Principles and Applications

Fifth Edition

Roger L. Tokheim

New York, New York Columbus, Ohio Woodland Hills, California Peoria, Illinois

Cover Credit: Astrid & Hanns-Frieder Michler/Science Photo Library/Photo Researchers, (bkgd) David McGlynn/FPG
Photo Credits: Page ix top Lou Jones/The Image Bank; page ix bottom © Cindy Lewis; page 4 Glencoe file photo; page 11 Microsoft Corporation; page 49 International Telecommunication Union and Inmarsat; page 72 © 1997 Intel Corporation and Sandia National Laboratories; by Randy Montoya; page 93 Hayes Microcomputer Products; page 131 Mrs. Engelbart; page 156 Photomicrograph by Leo Deriak, copyright © by Lucent Technologies Inc., All rights reserved; page 163 Alpine Electronics of America; page 181 Bryan Quintard, Lawrence Livermore National Laboratory; page 215 Photomicrograph by John Carnivale, © 1996 by Lucent Technologies, Inc., All rights reserved; page 237 Magellan Systems Corp.; page 284 The Phenom, GoldStar, LG Electronics; page 311 Photomicrograph, fiber-optic illumination, by Philip Harrington, copyright © 1996 by Lucent Technologies, Inc. All rights reserved; page 366 INTELSAT.

Library of Congress Cataloging-in-Publication Data
Tokheim, Roger L.
Digital electronics/Roger L. Tokheim.—5th ed.
p. cm.
Includes index.
ISBN 0-02-804161-5
1. Digital electronics. I. Title.
TK7868.D5T65 1998 98-8460
621.3815—dc21 CIP

Glencoe/McGraw-Hill

A Division of The ***McGraw-Hill*** *Companies*

Digital Electronics: Principles and Applications, Fifth Edition

Send all inquiries to:
Glencoe/McGraw-Hill
8787 Orion Place
Columbus, OH 43240-4027

5 6 7 9 071 06 05 04 03 02 01

ISBN 0-02-804161-5

Contents

The Glencoe *Basic Skills in Electricity and Electronics* series has been designed to provide entry-level competencies in a wide range of occupations in the electrical and electronic fields. The series consists of coordinated instructional materials designed especially for the career-oriented student. Each major subject area covered in the series is supported by a textbook, an experiments manual, and an instructor's productivity center. All the materials focus on the theory, practices, applications, and experiences necessary for those preparing to enter technical careers.

There are two fundamental considerations in the preparation of materials for such a series: the needs of the learner and needs of the employer. The materials in this series meet these needs in an expert fashion. The authors and editors have drawn upon their broad teaching and technical experiences to accurately interpret and meet the needs of the student. The needs of business and industry have been identified through personal interviews, industry publications, government occupational trend reports, and reports by industry associations.

The processes used to produce and refine the series have been ongoing. Technological change is rapid and the content has been revised to focus on current trends. Refinements in pedagogy have been defined and implemented based on classroom testing and feedback from students and instructors using the series. Every effort has been made to offer the best possible learning materials.

The widespread acceptance of the *Basic Skills in Electricity and Electronics* series and the positive responses from users confirm the basic soundness in content and design of these materials as well as their effectiveness as learning tools. Instructors will find the texts and manuals in each of the subject areas logically structured, well-paced, and developed around a framework of modern objectives. Students will find the materials to be readable, lucidly illustrated, and interesting. They will also find a generous amount of self-study and review materials and examples to help them determine their own progress.

Both the initial and on-going success of this series are due in large part to the wisdom and vision of Gordon Rockmaker who was a magical combination of editor, writer, teacher, electrical engineer and friend. Gordon has retired but he is still our friend. The publisher and editors welcome comments and suggestions from instructors and students using the materials in this series.

Charles A. Schuler,
Project Editor
and
Brian P. Mackin,
Editorial Director

Basic Skills in Electricity and Electronics

Charles A. Schuler, Project Editor

New Editions in This Series

Electricity: Principles and Applications, Fifth Edition, Richard J. Fowler
Electronics: Principles and Applications, Fifth Edition, Charles A. Schuler
Digital Electronics: Principles and Applications, Fifth Edition, Roger L. Tokheim

Other Series Titles Available:

Communication Electronics, Second Edition, Louis E. Frenzel
Microprocessors: Principles and Applications, Second Edition, Charles M. Gilmore
Industrial Electronics, Frank D. Petruzella
Mathematics for Electronics, Harry Forster, Jr.

Preface

Digital Electronics: Principles and Applications, Fifth Edition, is designed to be used as an introductory text for students who are new to the field of electronics. Prerequisites are general mathematics and basic DC circuits. Digital electronics can be studied before or concurrently with a course in basic electronics, since knowledge of active discrete components is not a prerequisite. Binary mathematics and Boolean concepts are introduced and explained in this book as needed.

Digital electronics is not a specialized field in electronics. Digital circuits were first used in computing devices, but they are now commonly found in a broad range of products. The advances in microelectronic design and manufacturing, computer technology, and information systems, have caused a rapid increase in the use of digital circuits.

For this edition, we sought the advice of instructors who have used the work for many years and instructors who have used the text for only a short period of time. They provided precise recommendations and consistent responses to questionnaires. Their collective information and suggestions are included in this new edition. The text interior was redesigned to use color more effectively, while retaining the popular marginal color strip. Students can quickly find their chapter-ending assignments with the red strip. Highlighted key terms now appear on each page across from their point of discussion in the text. Color strips are also used as a quick thumb reference to find the numerous illustrative examples within a chapter. The text design uses a consistent color-coding of circuit components. This edition also contains informative "Did You Know," "Job Tips," and "About Electronics" sidebars, which should spark interesting in-class discussions.

The fifth edition of *Digital Electronics: Principles and Applications* includes numerous changes and improvements that are a direct result of the reviewers' recommendations and the feedback from instructors of digital electronics. De Morgan's Theorems are now introduced in Chapter 4. Chapter 5 now contains coverage of optoisolators and stepper motors. Several subtle changes in the first chapter include information on the DT-1000 trainer display boards and questions related to using electronic simulation software. The chapter also discusses the use of simulation software. Additional objectives and coverage strengthens the coverage of Chapter 5, including interfacing circuits featuring an optoisolator and the operation of stepper motor driver circuits. Several of the changes in Chapter 6 relate to the Gray code commonly associated with *optical shaft encoding*. Instructors and students should both enjoy the new section in Chapter 8 on Counting Real World Events. Additional coverage of the *logic probe, DMM/Logic Probe, IC tester,* the *Portable IC Tester,* and the *Digital Real Time™ Oscilloscope* strengthens this chapter. Coverage of CPUs is improved in Chapter 10, along with coverage of the *arithmetic-logic unit (ALU).* A noteworthy addition to Chapter 11 is updated coverage of memories. Chapter 12 includes important new sections, including Section 12-18 on Programmable Logic Controllers (PLCs), and 12-19 on microcontrollers, along with several interesting examples. This edition retains the popular analysis techniques that help to develop the student's ability to troubleshoot. It also retains the electronic game circuits demonstrating the function of digital subsystems. In response to reviewer requests and recommendations received from the questionnaires, the Experiments Manual now contains nine experiments that can be performed using computer simulation software, such as Electronics Workbench®.

The material in this text is based on carefully selected and formulated performance objectives. Surveys, classroom testing, and feedback from students, teachers, and industry representatives were used in developing these objectives. The systems-subsystems approach is fundamental to digital electronics because of the extensive use of medium-, large-, and very large-scale integrated circuits. Small-scale integrated circuits are used when students learn the fundamentals. Most of the circuits in the text can be wired for classroom demonstration using off-the-shelf TTL or CMOS ICs. A companion volume, *Experiments Manual for Digital Electronics: Principles and Applications,* Fifth Edition, is closely correlated with this textbook. Digital design problems, troubleshooting problems, and chapter tests are also available in the companion experiments manual.

Appreciation should be given to the many instructors, students, and industry representatives who contributed to this book. I would like to give special thinks to family members—Marshall, Rachael, Dan, and Carrie for their work on this textbook.

Acknowledgements

The author, project editor, and publisher would like to thank the instructors listed below who provided general and detailed comments, and those who responded to the survey that was sent out while the book was being revised. Their comments and suggestions provided the valuable input necessary to make a good book even better.

John Baldwin
South Central Technical College
Faribault, MN

Stephen Brandt
Miami Valley Career Technology Center
Clayton, OH

Donin Custer
Western Iowa Tech Community College
Sioux City, IA

Thomas Day
ECPI College of Technology
Virginia Beach, VA

John Garry
RETS Institute
Nutley, NJ

Robert Gouveia
Greater Lowell Technical High School
Tyngsboro, MA

Michael Holcombe
EFTC Northwest Operations
Newberg, OR

Tony Jacobs
Ridley Lowell Technical Institute
New London, CT

Rod Minatra
Grayson College
Denison, TX

Alan B. Plemons
Metro Tech Aviation Career Center
Oklahoma City, OK

Mike Rybinski
Technology Center of Du Page
Addison, IL

Donald Shrum
Vatterott College
St. Louis, MO

Dan Siddall
SCP Global Technologies
Boise, ID

Tom Thomas
University of Alabama at Birmingham
Birmingham, AL

Robert Wheary
ECPI Technical College
Roanoke, VA

Daniel Wilson
Red River Vocational and Technical School
Duncan, OK

William R. Woodward
ECPI College of Technology
Virginia Beach, VA

Roger L. Tokheim

Electric and electronic circuits can be dangerous. Safe practices are necessary to prevent electrical shock, fires, explosions, mechanical damage, and injuries resulting from the improper use of tools.

Perhaps the greatest hazard is electrical shock. A current through the human body in excess of 10 milliamperes can paralyze the victim and make it impossible to let go of a "live" conductor or component. Ten milliamperes is a rather small amount of electrical flow: It is only *ten one-thousandths* of an ampere. An ordinary flashlight uses more than 100 times that amount of current!

Flashlight cells and batteries are safe to handle because the resistance of human skin is normally high enough to keep the current flow very small. For example, touching an ordinary 1.5-V cell produces a current flow in the microampere range (a microampere is one-millionth of an ampere). This amount of current is too small to be noticed.

High voltage, on the other hand, can force enough current through the skin to produce a shock. If the current approaches 100 milliamperes or more, the shock can be fatal. Thus, the danger of shock increases with voltage. Those who work with high voltage must be properly trained and equipped.

When human skin is moist or cut, its resistance to the flow of electricity can drop drastically. When this happens, even moderate voltages may cause a serious shock. Experienced technicians know this, and they also know that so-called low-voltage equipment may have a high-voltage section or two. In other words, they do not practice two methods of working with circuits: one for high voltage and one for low voltage. They follow safe procedures at all times. They do not assume protective devices are working. They do not assume a circuit is off even though the switch is in the OFF position. They know the switch could be defective.

As your knowledge and experience grow, you will learn many specific safe procedures for dealing with electricity and electronics. In the meantime:

1. Always follow procedures.
2. Use service manuals as often as possible. They often contain specific safety information. Read, and comply with, all appropriate material safety data sheets.
3. Investigate before you act.
4. When in doubt, *do not act*. Ask your instructor or supervisor.

General Safety Rules for Electricity and Electronics

Safe practices will protect you and your fellow workers. Study the following rules. Discuss them with others, and ask your instructor about any you do not understand.

1. Do not work when you are tired or taking medicine that makes you drowsy.
2. Do not work in poor light.
3. Do not work in damp areas or with wet shoes or clothing.
4. Use approved tools, equipment, and protective devices.
5. Avoid wearing rings, bracelets, and similar metal items when working around exposed electric circuits.
6. Never assume that a circuit is off. Double-check it with an instrument that you are sure is operational.
7. Some situations require a "buddy system" to guarantee that power will not be turned on while a technician is still working on a circuit.
8. Never tamper with or try to override safety devices such as an interlock (a type of switch that automatically removes power when a door is opened or a panel removed).
9. Keep tools and test equipment clean and in good working condition. Replace insulated probes and leads at the first sign of deterioration.
10. Some devices, such as capacitors, can store a *lethal* charge. They may store this charge for long periods of time. You must be certain these devices are discharged before working around them.
11. Do not remove grounds and do not use adaptors that defeat the equipment ground.
12. Use only an approved fire extinguisher for electrical and electronic equipment. Water can conduct electricity and may severely damage equipment. Carbon dioxide (CO_2) or halogenated-type extinguishers are usually preferred. Foam-type extin-

guishers may also be desired in *some* cases. Commercial fire extinguishers are rated for the type of fires for which they are effective. Use only those rated for the proper working conditions.

13. Follow directions when using solvents and other chemicals. They may be toxic, flammable, or may damage certain materials such as plastics. Always read and follow the appropriate material safety data sheets.
14. A few materials used in electronic equipment are toxic. Examples include tantalum capacitors and beryllium oxide transistor cases. These devices should not be crushed or abraded, and you should wash your hands thoroughly after handling them. Other materials (such as heat shrink tubing) may produce irritating fumes if overheated. Always read and follow the appropriate material safety data sheets.
15. Certain circuit components affect the safe performance of equipment and systems. Use only exact or approved replacement parts.
16. Use protective clothing and safety glasses when handling high-vacuum devices such as picture tubes and cathode-ray tubes.
17. Don't work on equipment before you know proper procedures and are aware of any potential safety hazards.
18. Many accidents have been caused by people rushing and cutting corners. Take the time required to protect yourself and others. Running, horseplay, and practical jokes are strictly forbidden in shops and laboratories.

Circuits and equipment must be treated with respect. Learn how they work and the proper way of working on them. Always practice safety: your health and life depend on it.

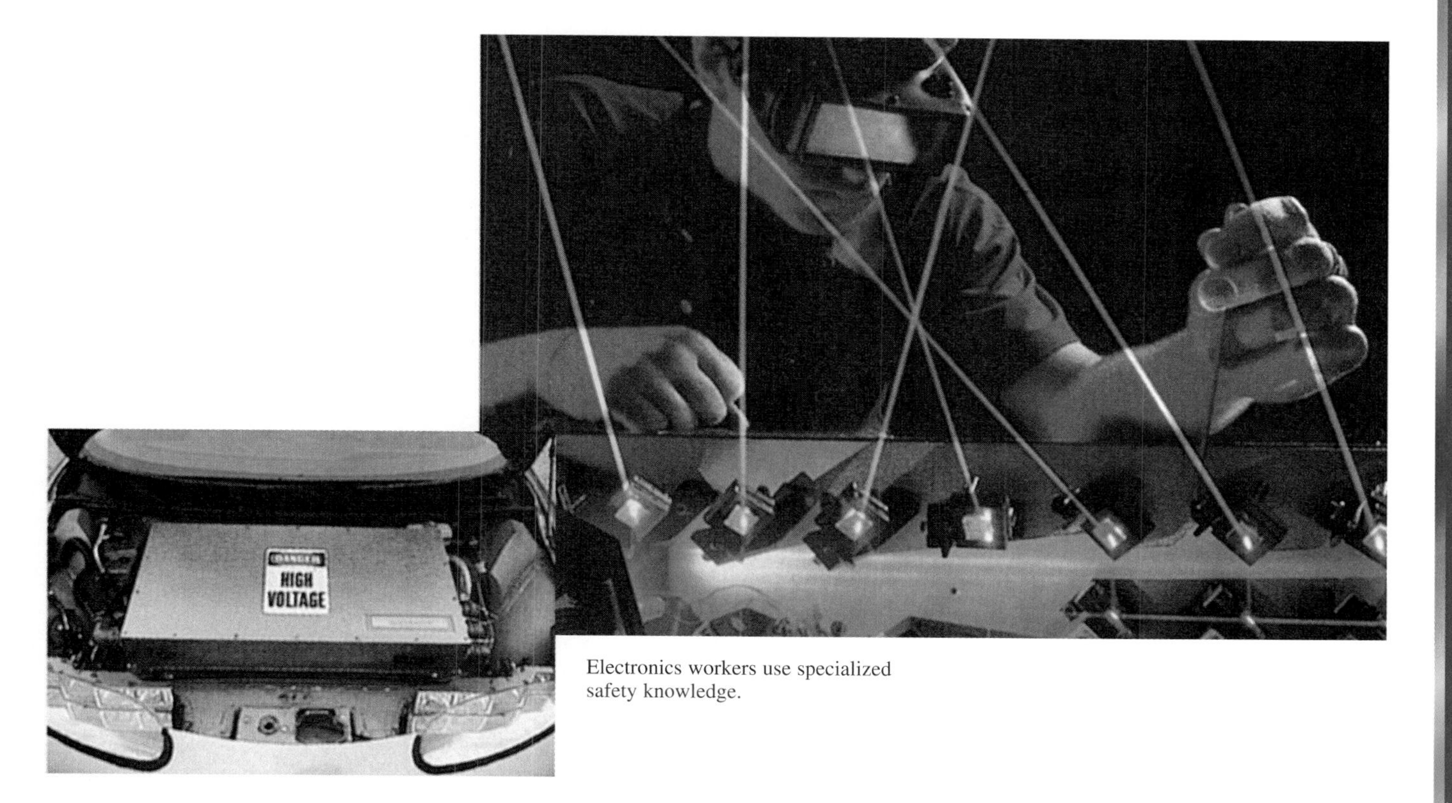

Electronics workers use specialized safety knowledge.

Digital Electronics

Chapter Objectives

This chapter will help you to:

1. *Identify* several characteristics of digital circuits as opposed to linear (analog) circuits.
2. *Classify* devices as using digital, analog, or a combination of technologies.
3. *Differentiate* between digital and analog signals and *identify* the HIGH and LOW portions of the digital waveform.
4. *List* three types of multivibrators and *describe* the general purpose of each type of circuit.
5. *Analyze* simple logic-level indicator circuits.
6. *Cite* several reasons for using digital circuits.
7. *Write* several limitations of digital circuits.

This chapter introduces you to digital electronics. Digital electronics is the world of the calculator, the computer, the integrated circuit, and the binary numbers 0 and 1. This is an exciting field within electronics because the uses for digital circuits are expanding so rapidly. One small integrated circuit can perform the task of thousands of transistors, diodes, and resistors. You see digital circuits in operation every day. At stores the cash registers read out digital displays. The tiny pocket calculators verge on becoming personal computers. All sizes of computers perform complicated tasks with fantastic speed and accuracy. Factory machines are controlled by digital circuits. Digital clocks and watches flash the time. Most automobiles use microprocessors to control engine functions. Technicians use digital voltmeters and frequency counters.

All persons working in electronics must now understand digital electronic circuits. The inexpensive integrated circuit has made the subject of digital electronics easy to study. You will use many integrated circuits to construct digital circuits.

Calculators

Integrated circuit

Digital clocks

Microprocessors

Personal computers

1-1 What Is a Digital Circuit?

In your experience with electricity and electronics you have probably used analog circuits. The circuit in Fig. 1-1(*a*) on the next page puts out an *analog* signal or voltage. As the wiper on the potentiometer is moved upward, the voltage from points *A* to *B* *gradually* increases. When the wiper is moved downward, the voltage gradually decreases from 5 to 0 volts (V). The waveform diagram in Fig. 1-1(*b*) is a graph of the analog output. On the left side the voltage from *A* to *B* is gradually increasing to 5 V; on the right side the voltage is gradually decreasing to 0 V. By stopping the potentiometer wiper at any midpoint, we can get an output voltage anywhere between 0 and 5 V. An analog device, then, is one that has a signal which *varies continuously* in step with the input.

Analog signal

A digital device operates with a digital signal. Figure 1-2(*a*) on the next page pictures a square-wave generator. The generator produces a square waveform that is displayed on the oscilloscope. The digital signal is only at +5 V *or* at 0 V, as diagramed in Fig. 1-2(*b*). The voltage at point *A* moves from 0 to 5 V. The voltage then stays at +5 V for a time. At point

Digital circuits

Logical 0s and 1s

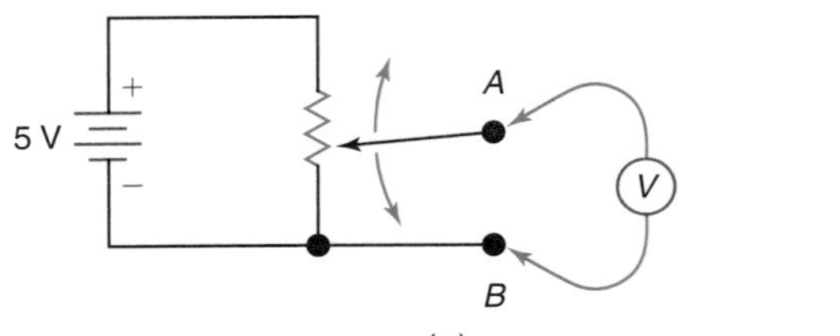

(*a*)

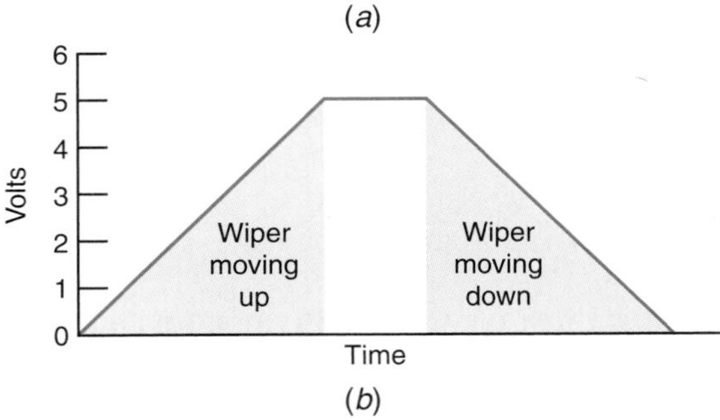

(*b*)

Fig. 1-1 (*a*) Analog output from a potentiometer. (*b*) Analog signal waveform.

Signal

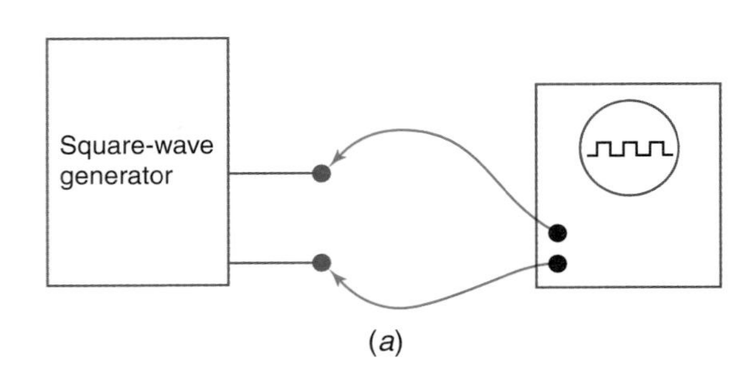

(*a*)

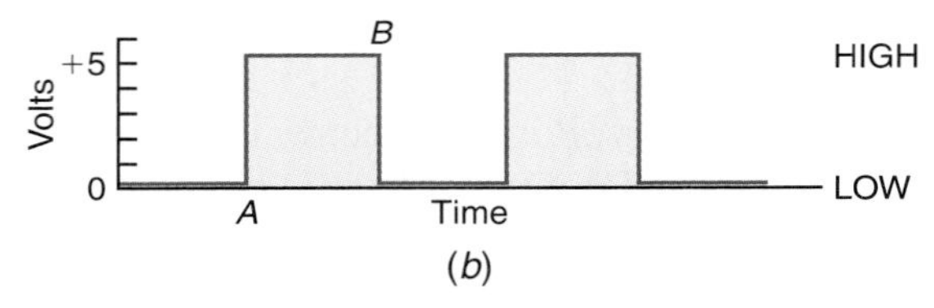

(*b*)

Fig. 1-2 (*a*) Digital signal displayed on scope. (*b*) Digital signal waveform.

Volt-ohm-millimeter

Digital multimeter

HIGH and LOW signals

Trend toward digital circuitry

B the voltage drops immediately from +5 to 0 V. The voltage then stays at 0 V for a time. Only two voltages are present in a digital electronic circuit. In the waveform diagram in Fig. 1-2(*b*) these voltages are labeled HIGH and LOW. The HIGH voltage is +5 V; the LOW voltage is 0 V. Later we shall call the HIGH voltage (+5 V) a logical 1 and the LOW voltage (0 V) a logical 0.

Circuits that handle only HIGH and LOW signals are called *digital circuits.* We mentioned that digital electronics is the world of logical 0s and 1s. The voltages in Fig. 1-2(*b*) are rather typical of the voltages you will be working with in digital electronics.

The digital signal in Fig. 1-2(*b*) could also be generated by a simple on-off switch. A digital signal could also be generated by a transistor turning on and off. In recent years digital electronic signals usually have been generated and processed by integrated circuits (ICs).

Both analog and digital signals are represented in graph form in Figs. 1-1 and 1-2. A *signal* can be defined as useful information transmitted within, to, or from electronics circuits. Signals are commonly represented as a voltage varying with time, as they are in Figs. 1-1 and 1-2. However, a signal could be an electrical current that either varies continuously (analog) or has an on-off characteristic (digital). Within most digital circuits, it is customary to represent signals in the voltage versus time format. When digital circuits are interfaced with nondigital devices such as lamps and motors, then the signal can be thought of as current versus time.

The standard volt-ohm-millimeter (VOM) shown in Fig. 1-3(*a*) is an example of an *analog* measuring device. As the voltage, resistance, or current being measured by the VOM increases, the needle *gradually and continuously moves* up the scale. A *digital multimeter* (DMM) is shown in Fig. 1-3(*b*). This is an example of a *digital* measuring device. As the current, resistance, or voltage being measured by the DMM increases, the display *jumps upward in small steps.* The DMM is an example of digital circuitry taking over tasks previously performed only by analog devices. This trend toward digital circuitry is growing. Currently, the modern technician's bench probably has both an analog VOM and a digital DMM.

About Electronics

A Changing Field. Electronics is among the most exciting areas of technical study. New developments are reported weekly. Interestingly, most developments are based on the fundamentals learned in the first classes in electricity, analog and digital circuits, computer technology and robotics, and communications.

TEST

Supply the missing word in each statement.

1. Refer to Fig. 1-2. The +5-V level of the ________ (analog, digital) signal could

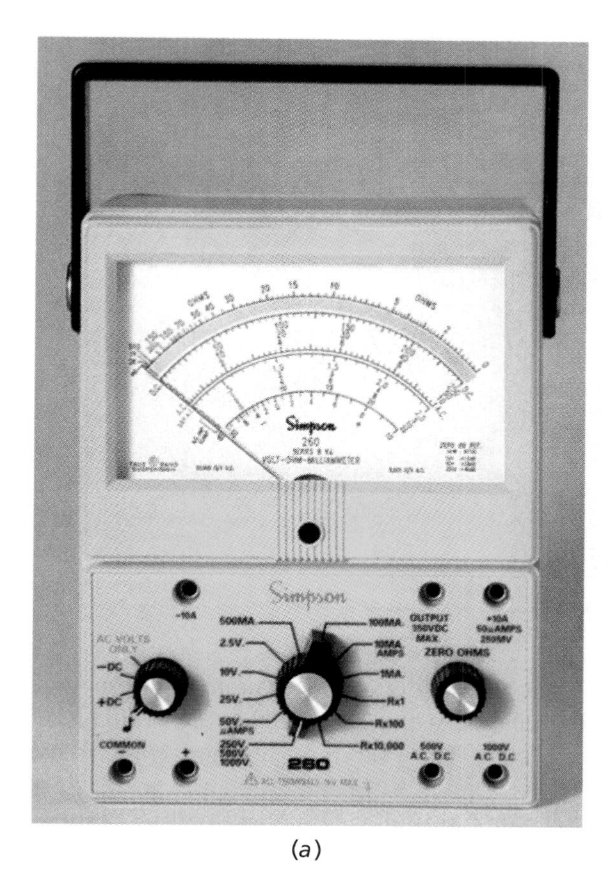

(a)

(b)

Fig. 1-3 (*a*) Analog meter. (*Courtesy of Simpson Electric Company.*) (*b*) Digital multimeter (DMM). (*Courtesy of Tektronix, Inc.*)

also be called a logical 1 or a _________ (HIGH, LOW).

2. A(n) _________ device is one that has a signal which varies continuously in step with the input.
3. Refer to Fig. 1-4. The *input* to the electronic block is classified as a(n) _________ (analog, digital) signal.
4. Refer to Fig. 1-4. The *output* from the electronic block is classified as a(n) _________ (analog, digital) signal.
5. An analog circuit is one that processes analog signals while a digital circuit processes _________ signals.

1-2 Where Are Digital Circuits Used?

Digital electronics is a fast-growing field, as witnessed by the widespread use of microcomputers. Microcomputers are only a few decades old; yet hundreds of millions of them are used in homes, schools, businesses, and governments. Microcomputers are extremely adaptable. At home a computer might be used for playing video games, managing a household budget, or controlling lights and appliances. At school students use the same computer to aid them in learning spelling, math, and writing. The staff uses the same computers for word processing, testing, and grading. In business

Microcomputer

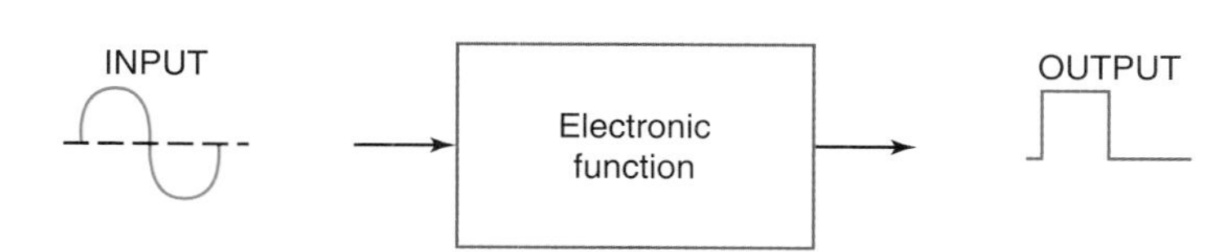

Fig. 1-4 Block diagram of electronic circuit shaping a sine wave into a square wave.

Did You Know?

The History of the Computer. One of the first computers was the Eniac, developed in the 1940s (pictured). The 1970s marked expanded use of the computer in businesses. At that time, expensive mainframe computers occupied special rooms in many offices. The 1980s ushered in the era of personal computers and individuals purchased them for home use. Today, personal computers can be taken anywhere as these devices continue to become both smaller and more powerful.

the same microcomputer might process payrolls, control inventory, and generate mailing lists. In the factories microcomputers are adapted for controlling machines, robots, and processes. In the military computers guide bombs and missiles, aim guns, and coordinate communications.

Microprocessors

Semiconductor memories

Microcomputers are designed around complex ICs called *microprocessors.* In addition, many IC semiconductor memories make up the microcomputer. Microcomputers with microprocessors, interface chips, and semiconductor memories have started the personal computer (PC) revolution. Small computers that used to cost tens of thousands of dollars now cost only hundreds. Digital circuits housed in ICs are used in both large and small computers.

Calculator

A handheld calculator is another example of a digital electronic device used by nearly everyone. Calculators range from the inexpensive models to the sophisticated versions used by engineers and scientists.

Peripheral devices

Only a few decades ago, even simple calculators would have cost hundreds of dollars. The more sophisticated programmable calculators can sometimes be connected to *peripheral devices* such as optical wands and printers. Programmable calculators verge on becoming very small computers. Scientists, engineers, and technicians have made great advances in producing digital ICs. As a result of these advances, the field of digital electronics has mushroomed.

Digital timepiece

The digital timepiece is a triumph of electronic technology. Very accurate multifunction digital clocks and wristwatches are available at low cost. The Marathon 50 "wristwatch" by Timex, shown in Fig. 1-5, is an example of a specialty digital watch that features a programmable electronic pedometer and a unique pulse calculation mode. It also includes a chronograph, countdown timer, light, and day and date indicator, and is water-resistant.

Fig. 1-5 A specialty wristwatch for runners. (*Courtesy of Timex Corporation.*)

(a)

(b)

Fig. 1-6 (*a*) RB5X educational robots are used in many schools. (*Courtesy of General Robotics Corporation.*) (*b*) Industrial robot arms are used to perform repetitive and hazardous tasks. (*Courtesy of Doug Martin.*)

Battery-operated wristwatches commonly use low-power liquid-crystal displays (LCDs).

Robots and other computer-controlled machines add to the mystique of electronic technology. Robots have captured the imagination of inventors, science fiction writers, and movie producers. A *robot* may be defined as a machine that can perform humanlike actions and functions. A computer serves as the control center for modern robots. The robot's computer can be reprogrammed to permit the machine to perform a different sequence of operations. The study of the science and technology of the robot is called *robotics.*

Robots are commonly classified as to their use. *Hobby robots* are used in education, advertising, and entertainment. *Industrial robots* usually take the form of arms and are used in manufacturing and materials handling. One popular educational robot is pictured in Fig. 1-6(*a*). The RB5X robot by General Robotics Corporation is controlled by a built-in microprocessor system (computer) which may be programmed using Tiny BASIC, factory-programmed PROMs, RCL (a "conversational robot control language"), or machine code. The RB5X is a mobile robot with tactile and sonar sensors. The RB5X robot is commonly equipped with a voice synthesis system and arm. The RB5X robot has been used in upper elementary through middle school science and technology programs to introduce students to concepts such as computer-controlled movement, voice generation, and programming. Older students may use the RB5X robot to study applied digital electronics, interfacing, programming, servo mechanisms, sensor/vision systems technology, and artificial intelligence.

Manufacturing robot arms are widely used in hazardous operations where fumes, radiation, sparks, repetitive motions, or high temperatures may be harmful to a human operator. One industrial robot is pictured in Fig. 1-6(*b*) performing welding operations.

Electronically synthesized voice

Robotics

Hobby robots

Industrial robots

DMM

The technician's bench has a new look. Digital multimeters read out resistance, voltage, and current values. The high-quality DMM pictured in Fig. 1-3(*b*) is extremely accurate and loaded with features. Besides measuring ac and dc voltage, ac and dc current, and resistance, it will measure capacitance, frequency, and temperature. The inexpensive DMM/logic probe pictured in Fig. 1-7(*a*) is a compact unit you might find in a student's toolkit. This unit is especially useful in digital circuits because it has a built-in *logic probe,* which is an easy-to-use test instrument that indicates logical HIGHs, LOWs, and detects pulses.

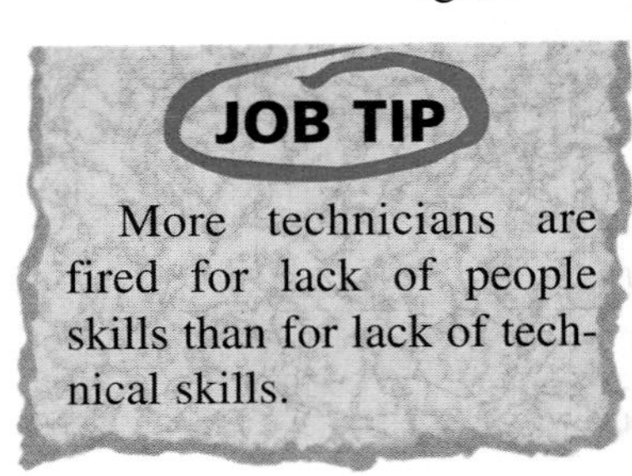

Logic probe

Digital capacitance meter

Frequency counters

Function generator

There is also a *digital capacitance meter* on most electronic workbenches. One such unit is pictured in Fig. 1-7(*b*). This handheld meter measures a wide range of capacitance. *Frequency counters* are also pieces of standard test equipment found in most school and industry labs. The digital test instrument accurately senses and displays the frequency. The high-quality unit pictured in Fig. 1-7(*c*) has a frequency range of 1 Hz to 100 MHz. Another popular digital instrument used in schools and labs is the *function generator.* The function generator shown in Fig. 1-7(*d*) can generate

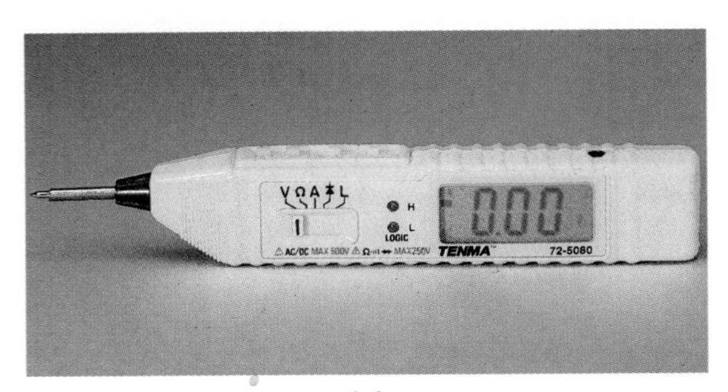

(*a*)

(*b*)

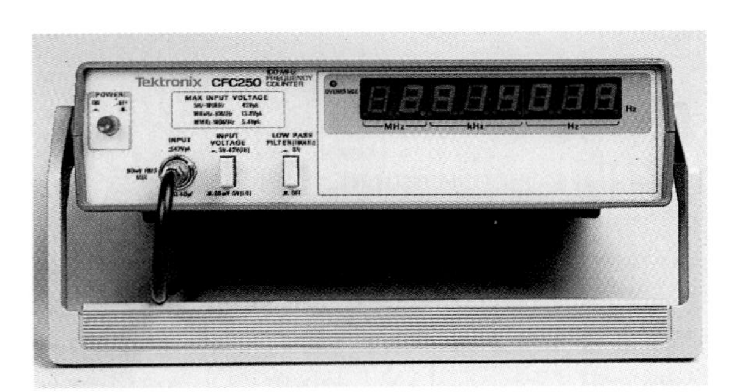

(*c*)

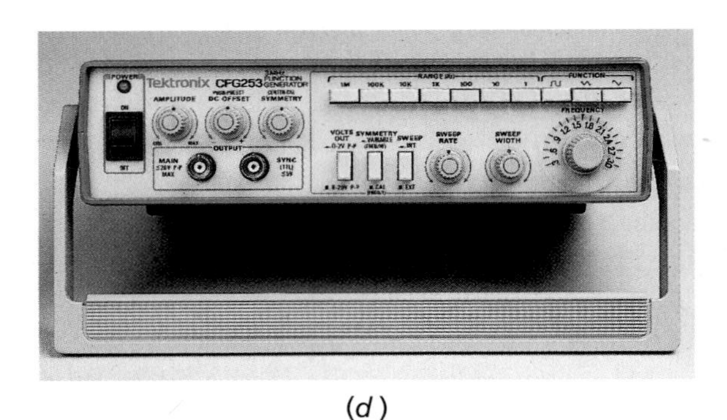

(*d*)

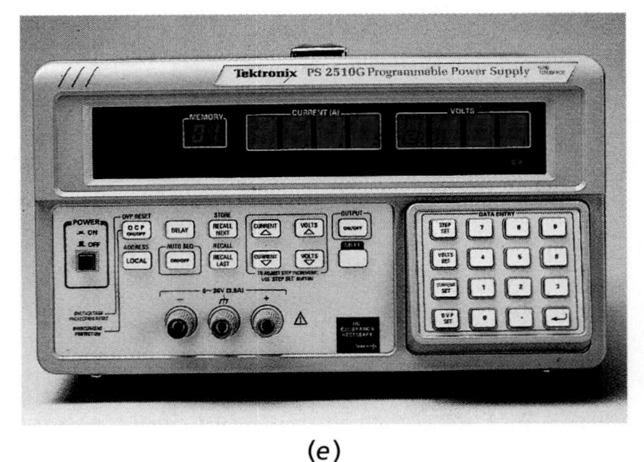

(*e*)

Fig. 1-7 (*a*) Combination DMM/logic probe. (*Courtesy of MCM Electronics.*) (*b*) Capacitance meter. (*Courtesy of MCM Electronics.*) (*c*) Frequency counter. (*Courtesy of Tektronix, Inc.*) (*d*) Function generator. (*Courtesy of Tektronix, Inc.*) (*e*) Programmable power supply. (*Courtesy of Tektronix, Inc.*)

Fig. 1-8 A digital picture-in-picture television. (*Courtesy of Phillips Consumer Electronics Company.*)

square-, triangular-, or sine-wave signals from less than 1 Hz to 3 MHz. Finally, the modern workbench features a high-tech version of a power supply. A *programmable power supply* from Tektronix, Inc. is shown in Fig. 1-7(*e*). Many modern test instruments make extensive use of digital circuitry.

Programmable power supply

Electronic products for home entertainment (such as televisions and stereos) have traditionally been designed using analog circuits. This is changing, with some digital circuitry used in both television and sound systems. The picture-in-picture television illustrated in Fig. 1-8 is one example of the use of digital circuits in the storage and generation of the smaller picture. Most modern consumer products used for home entertainment contain both analog and digital circuitry.

One high-quality entertainment product is Korg's 01/WFD Music Workstation pictured in Fig. 1-9. The 01/WFD music synthesizer uses full digital processing and its 48-M ROM stores 255 multisounds and 119 drum sounds. Like a computer, the music workstation has a built-in 3.5-in disk drive for long-term storage and a large RAM section for internal user storage. To aid programming, a backlit dot-matrix LCD (60 × 240 dot) graces the front of the synthesizer.

Digital synthesizer

Digital electronic circuits are at work in modern automobiles. There is a microprocessor in the "computer" of modern automobiles to control several ignition, fuel system, emissions, and transmission variables. Figure 1-10 gives an inside look at a powertrain control module by Delco Electronics Corporation. This powerful computer is featured in all Cadillac Northstar V8 engine-equipped vehicles. Notice the use of many large-scale ICs on the drivetrain control module PC board in Fig. 1-10.

Automotive electronics

From the driver's seat of some modern automobiles, a digital tachometer stares back at you from the instrument cluster. The glow of a digital display shows the time and tuning of your sound system. The same digital display identifies your selection on the compact disk player. A digital thermometer monitors the interior and exterior temperatures, while a digital compass points the way to your destination. Antilock braking and traction systems make driving less hazardous. Automobile manufacturers are spending large sums of money on research and development efforts in automotive electronics.

Many home appliances use electric motors. Federal regulations call on manufacturers to improve refrigerator, freezer, air conditioner, and

Fig. 1-9 Digital synthesizer with a floppy disk drive. (*Courtesy of Korg U.S.A., Inc.*)

Fig. 1-10 A powertrain control module is used in all Cadillac Northstar V8 engine–equipped vehicles. (*Courtesy of Delco Electronics Corporation.*)

heating equipment efficiencies by up to 35 percent. To save energy, General Electric Company has developed the electrically commutated motor (ECM) shown in Fig. 1-11. The ECM variable-speed motor used for air conditioning, heating, and refrigeration equipment is about 20 percent more efficient than conventional induction motors. The printed circuit board pictured contains a power supply and much digital circuitry. The PC board contains a microcomputer and an electrically erasable PROM that can be programmed by a personal computer for different appliances. The ECM is a direct-current brushless permanent-magnet motor.

Applications of digital circuits

In your home there are other pieces of digital equipment. Appliances such as microwave ovens, washers, and dryers may have microprocessor-controlled or microcontroller-based digital circuitry. Your bathroom may house a digital scale, digital thermometer, or digital blood pressure monitor. Your home's heating and cooling may be controlled by one of the newer "intelligent" setback thermostats. Electronic and video games make extensive use of digital electronics. Even your telephone equipment may now have digital readouts and memory characteristics.

Digital cameras

Digital imaging has been simplified recently with the introduction of *digital cameras.* Instead of using traditional film, these cameras record images as electronic data. The Sony Digital Mavica® camera is pictured in Fig. 1-12. This digital camera is somewhat unique in that the storage medium is a floppy disk which can be removed from the camera and inserted into a computer for a simple transfer of data. Computers with a preinstalled Joint Photographic Experts Group (JPEG) viewer (such as Windows® 95 or Adobe Photoshop) can view the pictures directly on either their PC or Macintosh systems. Viewing software also comes with the camera. The 3.5-in. floppy

Fig. 1-11 ECM (electrically commutated motor) variable-speed motor. (*Courtesy of General Electric Company.*)

disk used by the Sony camera in Fig. 1-12 can store up to 40 images at 640 × 480 24-bit videographics array (VGA) resolution. Digital cameras do not yet have the resolution of film cameras, but for some quick digitizing of color photos these units work well.

In recent years, manufacturers have produced unique products using digital technology. Some of these include portable electronic dictionaries and thesauruses, foreign language translators, handheld spelling checkers, electronic data organizer/schedulers, electronic Rolodex directories and telephone dialers, and specialized calculators. One shirt-pocket size product is the Digital Book System pictured in Fig. 1-13. This unit accepts two integrated circuit ROM cards at once allowing simultaneous and interactive access to both books. Two ROM cards are shown next to the compact Digital Book System unit in Fig. 1-13. Franklin's Digital Book System is a handheld alternative to the slower, more expensive CD-ROM.

One exotic robot in the news recently is the Mars Rover Sojourner. The six-wheeled semi-autonomous "rover" was landed on the planet Mars in 1997 as part of the Pathfinder mission.

Fig. 1-12 Mavica® digital camera with floppy disk storage. (*Courtesy of Sony Electronics, Inc.*)

Fig. 1-13 Digital Book System. (*Courtesy of Franklin Electronic Publishing.*)

Analog electronic system

The rover performed beyond expectations for several months. It steered autonomously to avoid obstacles as it inched toward its goal location. The rover is one of several exploration robots to be used in Mars missions.

Five large users of very advanced digital devices are the space, aviation, military, medical, and telecommunications fields. The space shuttle's robot arm is one example of the use of digital electronics in an exotic machine. The CAT/MRI machine permits physicians to "look inside" the human body to aid in diagnosis of various injuries or diseases. Formerly, digital circuits were used mainly in computers. Now these circuits are being applied in many other products because of their low cost and great accuracy. Because digital circuits are appearing in almost all electronic equipment, all well-trained technicians need to know how they operate.

TEST

Supply the missing word or words in each statement.

6. The DMM, or ________, in Fig. 1-3(*b*) uses a modern low-power ________ display.
7. The field of digital electronics has mushroomed as a result of advances in the making of ________ circuits.
8. Microcomputers are designed around a complex IC called a(n) ________.

Analog-to-digital (A/D) converter

Central processing unit (CPU)

1-3 Why Use Digital Circuits?

Electronic designers and technicians must have a working knowledge of both analog and digital systems. The designer must decide if the system will use analog or digital techniques or a combination of both. The technicians must build a prototype or troubleshoot and repair digital, analog, and combined systems.

Analog electronic systems have been more popular in the past. "Real-world" information dealing with time, speed, weight, pressure, light intensity, and position measurements are all *analog* in nature.

A simple analog electronic system for measuring the amount of liquid in a tank is illustrated in Fig. 1-14. The input to the system is a varying resistance. The processing proceeds according to the Ohm's law formula, $I = V/R$. The output indicator is an ammeter which is calibrated as a water tank gage. In the analog system in Fig. 1-14 as the water rises, the input resistance drops. Decreasing the resistance R causes an increase in current (I). Increased current causes the ammeter (water tank gage) to read higher.

The analog system in Fig. 1-14 is simple and efficient. The gage in Fig. 1-14 gives an indication of the water level in the tank. If more information is required about the water level, then a digital system such as the one shown in Fig. 1-15 might be used.

Digital systems are required when data must be stored, used for calculations, or displayed as numbers and/or letters. A somewhat more complex arrangement for measuring the amount of liquid in a water tank is the digital system shown in Fig. 1-15. The input is still a variable resistance as it was in the analog system. The resistance is converted into numbers by the analog-to-digital (A/D) converter. The central processing unit (CPU) of a computer can manipulate the input data, output the information, store the information, calculate things such as flow rates in and out, calculate the time until the tank is full (or empty) based on flow rates,

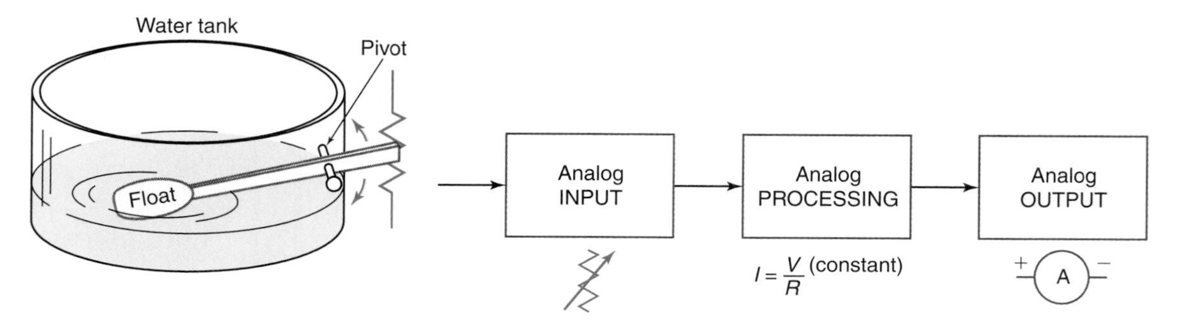

Fig. 1-14 Analog system used to interpret float level in water tank.

and so forth. Digital systems are valuable when calculations, data manipulations, and alphanumeric outputs are required.

Some of the advantages given for using digital circuitry instead of analog are as follows:

1. Inexpensive ICs can be used with few external components.
2. Information can be stored for short periods or indefinitely.
3. Data can be used for precise calculations.
4. Systems can be designed more easily using compatible digital logic families.
5. Systems can be programmed and show some manner of "intelligence."

The limitations of digital circuitry are as follows:

1. Most "real-world" events are analog in nature.
2. Analog processing is usually simpler and faster.

Digital circuits are appearing in more and more products primarily because of low-cost, reliable digital ICs. Other reasons for their growing popularity are accuracy, added stability, computer compatibility, memory, ease of use, simplicity of design, and compatibility with alphanumeric displays.

About Electronics

William (Bill) H. Gates III. The Chairman and CEO of Microsoft Corporation, Gates began programming computers at the age of 13. In 1974, while he was a college student, Gates developed BASIC, the programming language for the first microcomputer. Believing that the personal computer would eventually be found on every office desktop and in every home, Gates and Paul Allen formed Microsoft in 1975. Since then, Microsoft has been a leading developer of computer software.

Advantages of digital circuitry

Limitations of digital circuitry

TEST

Answer the following questions.

9. Generally, electronic circuits are classified as either analog or ________.
10. Measurements of time, speed, weight, pressure, light intensity, and position are ________ (analog, digital) in nature.
11. Refer to Fig. 1-14. As the water level drops, the input resistance increases. This causes the current *I* to ________ (decrease, increase) and the water level gage (ammeter) will read ________ (higher, lower).
12. Refer to Figs. 1-14 and 1-15. If this water tank were part of the city water system, where rates of water use are important, the system in Fig. ________ (1-14, 1-15) would be most appropriate.
13. True or false. The most important reason why digital circuitry is becoming more popular is because digital circuits are usually simpler and faster than analog circuits.

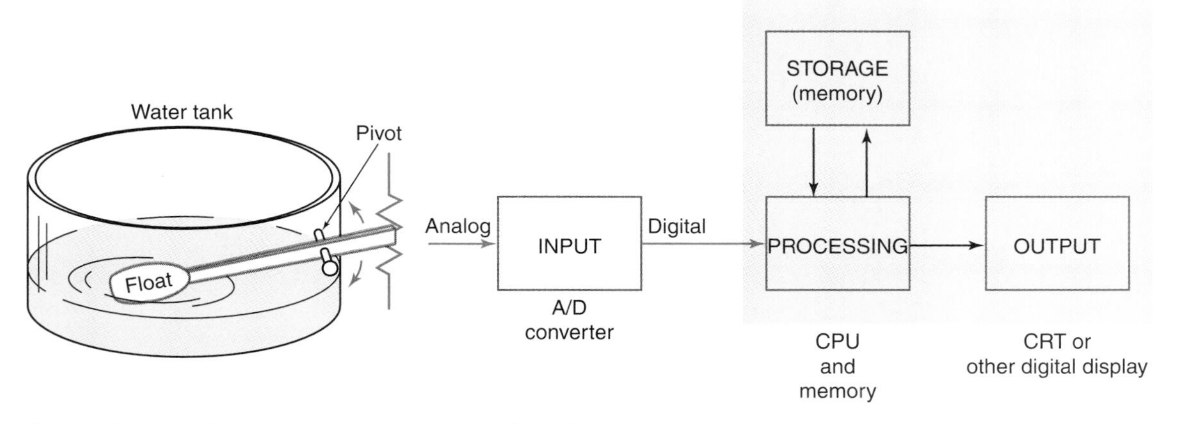

Fig. 1-15 Digital system used to interpret float level in water tank.

1-4 How Do You Make a Digital Signal?

TTL voltage levels

Transistor-transistor logic

Debounced logic switch

Digital waveform

Latch (flip-flop)

Contact bounce

Digital signals are composed of two well-defined voltage levels. For most of the circuits you will use, these voltage levels will be about 0 V (GND) and +3 to +5 V. These are called *TTL voltage levels* because they are used with the *transistor-transistor logic* family of ICs.

A TTL digital signal could be made manually by using a mechanical switch. Consider the simple circuit shown in Fig. 1-16(*a*). As the blade of the single-pole, double-throw (SPDT) switch is moved up and down, it produces the *digital waveform* shown at the right. At time period t_1, the voltage is 0 V, or LOW. At t_2 the voltage is +5 V, or HIGH. At t_3, the voltage is again 0 V, or LOW, and at t_4, it is again +5 V, or HIGH.

One problem with a mechanical switch is *contact bounce.* If we could look very carefully at a switch toggling from LOW to HIGH, it might look like the waveform in Fig. 1-16(*b*). The waveform first goes directly from LOW to HIGH (see *A*) but then, because of contact bounce, drops to LOW and then back to HIGH again. Although this happens in a very short time, digital circuits are fast enough to see this as a LOW, HIGH, LOW, HIGH waveform. Note that Fig. 1-16(*b*) shows that there is actually a range of voltages that are defined HIGH and LOW. The *undefined region* between HIGH and LOW causes trouble in digital circuits and should be avoided.

To cure the problem illustrated in Fig. 1-16(*b*), mechanical switches are sometimes *debounced.* A block diagram of a *debounced logic switch* is shown in Fig. 1-16(*c*). Note the use of the debouncing circuit, or latch. Many of the mechanical logic switches you will use on laboratory equipment will have been debounced with latch circuits. Latches are sometimes called *flip-flops* and will be studied in detail in a later chapter. Notice in Fig. 1-16(*c*) that the output of the latch during time period t_1 is LOW but not quite 0 V. During t_2 the output of the latch is HIGH even though it is something less than a full +5 V. Likewise t_3 is LOW and t_4 is HIGH in Fig. 1-16(*c*).

It might be suggested that a push-button switch be used to make a digital signal. If the button is pressed, a HIGH should be generated. If the push button is released, a LOW should be generated. Consider the simple circuit in

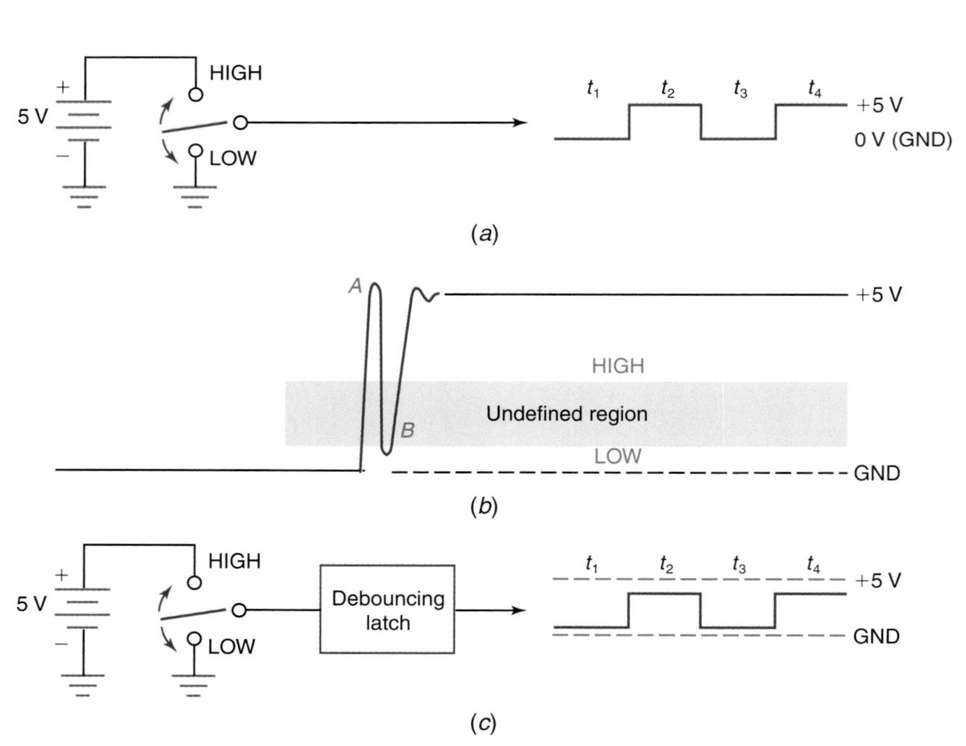

Fig. 1-16 (*a*) Generating a digital signal with a switch. (*b*) Waveform of contact bounce caused by a mechanical switch. (*c*) Adding a debouncing latch to a simple switch to condition the digital signal.

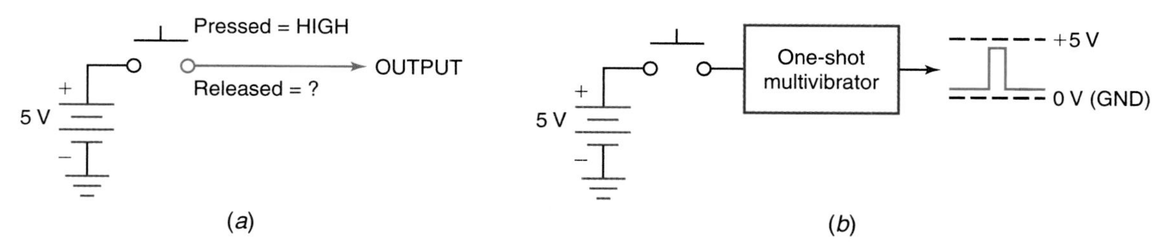

Fig. 1-17 (*a*) Push button will not generate a digital signal. (*b*) Push button used to trigger a one-shot multivibrator for a single-pulse digital signal.

Fig. 1-17(*a*). When the push button is pressed, a HIGH of about +5 V is generated at the output. When the push button is released, however, the voltage at the output is *undefined.* There is an open circuit between the power supply and the output. This would not work properly as a logic switch.

A normally open push-button switch can be used with a special circuit to generate a digital pulse. Figure 1-17(*b*) shows the push button connected to a *one-shot multivibrator* circuit. Now for each press of the push button, a *single short, positive pulse* is output from the one-shot circuit. The pulse width of the output is determined by the design of the multivibrator and *not* by how long you hold down the push button.

Both the latch circuit and the one-shot circuit were used earlier. Both are *multivibrator* (MV) circuits. The latch is also called a flip-flop or a *bistable multivibrator.* The one-shot is also called the *monostable multivibrator.* A third type of MV circuit is the *astable multivibrator.* This is also called a *free-running multivibrator.* In many digital circuits it may be referred to simply as the *clock.*

The free-running MV oscillates by itself without the need for external switching or an external signal. A block diagram of a free-running MV is shown in Fig. 1-18. The free-running MV generates a continuous series of TTL level pulses. The output in Fig. 1-18 alternately goes LOW and HIGH.

In the laboratory, you will need to generate digital signals. The equipment you will use will have slide switches, push buttons, and free-running clocks that will generate TTL level signals similar to those shown in Figs. 1-16, 1-17, and 1-18. In the laboratory, you will use *logic switches* which will have been debounced using a latch circuit as in Fig. 1-16(*c*). You will also use a *single-pulse clock* triggered by a push-button switch. The single-pulse clock push button will be connected to a one-shot multivibrator as shown in Fig. 1-17(*b*). Finally, your equipment will have a free-running clock. It will generate a continuous series of pulses, as shown in Fig. 1-18.

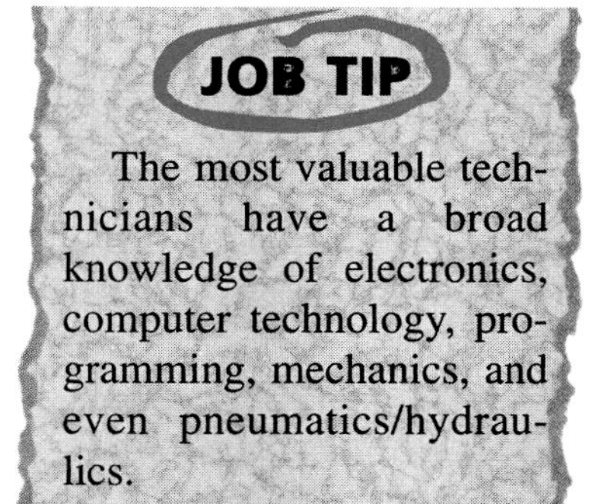

Astable, monostable, and bistable MVs can all be wired using discrete components (individual resistors, capacitors, and transistors) or purchased in IC form. Because of their superior performance, ease of use, and low cost, the IC forms of these circuits will be used in this course. A schematic diagram for a practical free-running clock circuit is shown in Fig. 1-19(*a*). This clock circuit produces a low-frequency (1- to 2-Hz) TTL level output. The heart of the free-running clock circuit is a common 555 timer IC. Note that several resistors, a capacitor, and a power supply must also be used in the circuit.

One-shot multivibrator

Multivibrator types: astable, bistable, and monostable

Free-running MV (clock)

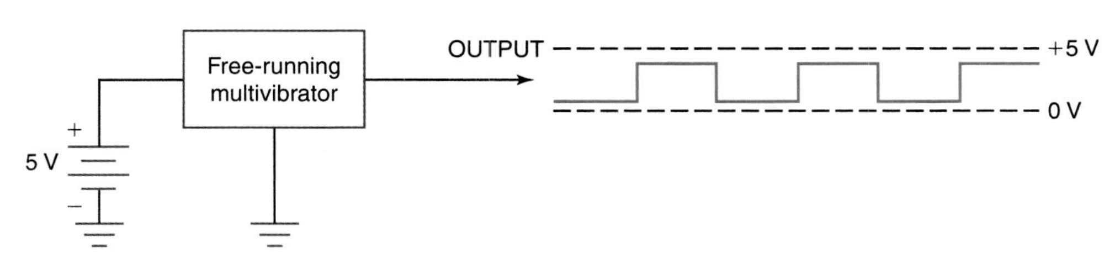

Fig. 1-18 Free-running multivibrator generates a string of digital pulses.

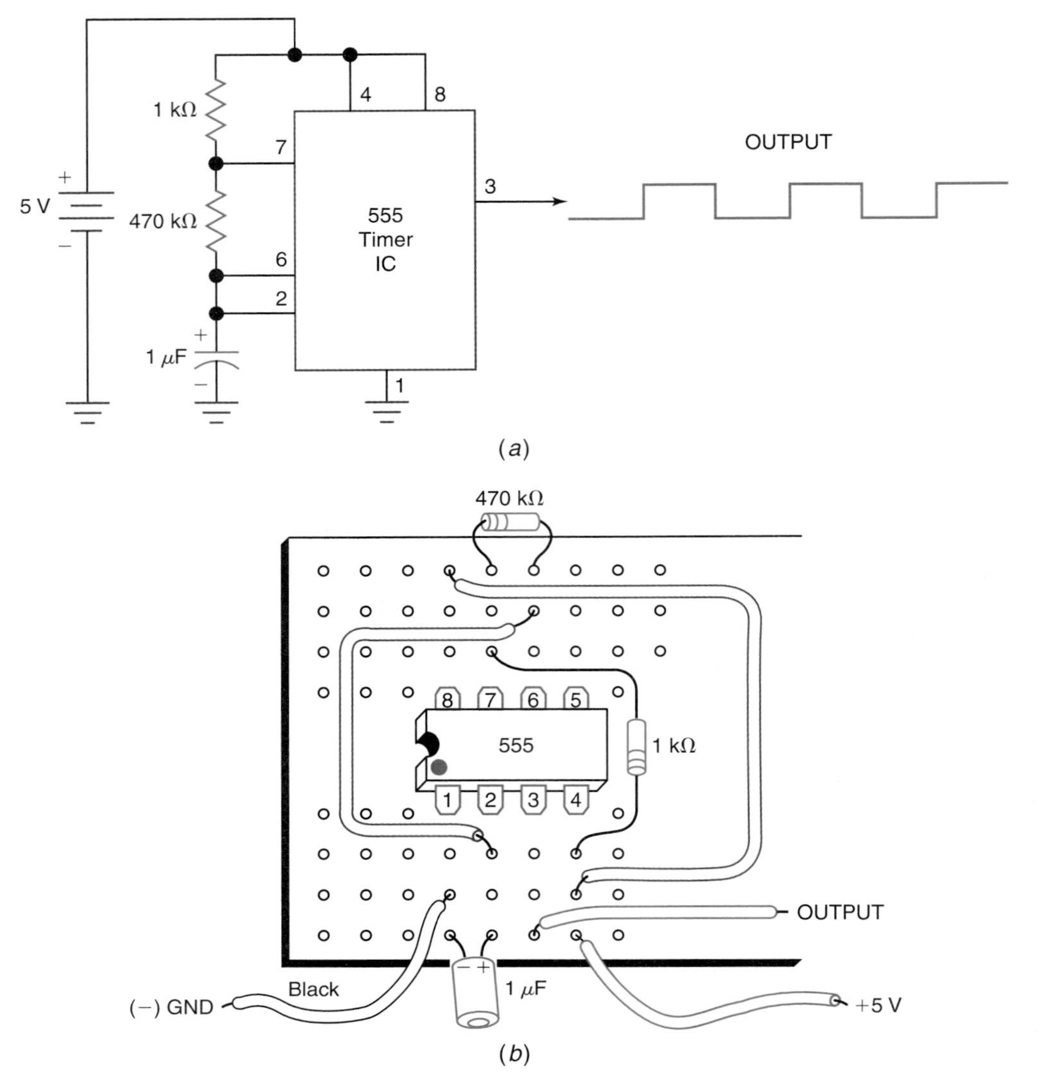

Fig. 1-19 (*a*) Schematic diagram of a free-running clock using a 555 timer IC. (*b*) Wiring the free-running clock circuit on a solderless breadboard.

A typical breadboard wiring of this free-running clock is sketched in Fig. 1-19(*b*). Notice the use of the IC type of mounting board. Also note that pin 1 on the IC is immediately counterclockwise from the notch or dot near the end of the eight-pin IC. The wiring diagram in Fig. 1-19(*b*) is shown for your convenience. You will normally have to wire circuits on solderless breadboards directly from the schematic diagram.

A typical digital trainer used during lab sessions is featured in Fig. 1-20. The photograph actually shows a pair of PC boards specifically designed to be used with this textbook's companion lab manual. Dynalogic's DT-1000 digital trainer board on the left includes a solderless breadboard for hooking up circuits. It also includes input devices such as 12 logic switches (two are debounced), a keypad, a one-shot MV, and a variable frequency clock (astable MV). Output devices mounted on the DT-1000 trainer board include 16 LED output indicators, a piezo buzzer, a relay, and a small dc motor. Power connections are available on the upper left of the DT-1000 digital trainer board. On the right in Fig. 1-20 is a second PC board which contains sophisticated LED, LCD, and VF displays. Dynalogic's DB-1000 display board is very useful when seven-segment displays are used as outputs. These boards along with individual ICs and other components could be used during your lab sessions to enable you to gain practical experience in digital electronics.

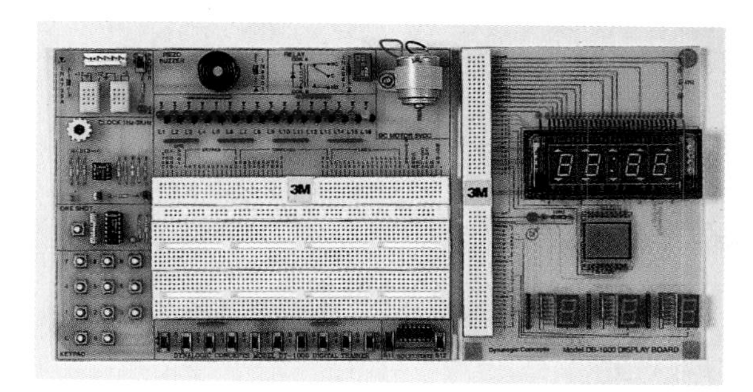

Fig. 1-20 Digital trainer and display boards used to set up lab experiments. (*Courtesy of Dynalogic Concepts.*)

TEST

Supply the missing word in each statement.

14. Refer to Fig. 1-16(*c*). The digital signal at t_2 ________ (HIGH, LOW), while it is ________ (HIGH, LOW) at t_3.
15. Refer to Fig. 1-17(*a*). When the push button is released (open), the output is ________.
16. Refer to Fig. 1-16(*c*). The debouncing latch is also called a flip-flop or ________ multivibrator.
17. Refer to Fig. 1-17(*b*). The one-shot multivibrator used for conditioning the digital signal is also called a(n) ________ multivibrator.
18. Refer to Fig. 1-19. A 555 ________ IC and several discrete components are being used to generate a continuous series of TTL level pulses. This free-running clock is also called a free-running multivibrator or ________ multivibrator.
19. Refer to Fig. 1-20. The one shot at the left on the DT-1000 board emits a single pulse each time the push-button switch is pressed. The one shot is also called a(n) ________ (astable, monostable) multivibrator.
20. Refer to Fig. 1-20. The clock at the left on the DT-1000 board generates a string of digital pulses. The clock is also called a(n) ________ (astable, bistable) multivibrator.
21. Refer to Fig. 1-20. The solid-state logic switches at the lower right on the DT-1000 board are debounced logic switches. The debouncing circuit is commonly called a ________ (bistable, monostable) multivibrator.
22. Refer to Fig. 1-20. The DT-1000 digital trainer board on the left features several output devices. List at least three of these output devices.
23. Refer to Fig. 1-20. The DB-1000 display board on the right features what three types of seven-segment displays?

interNET CONNECTION

Visit the website for Lucent Technologies for more information on digital circuits and components.

1-5 How Do You Test for a Digital Signal?

In the last section you generated digital signals using various MV circuits. These are the methods you will use in the laboratory to generate *input signals* for the digital circuits constructed. In this section, several simple methods of testing the *outputs* of digital circuits will be discussed.

Output indicators

Consider the circuit in Fig. 1-21(*a*). The *input* is provided by a simple SPDT switch and power supply. The *output indicator* is an LED. The 150-Ω resistor limits the current

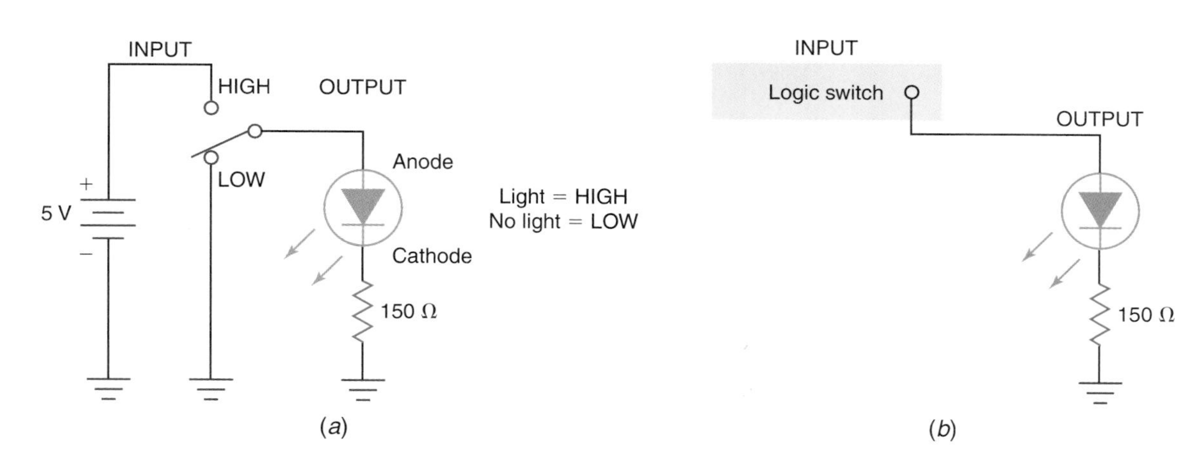

Fig. 1-21 (*a*) Simple LED output indicator. (*b*) Logic switch connected to simple LED output indicator.

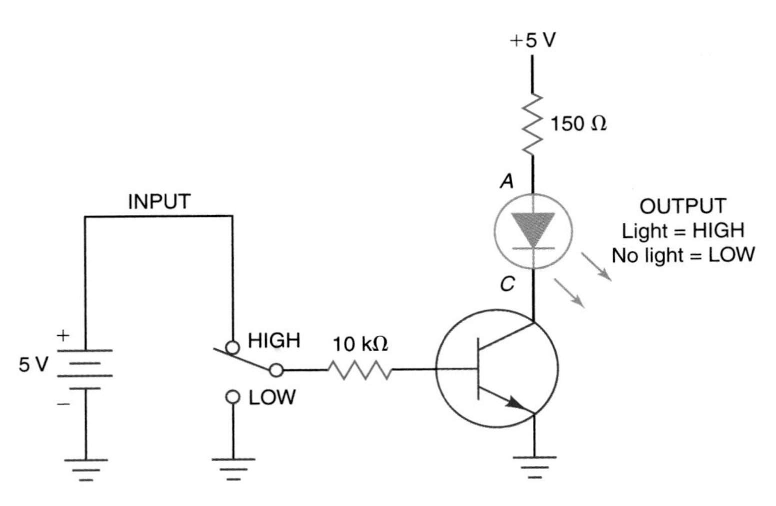

Fig. 1-22 Transistor-driven LED output indicator.

through the LED to a safe level. When the switch in Fig. 1-21(*a*) is in the HIGH (up) position, +5 V is applied to the anode end of the LED. The LED is forward-biased, current flows upward, and the LED lights. With the switch in the LOW (down) position, both the anode and cathode ends of the LED are grounded and it does not light. Using this indicator, a light means HIGH and no light generally means LOW.

The simple LED output indicator is shown again in Fig. 1-21(*b*). This time a simplified diagram of a logic switch forms the input. The logic switch acts like the switch in Fig. 1-21(*a*) except it is debounced. The output indicator is again the LED with a series-limiting resistor.

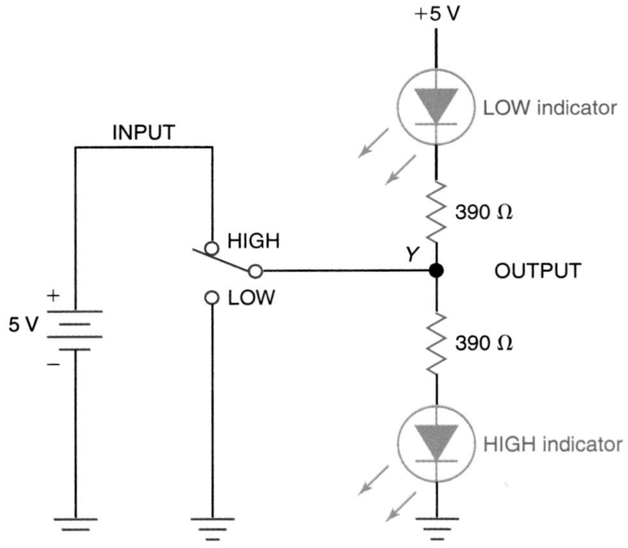

Fig. 1-23 LED output indicators that will show LOW, HIGH, and undefined logic levels.

When the input logic switch in Fig. 1-21(*b*) generates a LOW, the LED will not light. However, when the logic switch produces a HIGH, the LED will light.

Another LED output indicator is illustrated in Fig. 1-22. The LED acts exactly the same as the one shown previously. It lights to indicate a logical HIGH and does not light to indicate a LOW. The LED in Fig. 1-22 is driven by a transistor instead of directly by the input. The transistorized circuit in Fig. 1-22 holds an advantage over the direct-drive circuit in that it draws less current from the output of the digital circuit under test. Light-emitting diode output indicators wired like those in Fig. 1-22 may be found in your laboratory equipment.

Consider the output indicator circuit using two LEDs shown in Fig. 1-23. When the input is HIGH (+5 V), the bottom LED lights while the top LED does not light. When the input is LOW (GND), only the top LED lights. If point *Y* in the circuit in Fig. 1-23 enters the undefined region between HIGH and LOW or is not connected to a point in the circuit, both LEDs light.

Output voltages from a digital circuit can be measured with a standard voltmeter. With the TTL family of ICs, a voltage from 0 to 0.8 V is considered a LOW. A voltage from 2 to 5 V is considered a HIGH. Voltages between about 0.8 and 2 V are in the undefined region and signal trouble in TTL circuits.

Logic probe

A handy portable measuring device used to determine output logic levels is the *logic probe*. One such inexpensive, student-constructed logic probe is sketched in Fig. 1-24(*a*).

To use this unit for measuring TTL logic levels, follow this procedure:

1. Connect the red power lead to +5 V of the circuit under test.
2. Connect the black (GND) power lead to GND of the circuit under test.
3. Connect the third power lead (TTL) to GND of the circuit under test.
4. Touch the probe tip to the point in the digital circuit to be tested.
5. One or the other LED indicator shown in Fig. 1-24(*a*) should light. If both light, the tip is disconnected from the circuit or the point is in the undefined region between HIGH and LOW.

The logic probe in Fig. 1-24(*a*) can also be used with the CMOS logic family of ICs; CMOS is short for *complementary metal oxide semiconductor.* If the logic probe is to be used to measure *CMOS logic levels,* the TTL family power lead is left *disconnected.* Then the red power lead is connected to positive (+) of the power supply while the black lead (GND) goes to GND. Touch the probe tip to the test point in the CMOS digital circuit and the LED indicators tell its CMOS logic level, either HIGH or LOW.

A schematic diagram of the logic probe is shown in Fig. 1-24(*b*). The 555 timer IC is being used in this circuit. The 555 IC uses power supply voltages ranging from about 5 V up to

TTL logic levels

CMOS logic family

Complementary metal oxide semiconductor

CMOS logic levels

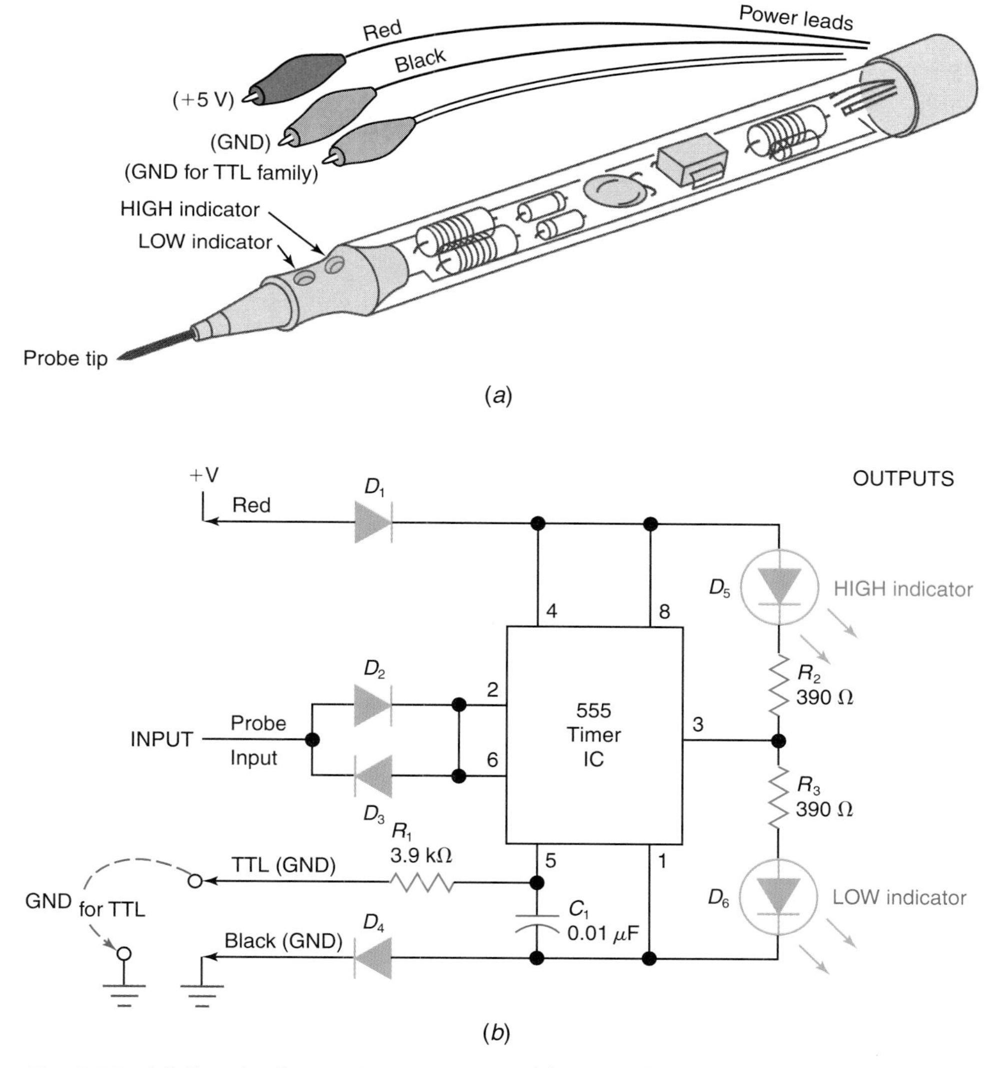

Fig. 1-24 (*a*) Sketch of a student-constructed logic probe. (*b*) Schematic diagram of the logic probe using the 555 timer IC. (*Courtesy of Electronic Kits International, Inc.*)

JOB TIP

Companies seek employees who are self-directed and good problem solvers.

18 V. TTL circuits always operate on 5 V, whereas some CMOS circuits operate on voltages as high as 15 V. The three power supply connections are shown at the left in Fig. 1-24(*b*). The red lead goes to the positive of the supply while the black lead goes to GND. If the circuit under test is TTL, the TTL (GND) lead is also connected to GND. If a CMOS circuit is being tested, the TTL (GND) power lead is left disconnected. The input to the logic probe is shown on the left, entering pins 2 and 6 of the 555 IC. If this voltage is LOW, the *bottom* LED (D_6) lights. If the input voltage is HIGH, the *top* LED (D_5) lights. With the input probe disconnected, both LEDs light. Note that pin 3 of the 555 IC (output of 555) always goes to the *opposite* logic level of the input. Therefore, if the input (pins 2 and 6) is HIGH, the output of the 555 IC (pin 3) goes LOW. This in turn activates and lights the top LED (the HIGH indicator).

Defining TTL and CMOS logic levels

The four silicon diodes (D_1 to D_4) in Fig. 1-24(*b*) protect the IC from reverse polarity. Capacitor C_1 prevents transient voltages from affecting the logic probe when the TTL (GND) lead is not connected. Pin 5 of the 555 IC is grounded through resistor R_1. This resistor "adjusts" for TTL instead of CMOS logic levels.

Logic probe

The logic probe in Fig. 1-24 responds to voltage levels differently in the CMOS and TTL modes. Figure 1-25 shows the TTL and CMOS logic levels in terms of percentage of total supply voltage. For TTL, which always uses a +5-V supply, this logic probe will indicate a HIGH for 2 V or greater. For TTL, it will indicate a LOW for 0.8 V or less.

Defining logic levels

In the laboratory, you may be asked to construct a logic probe like the one in Fig. 1-24 or your instructor might furnish you with a commercial logic probe for checking digital circuits. The operating instructions are different for each logic probe. Read the instruction manual on the unit you will be using.

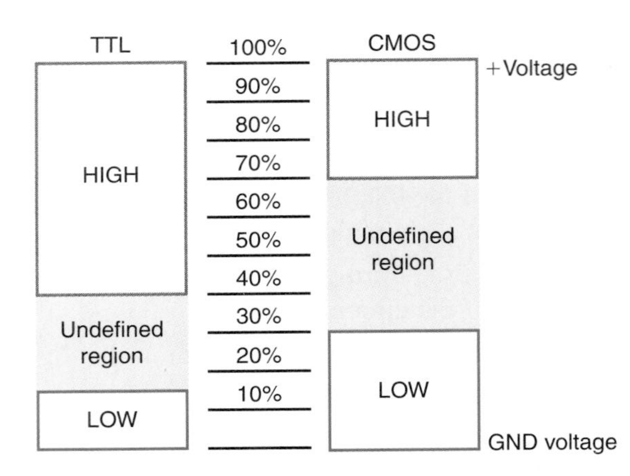

Fig. 1-25 Defining logic levels for the TTL and CMOS families of digital ICs.

TEST

Supply the missing word or words in each statement.

24. Refer to Fig. 1-21. If the input is HIGH, the LED will ________ (light, not light) because the diode is ________ (forward-, reverse-) biased.
25. Refer to Fig. 1-22. If the input is LOW, the transistor is turned ________ (off, on) and the LED ________ (does, does not) light.
26. Refer to Fig. 1-23. If the input is HIGH, the ________ (bottom, top) LED lights because its ________ (cathode, anode) has +5 V applied, forward biasing the diode.
27. The student-constructed ________ in Fig. 1-24 can be used to test either TTL of ________ digital circuits.
28. Refer to Fig. 1-25 and assume a 5-V power supply. In a TTL circuit, a voltage of 2.5 V would be considered a(n) ________ (HIGH, LOW, undefined) logic level.
29. Refer to Fig. 1-25 and assume a 10-V power supply. In a CMOS circuit, a voltage of 2 V would be considered a(n) ________ (HIGH, LOW, undefined) logic level.

Chapter 1 Review

Summary

1. Analog signals vary gradually and continuously, while digital signals produce discrete voltage levels commonly referred to as HIGH and LOW.
2. Computers, including the microcomputer, make extensive use of digital circuits. Calculators are also digital devices.
3. Most modern electronics equipment contains both analog and digital circuitry.
4. Logic levels are different for various digital logic families, such as TTL and CMOS. These logic levels are commonly referred to as HIGH, LOW, and undefined. Figure 1-25 details these TTL and CMOS logic levels.
5. Digital circuits have become very popular because of the availability of low-cost digital ICs. Other advantages of digital circuitry are computer compatibility, memory, ease of use, simplicity of design, accuracy, and stability.
6. Bistable, monostable, and astable multivibrators are used to generate digital signals. These are sometimes called latches, one-shot, and free-running multivibrators, respectively.
7. Logic level indicators may take the form of simple LED and resistor circuits, voltmeters, or logic probes. Light-emitting diode logic level indicators will probably be found on your laboratory equipment.

Chapter Review Questions

Answer the following questions.

1-1. Define the following:
  a. Analog signal
  b. Digital signal

1-2. Draw a square-wave digital signal. Label the bottom "0 V" and the top "+5 V." Label the HIGH and LOW on the waveform. Label the logical 1 and logical 0 on the waveform.

1-3. List two devices that contain digital circuits which do mathematical calculations.

1-4. List three test instruments that contain digital circuits and are used by electronic technicians.

1-5. Refer to Fig. 1-15. The processing, storage of data, and output in this system consist mostly of _________ (analog, digital) circuits.

1-6. Traditionally, most consumer electronics devices (TVs, radios) have used _________ (analog, digital) circuitry.

1-7. Refer to Fig. 1-16. When using a SPDT slide switch to produce a digital signal, a _________ latch is used to condition the output.

1-8. Refer to Fig. 1-17. A _________-_________, multivibrator is commonly used to condition the output of a push-button switch when generating a single digital pulse.

1-9. An astable, or _________-_________, multivibrator produces a string of digital pulses.

1-10. The circuit in Fig. 1-19 is classed as a(n) _________ (astable, bistable) multivibrator.

1-11. Refer to Fig. 1-20. The one-shot MV on the DT-1000 trainer produces a _________ (series of pulses, single pulse) when the push button is pressed once.

1-12. Refer to Fig. 1-20. The clock on the DT-1000 trainer produces a _________ (continuous series of pulses, single pulse) and the circuit can be referred to as a(n) _________ (astable, monostable) multivibrator.

1-13. Refer to Fig. 1-20. The two solid-state logic switches on the DT-1000 trainer are _________ (analog, debounced).

1-14. Refer to Fig. 1-20. The DB-1000 board features what *three* types of seven-segment displays?

1-15. The LED in Fig. 1-21(*b*) lights when the input logic switch is _________ (HIGH, LOW).

1-16. Refer to Fig. 1-23. The _________ (bottom, top) LED lights when the input switch is LOW.

1-17. The logic probe in Fig. 1-24 can be used with which two digital logic families?

1-18. Refer to Fig. 1-24(*b*). If the input is LOW, pin 3 of the 555 IC will be _________ (HIGH, LOW). This causes the _________ (bottom, top) LED to light.

1-19. Refer to Fig. 1-25 and assume a 5-V power supply. In a TTL circuit, a voltage of 1 V would be considered a(n) ________ (HIGH, LOW, undefined) logic level.

1-20. Refer to Fig. 1-25 and assume a 10-V power supply. In a CMOS circuit, a voltage of 8 V would be considered a(n) ________ (HIGH, LOW, undefined) logic level.

Critical Thinking Questions

1-1. List several advantages of digital over analog circuits.

1-2. When you are looking at electronic equipment, what are some clues that might indicate it contains at least some digital circuitry?

1-3. List at least ten devices that contain digital circuitry.

1-4. Refer to Fig. 1-16(*a*). What is the main drawback of this circuit for generating a digital signal?

1-5. Refer to Fig. 1-17(*a*). What is the difficulty with this circuit for generating a digital signal?

1-6. At the option of your instructor, use circuit simulation software to (1) draw a free-running clock circuit using a 555 timer IC such as that pictured in Fig. 1-26, (2) test the operation of the clock circuit, and (3) determine the approximate frequency of the clock using the time period measurements from the scope and the formula $f = 1/t$.

1-7. At the option of your instructor, use circuit simulation software to (1) draw the clock circuit in Fig. 1-26 as in question 1-6, (2) change the resistance of R_2 to 100 kΩ, (3) test the operation of the clock circuit, and (4) determine the approximate frequency of the clock using the time period measurements from the scope and the formula $f = 1/t$.

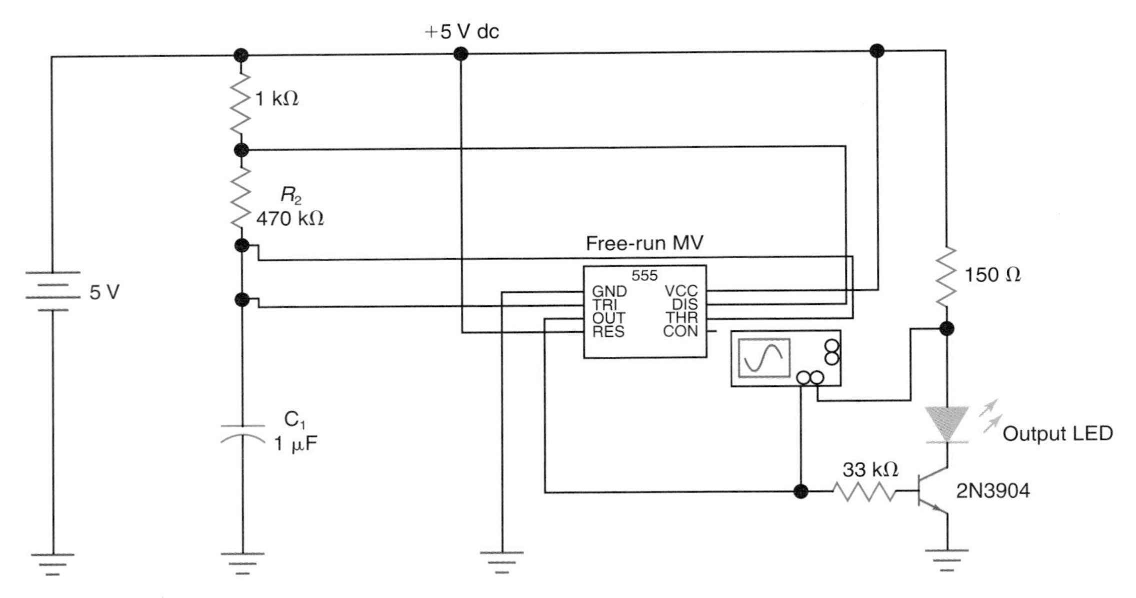

Fig. 1-26 Circuit simulation problem-clock circuit.

Answers to Tests

1. digital, HIGH
2. analog
3. analog
4. digital
5. digital
6. digital multimeter, liquid-crystal
7. integrated
8. microprocessor
9. digital
10. analog
11. decrease, lower
12. 1-15
13. F
14. HIGH, LOW
15. undefined
16. bistable
17. monostable
18. timer, astable
19. monostable
20. astable
21. bistable
22. LED (logic indicators), piezo buzzer, dc motor, relay
23. LED, LCD, VF
24. light, forward-
25. off, does not
26. bottom, anode
27. logic probe, CMOS
28. HIGH
29. LOW

Numbers We Use in Digital Electronics

Chapter Objectives

This chapter will help you to:

1. *Demonstrate* understanding of the idea of place value in the decimal, binary, octal, and hexadecimal number systems.
2. *Convert* binary numbers to decimal and decimal numbers to binary.
3. *Convert* hexadecimal numbers to binary, binary to hexadecimal, hexadecimal to decimal, and decimal numbers to hexadecimal.
4. *Convert* octal numbers to binary, binary to octal, octal to decimal, and decimal numbers to octal.

Most people understand us when we say we have nine pennies. The number 9 is part of the *decimal* number system we use every day. But digital electronic devices use a "strange" number system called *binary.* Digital computers and microprocessor-based systems use other strange number systems called *hexadecimal* and *octal.* Men and women who work in electronics must know how to convert numbers from the everyday decimal system to the binary, hexadecimal, and octal systems.

2-1 Counting in Decimal and Binary

A number system is a code that uses symbols to refer to a number of items. The decimal number system uses the symbols 0, 1, 2, 3, 4, 5, 6, 7, 8, and 9. The decimal number system contains 10 symbols and is sometimes called the *base 10 system.* The binary number system uses only the two symbols 0 and 1 and is sometimes called the *base 2 system.*

Decimal number system

Base 10 system

Binary number system

Base 2 system

Figure 2-1 compares a number of coins with the symbols we use for counting. The decimal symbols that we commonly use for counting from 0 to 9 are shown in the left column; the right column has the symbols we use to count nine coins in the binary system. Notice that the 0 and 1 count in binary is the same as in decimal counting. To represent two coins, the binary number 10 (say "one zero") is used. To represent three coins, the binary number 11 (say "one one") is used. To represent nine coins, the binary number 1001 (say "one zero zero one") is used.

COINS	DECIMAL SYMBOL	BINARY SYMBOL
No coins	0	0
●	1	1
●●	2	10
●●●	3	11
●●●●	4	100
●●●●●	5	101
●●●●●●	6	110
●●●●●●●	7	111
●●●●●●●●	8	1000
●●●●●●●●●	9	1001

Fig. 2-1 Symbols for counting.

For your work in digital electronics you should memorize the binary symbols used to count (at least up to 9).

TEST

Supply the missing word or number in each statement.

1. The binary number system is sometimes called the _________ system.
2. Decimal 8 equals _________ in binary.
3. The binary number 0110 equals _________ in decimal.
4. The binary number 0111 equals _________ in decimal.

2-2 Place Value

The clerk at the local store totals your bill and asks you for $2.43. We all know that this amount equals 243 cents. Instead of paying with 243 pennies, however, you could probably give the clerk the money shown in Fig. 2-2: two $1 dollar bills, four dimes, and three pennies. This money example illustrates the very important idea of *place value.*

JOB TIP

Interviewers look for interest in the company and its products. It pays to learn about the company before the interview.

Place value

Consider the decimal number 648 in Fig. 2-3. The digit 6 represents 600 because of its placement three positions left of the decimal point. The digit 4 represents 40 because of its placement two positions left of the decimal point. The digit 8 represents eight units because of its placement one position left of the decimal point. The total number 648, then, represents six hundred and forty-eight units. This is an example of place value in the decimal number system.

Decimal system

Binary number system

The binary number system also uses the idea of place value. What does the binary number 1101 (say "one one zero one") mean? Figure 2-4 shows that the digit 1 nearest the binary point is the units, or 1s, position, and so we add one item. The digit 0 in the 2s position tells us we have no 2s. The digit 1 in the 4s position tells us to add four items. The digit 1 in the 8s position tells us to add eight more items. When we count all the items, we find that the binary number 1101 represents 13 items.

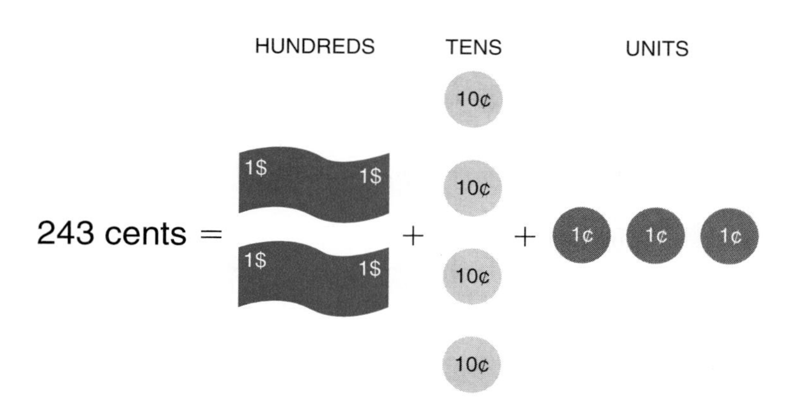

Fig. 2-2 An example of place value.

Fig. 2-3 Place value in the decimal system.

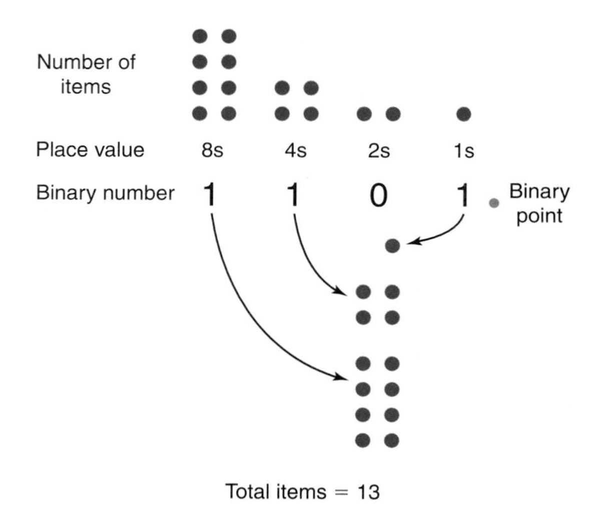

Fig. 2-4 Place value in the binary number system.

How about the binary number 1100 (say "one one zero zero")? Using the system from Fig. 2-4, we find that we have the following:

8s	4s	2s	1s	place value
yes	yes	no	no	binary
(1)	(1)	(0)	(0)	number
●●	●●			number of items
●●	●●			
●●				
●●				

The binary number 1100, then, represents 12 items.

Figure 2-5 shows each place in the value of the binary system. Notice that each place value is determined by multiplying the value to the right by 2. The term "base 2" for binary comes from this idea.

Many times the weight or value of each place in the binary number system is referred

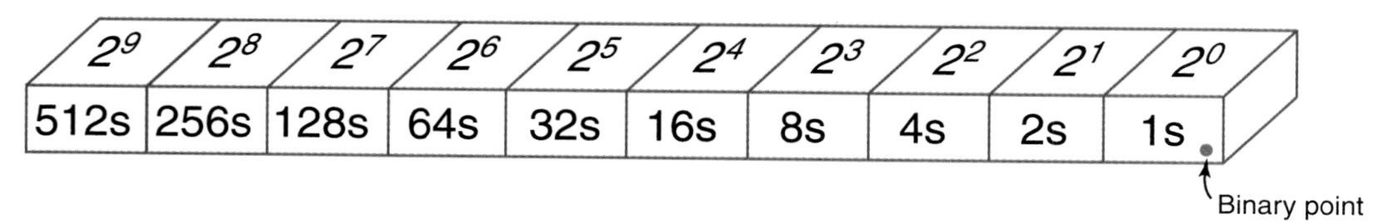

Fig. 2-5 Value of the places left of the binary point.

to as a power of 2. In Fig. 2-5, place values for a binary number are shown in decimal and also in powers of 2. For instance, the 8s place is the same as the 2^3 position, the 32s place the same as the 2^5 position and so on.

TEST

Supply the missing number in each statement.

5. The 1 in the binary number 1000 has a place value of ________ in decimal.
6. The binary number 1010 equals ________ in decimal.
7. The binary number 100000 equals ________ in decimal.
8. The number 2^7 equals ________ in decimal.

2-3 Binary to Decimal Conversion

While working with digital equipment you will have to convert from the binary code to decimal numbers. If you are given the binary number 110011, what would you say it equals in decimal? First write down the binary number as:

Binary	1	1	0	0	1	1	• binary point
Decimal	32 +	16	+		2 +	1	= 51

Start at the binary point and work to the left. For each binary 1, place the decimal value of that position (see Fig. 2-5) below the binary digit. Add the four decimal numbers to find the decimal equivalent. You will find that binary 110011 equals the decimal number 51.

Powers of 2

Another practical problem is to convert the binary number 101010 to a decimal number. Again write down the binary number as:

Binary	1	0	1	0	1	0	•
Decimal	32	+	8	+	2		= 42

Starting at the binary point, write the place value (see Fig. 2-5) for each binary 1 below the square in decimals. Add the three decimal numbers to get the decimal total. You will find that the binary number 101010 equals the decimal number 42.

Now try a long and difficult binary number: convert the binary number 1111101000 to a decimal number. Write down the binary number as:

Binary to decimal conversion

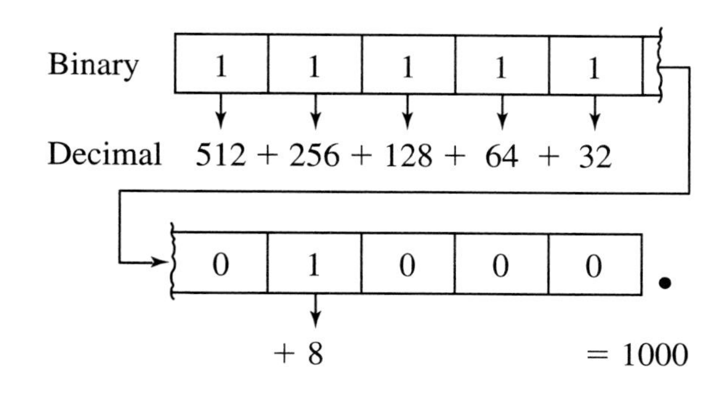

From Fig. 2-5, convert each binary 1 to its correct decimal value. Add the decimal values to get the decimal total. The binary num-

Binary point

Did You Know?

Television and the Worldwide Web. The word *intercasting* describes a television set which incorporates a Web browser. For example, you may watch a show on fly-fishing and click on something to see where to buy the equipment and where to fish locally.

ber 1111101000 equals the decimal number 1000.

■ TEST

Supply the missing number in each statement.

9. The binary number 1111 equals __________ in decimal.
10. The binary number 100010 equals __________ in decimal.
11. The binary number 1000001010 equals __________ in decimal.

2-4 Decimal to Binary Conversion

Decimal to binary conversion

Many times while working with digital electronic equipment you must be able to convert a decimal number to a binary number. We shall teach you a method that will help you to make this conversion.

Repeated divide-by-2 process

Suppose you want to convert the decimal number 13 to a binary number. One procedure you can use is the repeated divide-by-2 process shown below:

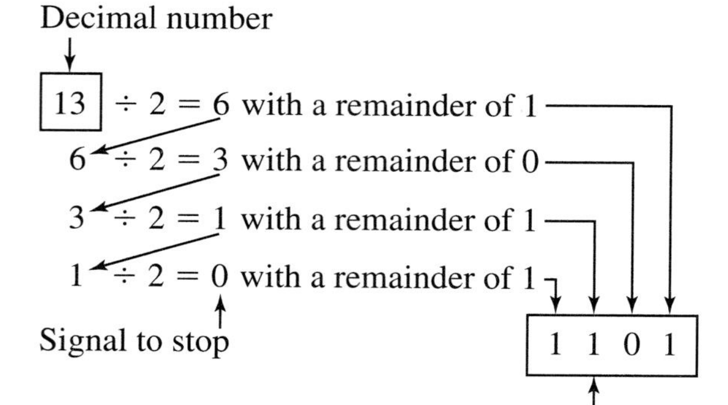

Notice that 13 is first divided by 2, giving a quotient of 6 with a remainder of 1. This remainder becomes the 1s place in the binary number. The 6 is then divided by 2, giving a quotient of 3 with a remainder of 0. This remainder becomes the 2s place in the binary number. The 3 is then divided by 2, giving a quotient of 1 with a remainder of 1. This remainder becomes the 4s place in the binary number. The 1 is then divided by 2, giving a quotient of 0 with a remainder of 1. This remainder becomes the 8s place in the binary number. The decimal number 13 has been converted to the binary number 1101.

Knowledge of computer technology is extremely important for almost all technicians.

Practice this procedure by converting the decimal number 37 to a binary number. Follow the procedure used before:

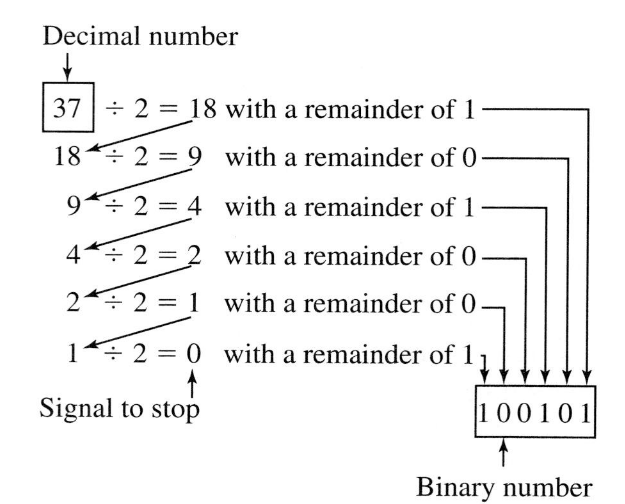

Notice that you stop dividing by 2 when the quotient becomes 0. According to this procedure, the decimal number 37 is equal to the binary number 100101.

■ TEST

Supply the missing number in each statement.

12. The decimal number 39 equals __________ in binary.
13. The decimal number 100 equals __________ in binary.
14. The decimal number 133 equals __________ in binary.

2-5 Electronic Translators

Electronic translators

If you were to try to communicate with a French-speaking person who did not know the English language, you would need someone to *translate* the English into French and then the French into English. A similar problem exists in digital electronics. Almost all digital circuits (calculators, computers) understand only binary numbers. But most people understand only decimal numbers. Thus we must have electronic devices that can translate from decimal to binary numbers and from binary to decimal numbers.

Figure 2-6 diagrams a typical system that might be used to translate from decimal to binary numbers and back to decimals. The de-

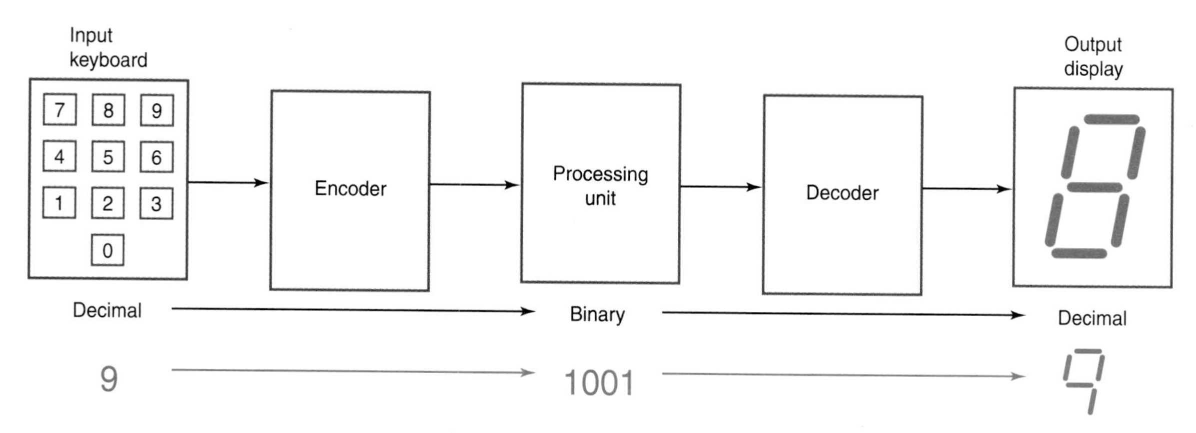

Fig. 2-6 A system using encoders and decoders.

vice that translates from the keyboard decimal numbers to binary is called an *encoder;* the device labeled *decoder* translates from binary to decimal numbers.

Encoders

Decoders

The bottom of Fig. 2-6 shows a typical conversion. If you press the decimal number 9 on the keyboard, the encoder will convert the 9 into the binary number 1001. The decoder will convert the binary 1001 into the decimal number 9 on the output display.

Encoders and decoders are very common electronic circuits in all digital devices. A pocket calculator, for instance, must have encoders and decoders to translate electronically from decimal to binary numbers and back to decimal. Figure 2-6 is a very basic block diagram of a pocket calculator. When you press the number 9 on the keyboard the number appears on the output display.

You can buy encoders and decoders that translate from any of the commonly used codes in digital electronics. Most of the encoders and decoders you will use will be packaged as single ICs.

TEST

Supply the missing word in each statement.

15. A(n) ________ is an electronic device that translates a decimal input number to binary.
16. The processing unit of a calculator outputs binary. This binary is converted to a decimal output display by an electronic device called a(n) ________.

2-6 Hexadecimal Numbers

Hexadecimal number system

The *hexadecimal number system* uses the 16 symbols 0, 1, 2, 3, 4, 5, 6, 7, 8, 9, A, B, C, D, E, and F and is referred to as the *base 16 system.* Figure 2-7 shows the equivalent binary and hexadecimal representations for the decimal numbers 0 through 17. The letter "A" stands for decimal 10, "B" for decimal 11, and so on. The

Base 16 system

Advantage of the hexadecimal system

Decimal	Binary	Hexadecimal
0	0000	0
1	0001	1
2	0010	2
3	0011	3
4	0100	4
5	0101	5
6	0110	6
7	0111	7
8	1000	8
9	1001	9
10	1010	A
11	1011	B
12	1100	C
13	1101	D
14	1110	E
15	1111	F
16	10000	10
17	10001	11

Fig. 2-7 Binary and hexadecimal equivalents to decimal numbers.

About Electronics

Growth of Microcontrollers. Motorola recently announced the shipment of its two billionth 68HC05 microcontroller. This microcontroller is one of perhaps hundreds of microcontrollers on the market. A recent listing of microcontrollers filled over 60 pages in the book, *IC Master*. Growth in the use of microcontrollers in "smart products" is expected to continue to rise rapidly.

advantage of the hexadecimal system is its usefulness in converting directly from a 4-bit binary number. For instance, hexadecimal F stands for the 4-bit binary number 1111. Hexadecimal notation is typically used to represent a binary number. For instance, the hexadecimal number A6 would represent the 8-bit binary number 10100110. Hexadecimal notation is widely used in microprocessor-based systems to represent 8-, 16-, or 32-bit binary numbers.

Hexadecimal notation

Microprocessor-based systems

Binary-to-hex conversion

Subscripts

Base 10

Base 2

Base 16

The number 10 represents how many objects? It can be observed from the table shown in Fig. 2-7 that the number 10 could mean ten objects, two objects, or sixteen objects depending on the base of the number. *Subscripts* are sometimes added to a number to indicate the base of the number. Using subscripts, the number 10_{10} represents ten objects. The subscript (10 in this example) indicates it is a base 10, or decimal number. Using subscripts, the number 10_2 represents two objects since this is in binary (base 2). Again using subscripts, the number 10_{16} represents sixteen objects since this is in hexadecimal (base 16).

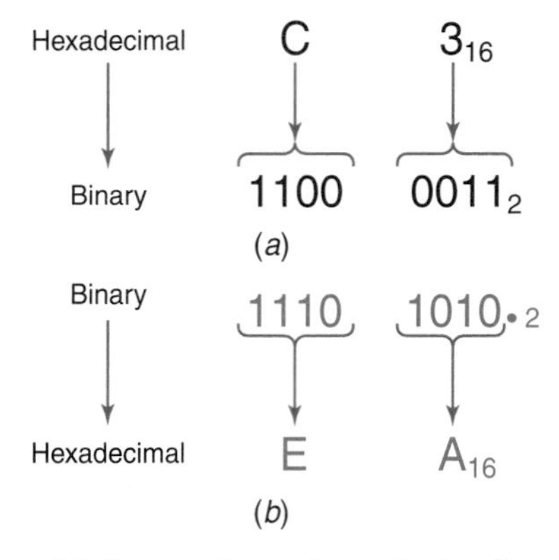

Fig. 2-8 (*a*) Converting a hexadecimal number to binary. (*b*) Converting a binary number to hexadecimal.

Decimal-to-hex conversion

Hex-to-binary conversion

Converting hexadecimal numbers to binary and binary numbers to hexadecimal is a common task when working with microprocessors and microcomputers. Consider converting $C3_{16}$ to its binary equivalent. In Fig. 2-8(*a*), each hexadecimal digit is converted to its 4-bit binary equivalent (see Fig. 2-7). Hexadecimal C equals the 4-bit binary number 1100, while 3_{16} equals 0011. Combining the binary groups yields $C3_{16} = 11000011_2$.

Now reverse the process and convert the binary number 11101010 to its hexadecimal equivalent. The simple process is detailed in Fig. 2-8(*b*). The binary number is divided into 4-bit groups starting at the binary point. Next, each 4-bit group is translated into its equivalent hexadecimal number with the help of the table shown in Fig. 2-7. The example in Fig. 2-8(*b*) shows $11101010_2 = EA_{16}$.

Consider converting hexadecimal $2DB_{16}$ to its decimal equivalent. The place values for the first three places in the hexadecimal number are shown across the top in Fig. 2-9 as 256s, 16s, and 1s. In Fig. 2-9 there are eleven 1s. There are thirteen 16s, which equals 208. There are two 256s, which equals 512. Adding 11 + 208 + 512 will equal 731_{10}. The example given in Fig. 2-9 shows that $2DB_{16} = 731_{10}$.

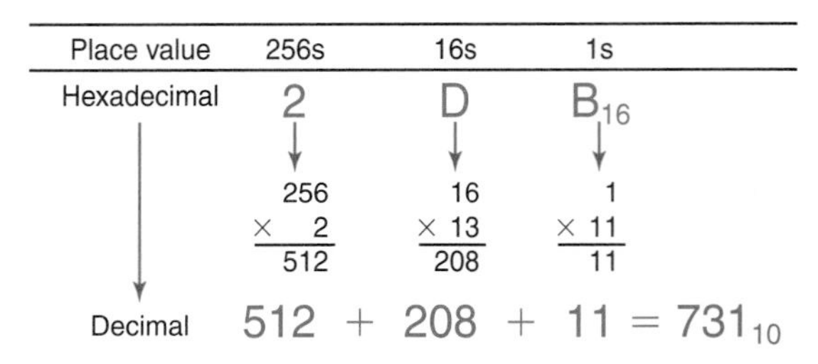

Fig. 2-9 Converting a hexadecimal number to decimal.

Now reverse the process and convert the decimal number 47 to its hexadecimal equivalent. Figure 2-10 details the repeated divide-by-16 process. The decimal number 47 is first divided by 16, resulting in a quotient of 2 with a remainder of 15. The remainder of 15 (F in hexadecimal) becomes the least significant

Did You Know?

Microprocessors Then and Now. The Intel 4004 4-bit microprocessor was unveiled in 1971. The 4004 processor contained about 2300 transistors. Today, an Intel Pentium II chip contains about 7.5 million transistors.

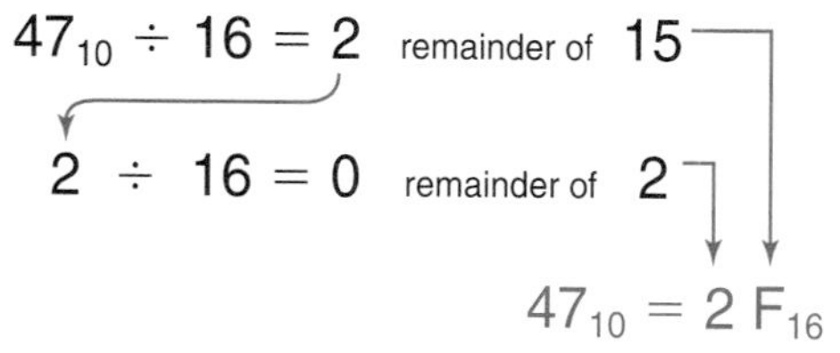

$47_{10} \div 16 = 2$ remainder of 15

$2 \div 16 = 0$ remainder of 2

$47_{10} = 2\text{F}_{16}$

Fig. 2-10 Converting a decimal number to hexadecimal using the repeated divide-by-16 process.

digit (LSD) of the hexadecimal number. The quotient (2 in this example) is transferred to the dividend position and divided by 16. This results in a quotient of 0 with a remainder of 2. The 2 becomes the next digit in the hexadecimal number. The process is complete because the integer part of the quotient is 0. The divide-by-16 process shown in Fig. 2-10 converts 47_{10} to its hexadecimal equivalent of 2F_{16}.

Repeated divide-by-16 process

TEST

Supply the missing number in each statement.

17. Decimal 15 equals ________ in hexadecimal.
18. Hexadecimal A6 equals ________ in binary.
19. Binary 11110 equals ________ in hexadecimal.
20. Hexadecimal 1F6 equals ________ in decimal.
21. Decimal 63 equals ________ in hexadecimal.

2-7 Octal Numbers

Octal number system

Some older computer systems use octal numbers to represent binary information. The *octal number system* uses the eight symbols 0, 1, 2, 3, 4, 5, 6, and 7. Octal numbers are also referred to as *base 8* numbers. The table shown in Fig. 2-11 gives the equivalent binary and octal representations for decimal numbers 0 through 17. The advantage of the octal system is its usefulness in converting directly from a 3-bit binary number. Octal notation is used to represent binary numbers.

Octal-to-binary conversion

Converting octal numbers to binary is a common operation when using some computer systems. Consider converting the octal number 67_8 (say "six seven base eight") to its binary equivalent. In Fig. 2-12(*a*), each octal digit is

Decimal	Binary	Octal
0	000	0
1	001	1
2	010	2
3	011	3
4	100	4
5	101	5
6	110	6
7	111	7
8	001 000	10
9	001 001	11
10	001 010	12
11	001 011	13
12	001 100	14
13	001 101	15
14	001 110	16
15	001 111	17
16	010 000	20
17	010 001	21

Fig. 2-11 Binary and octal equivalents to decimal numbers.

converted to its 3-bit binary equivalent. Octal 6 equals 110, while 7 equals 111. Combining the binary groups yields $67_8 = 110111_2$.

Binary-to-octal conversion

Now reverse the process and convert binary 100001101 to its octal equivalent. The simple process is detailed in Fig. 2-12(*b*). The binary number is divided into 3-bit groups (100 001 101) starting at the binary point. Next, each 3-bit group is translated into its equivalent octal number. The example in Fig. 2-12(*b*) shows $100\ 001\ 101_2 = 415_8$.

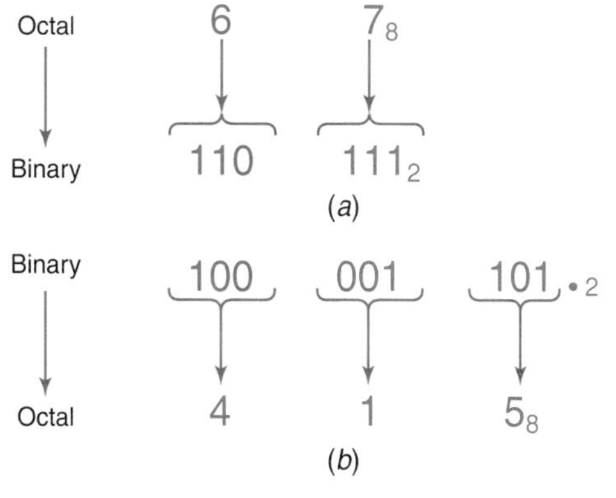

Fig. 2-12 (*a*) Converting an octal number to binary. (*b*) Converting a binary number to octal.

Octal-to-decimal conversion

Consider converting octal 415 (say "four one five base eight") to its decimal equivalent. The place values for the first three places in the octal number are shown across the top in Fig. 2-13 as 64s, 8s, and 1s. There are five 1s and one 8. There are four 64s, which equals 256. Adding $5 + 8 + 256 = 269_{10}$. The example in Fig. 2-13 shows that $415_8 = 269_{10}$.

Place value:	64s	8s	1s
Octal	4	1	5_8
	64 × 4 = 256	8 × 1 = 8	1 × 5 = 5
Decimal	256 +	8 +	5 = 269_{10}

Fig. 2-13 Converting an octal number to decimal.

Now reverse the process and convert the decimal number 498 to its octal equivalent. Figure 2-14 details the repeated divide-by-8 process. The decimal number 498 is first divided by 8, resulting in a quotient of 62 with a remainder of 2. The remainder (2) becomes the LSD of the octal number. The quotient (62 in this example) is transferred to the dividend position and divided by 8. This results in a quotient of 7 with a remainder of 6. The 6 becomes the next digit in the octal number. The last quotient (7 in this example) is transferred to the dividend position and divided by 8. The quotient is 0 with a remainder of 7. The 7 is the most significant digit (MSD) in the octal number. The divide-by-8 process shown

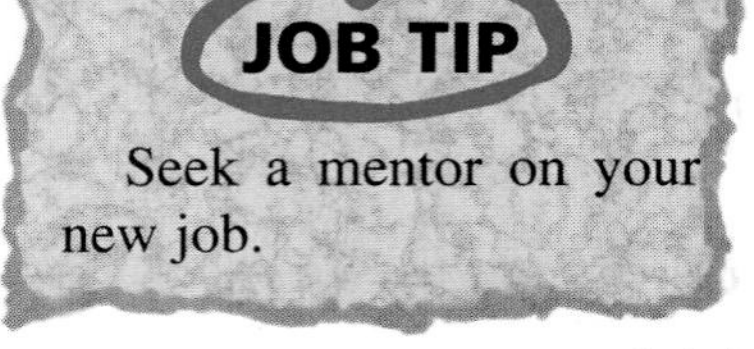

Decimal-to-octal conversion

Repeated divide-by-8 process

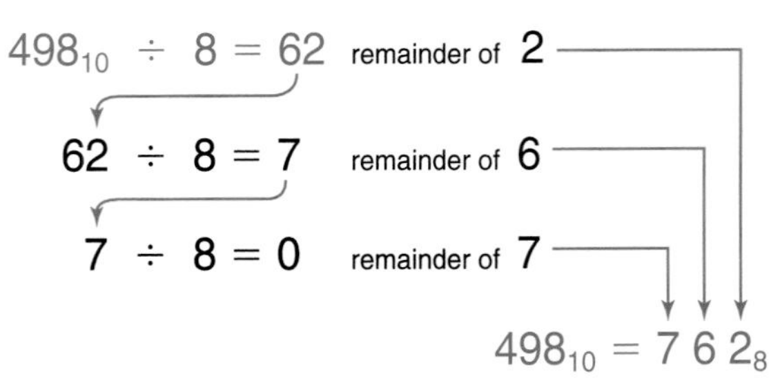

Fig. 2-14 Converting a decimal number to octal using the repeated divide-by-8 process.

in Fig. 2-14 converts 498_{10} to its octal equivalent of 762_8. Note that the signal to end the repeated divide-by-8 process is when the quotient becomes 0.

Technicians, engineers, and programmers must be able to convert between number systems. Many commercial calculators can aid in making binary, octal, hexadecimal, and decimal conversions. These calculators also perform arithmetic operations on binary, octal, and hexadecimal as well as decimal numbers.

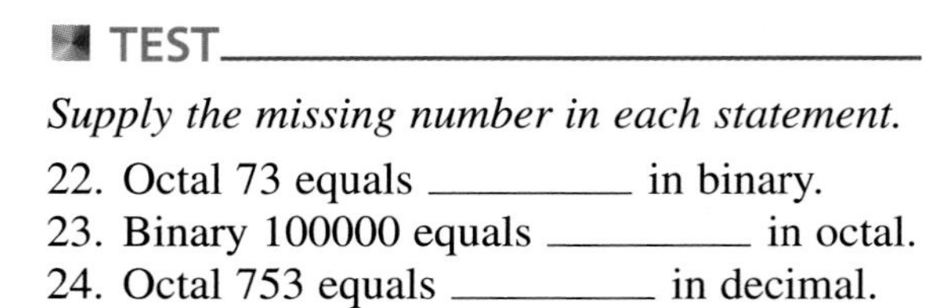

TEST

Supply the missing number in each statement.

22. Octal 73 equals ________ in binary.
23. Binary 100000 equals ________ in octal.
24. Octal 753 equals ________ in decimal.
25. Decimal 63 equals ________ in octal.

Chapter 2 Review

Summary

1. The decimal number system contains 10 symbols: 0, 1, 2, 3, 4, 5, 6, 7, 8, and 9.
2. The binary number system contains two symbols: 0 and 1.
3. The place values left on the binary point in binary are 64, 32, 16, 8, 4, 2, and 1.
4. All men and women in the field of digital electronics must be able to convert binary to decimal numbers and decimal to binary numbers.
5. Encoders are electronic circuits that convert decimal numbers to binary numbers.
6. Decoders are electronic circuits that convert binary numbers to decimal numbers.
7. The hexadecimal number system contains 16 symbols: 0, 1, 2, 3, 4, 5, 6, 7, 8, 9, A, B, C, D, E, and F.
8. Hexadecimal digits are widely used to represent binary numbers in the computer field.
9. The octal number system uses eight symbols: 0, 1, 2, 3, 4, 5, 6, and 7. Octal numbers are used to represent binary numbers in a few computer systems.

Chapter Review Questions

Answer the following questions.

2-1. How would you say the decimal number 1001?

2-2. How would you say the binary number 1001?

2-3. Convert the binary numbers in **a** to **j** to decimal numbers:

a. 1 f. 10000
b. 100 g. 10101
c. 101 h. 11111
d. 1011 i. 11001100
e. 1000 j. 11111111

2-4. Convert the decimal numbers in **a** to **j** to binary numbers:

a. 0 f. 64
b. 1 g. 69
c. 18 h. 128
d. 25 i. 145
e. 32 j. 1001

2-5. Encode the decimal numbers in **a** to **f** to binary numbers:

a. 9 d. 13
b. 3 e. 10
c. 15 f. 2

2-6. Decode the binary numbers in **a** to **f** to decimal numbers:

a. 0010 d. 0111
b. 1011 e. 0110
c. 1110 f. 0000

2-7. What is the job (function) of an encoder?

2-8. What is the job (function) of a decoder?

2-9. Write the decimal numbers from 0 to 15 in binary.

2-10. Convert the hexadecimal numbers in **a** to **d** to binary numbers:

a. 8A c. 6C
b. B7 d. FF

2-11. Convert the binary numbers in **a** to **d** to hexadecimal numbers:

a. 01011110 c. 11011011
b. 00011111 d. 00110000

2-12. Hexadecimal 3E6 = $\_\_\_\_\_\__{10}$.

2-13. Decimal 4095 = $\_\_\_\_\_\__{16}$.

2-14. Octal 156 = $\_\_\_\_\_\__{10}$.

2-15. Decimal 391 = $\_\_\_\_\_\__{8}$.

Critical Thinking Questions

2-1. If the digital circuits in a computer only respond to binary numbers, why are octal and hexadecimal numbers used extensively by computer specialists?

2-2. In a digital system such as a microcomputer, it is common to consider an 8-bit group (called a *byte*) as having a meaning. Predict some of the possible meanings of a byte (such as 11011011_2) in a microcomputer.

Answers to Tests

1. base 2
2. 1000
3. 6
4. 7
5. 8 or (2^3)
6. 10
7. 32
8. 128
9. 15
10. 34
11. 522
12. 100111
13. 1100100
14. 10000101
15. encoder
16. decoder
17. F
18. 10100110
19. 1E
20. 502
21. 3F
22. 111011
23. 40
24. 491
25. 77

Logic Gates

Chapter Objectives

This chapter will help you to:

1. *Memorize* the name, symbol, truth table, function, and Boolean expression for the eight basic logic gates.
2. *Draw* a logic diagram of any of the eight basic logic functions using only NAND gates.
3. *Convert* one type of basic gate to any other logic function using inverters.
4. *Sketch* logic diagrams illustrating how two-input gates could be used to create gates with more inputs.
5. *Memorize* the inverted input forms of the NAND and NOR gates.
6. *Identify* pin numbers and manufacturer's markings on both TTL and CMOS dual-in-line package ICs.
7. *Troubleshoot* simple logic gate circuits.
8. *Recognize* new logic gate symbols used in dependency notation (IEEE standard 91-1984).

Computers, calculators, and other digital devices are sometimes looked upon by the general public as being magical. Actually, digital electronic devices are extremely *logical* in their operation. The basic building block of any digital circuit is a *logic gate*. Persons working in digital electronics understand and use logic gates every day. Remember that logic gates are the building blocks for even the most complex computers. Logic gates can be constructed by using simple switches, relays, vacuum tubes, transistors and diodes, or ICs. Because of their availability, wide use, and low cost, ICs will be used to construct digital circuits. A variety of logic gates are available in all logic families including TTL and CMOS.

Logic gates

3-1 The AND Gate

AND gate

"All or nothing" gate

The AND gate is sometimes called the "all or nothing gate." Figure 3-1 shows the basic idea of the AND gate using simple switches.

What must be done in Fig. 3-1 to get the output lamp (L_1) to light? You must close *both* switches *A* and *B* to get the lamp to light. You could say that switch *A and* switch *B* must be closed to get the output to light. The AND gates you will operate most often are constructed of diodes and transistors and packaged inside an IC. To show the AND gate we use the *logic symbol* in Fig. 3-2. This standard AND gate symbol is used whether we are using relays, switches, pneumatic circuits, discrete diodes and transistors, or ICs. This is the symbol you will memorize and use from now on for AND gates.

AND gate logic symbol

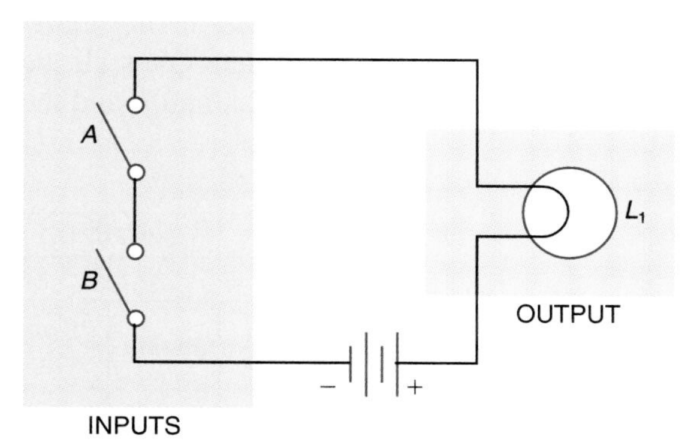

Fig. 3-1 AND circuit using switches.

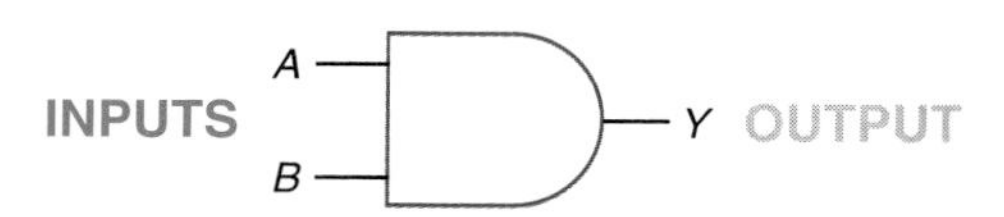

Fig. 3-2 AND gate logic symbol.

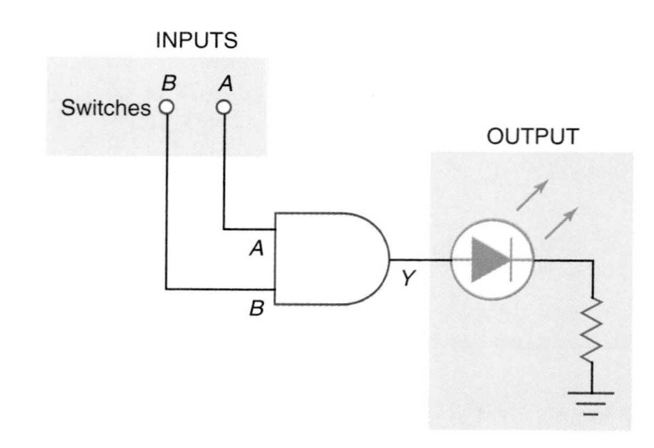

Fig. 3-3 Practical AND gate circuit.

The term "logic" is usually used to refer to a decision-making process. A logic gate, then, is a circuit that can decide to say yes or no at the output based upon the inputs. We already determined that the AND gate circuit in Fig. 3-1 says yes (light on) at the output only when we have a yes (switches closed) at *both* inputs.

AND function

Now let us consider an actual circuit similar to one you will set up in the laboratory. The AND gate in Fig. 3-3 is connected to input switches *A* and *B*. The output indicator is an LED. If a LOW voltage (GND) appears at inputs *A* and *B*, then the output LED is *not lit*. This situation is illustrated in line 1 of the truth table in Fig. 3-4. Notice also in line 1 that the inputs and outputs are given as *binary digits*. Line 1 indicates that if the inputs are binary 0 and 0, then the output will be a binary 0. Carefully look over the four combinations of switches *A* and *B* in Fig. 3-4. Notice that only binary 1s at both *A* and *B* will produce a binary 1 at the output (see the last line of the table).

Boolean expression

Boolean algebra

Positive logic

It is a +5 V compared to GND appearing at *A*, *B*, or *Y* that is called a binary 1 or a HIGH voltage. A binary 0, or LOW voltage, is defined as a GND voltage (near 0 V compared to GND) appearing at *A*, *B*, or *Y*. We are using *positive logic* because it takes a *positive* 5 V to produce what we call a binary 1. You will use positive logic in most of your work in digital electronics.

Truth table

The table in Fig. 3-4 is called a *truth table*. The truth table for the AND gate gives all the possible input combinations of *A* and *B* and the resulting outputs. Thus the truth table defines the exact operation of the AND gate. The truth table in Fig. 3-4 is said to describe the AND *function*. You should memorize the truth table for the AND function. The unique output from the AND gate is a HIGH only when all inputs are HIGH. The output column in Fig. 3-4 shows that *only* the last line in the AND truth table generates a 1 while the rest of the outputs are 0.

So far you have memorized the logic symbol and the truth table for the AND gate. Now you will learn a shorthand method of writing the statement "input *A* is ANDed with input *B* to get output *Y*." The short method used to represent this statement is called its *Boolean expression* ("Boolean" from Boolean algebra—the algebra of logic). The Boolean expression is a universal language used by engineers and technicians in digital electronics. Figure 3-5 shows the ways to express that input *A* is ANDed with input *B* to produce output *Y*. The top expression in Fig. 3-5 is how you would tell someone in the English language that you are ANDing inputs *A* and *B* to get output *Y*. Next in Fig. 3-5 you see the Boolean expression for ANDing inputs *A* and *B*. Note that a

INPUTS				OUTPUT	
B		*A*		*Y*	
Switch voltage	Binary	Switch voltage	Binary	Light	Binary
LOW	0	LOW	0	No	0
LOW	0	HIGH	1	No	0
HIGH	1	LOW	0	No	0
HIGH	1	HIGH	1	Yes	1

Fig. 3-4 AND truth table.

In the English language	Input A is ANDed with input B to get output Y.
As a Boolean expression	$A \cdot B = Y$ (AND symbol)
As a logic symbol	A, B → Y
As a truth table	see below

B	A	Y
0	0	0
0	1	0
1	0	0
1	1	1

Fig. 3-5 Four ways to express the logical ANDing of A and B.

multiplication dot (·) is used to symbolize the AND function in Boolean expressions. In common practice the Boolean expression $A \cdot B = Y$ can be simplified to $AB = Y$. Both $A \cdot B = Y$ or $AB = Y$ describe the two-input AND function.

In summary, Fig. 3-5 illustrates the four commonly used methods of describing the ANDing of inputs A and B. All these methods are widely used and must be learned by persons working in the electronics industry.

TEST

Answer the following questions.

1. The output side of an AND gate logic symbol is ________ (pointed, round).
2. Write the Boolean expression for a two-input AND gate.
3. Refer to Fig. 3-3. When both inputs are HIGH, output Y will be ________ and the LED will ________ (light, not light).

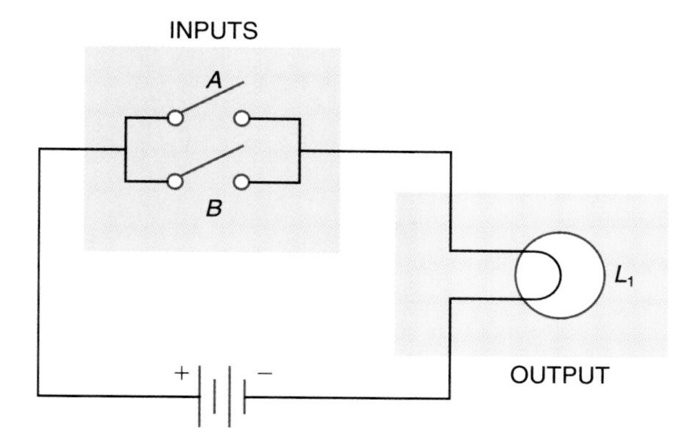

Fig. 3-6 OR circuit using switches.

3-2 The OR Gate

OR gate

"Any or all" gate

The OR gate is sometimes called the "any or all gate." Figure 3-6 illustrates the basic idea of the OR gate using simple switches. Looking at the circuit in Fig. 3-6, you can see that the output lamp will light when *either* or *both* of the input switches are closed but not when both are open. A truth table for the OR circuit is shown in Fig. 3-7. The truth table lists the switch and light conditions for the OR gate circuit in Fig. 3-6. The truth table in Fig. 3-7 is said to describe the *inclusive* OR *function.* The unique output from the OR gate is a LOW only when all inputs are LOW. The output column in Fig. 3-7 shows that *only* the first line in the OR truth table generates a 0 while all others are 1.

OR gate truth table

Inclusive OR function

The logic symbol for the OR gate is diagramed in Fig. 3-8 on the next page. Notice in the logic diagram that inputs A and B are being ORed to produce an output Y. The engineer's Boolean expression for the OR function is also illustrated in Fig. 3-8. Note that the plus (+) sign is the Boolean symbol for OR.

You should memorize the logic symbol, Boolean expression, and truth table for the OR gate.

INPUTS				OUTPUT	
B		A		Y	
Switch	Binary	Switch	Binary	Light	Binary
Open	0	Open	0	No	0
Open	0	Closed	1	Yes	1
Closed	1	Open	0	Yes	1
Closed	1	Closed	1	Yes	1

Fig. 3-7 OR gate truth table.

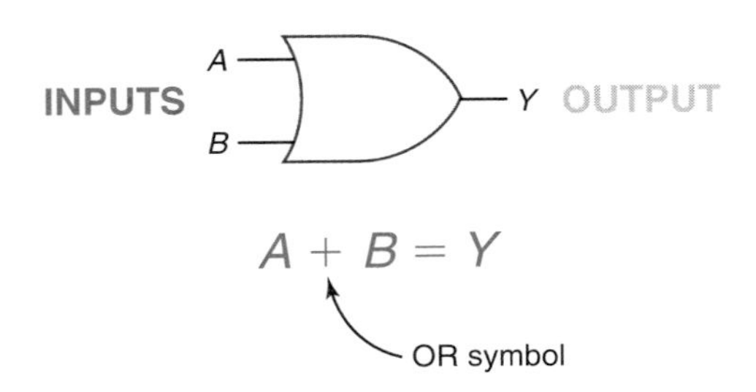

Fig. 3-8 OR logic gate symbol and Boolean expression.

TEST

Answer the following questions.

4. The output side of an OR gate logic symbol is ________ (pointed, round).
5. Write the Boolean expression for a two-input OR gate.
6. Refer to Fig. 3-8. If inputs A and B are both LOW, output Y will be ________.
7. Refer to Fig. 3-7. This truth table describes the ________ (exclusive, inclusive) OR logic function.

3-3 The Inverter and Buffer

All the gates so far have had at least two inputs and one output. The NOT circuit, however, has only one input and one output. The NOT circuit is often called an *inverter.* The job of the NOT circuit (inverter) is to give an output that is not the same as the input. The logic symbol for the inverter (NOT circuit) is shown in Fig. 3-9.

JOB TIP

Physical appearance and personal manner strongly influence the impression you make on others.

NOT circuit

Inverter

Negated

Complements

Inverts

Double inverting

If we were to put in a logic 1 at input A in Fig. 3-9, we would get out the opposite, or a logical 0, at output Y. We say that the inverter *complements* or *inverts* the input. Figure 3-9 also shows how we would write a Boolean expression for the NOT, or INVERT, function. Notice the use of the overbar ($^{-}$) symbol above the output to show that A has been inverted or complemented. We say that the Boolean term "$\overline{A}$" would be "not A."

An alternative NOT symbol used in Boolean expressions is also shown in Fig. 3-9. Notice the use of the apostrophe symbol to show that A has been inverted or complemented. We would say that the Boolean term A' would be "not A" or "A not." The use of the overbar is the preferred NOT symbol but the apostrophe

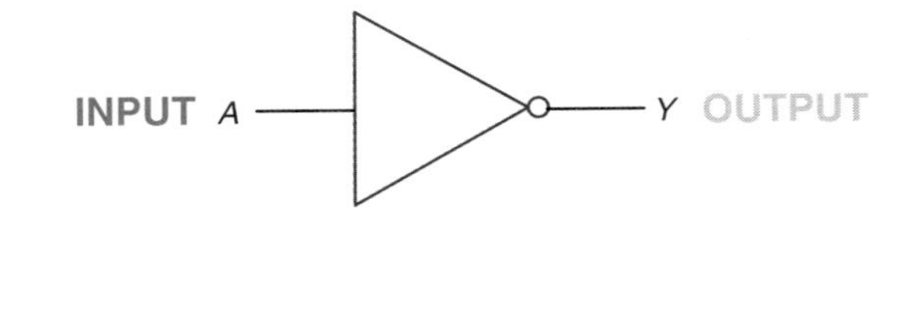

$Y = A'$ ← Alternate NOT symbol

Fig. 3-9 A logic symbol and Boolean expression for an inverter.

is common when Boolean expressions are shown on a computer screen as in circuit simulation programs.

The truth table for the inverter is shown in Fig. 3-10. If the voltage at the input of the inverter is LOW, then the output voltage is HIGH. However, if the input voltage is HIGH, then the output is LOW. As you learned, the output is always opposite the input. The truth table also gives the characteristics of the inverter in terms of binary 0s and 1s.

You learned that when a signal passes through an inverter, it can be said that the input is inverted or complemented. We can also say it is *negated.* The terms "negated," "complemented," and "inverted," then, mean the same thing.

The logic diagram in Fig. 3-11 shows an arrangement where input A is run through two inverters. Input A is first inverted to produce a "not A" ($\overline{A}$) and then inverted a second time for a "double not A" ($\overline{\overline{A}}$). In terms of binary digits, we find that when the input 1 is inverted twice, we end up with the original digit. Therefore we

INPUT		OUTPUT	
A		*Y*	
Voltages	Binary	Voltages	Binary
LOW	0	HIGH	1
HIGH	1	LOW	0

Fig. 3-10 Truth table for an inverter.

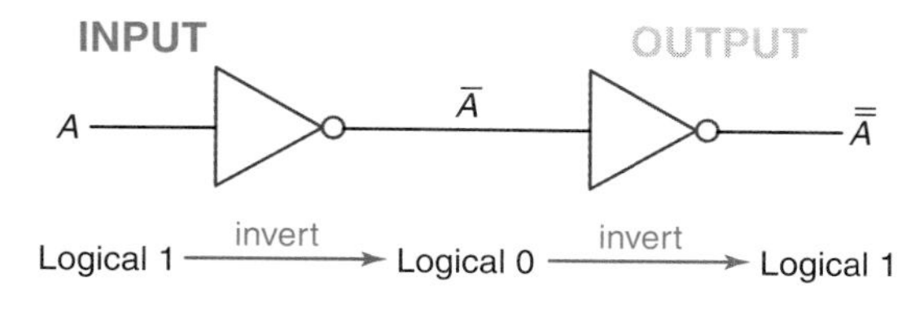

Therefore $\overline{\overline{A}} = A$

Fig. 3-11 Effect of double inverting.

find that $\overline{\overline{A}}$ equals A. Thus, a Boolean term with two bars over it is equal to the term under the two bars, as shown at the bottom of Fig. 3-11.

Two symbols found in logic diagrams that look something like an inverter symbol are pictured in Fig. 3-12. The logic symbol in Fig. 3-12(*a*) is an alternative symbol for an inverter and performs the NOT function. The placement of the "invert bubble" on the left side of the inverter symbol in Fig. 3-12(*a*) might suggest that this is an active LOW input.

The symbol in Fig. 3-12(*b*) is that of a *noninverting buffer/driver.* The noninverting buffer serves no logical purpose it does not invert, but is used to supply greater *drive current* at its output than is normal for a regular gate. Since regular digital ICs have limited drive current capabilities, the noninverting buffer/driver is very important when interfacing ICs with other devices such as LEDs, lamps, and others. Buffer/drivers are available in both noninverting and inverting form.

You now know the logic symbols, Boolean expression, and truth table for an inverter or NOT gate. You also recognize the symbol for a noninverting buffer/driver and know its purpose is to drive LEDs, lamps, and so forth. You will use inverters and buffers every day as you work in the field of digital electronics.

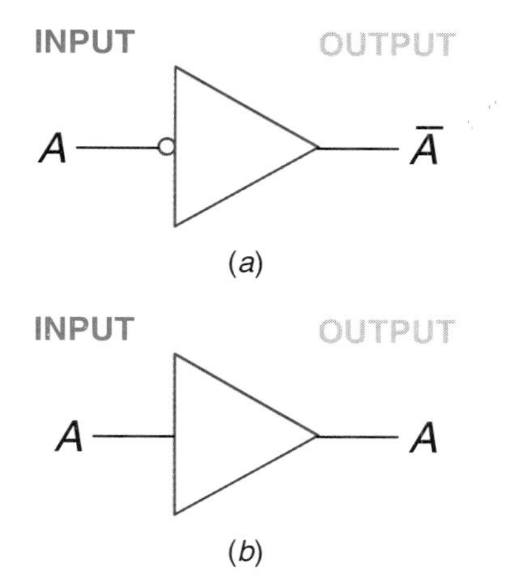

Fig. 3-12 (*a*) Alternative inverter logic symbol (note bubble at input). (*b*) Noninverting buffer/driver logic symbol.

TEST

Answer the following questions.

8. Refer to Fig. 3-9. If input A is HIGH, output Y from the inverter will be ________.
9. Refer to Fig. 3-11. If input A to the left inverter is LOW, the output from the right inverter will be ________.
10. Write the Boolean expression used to describe the action of an inverter.
11. List two words that are used to mean "inverted."
12. Refer to Fig. 3-12(*b*). If input A is LOW, the output from this buffer will be ________.

Alternative inverter logic symbol

Noninverting buffer/driver

3-4 The NAND Gate

NAND gate

NAND gate logic symbol

Invert bubble

NAND logic function

NAND Boolean expression

The AND, OR, and NOT gates are the three basic circuits that make up all digital circuits. The NAND gate is a NOT AND, or an inverted AND function. The standard logic symbol for the NAND gate is diagramed in Fig. 3-13(*a*). The little *invert bubble* (small circle) on the right end of the symbol means to invert the AND.

Figure 3-13(*b*) shows a separate AND gate and inverter being used to produce the NAND logic function. Also notice that the Boolean expressions for the AND gate ($A \cdot B$) and the NAND ($\overline{A \cdot B}$) are shown on the logic diagram in Fig. 3-13(*b*).

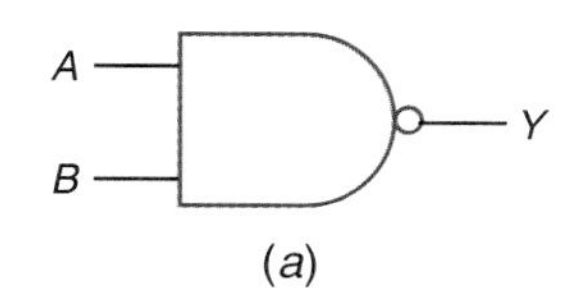

(*a*)

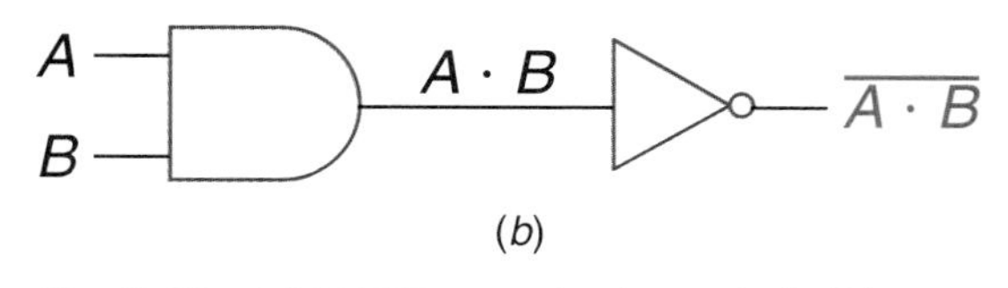

(*b*)

Fig. 3-13 (*a*) NAND gate logic symbol. (*b*) A Boolean expression for the output of a NAND gate.

INPUTS		OUTPUT	
B	A	AND	NAND
0	0	0	1
0	1	0	1
1	0	0	1
1	1	1	0

Fig. 3-14 Truth tables for AND and NAND gates.

NAND truth table

The truth table for the NAND gate is shown at the right in Fig. 3-14. Notice that the truth table for the NAND gate is developed by *inverting the outputs* of the AND gate. The AND gate outputs are also given in the table for reference.

NOR Boolean expression

NOR truth table

NAND gates are commonly employed in industrial practice and extensively used in all digital equipment. Do you know the logic symbol, Boolean expression, and truth table for the NAND gate? You must commit these to memory. The unique output from the NAND gate is a LOW only when all inputs are HIGH. The NAND output column in Fig. 3-14 shows that *only* the last line in the truth table generates a 0 while all other outputs are 1.

TEST

Answer the following questions.

13. The output side of a NAND gate logic symbol is ________ (flat with an added invert bubble, pointed with an added invert bubble, round with an added invert bubble).
14. Write the Boolean expression for a two-input NAND gate.
15. Refer to Fig. 3-13(*a*). When both inputs *A* and *B* are HIGH, output *Y* from the NAND gate will be ________. This ________ (is, is not) the unique output of the NAND gate.

3-5 The NOR Gate

NOR gate

NOR gate logic symbol

Invert bubble

The NOR gate is actually a NOT OR gate. In other words, the output of an OR gate is inverted to form a NOR gate. The logic symbol for the NOR gate is diagramed in Fig. 3-15(*a*). Note that the NOR symbol is an OR symbol with a small invert bubble (small circle) on the right side. The NOR function is being performed by an OR gate and an inverter in Fig.

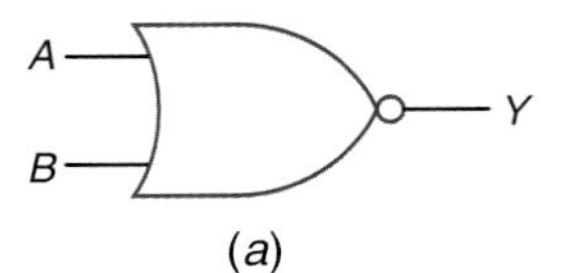

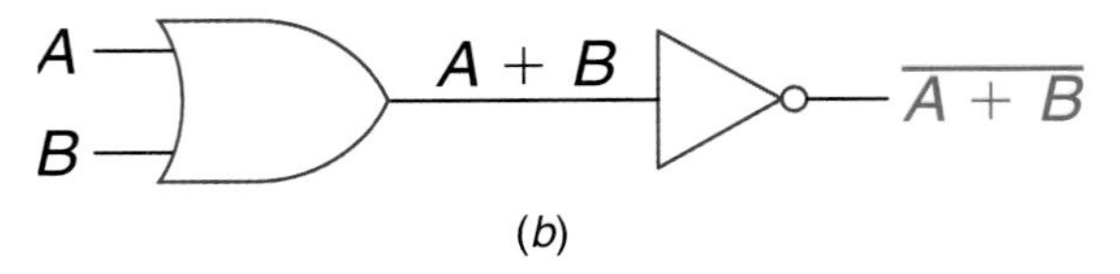

Fig. 3-15 (*a*) NOR gate logic symbol. (*b*) Boolean expression for the output of a NOR gate.

3-15(*b*). The Boolean expression for the OR function ($A + B$) is shown, The Boolean expression for the final NOR function is $\overline{A + B}$.

The truth table for the NOR gate is shown at the right in Fig. 3-16. Notice that the NOR gate truth table is just the complement of the output of the OR gate. The output of the OR gate is also included in the truth table in Fig. 3-16 for reference.

You now should memorize the symbol, Boolean expression, and truth table for the NOR gate. You will encounter these items often in your work in digital electronics. The unique output from the NOR gate is a HIGH only when all inputs are LOW. The output column in Fig. 3-16 shows that *only* the first line in the NOR truth table generates a 1 while all other outputs are 0.

TEST

Answer the following questions.

16. The output side of a NOR gate logic symbol is ________ (flat with an added invert bubble, pointed with an added invert bubble, round with an added invert bubble).

INPUTS		OUTPUT	
B	A	OR	NOR
0	0	0	1
0	1	1	0
1	0	1	0
1	1	1	0

Fig. 3-16 Truth table for OR and NOR gates.

17. Write the Boolean expression for a two-input NOR logic gate.
18. Refer to Fig. 3-15(*a*). If input *A* is LOW and input *B* is HIGH, output *Y* of the NOR gate will be ________. This ________ (is, is not) the unique output of the NOR gate.
19. Refer to Fig. 3-15(*a*). When both inputs are LOW, output *Y* of the NOR gate will be ________. This ________ (is, is not) the unique output of the NOR gate.

3-6 The Exclusive OR Gate

Exclusive OR gate

"Any but not all" gate

XOR gate

XOR function

XOR truth table

The exclusive OR gate is sometimes referred to as the "any but not all gate." The term "exclusive OR gate" is often shortened to "XOR gate." The logic symbol for the XOR gate is diagramed in Fig. 3-17(*a*); the Boolean expression for the XOR function is illustrated in Fig. 3-17(*b*). The symbol ⊕ means the terms are XORed together.

The output for the XOR gate is shown at the right in Fig. 3-18. Notice that if any but not all of the inputs are 1, then the output will be a binary, or logical, 1. The OR gate truth table is also given in Fig. 3-18, so that you may compare the OR gate truth table with the XOR gate truth table.

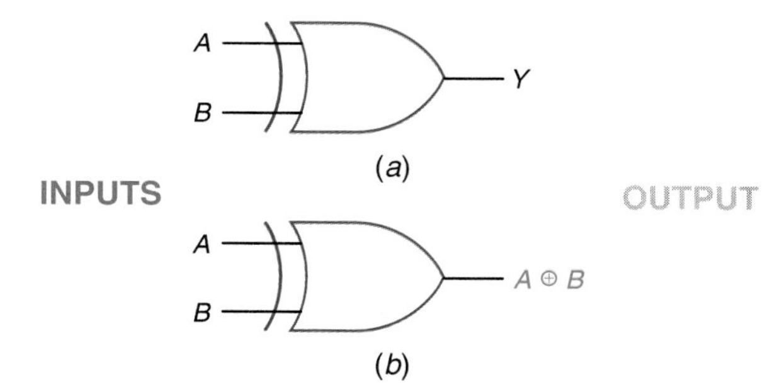

Fig. 3-17 (*a*) XOR gate logic symbol. (*b*) Boolean expression for the output of an XOR gate.

INPUTS		OUTPUT	
B	**A**	**OR**	**XOR**
0	0	0	0
0	1	1	1
1	0	1	1
1	1	1	0

Fig. 3-18 Truth table for OR and XOR gates.

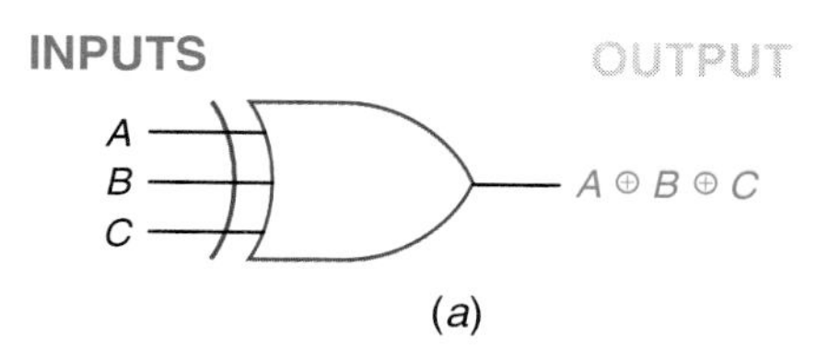

3-input **XOR**

INPUTS			OUTPUT
C	**B**	**A**	**Y**
0	0	0	0
0	0	1	1
0	1	0	1
0	1	1	0
1	0	0	1
1	0	1	0
1	1	0	0
1	1	1	1

(*b*)

Fig. 3-19 (*a*) Three-input XOR gate symbol and Boolean expression. (*b*) Truth table for three-input XOR gate.

The unique characteristic of the XOR gate is that it produces a HIGH output only when an *odd number of HIGH inputs are present.* To demonstrate this idea, Fig. 3-19 depicts a three-input XOR gate logic symbol, a Boolean expression, and a truth table. In Fig. 3-19(*b*) the three-input XOR function is described in the output column (*Y*). The HIGH outputs are generated only when an odd number of HIGH inputs are present (lines 2, 3, 5, and 8 in the truth table). If an even number of HIGH inputs to the XOR gate are present the output will be LOW (lines 1, 4, 6, and 7 in the truth table). The XOR gates are used in a variety of arithmetic circuits.

TEST

XOR gate logic symbol

XOR Boolean expression

Answer the following questions.

20. The exclusive OR gate is sometimes referred to as the ________ (five words).
21. Write the Boolean expression for a two-input XOR gate.
22. Refer to Fig. 3-17(*a*). If both inputs are HIGH, output *Y* from the XOR gate will be ________.
23. Refer to Fig. 3-19. If an odd number of inputs are HIGH, the output from an XOR gate will be ________.

3-7 The Exclusive NOR Gate

Exclusive NOR gate

XNOR gate logic symbol

XNOR function

XNOR truth table

XNOR Boolean expression

NAND gate as a universal gate

Gates with more than two inputs

XNOR gate

The term "exclusive NOR gate" is often shortened to "XNOR gate." The logic symbol for the XNOR gate is shown in Fig. 3-20(*a*). Notice that it is the XOR symbol with the added invert bubble on the output side. Figure 3-20(*b*) illustrates one of the Boolean expressions used for the XNOR function. Observe that the Boolean expression for the XNOR gate is $\overline{A \oplus B}$. The bar over the $A \oplus B$ expression tells us we have inverted the output of the XOR gate. Examine the truth table in Fig. 3-20(*c*). Notice that the output of the XNOR gate is the complement of the XOR truth table. The XOR gate output is also shown in the table in Fig. 3-20(*c*).

You now will have mastered the logic symbol, truth table, and Boolean expression for the XNOR gate.

TEST

Answer the following questions.

24. The logic symbol for an XNOR gate is formed by drawing a(n) ________ bubble at the output of the ________ logic symbol.
25. Write the Boolean expression for a two-input XNOR gate.

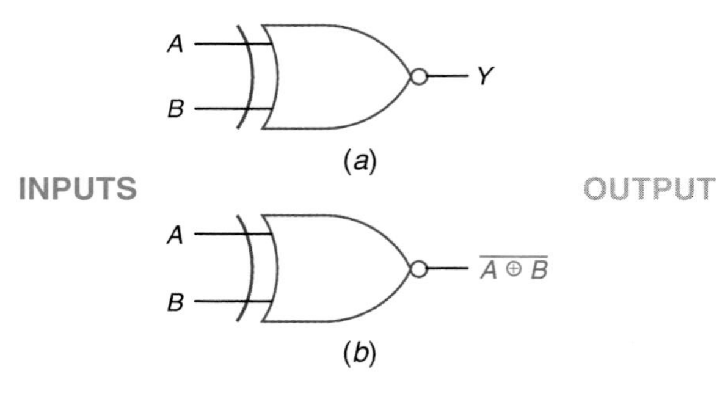

INPUTS		OUTPUT	
B	**A**	**XOR**	**XNOR**
0	0	0	1
0	1	1	0
1	0	1	0
1	1	0	1

(*c*)

Fig. 3-20 (*a*) XNOR gate logic symbol. (*b*) Boolean expression for the output of an XNOR gate. (*c*) Truth table for XOR and XNOR gates.

26. Refer to Fig. 3-20(*a*). If input *A* is LOW and *B* is HIGH, output *Y* of the XNOR gate will be ________.
27. Refer to Fig. 3-20(*a*). If an even number of inputs are HIGH, output *Y* of the XNOR gate will be ________.

3-8 The NAND Gate as a Universal Gate

So far in this chapter you have learned the basic building blocks used in all digital circuits. You also have learned about the seven types of gating circuits and now know the characteristics of the AND, OR, NAND, NOR, XOR, and XNOR gates and the inverter. You can buy ICs that perform any of these seven basic functions.

In looking through manufacturers' literature you will find that NAND gates seem to be more widely available than many other types of gates. Because of the NAND gate's wide use, we shall show how it can be used to make other types of gates. We shall be using the NAND gate as a "universal gate."

The chart in Fig. 3-21 shows how you would wire NAND gates to create any of the other basic logic functions. The logic function to be performed is listed in the left column of the table; the customary symbol for that function is listed in the center column. In the right column of Fig. 3-21 is a symbol diagram of how NAND gates would be wired to perform the logic function. The chart in Fig. 3-21 need *not* be memorized, but it may be referred to as needed in your future work in digital electronics.

TEST

Answer the following questions.

28. The NAND gate can perform the invert function if the inputs are ________ (connected together, left open).
29. How many two-input NAND gates must be used to produce the two-input OR function?

3-9 Gates with More Than Two Inputs

Figure 3-22(*a*) on page 40 shows a three-input AND gate. The Boolean expression for the three-input AND gate is $A \cdot B \cdot C = Y$, as

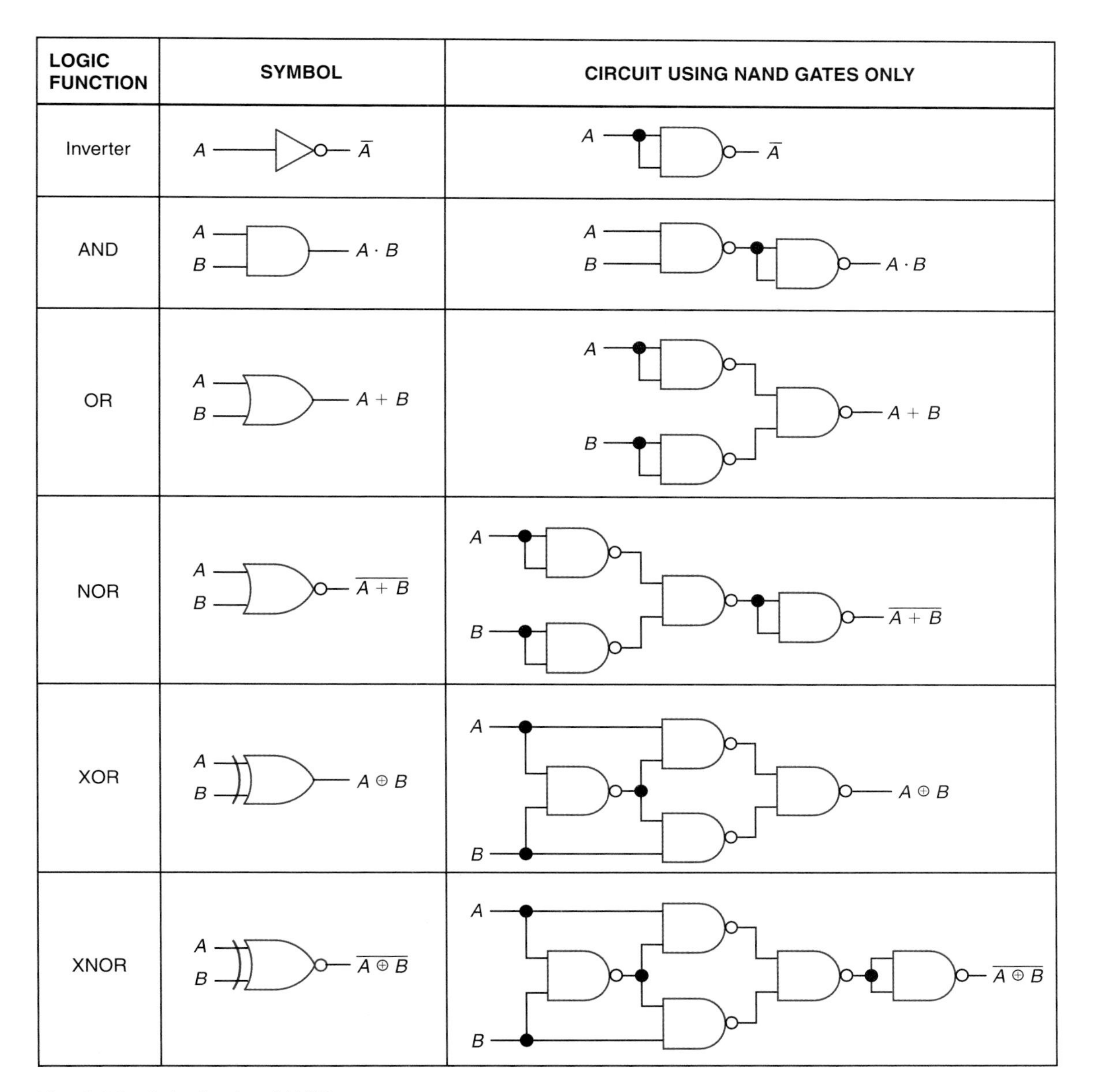

Fig. 3-21 Substituting NAND gates.

Three-input AND gate

Four-input AND gate

Four-input OR gate

illustrated in Fig. 3-22(*b*). All the possible combinations of inputs *A*, *B*, and *C* are given in the truth table in Fig. 3-22(*c*); the outputs for the three-input AND gate are tabulated in the right column of the truth table. Notice that with three inputs the possible combinations in the truth table have increased to eight (2^3).

How could you produce a three-input AND gate as illustrated in Fig. 3-22 if you have only two-input AND gates available? The solution is given in Fig. 3-23(*a*). Note the wiring of the two-input AND gates on the right side of the diagram to form a three-input AND gate. Figure 3-23(*b*) illustrates how a four-input AND gate could be wired by using available two-input AND gates.

The logic symbol for a four-input OR gate is illustrated in Fig. 3-24(*a*) on page 40. The Boolean expression for the four-input OR gate is $A + B + C + D = Y$. This Boolean expression is written in Fig. 3-24(*b*). Read the Boolean expression $A + B + C + D = Y$ as "input *A* or input *B* or input *C* or input *D* will equal output *Y*." Remember that the + symbol means the logic function OR in Boolean expressions. The truth table for the four-input OR gate is shown in

If in doubt, read the directions or manual.

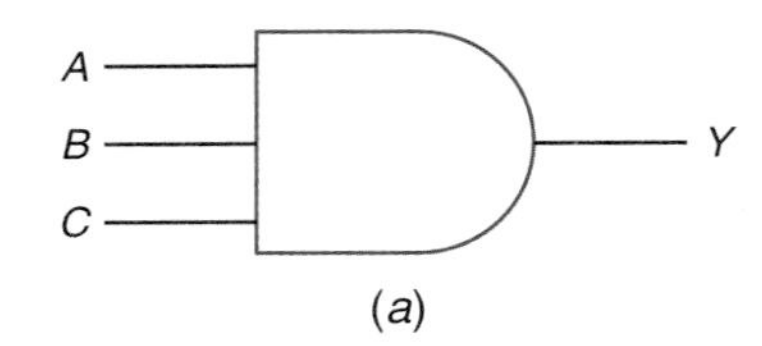

(*a*)

$$A \cdot B \cdot C = Y$$

(*b*)

INPUTS			OUTPUT
C	**B**	**A**	**Y**
0	0	0	0
0	0	1	0
0	1	0	0
0	1	1	0
1	0	0	0
1	0	1	0
1	1	0	0
1	1	1	1

(*c*)

Fig. 3-22 Three-input AND gate. (*a*) Logic symbol. (*b*) Boolean expression. (*c*) Truth table.

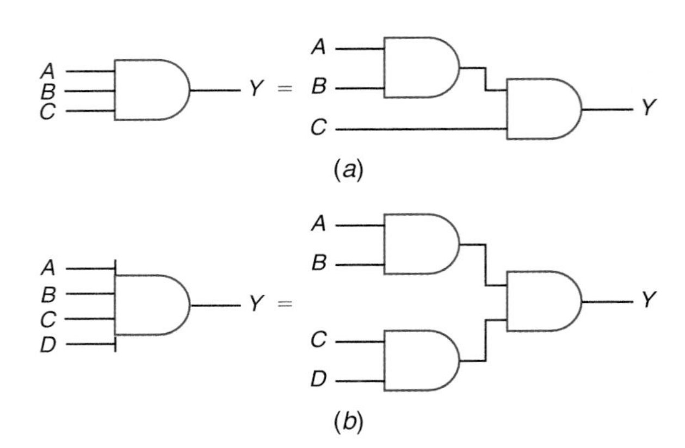

Fig. 3-23 Expanding the number of inputs. (*a*) Using two AND gates to wire three-input AND. (*b*) Using three AND gates to wire four-input AND.

Fig. 3-24(*c*). Notice that because of the four inputs there are 16 possible combinations (2^4) of *A, B, C,* and *D*. To wire the four-input OR gate, you could buy the correct gate from a manufacturer of digital logic circuits or you could use two-input OR gates to wire the four-input OR gate. Figure 3-25(*a*) diagrams how you could wire a four-input OR gate using two-input OR gates. Figure 3-25(*b*) shows how to convert two-input OR gates into a three-input OR gate. Notice that the *pattern* of connecting both OR and AND gates to expand the number of inputs is the same (compare Figs. 3-23 and 3-25).

Four-input NAND gate

Four-input OR gate

Expanding the number of inputs of a NAND gate is somewhat more difficult than expanding AND and OR gates. Figure 3-26 shows how a four-input NAND gate can be wired us-

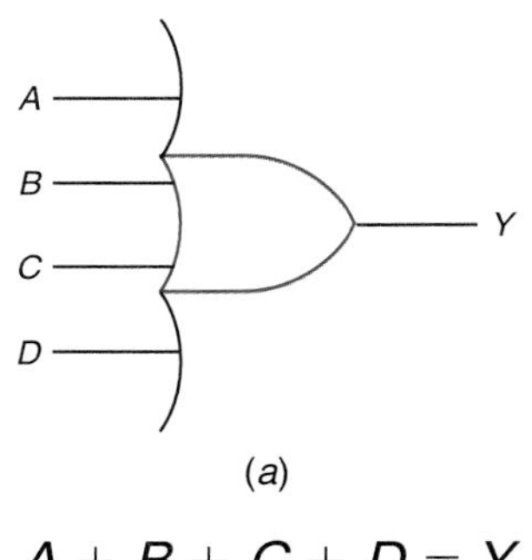

(*a*)

$$A + B + C + D = Y$$

(*b*)

INPUTS				OUTPUT
D	**C**	**B**	**A**	**Y**
0	0	0	0	0
0	0	0	1	1
0	0	1	0	1
0	0	1	1	1
0	1	0	0	1
0	1	0	1	1
0	1	1	0	1
0	1	1	1	1
1	0	0	0	1
1	0	0	1	1
1	0	1	0	1
1	0	1	1	1
1	1	0	0	1
1	1	0	1	1
1	1	1	0	1
1	1	1	1	1

(*c*)

Fig. 3-24 Four-input OR gate. (*a*) Logic symbol illustrating the method used to show extra inputs beyond the width of the symbol. (*b*) Boolean expression. (*c*) Truth table.

About Electronics

Superconductor Sleuth. A Dutch physicist, H.K. Onnes, found the first superconductor. In 1911, Onnes realized that mercury had no resistance at 4.3° above absolute zero.

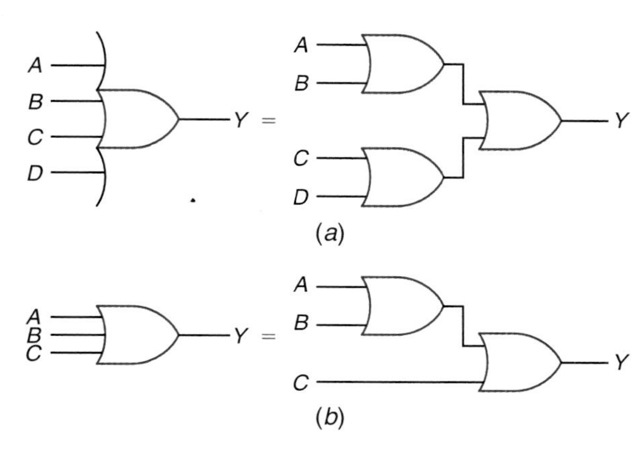

Fig. 3-25 Expanding the number of inputs.

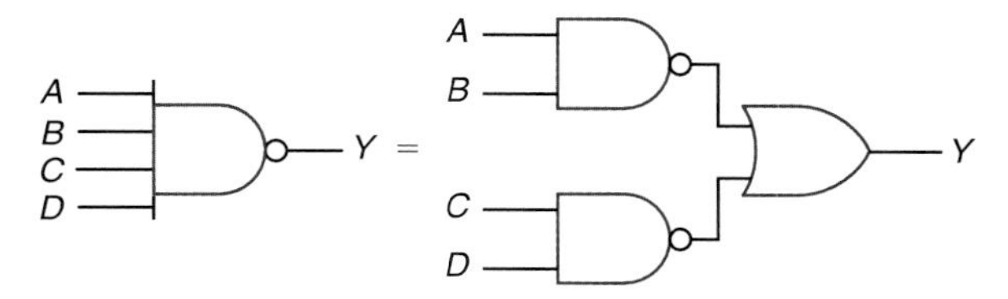

Fig. 3-26 Expanding the number of inputs.

ing 2 two-input NAND gates and 1 two-input OR gate.

You frequently will run into gates that have from two to as many as eight and more inputs. The basics covered in this section are a handy reference when you need to expand the number of inputs to a gate.

Expanding the number of inputs

TEST

Answer the following questions.

30. Write the Boolean expression for a three-input NAND gate.
31. The truth table for a three-input NAND gate would have _________ lines to include all the possible input combinations.
32. Write the Boolean expression for a four-input NOR gate.
33. The truth table for a five-input NOR gate would contain _________ lines to include all the possible input combinations.

3-10 Using Inverters to Convert Gates

Using inverters to convert gates

Frequently it is convenient to convert a basic gate such as an AND, OR, NAND, or NOR to another logic function. This can be done easily with the use of inverters. The chart in Fig. 3-27 on page 42 is a handy guide for converting any given gate to any other logic function. Look over the chart: notice that in the top section only the outputs are inverted. Inverting the outputs leads to rather predictable results, shown on the right side of the chart.

The center section of the chart shows only the gate inputs being inverted. For instance, if you invert both inputs of an OR gate, the gate becomes a NAND gate. This fact is emphasized in Fig. 3-28(*a*) on page 43. Notice that the invert bubbles have been added to the inputs of the OR gate in Fig. 3-28(*a*), which converts the OR gate to a NAND function. Also, in the center section of the chart the inputs of the AND gate are being inverted. This is redrawn in Fig. 3-28(*b*). Notice that the invert bubbles at the input of the AND gate convert it into a NOR gate. The new symbols at the left (with the input invert bubbles) in Fig. 3-28 are used in some logic diagrams in place of the more standard NAND and NOR logic symbols at the right. Be aware of these new symbols because you will run into them in your future work in digital electronics.

Invert bubbles

Alternate NAND symbol

Alternate NOR symbol

Figure 3-29 on page 43 illustrates how adding inverters (invert bubbles) to a logic symbol is described in Boolean expression form. Consider the NAND symbol at the left in Fig. 3-29(*a*) as an AND with an inverter attached to the output. The Boolean expression for the AND gate alone is $A \cdot B = Y$. Adding the inverter to the *output* of the AND gate in Fig. 3-29(*a*) is symbolized in the Boolean expression as a *long overbar* as $\overline{A \cdot B} = Y$. At the right in Fig. 3-29(*a*) is a simple truth table describing the NAND logic function.

Next consider the alternative NAND symbol in Fig. 3-29(*b*). Notice that the inverters (invert bubbles) are attached to the inputs of the OR symbol. Inverters at an input are symbolized with a short overbar as shown in Fig. 3-29(*b*). The $\overline{A} + \overline{B} = Y$ expression describes the alternative NAND logic symbol with its logic function shown in the NAND truth table at the right. The two Boolean expressions $\overline{A \cdot B} = Y$ and $\overline{A} + \overline{B} = Y$ both describe the NAND logic function. The two logic symbols at the left in Fig. 3-29 both produce the NAND truth table. Applying *DeMorgan's theorem* (part of Boolean algebra) is a systematic way of converting simple logic functions to fundamental AND or OR circuits. DeMorgan's theorem will be covered in some detail in Chap. 4.

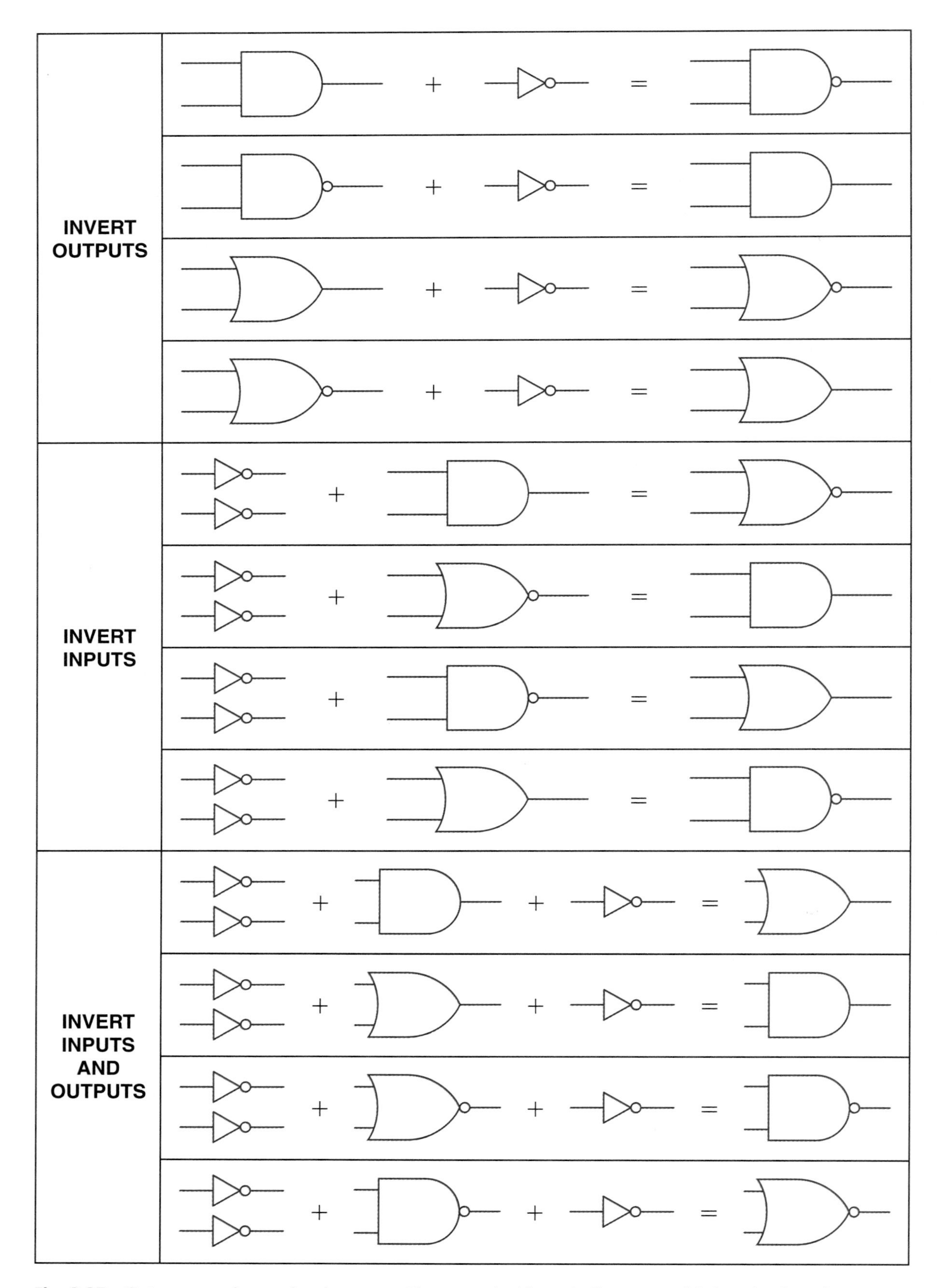

Fig. 3-27 Gate conversions using inverters. The + symbol here indicates combining the functions.

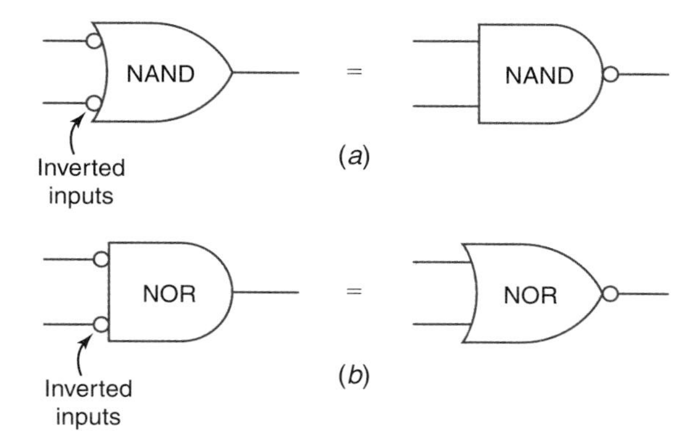

Fig. 3-28 Common alternative logic symbols. (*a*) NAND symbols. (*b*) NOR symbols. *Note:* Invert bubbles at inputs commonly mean an active-LOW input.

Did You Know?

Electronic Highways. *Intelligent Vehicle Highway Systems* (IVHS) are currently being developed. These electronic highways are able to react quickly to changing traffic conditions. In Japan, drivers can access roadside beacons to get traffic information, individual communications, and directions. Monitoring systems that allow controllers to quickly reroute traffic from backed-up areas are also being developed.

The bottom section of the chart in Fig. 3-27 shows both the inputs and the outputs being inverted. Notice that by using inverters at the inputs and outputs you can convert back and forth from AND to OR and from NAND to NOR.

With the 12 conversions shown in the chart in Fig. 3-27 you can convert any basic gate (AND, OR, NAND, and NOR) to any other gate with just the use of inverters. You will *not* need to memorize the chart in Fig. 3-27, but remember it for reference.

TEST

Supply the missing word in each sentence.

34. The OR gate can be converted to the NAND function by adding ________ to the inputs of the OR gate.
35. Adding inverters to the inputs of the AND gate produces the ________ logic function.
36. Adding an inverter to the output of an AND gate produces the ________ logic function.
37. Adding inverters to all inputs and outputs of an AND gate produces the ________ logic function.
38. Write the Boolean expression for the standard NOR logic symbol shown in Fig. 3-30(*a*) on page 44 (use long overbar).
39. Write the Boolean expression that best describes the alternative NOR logic symbol shown in Fig. 3-30(*b*) (use two short overbars).

3-11 Practical TTL Logic Gates

The popularity of digital circuits is due partly to the availability of inexpensive ICs. Manufacturers have developed many *families*

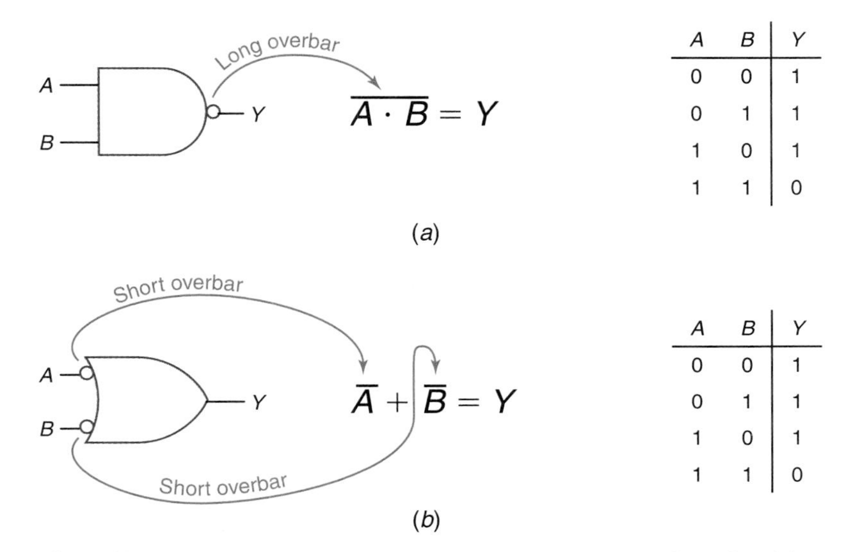

A	B	Y
0	0	1
0	1	1
1	0	1
1	1	0

A	B	Y
0	0	1
0	1	1
1	0	1
1	1	0

Fig. 3-29 NAND logic symbols, Boolean expressions, and truth tables.

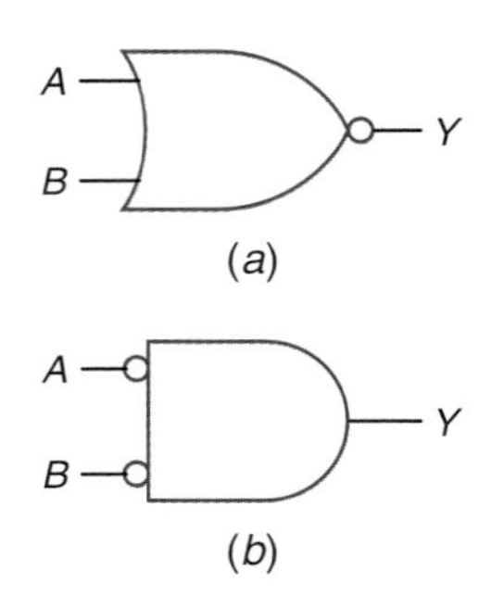

Fig. 3-30 NOR logic symbols.

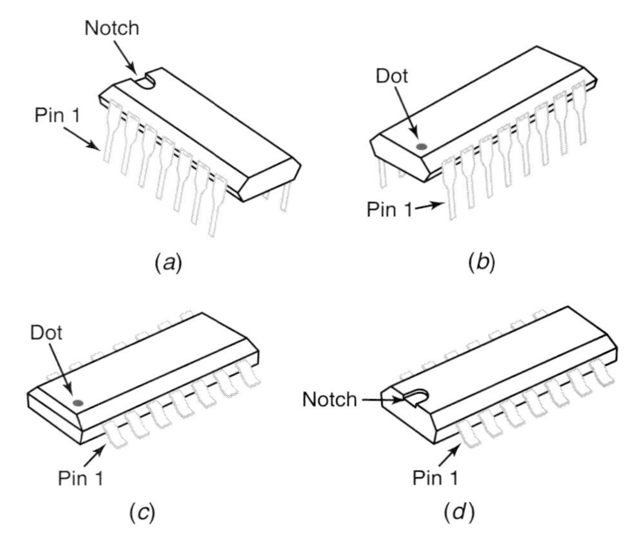

Fig. 3-31 Dual- in-line package (DIP) ICs. Regular and micro size. (*a*) Regular-size DIP- locating pin 1 using a notch. (*b*) Regular-size DIP- locating pin 1 using a dot. (*c*) Micro-size DIP surface mount IC locating pin 1 using a dot. (*d*) Micro-size DIP surface mount IC locating pin 1 using a notch.

Families of digital ICs

of digital ICs. These families are groups of devices that can be used together. The ICs in a family are said to be compatible and can be easily connected to one another.

Bipolar technology

Metal-oxide semiconductor (MOS) technology

CMOS family

One group of families is manufactured using bipolar technology. These ICs contain parts comparable to discrete bipolar transistors, diodes, and resistors. Another group of digital IC families uses metal oxide semiconductor (MOS) technology. In the laboratory, you will probably have an opportunity to use both TTL and CMOS ICs. The CMOS family is a very low power and widely used family using MOS technology. The CMOS ICs contain parts comparable to insulated-gate field-effect transistors (IGFETs).

Dual-in-line package (DIP)

A popular type of IC is illustrated in Fig. 3-31(*a*). This case style is referred to as a *dual in-line package* (DIP) by IC manufacturers. This particular IC is called a 14-pin DIP IC.

7400 series of TTL ICs

Just *counterclockwise* from the notch on the IC in Fig. 3-31(*a*) is pin 1. The pins are numbered counterclockwise from 1 to 14 when viewed from the top of the IC. A dot on the top of the DIP IC as in Fig. 3-31(*b*) is another method used to locate pin 1.

SMT

The ICs in Fig. 3-31(*a*) and (*b*) have longer pins that are commonly inserted through holes drilled in a printed circuit board and soldered to traces on the *bottom.* The two ICs in Fig. 3-31(*c*) and (*d*) are much smaller and have shorter pins bent to be soldered to the traces on the *top* of the PC board. The smaller micropackages in Fig. 3-31(*c*) and (*d*) are commonly called *SMT (surface-mount technology)* packages. The SMT packages are typically much smaller in order to save PC board space and are easier to align when being positioned and soldered using automated manufacturing equipment. Two methods of locating pin 1 on the small SMT packages are illustrated in Fig. 3-31(*c*) and (*d*). In your school lab you will probably use the larger DIP IC, which is shown in Fig. 3-31(*a*) with long pins because they can be inserted into a solderless breadboard.

Pin diagram

Manufacturers of ICs provide pin diagrams similar to the one shown in Fig. 3-32. This IC contains 4 two-input AND gates. Thus, it is called a *quadruple two-input AND gate.* This 7408 unit is one of many of the 7400 series of TTL ICs available. The power connections to

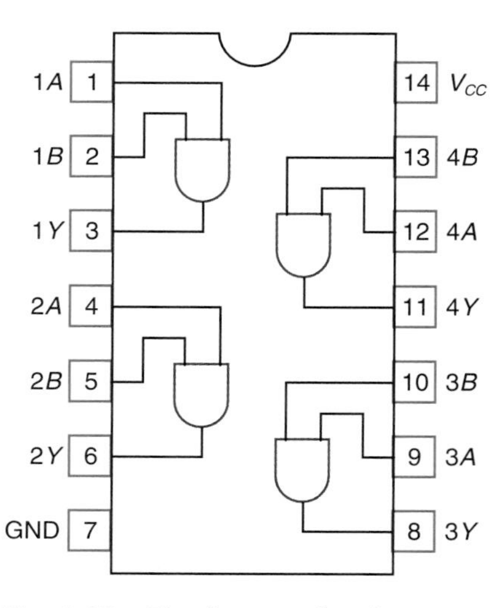

Fig. 3-32 Pin diagram for the 7408 digital IC.

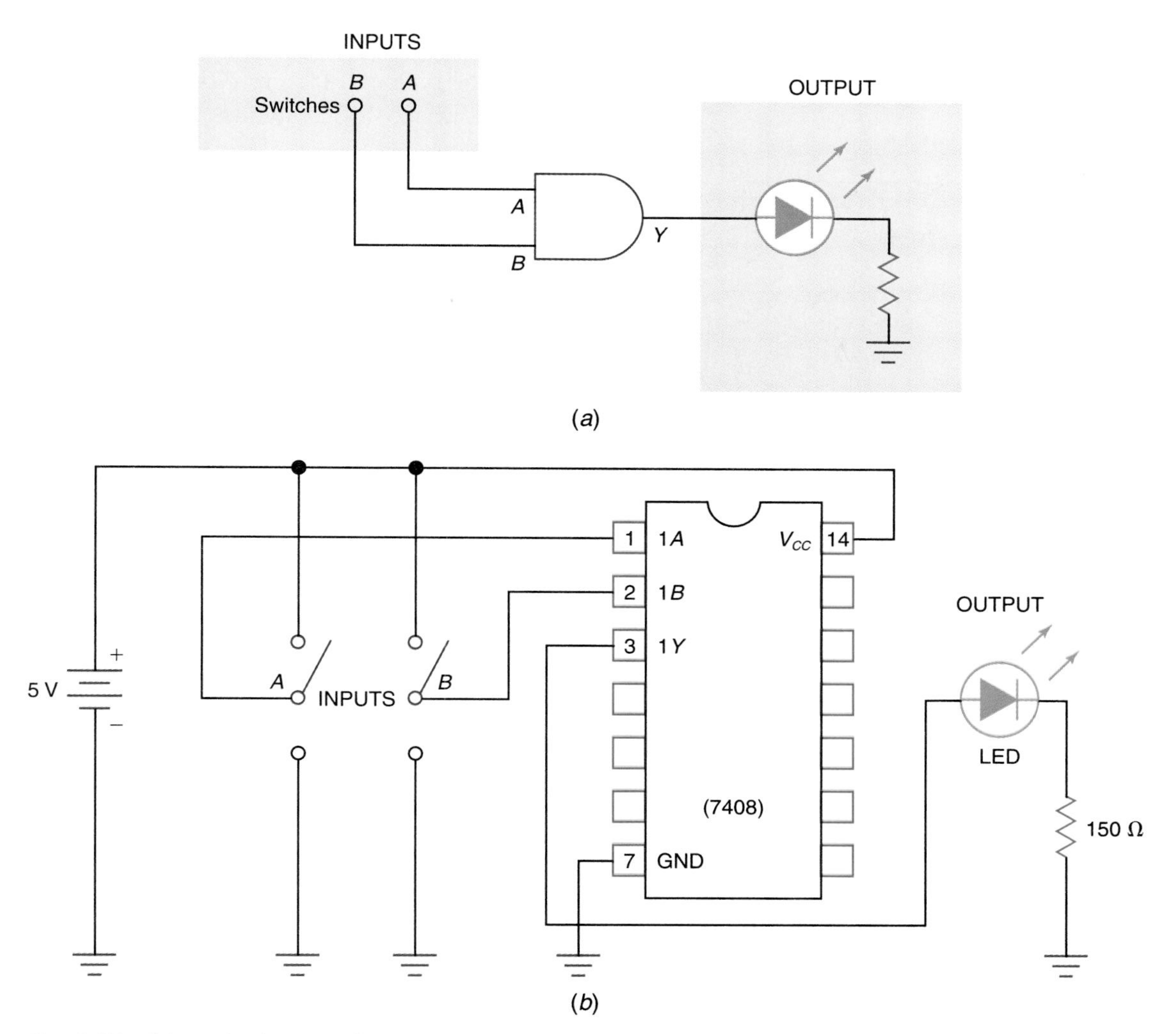

Fig. 3-33 (*a*) Logic diagram for a two-input AND gate circuit. (*b*) Wiring diagram to implement the two-input AND function.

the IC are the GND (pin 7) and V_{CC} (pin 14) pins. All other pins are the inputs and outputs to the four TTL AND gates.

Given the logic diagram in Fig. 3-33(*a*), wire this circuit using the 7408 TTL IC. A wiring diagram for the circuit is shown in Fig. 3-33(*b*). A 5-V dc regulated power supply is typically used with all TTL devices. The positive (V_{CC}) and negative (GND) power connections are made to pins 14 and 7 of the IC. Input switches (*A* and *B*) are wired to pins 1 and 2 of the 7408 IC. Notice that if a switch is in the *up position,* a logical 1 (+5 V) is applied to the input of the AND gate. If a switch is in the down position, however, a logical 0 is applied to the input. At the right in Fig. 3-33(*b*), an LED and 150-Ω limiting resistor are connected to GND. If the output at pin 3 is HIGH (near +5 V), current will flow through the LED. When the LED is lit, it indicates a HIGH output from the AND gate.

The top of a typical TTL digital IC is shown in Fig. 3-34(*a*) on page 46. The block form of the letters "NS" on the top of the IC shows the manufacturer as National Semi-conductor. The DM7408N part number can be divided into sections as shown in Fig. 3-34(*b*). The prefix "DM" is the manufacturer's code (National Semiconductor uses the letters "DM" as a prefix). The core part number is 7408, which is a quadruple two-input AND gate TTL IC. This core part number is the same from manufacturer to manufacturer. The trailing letter "N" (the suffix) is a code used by several manufacturers to designate the DIP.

The top of another digital IC is shown in Fig. 3-35(*a*) on page 46. The letters "SN" on this IC stand for the manufacturer, Texas

TTL digital IC

For related information, visit the website for Fairchild Semiconductor.

Low-power Schottky

Core part number

Logic families or subfamilies

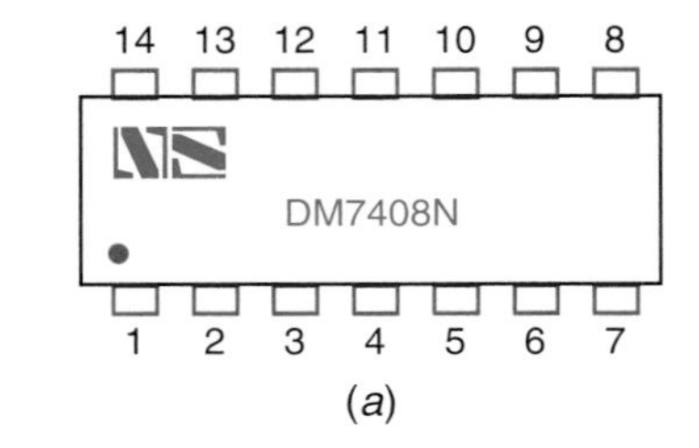

(a)

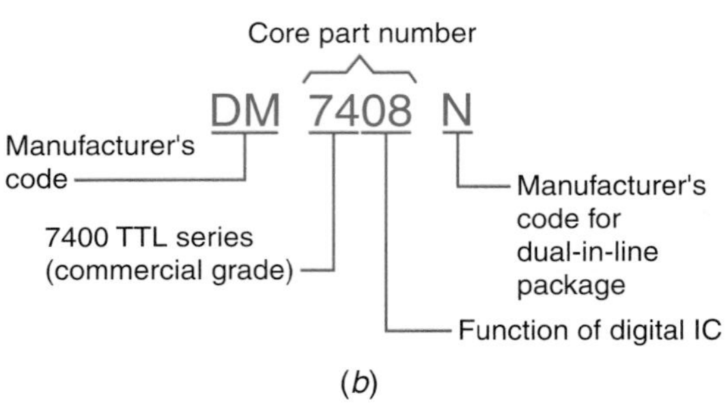

(b)

Marking on a typical digital IC

Decoding the part number on a typical IC

Fig. 3-34 (a) Marking on a typical digital IC. *(Courtesy of National Semiconductor Corporation.)* (b) Decoding the part number on a typical IC.

Commercial grade

Instruments. On this unit, the suffix "J" stands for a ceramic DIP packaging. This is typically referred to as the *commercial grade*. The core part number of the IC in Fig. 3-35 is 74LS08. This is similar to the 7408 quadruple two-input AND gate IC discussed earlier. The letters "LS" in the center of the core number designate the type of TTL circuitry used in the IC. In this case "LS" stands for *low-power Schottky*.

The internal letter(s) in a core part number of a 7400 series IC tell something about the *logic* family or *subfamily*. Typical internal letters used are:

AC = FACT Fairchild Advanced CMOS Technology logic (a newer advanced family of CMOS)
ACT = FACT Fairchild Advanced CMOS Technology logic (a newer family of CMOS with TTL logic levels)
ALS = advanced low-power Schottky TTL logic (a subfamily of TTL)
AS = advanced Schottky TTL logic (a subfamily of TTL)
C = CMOS logic (an early family of CMOS)
F = FAST Fairchild Advanced Schottky TTL logic (a new subfamily of TTL)
FCT = FACT Fairchild Advanced CMOS Technology logic (a family of CMOS with TTL logic levels)
H = high-speed TTL logic (a subfamily of TTL)
HC = high-speed CMOS logic (a family of CMOS)
HCT = high-speed CMOS logic (a family of CMOS with TTL inputs)
L = low-power TTL logic (a subfamily of TTL)
LS = low-power Schottky TTL logic (a subfamily of TTL)
S = Schottky TTL logic (a subfamily of TTL)

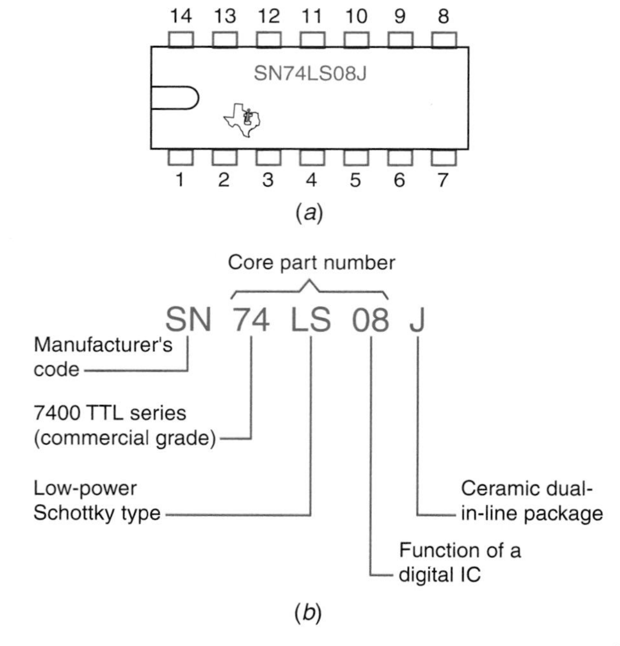

(a)

(b)

CMOS ICs

Fig. 3-35 (a) Markings on a Texas Instruments digital IC. (b) Decoding the part number of a typical low-power Schottky IC.

The internal letters give information about the speed, power consumption, and process technology of digital ICs. Because of these speed and power consumption differences, manufacturers usually recommend that exact part numbers be used when replacing digital ICs. When the letter "C" is used inside a 7400 series part number, it designates a CMOS and not a TTL digital IC. The internal letters "HC," "HCT," "AC," "ACT," and "FCT" also designate CMOS ICs.

Data manuals from manufacturers contain much valuable information on digital ICs. They contain pin diagrams and packaging information. Data manuals also contain details on part numbering and other valuable data for the technician, student, or engineer.

TEST

Answer the following questions.

40. List two popular digital IC families.
41. Refer to Fig. 3-31(*a*). This IC uses the popular case style called the ________ package.
42. A ________ -V dc power supply is used with TTL ICs. The V_{CC} pin is connected to the ________ ($^-$, $^+$) of the supply.
43. Refer to Fig. 3-33(*b*). How is the 7408 IC described by the manufacturer?
44. What can you tell about a digital IC that has the marking "DM74LS08N" printed on the top?

3-12 Practical CMOS Logic Gates

Practical CMOS logic gates

The older 7400 series of TTL logic devices has been extremely popular for many decades. One of its disadvantages is its higher power consumption. In the late 1960s, manufacturers developed CMOS digital ICs which consume little power and were perfect for battery-operated electronic devices. CMOS stands for *complementary metal oxide semiconductor.*

Complementary metal oxide semiconductor

Several families of compatible CMOS ICs have been developed. The first was the 4000 series. Next came the 74C00 series and more recently the 74HC00 series of CMOS digital ICs. In 1985, the FACT (Fairchild Advanced CMOS Technology) 74AC00 series, 74ACT00 series, and 74FCT00 series of extremely fast, low-power CMOS digital ICs were introduced by Fairchild. Fairchild Camera and Instrument Corporation has since been purchased by National Semiconductor Corporation. FACT is currently a trademark of National Semiconductor Corporation. Many large-scale integrated (LSI) circuits such as digital wristwatch and calculator chips are also manufactured using the CMOS technology.

4000 series

74C00 series

74HC00 series

FACT series

A typical 4000 series CMOS IC is pictured in Fig. 3-36(*a*). Note that pin 1 is marked as such on the top of the IC immediately counterclockwise from the notch. The CD4081BE part number can be divided into sections as shown in Fig. 3-36(*b*). The prefix CD is the manufacturer's code for CMOS digital ICs. The core part number is 4081B, which stands for a CMOS quadruple two-input AND gate IC. This core part number is almost always the same from manufacturer to manufacturer. The trailing letter "E" is the manufacturer's packaging code for a plastic DIP IC. The letter "B" is a "buffered version" of the original 4000A series. The buffering provides the 4000B series devices with greater output drive and some protection from static electricity.

Figure 3-36(*c*) is a pin diagram for the CD4081BE CMOS quad two-input AND gate

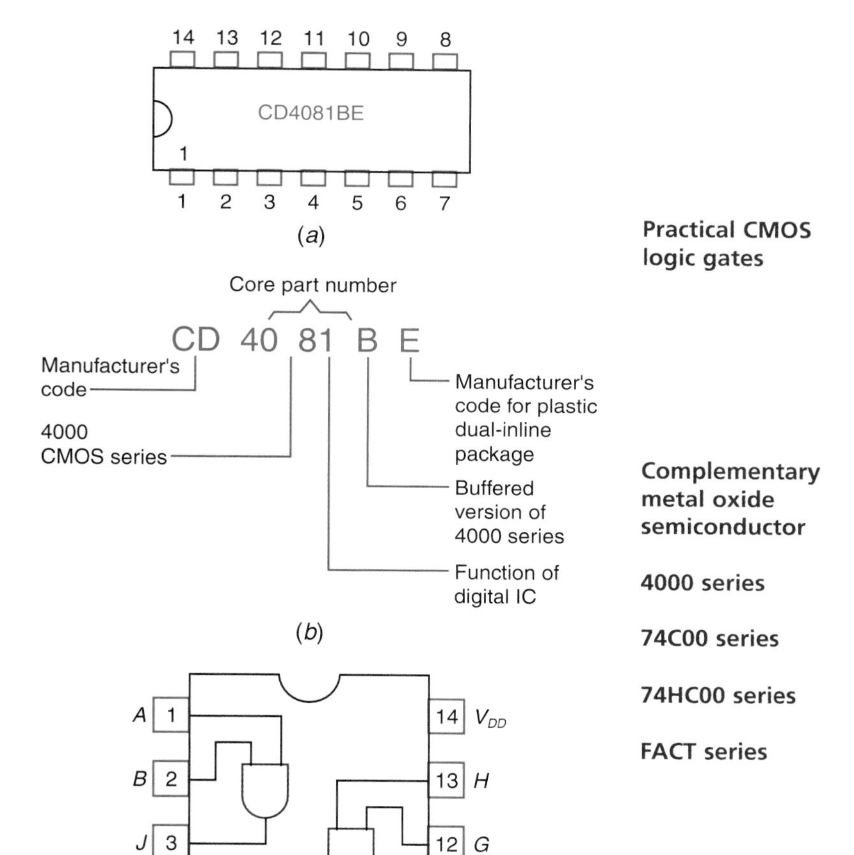

Fig. 3-36 (*a*) Markings on CMOS digital IC. (*b*) Decoding the part number on a typical 4000B series CMOS IC. (*c*) Pin diagram for the 4081B CMOS IC.

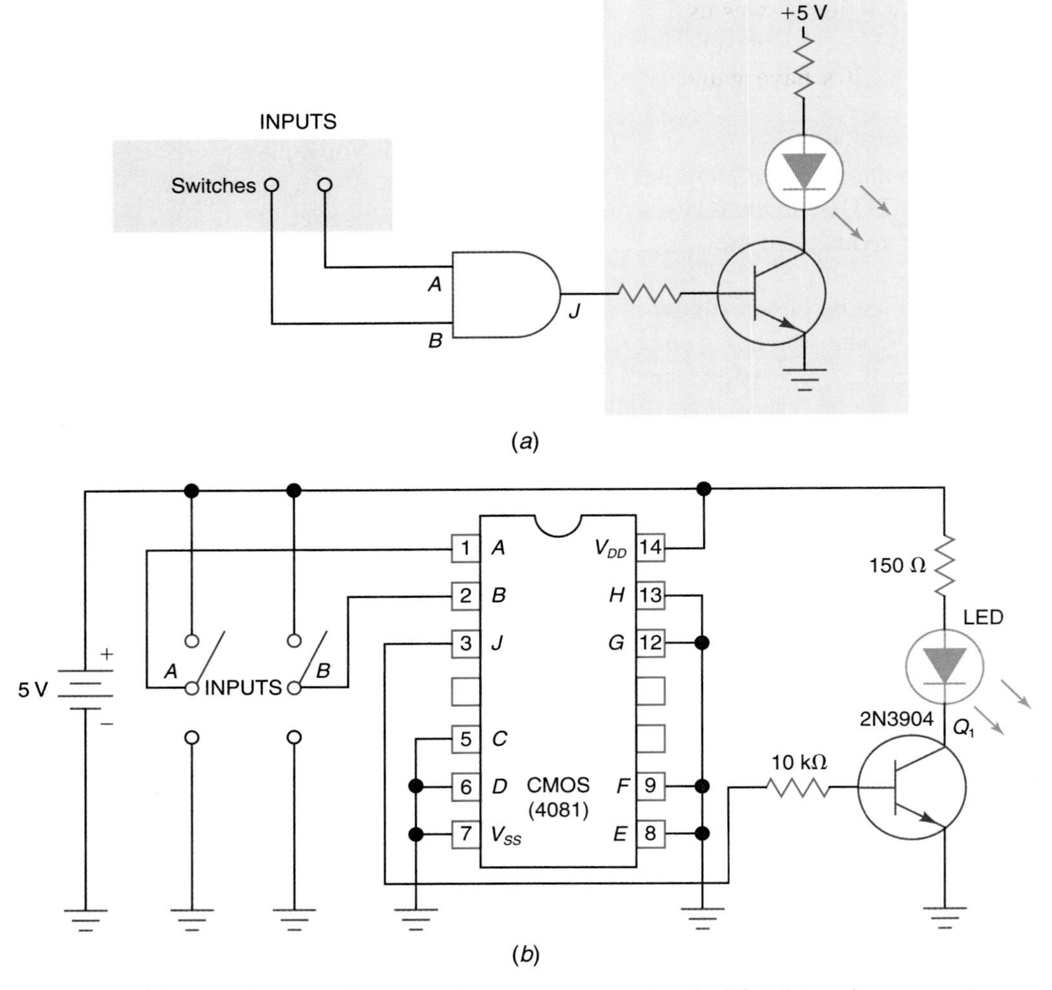

Fig. 3-37 (*a*) Logic diagram for a two-input AND gate circuit. (*b*) Wiring diagram using the 4081 CMOS IC to implement the two-input AND function.

Unused inputs

IC. The power connections are V_{DD} (positive voltage) and V_{SS} (GND or negative voltage). The labeling of the power connections on TTL and 4000 series CMOS ICs is different. This difference can be observed by comparing Figs. 3-32 and 3-36(*c*).

Given the schematic diagram in Fig. 3-37(*a*), wire this circuit using the 4081B CMOS IC. A wiring diagram for the circuit is shown in Fig. 3-37(*b*). A 5-V dc power supply is shown but the 4000 series CMOS ICs can use voltages from 3 to 18 V dc. Care is taken in removing the 4081 from its conductive foam storage because CMOS ICs can be damaged by static charges. Do not touch the pins when inserting the 4081 CMOS IC in a socket or mounting board. V_{DD} and V_{SS} power connections are made with the power off. *When using CMOS, all unused inputs are tied to GND or* V_{DD}. In this example, unused inputs (*C, D, E, F, H, G*) are grounded. The output of the AND gate (pin 3) is connected to the driver transistor. The transistor turns the LED on when pin 3 is HIGH or off when the output is LOW. Finally, inputs *A* and *B* are connected to input switches.

When the input switches in Fig. 3-37(*b*) are in the up position, they generate a HIGH input. A LOW input is generated when the switches are in the down position. Two LOW inputs to the AND gate produce a LOW output at pin 3 of the IC. The LOW output turns off the transistor and the LED does not light. Two HIGH inputs to the AND gate produce a HIGH output at pin 3. The HIGH (about +5 V) output at the base of Q1 turns on the transistor and the LED lights. The 4081 CMOS IC will generate a two-input AND truth table like the one in Fig. 3-4 on page 32.

Several families of CMOS digital ICs are available. A 4000 series IC was used as an example in this section. The newer 74HC00 series CMOS digital ICs have gained favor because they are somewhat more compatible with the popular TTL logic. The 74HC00 series of CMOS ICs also has more drive capabilities than the older 4000 and 74C00 series units and operates at high frequencies. The "HC" in the 74HC00 series part number stands for high-speed CMOS.

The FACT Fairchild Advanced CMOS Technology logic series is the latest CMOS family of ICs. It includes the 74AC00-, 74ACT00-, 74ACTQ00-, 74FCT00-, and 74FCTA00-series subfamilies of CMOS digital ICs. The FACT logic family has outstanding operating characteristics exceeding all CMOS and most TTL subfamilies. For direct replacement of 74LS00 and 74ALS00 TTL series ICs, the 74ACT00-, 74ACTQ00-, 74FCT00-, and 74FCTA00-series circuits with TTL-type input voltage characteristics are included in the FACT CMOS family. FACT logic devices are ideal for portable systems such as laptop computers because of their extremely low power consumption and excellent high-speed characteristics.

Caution must be used so that static charges do not damage CMOS ICs. All unused inputs to CMOS gates must be grounded or connected to V_{DD}. Most important, *input voltages must not exceed the GND (V_{SS}) to V_{DD} voltage.*

Cautions when using CMOS ICs

TEST

Answer the following questions.

45. The primary advantage of CMOS digital ICs is their _________ (high, low) power consumption.
46. While TTL must use a 5-V power supply, 4000 series CMOS ICs can operate on dc voltages from _________ to _________ V.
47. Refer to Fig. 3-36. How is this 4081B IC described by the manufacturer?
48. What rule (dealing with unused inputs) must be followed when wiring CMOS ICs?

3-13 Troubleshooting Simple Gate Circuits

The most basic piece of test equipment used in digital troubleshooting is the *logic probe.* A simple logic probe is pictured in Fig. 3-38 on page 50. The slide switch on the unit is used to select the type of logic family under test, either TTL or CMOS. The logic probe in Fig. 3-38 is set to test a TTL type of digital circuit in this example. Typically, two leads provide power

Logic probe

Did You Know?

Communications in Orbit. Satellites in *geostationary earth orbit* (GEO) enable fax, videoconferencing, Internet, long-distance fixed phone service, television, and broadband multimedia to be provided in developing areas of the world. Satellites in *medium-earth orbit* (MEO) are used for mobile cell phones, fixed phones, and other personal communications. Satellites in *low-earth orbit* (LEO) are used for handheld mobile phones, paging, fax, ship or truck tracking, fixed ordinary phones, broadband multimedia, and monitoring of remote industrial spots. In the aftermath of the earthquake pictured, search and rescue teams kept in touch with one another using satellite technology and were also able to maintain international communications.

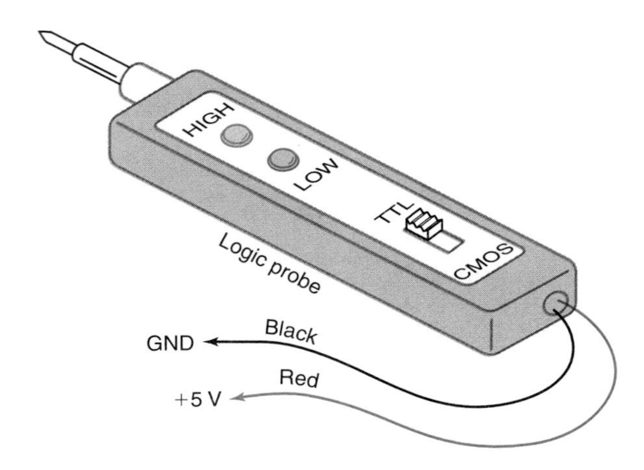

Fig. 3-38 Logic probe.

to the logic probe. The red lead is connected to the positive (+) of the power supply while the black lead of the logic probe is connected to the negative (−) or GND of the power supply. After powering the logic probe, the needlelike probe is touched to a test point or node in the circuit. The logic probe will light either the HIGH or the LOW indicator. If neither indicator lights, it usually means that the voltage is somewhere between the HIGH and LOW.

Printed circuit (PC) board

In practical electronic equipment most digital ICs are mounted on a printed circuit (PC) board. An example is shown in Fig. 3-39(*a*). Also available to the student or technician is a circuit wiring diagram or schematic similar to that in Fig. 3-39(*b*). Many times the +5-V (V_{CC}) and GND connections to the ICs are not shown on the wiring diagram. However, they are always understood to be present. Pin numbers are usually given in a wiring diagram. The IC type may not be given on the schematic but is usually listed on a parts list in the equipment manual.

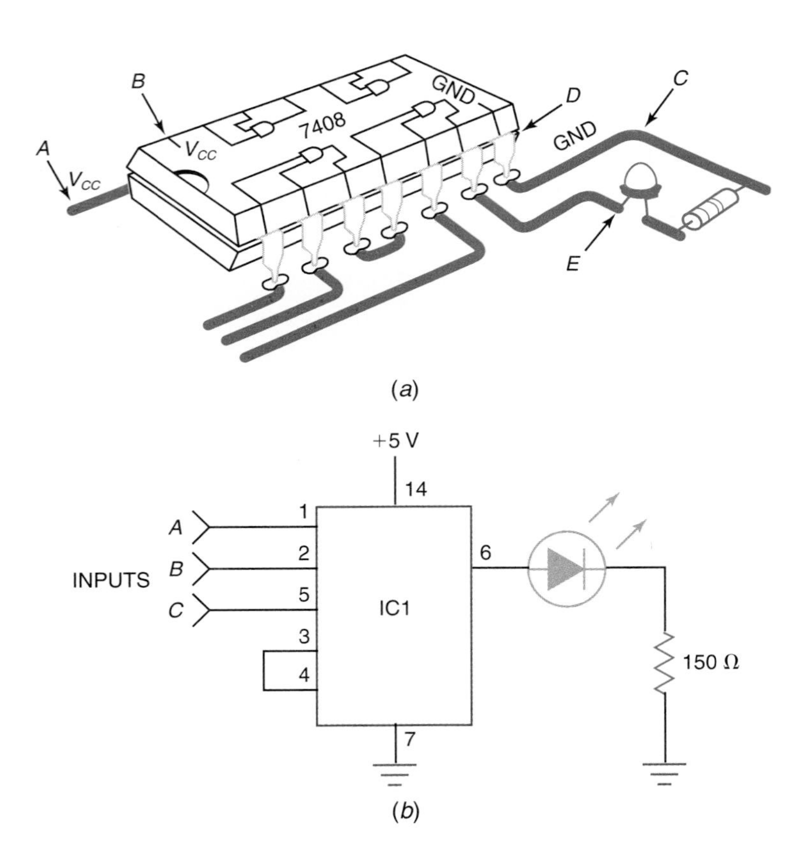

Fig. 3-39 (*a*) Digital IC mounted on a printed circuit (PC) board. (*b*) Wiring or schematic diagram of a digital gating circuit.

The first step in troubleshooting is to use your senses. *Feel* the flat top of the ICs to determine if they are hot. Some ICs operate cool, others run slightly warm. CMOS ICs should always be cool. *Look* for broken connections, solder bridges, broken PC board traces, and bent IC pins. *Smell* for possible overheating. *Look* for signs of excessive heat, such as discoloration or charring.

The next step in troubleshooting might involve checking whether each IC has power. With the logic probe connected to power, check at the points labeled *A*, *B* (the V_{CC} pin), *C*, and *D* in Fig. 3-39(*a*). Nodes *A* and *B* should give a bright HIGH light on the logic probe. Nodes *C* and *D* should give a bright LOW light on the logic probe.

The next step might be to trace the path of logic through the circuit. The circuit is equal to a three-input AND gate in this example (Fig. 3-39). Its unique state is a HIGH when all inputs are HIGH. Check pins 1, 2, and 5 of the IC in Fig. 3-39(*a*) with the logic probe. Manipulate the equipment to get all inputs HIGH. When all inputs are HIGH, the output (pin 6 of the IC) should be HIGH and the circuit LED should light. If the unique state works, try several other input combinations and verify their proper operation.

Refer to Fig. 3-39(*a*). Assume a HIGH reading at node *A* and a LOW reading on the logic probe at node *B* (pin 14 of the IC). This probably means an open circuit in the PC board trace or a faulty solder joint between points *A* and *B*. If DIP IC sockets are used, the thin part of the IC pin can be bent. This common difficulty causes an open between the IC pin and the socket and PC board trace.

Refer to Fig. 3-39(*a*). Assume LOW readings at pins 1, 2, and 3 with no reading (neither LED on the logic probe lit) at pin 4. No reading on most logic probes means a voltage between LOW and HIGH (perhaps 1 to 2 V in TTL). This input (pin 4) is floating (not connected) and is considered to be a HIGH by the TTL circuitry inside the 7408 IC. The output of the first AND gate (pin 3) is supposed to pull the input to the second AND gate (pin 4) LOW. If it does not, the fault could be in the PC board trace, solder connections, or a bent IC pin. Internal opens and shorts also occur in digital ICs.

Floating inputs

Troubleshooting a comparable CMOS circuit proceeds in the same way with a few exceptions. The logic probe must be set to CMOS instead of TTL. Floating inputs on CMOS ICs can harm the IC. A LOW in CMOS is defined as approximately 0 to 20 percent of the power supply voltage. A HIGH in CMOS is defined as approximately 80 to 100 percent of supply voltage.

Steps in troubleshooting

In summary, troubleshooting *first* involves using your senses. *Second,* check with a logic probe to see if each IC has power. *Third,* determine the exact job of the gating circuit and test for the unique output conditions. *Finally,* check other input and output conditions. Open and short-circuit conditions can occur inside ICs as well as in the wiring. Digital ICs should be replaced with exact subfamily replacements when possible.

TEST

Answer the following questions.

49. Refer to Fig. 3-38. With what two logic families can this logic probe be used?
50. What is the first step in troubleshooting gating circuits using TTL ICs?
51. What is the second step in troubleshooting?
52. Floating inputs to CMOS ICs are ________ (allowed, not allowed).

3-14 IEEE Logic Symbols

IEEE symbols

The logic gate symbols you have memorized are the traditional ones recognized by all workers in the electronics industry. These symbols are very useful in that they have distinctive shapes. Manufacturers' data manuals include traditional logic symbols and are recently including the newer *IEEE functional logic symbols.* These newer symbols are in accordance with ANSI/IEEE Standard 91-1984 and IEC Publication 617-12. These newer IEEE symbols are commonly referred to as *"dependency notation."* For simple gating circuits, the traditional logic symbols are probably preferred, but the IEEE standard symbols have advantages as ICs become more complicated. Most military contracts call for the use of IEEE standard symbols.

IEEE logic gate symbols

LOGIC FUNCTION	TRADITIONAL LOGIC SYMBOL	IEEE LOGIC SYMBOL*
AND	A, B — Y	A, B — & — Y
OR	A, B — Y	A, B — ≥ 1 — Y
NOT	A — Y	A — 1 — Y
NAND	A, B — Y	A, B — & — Y
NOR	A, B — Y	A, B — ≥ 1 — Y
XOR	A, B — Y	A, B — = — Y
XNOR	A, B — Y	A, B — = — Y

*ANSI/IEEE Standard 91-1984 and IEC Publication 617-12.

Fig. 3-40 Comparing traditional and IEEE logic gate symbols.

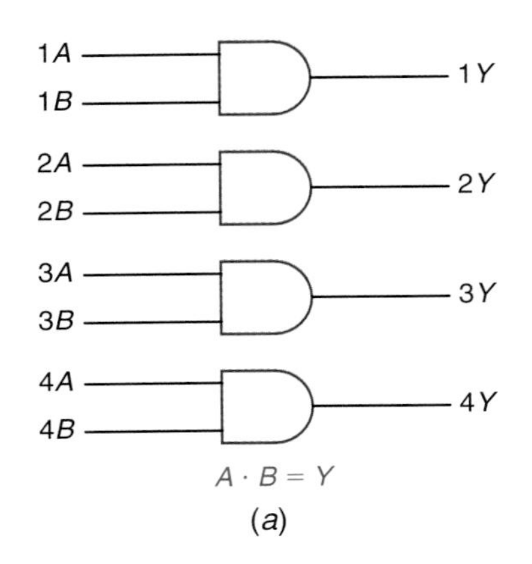

(a)

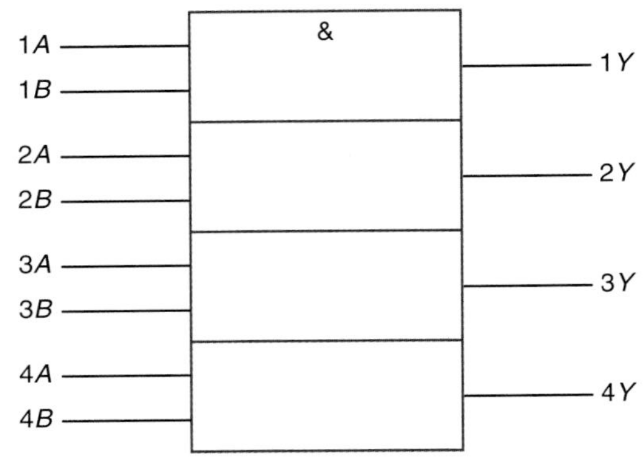

This symbol is in accordance with ANSI/IEEE Standard 91-1984 and IEC Publication 617-12.

(b)

Fig. 3-41 Logic symbol for 7408 quadruple two-input AND gate. (*a*) Traditional symbol (most common). (*b*) IEEE functional logic symbol (newer method).

Figure 3-40 on page 51 shows the traditional logic symbols and their IEEE counterparts. All IEEE logic symbols are rectangular. There is an identifying character or symbol inside the rectangle. For instance, notice in Fig. 3-40 that the ampersand (&) character is printed inside the IEEE standard AND gate symbol. Characters *outside* the rectangle are not part of the standard symbol and may vary from manufacturer to manufacturer. The invert bubble on traditional logic symbols (NOT, NAND, NOR, and XOR) is replaced with a right triangle on corresponding IEEE standard symbols. The IEEE right triangle can also be used on inputs to signify an active-low input. You have memorized the traditional logic gate symbols. You will not have to memorize the IEEE logic gate symbols but should be aware they exist.

Recent manufacturers' data manuals will probably give both the traditional and IEEE functional logic symbols for a particular IC. For instance, logic symbols for the 7408 quadruple two-input AND gate are illustrated in Fig. 3-41. The traditional logic diagram for the 7408 IC is shown in Fig. 3-41(*a*). The IEEE logic diagram for the 7408 IC is reproduced in Fig. 3-41(*b*). Note in the IEEE symbol for the 7408 IC that only the top AND gate contains the & symbol, but it is understood that the lower three rectangles also represent two-input AND gates.

TEST

Answer the following questions.

53. Draw the IEEE standard logic symbol for a three-input AND gate.
54. Draw the IEEE standard logic symbol for a three-input OR gate.
55. Draw the IEEE standard logic symbol for a three-input NAND gate.
56. The right triangle on IEEE symbols replaces the invert ________ on traditional logic symbols.
57. For simple gating circuits, the ________ (IEEE standard, traditional) logic symbols are probably preferred because of their distinctive shapes.

Chapter 3 Review

Summary

1. Binary logic gates are the basic building blocks for all digital circuits.
2. Figure 3-42 shows a summary of the seven basic logic gates. This information should be memorized.
3. NAND gates are widely employed and can be used to make other logic gates.
4. Logic gates are often needed with 2 to 10 inputs. Several gates may be connected in the proper manner to get more inputs.
5. AND, OR, NAND, and NOR gates can be converted back and forth by using inverters.
6. Logic gates are commonly packaged in DIP ICs. Both TTL and CMOS digital ICs are very widely used in small systems.
7. Very low power consumption is an advantage of CMOS digital ICs.

LOGIC FUNCTION	LOGIC SYMBOL	BOOLEAN EXPRESSION	TRUTH TABLE		
			INPUTS		OUTPUT
			B	*A*	*Y*
AND	*A*, *B* — *Y*	$A \cdot B = Y$	0	0	0
			0	1	0
			1	0	0
			1	1	1
OR	*A*, *B* — *Y*	$A + B = Y$	0	0	0
			0	1	1
			1	0	1
			1	1	1
Inverter	*A* — $\overline{A}$	$A = \overline{A}$		0	1
				1	0
NAND	*A*, *B* — *Y*	$\overline{A \cdot B} = Y$	0	0	1
			0	1	1
			1	0	1
			1	1	0
NOR	*A*, *B* — *Y*	$\overline{A + B} = Y$	0	0	1
			0	1	0
			1	0	0
			1	1	0
XOR	*A*, *B* — *Y*	$A \oplus B = Y$	0	0	0
			0	1	1
			1	0	1
			1	1	0
XNOR	*A*, *B* — *Y*	$\overline{A \oplus B} = Y$	0	0	1
			0	1	0
			1	0	0
			1	1	1

Fig. 3-42 Summary of basic logic gates.

8. When using CMOS ICs, all unused inputs must go to V_{DD} or GND. Care must be exercised in storing and handling CMOS ICs to avoid static electricity. Input voltages to a CMOS IC must never exceed the power supply voltages.
9. The logic probe, knowledge of the circuit, and your senses of sight, smell, and touch are basic tools used in troubleshooting gating circuits.
10. Figure 3-40 compares the traditional logic gate symbols with the newer IEEE standard logic symbols.

Chapter Review Questions

Answer the following questions.

3-1. Draw the traditional logic symbols for **a** to **j** (label inputs *A, B, C, D* and outputs *Y*):
 a. Two-input AND gate
 b. Three-input OR gate
 c. Inverter (two symbols)
 d. Two-input XOR gate
 e. Four-input NAND gate
 f. Two-input NOR gate
 g. Two-input XNOR gate
 h. Two-input NAND gate (special symbol)
 i. Two-input NOR gate (special symbol)
 j. Buffer (noninverting)

3-2. Write the Boolean expression for each gate in question 1.

3-3. Draw the truth table for each gate in question 1.

3-4. Look at the chart in Fig. 3-42. Which logic gate always responds with an output of logical 0 only when all inputs are HIGH?

3-5. Which logic gate might be called the "all or nothing gate"?

3-6. Which logic gate might be called the "any or all gate"?

3-7. Which logic circuit complements the input?

3-8. Which logic gate might be called the "any but not all gate"?

3-9. Given an AND gate and inverters, draw how you would produce a NOR function.

3-10. Given a NAND gate and inverters, draw how you would produce an OR function.

3-11. Given a NAND gate and inverters, draw how you would produce an AND function.

3-12. Given 4 two-input AND gates, draw how you would produce a five-input AND gate.

3-13. Given several two-input NAND and OR gates, draw how you would produce a four-input NAND gate.

3-14. Switches arranged in series (see Fig. 3-1) act like what type of logic gate?

3-15. Switches arranged in parallel (see Fig. 3-6) act like what type of logic gate?

3-16. Figure 3-31(*b*) illustrates a(n) ________ (8, 16) pin ________ (three letters) IC.

3-17. Draw a wiring diagram similar to Fig. 3-33(*b*) for a circuit that will perform the three-input AND function. Use a 7408 IC, a 5-V dc power supply, three input switches, and an output indicator.

3-18. The PC board pad labeled ________ (*A, C*) is pin 1 of the IC in Fig. 3-43.

3-19. The PC board pad labeled ________ (letter) is the GND pin on the 7408 IC in Fig. 3-43.

3-20. The PC board pad labeled ________ (letter) is the V_{CC} pin on the 7408 IC in Fig. 3-43.

3-21. The IC shown in Fig. 3-44 was manufactured by ________ (Motorola, Texas Instruments).

3-22. What does the prefix "SN" on the IC in Fig. 3-44 represent?

3-23. The unit shown in Fig. 3-44 is a ________ (low-power, standard) TTL 14-pin DIP IC.

3-24. Pin 1 of the IC shown in Fig. 3-44 is labeled with the letter ________.

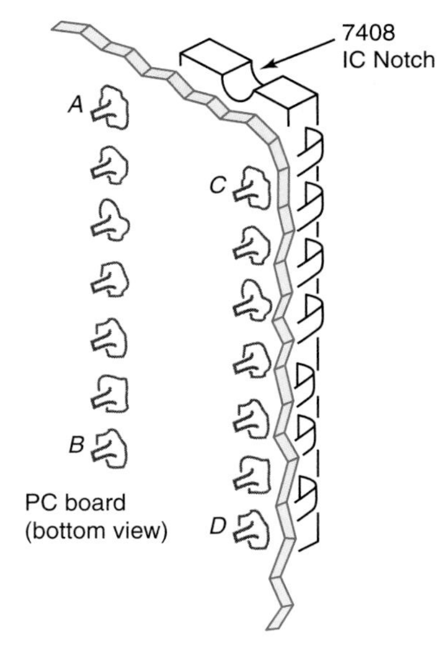

Fig. 3-43 An IC soldered to a PC board.

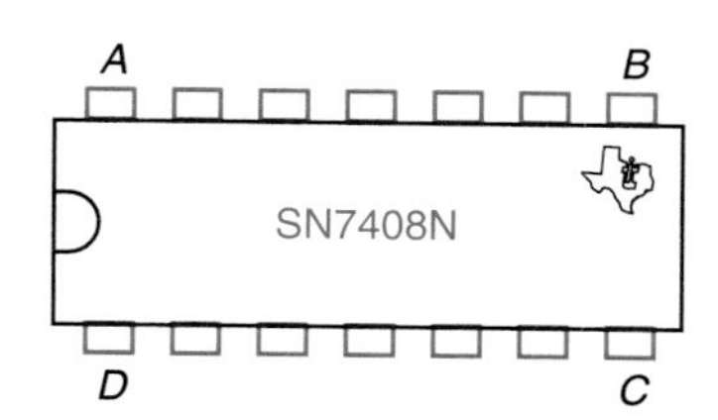

Fig. 3-44 Top view of a typical digital IC.

3-25. The pin labeled with a "C" on the IC in Fig. 3-44 is pin number _________.

3-26. Refer to Fig. 3-39. Draw a logic diagram of this circuit using traditional logic symbols.

3-27. Figure 3-39(*b*) is an example of a _________ (logic, wiring) diagram that might be used by service personnel.

3-28. Refer to Fig. 3-39(*a*). If all input pins (1, 2, and 5) are HIGH and output pin 6 is HIGH but point *E* is LOW, the LED _________ (will, will not) light and the circuit _________ (is, is not) working properly.

3-29. Refer to Fig. 3-39(*a*). List several possible problems if pin 6 is HIGH but point *E* is LOW.

3-30. Refer to Fig. 3-39(*a*). An *internal open* between the output of the first AND gate and pin 3 might give neither a HIGH nor LOW indication on the logic probe. This means that both pins 3 and 4 are floating _________ (HIGH, LOW).

3-31. Refer to Fig. 3-45. The IC shown in Fig. 3-45 was manufactured by _________ (National Semiconductor, RCA).

3-32. Refer to Fig. 3-45. The core number of this IC is _________, which means it is a _________ (CMOS, TTL) logic device.

3-33. Pin 1 of the IC shown in Fig. 3-45 is labeled with the letter _________.

3-34. What precaution should be taken when storing the IC in Fig. 3-45?

3-35. Draw the IEEE standard logic symbol for a three-input NOR gate.

3-36. Draw the IEEE standard logic symbol for a three-input XNOR gate.

3-37. The right _________ (circle, triangle) at the output of an IEEE standard NAND logic symbol signifies to invert the output of the AND function.

3-38. The IEEE standard AND logic symbol uses the _________ sign to signify the AND function.

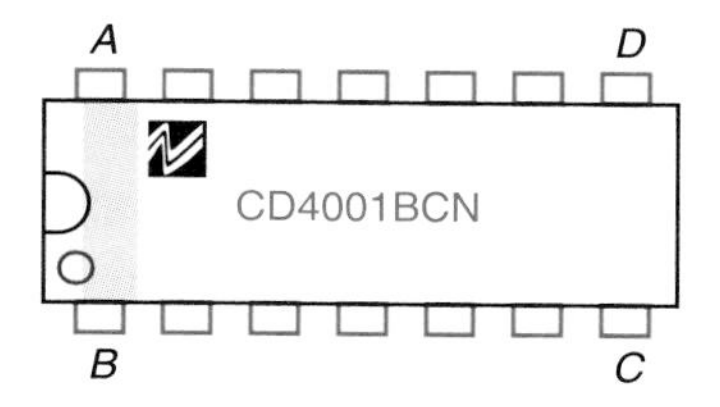

Fig. 3-45 Top view of a typical digital IC.

Critical Thinking Questions

3-1. What three-input logic gate would you use in your design if you require a HIGH output *only* when all three input switches go HIGH?

3-2. What four-input logic gate would you use in your design if you require a HIGH output only when an *odd* number of input switches are HIGH.

3-3. Refer to Fig. 3-28(*a*). Explain why the OR gate with inverted inputs produces the NAND function.

3-4. Inverting both inputs of a two-input NAND gate produces a circuit that generates the _________ logic function.

3-5. Inverting both inputs and the output of a two-input OR gate produces a circuit that generates the _________ logic function.

3-6. Refer to Fig. 3-37. If input *A* is HIGH and input *B* is LOW, output *J* (pin 3) will be _________ (HIGH, LOW). Transistor Q_1 is turned _________ (off, on) and the LED _________ (does, does not) light.

3-7. Refer to Fig. 3-37. Why are pins 5, 6, 8, 9, 12, and 13 grounded in this circuit?

3-8. Refer to Fig. 3-39. If the 7408 TTL IC developed an internal "short circuit," the top of the IC would probably feel _________ (hot, cool) to the touch.

3-9. Draw a logic diagram (use AND and inverter symbols) for the Boolean expression $\overline{A} \cdot \overline{B} = Y$.

3-10. The Boolean expression $\overline{A} \cdot \overline{B} = Y$ is one representation of the _________ (NAND, NOR) logic function.

3-11. Draw a waveform to represent the logic levels (*H* and *L*) at output *Y* of the AND gate in Fig. 3-46.

3-12. Draw a waveform to represent the logic levels (*H* and *L*) at output *Y* of the NOR gating circuit in Fig. 3-47.

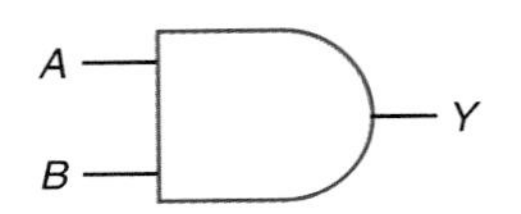

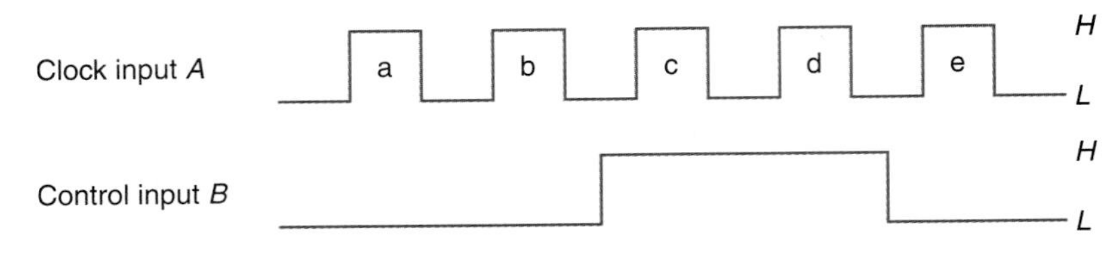

Fig. 3-46 Pulse train problem.

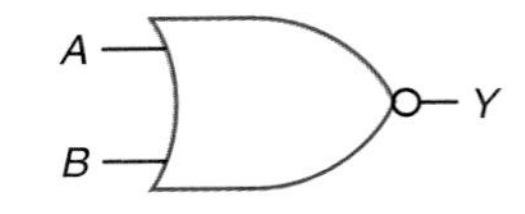

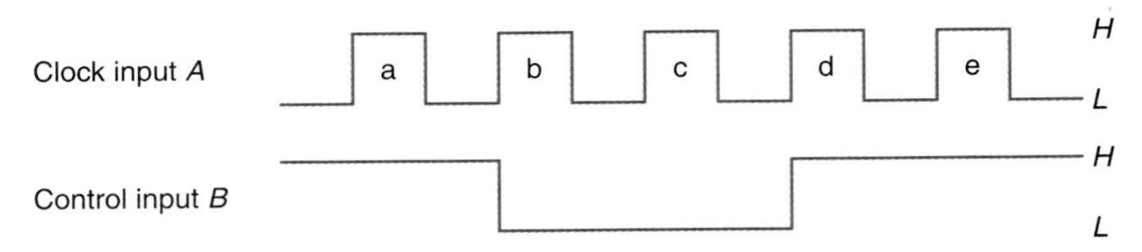

Fig. 3-47 Pulse train problem.

Answers to Tests

1. round
2. $A \cdot B = Y$ or $AB = Y$
3. HIGH, light
4. pointed
5. $A + B = Y$
6. LOW
7. inclusive
8. LOW
9. LOW
10. $Y = \overline{A}$ or $Y = A'$
11. negated, complemented
12. LOW
13. round with an added invert bubble
14. $\overline{A \cdot B} = Y$
15. LOW, is
16. pointed with an added invert bubble
17. $\overline{A + B} = Y$
18. LOW, is not
19. HIGH, is
20. any but not all gate
21. $A \oplus B = Y$
22. LOW
23. HIGH
24. invert, XOR
25. $\overline{A \oplus B} = Y$
26. LOW
27. HIGH
28. connected together
29. three
30. $\overline{A \cdot B \cdot C} = Y$
31. eight
32. $\overline{A + B + C + D} = Y$
33. 32 (2^5)
34. inverters
35. NOR
36. NAND
37. OR
38. $\overline{A + B} = Y$
39. $\overline{A} \cdot \overline{B} = Y$
40. TTL, CMOS
41. dual-inline (DIP)
42. 5, +, −
43. TTL quad two-input AND gate
44. manufacturer—National Semiconductor; package—dual-inline (DIP); family—low-power Schottky; function—quad two-input AND gate
45. low
46. 3, 18
47. CMOS quad two-input AND gate
48. All unused CMOS inputs must be connected to either GND (−) or positive (+) of the power supply.
49. TTL and CMOS
50. Use your senses to locate possible open or short circuits or possible overheating.
51. Check that each IC has power.
52. not allowed
53. A B C & Y
54. A B C ≥1 Y
55. A B C & Y
56. bubble
57. traditional

Chapter 4 Using Logic Gates

Chapter Objectives

This chapter will help you to:

1. *Draw* logic diagrams from minterm and maxterm Boolean expressions.
2. *Design* a logic diagram from a truth table by first developing a minterm Boolean expression and then drawing the AND-OR logic diagram.
3. *Reduce* a minterm Boolean expression to its simplest form using two-, three-, four-, and five-variable Karnaugh maps.
4. *Simplify* AND-OR logic circuits using NAND gates.
5. *Solve* logic problems using data selectors.
6. *Convert* minterm-to-maxterm and maxterm-to-minterm Boolean expressions using De Morgan's theorems.
7. *Use* a "keyboard version" of Boolean expressions.

In Chap. 3 you memorized the symbol, truth table, and Boolean expression for each of the binary logic gates. These gates are the basic building blocks for *all* digital systems. In this chapter you will see how your knowledge of gate symbols, truth tables, and Boolean expressions can be used to solve real world problems in electronics. You will be connecting gates to form what engineers refer to as *combinational logic circuits,* and you will be combining gates (ANDs, ORs) and inverters and so on to solve logic problems. There are three "tools of the trade" in solving logic problems: *gate symbols, truth tables,* and *Boolean expressions.* Do you have these tools of the trade? Do you know your gate symbols, truth tables, and Boolean expressions? If you need review on logic gates, please refer to the summary of Chap. 3; Fig. 3-42 especially will be helpful. Your understanding of using logic gates is important because to be successful as a technician, troubleshooter, engineer, or hobbyist in digital electronics you must master combining gates. It is strongly suggested that you try out your *combinational logic circuits* in the shop or laboratory. Logic gates are packaged in inexpensive, easy-to-use ICs. Combinational logic circuits can be implemented using either TTL or CMOS ICs. Combinational logic circuits can also be implemented using circuit simulation software on your computer.

Combinational logic circuits

4-1 Constructing Circuits from Boolean Expressions

We use Boolean expressions to guide us in building logic circuits. Suppose you are given the Boolean expression $A + B + C = Y$ (read as "A or B or C equals output Y") and told to build a circuit that will perform this logic function. Looking at the expression, notice that each input must be ORed to get output Y. Figure 4-1 illustrates the *gate* needed to do the job.

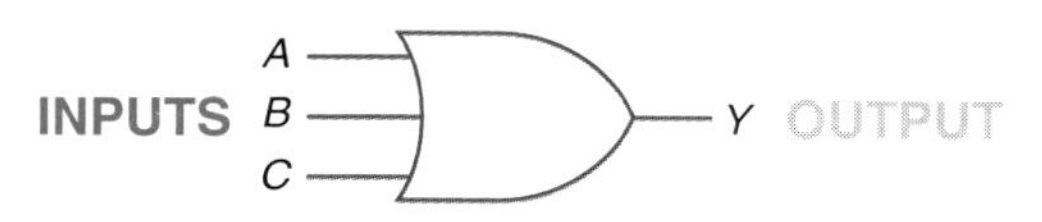

Fig. 4-1 Logic diagram for Boolean expression $A + B + C = Y$

Constructing logic circuit from a Boolean expression

Now suppose you are given the Boolean expression $\overline{A} \cdot B + A \cdot \overline{B} + \overline{B} \cdot C = Y$ (read as "not A and B, or A and not B, or not B and C equals output Y"). How would you construct a circuit that will do the job of this expression? The first step is to look at the Boolean expression and note that you must OR $\overline{A} \cdot B$ with $A \cdot \overline{B}$ with $\overline{B} \cdot C$. Figure 4-2(a) shows that a three-input OR gate will form the output Y. This may be redrawn as in Fig. 4-2(b).

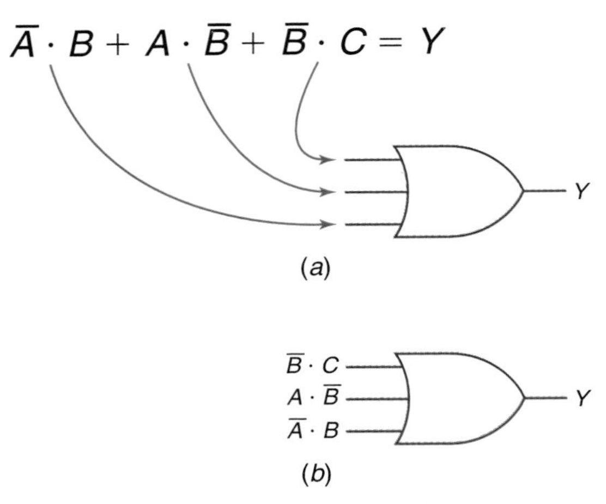

Fig. 4-2 Step 1 in constructing a logic circuit.

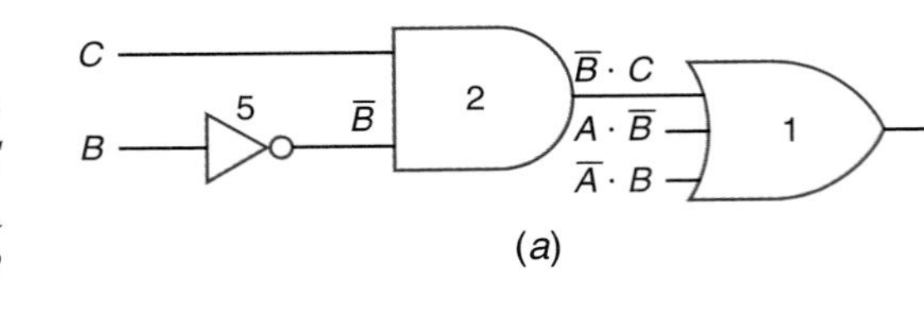

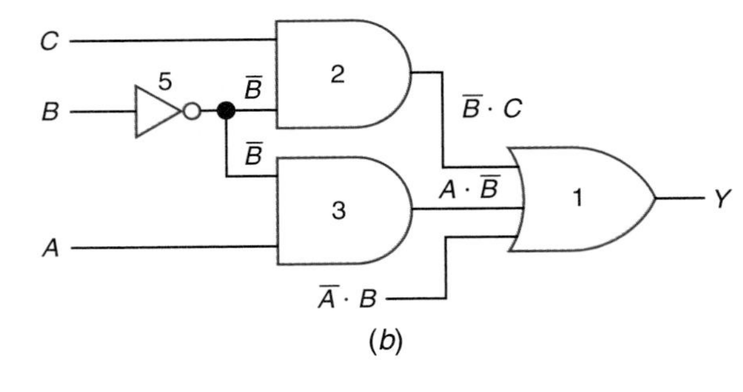

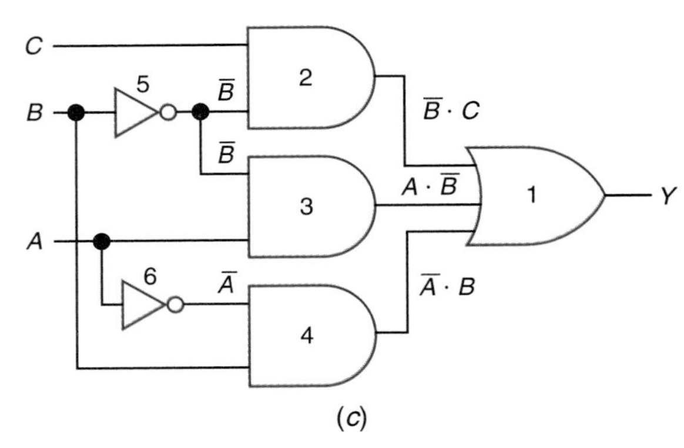

Fig. 4-3 Step 2 in constructing a logic circuit.

Maxterm form

Circuit simulation software

The second step used in constructing a logic circuit from the given Boolean expression $\overline{A} \cdot B + A \cdot \overline{B} + \overline{B} \cdot C = Y$ is shown in Fig. 4-3. Notice in Fig. 4-3(a) that an AND gate has been added to feed the $\overline{B} \cdot C$ to the OR gate and an inverter has been added to form the $\overline{B}$ for the input to AND gate 2. Figure 4-3(b) adds AND gate 3 to form the $A \cdot \overline{B}$ input to the OR gate. Finally, Fig. 4-3(c) adds AND gate 4 and inverter 6 to form the $\overline{A} \cdot B$ input to the OR gate. Figure 4-3(c) is the circuit that would be constructed to perform the required logic given in the Boolean expression $\overline{A} \cdot B + A \cdot \overline{B} + \overline{B} \cdot C = Y$.

Notice that we started at the output of the logic circuit and worked toward the inputs. You have now experienced how combinational logic circuits are constructed from Boolean expressions.

Sum-of-products form

Product-of-sums form

Minterm form

Boolean expressions come in two forms. The *sum-of-products* form is the type we saw in Fig. 4-2. Another example of this form is $A \cdot B + B \cdot C = Y$. The other Boolean expression form is the *product-of-sums;* an example is $(D + E) \cdot (E + F) = Y$. The sum-of-products form is called the *minterm form* in engineering texts. The product-of-sums form is called the *maxterm form* by engineers, technicians, and scientists.

Computer *circuit simulation software,* such as Electronics Workbench®, will draw a logic diagram from a Boolean expression. This software can draw logic diagrams from either minterm or maxterm Boolean expressions. Professionals in digital design will commonly use computer circuit simulations. Your instructor may have you use circuit simulation software in the lab.

TEST

Answer the following questions.

1. Construct logic circuits using AND, OR, and NOT gates for the following minterm Boolean expressions:
   a. $\overline{A} \cdot \overline{B} + A \cdot B = Y$
   b. $\overline{A} \cdot \overline{C} + A \cdot B \cdot C = Y$
   c. $A \cdot D + \overline{B} \cdot \overline{D} + C \cdot \overline{D} = Y$
2. A minterm Boolean expression is also called the ________ form.

3. A maxterm Boolean expression is also called the ________ form.
4. The minterm Boolean expression $A \cdot D + \overline{B} \cdot \overline{D} + C \cdot \overline{D} = Y$ has a pattern that is called the ________ (product-of-sums, sum-of-products) form.
5. The maxterm Boolean expression $(A + D) \cdot (B + \overline{C}) \cdot (A + C) = Y$ has a pattern that is called the ________ (product-of-sums, sum-of-products) form.

4-2 Drawing a Circuit from a Maxterm Boolean Expression

Suppose you are given the maxterm Boolean expression $(A + B + C) \cdot (\overline{A} + \overline{B}) = Y$. The first step in constructing a logic circuit for this Boolean expression is shown in Fig. 4-4(*a*). Notice that the terms $(A + B + C)$ and $(\overline{A} + \overline{B})$ are ANDed together to form output Y. Figure 4-4(*b*) shows the logic circuit redrawn. The second step in drawing the logic circuit is shown in Fig. 4-5. The $(\overline{A} + \overline{B})$ part of the expression is produced by adding OR gate 2 and inverters 3 and 4, as illustrated in Fig. 4-5(*a*). Then, the expression $(A + B + C)$ is delivered to the AND gate by OR gate 5 in Fig. 4-5(*b*). The logic circuit shown in Fig. 4-5(*b*) is the complete logic circuit for the maxterm Boolean expression $(A + B + C) \cdot (\overline{A} + \overline{B}) = Y$.

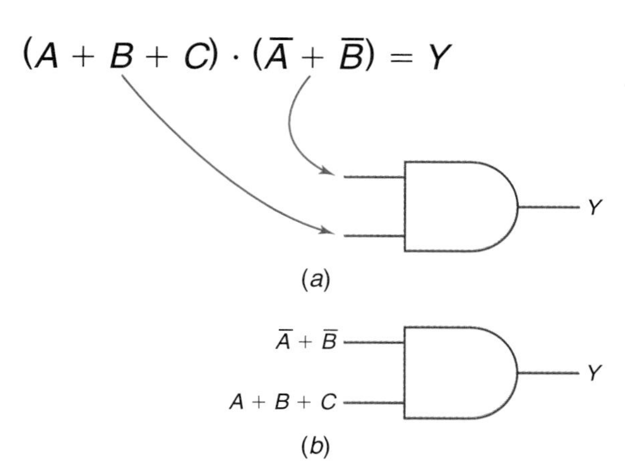

Fig. 4-4 Step 1 in constructing a product-of-sums logic circuit.

In summary, we work from right to left (from output to input) when converting a Boolean expression to a logic circuit. Notice that we use only AND, OR, and NOT gates when constructing combinational logic circuits. Maxterm and minterm Boolean expressions both can be converted to logic circuits. Minterm expressions create AND-OR logic circuits similar to that in Fig. 4-3(*c*), whereas maxterm expressions create OR-AND logic circuits similar to that in Fig. 4-5(*b*).

You now should be able to identify minterm and maxterm Boolean expressions, and you should be able to convert Boolean expressions to combinational logic circuits by using AND, OR, and NOT gates.

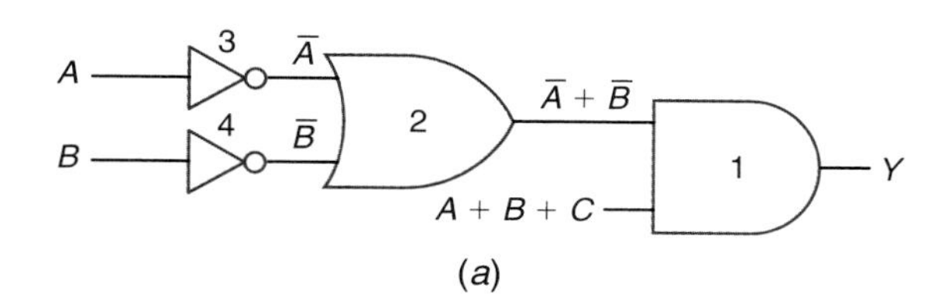

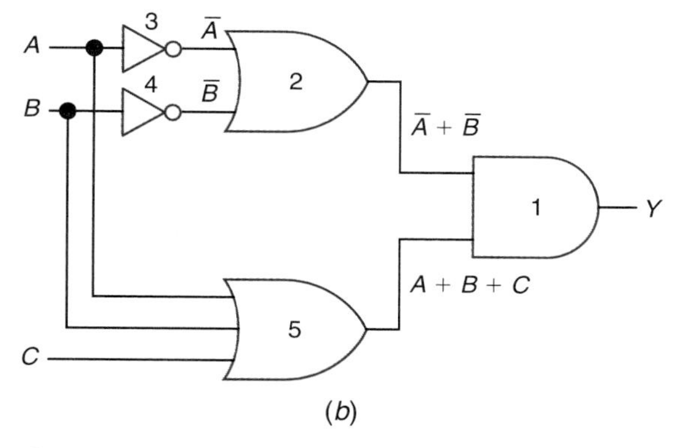

Fig. 4-5 Step 2 in constructing a product-of-sums logic circuit.

TEST

Answer the following questions.

6. Construct a logic circuit using AND, OR, and NOT gates from the following Boolean expressions:
   a. $(A + B) \cdot (\overline{A} + \overline{B}) = Y$
   b. $(\overline{A} + B) \cdot \overline{C} = Y.$
   c. $(A + B) \cdot (\overline{C} + \overline{D}) \cdot (\overline{A} + C) = Y$
7. Refer to question 6. These Boolean expressions are ________ (maxterm, minterm).
8. Refer to question 6. These Boolean expressions are in ________ (product-of-sums, sum-of-products) form.
9. Maxterm Boolean expressions are used to create ________ (AND-OR, OR-AND) logic circuits.

4-3 Truth Tables and Boolean Expressions

Boolean expressions are a convenient method of describing how a logic circuit operates. The *truth table* is another precise method of describing how a logic circuit works. As you work in digital electronics, you will have to convert information from truth-table form to a Boolean expression.

Truth Table to Boolean Expression

Forming a Boolean expression from a truth table

Look at the truth table in Fig. 4-6(*a*). Notice that only two of the eight possible combinations of inputs *A*, *B*, and *C* generate a logical 1 at the output. The two combinations that generate a 1 output are shown as $\overline{C} \cdot B \cdot A$ (read as "not *C* and *B* and *A*") and $C \cdot \overline{B} \cdot \overline{A}$ (read as "*C* and not *B* and not *A*"). Figure 4-6(*b*) shows how the combinations are ORed together to form the Boolean expression for the truth table. Both the truth table in Fig. 4-6(*a*) and the Boolean expression in Fig. 4-6(*b*) describe how the logic circuit should work.

The truth table is the origin of most logic circuits. You must be able to convert the truth-table information into a Boolean expression as in this section. Remember to look for combinations of variables that generate a logical 1 in the truth table.

Truth table

INPUTS			OUTPUT	
C	*B*	*A*	*Y*	
0	0	0	0	
0	0	1	0	
0	1	0	0	
0	1	1	1	← $\overline{C} \cdot B \cdot A = 1$
1	0	0	1	← $C \cdot \overline{B} \cdot \overline{A} = 1$
1	0	1	0	
1	1	0	0	
1	1	1	0	

(*a*)

(*b*) Boolean expression

$$\overline{C} \cdot B \cdot A + C \cdot \overline{B} \cdot \overline{A} = Y$$

Fig. 4-6 Forming a minterm Boolean expression from a truth table.

Boolean Expression to Truth Table

Constructing a truth table from a Boolean expression

Occasionally you must reverse the procedure you have just learned. That is, you must take a Boolean expression and from it construct a truth table. Consider the Boolean expression in Fig. 4-7(*a*). It appears that two combinations of inputs *A*, *B*, and *C* generate a logical 1 at the output. In Fig. 4-7(*b*) we find the correct combinations of *A*, *B*, and *C* that are given in the Boolean expression and mark a 1 in the output column. All other outputs in the truth table are 0. The Boolean expression in Fig. 4-7(*a*) and the truth table in Fig. 4-7(*b*) both accurately describe the operation of the same logic circuit.

Suppose you are given the Boolean expression in Fig. 4-8(*a*). At first glance it seems that this would produce two outputs with a logical 1. However, if you look closely at Fig. 4-8(*b*) you will see that the Boolean expression $\overline{C} \cdot \overline{A} + C \cdot B \cdot A = Y$ actually generates three

(*a*) Boolean expression

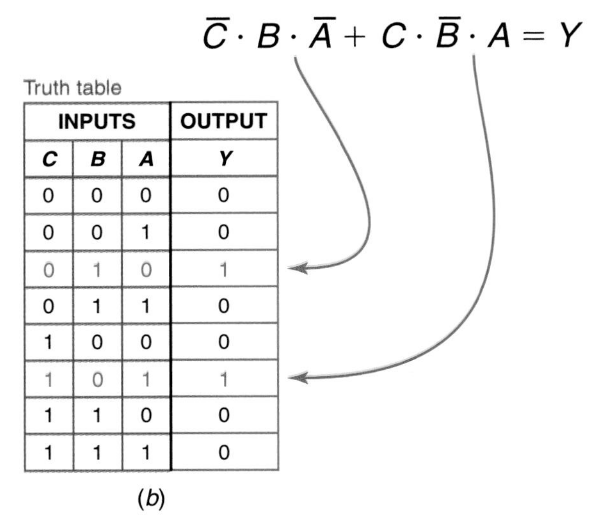

$$\overline{C} \cdot B \cdot \overline{A} + C \cdot \overline{B} \cdot A = Y$$

Truth table

INPUTS			OUTPUT
C	*B*	*A*	*Y*
0	0	0	0
0	0	1	0
0	1	0	1
0	1	1	0
1	0	0	0
1	0	1	1
1	1	0	0
1	1	1	0

(*b*)

Fig. 4-7 Constructing a truth table from a minterm Boolean expression.

Did You Know?

Electronic Thermometers. Today, taking a child's temperature is not the challenge it was for previous generations. Thermoscan, an electronic thermometer, takes a reading from inside the ear in one second. This thermometer collects infrared energy from the ear's tissues, sends it to a pyroelectric sensor, and provides a digital readout.

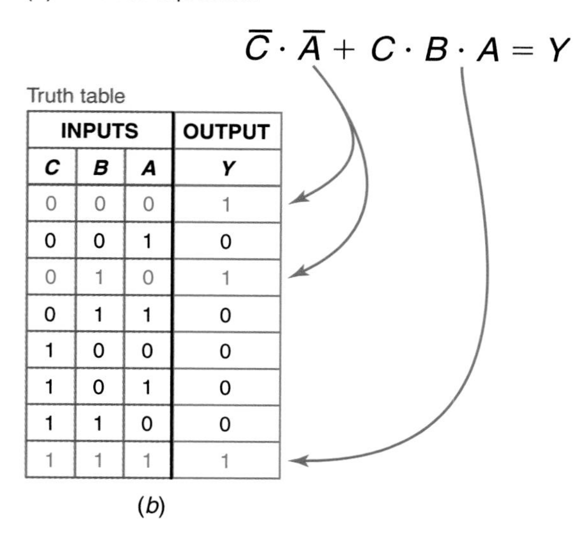

INPUTS			OUTPUT
C	B	A	Y
0	0	0	1
0	0	1	0
0	1	0	1
0	1	1	0
1	0	0	0
1	0	1	0
1	1	0	0
1	1	1	1

Fig. 4-8 Constructing a truth table from a minterm Boolean expression.

logical 1s in the output column. The "trick" illustrated in Fig. 4-8 should make you very cautious. Make sure you have all the combinations that generate a logical 1 in the truth table. The Boolean expression in Fig. 4-8(*a*) and the truth table in Fig. 4-8(*b*) both describe the same logic circuit or logic function.

You have now converted truth tables to Boolean expressions and Boolean expressions to truth tables. You were reminded that the Boolean expressions you worked with were minterm Boolean expressions. The procedure for producing maxterm Boolean expressions from a truth table is quite different.

Circuit Simulation Conversions

Circuit simulation software running on modern computers can accurately convert Boolean expressions to truth tables or truth tables to Boolean expressions. We will demonstrate the use of one popular electronic circuit simulator.

One easy-to-use circuit simulator is Electronics Workbench ® (EWB) by Interactive Image Technologies, Ltd. The EWB software contains an instrument called a *logic converter,* shown in Fig. 4-9(*a*) on page 62. To use this EWB instrument to convert a Boolean expression to a truth table, you would take the following steps:

Step 1. Type the expression in the bottom section (see Fig. 4-9(*b*)).

Step 2. Activate the Boolean expression to a truth-table option (see Fig. 4-9(*b*)).

Step 3. View the resulting truth table on the computer monitor (see Fig. 4-9(*b*)).

The Boolean expression $A'B'C + ABC'$ entered in step 1, Fig. 4-9(*b*), is a shortened "keyboard version" of the $C \cdot \overline{B} \cdot \overline{A} + \overline{C} \cdot B \cdot A = Y$ in Fig. 4-6(*b*). It is important to recognize that $A'B'C + ABC'$ equals $C \cdot \overline{B} \cdot \overline{A} + \overline{C} \cdot B \cdot A = Y$. The apostrophe in the "keyboard version" of a Boolean expression means the same as an overbar over that letter. Therefore A' (say A not) means the same as $\overline{A}$ (say A not). Notice that the *order* that the variables appear in the Boolean expression are reversed. This difference in order has no effect on the logic function. Therefore, ABC means the same as CBA. Also notice that the AND dot between variables has been eliminated so that $A \cdot B \cdot C$ can be shortened to ABC.

Compare the output columns in Figs. 4-6(*a*) and 4-9(*b*). Both these truth tables describe the same logic function although the output columns seem different. This is because the order of the input variables are listed as CBA in Fig. 4-6(*a*) whereas they appear as ABC in Fig. 4-9(*b*). Truth table line 5 (100) in Fig. 4-6(*a*) is the same as line 2 (001) in Fig. 4-9(*b*). This demonstrates that the headings on truth tables and Boolean expression representations vary. Workers in electronics will become familiar with several methods of labeling truth tables and variations in Boolean expressions.

Electronic circuit simulators such as EWB can commonly handle either minterm or maxterm Boolean expressions. Observe from Fig. 4-9(*a*) that five other logic conversions are available using this version of EWB. Your instructor may have you use the many features available on your electronic circuit simulation software.

TEST

Answer the following questions.

10. Refer to Fig. 4-6(*a*). Assume that only the bottom two lines of the truth table produce an output of 1 (all other outputs = 0). Write the sum-of-products Boolean expression for this situation.
11. Refer to Fig. 4-6(*a*). The Boolean expression $\overline{C} \cdot \overline{B} \cdot \overline{A} + \overline{C} \cdot \overline{B} \cdot A = Y$ produces a truth table that has HIGH outputs in what two lines?

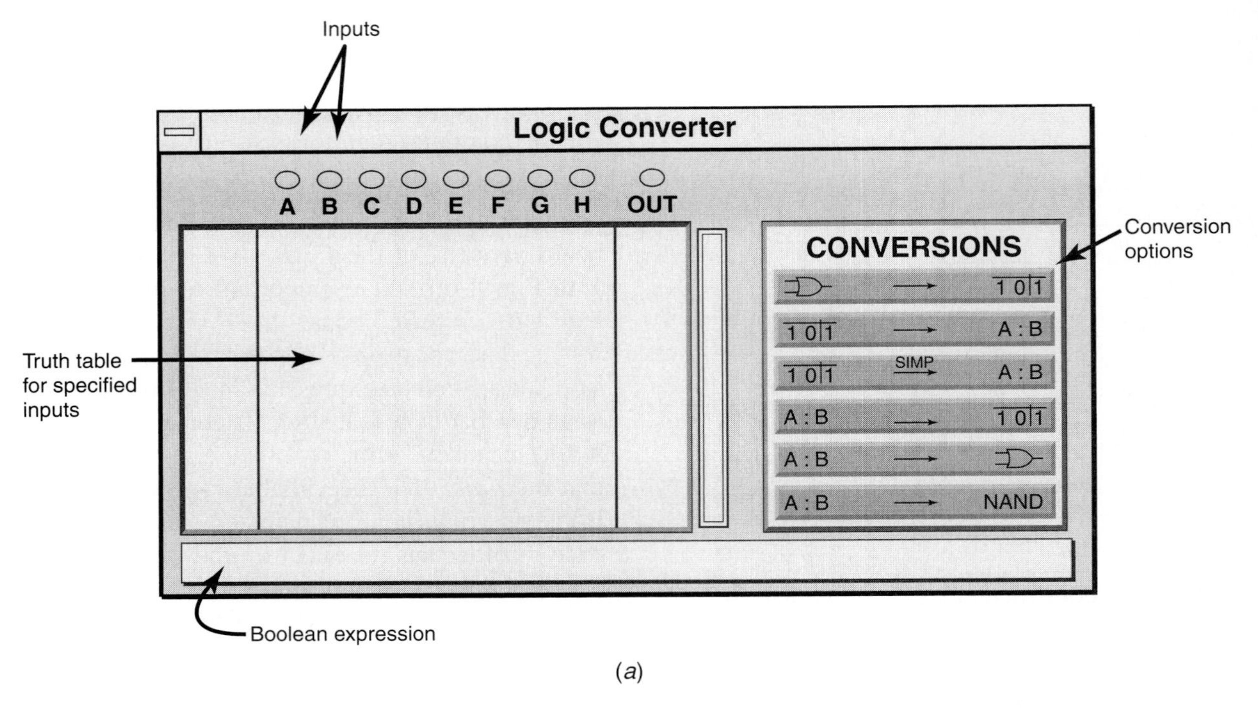

(a)

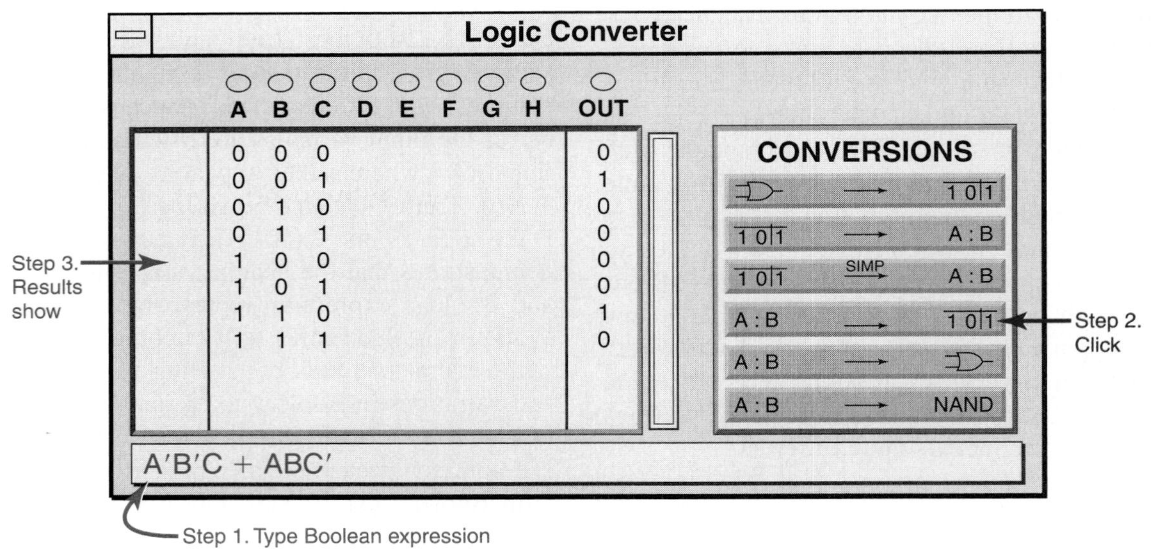

(b)

Fig. 4-9 Logic converter instrument from an electronic circuit simulator. (*From Electronics Workbench, courtesy of Interactive Image Technologies, Ltd.*) (*a*) Logic converter instrument layout. (*b*) The three steps in converting a Boolean expression to a truth table.

12. Construct a truth table for the Boolean expression $C \cdot B \cdot \overline{A} + C \cdot \overline{B} \cdot A = Y$.
13. The procedure illustrated in Fig. 4-6 converts a truth table to a ________ (maxterm, minterm) Boolean expression.
14. The procedure illustrated in Figs. 4-7 and 4-8 converts a ________ (maxterm, minterm) Boolean expression to a truth table.
15. Write the "keyboard version" of the Boolean expression $\overline{C} \cdot \overline{B} \cdot A + B \cdot \overline{A} = Y$.

4-4 Sample Problem

The procedures in Secs. 4-1 to 4-3 are useful skills as you work in digital electronics. To assist you in developing your skills, we shall take an everyday logic problem and work from truth

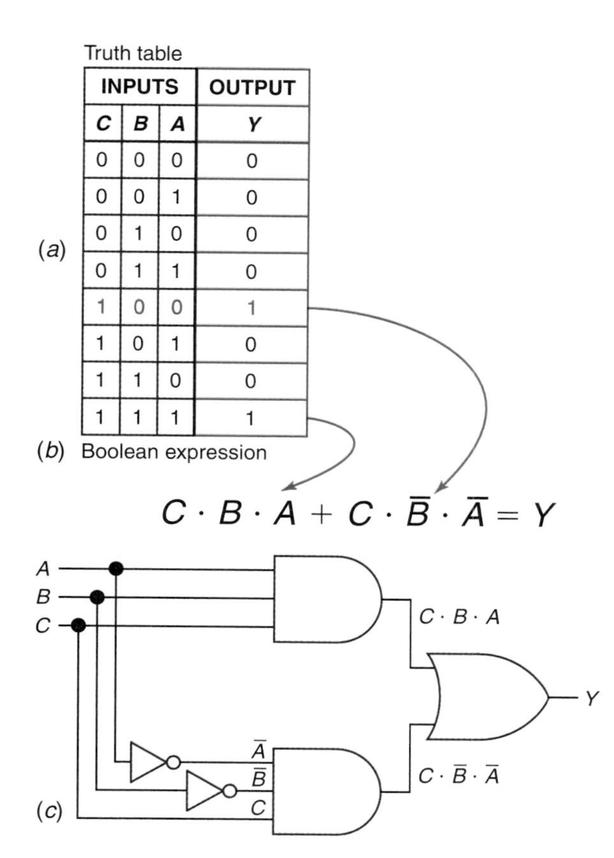

C	B	A	Y
0	0	0	0
0	0	1	0
0	1	0	0
0	1	1	0
1	0	0	1
1	0	1	0
1	1	0	0
1	1	1	1

Fig. 4-10 Electronic lock problem. (*a*) Truth table. (*b*) Boolean expression. (*c*) Logic circuit.

table to Boolean expression to logic circuit as shown in Fig. 4-10.

Let us assume that we are designing a simple *electronic lock.* The lock will open only when certain switches are activated. Figure 4-10(*a*) is the truth table for the electronic lock. Notice that the two combinations of input switches, *A*, *B*, and *C* generate a 1 at the output. A 1 output will open the lock. Figure 4-10(*b*) shows how we form the minterm Boolean expression for the electronic lock circuit. The logic circuit in Fig. 4-10(*c*) is then drawn from the Boolean expression. Look over the sample problem in Fig. 4-10 and be sure you can follow how we converted from the truth table to the Boolean expression and then to the logic circuit.

Many electronic circuit simulation programs can handle these conversions. For instance, the logic converter instrument in Electronics Workbench® could make these conversions. As such, EWB (see Fig. 4-9) could be used to solve this problem by doing the following:

Step 1. Fill out the truth table (see Fig. 4-9).

Step 2. Activate the truth table to Boolean expression button (second button from top in Fig. 4-9).

Step 3. Activate the Boolean expression to logic diagram button (second button from bottom in Fig. 4-9).

The Boolean expression and logic diagram to solve this problem then appear on the computer monitor.

You now should be able to solve a problem such as the one in Fig. 4-10. The following test will give you some practice in solving problems dealing with truth tables, Boolean expressions, and combinational logic circuits.

TEST

Answer the following questions.

16. Using the truth table below for an electronic lock, write the minterm Boolean expression for this truth table.

INPUT SWITCHES			OUTPUT
C	B	A	Y
0	0	0	0
0	0	1	0
0	1	0	1
0	1	1	0
1	0	0	0
1	0	1	1
1	1	0	0
1	1	1	0

Truth Table for Question 16—Lock Problem

Electronic lock

17. From the Boolean expression developed in question 16, draw a logic symbol diagram for the electronic lock problem.

4-5 Simplifying Boolean Expressions

Consider the Boolean expression $\overline{A} \cdot B + A \cdot \overline{B} + A \cdot B = Y$ in Fig. 4-11(*a*). In constructing a logic circuit for this Boolean expression, we find that we need three AND gates, two inverters, and one 3-input OR gate. Figure 4-11(*b*) is a logic circuit that would perform the logic of the Boolean expression $\overline{A} \cdot B + A \cdot \overline{B} + A \cdot B = Y$. Figure 4-11(*c*) details the truth table for the Boolean expression and logic circuit in Fig. 4-11(*a*) and (*b*). Immediately you recognize the truth table in Fig. 4-11(*c*) as the truth table for a two-input OR gate. The simple Boolean expression for a two-input OR gate is $A + B = Y$, as shown in Fig. 4-11(*d*). The logic circuit for a two-input OR gate in its simplest form is diagramed in Fig. 4-11(*e*).

JOB TIP

Listening, writing, and speaking skills are important for all employees.

Karnaugh mapping

Boolean algebra

The example summarized in Fig. 4-11 shows how we must try to simplify our original Boolean expression to get a simple, inexpensive logic circuit. In this case we were lucky enough to notice that the truth table belonged to an OR gate. However, usually we must use more systematic methods of simplifying our Boolean expression. Such methods include applying Boolean algebra and *Karnaugh mapping.*

Tabular method of simplification

Quine-McCluskey method

(*a*) Original Boolean expression

$$\overline{A} \cdot B + A \cdot \overline{B} + A \cdot B = Y$$

(*b*)

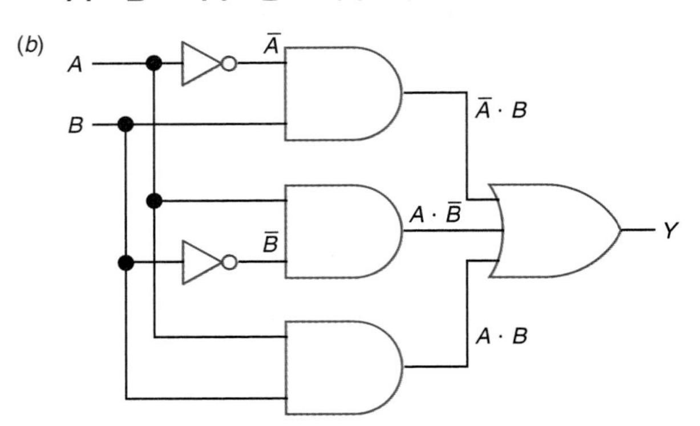

(*c*) Truth table

INPUTS		OUTPUT
B	*A*	*Y*
0	0	0
0	1	1
1	0	1
1	1	1

Maurice Karnaugh

(*d*) Simplified Boolean expression

$$A + B = Y$$

(*e*)

A

B

Y

Simplifying Boolean expressions

Fig. 4-11 Simplifying Boolean expressions. (*a*) Unsimplified Boolean expression. (*b*) Complex logic diagram. (*c*) Truth table. (*d*) Simplified Boolean expression: two-input OR by inspection. (*e*) Simple logic diagram.

Boolean algebra was originated by George Boole (1815–1864). Boole's algebra was adapted in the 1930s for use in digital logic circuits; it is the basis for the tricks we shall use to simplify Boolean expressions. We do not deal directly with Boolean algebra in this text. Many of you who continue on in digital electronics and engineering will study Boolean algebra in detail.

Karnaugh mapping, an easy-to-use graphic method of simplifying Boolean expressions, is covered in detail in Secs. 4-6 to 4-10. Several other simplification methods are available, including Veitch diagrams, Venn diagrams, and the tabular method of simplification. The tabular method used by computer software such as the Electronics Workbench® is called the *Quine-McCluskey method.*

TEST

Supply the missing word or words in each statement.

18. The logic circuits in Fig. 4-11(*b*) and (*e*) produce ________ (different, identical) truth tables.
19. Boolean expressions can many times be simplified by inspection or by using methods which include ________ algebra or ________ mapping.
20. Karnaugh mapping is a systematic graphic method of logic circuit simplification but the ________-________ method is better suited for computer simplification.

4-6 Karnaugh Maps

In 1953 Maurice Karnaugh published an article about his system of mapping and thus simplifying Boolean expressions. Figure 4-12 illustrates a Karnaugh map. The four squares (1, 2, 3, 4) represent the four possible combinations of *A* and *B* in a two-variable truth table. Square 1 in the Karnaugh map, then, stands for $\overline{A} \cdot \overline{B}$, square 2 for $\overline{A} \cdot B$, and so forth.

Let us map the familiar problem from Fig. 4-11. The original Boolean expression $\overline{A} \cdot B + A \cdot \overline{B} + A \cdot B = Y$ is rewritten in Fig. 4-13(*a*) for your convenience. Next, a 1 is placed in

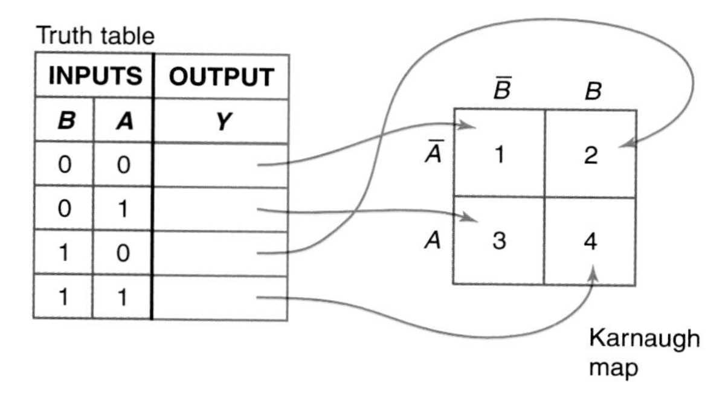

Fig. 4-12 The meaning of squares in a Karnaugh map.

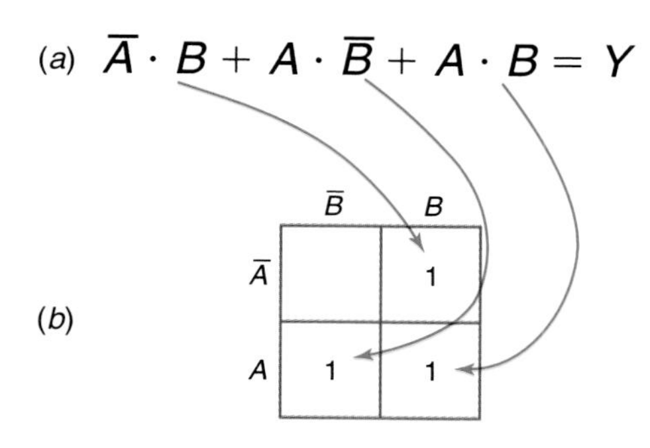

Fig. 4-13 Marking 1s on a Karnaugh map.

each square of the Karnaugh map, as shown in Fig. 4-13(*b*). The filled-in Karnaugh map is now ready for *looping*. The looping technique is shown in Fig. 4-14. *Adjacent 1s* are *looped together* in groups of two, four, or eight. Looping continues until all 1s are included inside a loop. Each loop is a new term in the simplified Boolean expression. Notice that we have two loops in Fig. 4-14. These two loops mean that we shall have two terms ORed together in our new simplified Boolean expression.

Now let us simplify the Boolean expression based upon the two loops that are redrawn in Fig. 4-15. First the bottom loop: notice that an A is included along with a B and a $\overline{B}$. The B and $\overline{B}$ terms can be *eliminated* according to the rules of Boolean algebra. This leaves the A term in the bottom loop. Likewise, the vertical loop contains an A and a $\overline{A}$, which are eliminated, leaving only a B term. The leftover A and B terms are then ORed together, giving the simplified Boolean expression $A + B = Y$.

The procedure for simplifying a Boolean expression sounds complicated. Actually, this procedure is quite easy after some practice. Here is a summary of the six steps:

1. Start with a minterm Boolean expression.
2. Record 1s on a Karnaugh map.
3. Loop adjacent 1s (loops of two, four, or eight squares).
4. Simplify by dropping terms that contain a term and its complement within a loop.
5. OR the remaining terms (one term per loop).
6. Write the simplified minterm Boolean expression.

TEST

Answer the following questions.

21. The map shown in Fig. 4-13 was developed by ________.
22. List the six steps used in simplifying a Boolean expression using a Karnaugh map.

4-7 Karnaugh Maps with Three Variables

Consider the unsimplified Boolean expression $A \cdot \overline{B} \cdot \overline{C} + \overline{A} \cdot \overline{B} \cdot \overline{C} + \overline{A} \cdot \overline{B} \cdot C + A \cdot B \cdot \overline{C} = Y$, as given in Fig. 4-16(*a*) on page 66. A three-variable Karnaugh map is illustrated in Fig. 4-16(*b*). Notice the eight possible combinations of A, B, and C, which are represented by the eight squares in the map. Tabulated on the map are four 1s, which represent each of the four terms in the original Boolean expres-

Three-variable Karnaugh map

Karnaugh map

Looping

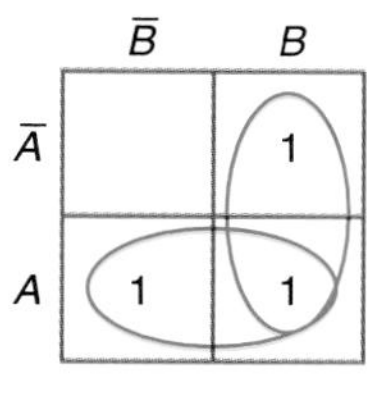

Fig. 4-14 Looping 1s together on a Karnaugh map.

Simplifying Boolean expressions

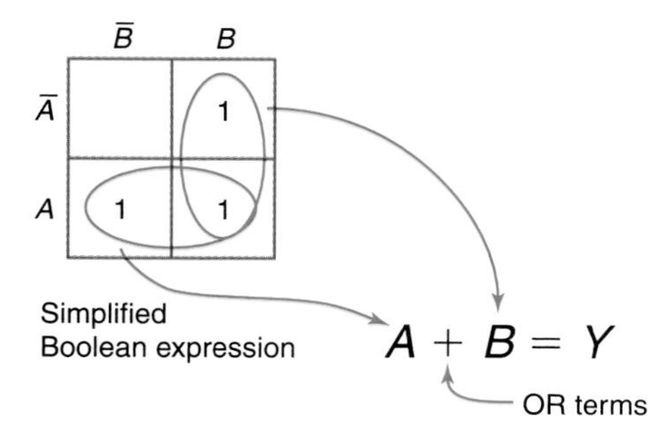

Fig. 4-15 Simplifying a Boolean expression from a Karnaugh map.

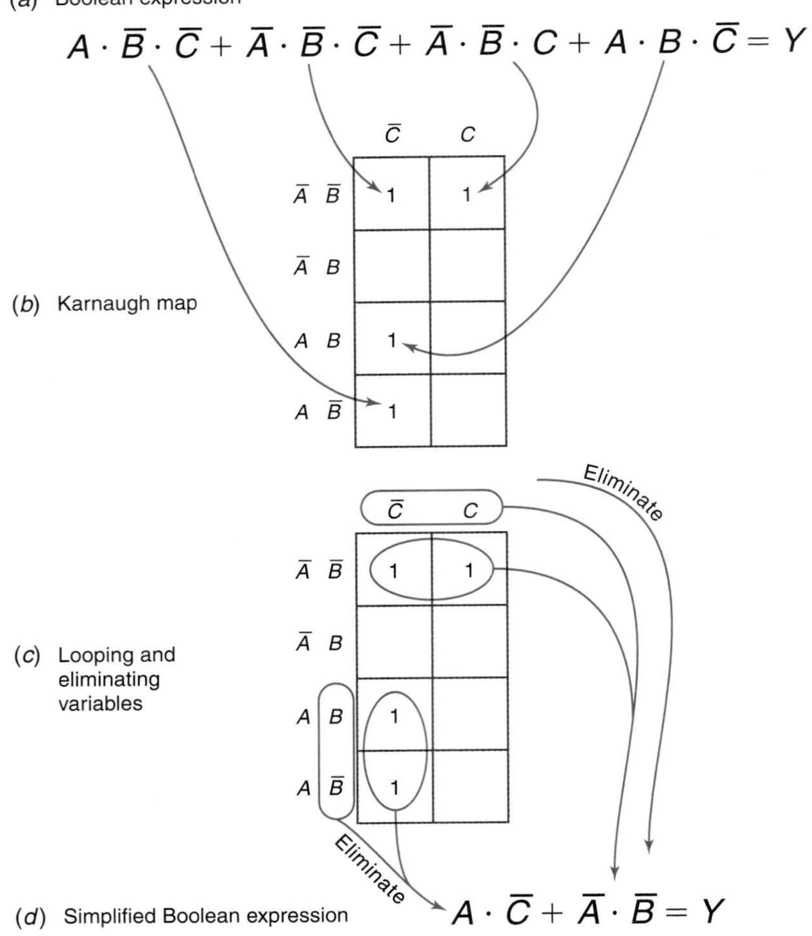

Simplifying a Boolean expression

Fig. 4-16 Simplifying a Boolean expression using a Karnaugh map. (*a*) Unsimplified expression. (*b*) Mapping 1s. (*c*) Looping 1s and eliminating variables. (*d*) Forming simplified minterm expression.

sion. The Karnaugh map with loops is redrawn in Fig. 4-16(*c*). Adjacent groups of two 1s are looped. The bottom loop contains both a B and a $\overline{B}$. The B and $\overline{B}$ terms are eliminated. The bottom loop still contains the A and $\overline{C}$, giving the $A \cdot \overline{C}$ term. The upper loop contains both a C and a $\overline{C}$. The C and $\overline{C}$ terms are eliminated, leaving the $\overline{A} \cdot \overline{B}$ term. A minterm Boolean expression is formed by adding the OR symbol. The simplified Boolean expression is written in Fig. 4-16(*d*) as $A \cdot \overline{C} + \overline{A} \cdot \overline{B} = Y$.

JOB TIP

Be on time — to interviews and to work!!

You can see that the simplified Boolean expression in Fig. 4-16 would take fewer electronic parts than the original expression. Remember that the much different looking simplified Boolean expression produces the same truth table as the original Boolean expression.

It is critical that the Karnaugh map be prepared just as the one shown in Fig. 4-16 was. Note that as you progress downward on the left side of the map, only one variable changes for each step. At the top left $\overline{A}\,\overline{B}$ is listed, while directly below is $\overline{A}\,B$ ($\overline{B}$ changed to B). Then, progressing downward from $\overline{A}B$ to AB the $\overline{A}$ term is changed to A. Finally, moving downward from AB to $A\overline{B}$ the B term is changed to $\overline{B}$. The Karnaugh map will not work properly if it is not laid out correctly.

TEST

Answer the following questions.

23. Simplify the Boolean expression $\overline{A} \cdot \overline{B} \cdot C + \overline{A} \cdot B \cdot C + A \cdot \overline{B} \cdot \overline{C} + A \cdot \overline{B} \cdot C = Y$ by:

a. Plotting 1s on a three-variable Karnaugh map
b. Looping groups of two or four 1s
c. Eliminating variables whose complement appears within the loop(s)
d. Writing the simplified minterm Boolean expression

24. Simplify the Boolean expression $\overline{A} \cdot B \cdot \overline{C} + \overline{A} \cdot B \cdot C + A \cdot B \cdot \overline{C} + A \cdot B \cdot C = Y$ by:
 a. Plotting 1s on a three-variable Karnaugh map
 b. Looping groups of two or four 1s
 c. Eliminating variables whose complement appears within the loop(s)
 d. Writing the simplified minterm Boolean expression

4-8 Karnaugh Maps with Four Variables

The truth table for four variables has 16 (2^4) possible combinations. Simplifying a Boolean expression that has four variables sounds complicated, but a Karnaugh map makes the job of simplifying easy.

Karnaugh maps with four variables

Consider the Boolean expression $A \cdot \overline{B} \cdot \overline{C} \cdot \overline{D} + \overline{A} \cdot B \cdot \overline{C} \cdot D + \overline{A} \cdot \overline{B} \cdot \overline{C} \cdot D + \overline{A} \cdot \overline{B} \cdot C \cdot D + \overline{A} \cdot B \cdot C \cdot D + A \cdot \overline{B} \cdot \overline{C} \cdot D = Y$, as in Fig. 4-17(*a*). The four-variable Karnaugh map in Fig. 4-17(*b*) gives the 16 possible combinations of *A*, *B*, *C*, and *D*. These are represented in the 16 squares of the map. Tabulated on the map are six 1s, which represent the six terms in the original Boolean expression. The Karnaugh map is redrawn in Fig. 4-17(*c*).

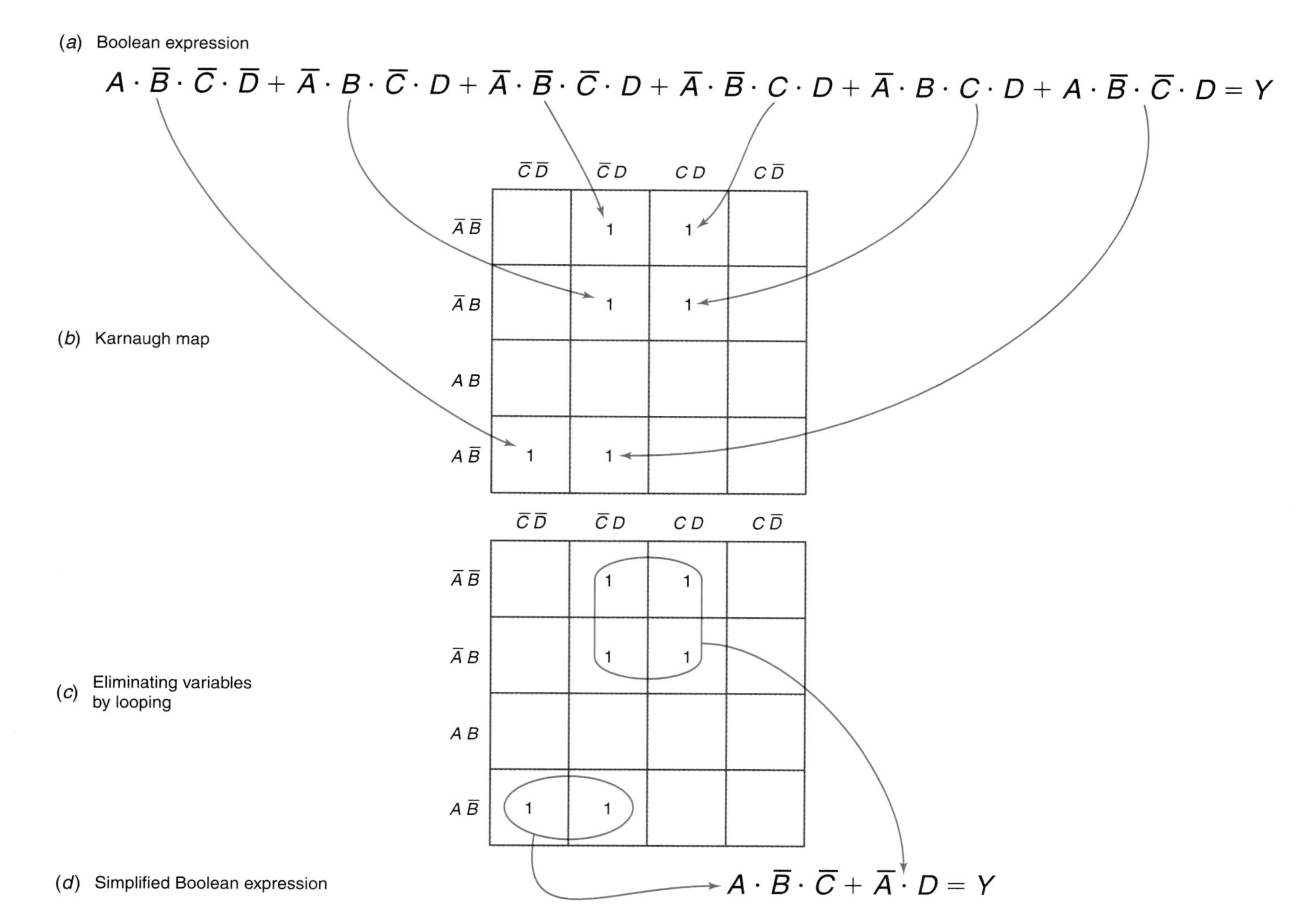

Fig. 4-17 Simplifying a four-variable Boolean expression using a Karnaugh map.

Adjacent groups of two 1s and four 1s are looped. The bottom loop of two 1s eliminates the D and $\overline{D}$ terms. The bottom loop then produces the $A \cdot \overline{B} \cdot \overline{C}$ term. The upper loop of four 1s eliminates the C and $\overline{C}$ *and* B *and* $\overline{B}$ terms. The upper loop then produces the $\overline{A} \cdot D$ term. The $A \cdot \overline{B} \cdot \overline{C}$ and $\overline{A} \cdot D$ terms are then ORed together. The simplified minterm Boolean expression is written in Fig. 4-17(*d*) as $A \cdot \overline{B} \cdot \overline{C} + \overline{A} \cdot D = Y$.

Observe that the same procedure and rules are used for simplifying Boolean expressions with two, three, or four variables and that larger loops in a Karnaugh map eliminate more variables. You must take care to make sure that the maps look just like the ones in Figs. 4-16 and 4-17.

TEST

Answer the following questions.

25. Simplify the Boolean expression $\overline{A} \cdot B \cdot \overline{C} \cdot \overline{D} + A \cdot B \cdot \overline{C} \cdot \overline{D} + \overline{A} \cdot B \cdot \overline{C} \cdot D + A \cdot B \cdot \overline{C} \cdot D + A \cdot \overline{B} \cdot C \cdot D + A \cdot \overline{B} \cdot C \cdot \overline{D} = Y$ by:
    a. Plotting 1s on a four-variable Karnaugh map
    b. Looping groups of two or four 1s
    c. Eliminating variables whose complements appear within loops
    d. Writing the simplified minterm Boolean expression
26. Simplify the Boolean expression $\overline{A} \cdot \overline{B} \cdot \overline{C} \cdot \overline{D} + \overline{A} \cdot \overline{B} \cdot \overline{C} \cdot D + \overline{A} \cdot B \cdot \overline{C} \cdot \overline{D} + \overline{A} \cdot B \cdot \overline{C} \cdot D + A \cdot B \cdot C \cdot D + A \cdot B \cdot C \cdot \overline{D} = Y$ by:
    a. Plotting 1s on a four-variable Karnaugh map
    b. Looping groups of two or four 1s
    c. Eliminating variables whose complement appears within the loop(s)
    d. Writing the simplified minterm Boolean expression

4-9 More Karnaugh Maps

K map looping variations

This section presents some sample Karnaugh maps. Notice the unusual looping procedures used on most maps in this section.

Consider the Boolean expression in Fig. 4-18(*a*). The four terms are shown as four 1s on the Karnaugh map in Fig. 4-18(*b*). The correct looping procedure is shown. Notice that the Karnaugh map is considered to be wrapped in a cylinder, with the left side adjacent to the right side. Also notice the elimination of the A and $\overline{A}$ and C and $\overline{C}$ terms. The simplified Boolean expression of $B \cdot \overline{D} = Y$ is shown in Fig. 4-18(*c*).

Another unusual looping variation is illustrated in Fig. 4-19(*a*). Notice that, while looping, the top and bottom of the map are adjacent

(*a*) Boolean expression

$$A \cdot B \cdot \overline{C} \cdot \overline{D} + \overline{A} \cdot B \cdot \overline{C} \cdot \overline{D} + \overline{A} \cdot B \cdot C \cdot \overline{D} + A \cdot B \cdot C \cdot \overline{D} = Y$$

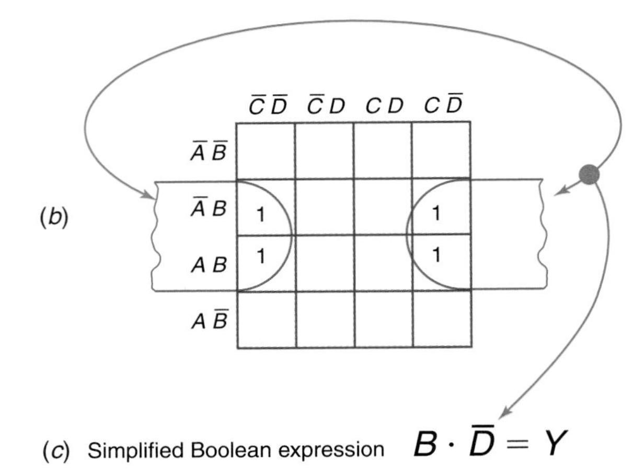

(*c*) Simplified Boolean expression $B \cdot \overline{D} = Y$

Fig. 4-18 Simplifying a Boolean expression by considering the map as a vertical cylinder. In this way, the four 1s can be looped.

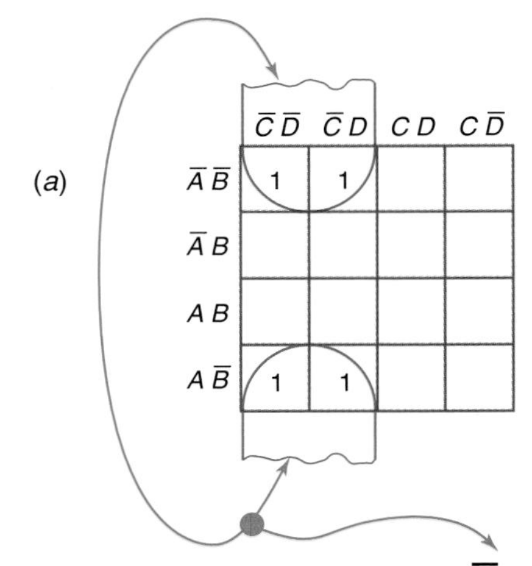

(*b*) Simplified Boolean expression $\overline{B} \cdot \overline{C} = Y$

Fig. 4-19 Simplifying a Boolean expression by considering the map as a horizontal cylinder. In this way, the four 1s can be looped.

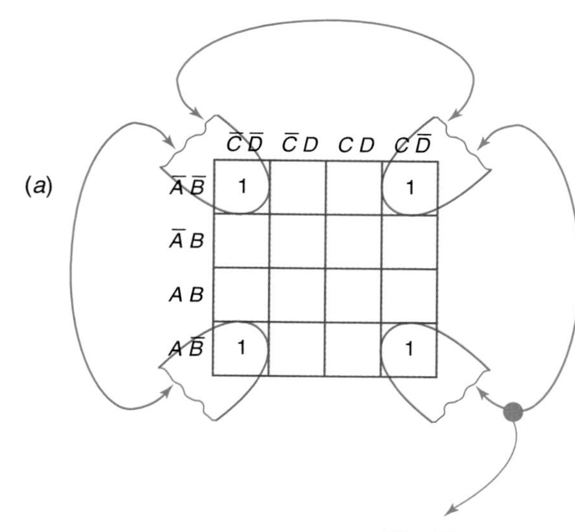

(b) Simplified Boolean expression $\overline{B} \cdot \overline{D} = Y$

Fig. 4-20 Simplifying a Boolean expression by thinking of the Karnaugh map as a ball. In this way, the 1s at the four corners can be enclosed in a single loop.

to one another, as if rolled into a cylinder. The simplified Boolean expression for this map is given as $\overline{B} \cdot \overline{C} = Y$ in Fig. 4-19(*b*). The A and $\overline{A}$ as well as the D and $\overline{D}$ terms have been eliminated in Fig. 4-19.

Figure 4-20(*a*) shows still another unusual looping pattern. The four corners of the Karnaugh map are considered connected, as if the map were formed into a ball. The four corners are then adjacent and may be formed into one loop as shown. The simplified Boolean expression is $\overline{B} \cdot \overline{D} = Y$, given in Fig. 4-20(*b*). In this example, the A and $\overline{A}$ as well as the C and $\overline{C}$ terms have been eliminated.

TEST

Answer the following questions.

27. Simplify the following Boolean expression $\overline{A} \cdot B \cdot \overline{C} \cdot \overline{D} + \overline{A} \cdot \overline{B} \cdot \overline{C} \cdot D + \overline{A} \cdot \overline{B} \cdot C \cdot D + \overline{A} \cdot B \cdot C \cdot \overline{D} + A \cdot \overline{B} \cdot \overline{C} \cdot D + A \cdot \overline{B} \cdot C \cdot D = Y$ by:
    a. Plotting 1s on a four-variable Karnaugh map
    b. Looping groups of two or four 1s
    c. Eliminating variables whose complements appear within loops
    d. Writing the simplified minterm Boolean expression
28. Simplify the following Boolean expression $\overline{A} \cdot \overline{B} \cdot \overline{C} + \overline{A} \cdot \overline{B} \cdot C + A \cdot \overline{B} \cdot \overline{C} + A \cdot \overline{B} \cdot C + A \cdot B \cdot C = Y$ by:
    a. Plotting 1s on a three-variable Karnaugh map
    b. Looping groups of two or four 1s
    c. Eliminating variables whose complements appear within loops
    d. Writing the simplified minterm Boolean expression

4-10 A Five-Variable Karnaugh Map

Five variable Karnaugh map

The Karnaugh map becomes three-dimensional when solving logic problems with more than four variables. A three-dimensional Karnaugh map will be used in this section.

Three-dimensional Karnaugh map

A five-variable unsimplified Boolean expression is given in Fig. 4-21(*a*) on page 70. A five-variable Karnaugh map is drawn in Fig. 4-21(*b*). Notice that it has 2 four-variable Karnaugh maps stacked to make it three-dimensional. The top map is the E plane while the bottom is the $\overline{E}$ (E not) plane.

Each of the nine terms in the unsimplified Boolean expression is plotted as a 1 on the Karnaugh map in Fig. 4-21(*b*). Adjacent groups of two, four, and eight are looped. The four 1s on the E and $\overline{E}$ planes are also adjacent so that the entire group is enclosed in a cylinder and is considered a single group of eight 1s.

The next step is the conversion of the looped 1s on the Karnaugh map to a simplified minterm Boolean expression. The lone 1 on the $\overline{E}$ plane of the map in Fig. 4-21(*b*) cannot be simplified and is written as $A \cdot \overline{B} \cdot \overline{C} \cdot \overline{D} \cdot \overline{E}$ in Fig. 4-21(*c*). The eight 1s enclosed in the looping cylinder can be simplified. The E and $\overline{E}$, the C and $\overline{C}$, and the B and $\overline{B}$ variables are eliminated leaving the term $\overline{A} \cdot D$. The terms $A \cdot \overline{B} \cdot \overline{C} \cdot \overline{D} \cdot \overline{E}$ and $\overline{A} \cdot D$ are ORed yielding the simplified minterm Boolean expression shown in Fig. 4-21(*c*) as $A \cdot \overline{B} \cdot \overline{C} \cdot \overline{D} \cdot \overline{E} + \overline{A} \cdot D = Y$.

Looping cylinder

TEST

Answer the following question.

29. Simplify the Boolean expression $A \cdot \overline{B} \cdot \overline{C} \cdot \overline{D} \cdot \overline{E} + A \cdot \overline{B} \cdot \overline{C} \cdot D \cdot \overline{E} + A \cdot \overline{B} \cdot \overline{C} \cdot \overline{D} \cdot E + A \cdot \overline{B} \cdot \overline{C} \cdot D \cdot E + \overline{A} \cdot B \cdot C \cdot D \cdot E + \overline{A} \cdot \overline{B} \cdot C \cdot D \cdot E = Y$ by:
    a. Plotting 1s on a five-variable Karnaugh map
    b. Looping groups of two, four, or eight adjacent 1s

$$A \cdot \overline{B} \cdot \overline{C} \cdot \overline{D} \cdot \overline{E} + \overline{A} \cdot \overline{B} \cdot \overline{C} \cdot D \cdot \overline{E} + \overline{A} \cdot B \cdot \overline{C} \cdot D \cdot \overline{E} +$$
$$\overline{A} \cdot \overline{B} \cdot C \cdot D \cdot \overline{E} + \overline{A} \cdot B \cdot C \cdot D \cdot \overline{E} + \overline{A} \cdot \overline{B} \cdot \overline{C} \cdot D \cdot E +$$
$$\overline{A} \cdot B \cdot \overline{C} \cdot D \cdot E + \overline{A} \cdot \overline{B} \cdot C \cdot D \cdot E + \overline{A} \cdot B \cdot C \cdot D \cdot E = Y$$

(*a*) Unsimplified Boolean expression

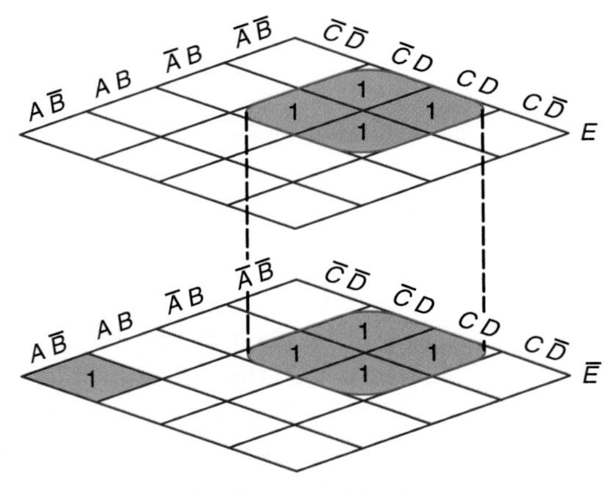

(*b*) Karnaugh map. Plotting 1s and looping

$$A \cdot \overline{B} \cdot \overline{C} \cdot \overline{D} \cdot \overline{E} + \overline{A} \cdot D = Y$$

(*c*) Simplified Boolean expression

Fig. 4-21 Using a five-variable Karnaugh map to simplify a Boolean expression.

c. Eliminating variables whose complements appear within loops or cylinders
d. Writing the simplified minterm Boolean expression

4-11 Using NAND Logic

Section 3-8 explained how NAND gates could be wired to form other gates or inverters (see Fig. 3-21). We mentioned that the NAND gate can be used as a universal gate. In this section, you will see how NAND gates are used in wiring combinational logic circuits. NAND gates are widely employed in industry because they are easy to use and readily available.

Suppose your supervisor gives you the Boolean expression $A \cdot B + A \cdot \overline{C} = Y$, as shown in Fig. 4-22(*a*). You are told to solve this logic problem at the least cost. You first draw the logic circuit for the Boolean expression shown in Fig. 4-22(*b*), using AND gates, an OR gate, and an inverter. Checking a manufacturer's catalog, you determine that you must use three different ICs to do the job.

Using NAND logic

Your supervisor suggests that you try using NAND logic. You redraw your logic circuit to look like the NAND-NAND logic circuit in Fig. 4-22(*c*). Upon checking a catalog, you find you need only one IC that contains the four NAND gates to do the job. Recall from Chap. 3 that the OR symbol with invert bubbles at the inputs is another symbol for a NAND gate. You finally test the circuit in Fig. 4-22(*c*) and find that it performs the logic $A \cdot B + A \cdot \overline{C} = Y$. Your supervisor is pleased you have found a circuit that requires only one IC, as compared to the circuit in Fig. 4-22(*b*), which uses three ICs.

Remembering this trick will help you appreciate *why* NAND gates are used in many logic circuits. If your future job is in digital circuit design, this can be a useful tool for making your final circuit the best for the least cost.

You may have questioned why the NAND gates in Fig. 4-22(*c*) could be substituted for the AND and OR gates in Fig. 4-22(*b*). If you look carefully at 4-22(*c*), you will see two AND symbols feeding into an OR symbol. From previous experience we know that if we invert twice we have the original logic state. Hence the two invert bubbles in Fig. 4-22(*c*) between the AND and OR symbols cancel one another. Because the two invert bubbles cancel one another, we end up with two AND gates feeding an OR gate.

In summary, using NAND gates involves these steps:

1. Start with a minterm (sum-of-products) Boolean expression.
2. Draw the AND-OR logic diagram using AND, OR, and NOT symbols.
3. Substitute NAND symbols for each AND and OR symbol, keeping all connections the same.
4. Substitute NAND symbols with all inputs tied together for each inverter.
5. Test the logic circuit containing all NAND gates to determine if it generates the proper truth table.

TEST

Answer the following questions.

30. The logic circuit in Fig. 4-22(*b*) is called a(n) ________ (AND-OR, NAND-NAND) circuit.
31. The logic circuits in Fig. 4-22(*b*) and (*c*) generate ________ (different, identical) truth tables.
32. List five steps in converting a sum-of-products Boolean expression to a NAND-NAND logic circuit.
33. Convert the minterm Boolean expression $\overline{A} \cdot \overline{B} + A \cdot B = Y$ to NAND logic by:
    a. Drawing an AND-OR logic diagram of this expression
    b. Redrawing the AND-OR diagram as a NAND-NAND logic diagram

4-12 Solving Logic Problems the Easy Way

Manufacturers of ICs have simplified the job of solving many combinational logic problems by producing *data selectors.* A data selector is often a *one-package solution* to a complicated logic problem. The data selector actually contains a rather large number of gates packaged inside a single IC. In this chapter the data selector will be used as a "universal package" for solving combinational logic problems.

Data selector

A *1-of-8 data selector* is illustrated in Fig. 4-23 on page 72. Notice the eight *data inputs*

1-of-8 selector

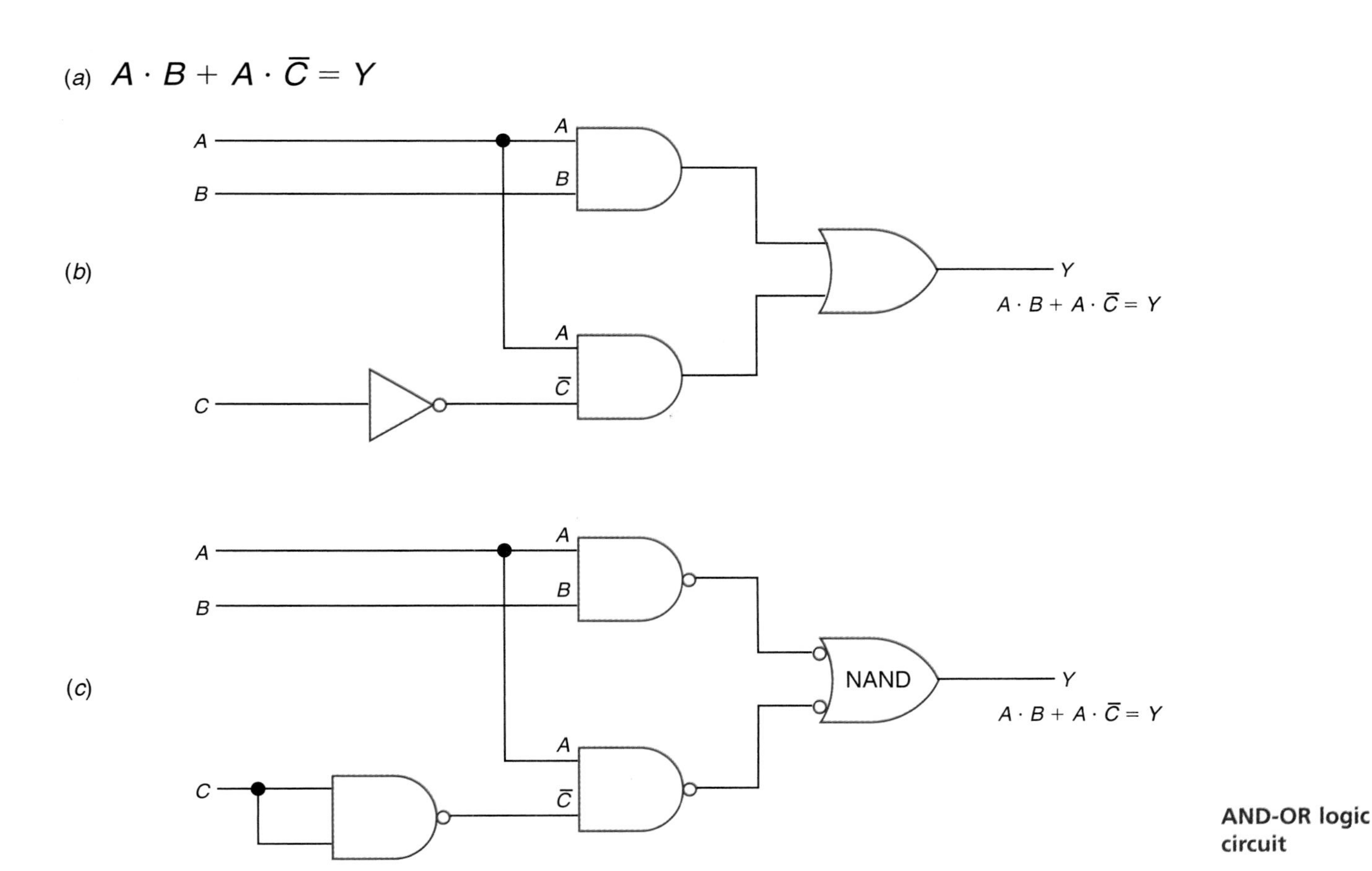

AND-OR logic circuit

NAND-NAND logic circuit

Fig. 4-22 Using NAND gates in logic circuits. (*a*) Boolean expression. (*b*) AND-OR logic circuit. (*c*) Equivalent NAND-NAND logic circuit.

About Electronics

How Fast Is Your Math? Teraflops—the ability to perform one trillion mathematical operations per second—is a mathematical boundary recently surpassed by Intel. In order to achieve this, Intel developed a computer with 4536 pairs of microprocessors and newly designed interface chips for each node and programming software. The U.S. government purchased one of these computers for $55 million.

numbered from 0 to 7 on the left. Also notice the three *data selector inputs* labeled *A*, *B*, and *C* at the bottom of the data selector. The output of the data selector is labeled *W*.

The basic job the data selector performs is transferring data from a *given* data input (0 to 7) to the output (*W*). Which data input is selected is determined by which binary number you place on the data selector inputs at the bottom (see Fig. 4-23). The data selector in Fig. 4-23 functions in the same manner as a rotary switch. Figure 4-24 shows the data at input 3 being transferred to the output by the rotary switch contacts. In like manner the data from data input 3 in Fig. 4-23 is being transferred to output *W* of the data selector. In the rotary switch you must mechanically change the switch position to transfer data from another input. In the 1-of-8 data selector in Fig. 4-23 you need only change the binary input at the data selector inputs to transfer data from another data input to the output. Remember that the data selector operates somewhat as a rotary switch in transferring logical 0s or 1s from a given input to the single output.

Now you will learn how data selectors can be used to solve logic problems. Consider the *simplified* Boolean expression shown in Fig. 4-25(*a*). For your convenience a logic circuit for this complicated Boolean expression is drawn in Fig. 4-25(*b*). Using standard ICs, we probably would have to use from six to nine IC packages to solve this problem. This would be quite expensive because of the cost of the ICs and PC board space.

1-of-16 data selector

A less costly solution to the logic problem is to use a data selector. The Boolean expression from Fig. 4-25(*a*) is repeated in truth-table form in Fig. 4-26 on page 74. A *1-of-16 data selector* is added in Fig. 4-26. Notice that logical 0s and 1s are placed at the 16 data inputs of the data selector corresponding to the truth-table output column *Y*. These are *permanently* connected for this truth table. Data selector inputs (*D, C, B,* and *A*) are switched to the binary numbers on the input side of the truth table. If the data selector inputs *D, C, B,* and *A* are at binary

Rotary switch

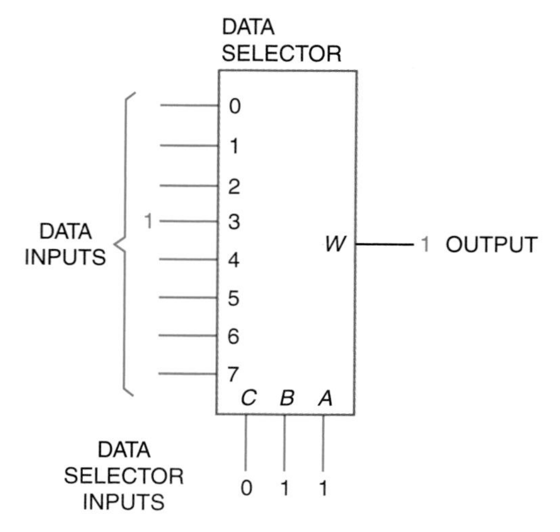

Fig. 4-23 Logic symbol for a 1-of-8 data selector.

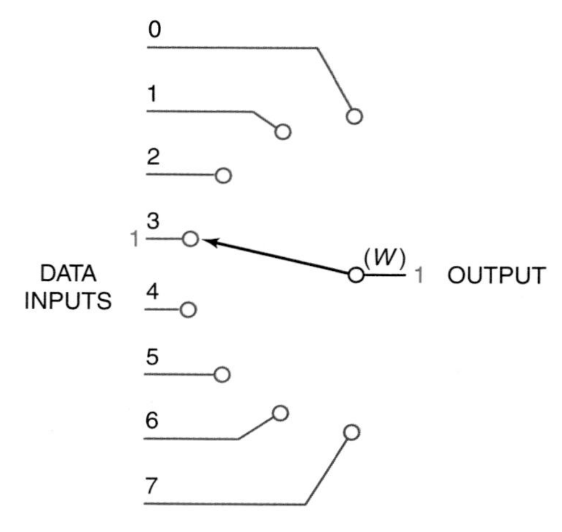

Fig. 4-24 Single-pole eight-position rotary switch works as a data selector.

(*a*) Simplified Boolean expression

$$A \cdot B \cdot C \cdot D + \bar{A} \cdot \bar{B} \cdot \bar{C} \cdot \bar{D} + A \cdot \bar{B} \cdot \bar{C} \cdot D + A \cdot B \cdot \bar{C} \cdot \bar{D} + \bar{A} \cdot B \cdot C \cdot \bar{D} + \bar{A} \cdot B \cdot \bar{C} \cdot D + \bar{A} \cdot \bar{B} \cdot C \cdot D = Y$$

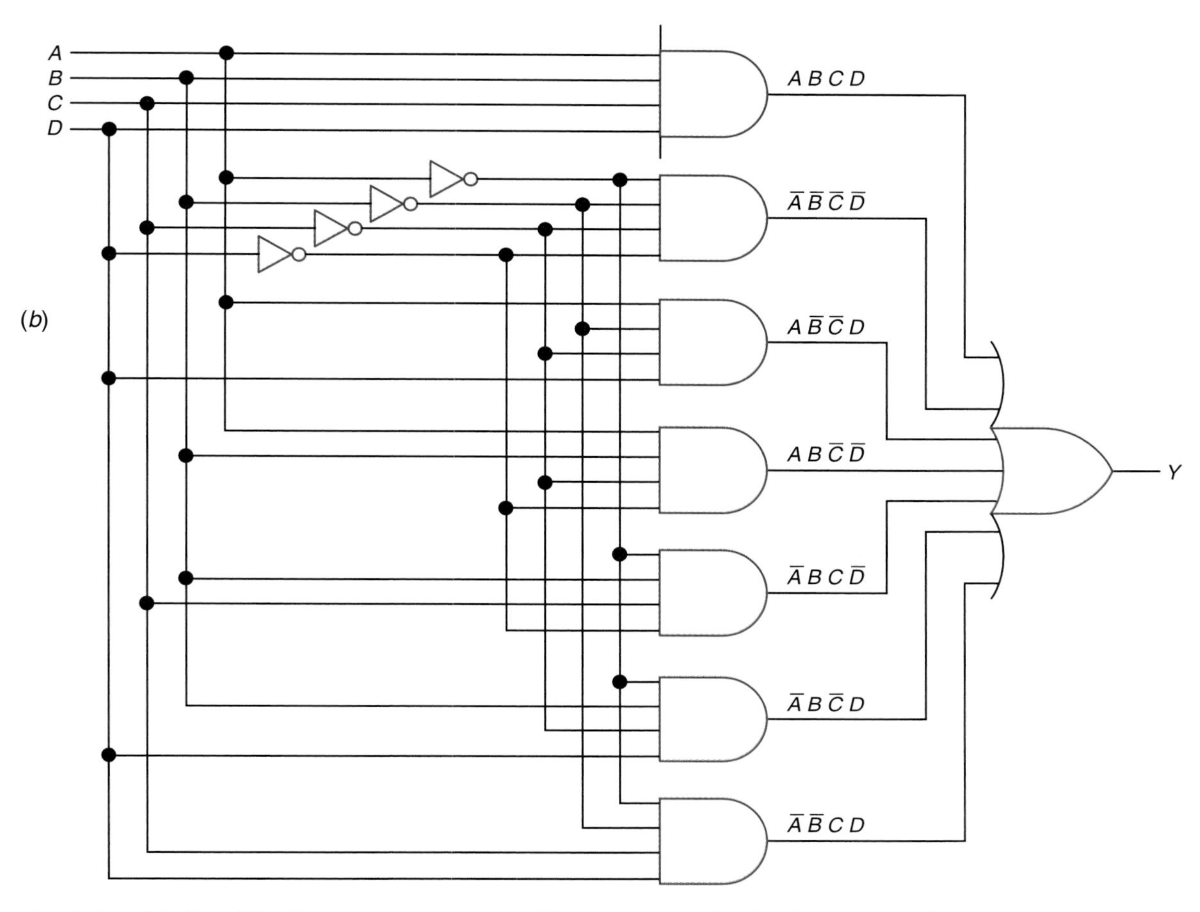

Fig. 4-25 (*a*) Simplified Boolean expression. (*b*) Logic circuit for Boolean expression.

0000, then a logical 1 is transferred to output *W* of the data selector. The first line of the truth table requires that a logical 1 appear at output *W* when *D, C, B,* and *A* are all 0s. If data selector inputs *D, C, B,* and *A* are at binary 0001, a logical 0 appears at output *W*, as required by the truth table. Any combination of *D, C, B,* and *A* generates the proper output according to the truth table.

We used the data selector to solve a complicated logic problem. In Fig. 4-25 we found we needed at least six ICs to solve this logic problem. Using the data selector in Fig. 4-26, we solved this problem by using only one IC.

The data selector seems to be an easy-to-use and efficient way to solve combinational logic problems. Commonly available data selectors can solve logic problems with three, four, or five variables. When using manufacturers' data manuals, you will notice that data selectors are also called *multiplexers.*

TEST

Supply the missing word, letter, or number in each statement.

34. Figure 4-23 illustrates the logic symbol for a 1-of-8 ________.
35. Refer to Fig. 4-23. If all data select inputs are HIGH, data at input ________ (number) is selected and transferred to output ________ (letter) of the data selector.
36. The action of a data selector is many times compared to that of a mechanical ________ switch.

Multiplexers

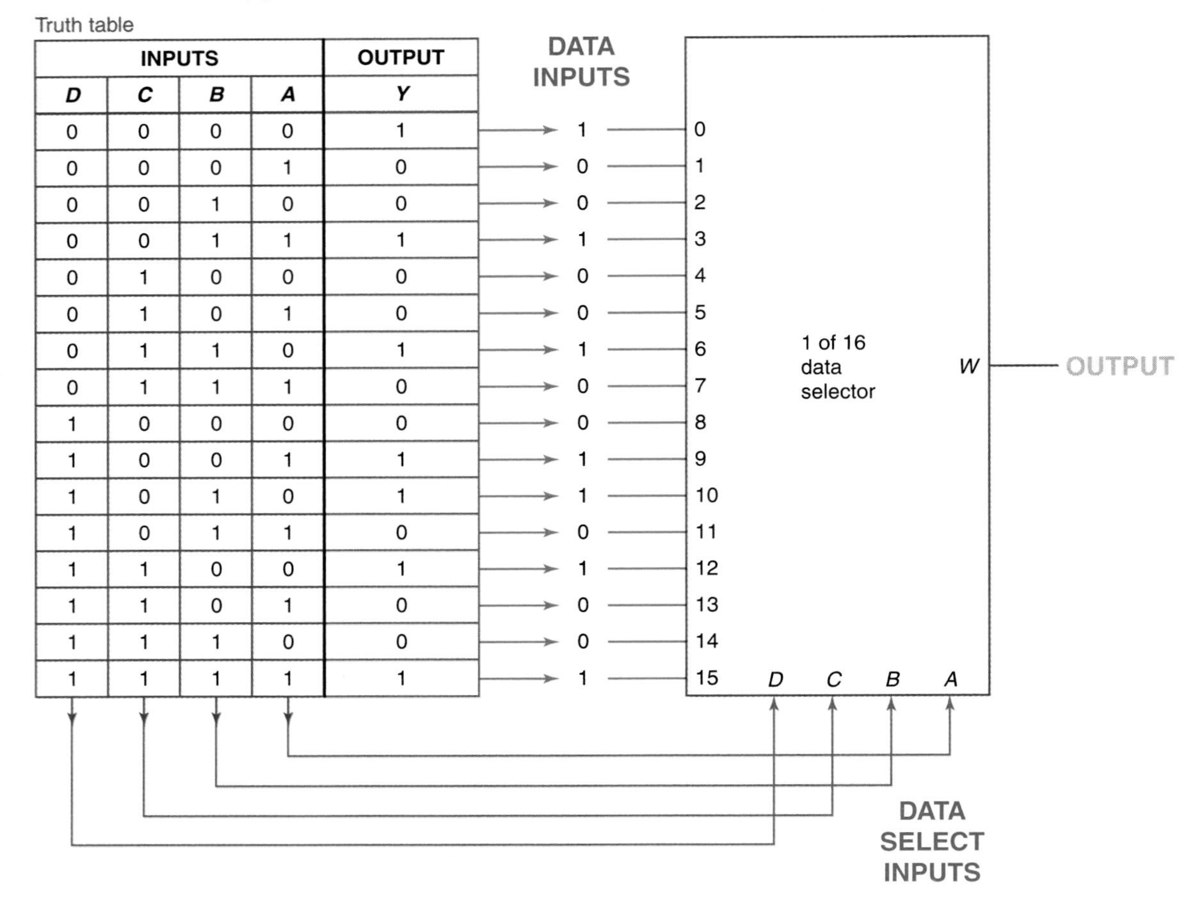

INPUTS				OUTPUT
D	C	B	A	Y
0	0	0	0	1
0	0	0	1	0
0	0	1	0	0
0	0	1	1	1
0	1	0	0	0
0	1	0	1	0
0	1	1	0	1
0	1	1	1	0
1	0	0	0	0
1	0	0	1	1
1	0	1	0	1
1	0	1	1	0
1	1	0	0	1
1	1	0	1	0
1	1	1	0	0
1	1	1	1	1

Data input	1	0	0	1	0	0	1	0	0	1	1	0	1	0	0	1
Input no.	0	1	2	3	4	5	6	7	8	9	10	11	12	13	14	15

Fig. 4-26 Solving logic problem with a data selector IC.

Multiplexers

37. Refer to Fig. 4-26. If all data select inputs are HIGH, data from input _________ (number) will be transferred to output *W*. Under these conditions, output *W* will be _________ (HIGH, LOW).
38. Logic problems can many times be solved using a single _________ IC.

Solving logic problem with a data selector

Folding technique

4-13 More Data Selector Problems

The previous section used a 1-of-16 data selector to solve a four-variable logic problem. A similar logic problem can be solved using a less expensive, 1-of-8 data selector. This is done by using what is sometimes called the "folding" technique.

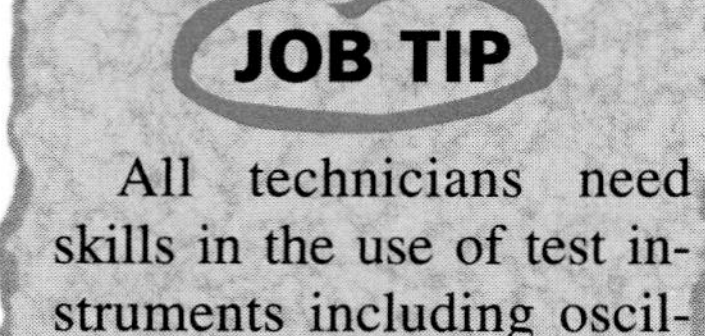

Consider the four-variable truth table shown in Fig. 4-27. Note that the pattern of inputs *C*, *B*, and *A* is the same in lines 0 through 7 as in lines 8 through 15. These areas are circled on the truth table in Fig. 4-27. To solve this logic problem using a 1-of-8 data selector, inputs *C*, *B*, and *A* are connected to the data select inputs of the unit. This is shown in the lower part of Fig. 4-27.

The eight data inputs (D_0 to D_7) shown in Fig. 4-28(*i*), on page 76, must now be determined one by one. The input to D_0 on the 74151 1-of-8 data selector IC is determined in Fig. 4-28(*a*). The truth table from Fig. 4-27 is folded over to compare lines 0 and 8. Inputs *C*, *B*, and *A* (which connect to the 74151 IC's data select or inputs) are each 000. Whether input *D* is 0 or 1, output *Y* is always 0 according to Fig. 4-28(*a*). Therefore, a logical 0 (GND) is applied to the D_0 input to the 74151 1-of-8 data selector IC. This is shown in Fig. 4-28(*i*).

The input to D_1 of the 74151 IC is determined in Fig. 4-28(*b*). The folding technique is used to compare lines 1 and 9 of the truth table.

Line Number	INPUTS				OUTPUT
	D	C	B	A	Y
0	0	0	0	0	0
1	0	0	0	1	1
2	0	0	1	0	0
3	0	0	1	1	1
4	0	1	0	0	0
5	0	1	0	1	0
6	0	1	1	0	1
7	0	1	1	1	0
8	1	0	0	0	0
9	1	0	0	1	1
10	1	0	1	0	1
11	1	0	1	1	0
12	1	1	0	0	1
13	1	1	0	1	0
14	1	1	1	0	0
15	1	1	1	1	0

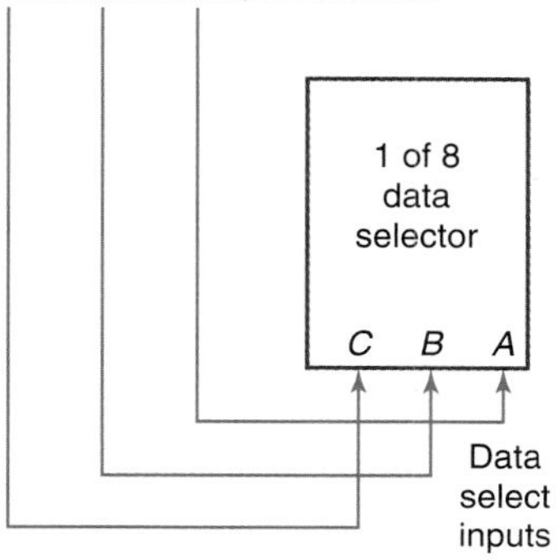

Fig. 4-27 First step in using a 1-of-8 data selector to solve a four-variable logic problem.

Inputs C, B, and A must be the same. Whether input D is 0 or 1, output Y is always 1. Therefore, a logical 1 (+5 V) is applied to the D_1 input to the 74151 data selector IC. This is shown in Fig. 4-28(*i*).

The input to D_2 of the 74151 IC is determined in Fig. 4-28(*c*). Folding the truth table compares lines 2 and 10. Inputs C, B, and A are the same. The outputs are different. In each case, the output is the same as the D input. Therefore, data input D_2 to the 74151 IC is equal to input D from the truth table. The logic symbol for the 74151 IC shows a D written to the left of input D_2 [Fig. 4-28(*i*)].

The input to D_3 of the 74151 data selector IC is determined in Fig. 4-28(*d*). Folding the truth table compares lines 3 and 11. Inputs C, B, and A are the same. The outputs are different. In each case, the output is the complement of the D input. Therefore, data input D_3 to the 74151 IC is equal to a not D ($\overline{D}$). The logic symbol for the 74151 IC in Fig. 4-28(*i*) shows a $\overline{D}$ written to the left of input D_3.

In like manner, the input to D_4 is determined in Fig. 4-28(*e*). Data input D_4 to the 1-of-8 data selector is equal to D.

The input to D_5 is determined in Fig. 4-28(*f*). Data input D_5 to the 74151 IC is equal to 0 (GND).

The input to D_6 is determined in Fig. 4-28(*g*). Data input D_6 to the 1-of-8 data selector is equal to $\overline{D}$ (not D).

Finally, the input to D_7 is determined in Fig. 4-28(*h*). Data input D_7 of the 74151 data selector IC is equal to 0 (GND).

Note in Fig. 4-28(*i*) that data inputs D_0, D_5, and D_7 of the 74151 IC are permanently grounded. Data input D_1 is permanently connected to +5 V. Data inputs D_2 and D_4 are connected directly to input D from the truth table. Data inputs D_3 and D_6 are connected through an inverter to the complement of input D. The enable, or strobe, input to the 74151 1-of-8 data selector must be held LOW (at logical 0) for the unit to operate. The small bubble on the logic symbol in Fig. 4-28(*i*) means that the enable input is an active LOW input.

Universal logic element

The data selector (multiplexer) has been used as a universal logic element in the last two sections. It is a simple, low-cost solution to many logic problems with from three to five input variables.

Data selectors

Simplified gate circuits and data selector ICs have been used to implement logic problems. More complex logic problems are created when there are more variables or when the logic circuit has several outputs. For these problems, designers can use a programmable array of logic gates within a single IC. This device is called *programmable array logic* or *PAL*. The PAL is based on programmable AND/OR architecture. These programmable logic devices are available in both TTL and CMOS. These devices are user-programmable. A typical PAL may have 16 inputs and 8 outputs.

Very complicated logic problems can be solved using either factory-programmed *gate-arrays* or *read-only memories (ROMs)*. User-programmable gate arrays and ROMs are also available in the form of *PROMs (programmable ROMs)*, *EPROMs (erasable PROMs)*, and *programmable gate arrays*.

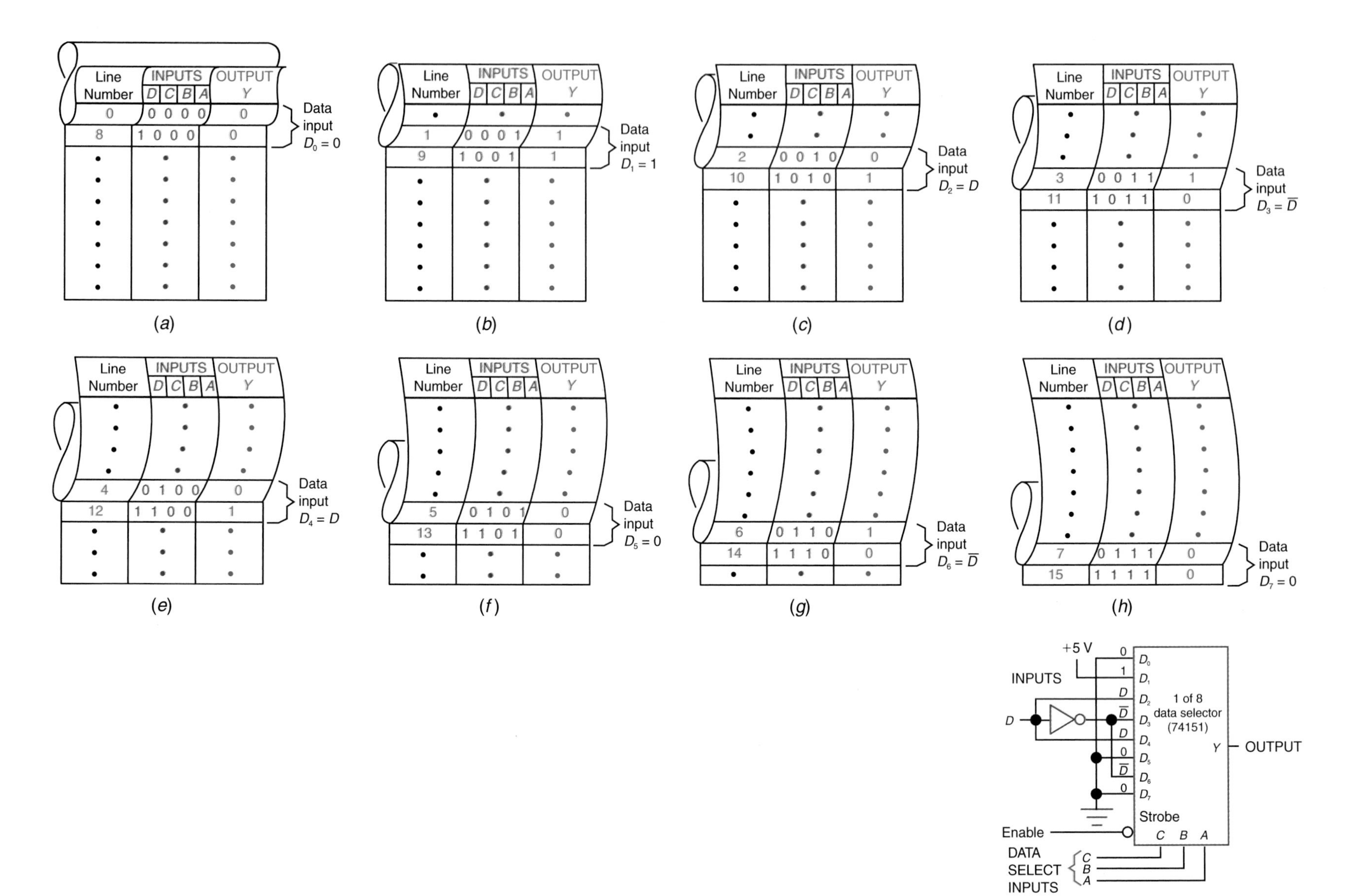

Fig. 4-28 Second step in using a 1-of-8 data selector to solve a four-variable logic problem using the "folding" technique. (*a*) Determining data to be placed at input D_0. (*b*) Determining data to be placed at input D_1. (*c*) Determining data to be placed at input D_2. (*d*) Determining data to be placed at input D_3. (*e*) Determining data to be placed at input D_4. (*f*) Determining data to be placed at input D_5. (*g*) Determining data to be placed at input D_6. (*h*) Determining data to be placed at input D_7. (*i*) Solution to four-variable logic problem posed in truth table.

TEST

Select the correct word in each statement.

39. Refer to Fig. 4-28(*i*). With inputs of $D = 1$, $C = 0$, $B = 0$, $A = 0$, and enable $= 0$, output Y of the data selector will be ________ (HIGH, LOW).
40. Refer to Fig. 4-28(*i*). With inputs of $D = 1$, $C = 1$, $B = 0$, $A = 0$, and enable $= 0$, output Y of the data selector will be ________ (HIGH, LOW).
41. Refer to Fig. 4-28(*i*). With inputs of $D = 1$, $C = 0$, $B = 1$, $A = 1$, and enable $= 0$, output Y of the data selector will be ________ (HIGH, LOW).
42. List several devices that can be used to solve very complicated logic problems.

4-14 Using De Morgan's Theorems

Boolean algebra, the algebra of logic circuits, has many laws or theorems. *De Morgan's theorems* are very useful. They allow us to convert back and forth from minterm to maxterm forms of Boolean expressions. They also allow us to eliminate long overbars that cover several variables.

De Morgan's theorems can be stated in the form shown in Fig. 4-29. The first theorem ($\overline{A + B} = \overline{A} \cdot \overline{B}$) shows that the long overbar covering the $\overline{A + B}$ term can be eliminated. A simple example of the first theorem is shown in the example section when the customary NOR logic symbol ($\overline{A + B} = Y$) is shown as equivalent to the alternative NOR symbol ($\overline{A} + \overline{B} = Y$).

De Morgan's second theorem is stated in Fig. 4-29 as $\overline{A \cdot B} = \overline{A} + \overline{B}$. A simple example of the second theorem is shown in the example section where the customary NAND logic symbol $\overline{A \cdot B}$) is shown as equivalent to the alternative NAND symbol ($\overline{A} + \overline{B} = Y$).

Boolean Expressions—Keyboard Version

The long overbars in Boolean expressions (for example, $\overline{A \cdot B}$) are somewhat more difficult to show in the keyboard versions of an expression. For instance, the keyboard version of $\overline{A \cdot B}$ would be (*AB*)'. The apostrophe *outside* the parenthesis means a long overbar. Next consider the Boolean expression $\overline{A \cdot B \cdot \overline{C} + \overline{A} \cdot \overline{B} \cdot \overline{C}} = Y$. The keyboard version of this expression would be ((*ABC*') + (*A*'*B*'*C*'))'. The customary Boolean expression for the NOR function is $\overline{A + B}$ while the keyboard version could be typed as (*A* + *B*)'. Do not be surprised when working with some circuit simulation programs if the software converts a minterm to a maxterm or maxterm to a minterm type expression. For instance, it might convert the conventional NAND notation ($\overline{A \cdot B}$) to alternative NAND notation (*A*' + *B*'). Computer circuit simulation programs use De Morgan's theorems to make these conversions.

Minterm-to-Maxterms or Maxterm-to-Minterms

Four steps are need to convert a maxterm to minterm Boolean expression or from minterm to maxterm form. The four steps, which are

(*a*) First theorem

$$\overline{A + B} = \overline{A} \cdot \overline{B}$$

(*b*) Example—first theorem

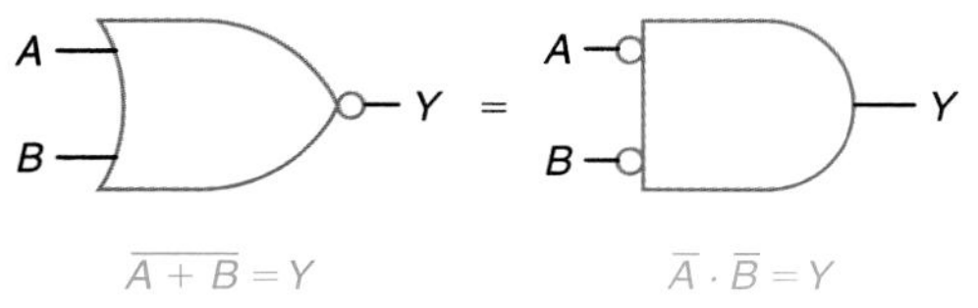

A, B, Y, =, A, B, Y

$\overline{A + B} = Y$ $\overline{A} \cdot \overline{B} = Y$

(*c*) Second theorem

$$\overline{A \cdot B} = \overline{A} + \overline{B}$$

(*d*) Example—second theorem

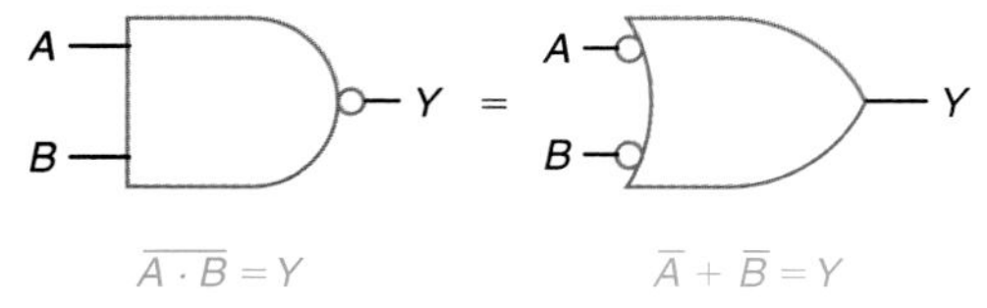

A, B, Y, =, A, B, Y

$\overline{A \cdot B} = Y$ $\overline{A} + \overline{B} = Y$

Fig. 4-29 De Morgan's theorems and practical examples.

based on De Morgan's theorems, are as follows:

> Step 1. Change all ORs to ANDs and all ANDs to ORs.
> Step 2. Complement each individual variable (add short overbars to each).
> Step 3. Complement the entire function (add long overbar to entire function).
> Step 4. Eliminate all groups of double overbars.

As an example, consider converting the customary NAND expression ($\overline{A \cdot B} = Y$) to its alternative NAND form ($\overline{A} + \overline{B} = Y$). Follow the four-step process in Fig. 4-30 to get familiar with the procedure. At the end of the procedure, the alternative NAND expression is shown as $\overline{A} + \overline{B} = Y$, but on the computer it would be represented as *A' + B' = Y*.

Now we will use the four-step procedure in converting a more complicated maxterm expression to its minterm form. Conversions from maxterm-to-minterm or minterm-to-maxterm form are commonly undertaken to get rid of long overbars in the Boolean expression. The new example illustrated in Fig. 4-31 will change the maxterm expression $\overline{(\overline{A} + \overline{B} + \overline{C}) \cdot (A + B + \overline{C})} = Y$ to its minterm equivalent and eliminate the long overbar. Carefully follow the conversion process in Fig. 4-31. The result of this conversion is that the minterm form $A \cdot B \cdot C + \overline{A} \cdot \overline{B} \cdot C = Y$, which performs exactly the same logic function as the maxterm expression $\overline{(\overline{A} + \overline{B} + \overline{C}) \cdot (A + B + \overline{C})} = Y$. The resulting minterm expression can be written in conventional form as $A \cdot B \cdot C + \overline{A} \cdot \overline{B} \cdot C = Y$ using overbars or in the shortened keyboard version *ABC + A'B'C = Y* using apostrophes.

It must be understood that the logic diagrams that would be wired using the maxterm expres-

Begin. Customary NAND expression.

$$\overline{A \cdot B} = Y$$

Step 1. Change all ORs to ANDs and all ANDs to ORs.

$$\overline{A + B} = Y$$

Step 2. Complement each individual variable (short overbar).

$$\overline{\overline{A} + \overline{B}} = Y$$

Step 3. Complement the entire function (long overbar).

$$\overline{\overline{\overline{A} + \overline{B}}} = Y$$

Step 4. Eliminate all groups of double overbars.

$$\overline{A} + \overline{B} = Y$$

End. Alternative NAND expression.

$$\overline{A} + \overline{B} = Y$$

Fig. 4-30 Four-step process using De Morgan's second theorem to convert conventional NAND to alternative NAND. Note that the long overbar is eliminated.

Begin. Maxterm expression.

$$\overline{(\overline{A} + \overline{B} + \overline{C}) \cdot (A + B + \overline{C})} = Y$$

Step 1. Change all ORs to ANDs and all ANDs to ORs.

$$\overline{\overline{A} \cdot \overline{B} \cdot \overline{C} + A \cdot B \cdot \overline{C}} = Y$$

Step 2. Complement each individual variable (short overbars).

$$\overline{\overline{\overline{A}} \cdot \overline{\overline{B}} \cdot \overline{\overline{C}} + \overline{A} \cdot \overline{B} \cdot \overline{\overline{C}}} = Y$$

Step 3. Complement the entire function (long overbar).

$$\overline{\overline{\overline{\overline{A}} \cdot \overline{\overline{B}} \cdot \overline{\overline{C}} + \overline{A} \cdot \overline{B} \cdot \overline{\overline{C}}}} = Y$$

Step 4. Eliminate all groups of double overbars.

$$A \cdot B \cdot C + \overline{A} \cdot \overline{B} \cdot C = Y$$

End. Minterm expression.

$$A \cdot B \cdot C + \overline{A} \cdot \overline{B} \cdot C = Y$$

Fig. 4-31 Four-step process using De Morgan's theorems to convert from maxterm-to-minterm form. Note that the long overbar is eliminated.

sion $\overline{(\overline{A} + \overline{B} + \overline{C}) \cdot (A + B + \overline{C})} = Y$ or its equivalent minterm form $A \cdot B \cdot C + \overline{A} \cdot \overline{B} \cdot C = Y$ from Fig. 4-31 would *look different,* but they would generate the same truth table. It is said that they generate the same logic function.

In summary, De Morgan's theorems are useful for converting from maxterm-to-minterm or minterm-to-maxterm form of Boolean expressions. We commonly make this conversion to eliminate long overbars in a Boolean expression. A second reason to use De Morgan's theorem might be to examine two different logic diagrams that perform the same logic function. One logic diagram might be simpler than the other.

TEST

Answer the following questions.

43. State two of De Morgan's theorems.
44. Convert the maxterm Boolean expression $\overline{(A + \overline{B} + \overline{C}) \cdot (\overline{A} + B + \overline{C})} = Y$ to its minterm form. Show each step as is done in Fig. 4-31.
45. Convert the minterm Boolean expression $\overline{\overline{A} \cdot B \cdot C + \overline{A} \cdot \overline{B} \cdot \overline{C}} = Y$ to its maxterm form. Show each step as is done in Fig. 4-31.
46. Write the Boolean expression $\overline{\overline{A} \cdot B \cdot C + \overline{A} \cdot \overline{B} \cdot \overline{C}} = Y$ in the keyboard version using apostrophes instead of overbars.
47. Draw a logic symbol diagram for the Boolean expression $(A'BC + A'B'C')' = Y$. HINT: Use a two-input NOR gate nearest output.
48. Draw a logic symbol diagram for the Boolean expression $((A + B + C + D)(A' + D)(A' + B' + C'))' = Y$. HINT: Use three-input NAND gate nearest output.

Chapter 4 Review

Summary

1. Combining gates in combinational logic circuits from Boolean expressions is a necessary skill for most competent technicians and engineers.
2. Workers in digital electronics must have an excellent knowledge of gate symbols, truth tables, and Boolean expressions and know how to convert from one form to another.
3. The minterm Boolean expression (sum-of-products form) might look like the expression in Fig. 4-32(*a*). The Boolean expression $A \cdot B + \overline{A} \cdot \overline{C} = Y$ would be wired as shown in Fig. 4-32(*b*).
4. The pattern of gates shown in Fig. 4-32(*b*) is called an AND-OR logic circuit.
5. The maxterm Boolean expression (product-of-sums form) might look like the expression in Fig. 4-32(*c*). The Boolean expression $(A + \overline{C}) \cdot (\overline{A} + B) = Y$ would be wired as shown in Fig. 4-32(*d*). This is an OR-AND logic circuit.
6. A Karnaugh map is a convenient method of simplifying Boolean expressions.
7. AND-OR logic circuits can be wired easily by using only NAND gates, as shown in Fig. 4-33.
8. Data selectors are a simple, one-package method of solving many gating problems. Less expensive data selectors can be used when the folding design technique is utilized.
9. Very complex logic problems can be solved using PALs, ROMs, PROMS, gate arrays, or user-programmable gate arrays.
10. De Morgan's theorems are useful in converting maxterm-to-minterm and minterm-to-maxterm Boolean expressions.
11. A keyboard version of a Boolean expression is used with computer systems. An example would be $\overline{A \cdot B} = Y$ is equivalent to $(A'B)' = Y$.

(*a*) Minterm Boolean expression

$$A \cdot B + \overline{A} \cdot \overline{C} = Y$$

(*b*)

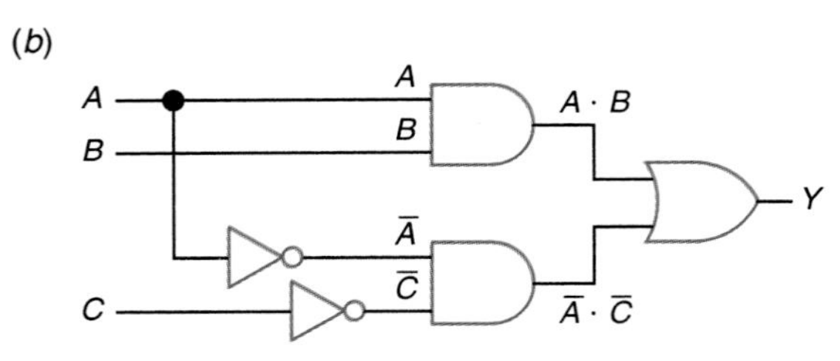

(*c*) Maxterm Boolean expression

$$(A + \overline{C}) \cdot (\overline{A} + B) = Y$$

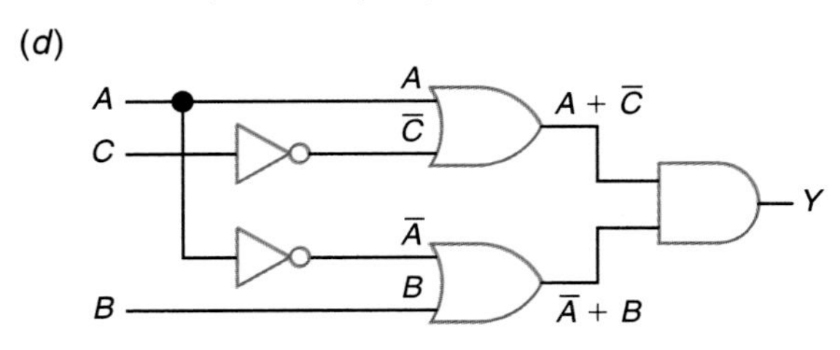

Fig. 4-32 (*a*) Minterm expression. (*b*) AND-OR logic circuit. (*c*) Maxterm expression. (*d*) OR-AND logic circuit.

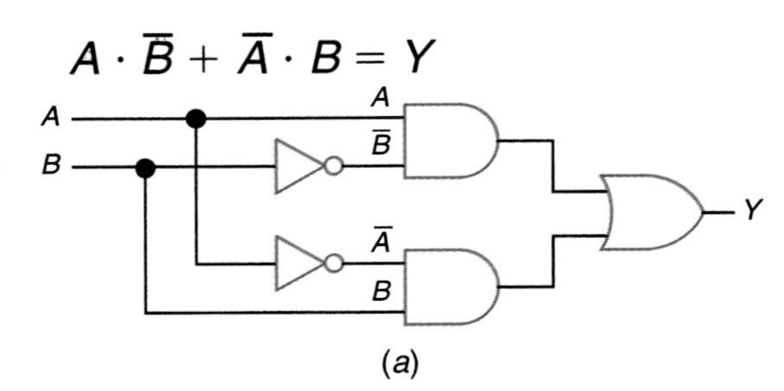

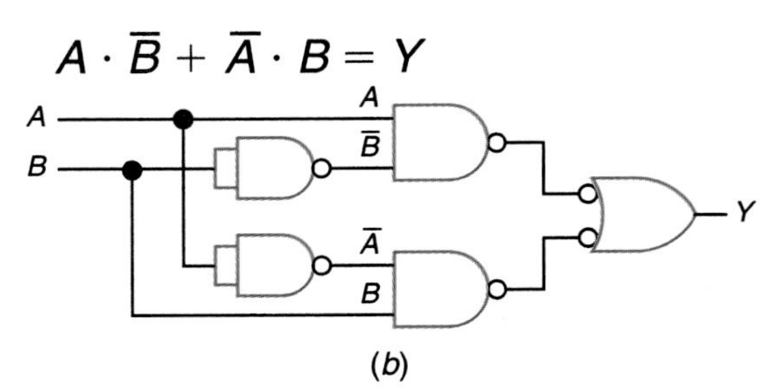

Fig. 4-33 (*a*) AND-OR logic circuit. (*b*) Equivalent NAND-NAND logic circuit.

Chapter Review Questions

Answer the following questions.

4-1. Engineers and technicians refer to circuits that are a combination of different gates as ________ logic circuits.

4-2. Draw a logic diagram for the Boolean expression $\overline{A} \cdot \overline{B} + B \cdot C = Y$. Use one OR gate, two AND gates, and two inverters.

4-3. The Boolean expression $\overline{A} \cdot \overline{B} + B \cdot C = Y$ is in ________ (product-of-sums, sum-of-products) form.

4-4. The Boolean expression $(A + B) \cdot (C + D) = Y$ is in ________ (product-of-sums, sum-of-products) form.

4-5. A Boolean expression in product-of-sums form is also called a ________ expression.

4-6. A Boolean expression in sum-of-products form is also called a ________ expression.

4-7. Write the minterm Boolean expression that would describe the truth table in Fig. 4-34. Do not simplify the Boolean expression.

4-8. Draw a truth table (three-variable) that represents the Boolean expression $\overline{C} \cdot \overline{B} + C \cdot \overline{B} \cdot A = Y$.

4-9. The truth table in Fig. 4-35 is for an electronic lock. The lock will open only when a logical 1 appears at the output. First, write the minterm Boolean expression for the lock. Second, draw the logic circuit for the lock (use AND, OR, and NOT gates).

4-10. List the six steps for simplifying a Boolean expression using a Karnaugh map as discussed in Sec. 4-6.

4-11. Use a Karnaugh map to simplify the Boolean expression $\overline{A} \cdot \overline{B} \cdot \overline{C} + \overline{A} \cdot \overline{B} \cdot C + A \cdot B \cdot \overline{C} + A \cdot \overline{B} \cdot \overline{C} = Y$. Write the simplified Boolean expression in minterm form.

4-12. Use a Karnaugh map to simplify the Boolean expression $A \cdot \overline{B} \cdot \overline{C} \cdot \overline{D} + A \cdot \overline{B} \cdot \overline{C} \cdot D + A \cdot \overline{B} \cdot C \cdot D + A \cdot \overline{B} \cdot C \cdot \overline{D} = Y$.

4-13. From the truth table in Fig. 4-34 do the following:

a. Write the unsimplified Boolean expression.

b. Use a Karnaugh map to simplify the Boolean expression from a.

c. Write the simplified minterm Boolean expression for the truth table.

d. Draw a logic circuit from the simplified Boolean expression (use AND, OR, and NOT gates).

e. Redraw the logic circuit from d using only NAND gates.

4-14. Use a Karnaugh map to simplify the Boolean expression $\overline{A} \cdot \overline{B} \cdot C \cdot D + A \cdot B \cdot \overline{C} \cdot \overline{D} + A \cdot B \cdot C \cdot \overline{D} + A \cdot \overline{B} \cdot C \cdot D = Y$. Write the answer as a minterm Boolean expression.

4-15. From the Boolean expression $\overline{A} \cdot \overline{B} \cdot \overline{C} \cdot \overline{D} + \overline{A} \cdot \overline{B} \cdot C \cdot D + \overline{A} \cdot B \cdot \overline{C} \cdot D + A \cdot B \cdot C \cdot D + A \cdot B \cdot C \cdot \overline{D} + A \cdot \overline{B} \cdot \overline{C} \cdot \overline{D} = Y$ do the following:

a. Draw a truth table for the expression.

b. Use a Karnaugh map to simplify.

c. Draw a logic circuit of the simplified Boolean expression (use AND, OR, and NOT gates).

d. Draw a circuit to solve this problem using a 1-of-16 data selector.

e. Draw a circuit to solve this problem using the folding technique and a 1-of-8 data selector.

INPUTS			OUTPUT
C	B	A	Y
0	0	0	1
0	0	1	0
0	1	0	1
0	1	1	0
1	0	0	0
1	0	1	1
1	1	0	0
1	1	1	1

Fig. 4-34 Truth table.

INPUTS			OUTPUT
C	B	A	Y
0	0	0	0
0	0	1	0
0	1	0	0
0	1	1	1
1	0	0	1
1	0	1	0
1	1	0	0
1	1	1	0

Fig. 4-35 Truth table.

4-16. From the Boolean expression $\overline{A}\cdot\overline{B}\cdot\overline{C}\cdot D\cdot E + \overline{A}\cdot B\cdot\overline{C}\cdot D\cdot E + A\cdot B\cdot\overline{C}\cdot D\cdot E + A\cdot\overline{B}\cdot\overline{C}\cdot D\cdot E + A\cdot B\cdot\overline{C}\cdot D\cdot\overline{E} + \overline{A}\cdot\overline{B}\cdot C\cdot D\cdot\overline{E} + \overline{A}\cdot\overline{B}\cdot C\cdot\overline{D}\cdot\overline{E} = Y$ do the following:
   a. Use a Karnaugh map to simplify.
   b. Write the simplified minterm Boolean expression.
   c. Draw a logic circuit from the simplified Boolean expression (use AND, OR, and NOT gates).

4-17. The Boolean algebra laws that allow us to convert from minterm-to-maxterm or maxterm-to-minterm forms of expressions are called ________ ________.

4-18. Based on De Morgan's first theorem, $\overline{A + B} =$ ________.

4-19. Based on De Morgan's second theorem, $\overline{\overline{A}\cdot\overline{B}} =$ ________.

4-20. Using De Morgan's theorems, convert the maxterm Boolean expression $\overline{(A + \overline{B} + C)\cdot(\overline{A} + \overline{B} + \overline{C})} = Y$ to its minterm form. This will remove the long overbar.

4-21. Using De Morgan's theorems, convert the minterm Boolean expression $\overline{\overline{A}\cdot\overline{B}\cdot C + A\cdot B\cdot C} = Y$ to its maxterm form. This will remove the long overbar.

Critical Thinking Questions

4-1. When implemented, a minterm Boolean expression produces what pattern of logic gates?

4-2. When implemented, a maxterm Boolean expression produces what pattern of logic gates?

4-3. Simplify the Boolean expression $\overline{A}\cdot\overline{B}\cdot\overline{C}\cdot\overline{D} + \overline{A}\cdot\overline{B}\cdot\overline{C}\cdot D + \overline{A}\cdot B\cdot\overline{C}\cdot\overline{D} + \overline{A}\cdot B\cdot\overline{C}\cdot D + A\cdot\overline{B}\cdot\overline{C}\cdot\overline{D} + A\cdot\overline{B}\cdot\overline{C}\cdot D + A\cdot\overline{B}\cdot C\cdot D = Y$.

4-4. Do you think it is possible to develop a maxterm (product-of-sums) Boolean expression from a truth table?

4-5. Do you think the Karnaugh map shown in Fig. 4-17(*b*) can be used to simplify either minterm or maxterm Boolean expressions?

4-6. A five-variable logic problem could be solved using a 1-of-16 data selector employing the ________ technique.

4-7. A six-variable truth table would have how many combinations?

4-8. Write the keyboard version of the Boolean expression $\overline{A}\cdot\overline{B}\cdot C + A\cdot B\cdot\overline{C} + A\cdot\overline{B}\cdot C = Y$ as you may have to do when entering information into a electronic circuit simulator.

4-9. Write a *maxterm* Boolean expression for the logic diagram shown in Fig. 4-36.

4-10. Using De Morgan's theorems (or circuit simulator software if available) write the *minterm* Boolean expression that would describe the logic function of the circuit in Fig. 4-36. HINT: Use the maxterm expression developed in question 4-9.

4-11. Draw a three-variable truth table that would describe the logic function of the circuit in Fig. 4-36. HINT: Work from the minterm expression developed in question 4-10.

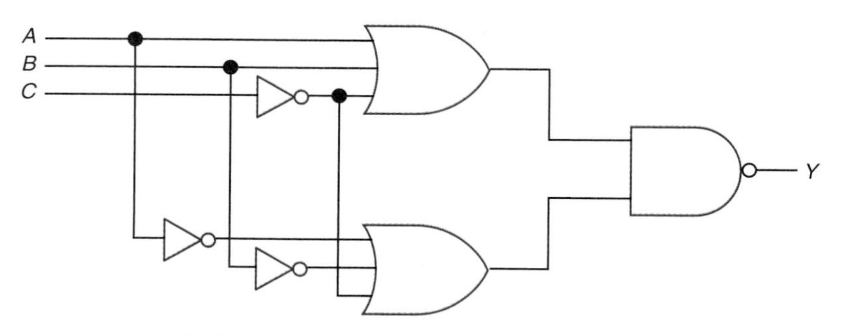

Fig. 4-36 Logic circuit.

Answers to Tests

1. See logic circuits a, b, and c below.

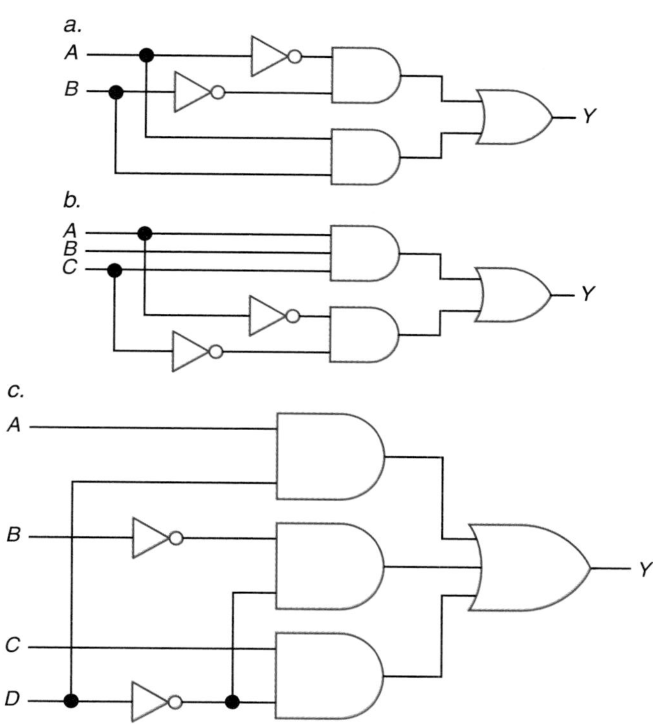

2. sum-of-products
3. product-of-sums
4. sum-of-products
5. product-of-sums
6. See logic circuits a, b, and c below.

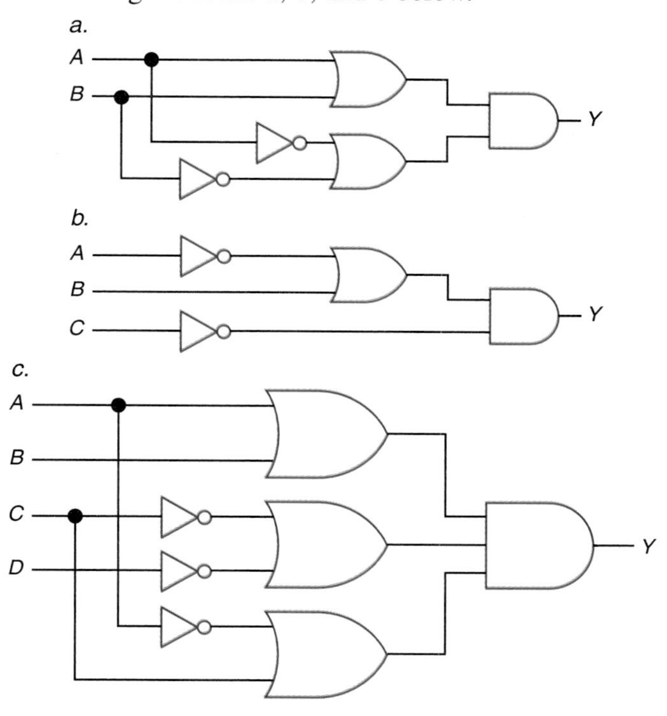

7. maxterm
8. product-of-sums
9. OR-AND
10. $C \cdot B \cdot \overline{A} + C \cdot B \cdot A = Y$
11. lines 1 and 2
12. See table below.

INPUTS			OUTPUT
C	**B**	**A**	**Y**
0	0	0	0
0	0	1	0
0	1	0	0
0	1	1	0
1	0	0	0
1	0	1	1
1	1	0	1
1	1	1	0

13. minterm
14. minterm
15. $C'B'A + BA' = Y$
16. $\overline{C} \cdot B \cdot \overline{A} + C \cdot \overline{B} \cdot A = Y$
17. See figure below.

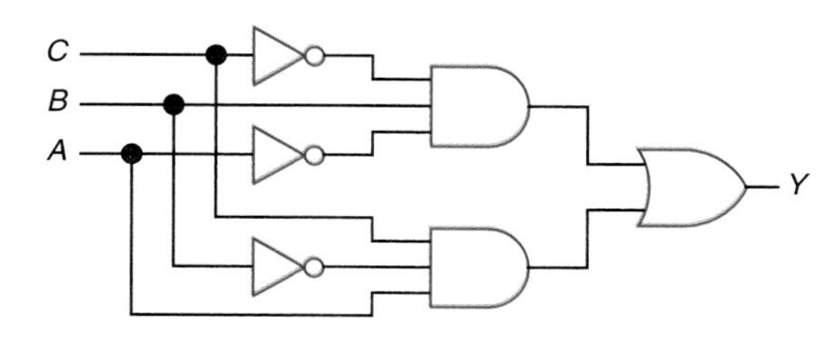

18. identical
19. Boolean, Karnaugh
20. Quine-McCluskey or tabular method
21. Maurice Karnaugh
22. 1. Start with a minterm Boolean expression.
    2. Record 1s on a Karnaugh map.
    3. Loop adjacent 1s (loops of two, four, or eight squares).
    4. Simplify by dropping terms that contain a term and its complement within a loop.
    5. OR the remaining terms (one term per loop).
    6. Write the simplified minterm Boolean expression.

23. See a–c below.

	$\bar{C}$	C
$\bar{A}\,\bar{B}$		1
$\bar{A}\,B$		1
$A\,B$		
$A\,\bar{B}$	1	1

d. $\bar{A} \cdot C + A \cdot \bar{B} = Y$

24. See a–c below.

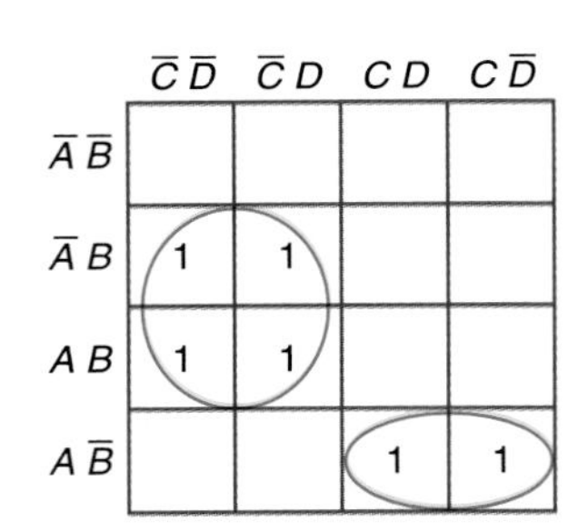

d. $B = Y$

25. See a–c below.

	$\bar{C}\,\bar{D}$	$\bar{C}\,D$	$C\,D$	$C\,\bar{D}$
$\bar{A}\,\bar{B}$				
$\bar{A}\,B$	1	1		
$A\,B$	1	1		
$A\,\bar{B}$			1	1

d. $B \cdot \bar{C} + A \cdot \bar{B} \cdot C = Y$

26. See a–c below.

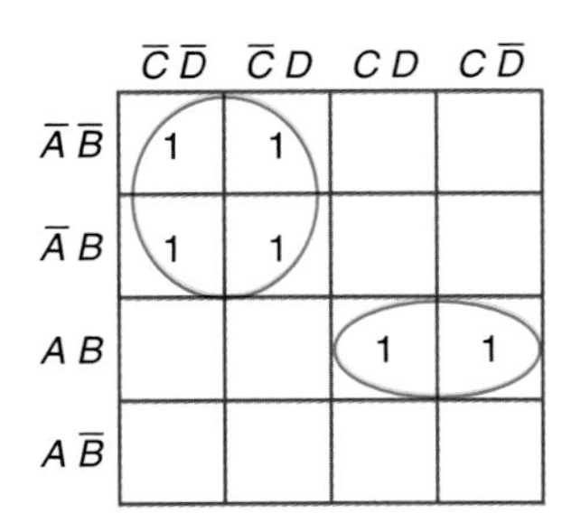

d. $\bar{A} \cdot \bar{C} + A \cdot B \cdot C = Y$

27. See a–c below.

	$\bar{C}\,\bar{D}$	$\bar{C}\,D$	$C\,D$	$C\,\bar{D}$
$\bar{A}\,\bar{B}$		1	1	
$\bar{A}\,B$	1			1
$A\,B$				
$A\,\bar{B}$		1	1	

d. $\bar{A} \cdot B \cdot \bar{D} + \bar{B} \cdot D = Y$

28. See a–c below.

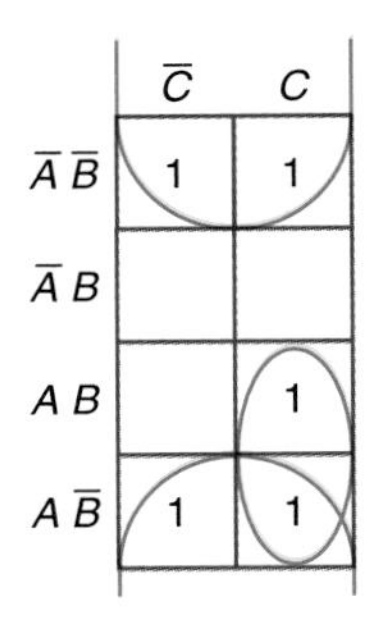

d. $\bar{B} + A \cdot C = Y$

29. See a–c below.

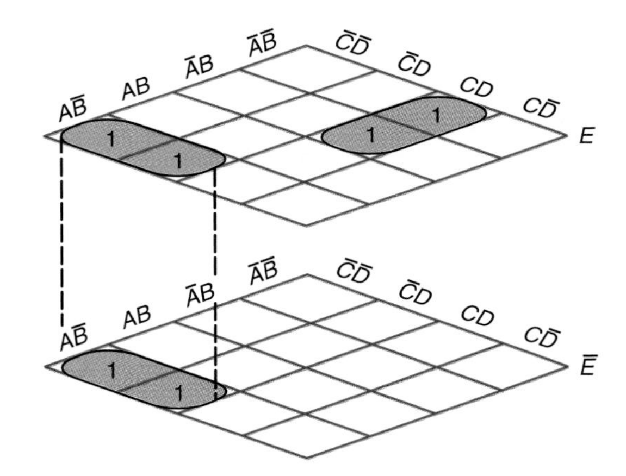

d. $A \cdot \bar{B} \cdot \bar{C} + \bar{A} \cdot C \cdot D \cdot E = Y$

30. AND-OR
31. identical
32. 1. Start with a minterm Boolean expression.
    2. Draw the AND-OR logic diagram using AND, OR, and NOT symbols.
    3. Substitute NAND symbols for each AND and OR symbol, keeping all connections the same.
    4. Substitute NAND symbols with all inputs tied together for each inverter.
    5. Test the logic circuit containing all NAND gates to determine that it generates the proper truth table.

33. a.

b.

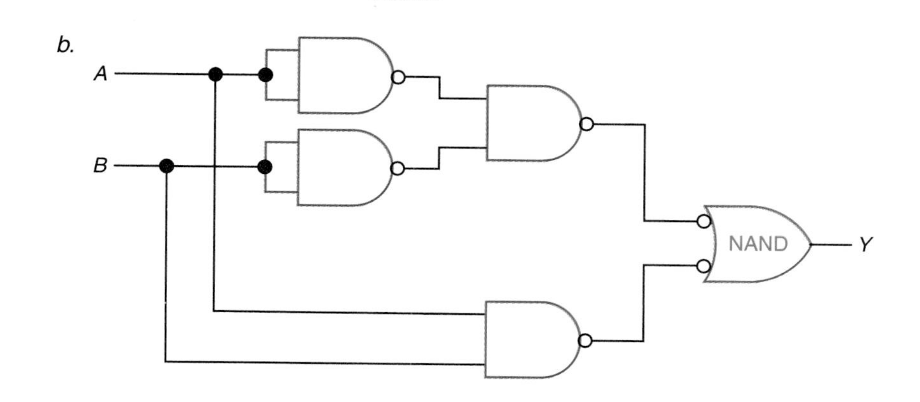

34. data selector
35. 7, *W*.
36. rotary
37. 15, HIGH
38. data selector
39. LOW
40. HIGH
41. LOW
42. ROMs, PROMs, PALs, gate arrays, programmable gate arrays
43. $\overline{A + B} = \overline{A} \cdot \overline{B}$

    $\overline{A \cdot B} = \overline{A} + \overline{B}$
44. Begin. $\overline{(A + \overline{B} + \overline{C}) \cdot (\overline{A} + B + \overline{C})} = Y$

    Step 1. $\overline{(A \cdot \overline{B} \cdot \overline{C}) + (\overline{A} \cdot B \cdot \overline{C})} = Y$

    Step 2. $\overline{(\overline{A} \cdot \overline{\overline{\overline{B}}} \cdot \overline{C}) + (\overline{\overline{\overline{A}}} \cdot \overline{B} \cdot \overline{\overline{\overline{C}}})} = Y$

    Step 3. $\overline{\overline{(\overline{A} \cdot \overline{\overline{\overline{B}}} \cdot \overline{C}) + (\overline{\overline{\overline{A}}} \cdot \overline{B} \cdot \overline{\overline{\overline{C}}})}} = Y$

    Step 4. Eliminate double overbars.

    End. $\overline{A} \cdot B \cdot C + A \cdot \overline{B} \cdot C = Y$
45. Begin. $\overline{\overline{\overline{A} \cdot B \cdot C + \overline{A} \cdot \overline{B} \cdot \overline{C}}} = Y$

    Step 1. $\overline{(\overline{A} + B + C) \cdot (\overline{A} + \overline{B} + \overline{C})} = Y$

    Step 2. $\overline{(\overline{\overline{\overline{A}}} + \overline{B} + \overline{C}) \cdot (\overline{\overline{\overline{A}}} + \overline{\overline{\overline{B}}} + \overline{\overline{\overline{C}}})} = Y$

    Step 3. $\overline{\overline{(\overline{\overline{\overline{A}}} + \overline{B} + \overline{C}) \cdot (\overline{\overline{\overline{A}}} + \overline{\overline{\overline{B}}} + \overline{\overline{\overline{C}}})}} = Y$

    Step 4. Eliminate double overbars.

    End. $(A + \overline{B} + \overline{C}) \cdot (A + B + C) = Y$
46. $(A'BC + A'B'C')'$

47.

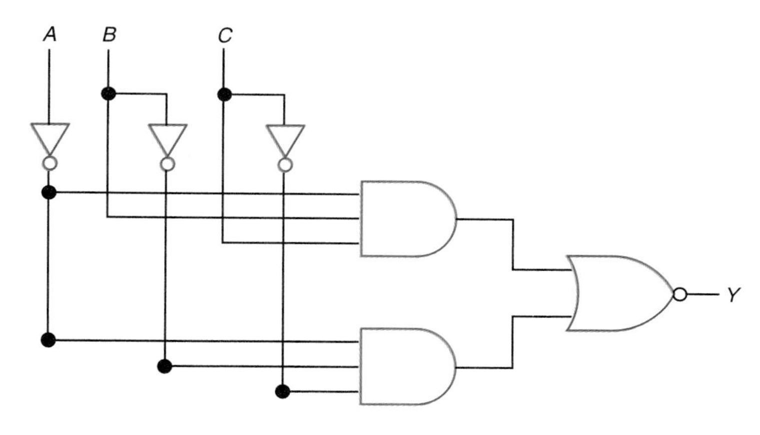

48.

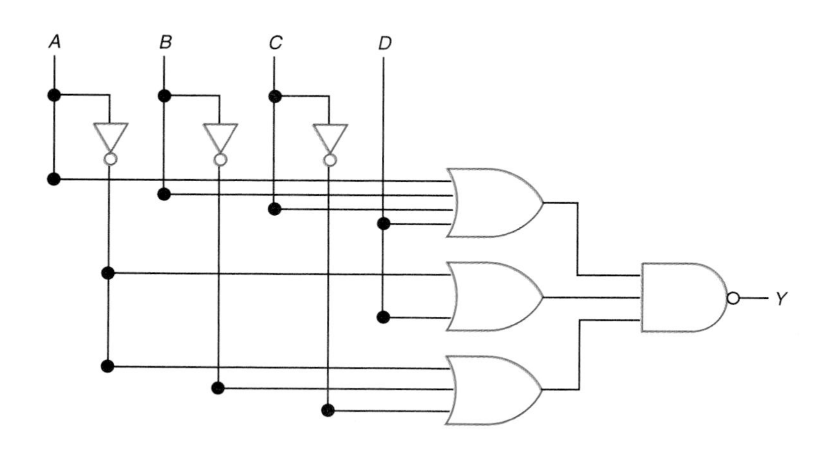

IC Specifications and Simple Interfacing

Chapter Objectives

This chapter will help you to:

1. *Determine* logic levels using TTL and CMOS voltage profile diagrams.
2. *Discuss* selected TTL and CMOS IC specifications such as input and output voltages, noise margin, drive capability, fan-in, fan-out, propagation delay, and power consumption.
3. *List* several safety precautions for handling and designing with CMOS ICs.
4. *Recognize* several simple switch interface and debounce circuits using both TTL and CMOS ICs.
5. *Analyze* interfacing circuits for LEDs and incandescent lamps using both TTL and CMOS ICs.
6. *Draw* TTL-to-CMOS and CMOS-to-TTL interface circuits.
7. *Describe* the operation of interface circuits for buzzers, relays, motors, and solenoids using both TTL and CMOS ICs.
8. *Analyze* interfacing circuits featuring an optoisolator.
9. *List* the primary characteristics and features of a stepper motor.
10. *Describe* the operation of stepper motor driver circuits.
11. *Troubleshoot* a simple logic circuit.

Logic families: TTL and CMOS

Interfacing

The driving force behind the increased use of digital circuits has been the availability of a variety of *logic families.* Integrated circuits within a logic family are designed to *interface* easily with one another. For instance, in the TTL logic family you may connect an output directly into the input of several other TTL inputs with no extra parts. The designer can have confidence that ICs from the same logic family will interface properly. Interfacing *between* logic families and between digital ICs and the outside world is a bit more complicated. *Interfacing* can be defined as the design of the interconnections between circuits that shift the levels of voltage and current to make them compatible. A fundamental knowledge of simple interfacing techniques is required of technicians and engineers who work with digital circuits. Most logic circuits are of no value if they are not interfaced with "real world" devices.

5-1 Logic Levels and Noise Margin

In any field of electronics most technicians and engineers start investigating a new device in terms of voltage, current, and resistance or impedance. In this section just the *voltage characteristics* of both TTL and CMOS ICs will be studied.

Logic Levels

How is a logical 0 (LOW) or logical 1 (HIGH) defined? Figure 5-1 shows an inverter (such as the 7404 IC) from the bipolar TTL logic family. Manufacturers specify that for correct operation, a LOW *input* must range from GND to 0.8 V. Also, a HIGH *input* must be in the range from 2.0 to 5.5 V. The unshaded section from

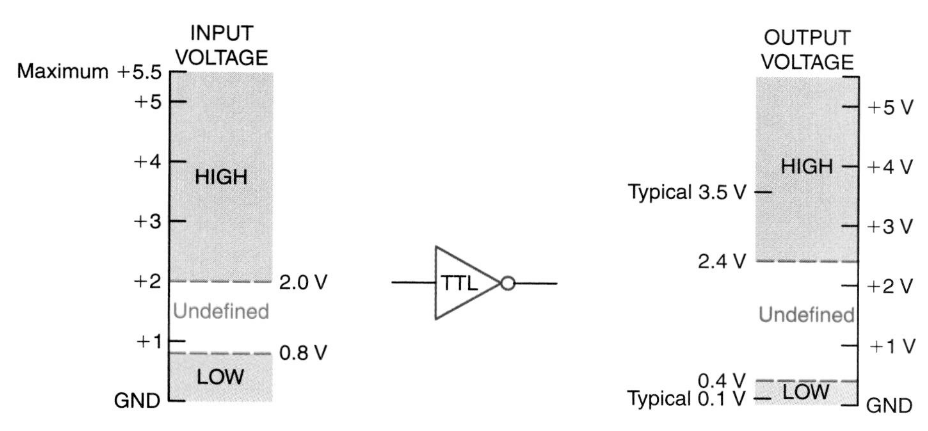

Fig. 5-1 Defining TTL input and output voltage levels.

0.8 to 2.0 V on the input side is the undefined area, or indeterminate region. Therefore, an input of 3.2 is a HIGH input. An input of 0.5 V is considered a LOW input. An input of 1.6 V is in the undefined region and should be avoided. Inputs in the undefined region yield unpredictable results at the output. Expected outputs from the TTL inverter are shown on the right in Fig. 5-1. A typical LOW output is about 0.1 V. A typical HIGH output might be about 3.5 V. However, a HIGH output could be as low as 2.4 V according to the voltage profile diagram in Fig. 5-1. The HIGH output depends on the resistance value of the load placed at the output. The greater the load current, the lower the HIGH output voltage. The unshaded section on the output voltage side in Fig. 5-1 is the undefined region. Suspect trouble if the output voltage is in the undefined region (0.4 to 2.4 V).

The voltages given for LOW and HIGH logic levels in Fig. 5-1 are for a TTL device. These voltages are different for other logic families.

The popular 4000 and 74C00 series CMOS logic families of ICs operate on a wide range of power supply voltages (from +3 to +15 V). The definition of a HIGH and LOW logic level for a typical CMOS inverter is illustrated in Fig. 5-2(*a*) on page 88. A 10-V power supply is being used in this voltage profile diagram.

The CMOS inverter shown in Fig. 5-2(*a*) will respond to any input voltage within 70–100 percent of V_{DD} (+10 V in this example) as a HIGH. Likewise, any voltage within 0 to 30 percent of V_{DD} is regarded as a LOW input to ICs in the 4000 and 74C00 series.

Typical output voltages for CMOS ICs are shown in Fig. 5-2(*a*). Output voltages are normally almost at the *voltage rails* of the power supply. In this example, a HIGH output would be about +10 V while a LOW output would be about 0 V or GND.

The 74HC00 series and the newer 74AC00 and 74ACQ00 series operate on a lower-voltage power supply (from +2 to +6 V) than the older 4000 and 74C00 series CMOS ICs. The input and output voltage characteristics are summarized in the voltage profile diagram in Fig. 5-2(*b*). The definition of HIGH and LOW for both input and output on the 74HC00, 74AC00, and 74ACQ00 series is approximately the same as for the 4000 and 74C00 series CMOS ICs. This can be seen in a comparison of the two voltage profiles in Figs. 5-2(*a*) and (*b*).

The 74HCT00 series and newer 74ACT00, 74ACTQ00, 74FCT00, and 74FCTA00 series of CMOS ICs are designed to operate on a 5-V power supply like TTL ICs. The function of the 74HCT00, 74ACT00, 74ACTQ00, 74FCT00, and 74FCTA00 series of ICs is to interface between TTL and CMOS devices. These CMOS ICs with a "T" designator can serve as direct replacements for many TTL ICs.

4000 and 74C00 series CMOS voltage profile

TTL voltage profile

74HC00 and 74AC00 series CMOS voltage profile

"T" series CMOS profile

A voltage profile diagram for the 74HCT00, 74ACT00, 74ACTQ00, 74FCT00, and 74FCTA00 series CMOS ICs is drawn in Fig. 5-2(*c*). Notice that the definition of LOW and HIGH at the *input* is the same for these "T" CMOS ICs as it is for regular bipolar TTL ICs. This can be seen in a comparison of the input side of the voltage profiles of TTL and the "T" CMOS ICs (see Figs. 5-1 and 5-2(*c*)). The output voltage profiles for all the CMOS ICs are similar. In summary, the "T" series CMOS ICs

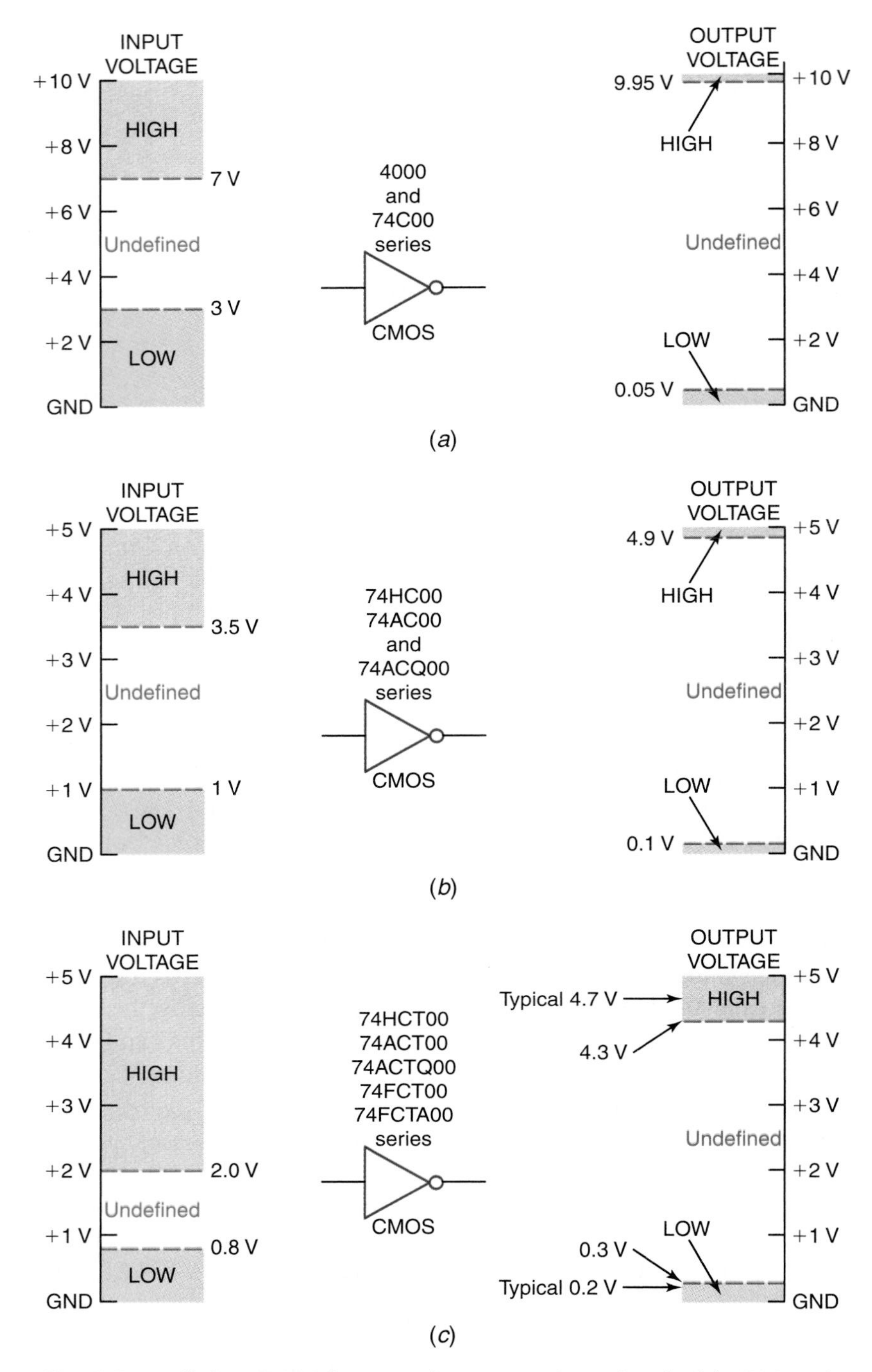

CMOS voltage profiles

Fig. 5-2 Defining CMOS input and output voltage levels. (*a*) 4000 and 74C00 series voltage profile. (*b*) 74HC00, 74AC00, and 74ACQ00 series voltage profile. (*c*) 74HCT00, 74ACT00, 74ACTQ00, 74FCT00, 74FCTA00 series voltage profile.

have typical TTL input voltage characteristics with CMOS outputs.

Noise Margin

Advantages of CMOS

The most often cited advantages of CMOS are its low power requirements and good noise immunity. *Noise immunity* is a circuit's insensitivity or resistance to undesired voltages or noise. It is also called *noise margin* in digital circuits.

Noise immunity

Noise margin

The noise margins for typical TTL and CMOS families are compared in Fig. 5-3. The noise margin is much better for the CMOS

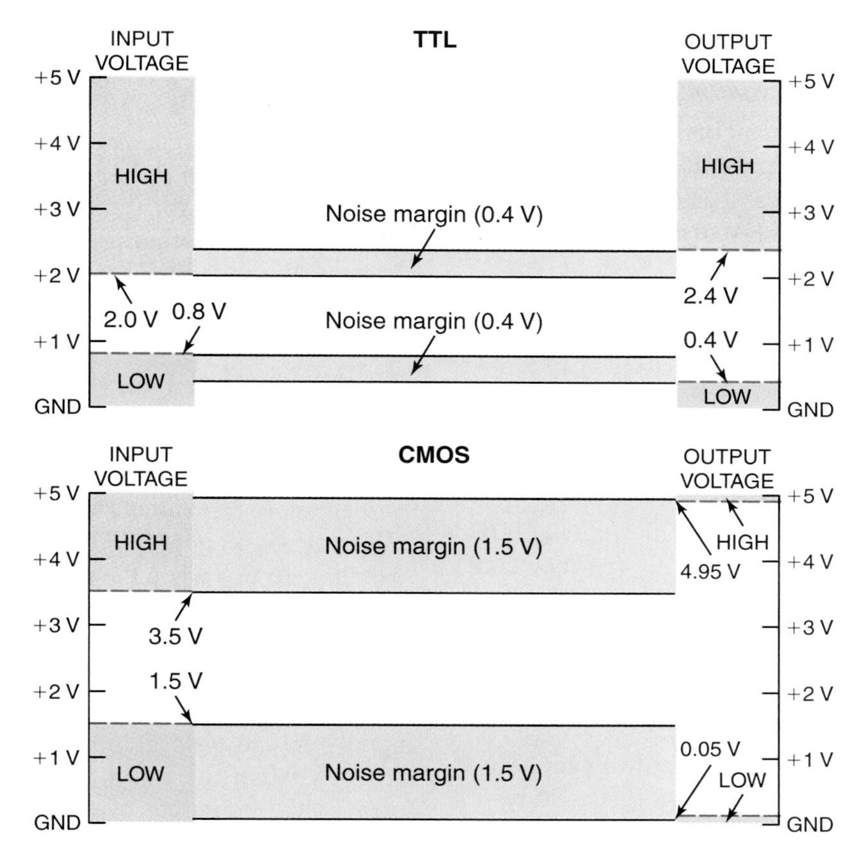

Fig. 5-3 Defining and comparing TTL and CMOS noise margins.

Comparing TTL and CMOS noise margin

than for the TTL family. You may introduce almost 1.5 V of unwanted noise into the CMOS input before getting unpredictable results.

Noise in a digital system is *unwanted voltages* induced in the connecting wires and printed circuit board traces that might affect the input logic levels, thereby causing faulty output indications.

Consider the diagram in Fig. 5-4. The LOW, HIGH, and undefined regions are defined for TTL inputs. If the actual input voltage is 0.2 V, then the margin of safety between it and the undefined region is 0.6 V (0.8 − 0.2 = 0.6 V). This is the *noise margin.* In other words, it would take more than +0.6 V added to the LOW voltage (0.2 V in this example) to move the input into the undefined region.

Noise

In actual practice, the noise margin is even greater because the voltage must increase to the *switching threshold,* which is shown as 1.2 V in Fig. 5-4. With the actual LOW input at +0.2 V and the switching threshold at about +1.2 V, the actual noise margin is 1 V (1.2 − 0.2 = 1 V).

Switching threshold

The switching threshold is *not* an absolute voltage. It does occur within the undefined region but varies widely because of manufacturer, temperature, and the quality of the components. However, the logic levels are guaranteed by the manufacturer.

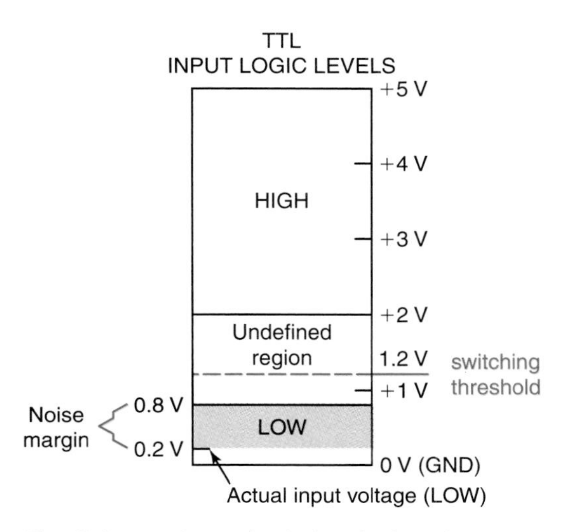

Fig. 5-4 TTL input logic levels showing noise margin.

TEST

Supply the missing word in each statement.

1. The design of interconnections between two circuits to make them compatible is called ________.
2. An input of +3.1 V to a TTL IC would be considered ________ (HIGH, LOW, undefined).
3. An input of +0.7 V to a TTL IC would be considered ________ (HIGH, LOW, undefined).
4. An output of +2.0 V from a TTL IC would be considered ________ (HIGH, LOW, undefined).
5. An input of +6 V (10-V power supply) to a 4000 series CMOS IC would be considered ________ (HIGH, LOW, undefined).
6. A typical HIGH output (10-V power supply) from a CMOS IC would be about ________ V.
7. An input of +3 V (5-V power supply) to a 74HCT00 series CMOS IC would be considered ________ (HIGH, LOW, undefined).
8. The ________ (CMOS, TTL) family of ICs has better noise immunity.
9. The switching threshold of a digital IC is the *input* voltage at which the output logic level switches from HIGH-to-LOW or LOW-to-HIGH (T or F).
10. The 74FCT00 series of ________ (CMOS, TTL) ICs has an input voltage profile that looks like that of a TTL IC.

5-2 Other Digital IC Specifications

Digital IC specifications

Digital logic voltage levels and noise margins were studied in the last section. In this section, other important specifications of digital ICs will be introduced. These include drive capabilities, fan-out and fan-in, propagation delay, and power dissipation.

Drive Capabilities

Drive capabilities

Fan-out

Fan-in

A bipolar transistor has its maximum wattage and collector current ratings. These ratings determine its *drive capabilities.* One indication of output drive capability of a digital IC is called its fan-out. The *fan-out* of a digital IC is the number of "standard" inputs that can be driven by the gate's output. If the fan-out for standard TTL gates is 10, this means that the output of a single gate can drive up to 10 inputs of the gates *in the same subfamily.* A typical fan-out value for standard TTL ICs is 10. The fan-out for low-power Schottky TTL (LS-TTL) is 20 and for the 4000 series CMOS it is considered to be about 50.

Another way to look at the current characteristics of gates is to examine their output drive and input loading parameters. The diagram in Fig. 5-5(*a*), on the next page, is a simplified view of the output drive capabilities and input load characteristics of a standard TTL gate. A standard TTL gate is capable of handling 16 mA when the output is LOW (I_{OL}) and 400 μA when the output is HIGH (I_{OH}). This seems like a mismatch until you examine the input loading profile for a standard TTL gate. The input loading (worst-case conditions) is only 40 μA with the input HIGH (I_{IH}) and 1.6 mA when the input is LOW (I_{IL}). This means that the output of a standard TTL gate can drive 10 inputs (16 mA/1.6 mA = 10). Remember, these are *worst-case conditions* and in actual bench tests under static conditions these input load currents are much less than specified.

A summary of the *output drive* and *input loading* characteristics of several popular families of digital ICs is detailed in Fig. 5-5(*b*). Look over this chart of very useful information. You will need this data later when interfacing TTL and CMOS ICs.

Notice the outstanding output drive capabilities of the FACT series of CMOS ICs (see Fig. 5-5(*b*)). The superior drive capabilities, low power consumption, excellent speed, and great noise immunity make the FACT series of CMOS ICs one of the preferred logic families for new designs. The newer FAST TTL logic series also has many desirable characteristics that make it suitable for new designs.

The load represented by a single gate is called the *fan-in* of that family of ICs. The input loading column in Fig. 5-5(*b*) can be thought of as the fan-in of these IC families. Notice that the fan-in or input loading characteristics are different for each family of ICs.

Suppose you are given the interfacing problem in Fig. 5-6(*a*) on page 92. You are asked if the 74LS04 inverter has enough fan-out to

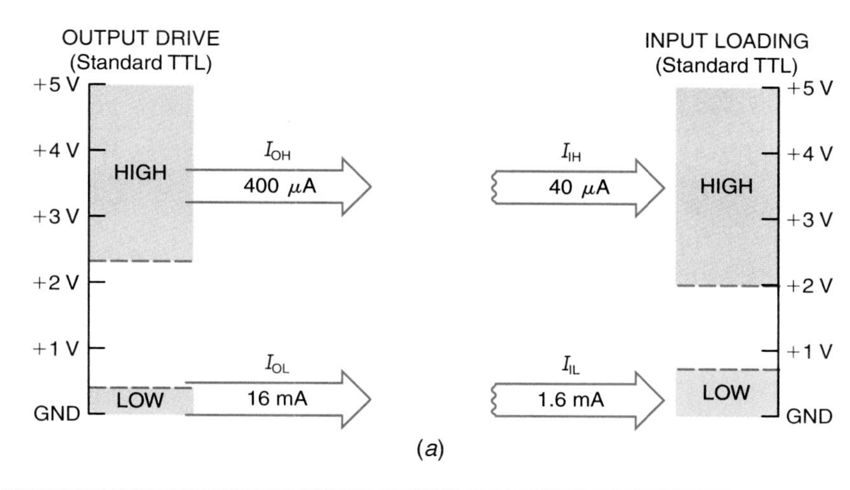

	Device Family	Output Drive*	Input Loading
TTL	Standard TTL	I_{OH} = 400 μA I_{OL} = 16 mA	I_{IH} = 40 μA I_{IL} = 1.6 mA
	Low-Power Schottky	I_{OH} = 400 μA I_{OL} = 8 mA	I_{IH} = 20 μA I_{IL} = 400 μA
	Advanced Low-Power Schottky	I_{OH} = 400 μA I_{OL} = 8 mA	I_{IH} = 20 μA I_{IL} = 100 μA
	FAST Fairchild Advanced Schottky TTL	I_{OH} = 1 mA I_{OL} = 20 mA	I_{IH} = 20 μA I_{IL} = 0.6 mA
CMOS	4000 Series	I_{OH} = 400 μA I_{OL} = 400 μA	I_{in} = 1 μA
	74HC00 Series	I_{OH} = 4 mA I_{OL} = 4 mA	I_{in} = 1 μA
	FACT Fairchild Advanced CMOS Technology Series (AC/ACT/ACQ/ACTQ)	I_{OH} = 24 mA I_{OL} = 24 mA	I_{in} = 1 μA
	FACT Fairchild Advanced CMOS Technology Series (FCT/FCTA)	I_{OH} = 15 mA I_{OL} = 64 mA	I_{in} = 1 μA

*Buffers and drivers may have more output drive.

(*b*)

Fig. 5-5 (*a*) Standard TTL voltage and current profiles. (*b*) Output drive and input loading characteristics for selected TTL and CMOS logic families.

drive the four standard TTL NAND gates on the right.

The voltage and current profiles for LS-TTL and standard TTL gates are sketched in Fig. 5-6(*b*). The voltage characteristics of all TTL families are compatible. The LS-TTL gate can drive 10 standard TTL gates when its output is HIGH (400 μA/40 μA = 10). However, the LS-TTL gate can drive only five standard TTL gates when it is LOW (8 mA/1.6 mA = 5). We could say that the fan-out of LS-TTL gates is only 5 when driving standard TTL gates. It is true that the LS-TTL inverter can drive four standard TTL inputs in Fig. 5-6(*a*).

Propagation Delay

Speed, or quickness of response to a change at the inputs, is an important consideration in high-speed applications of digital ICs. Consider the waveforms in Fig. 5-7(*a*) on page 93. The top waveform shows the input to an inverter going from LOW to HIGH and then from HIGH to LOW. The bottom waveform shows the output response to the changes at the input. The slight delay between the time the input changes and the time the output changes is called the *propagation delay* of the inverter. Propagation delay is measured in seconds. The

Propagation delay

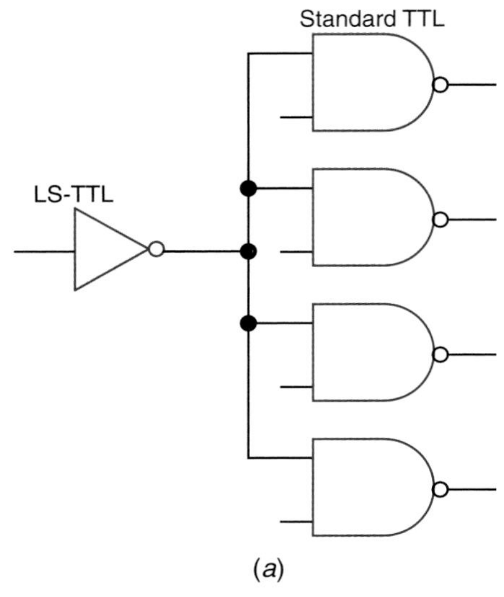

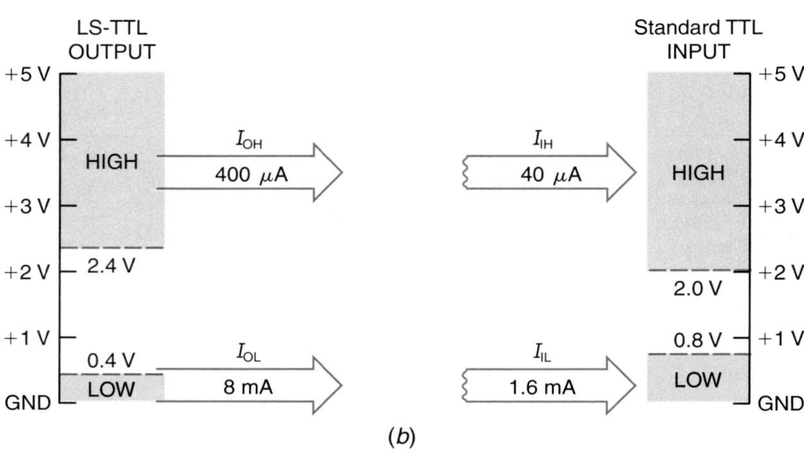

Fig. 5-6 Interfacing LS-TTL to standard TTL problem. (*a*) Logic diagram of interfacing problem. (*b*) Voltage and current profiles for visualizing the solution to the problem.

propagation delay for the LOW-to-HIGH transition of the input to the inverter is different from the HIGH-to-LOW delay. Propagation delays are shown in Fig. 5-7(*a*) for a standard TTL 7407 inverter IC.

The typical propagation delay for a standard TTL inverter (such as the 7404 IC) is about 12 ns for the LOW-to-HIGH change while only 7 ns for the HIGH-to-LOW transition of the input.

FACT series CMOS ICs

Emitter coupled logic (ECL)

Representative minimum propagation delays are summarized on the graph in Fig. 5-7(*b*). The lower the propagation delay specification for an IC, the higher the speed. Notice that the AS-TTL (advanced Schottky TTL) and AC-CMOS (FACT Fair-child Advanced CMOS Technology—AC subfamily) are the fastest with minimum propagation delays of about 1 ns for a simple inverter. The older 4000 and 74C00 series CMOS families are the slowest families (highest propagation delays). Some 4000 series ICs have propagation delays of over 100 ns. In the past, TTL ICs were considered faster than those manufactured using the CMOS technology. Currently, however, the FACT CMOS series rival the best TTL ICs in low propagation delays (high speed). For extremely high-speed operation, the ECL (emitter coupled logic) and the developing gallium arsenide families are required.

Power dissipation

Power Dissipation

Generally, as propagation delays decrease (increased speed), the power consumption and related heat generation increase. Historically, a

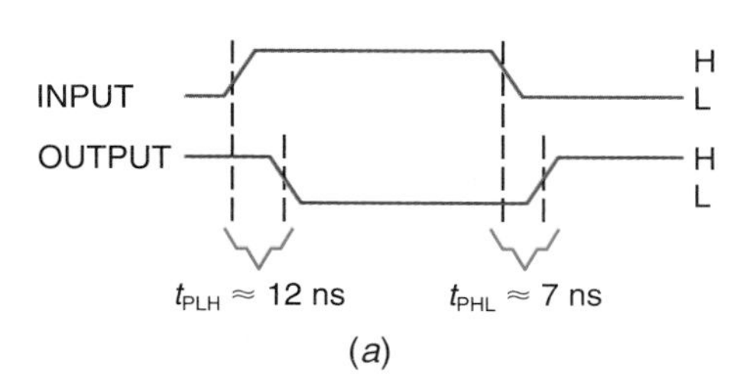

(*a*)

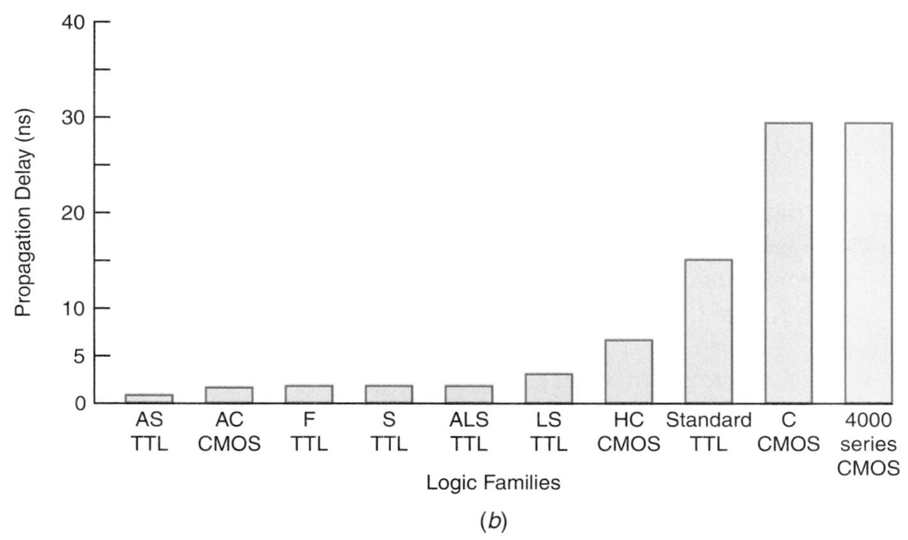

(*b*)

Fig. 5-7 (*a*) Waveforms showing propagation delays for a standard TTL inverter. (*b*) Graph of propagation delays for selected TTL and CMOS families.

Propagation delays

standard TTL IC might have a propagation delay of about 10 ns compared with a propagation delay of about 30 to 50 ns for a 4000 series CMOS IC. The 4000 CMOS IC, however, would consume only 0.001 mW, while the standard TTL gate might consume 10 mW of power. The power dissipation of CMOS increases with frequency. So at 100 kHz, the 4000 series gate may consume 0.1 mW of power.

The speed versus power table in Fig. 5-8 on page 94 compares in graphic form several of the modern TTL and CMOS families. The vertical axis on the graph represents the propagation delay (speed) in nanoseconds while the horizontal axis depicts the power consumption (in milliwatts) of each gate. Families with the most desirable combination of both speed and power are those near the lower left corner of the table. A few years ago many designers suggested that the ALS (advanced low-power Schottky TTL) family was the best compromise between speed and power dissipation. With the introduction of new families, it appears that the FACT (Fairchild Advanced CMOS Technology) series is now one of the best compromise logic families. Both the ALS and the FAST (Fairchild Advanced Schottky

Best logic families

About Electronics

Dennis C. Hayes In 1978, 28-year-old Dennis C. Hayes formed what became Hayes Microcomputer Products, Inc. When he began, Hayes hand-assembled and soldered products on a borrowed dining room table in his home. Then, in 1981 the Hayes™ Smartmodem™, which was easily integrated into the computer environment, started a communications revolution.

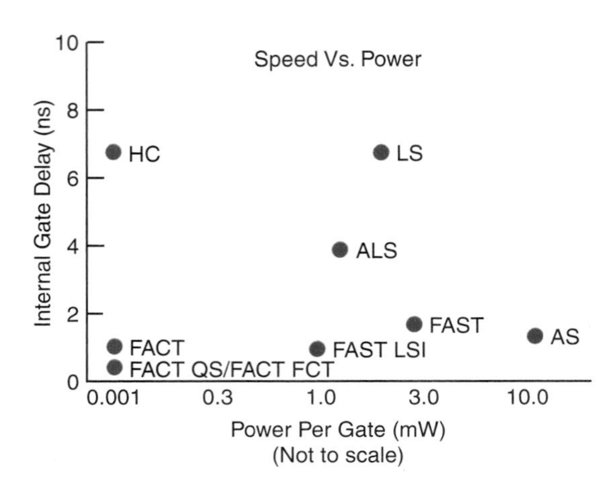

Fig. 5-8 Speed versus power for selected TTL and CMOS families. (*Courtesy of National Semiconductor Corporation.*)

Speed versus power chart

TTL) families are also excellent choices for new designs.

TEST

Supply the missing word in each statement.

11. The number of "standard" input loads that can be driven by an IC is called its ________ (fan-in, fan-out).
12. The ________ (4000 series CMOS, FAST TTL series) gates have more output drive capabilities.
13. Refer to Fig. 5-5(*b*). The calculated fan-out when interfacing LS-TTL to LS-TTL is ________.
14. The 4000 series CMOS gates have very low power dissipation, good noise immunity, and ________ (long, short) propagation delays.
15. Refer to Fig. 5-7(*b*). The fastest CMOS subfamily is ________.
16. All TTL subfamilies have ________ (different, the same) voltage and different output drive and input loading characteristics.

5-3 MOS and CMOS ICs

MOS ICs

PMOS

NMOS

Advantages and disadvantages of MOS devices

The enhancement type of metal oxide semiconductor field-effect transistor (MOSFET) forms the primary component in MOS ICs. Because of their simplicity, MOS devices use less space on a silicon chip. Therefore, more functions per chip are typical in MOS devices than in bipolar ICs (such as TTL). Metal oxide semiconductor technology is widely used in large scale integration (LSI) and very large scale integration (VLSI) devices because of this packing density on the chip. Microprocessors, memory, and clock chips are typically fabricated using MOS technology. Metal oxide semiconductor circuits are typically of either the PMOS (P-channel MOS) or the newer, faster NMOS (N-channel MOS) type. Metal oxide semiconductor chips are smaller, consume less power, and have a better noise margin and higher fan-out than bipolar ICs. The main disadvantage of MOS devices is their relative lack of speed.

CMOS ICs

Complementary symmetry metal oxide semiconductor ICs

Advantages and disadvantages of CMOS devices

Complementary symmetry metal oxide semiconductor (CMOS) devices use both P-channel and N-channel MOS devices connected end to end. Complementary symmetry metal oxide semiconductor ICs are noted for their *exceptionally low power consumption.* The CMOS family of ICs also has the advantages of low cost, simplicity of design, low heat dissipation, good fan-out, wide logic swings, and good noise-margin performance. Most CMOS families of digital ICs operate on a wide range of voltages.

The main disadvantage of many CMOS ICs is that they are somewhat slower than bipolar digital ICs such as TTL devices. Also, extra care must be taken when handling CMOS ICs because they must be protected from static discharges. A static charge or transient voltage in a circuit can damage the very thin silicon dioxide layers inside the CMOS chip. The silicon dioxide layer acts like the dielectric in a capacitor and can be punctured by static discharge and transient voltages.

Cautions when using CMOS

If you do work with CMOS ICs, manufacturers suggest preventing damage from static discharge and transient voltages by:

Conductive foam

1. Storing CMOS ICs in special *conductive foam* or static shielding bags or containers
2. Using battery-powered soldering irons when working on CMOS chips or grounding the tips or ac-operated units
3. Changing connections or removing CMOS ICs only when the power is turned off

4. Ensuring that input signals do not exceed power supply voltages
5. Always turning off input signals before circuit power is turned off
6. Connecting *all unused input leads* to either the positive supply voltage or GND, whichever is appropriate (only unused CMOS *outputs* may be left unconnected)

FACT CMOS ICs are much more tolerant of static discharge.

The extremely low power consumption of CMOS ICs makes them ideal for battery-operated portable devices. Complementary symmetry metal oxide semiconductor ICs are widely used in electronic wristwatches, calculators, portable computers, and space vehicles.

A typical CMOS device is shown in Fig. 5-9. The top half is a P-channel MOSFET, while the bottom half is an N-channel MOSFET. Both are enhancement-mode MOSFETs. When the input voltage (V_{in}) is LOW, the top MOSFET is on and the bottom unit is off. The output voltage (V_{out}) is then HIGH. However, if (V_{in}) is HIGH, the bottom device is on and the top MOSFET is off. Therefore, (V_{out}) is LOW. The device in Fig. 5-9 acts as an inverter.

Notice in Fig. 5-9 that the V_{DD} of the CMOS unit goes to the positive supply voltage. The V_{DD} lead is labeled V_{CC} (as in TTL) by some manufacturers. The "*D*" in V_{DD} stands for the *drain* supply in MOSFET. The V_{SS} lead of the CMOS unit is connected to the negative of the power supply. This connection is called GND (as in TTL) by some manufacturers. The "*S*" in V_{SS} stands for *source* supply in a MOSFET. CMOS ICs typically operate on 5-, 6-, 9-, or 12-V power supplies.

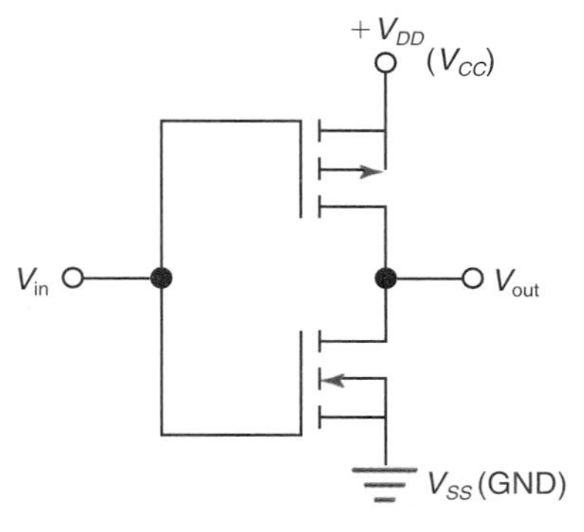

Fig. 5-9 CMOS structure using P-channel and N-channel MOSFETs in series.

The CMOS technology is used in making several families of digital ICs. The most popular are the 4000, 74C00, 74HC00, and FACT series ICs. The 4000 series is the oldest and is still widely used. This family has all the customary logic functions plus a few devices that have no equivalent in TTL families. For instance, in CMOS it is possible to produce *transmission gates* or *bilateral switches.* These gates can conduct or allow a signal to pass in either direction like relay contacts.

4000 series

Transmission gates

Bilateral switches

The 74C00 series is an older CMOS logic family that is the pin-for-pin, function-for-function equivalent of the 7400 series of TTL ICs. As an example, a 7400 TTL IC is designated as a quadruple ("quad") two-input NAND gate as is the 74C00 CMOS ICs.

74C00 series

The 74HC00 series CMOS logic family is designed to replace the 74C00 series and many 4000 series ICs. It has pin-for-pin, function-for-function equivalents for both 7400 and 4000 series ICs. It is a high-speed CMOS family with good drive capabilities. It operates on a 2- to 6-V power supply.

74HC00 series

The FACT (Fairchild Advanced CMOS Technology) logic IC series includes the 74AC00, 74ACQ00, 74ACT00, 74ACTQ00, 74FCT00, and 74FCTA00 subfamilies. The FACT family provides pin-for-pin, function-for-function equivalents for 7400 TTL ICs. The FACT series was designed to outperform existing CMOS and bipolar logic families. As noted before, the FACT series of CMOS ICs may be the best overall logic family currently available to the designer. It features low power consumption even at modest frequencies (0.1 mW/fate at 1 MHz). Power consumption does however increase at higher frequencies (>50 mW at 40 MHz). It has outstanding noise immunity, with the "Q" devices having patented noise-suppression circuitry. The "T" devices have TTL voltage level inputs. The propagation delays for the FACT series are outstand-

FACT series CMOS ICs

CMOS structure

Did You Know?

Locks with Memory. Intel chips used in safe locks offer 500 billion possible combinations. But in case someone still wants to try, the lock records the number of unsuccessful attempts to open it.

ing (see Fig. 5-7(*b*)). FACT ICs show excellent resistance to static electricity. The series is also radiation-tolerant making it good in space, medical, and military applications. The output drive capabilities of the FACT family are outstanding (see Fig. 5-5(*b*)).

TEST

Supply the missing word in each statement.

17. Large-scale integration (LSI) and very large-scale integration (VLSI) devices make extensive use of ________ (bipolar, MOS) technology.
18. The letters CMOS stand for ________.
19. The most important advantage of using CMOS is its ________.
20. The V_{SS} pin on a CMOS IC is connected to ________ (+5 V, GND) of the power supply.
21. The V_{DD} pin on a CMOS IC is connected to ________ (positive, GND) of the power supply.

5-4 Interfacing TTL and CMOS with Switches

One of the most common means of entering information into a digital system is the use of switches or a keyboard. Examples might be the switches on a digital clock, the keys on a calculator, or the large keyboard on a microcomputer. This section will detail several methods of using a switch to enter data into either TTL or CMOS digital circuits.

Three simple switch interface circuits are depicted in Fig. 5-10. Pressing the push-button switch in Fig. 5-10(*a*) will drop the input of the TTL inverter to ground level or LOW. Releasing the push-button switch in Fig. 5-10(*a*) opens the switch. The input to the TTL inverter now is allowed to "float." In TTL, inputs usually float at a HIGH logic level.

Switch-to-TTL interfaces

Floating inputs on TTL are not dependable. Figure 5-10(*b*) is a slight refinement of the switch input circuit in Fig. 5-10(*a*). The 10-kΩ resistor has been added to make sure the input to the TTL inverter goes HIGH when the switch is open. The 10-kΩ resistor is called a *pull-up resistor*. Its purpose is to pull the input voltage up to +5 V. Both circuits in Figs. 5-10(*a*) and (*b*) illustrate active LOW switches. They are called active LOW switches because the inputs go LOW only when the switch is activated.

Pull-up resistor

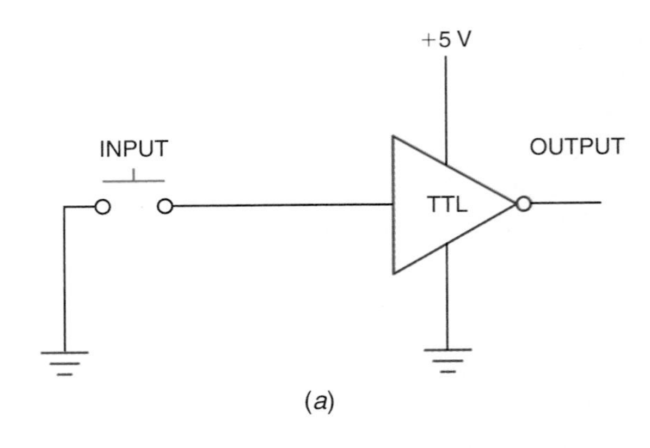

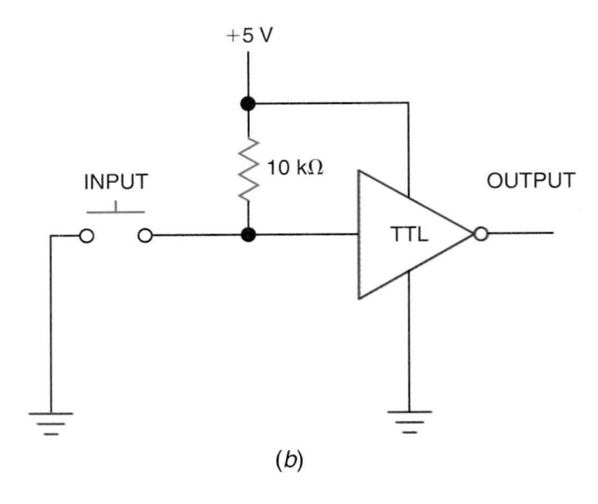

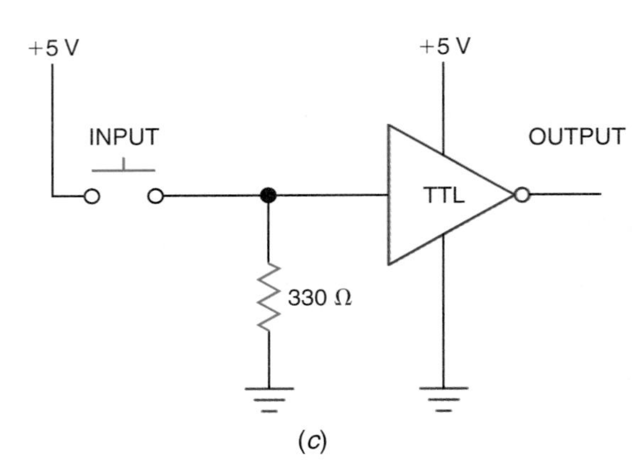

Fig. 5-10 Switch-to-TTL interfaces. (*a*) Simple active-LOW switch interface. (*b*) Active-LOW switch interface using pull-up resistor. (*c*) Active-HIGH switch interface using pull-down resistor.

An active HIGH input switch is sketched in Fig. 5-10(*c*). When the input switch is activated, the +5 V is connected directly to the input of the TTL inverter. When the switch is released (opened) the input is pulled LOW by the pull-down resistor. The value of the pull-down resistor is relatively low because the input current required by a standard TTL gate may be as high as 1.6 mA (see Fig. 5-5(*b*)).

Two switch-to-CMOS interface circuits are drawn in Fig. 5-11. An active LOW input switch is drawn in Fig. 5-11(*a*). The 100-kΩ pull-up resistor pulls the voltage to +5 V when the input switch is open. Figure 5-11(*b*) illustrates an active HIGH switch feeding a CMOS inverter. The 100-kΩ pull-down resistor makes sure the input to the CMOS inverter is near ground when the input switch is open. The resistance value of the pull-up and pull-down resistors is much greater than those in TTL interface circuits. This is because the input loading currents are much greater in TTL than in CMOS. The CMOS inverter illustrated in Fig. 5-11 could be from the 4000, 74C00, 74HC00, or the FACT series of CMOS ICs.

Pull-down resistor

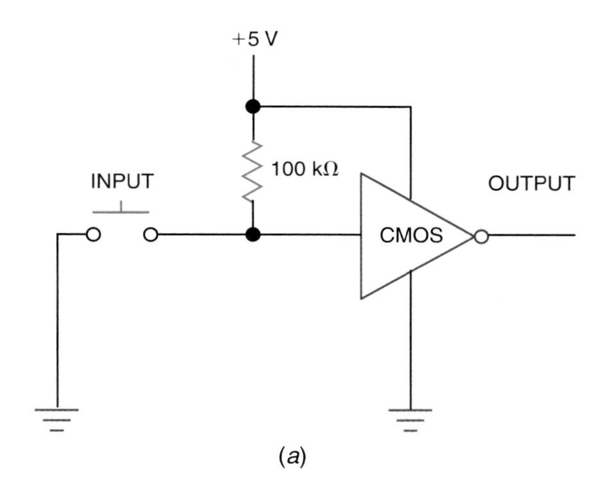

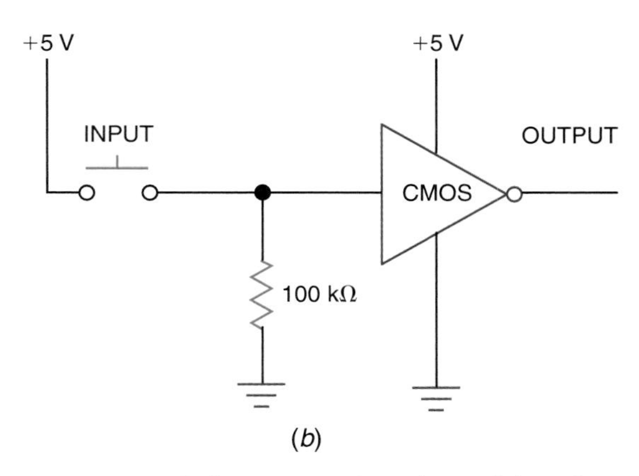

Fig. 5-11 Switch-to-CMOS interfaces. (*a*) Active-LOW switch interface with pull-up resistor. (*b*) Active-HIGH switch interface with pull-down resistor.

Switch Debouncing

The switch interface circuits in Figs. 5-10 and 5-11 work well for some applications. However, none of the switches in Figs. 5-10 and 5-11 were *debounced.* The lack of a debouncing circuit can be demonstrated by operating the counter shown in Fig. 5-12(*a*) on page 98. Each press of the input switch should cause the decade (0–9) counter to increase by 1. However, in practice each press of the switch increases the count by 1, 2, 3, or sometimes more. This means that several pulses are being fed into the clock (CLK) input of the counter each time the switch is pressed. This is caused by switch bounce.

Switch debouncing circuit

A switch *debouncing circuit* has been added to the counting circuit in Fig. 5-12(*b*). The decade counter will now count each HIGH–LOW cycle of the input switch. The cross-coupled NAND gates in the debouncing circuit are sometimes called an RS flip-flop or latch. Flip-flops will be studied in greater detail in Chap. 7.

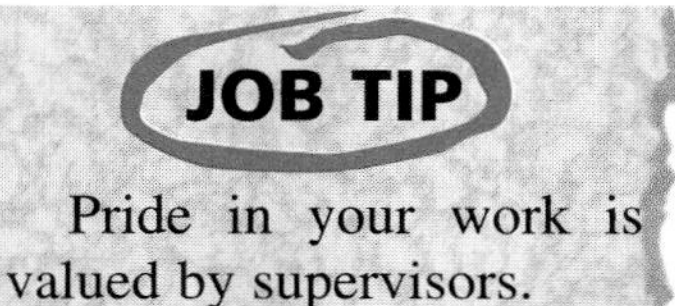

RS flip-flop latch

Several other switch debouncing circuits are illustrated in Fig. 5-13 on page 99. The simple debouncing circuit drawn in Fig. 5-13(*a*) will only work on the slower 4000 series CMOS IC. The 40106 CMOS IC is a special inverter. The 40106 is a Schmitt trigger inverter, which means it has a "snap action" when changing to either HIGH or LOW. A Schmitt trigger can also change a slow-rising signal (such as a sine wave) into a square wave.

Schmitt trigger inverter

The switch debouncing circuit in Fig. 5-13(*b*) will drive 4000, 74HC00, or FACT series CMOS or TTL ICs. Another general-purpose switch debouncing circuit is illustrated in Fig. 5-13(*c*). This debouncing circuit can drive either CMOS or TTL inputs. The 7403 is an *open-collector* NAND TTL IC and needs pull-up resistors as shown in Fig. 5-13(*c*). The external pull-up resistors make it possible to have an output voltage of just about +5 V for a HIGH. Open-collector TTL gates with external pull-up resistors are useful when driving CMOS with TTL.

Open-collector TTL output

Switch-to-CMOS interfaces

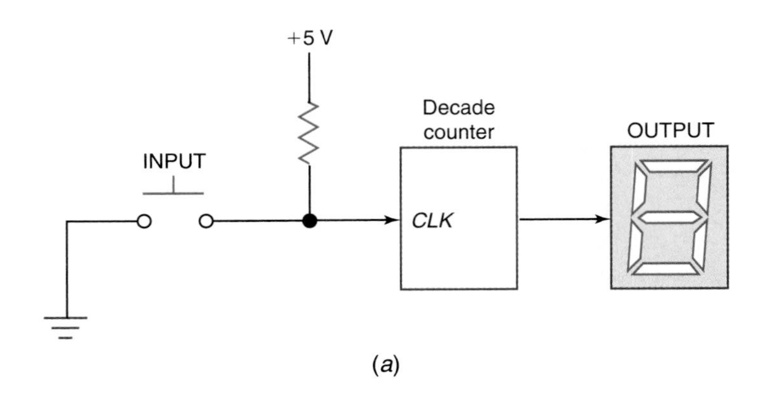

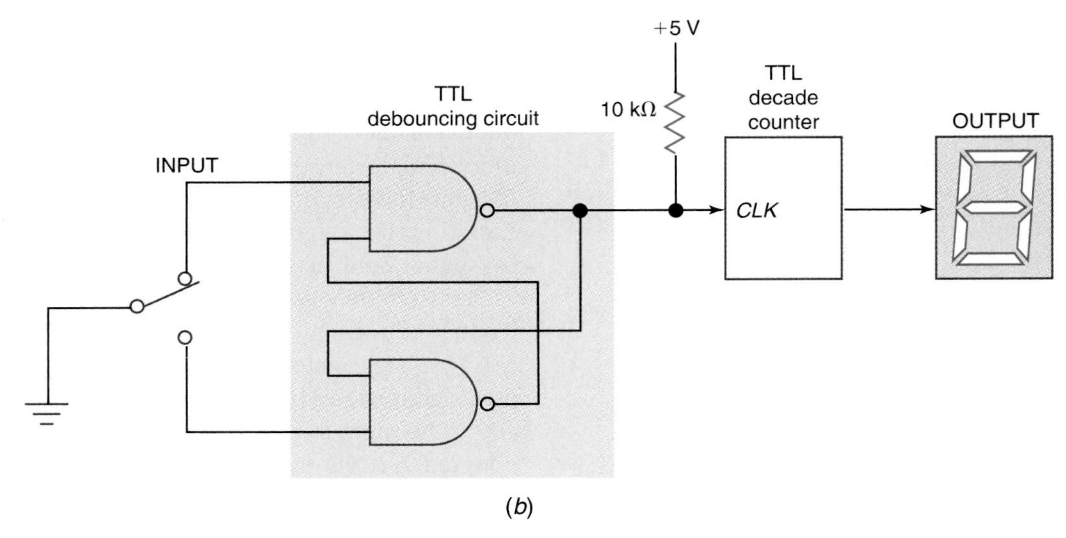

Fig. 5-12 (*a*) Block diagram of switch interfaced to a decimal counter system. (*b*) Adding a debouncing circuit to make the decimal counter work properly.

TEST

Answer the following questions.

22. Refer to Fig. 5-10(*a*). The input to the TTL inverter goes _________ (HIGH, LOW) when the switch is pressed (closed) but _________ (floats HIGH, goes LOW) when the input push button is open.
23. Refer to Fig. 5-10(*b*). The 10-kΩ resistor, which assures the input of the TTL inverter, will go HIGH when the switch that is open is called a _________ (filter, pull-up) resistor.
24. Refer to Fig. 5-12(*b*). The cross-coupled NAND gates that function as a debouncing circuit are sometimes called a(n) _________ or latch.
25. Refer to Fig. 5-10(*c*). Pressing the switch causes the input of the inverter to go _________ (HIGH, LOW) while the output goes _________ (HIGH, LOW).
26. Refer to Fig. 5-11. The inverters and associated resistors form switch debouncing circuits (T or F).
27. Refer to Fig. 5-12(*a*). This decade counter circuit lacks what circuit?
28. Refer to Fig. 5-13(*c*). The 7403 is a TTL inverter with a(n) _________ output.
    a. Open collector
    b. Totem pole
    c. Tri state

Light-emitting diode (LED)

5-5 Interfacing TTL and CMOS with LEDs

Many of the lab experiments you will perform using digital ICs require an output indicator. The LED (light-emitting diode) is perfect for this job because it operates at low currents and voltages. The maximum current required by many LEDs is about 20 to 30 mA with about

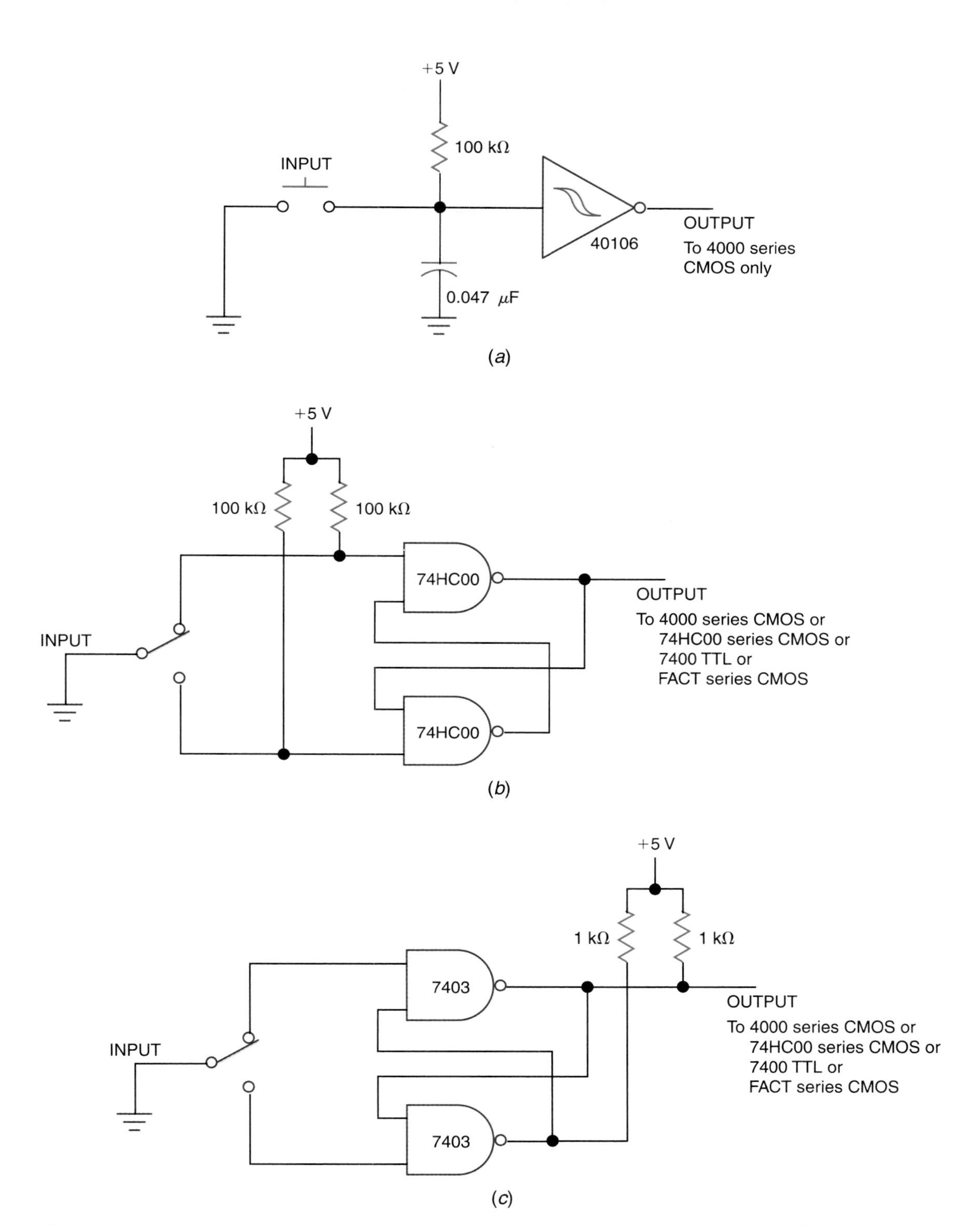

Fig. 5-13 Switch debouncing circuits. (*a*) A 4000 series switch debouncing circuit. (*b*) General purpose switch debouncing circuit that will drive CMOS or TTL inputs. (*c*) Another general-purpose switch debouncing circuit that will drive CMOS or TTL inputs.

Switch debouncing circuits

2 V applied. An LED will light dimly on only 1.7 to 1.8 V and 2 mA.

Interfacing 4000 series CMOS devices with simple LED indicator lamps is easy. Figure 5-14(*a–f*) shows six examples of CMOS ICs driving LED indicators. Figures 5-14(*a*) and (*b*) show the CMOS supply voltage at +5 V. At this low voltage, no limiting resistors are

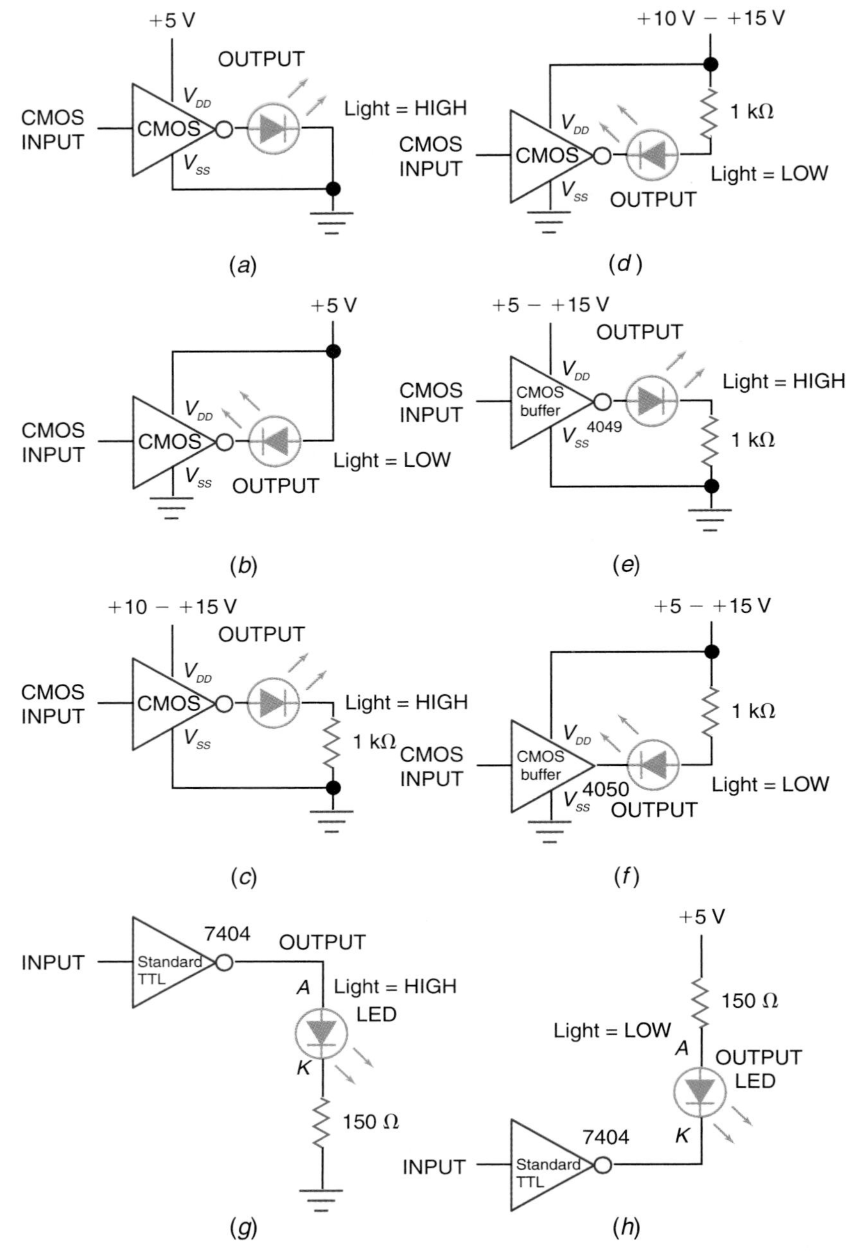

Simple CMOS-to-LED interfacing

Simple TTL-to-LED interfacing

Fig. 5-14 Simple CMOS- and TTL-to-LED interfacing. (*a*) CMOS active HIGH. (*b*) CMOS active-LOW. (*c*) CMOS active HIGH, supply voltage = 10 to 15 V. (*d*) CMOS active LOW, supply voltage = 10 to 15 V. (*e*) CMOS inverting buffer to LED interfacing. (*f*) CMOS noninverting buffer to LED interfacing. (*g*) TTL active-HIGH. (*h*) TTL active-LOW.

needed in series with the LEDs. In Fig. 5-14(*a*), when the output of the CMOS inverter goes HIGH, the LED output indicator lights. The opposite is true in Fig. 5-14(*b*): when the CMOS output goes LOW, the LED indicator lights.

Figures 5-14(*c*) and (*d*) show the 4000 series CMOS ICs being operated on a higher supply voltage (+10 to +15 V). Because of the higher voltage, a 1-kΩ limiting resistor is placed in series with the LED output indicator lights. When the output of the CMOS inverter in Fig. 5-14(*c*) goes HIGH, the LED output indicator lights. In Fig. 5-14(*d*), however, the LED indicator is activated by a LOW at the CMOS output.

Figures 5-14(*e*) and (*f*) show CMOS buffers being used to drive LED indicators. The circuits may operate on voltages from +5 to +15 V. Figure 5-14(*e*) shows the use of an inverting CMOS buffer (like the 4049 IC), while Fig. 5-14(*f*) uses the noninverting buffer (like the 4050 IC). In both cases, a 1-kΩ limiting resistor must be used in series with the LED output indicator.

Standard TTL gates are sometimes used to drive LEDs directly. Two examples are illustrated in Figs. 5-14(*g*) and (*h*). When the output of the inverter in Fig. 5-14(*g*) goes HIGH, current will flow through the LED causing it to light. The indicator light in Fig. 5-14(*h*) only lights when the output of the 7404 inverter goes LOW. The circuits in Fig. 5-14 are not recommended for critical uses because they exceed the output current ratings of the ICs. However, the circuits in Fig. 5-14 have been tested and work properly as simple output indicators.

Three improved LED output indicator designs are diagramed in Fig. 5-15. Each of the circuits uses transistor drivers and can be used with either CMOS or TTL. The LED in Fig. 5-15(*a*) lights when the output of the inverter goes HIGH. The LED in Fig. 5-15(*b*) lights when the output of the inverter goes LOW. Notice that the indicator in Fig. 5-15(*b*) uses a PNP instead of an NPN transistor.

The LED indicator circuits in Figs. 5-15(*a*) and (*b*) are combined in Fig. 5-15(*c*). The red light (LED1) will light when the inverter's output is HIGH. During this time LED2 will be off. When the output of the inverter goes LOW, transistor Q_1 turns off while Q_2 turns on. The green light (LED2) lights when the output of the inverter is LOW.

The circuit is Fig. 5-15(*c*) is a very basic logic probe. However, its accuracy is less than most logic probes.

The indicator light shown in Fig. 5-16 on page 103 uses an incandescent lamp. When the output of the inverter goes HIGH, the transistor is turned on and the lamp lights. When the inverter's output is LOW, the lamp does not light.

TEST

Supply the missing word(s) in each statement.

29. Refer to Fig. 5-14(*a–f*). The _________ (4000, FAST) series CMOS ICs are being used to drive the LEDs in these circuits.
30. Refer to Fig. 5-14(*h*). When the output of the inverter goes HIGH, the LED _________ (goes out, lights).
31. Refer to Fig. 5-15(*a*). When the output of the inverter goes LOW, the transistor is turned _________ (off, on) and the LED _________ (does not light, lights).
32. Refer to Fig. 5-15(*c*). When the output of the inverter goes HIGH, transistor _________ (Q_1, Q_2) is turned on and the _________ (green, red) LED lights.

5-6 Interfacing TTL and CMOS ICs

Interfacing TTL and CMOS ICs

CMOS and TTL logic levels (voltages) are defined differently. These differences are illustrated in the voltage profiles for TTL and CMOS shown in Fig. 5-17(*a*). Because of the differences in voltage levels, CMOS and TTL ICs usually cannot simply be connected together. Just as important, current requirements for CMOS and TTL ICs are different.

Transistor driver circuit

Look at the voltage and current profile in Fig. 5-17(*a*). Note that the output drive currents for the standard TTL are more than adequate to drive CMOS inputs. However, the voltage profiles do not match. The LOW outputs from the TTL are compatible because they fit within the wider LOW input band on the CMOS IC. There is a range of possible HIGH outputs from the TTL IC (2.4 to 3.5 V) that do not fit within the HIGH range of the CMOS IC. This incompatibility could cause problems. These problems can be solved by using a pull-up resistor between gates to pull the HIGH output of the standard TTL up closer to +5 V. A completed circuit for interfacing standard TTL to CMOS is shown in Fig. 5-17(*b*). Note the use of the 1-kΩ pull-up resistor. This circuit works for driving either 4000 series, 74HC00, or FACT series CMOS ICs.

During interviews, show an interest in continuing education opportunities.

Pull-up resistor

TTL-CMOS interfacing

Several other examples of TTL-to-CMOS and CMOS-to-TTL interfacing using a com-

CMOS-to-TTL interfacing

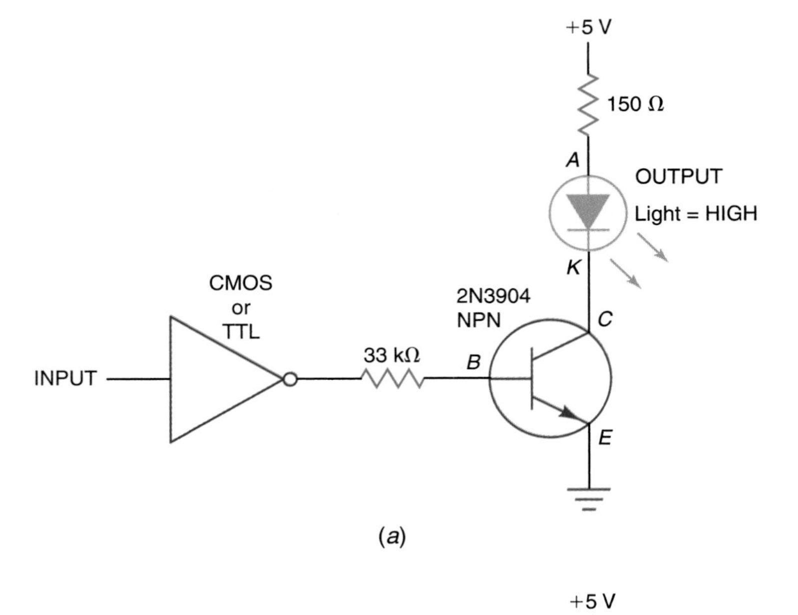

Interfacing to LEDs using a transistor drive circuit

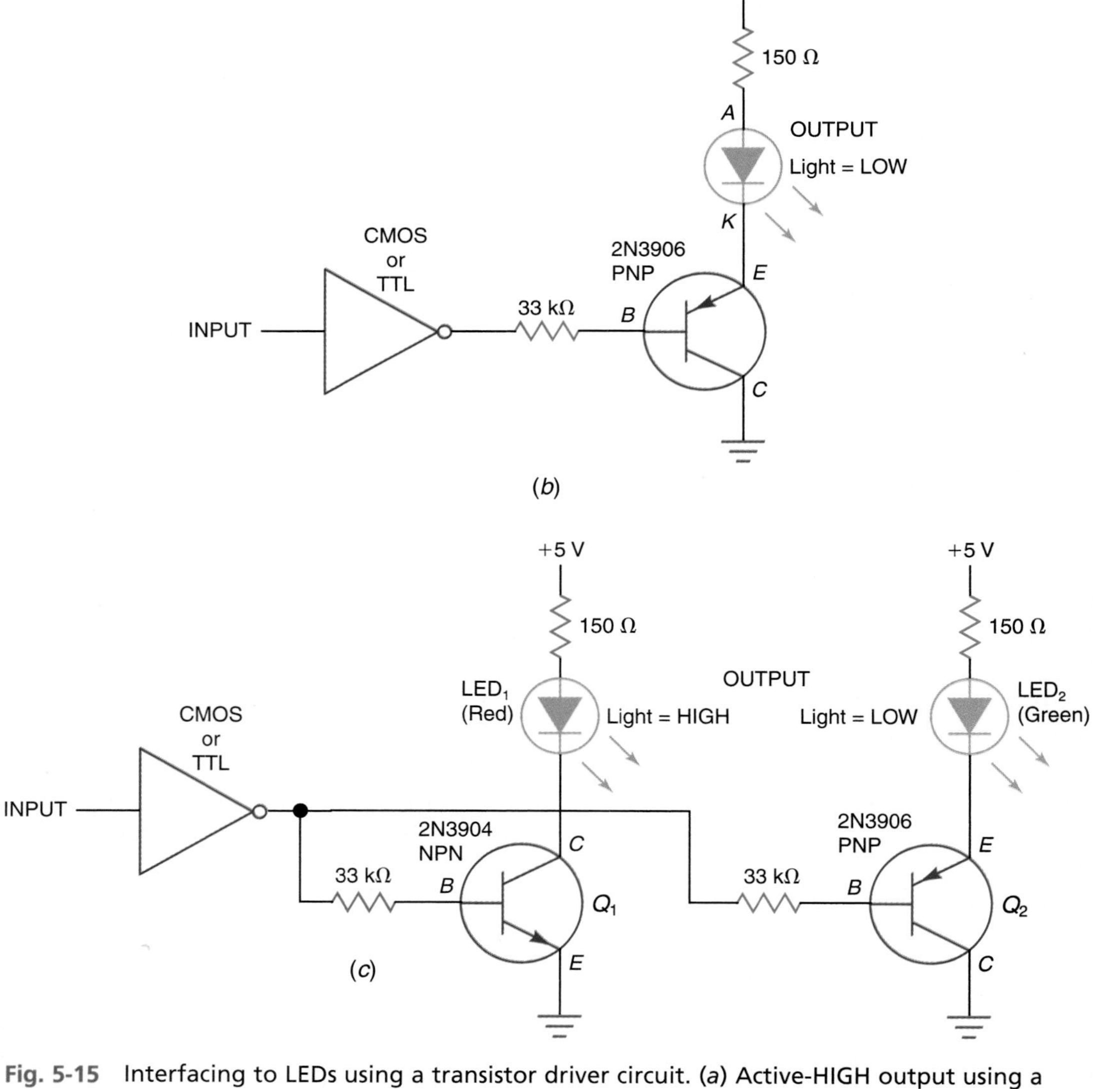

Fig. 5-15 Interfacing to LEDs using a transistor driver circuit. (*a*) Active-HIGH output using a NPN transistor driver. (*b*) Active-LOW output using a PNP transistor driver. (*c*) HIGH-LOW indicator circuit (simplified logic probe).

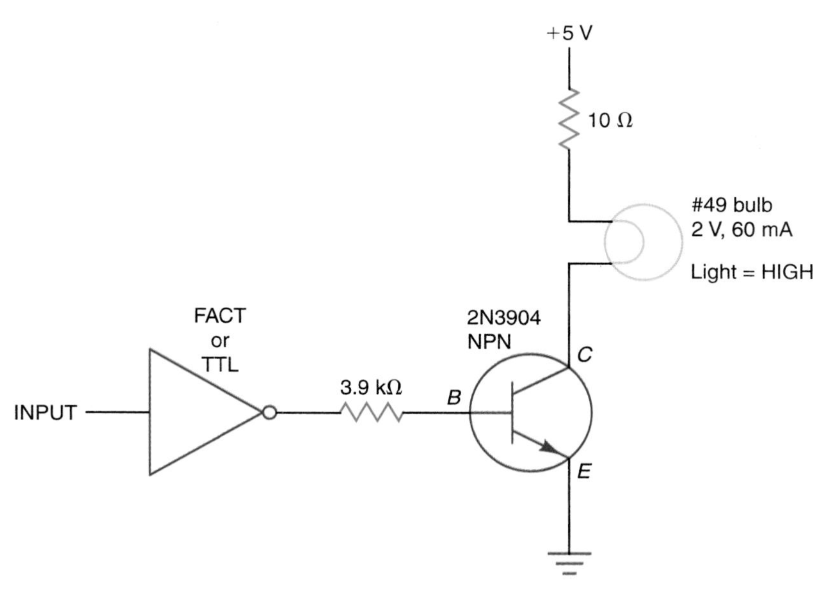

Fig. 5-16 Interfacing to an incandescent lamp using a transistor driver circuit.

Interfacing to an incandescent lamp

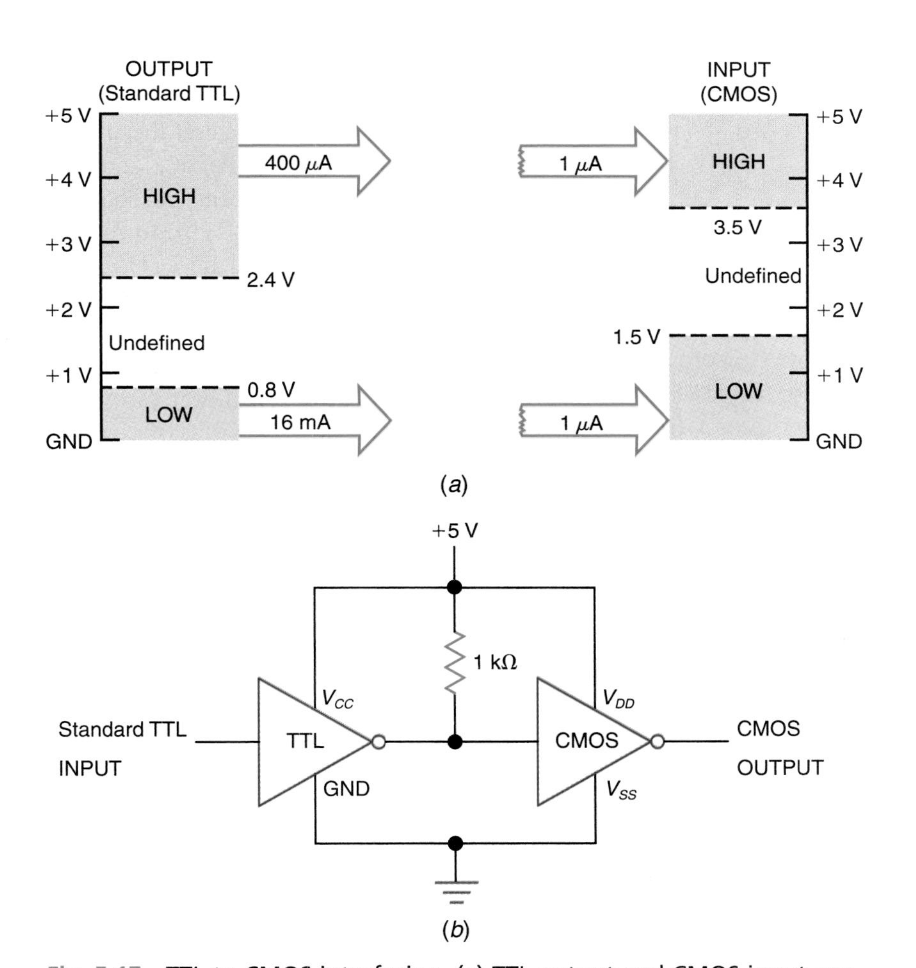

Fig. 5-17 TTL-to-CMOS interfacing. (*a*) TTL output and CMOS input profiles for visualizing compatibility. (*b*) TTL-to-CMOS interfacing using a pull-up resistor.

TTL-to-CMOS interfacing

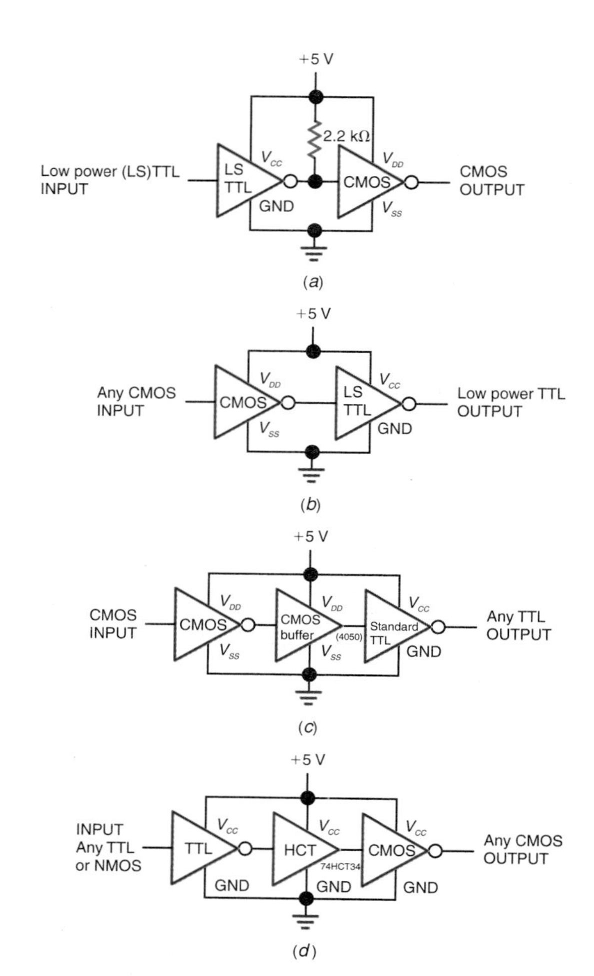

Fig. 5-18 Interfacing TTL and CMOS when both use a common +5-V power supply. (*a*) Low-power Schottky TTL to CMOS interfacing using a pull-up resistor. (*b*) CMOS to low-power Schottky TTL interfacing. (*c*) CMOS to standard TTL interfacing using a CMOS buffer IC. (*d*) TTL to CMOS interfacing using a 74HCT00 series IC.

74HCT00 series of CMOS ICs

FACT series of CMOS ICs

TTL-to-CMOS interfacing

CMOS-to-TTL interfacing

mon 5-V power supply are detailed in Fig. 5-18. Figure 5-18(*a*) shows the popular LS-TTL driving any CMOS gate. Notice the use of a 2.2-kΩ pull-up resistor. The pull-up resistor is being used to pull the TTL HIGH up near +5 V so that it will be compatible with the input voltage characteristics of CMOS ICs.

In Fig. 5-18(*b*), a CMOS inverter (any series) is driving an LS-TTL inverter *directly*. Complementary symmetry metal oxide semiconductor ICs can drive LS-TTL and ALS-TTL (advanced low-power Schottky) inputs: most CMOS ICs cannot drive standard TTL inputs without special interfacing.

Manufacturers have made interfacing easier by designed special buffers and other interface chips for designers. One example is the use of the 4050 noninverting buffer in Fig. 5-18(*c*). The 4050 buffer allows the CMOS inverter to have enough drive current to operate up to two standard TTL inputs.

The problem of voltage incompatibility from TTL (or NMOS) to CMOS was solved in Fig. 5-17 using a pull-up resistor. Another method of solving this problem is illustrated in Fig. 5-18(*d*). The 74HCT00 series of CMOS ICs is specifically designed as a convenient interface between TTL (or NMOS) and CMOS. Such an interface is implemented in Fig. 5-18(*d*) using the 74HCT34 noninverting IC.

The 74HCT00 series of CMOS ICs is widely used when interfacing between NMOS devices and CMOS. The NMOS output characteristics are almost the same as for LS-TTL.

The modern FACT series of CMOS ICs has excellent output drive capabilities. For this reason FACT series chips can drive TTL, CMOS, NMOS, or PMOS ICs directly as illustrated in Fig. 5-19(*a*). The output voltage characteristics of TTL do not match the input voltage profile of 74HC00, 74AC00, and 74ACQ00 series CMOS ICs. For this reason, a pull-up resistor is used in Fig. 5-19(*b*) to make sure the HIGH output voltage of the TTL gate is pulled up near the +5-V rail of the power supply. Manufacturers produce "T"-type CMOS gates that have the input voltage profile of a TTL IC. TTL gates can directly drive any 74HCT00, 74ACT00, 74FCT00, 74FCTA00, or 74ACTQ00 series CMOS IC, as summarized in Fig. 5-19(*c*).

Interfacing CMOS devices with TTL devices takes some added components when each operates on a *different voltage power supply*. Figure 5-20 shows three examples of TTL-to-CMOS and CMOS-to-TTL interfacing. Figure 5-20(*a*) shows the TTL inverter driving a general-purpose NPN transistor. The transistor and associated resistors translate the lower-voltage TTL outputs to the higher-voltage inputs needed to operate the CMOS inverter. The CMOS output has a voltage swing from about 0 to almost +10 V. Figure 5-20(*b*) shows an open-collector TTL buffer and a 10-kΩ pull-up resistor being used to translate the lower TTL to the higher

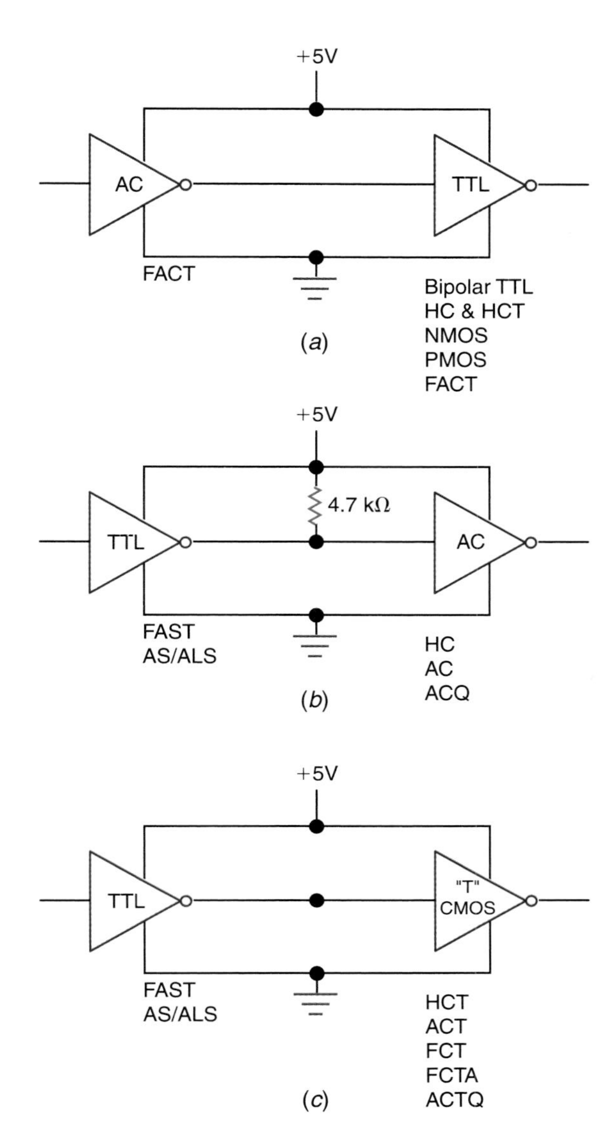

Fig. 5-19 Interfacing FACT with other families. (*a*) FACT driving most other TTL and CMOS families. (*b*) TTL-to-FACT interfacing using a pull-up resistor. (*c*) TTL-to-"T" CMOS ICs.

CMOS voltages. The 7406 and 7416 TTL ICs are two inverting, open-collector (OC) buffers.

Interfacing between a higher-voltage CMOS inverter and a lower-voltage TTL inverter is shown in Fig. 5-20(*c*). The 4049 CMOS buffer is used between the higher-voltage CMOS inverter and the lower-voltage TTL IC. Note that the CMOS buffer is powered by the lower-voltage (+5 V) power supply in Fig. 5-20(*c*).

Looking at the voltage and current profiles (such as in Fig. 5-17(*a*)) is a good starting point when learning about or designing an interface. Manufacturers' manuals are also very helpful. Several techniques are used to interface between different logic families. These include the use of pull-up resistors and special interface ICs. Sometimes no extra parts are needed.

TEST

Supply the missing word in each statement.

33. Refer to Fig. 5-17(*a*). According to this profile of TTL output and CMOS input characteristics, the logic devices ________ (are, are not) voltage compatible.
34. Refer to Fig. 5-18(*a*). The 2.2-kΩ resistor in this circuit is called a ________ resistor.
35. Refer to Fig. 5-18(*c*). The 4050 buffer is a special interface IC that solves the ________ (current drive, voltage) incompatibility between the logic families.
36. Refer to Fig. 5-20(*a*). The ________ (NMOS IC, transistor) translates the TTL logic levels to the higher-voltage CMOS logic levels.

5-7 Interfacing with Buzzers, Relays, Motors, and Solenoids

The objective of many electromechanical systems is to control a simple output device. This device might be as simple as a light, buzzer, relay, electric motor, stepper motor, or solenoid. Interfacing to LEDs and lamps has been explored. Simple interfacing between logic elements and buzzers, relays, motors, and solenoids will be investigated in this section.

Piezo buzzer

The piezo buzzer is a modern signaling device drawing much less current than older buzzers and bells. The circuit in Fig. 5-21 on page 107 shows the interfacing necessary to drive a piezo buzzer with digital logic elements. A standard TTL or FACT CMOS inverter is shown driving a piezo buzzer *directly.* The standard TTL output can sink up to 16 mA while a FACT output has 24 mA of drive current. The piezo buzzer draws about 3 to 5 mA when sounding. Notice that the piezo buzzer has polarity markings. The diode across the buzzer is to suppress any transient voltages

CMOS buffer

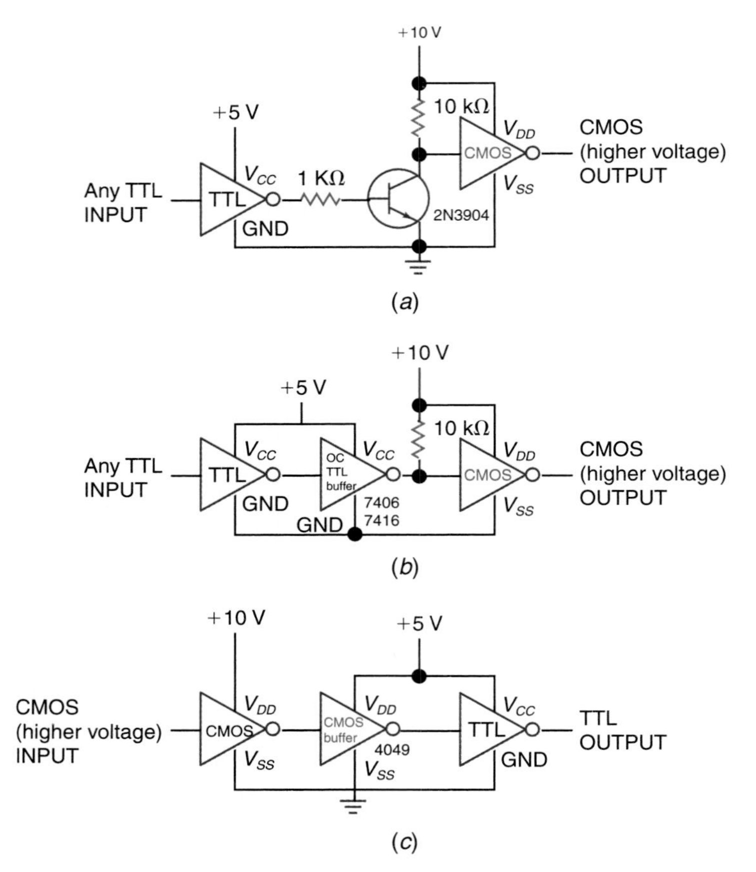

Fig. 5-20 Interfacing TTL and CMOS when each use a different power supply voltage. (*a*) TTL-to-CMOS interfacing using a driver transistor. (*b*) TTL-to-CMOS interfacing using an open collector TTL buffer IC. (*c*) CMOS-to-TTL interfacing using a CMOS buffer IC.

that might be induced in the system by the buzzer.

Logic device to buzzer interface

Most logic families do not have the current capacity to drive a buzzer directly. A transistor has been added to the output of the inverter in Fig. 5-21(*b*) to drive the piezo buzzer. When the output of the inverter goes HIGH, the transistor is turned on and the buzzer sounds. A LOW at the output of the inverter turns the transistor off, switching the buzzer off. The diode protects against transient voltages. The interface circuit sketched in Fig. 5-21(*b*) will work for both TTL and CMOS.

Clamp diode

Logic device to relay interface

Relay

A relay is an excellent method of isolating a logic device from a high-voltage circuit. Figure 5-22 shows how a TTL or CMOS inverter could be interfaced with a relay. When the output of the inverter goes HIGH, the transistor is turned on and the relay is activated. When activated, the normally open (NO) contacts of the relay close as the armature clicks downward. When the output of the inverter in Fig. 5-22 goes LOW, the transistor stops conducting and the relay is deactivated. The armature springs upward to its normally closed (NC) position. The clamp diode across the relay coil prevents voltage spikes which might be induced in the system.

The circuit in Fig. 5-23(*a*) on page 108 uses a relay to isolate an electric motor from the logic devices. Notice that the logic circuit and dc motor have separate power supplies. When the output of the inverter goes HIGH, the transistor is turned on and the NO contacts of the relay snap closed. The dc motor operates. When the output of the inverter goes lOW, the transistor stops conducting and the relay contacts spring back to their NC position. This turns off the motor. The electric motor in Fig. 5-23(*a*) produces rotary motion. A solenoid is an elec-

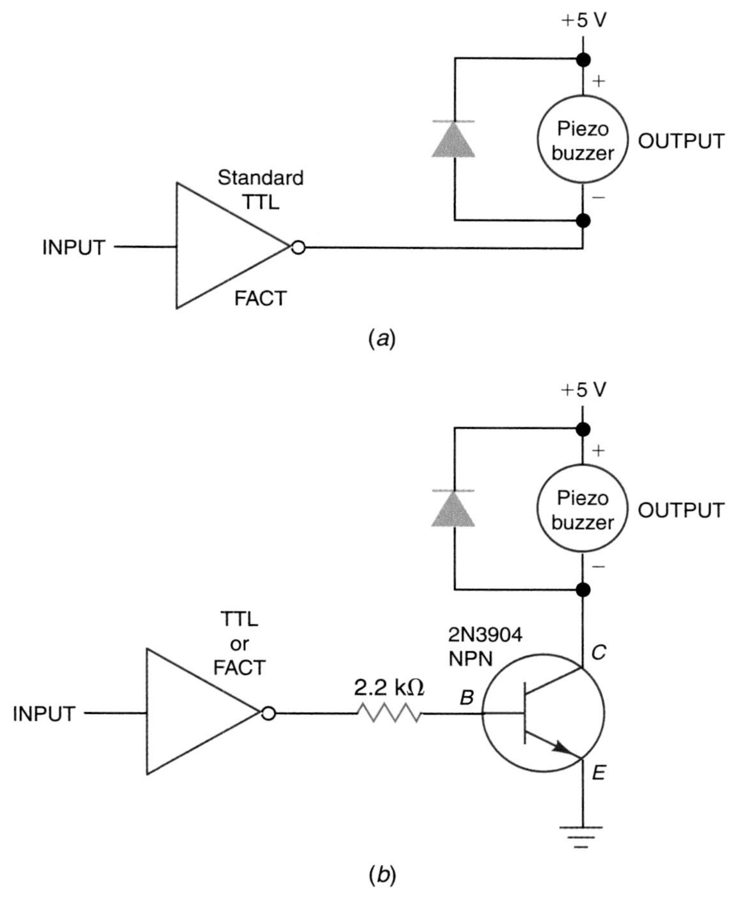

Fig. 5-21 Logic device to buzzer interfacing. (*a*) Standard TTL or FACT CMOS inverter driving a piezo buzzer directly. (*b*) TTL or CMOS interfaced with buzzer using a transistor driver.

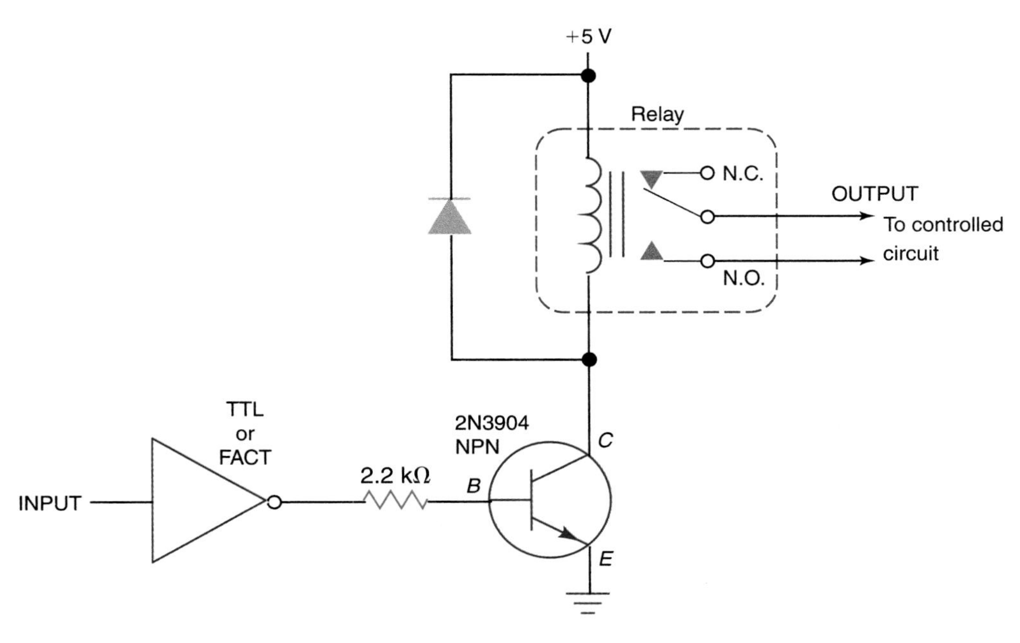

Fig. 5-22 TTL or CMOS interfaced with a relay using a transistor driver circuit.

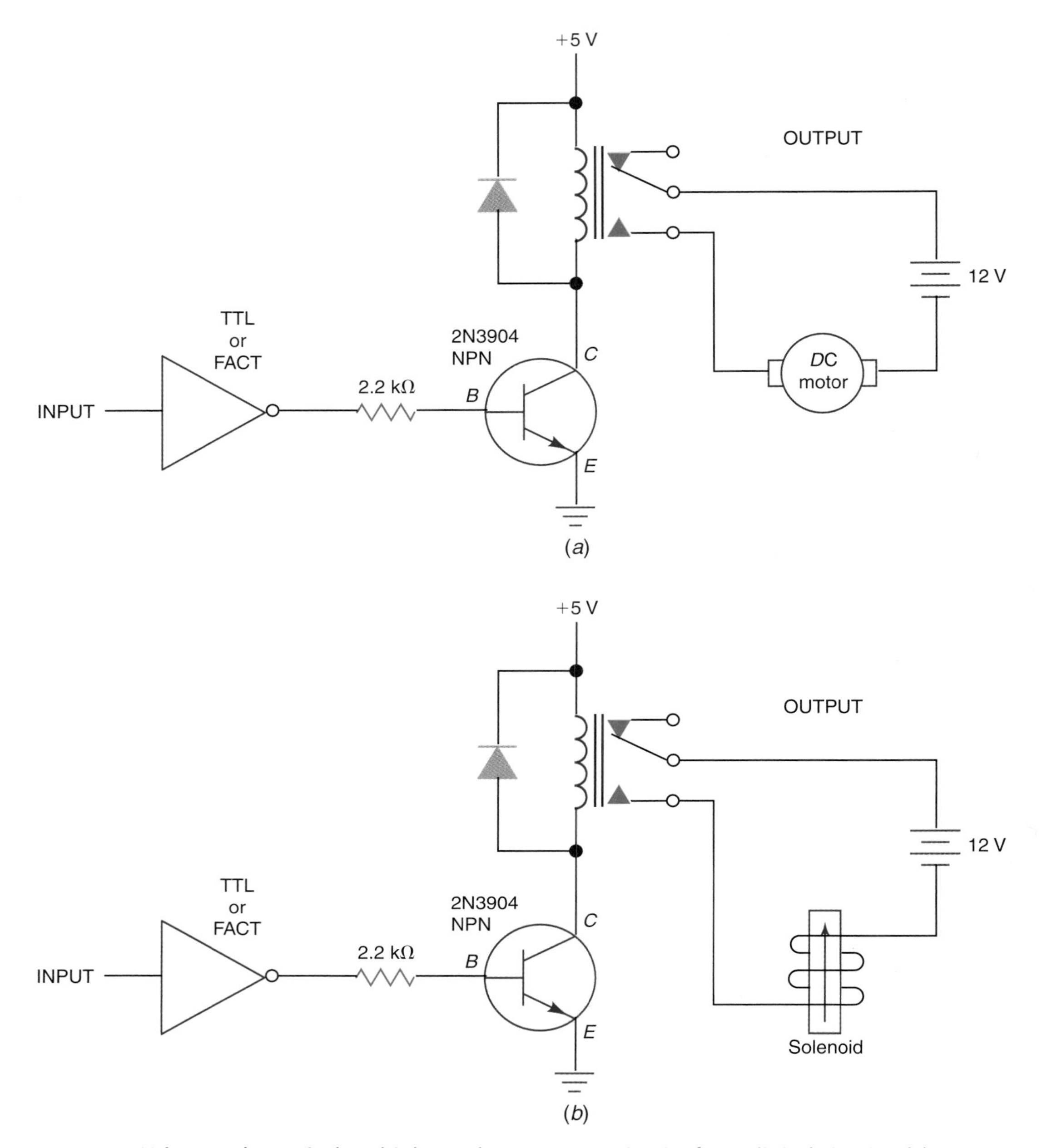

Fig. 5-23 Using a relay to isolate higher voltage/current circuits from digital circuits. (*a*) Interfacing TTL or CMOS with an electric motor. (*b*) Interfacing TTL or CMOS with a solenoid.

trical device that can produce linear motion. A solenoid is being driven by a logic gate in Fig. 5-23(*b*). Note the separate power supplies. This circuit works the same as the motor interface circuit in Fig. 5-23(*a*).

Logic device to motor interfacing

Logic device to solenoid interfacing

Voltage and current characteristics of most buzzers, relays, electric motors, and solenoids are radically different from those of logic circuits. Most of these electrical devices need special interfacing circuits to drive and isolate the devices from the logic circuits.

TEST

Supply the missing word(s) in each statement

37. Refer to Fig. 5-21(*a*). If the piezo buzzer draws only 6 mA, it ________ (is, is not) possible for a 4000 series CMOS IC to drive the buzzer directly (see Fig. 5-5(*b*) for 4000 series data).
38. Refer to Fig. 5-21(*b*). When the input to the inverter goes LOW, the transistor turns ________ (off, on) and the buzzer ________ (does not sound, sounds).

39. Refer to Fig. 5-22. The purpose of the diode across the coil of the relay is to suppress ________ (sound, transient voltages) induced in the circuit.
40. Refer to Fig. 5-23(*a*). The dc motor will run only when a ________ (HIGH, LOW) appears at the output of the inverter.

5-8 Optoisolators

The relay featured in Fig. 5-23 isolated the lower voltage digital circuitry from the high voltage/current devices such as a solenoid and electric motor. *Electromechanical relays* are relatively large and expensive but are a widely used method of control and isolation. Electromechanical relays can cause unwanted voltage spikes and noise due to the coil windings and opening and closing of contact points. A useful alternative to an electromagnetic relay when interfacing with digital circuits is the *optoisolator* or *optocoupler.* One close relative of the optoisolator is the *solid-state relay.*

One economical optoisolator is featured in Fig. 5-24. The *4N25 optoisolator* consists of a *gallium arsenide infrared-emitting diode* optically coupled to a silicon *phototransistor detector* enclosed in a six-pin dual in-line package (DIP). Figure 5-24(*a*) details the pin diagram for the 4N25 optoisolator with the names of the pins. On the input side, the LED is typically activated with a current of about 10 to 30 mA. When the input LED is activated, the light activates (turns on) the phototransistor. With no current through the LED the output phototransistor of the optoisolator is turned off (high resistance from emitter-to-collector).

A simple test circuit using the 4N25 optoisolator is shown in Fig. 5-24(*b*). The digital signal from the output of a TTL or FACT inverter directly drives the infrared-emitting diode. The circuit is designed so the LED is activated when the output of the inverter goes LOW, which allows the inverter to sink the 10 to 20 mA LED current to ground. When the LED is activated, infrared light shines (inside the package) activating the phototransistor. The transistor is turned on (low resistance from emitter-to-collector) dropping the voltage at the collector (OUTPUT) to near 0 V. If the output of the inverter goes HIGH, the LED does not light and the NPN phototransistor turns off (high resistance from emitter-to-collector). The output (at the collector) is pulled to about +12 V (HIGH) by the 10-kΩ pull-up resistor. In this example, notice that the input side of the circuit operates on +5 V while the output side in this example uses a separate +12 V power supply. In summary, the input and output sides of the circuit are isolated from one another. When pin 2 of the optoisolator goes LOW, the output at the collector of the transistor goes LOW. The grounds of the separate power supplies should not be connected to complete the isolation between the low- and high-voltage sides of the optoisolator.

A simple application of the optoisolator being used to interface between TTL circuitry and a piezo buzzer is diagramed in Fig. 5-24(*c*). In this example the pull-up resistor is removed because we are using the NPN phototransistor in the optoisolator to sink the 2 to 4 mA of current when the transistor is activated. A LOW at the output of the inverter (pin 2 of optoisolator) activates the LED, which in turn activates the phototransistor.

Optoisolator

To control heavier loads using the optoisolator we could attach a power transistor to the output as is done in Fig. 5-24(*d*). In this example, if the LED is activated it activates the phototransistor. The output of the optoisolator (pin 5) drops LOW, which turns off the power transistor. The emitter-to-collector resistance of the power transistor is high, turning off the dc motor. When the output of the TTL inverter goes HIGH it turns off the LED and the phototransistor in the optoisolator. The voltage at output pin 5 goes positive, which turns on the power transistor and operates the dc motor.

If the power transistor (or other power-handling device such as a triac) in Fig. 5-24(*d*) were housed in the isolation unit the entire device is sometimes a *solid-state relay.* Solid-state relays can be purchased to handle a variety of outputs included in either ac or dc loads. The output circuitry in a solid-state relay may be more complicated than that shown in Fig. 5-24(*d*).

Solid-state relay

Several examples of solid-state relay packages are shown in Fig. 5-25 on page 111. The unit in Fig. 5-25(*a*) is a smaller PC-mounted unit. The larger bolted-on solid-state relay has screw terminals and can handle greater ac currents and voltages.

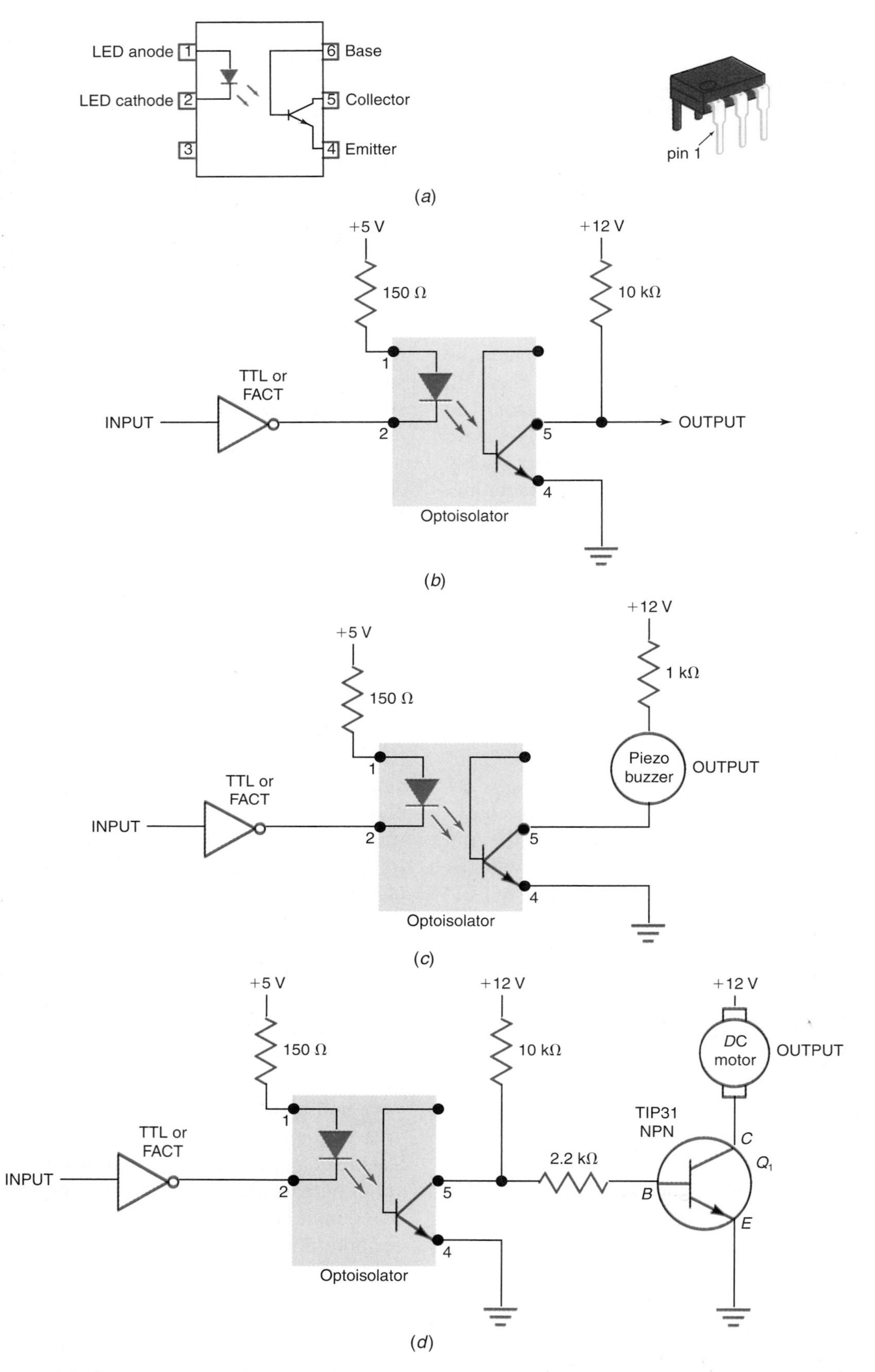

Fig. 5-24 (*a*) The 4N25 optoisolator pin out and six-pin DIP. (*b*) Basic optoisolator circuit separates 5 V and 12 V circuits. (*c*) Optoisolator driving piezo buzzer. (*d*) Optoisolator isolating low-voltage digital circuit from high voltage/current motor circuit.

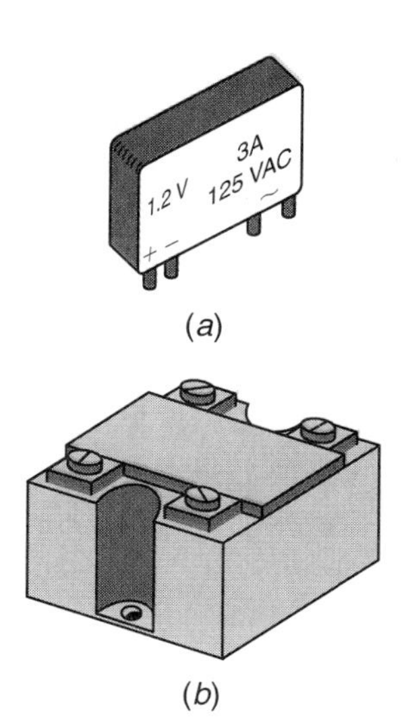

(a)

(b)

Fig. 5-25 (*a*) Solid-state relay—small PC-mounted package. (*b*) Solid-state relay—heavy duty package.

In summary, it is common to isolate digital circuitry from some devices because of high operating voltages and currents or because of dangerous feedback in the form of voltage spikes and noise. Traditionally, electromagnetic relays have been used for isolation, but optoisolators and solid-state relays are an inexpensive and effective alternative when interfacing with digital circuits. A typical optoisolator, shown in Fig. 5-24(*a*), contains an infrared-emitting diode that activates a phototransistor. If you are building an interface project using the parallel port from an IBM-compatible or PC, you will want to use optoisolators between your circuits and the computer. The PC parallel-port outputs and inputs operate with TTL level signals. Good isolation protects your computer from voltage spikes and noise.

TEST

Supply the missing word(s) in each statement.

41. Refer to Fig. 5-23(*a*). The ________ (relay, transistor) isolates the digital circuitry from the higher voltage and noisy dc motor circuit.
42. The 4N25 optoisolator device contains an infrared-emitting diode optically coupled to a ________ (phototransistor, triac) detector enclosed in a six-pin DIP.
43. Refer to Fig. 5-24(*b*). If the output of the TTL inverter goes LOW, the infrared LED ________ (does not light, lights), which ________ (activates, deactivates) the phototransistor and the voltage at pin 5 (output) goes ________ (HIGH, LOW).
44. Refer to Fig. 5-24(*b*). The 10-kΩ resistor connecting the collector of the phototransistor to +12 V is called a(n) ________ resistor.
45. Refer to Fig. 5-24(*c*). If the output of the TTL inverter goes HIGH the LED ________ (does not light, lights), which ________ (activates, deactivates) the phototransistor and the voltage at pin 5 (output) goes ________ (HIGH, LOW) and the buzzer ________ (does not sound, sounds).
46. Refer to Fig. 5-24(*d*). If the output of the TTL inverter goes HIGH the LED does not light, which deactivates (turns off) the phototransistor and the voltage at pin 5 (output) goes more positive. This positive-going voltage at the base of the power transistor ________ (turns on, turns off) Q_1 and the dc motor ________ (does not run, runs).

For information on industry standards, visit the website for the National institute of standards and technology (NIST).

5-9 Interfacing with Stepper Motors

The dc motor mentioned previously in this chapter is a device that rotates continuously when power is applied. The control over the dc motor is limited to ON-OFF, or if you reverse the direction of current flow through the rotor the direction of rotation reverses. A simple dc motor does not facilitate good speed control and it will not rotate a given number of degrees to stop for angular positioning. Where precision positioning or exacting speed are required, a regular dc motor does not do the job.

Both the servo and the stepper motor can rotate to a given position and stop and also reverse direction. The word "servo" is short for *servomotor.* "Servo" is a general term for a motor in which either the angular position or speed can be controlled precisely by a servo loop which uses feedback from the output back to input for control. The most common servos are the inexpensive units used in model aircraft, model cars, and some educational robot kits. These servos are geared-down dc motors with built-in electronics that respond to different pulse widths. These servos use feedback to ensure the device rotates to and stays at the current angular position. These servos are popular in remote-control models and toys. They commonly have three wires (one wire for input and two wires for power) and are not used for continuous rotation.

Stepper Motor

Bipolar stepper motor

The *stepper motor* can rotate a *fixed angle* with each input signal. A common four-wire stepper motor is sketched in Fig. 5-26(*a*). From the label you can see some of the important characteristics of the stepper motor. This stepper motor is designed to operate on 5 V dc. Each of the two coils ($L1$ and $L2$) has a resistance of 20 Ω. Using Ohm's law we calculate that the dc current through each coil is 0.25 A or 250 mA ($I = V/R$, substituting $I = 5/20$, then $I = 0.25$ A). The 2 ph means this is a *two-phase* or *bipolar* (as opposed to unipolar) stepper motor. Bipolar stepper motors typically have four wires coming from the case as is shown in Fig. 5-26(*a*). Unipolar stepper motors can have five to eight wires coming from the unit. The label on the stepper motor in Fig. 5-26(*a*) indicates that each step of the motor is 18° (meaning each signal rotates the shaft of the stepper motor an angle of 18°). Other important characteristics that might be given in a catalog or

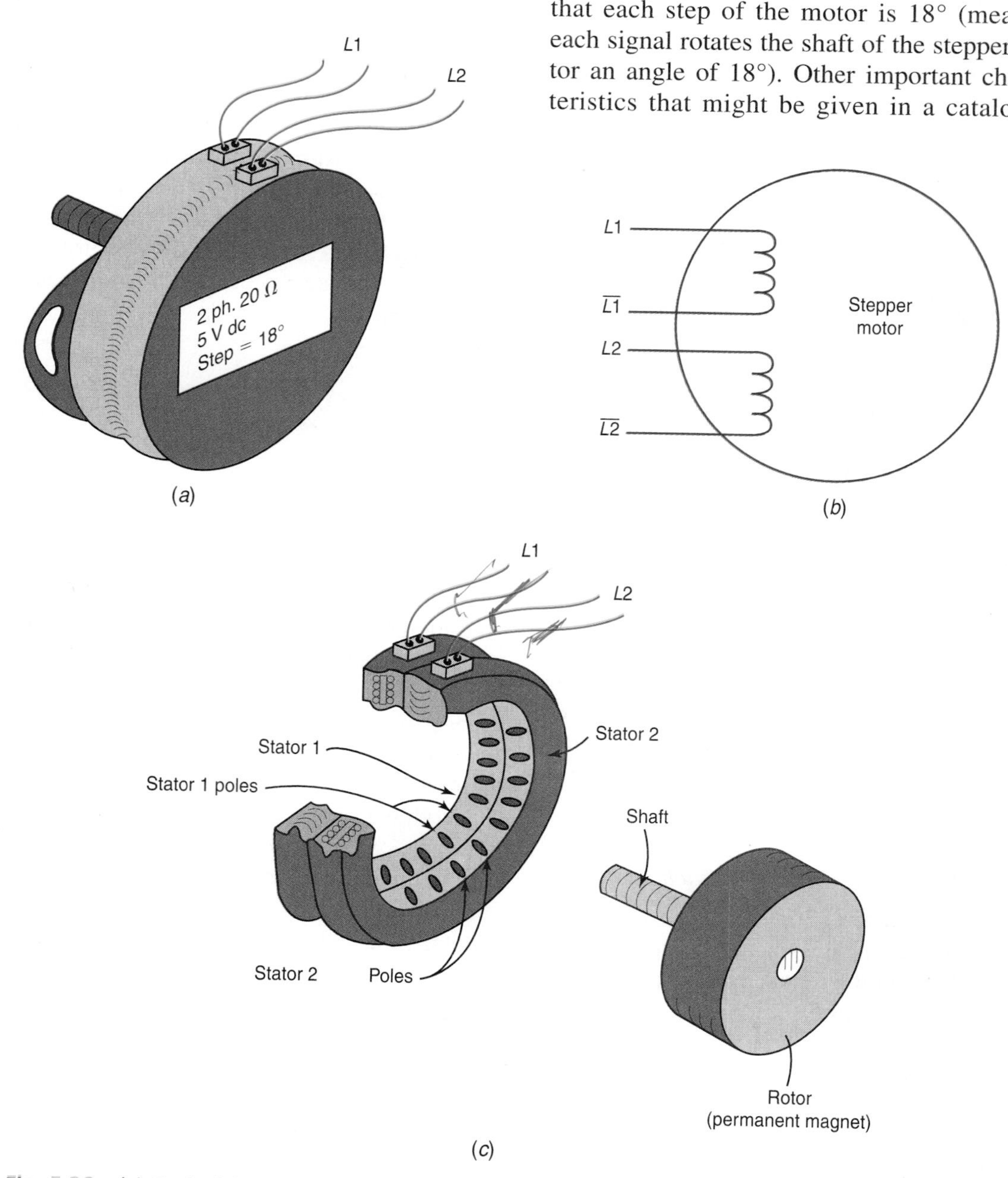

Fig. 5-26 (*a*) Typical four-wire stepper motor. (*b*) Schematic of four-wire bipolar stepper motor. (*c*) Simple exploded view of typical permanent magnet type stepper motor.

manufacturer's data sheet are physical size, inductance of coils, holding torque, and detent torque of motor. A schematic of the stepper motor's coils would probably be included such as the one with the wire colors shown in Fig. 5-26(*b*). Notice that there are two coils in the schematic diagram of this stepper motor. A control sequence is also usually given for a stepper motor.

A simplified exploded view of a stepper motor is drawn in Fig. 5-26(*c*). Of interest is the *permanent magnet rotor* attached to the output shaft. Some stepper motors have a gearlike soft-iron rotor with the number of poles unequal to the number of poles in the stator. These are referred to as *variable reluctance stepper motors.* There are two stators as shown in Fig. 5-26(*c*). A series of poles are visible on both stator 1 and 2. The number of poles on a single stator are the number of steps required to complete one revolution of the stepper motor. For instance, if a stepper motor has a single step angle of 18°, you can calculate the number of steps in a revolution as:

Degs. in circle/single-step angle = steps per revolution
360°/18° = 20 steps per revolution

In this example, each stator has 20 visible poles. Notice that the poles of stator 1 and 2 are not aligned but are one-half the single-step angle, or 9° different. Common stepper motors are available in step angles of 0.9°, 1.8°, 3.6°, 7.5°, 15°, and 18°.

The stepper motor responds to a standard control sequence. That control sequence for a sample bipolar stepper motor is charted in Fig. 5-27(*a*). Step 1 on the chart shows coil lead $L1$ at about +5 V, while the other end of the coil ($\overline{L1}$) is grounded. Likewise, step 1 also shows coil lead $L2$ at about +5 V while the other end of the coil ($\overline{L2}$) is grounded. In step 2, note that the polarity of coil $L1/\overline{L1}$ is reversed while $L2/\overline{L2}$ stays the same, causing a clockwise (CW) rotation of one step (18° for the sample stepper motor). In step 3, only the polarity of coil $L2/\overline{L2}$ is reversed, causing a second CW rotation of one step. In step 4, only the polarity of coil $L1/\overline{L1}$ is reversed, which causes a third CW rotation of a single step. In step 1, only the polarity of $L2/\overline{L2}$ has been reversed, causing a fourth CW rotation of a single step. Continuing the sequence of steps 2, 3, 4, 1, 2, 3, and so on would cause the stepper motor to continue rotating in a CW direction 18° at each step.

To reverse the stepper motor's direction of rotation, move upward on the control sequence chart in Fig. 5-27(*a*). Suppose we are at step 2 at the bottom of the chart. Moving upward to step 1, the polarity of only coil $L1/\overline{L1}$ changes and the motor rotates one step counterclockwise (CCW). Moving upward again to step 4, the polarity of only coil $L2/\overline{L2}$ changes and the motor rotates a second step CCW. In step 3, the polarity of only coil $L1/\overline{L1}$ changes as the motor rotates a third step CCW. CCW rotation continues as long as the sequence 2, 1, 4, 3, 2, 1, 4, 3, and so forth from the control sequence is followed.

Variable reluctance stepper motor

In summary, CW rotation occurs when you progress downward on the control sequence chart in Fig. 5-27(*a*). Counterclockwise rotation occurs when you stop at any step on the chart in Fig. 5-27(*a*) and then progress upward. The stepper motor is excellent at exact angular positioning, which is important in computer disk drives and printers, robotics and all types of automated machinery, and NC machine tools. The stepper motor can also be used for continuous rotation applications where the exact speed of rotation is important. Continuous rotation of a stepper motor can be accomplished by sequencing through the control sequence quickly. For instance, suppose you want the motor from Fig. 5-26(*a*) to rotate at 600 rpm. This means that the motor rotates 10 revolutions per second (600 rpm/60 sec = 10 rev/sec). You would have to send the code from the control sequence in Fig. 5-27(*a*) to the stepper motor at a frequency of 200 Hz (10 rev/sec × 20 steps per rev = 200 Hz).

Control sequence for stepper motor

Interfacing

Consider the simple test circuit in Fig. 5-27(*b*) which could be used to check a bipolar stepper motor. The single-pole double-throw (SPDT) switches are currently set to deliver the voltages defined by step 1 on the control sequence chart in Fig. 5-27(*a*). As you change the voltage inputs to the coils as specified by step 2, then step 3, and then step 4, and so on, the mo-

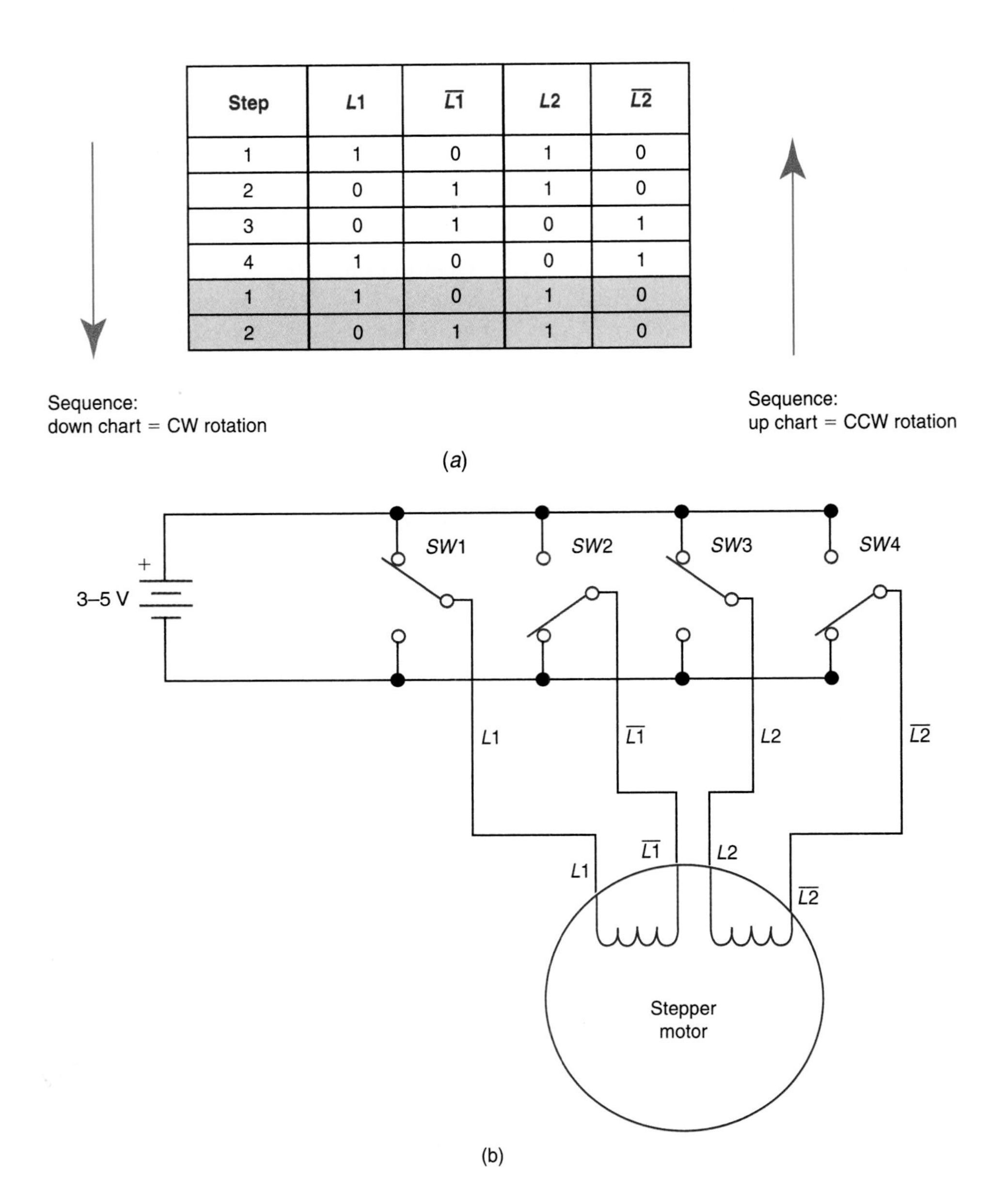

Step	L1	$\overline{L1}$	L2	$\overline{L2}$
1	1	0	1	0
2	0	1	1	0
3	0	1	0	1
4	1	0	0	1
1	1	0	1	0
2	0	1	1	0

Fig. 5-27 (*a*) Bipolar control sequence chart. (*b*) Test circuit for hand checking a four-wire bipolar stepper motor.

tor rotates by stepping in a CW direction. If you reverse the order and sequence upward on the control sequence chart in Fig. 5-27(*a*) the motor reverses and rotates by stepping in a CCW direction. The circuit in Fig. 5-27(*b*) is an impractical interface circuit but can be used for hand testing a stepper motor.

A practical bipolar stepper motor interface is based on the *MC3479 stepper motor driver IC* from Motorola. The schematic diagram in Fig. 5-28(*a*) details how you might wire the MC3479 driver IC to a bipolar stepper motor. The MC3479 IC has a logic section that generates the proper control sequence to drive a bipolar stepper motor. The motor driver section has a drive capability of 350 mA per coil. Each step of the motor is triggered by a single positive-going clock pulse entering the CLK input (pin 7) of the IC. One input control sets the direction of rotation of the stepper motor. A logic 0 at the CW/$\overline{\text{CCW}}$ input to the MC3479 allows CW rotation while a logic 1 input at pin 10 changes to CCW rotation of the stepper motor.

The MC3479 IC also has a $\overline{\text{Full}}$/Half input (pin 9) which can change the operation of the IC from stepping by full steps or half steps. In the *full-step mode,* the stepper motor featured

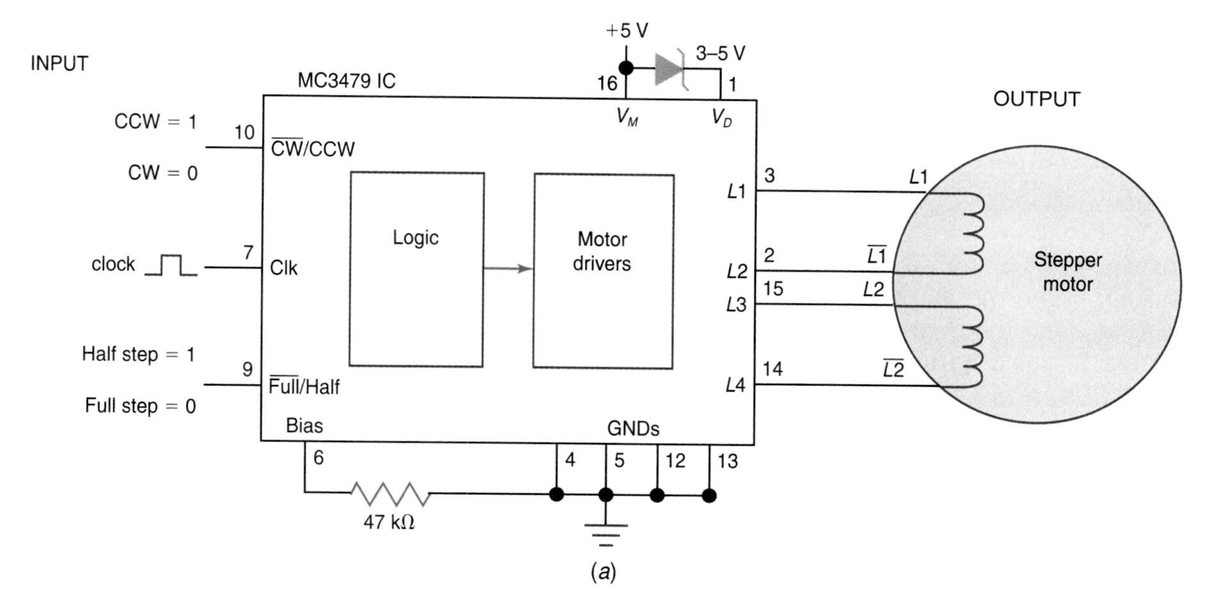

Step	L1	L2	L3	L4
1	1	0	1	0
2	0	1	1	0
3	0	1	0	1
4	1	0	0	1
1	1	0	1	0
2	0	1	1	0

(*b*)

Step	L1	L2	L3	L4
1	1	0	1	0
2	1	1	1	0
3	0	1	1	0
4	0	1	1	1
5	0	1	0	1
6	1	1	0	1
7	1	0	0	1
8	1	0	1	1
1	1	0	1	0
2	1	1	1	0
3	0	1	1	0
4	0	1	1	1

(*c*)

Fig. 5-28 (*a*) Using the MC3479 Stepper Motor Driver IC to interface with a bipolar stepper motor. (*b*) Control sequence of the MC3479 IC in the full-step mode. (*c*) Control sequence for the MC3479 IC in the half-step mode.

in Fig. 5-26 rotates 18° for each clock pulse (each single step). In the *half-step mode,* the stepper motor rotates half of a regular step or only 9° per clock pulse. The *control sequence* used by the MC3479 IC in the full-step mode is shown in chart form in Fig. 5-28(*b*). Note that this is the same control sequence used in Fig. 5-27(*a*). The control sequence used by the MC3479 IC in the half-step mode is detailed in chart form in Fig. 5-28(*c*). These control sequences are standard for bipolar or two-phase stepper motors and are built into the logic block of the MC3479 stepper motor driver IC. Specialized ICs such as the MC3479 stepper motor driver are usually the simplest and least expensive method of solving the problem of generating the correct control sequences, allowing for either CW or CCW rotation, and allowing the stepper motor to operate in either the full-step or half-step mode. The motor driver circuitry of the MC3479 is included inside the IC so lower power stepper motors can be driven directly by the IC as illustrated in Fig. 5-28(*a*).

MC 3479 Stepper motor driver IC

Unipolar or four-phase stepper motors have five or more leads exiting the motor. Specialized ICs are also available for generating the correct control sequence for these four-phase motors. One such product is the *EDE1200 unipolar stepper motor IC* by E-LAB Engineering. The EDE1200 has many of the same features of the Motorola MC3479 except it does not have the motor drivers inside the IC. External driver transistors or a driver IC must be used in conjunction with the EDE1200 unipolar stepper motor IC. The control sequence for four-phase (unipolar) and two-phase (bipolar) stepper motors is different.

Unipolar Stepper motor

Troubleshooting using a logic probe

TEST

Answer the following questions.

47. The ________ (dc motor, servo motor) is a good choice for continuous rotation applications not requiring speed control.
48. The ________ (dc motor, stepper motor) is a good choice for applications that require exact angular positioning of a shaft.
49. Both the servo and stepper motors can be used in applications that require exact angular positioning (T or F).
50. The device featured in Fig. 5-26 is a ________ (bipolar, unipolar) stepper motor.
51. The chart in Fig. 5-27(*a*) shows the ________ sequence for a ________ (bipolar, unipolar) stepper motor.
52. Refer to Fig. 5-27(*a*). If we are at step 4 and progress upward on the control sequence chart to step 3, the stepper motor rotates in a ________ (CCW, CW) direction.
53. Refer to Fig. 5-28(*a*). The ________ (logic, motor drive) block inside the MC3479 IC assures that the control sequence for driving a bipolar stepper motor is followed.
54. Refer to Fig. 5-28(*a*). The maximum drive current for each coil available using the MC3479 is ________ (10, 350) mA, which allows it to drive many smaller stepper motors directly.
55. Refer to Fig. 5-28(*a*) and assume input pins 9 and 10 are HIGH. When a clock pulse enters pin 7, the attached stepper motor rotates ________ (CCW, CW) a ________ (full step, half step).

5-10 Troubleshooting Simple Logic Circuits

One test equipment manufacturer suggests that about three-quarters of all faults in digital circuits occur because of open input or output circuits. Many of these faults can be isolated in a logic circuit using a logic probe.

Consider the combinational logic circuit mounted on a printed circuit board in Fig. 5-29(*a*). The equipment manual might include a schematic similar to the one shown in Fig. 5-29(*b*). Look at the circuit and schematic and determine the logic diagram. From that you can determine the Boolean expression and truth table. You will find that in this example, two NAND gates are feeding an OR gate. This is equivalent to the four-input NAND function. Its logic symbol diagram is the same as the one on the right side in Fig. 3-26.

The fault in the circuit in Fig. 5-29(*a*) is shown as an open circuit in the input to the OR gate. Now let's troubleshoot the circuit to see how we find this fault.

1. Set the logic probe to TTL and connect the power.
2. Test nodes 1 and 2 (see Fig. 5-29(*a*)). *Result:* Both are HIGH.
3. Test nodes 3 and 4. *Result:* Both are LOW. *Conclusion:* Both ICs have power.
4. Test the four-input NAND circuit's unique state (inputs *A, B, C,* and *D* are all HIGH). Test at pins 1, 2, 4, and 5 of the 7400 IC. *Results:* All inputs are HIGH but the LED still glows and indicates a HIGH output. *Conclusion:* The unique state of the four-input NAND circuit is faulty.
5. Test the outputs of the NAND gates at pins 3 and 6 of the 7400 IC. *Results:* Both outputs are LOW. *Conclusions:* The NAND gates are working.
6. Test the inputs to the OR gate at pins 1 and 2 of the 7432 IC. *Results:* Both inputs are LOW. *Conclusions:* The OR gate inputs at pins 1 and 2 are correct but the output is still incorrect. Therefore the OR gate is faulty and the 7432 IC needs to be replaced.

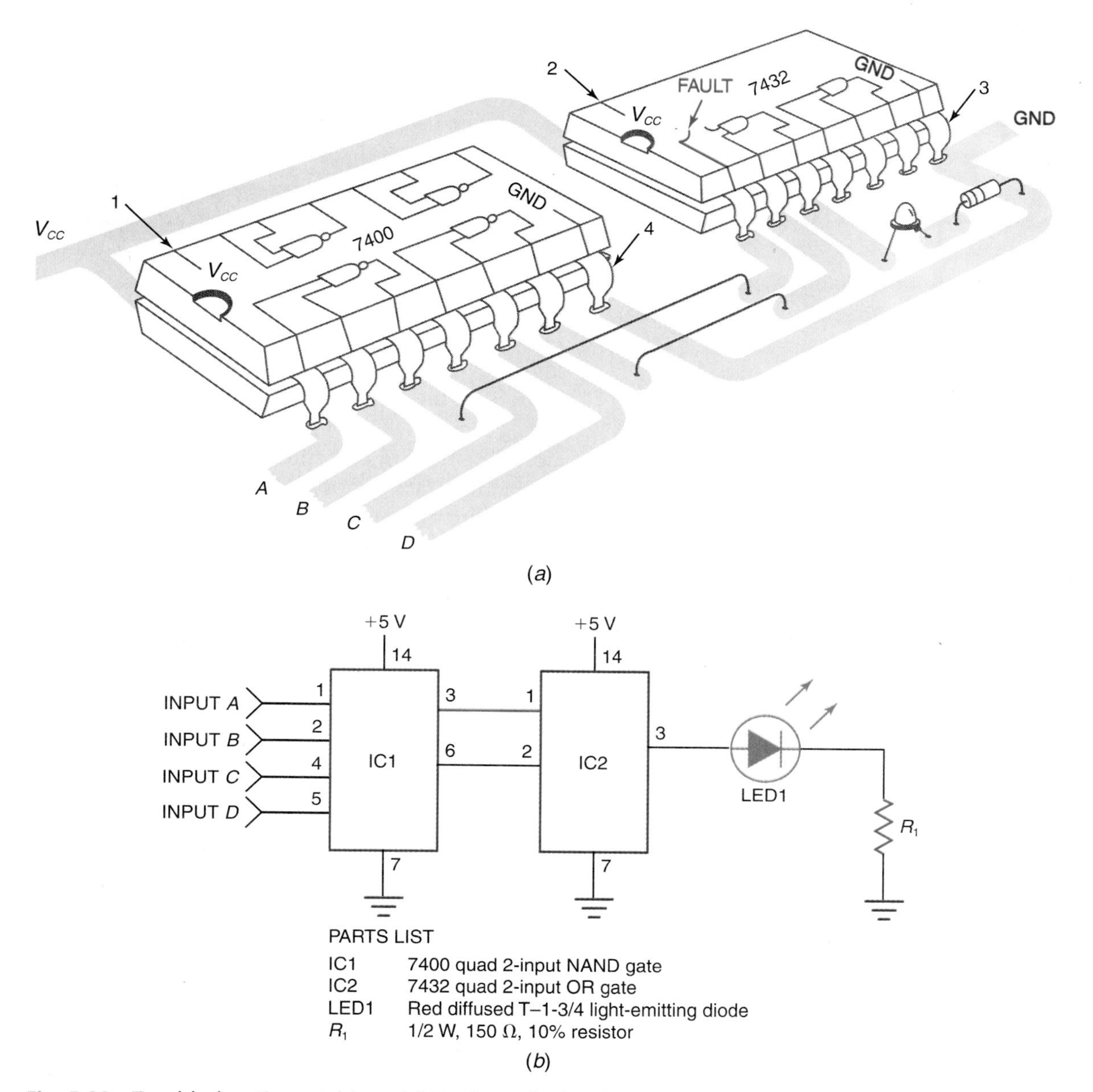

Fig. 5-29 Troubleshooting problem. (*a*) Testing a faulty circuit mounted on a PC board. (*b*) Schematic diagram of four-input NAND circuit.

TEST

Supply the missing word in each statement.

56. Most faults in digital circuits occur because of ________ (open, short) circuits in the inputs and outputs.
57. A simple piece of test equipment, such as a(n)________ , can be used for checking a digital logic circuit for open circuits in the inputs and outputs.
58. Refer to Fig. 5-29. With inputs *A, B, C,* and *D* all HIGH, the output (pin 3 of IC2) should be ________ (HIGH, LOW).

Chapter 5 Review

Summary

1. Interfacing is the design of circuitry between devices that shifts voltage and current levels to make them compatible.
2. Interfacing between members of the same logic family is usually as simple as connecting one gate's output to the next input, etc.
3. In interfacing between logic families or between logic devices and the "outside world," the voltage and current characteristics are very important factors.
4. Noise margin is the amount of unwanted induced voltage that can be tolerated by a logic family. Complementary symmetry metal oxide semiconductor ICs have better noise margins than TTL families.
5. The fan-out and fan-in characteristics of a digital IC are determined by its output drive and input loading specifications.
6. Propagation delay (or speed) and power dissipation are important IC family characteristics.
7. The ALS-TTL, FAST (Fairchild Advanced Schottky TTL), and FACT (Fairchild Advanced CMOS Technology) logic families are very popular owing to a combination of low power consumption, high speed, and good drive capabilities. Earlier TTL and CMOS families are still in wide use in existing equipment.
8. Most CMOS ICs are sensitive to static electricity and must be stored and handled properly. Other precautions to be observed include turning off an input signal before circuit power and connecting all unused inputs.
9. Simple switches can drive logic circuits using pull-up and pull-down resistors. Switch debouncing is usually accomplished using latch circuits.
10. Driving LEDs and incandescent lamps with logic devices usually requires a driver transistor.
11. Most TTL-to-CMOS and CMOS-to-TTL interfacing requires some additional circuitry. This can take the form of a simple pull-up resistor, special interface IC, or transistor driver.
12. Interfacing digital logic devices with buzzers and relays usually requires a transistor driver circuit. Electric motors and solenoids can be controlled by logic elements using a relay to isolate them from the logic circuit.
13. Optoisolators are also called optocouplers. Solid-state relays are a variation of the optoisolator. Optoisolators are used to electrically isolate digital circuitry from circuits that contain motors or other high voltage/current devices that might cause voltage spikes and noise.
14. Stepper motors operate on dc and are useful in applications where precise angular positioning or speed of an output shaft is important.
15. Stepper motors are classified as either bipolar (two-phase) or unipolar (four-phase). Other important characteristics are step angle, voltage, current, coil resistance, and torque.
16. Specialized ICs are useful for interfacing and driving stepper motors. The logic section of the IC generates the correct control sequence to step the motor.
17. Each logic family has its own definition of logical HIGH and LOW. Logic probes test for these levels.

Chapter Review Questions

Answer the following questions.

5-1. Applying 3.1 V to a TTL input is interpreted by the IC as a(n) _________ (HIGH, LOW, undefined) logic level.

5-2. A TTL output of 2.0 V is considered a(n) _________ (HIGH, LOW, undefined) output.

5-3. Applying 2.4 V to a CMOS input (10-V power supply) is interpreted by the IC as a(n) _________ (HIGH, LOW, undefined) logic level.

5-4. Applying 3.0 V to a 74HC00 series CMOS input (5-V power supply) is interpreted by the IC as a(n) _________ (HIGH, LOW, undefined) logic level.

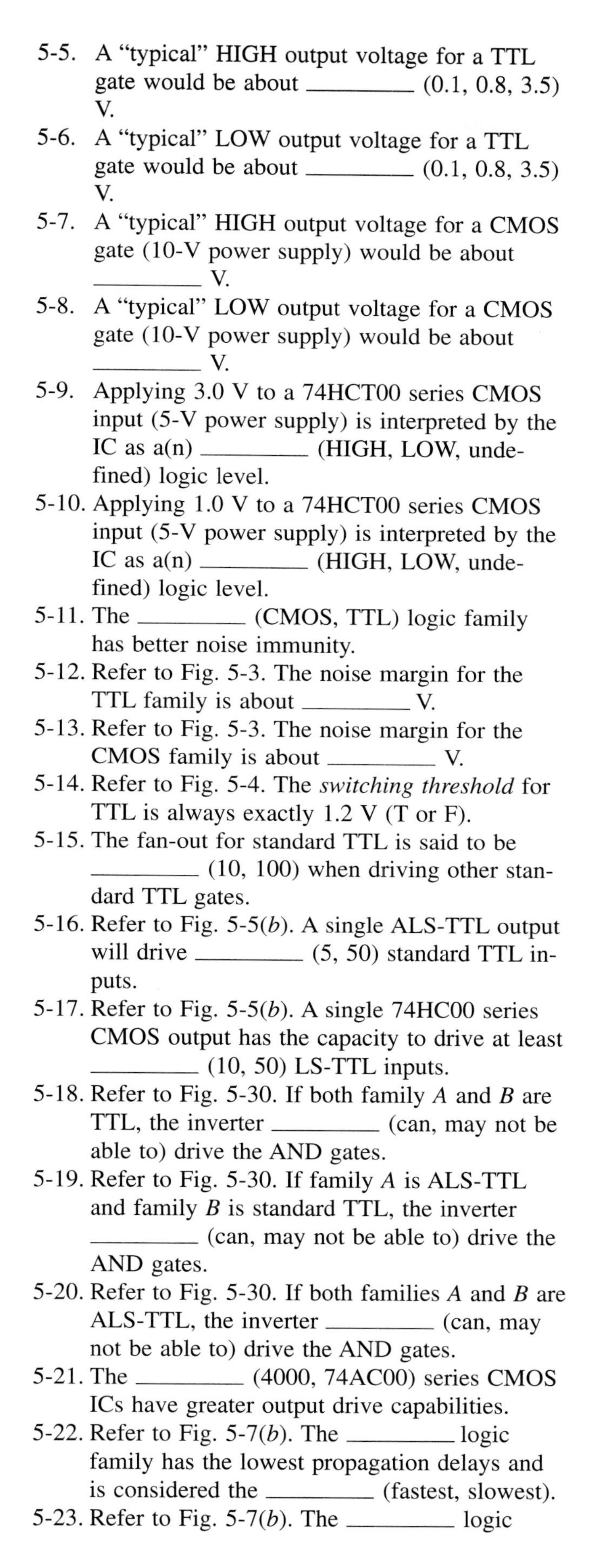

5-5. A "typical" HIGH output voltage for a TTL gate would be about ________ (0.1, 0.8, 3.5) V.

5-6. A "typical" LOW output voltage for a TTL gate would be about ________ (0.1, 0.8, 3.5) V.

5-7. A "typical" HIGH output voltage for a CMOS gate (10-V power supply) would be about ________ V.

5-8. A "typical" LOW output voltage for a CMOS gate (10-V power supply) would be about ________ V.

5-9. Applying 3.0 V to a 74HCT00 series CMOS input (5-V power supply) is interpreted by the IC as a(n) ________ (HIGH, LOW, undefined) logic level.

5-10. Applying 1.0 V to a 74HCT00 series CMOS input (5-V power supply) is interpreted by the IC as a(n) ________ (HIGH, LOW, undefined) logic level.

5-11. The ________ (CMOS, TTL) logic family has better noise immunity.

5-12. Refer to Fig. 5-3. The noise margin for the TTL family is about ________ V.

5-13. Refer to Fig. 5-3. The noise margin for the CMOS family is about ________ V.

5-14. Refer to Fig. 5-4. The *switching threshold* for TTL is always exactly 1.2 V (T or F).

5-15. The fan-out for standard TTL is said to be ________ (10, 100) when driving other standard TTL gates.

5-16. Refer to Fig. 5-5(*b*). A single ALS-TTL output will drive ________ (5, 50) standard TTL inputs.

5-17. Refer to Fig. 5-5(*b*). A single 74HC00 series CMOS output has the capacity to drive at least ________ (10, 50) LS-TTL inputs.

5-18. Refer to Fig. 5-30. If both family *A* and *B* are TTL, the inverter ________ (can, may not be able to) drive the AND gates.

5-19. Refer to Fig. 5-30. If family *A* is ALS-TTL and family *B* is standard TTL, the inverter ________ (can, may not be able to) drive the AND gates.

5-20. Refer to Fig. 5-30. If both families *A* and *B* are ALS-TTL, the inverter ________ (can, may not be able to) drive the AND gates.

5-21. The ________ (4000, 74AC00) series CMOS ICs have greater output drive capabilities.

5-22. Refer to Fig. 5-7(*b*). The ________ logic family has the lowest propagation delays and is considered the ________ (fastest, slowest).

5-23. Refer to Fig. 5-7(*b*). The ________ logic family has the highest propagation delays and is considered the ________ (fastest, slowest).

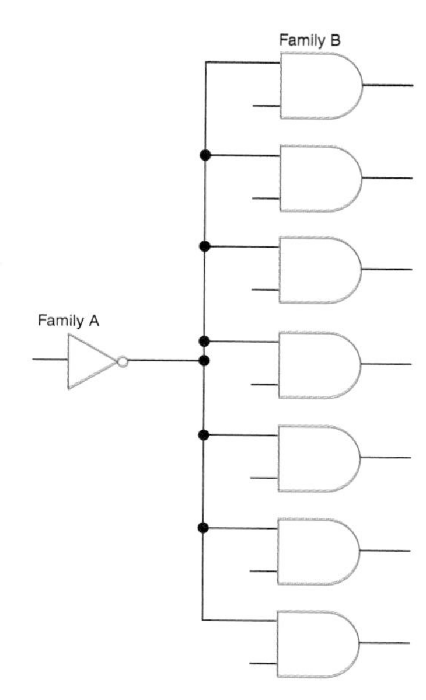

Fig. 5-30 Interfacing problem.

5-24. Refer to Fig. 5-7(*b*). The ________ is the fastest CMOS family.

5-25. Generally, ________ (CMOS, TTL) ICs consume the least power.

5-26. List several precautions that should be observed when working with CMOS ICs.

5-27. The V_{DD} pin on a 4000 series CMOS IC is connected to ________ (ground, positive) of the dc power supply.

5-28. Refer to Fig. 5-10(*b*). With the switch open, the inverter's input is ________ (HIGH, LOW) while the output is ________ (HIGH, LOW).

5-29. Refer to Fig. 5-11(*a*). When the switch is open, the ________ resistor causes the input of the CMOS inverter to be pulled HIGH.

5-30. Refer to Fig. 5-31. Component R_1 is called a ________ resistor.

5-31. Refer to Fig. 5-31. Closing SW_1 causes the input to the inverter to go ________ (HIGH, LOW) and the LED ________ (goes out, lights).

5-32. Refer to Fig. 5-31. With SW_1 open, a ________ (HIGH, LOW) appears at the input of the inverter causing the output LED to ________ (go out, light).

5-33. The common switch debouncing circuits in Figs. 5-13(*b*) and (*c*) are called RS flip-flops or ________.

5-34. A TTL output can drive a regular CMOS input with the addition of a(n) ________ resistor.

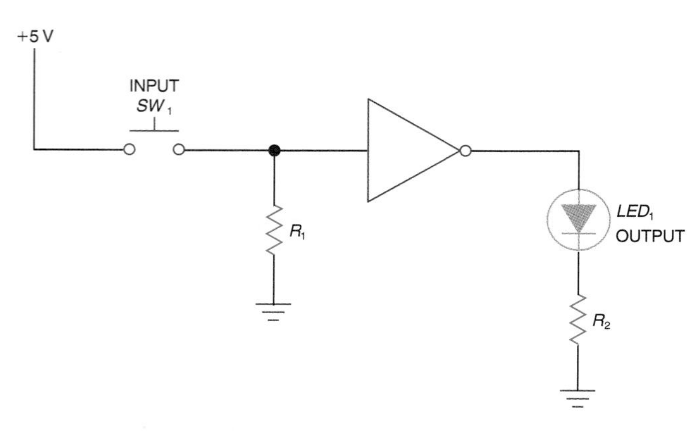

Fig. 5-31 Interfacing problem.

5-35. Any CMOS gate can drive at least one LS-TTL input (T or F).

5-36. A 4000 series CMOS output can drive a standard TTL input with the addition of a(n) ________.

5-37. Open-collector TTL gates require the use of ________ resistors at the outputs.

5-38. Refer to Fig. 5-21. The transistor functions as a(n) ________ (AND gate, driver) in this circuit.

5-39. Refer to Fig. 5-21. When the input to the inverter goes LOW, its output goes ________ (HIGH, LOW) which ________ (turns off, turns on) the transistor allowing current to flow through the transistor and piezo buzzer to sound the buzzer.

5-40. Refer to Fig. 5-23(*a*). When the input to the inverter goes LOW, its output goes HIGH which ________ (turns off, turns on) the NPN transistor; the coil of the relay is ________ (activated, deactivated), the relay armature clicks downward, and the dc motor ________ (rotates, will not rotate).

5-41. Refer to Fig. 5-23(*b*). When the input to the inverter goes HIGH, its output goes LOW which ________ (turns off, turns on) the NPN transistor; the coil of the relay is ________ (activated, deactivated), the armature of the relay ________ (clicks, will not click) downward, and the solenoid ________ (is, will not be) activated.

5-42. Refer to Fig. 5-24. The 4N25 optoisolator contains a gallium arsenide ________ (infrared-emitting diode, incandescent lamp) optically coupled to a phototransistor output.

5-43. Refer to Fig. 5-24(*b*). If the input to the inverter goes HIGH, its output goes LOW which ________ (activates, deactivates) the LED, the phototransistor is ________ (turned off, turned on), and the output voltage goes ________ (HIGH, LOW).

5-44. Refer to Fig. 5-24(*c*). The piezo buzzer sounds when the input to the inverter goes ________ (HIGH, LOW).

5-45. Refer to Fig. 5-24(*d*). This is an example of good design practice by using an optoisolator to isolate the low-voltage digital circuit from the higher-voltage noisy motor circuit (T or F).

5-46. Refer to Fig. 5-24(*d*). The dc motor turns on when a ________ (HIGH, LOW) logic level appears at the input of the inverter.

5-47. A solid-state relay is a close relative of the optoisolator (T or F).

5-48. A ________ (dc motor, stepper motor) should be used when the application calls for exact angular positioning of a shaft (as in a robot wrist).

5-49. The stepper motor sketched in Fig. 5-26(*a*) is classified as a unipolar or four-phase unit (T or F).

5-50. The device featured in Fig. 5-26 is a ________ (permanent magnet, variable reluctance) type stepper motor.

5-51. The step angle for the stepper motor in Fig. 5-26(*a*) is ________ degrees.

5-52. The control sequence shown in Fig. 5-27(*a*) is for a ________ (bipolar, unipolar) stepper motor.

5-53. Refer to Fig. 5-28(*a*). How is the MC3479 IC described by its manufacturer?

5-54. Refer to Fig. 5-28(*a*) and assume pins 9 and 10 of the MC3479 IC are LOW. When a single clock pulse enters the CLK input (pin 7) the stepper motor rotates a ________ (full step, half step) in the ________ (CCW, CW) direction.

5-55. Refer to Fig. 5-28(*a*) and assume pins 9 and 10 of the MC3479 IC are HIGH and the stepper motor has a step angle of 18°. Under these conditions, how many clock pulses must enter the CLK input to cause the stepper motor to rotate one revolution?

Critical Thinking Questions

5-1. How would you define *interfacing*?
5-2. How do you define *noise* in a digital system?
5-3. What is the propagation delay of a logic gate?
5-4. List several advantages of CMOS logic elements.
5-5. Refer to Fig. 5-5(*b*). A single 4000 series CMOS output has the capacity to drive at least ________ LS-TTL input(s).
5-6. Refer to Fig. 5-25. If family *A* is standard TTL and family *B* is ACT-CMOS, the inverter ________ (can, may not be able to) drive the AND gates.
5-7. Refer to Fig. 5-15(*c*). Explain the operation of this HIGH-LOW indicator circuit.
5-8. What is the purpose of "T"-type CMOS ICs (HCT, ACT, etc.)?
5-9. What electromechanical device could be used to isolate higher voltage equipment (such as motors or solenoids) from a logic circuit?
5-10. An electric motor converts electric energy into ________ motion.
5-11. A(n) ________ is an electromechanical device that converts electric energy into linear motion.
5-12. Why is the FACT-CMOS series considered by many engineers to be one of the best logic families for new designs?
5-13. Refer to Fig. 5-22. Explain the circuit action when the inverter input is LOW.
5-14. Refer to Fig. 5-23(*a*). Explain the circuit action when the inverter input is HIGH.
5-15. An optoisolator prevents the transmission of ________ (the signal, unwanted noise) from one electronic system to another that operates on a different voltage.
5-16. Refer to Fig. 5-24(*d*). Explain the circuit action when the inverter input is LOW.
5-17. If the coil resistance of a 12-V stepper motor is 40Ω, what is the current draw for the coil?
5-18. If a stepper motor is designed with a step angle of 3.6°, how many steps are required for one revolution of the motor?

Answers to Tests

1. interfacing
2. HIGH
3. LOW
4. undefined
5. undefined
6. +10
7. HIGH
8. CMOS
9. T
10. CMOS
11. fan-out
12. FAST TLL series
13. 20 (8 mA/400 μA = 20)
14. long
15. FACT series CMOS
16. the same
17. MOS
18. complementary symmetry metal oxide semiconductor
19. low-power consumption
20. GND
21. positive
22. LOW, floats HIGH
23. pull-up
24. RS flip-flop
25. HIGH, LOW
26. F
27. switch debouncing circuit
28. open collector
29. 4000
30. goes out
31. off, does not light
32. Q_1, red
33. are not
34. pull-up
35. current drive
36. transistor
37. is not
38. on, sounds
39. transient voltages
40. HIGH
41. relay
42. phototransistor
43. lights, activates, LOW
44. pull-up
45. does not light, deactivates, HIGH, does not sound
46. turns on, runs
47. dc motor
48. stepper motor
49. T
50. bipolar
51. control, bipolar
52. CCW
53. logic
54. 350
55. CCW, half step
56. open
57. logic probe or voltmeter
58. LOW

Encoding, Decoding, and Seven-Segment Displays

Chapter Objectives

This chapter will help you to:

1. *Identify* the characteristics and applications of several commonly used codes.
2. *Convert* decimal numbers to BCD code and BCD to decimal numbers.
3. *Compare* decimal numbers with excess-3 code, Gray code, and 8421 BCD code.
4. *Convert* ASCII code to letters and numbers, and characters to ASCII code.
5. *Demonstrate* the coding of a seven-segment display.
6. *Describe* the construction and important characteristics of LCD, LED, and vacuum fluorescent (VF) seven-segment displays.
7. *Demonstrate* the operation of several TTL and CMOS BCD-to-seven segment decoder/driver ICs used for driving LED, LCD, and VF seven-segment displays.
8. *Troubleshoot* a faulty decoder/driver seven-segment display circuit.

We humans use the decimal code to represent numbers. Digital electronic circuits in computers and calculators use mostly the *binary* code to represent numbers. Many other special codes are used in digital electronics to represent numbers, letters, punctuation marks, and control characters. This chapter covers several common codes used in digital electronic equipment. Electronic translators, which convert from one code to another, are widely used in digital electronics. In Chap. 2 we used an *encoder* to translate from decimal to binary numbers and a *decoder* to translate back from binary to decimal numbers. This chapter introduces you to several very common encoders and decoders used for translating from code to code.

6-1 The 8421 BCD Code

How would you represent the decimal number 926 in binary form? In other words, how would you convert 926 to the binary number 1110011110? The decimal-to-binary conversion would be done by using the method from Chap. 2 and illustrated in Fig. 6-1.

The binary number 1110011110 does not make much sense to most of us. A code that uses binary in a different way than in the preceding example is called the *8421 binary-coded decimal code.* This code is frequently referred to as just the *BCD code.*

BCD code

The decimal number 926 is converted to the BCD (8421) code in Fig. 6-2(*a*). The result is that the decimal number 926 equals 1001 0010 0110 in the 8421 BCD code. Notice from Fig. 6-2(*a*) that each group of four binary digits represents a decimal digit. The right group (0110) represents the 1s place value in the decimal number. The middle group (0010) represents the 10s place value in the decimal number. The left group (1001) represents the 100s place value in the decimal number.

Suppose you are given the 8421 BCD number 0001 1000 0111 0001. What decimal num-

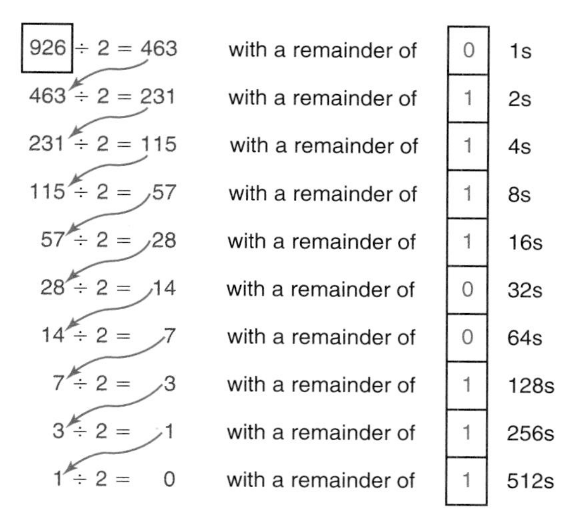

Fig. 6-1 Converting decimal to binary numbers.

	HUNDREDS	TENS	ONES
Decimal number	9	2	6
	↓	↓	↓
8421 BCD coded number	1001	0010	0110

(*a*)

	THOUSANDS	HUNDREDS	TENS	ONES
8421 BCD coded number	0001	1000	0111	0001
	↓	↓	↓	↓
Decimal number	1	8	7	1

(*b*)

Fig. 6-2 (*a*) Converting from decimal to 8421 BCD code. (*b*) Converting from BCD to decimal.

ber does this represent? Figure 6-2(*b*) shows how you translate from the BCD code to a decimal number. We find that the BCD number 0001 1000 0111 0001 is equal to the decimal number 1871. The 8421 BCD code does not use the numbers 1010 1011 1100 1101 1110 1111. These are considered invalid numbers.

The 8421 BCD code is very widely used in digital systems. As pointed out, it is common practice to substitute the term "BCD code" to mean the 8421 BCD code. A word of caution, however: some BCD codes do have different weightings of the place values, such as the 4221 code and the excess-3 code.

TEST

Supply the missing number in each statement.

Converting decimal to binary numbers

1. The decimal number 29 is the same as ________ in binary.
2. The decimal number 29 is the same as ________ in the 8421 BCD code.
3. The 8421 BCD number 1000 0111 0110 0101 equals ________ in decimal.

6-2 The Excess-3 Code

Excess-3 code

Converting a decimal number to the excess-3 code

The term "BCD" is a general term, usually referring to an 8421 code. Another code that is really a BCD code is the *excess-3 code*. To convert a decimal number to the excess-3 form we *add 3 to each digit of the decimal number* and convert to binary form. Figure 6-3 shows how the decimal number 4 is converted to the excess-3 code number 0111. Some decimal numbers are converted to excess-3 code in Table 6-1 on page 124. You probably have noticed that the excess-3 code for decimal numbers is rather difficult to figure out. This is because the binary digits are not weighted as they are in regular binary numbers and in the 8421 BCD code. The excess-3 code is used in some arithmetic circuits because it is self-complementing.

The 8421 and excess-3 codes are but two of many BCD codes used in digital electronics. The 8421 code is by far the most widely used BCD code.

JOB TIP

Networking leads to many job offers and promotions. Start this while still in school.

TEST

Converting decimal to 8421 BCD

Converting from BCD to decimal numbers

Supply the missing number in each statement.

4. The decimal number 18 equals ________ in excess-3 code.
5. The excess-3 code number 1001 0011 equals ________ in decimal.

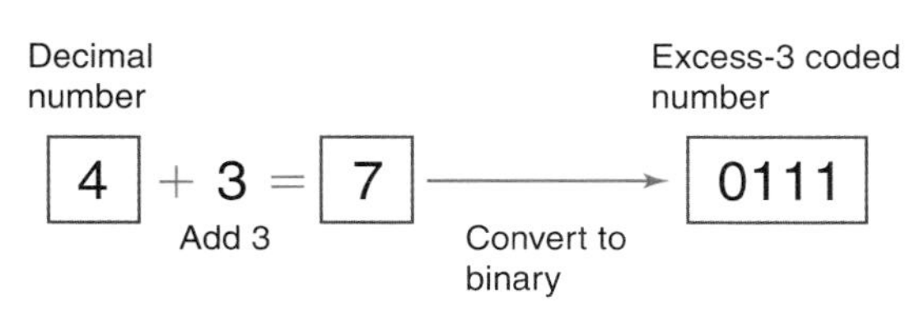

Fig. 6-3 Converting a decimal number to the excess-3 code.

TABLE 6-1 The Excess-3 Code

Decimal Number	Excess-3-Coded Number		
0			0011
1			0100
2			0101
3			0110
4			0111
5			1000
6			1001
7			1010
8			1011
9			1100
14		0100	0111
27		0101	1010
38		0110	1011
459	0111	1000	1100
606	1001	0011	1001
	Hundreds	Tens	Ones

6-3 The Gray Code

Gray code

Table 6-2 compares the *Gray code* with some codes you already know. The important characteristic of the Gray code is that only *one digit changes* as you *count* from top to bottom, as shown in Table 6-2. The Gray code cannot be used in arithmetic circuits. The Gray code is used for input and output devices in digital systems. You can see from Table 6-2 that the Gray code is not classed as one of the many BCD codes. Also notice that it is quite difficult to translate from decimal numbers to the Gray code and back to decimals again. There is a method for making this conversion, but we usually use electronic decoders to do the job for us.

TABLE 6-2 The Gray Code

Decimal Number	Binary Number	8421 BCD Coded Number		Gray-Code Number
0	0000		0000	0000
1	0001		0001	0001
2	0010		0010	0011
3	0011		0011	0010
4	0100		0100	0110
5	0101		0101	0111
6	0110		0110	0101
7	0111		0111	0100
8	1000		1000	1100
9	1001		1001	1101
10	1010	0001	0000	1111
11	1011	0001	0001	1110
12	1100	0001	0010	1010
13	1101	0001	0011	1011
14	1110	0001	0100	1001
15	1111	0001	0101	1000
16	10000	0001	0110	11000
17	10001	0001	0111	11001

Optical encoding of shaft position

The Gray code, which was invented by Frank Gray of Bell Labs, is commonly associated with *optical encoding* of a shaft's angular position. A simple example of this idea is sketched in Fig. 6-4. The encoder disk is attached to a shaft. The lighter areas of the disk represent transparent areas; the darker areas are opaque. A light source (usually infrared) shines from above the disk and light detectors are positioned below. The disk is free to rotate while the light sources and detectors stay in their position.

In the example shown in Fig. 6-4, light passes through all three transparent areas activating all three light detectors. In this example the detectors send the Gray code 111 to the Gray code to binary decoder. The decoder translates Gray code 111 to binary 101. As this is only a 3-bit shaft position encoder disk, the resolution is only 1 of 8. It can only detect a change in angular shaft position each 45° (360°/8 = 45°). The encoder disk in Fig. 6-4 is not very practical but serves to show how shaft positioning might be accomplished using the Gray code.

The Gray and excess-3 codes are not used extensively today. The purpose of mentioning them briefly is to make you aware that many codes exist in digital equipment. The codes you will probably encounter most often are binary, BCD (8421), and ASCII.

TEST

Answer the following questions.

6. The Gray code ________ (is, is not) a BCD-type code.
7. What characteristic is most important about the Gray code?
8. The inventor of the Gray code was ________ of Bell Labs.
9. The Gray code is most commonly associated with ________ of a shaft's angular position using an encoder disk.

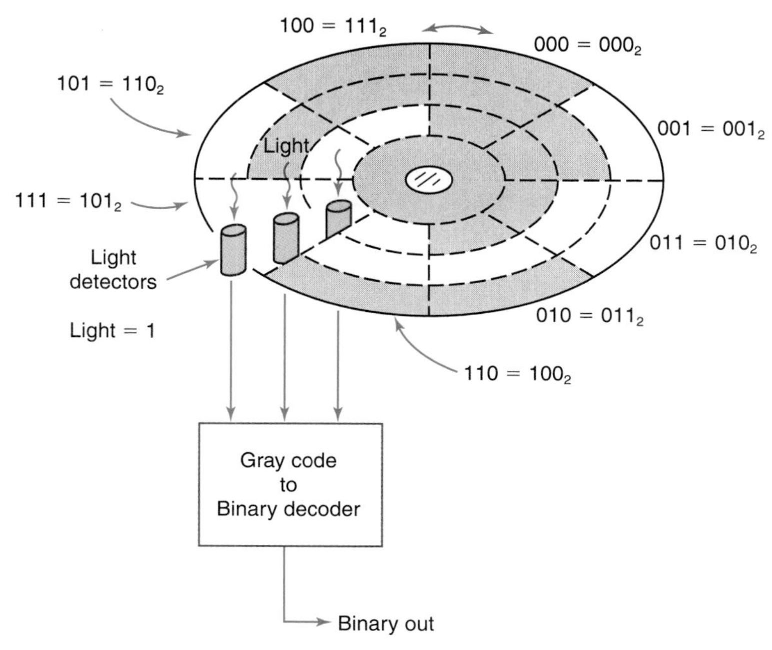

Fig. 6-4 Gray code used on shaft encoder disk to determine angular position.

6-4 The ASCII Code

The ASCII code is widely used to send information to and from microcomputers. The standard ASCII code is a 7-bit code used in transferring coded information from keyboards and to computer displays and printers. The abbreviation ASCII (pronounced "ask-ee") stands for the *American Standard Code for Information Interchange.*

Table 6-3 is a summary of the ASCII code. The ASCII code is used to represent numbers, letters, punctuation marks, as well as control characters. For instance, the 7-bit ASCII code 111 1111 stands for DEL from the top chart. From the bottom chart we see that DEL means delete.

What is the coding for "A" in ASCII? Locate A on the top chart in Table 6-3. Assembling the 7-bit code gives 100 0001 = A. This is the code you would expect to be sent to a microcomputer's CPU if you pressed the A key on the keyboard.

Some care must be used in applying Table 6-3 to specific equipment. Be aware that the shaded control characters may have other meanings on specific computers or other equipment. However, common control characters such as BEL (bell), BS (backspace), LF (line feed), CR (carriage return), DEL (delete), and SP (space) are used on most computers. The exact meaning of the ASCII control codes should be looked up in your equipment manual.

The ASCII code is an *alphanumeric code.* It can represent both letters and numbers. Several other alphanumeric codes are *EBCDIC* (extended binary-coded decimal interchange code), *Baudot,* and *Hollerith.*

ASCII code

Alphanumeric code

EBCDIC

Baudot

Hollerith

American Standard Code for Information Interchange

TEST

Answer the following questions.

10. ASCII is classified as a(n) ________ code because it can represent both numbers and letters.
11. The letters ASCII stand for ________.
12. The letter R is represented by the 7-bit ASCII code ________.
13. The ASCII code 010 0100 represents what character?

6-5 Encoders

A digital system using an *encoder* is shown in Fig. 6-5 on page 127. The encoder in this system must translate the decimal input from the keyboard to an 8421 BCD code. You used an encoder of this type in the experiments in Chap. 2. This encoder is called a *10-line-to-4-line priority encoder* by the manufacturer. Figure 6-6(*a*)

10-line-to-4-line priority encoder

JOB TIP

Companies use internships to identify outstanding employees.

ASCII code

TABLE 6-3 The ASCII Code

Bit 7 →					0	0	0	0	1	1	1	1
Bit 6 →					0	0	1	1	0	0	1	1
Bit 5 →					0	1	0	1	0	1	0	1
	Bit 4	Bit 3	Bit 2	Bit 1								
	0	0	0	0	NUL	DLE	SP	0	@	P	\	P
	0	0	0	1	SOH	DC1	!	1	A	Q	a	q
	0	0	1	0	STX	DC2	"	2	B	R	b	r
	0	0	1	1	ETX	DC3	#	3	C	S	c	s
	0	1	0	0	EOT	DC4	$	4	D	T	d	t
	0	1	0	1	ENQ	NAK	%	5	E	U	e	u
	0	1	1	0	ACK	SYN	&	6	F	V	f	v
	0	1	1	1	BEL	ETB	'	7	G	W	g	w
	1	0	0	0	BS	CAN	(	8	H	X	h	x
	1	0	0	1	HT	EM	)	9	I	Y	i	y
	1	0	1	0	LF	SUB	*	:	J	Z	j	z
	1	0	1	1	VT	ESC	+	;	K	[	k	l
	1	1	0	0	FF	FS	,	<	L	\	l	|
	1	1	0	1	CR	GS	—	=	M	]	m	}
	1	1	1	0	SO	RS	.	>	N	∧	n	~
	1	1	1	1	S1	US	/	?	O	—	o	DEL

Control Functions

NUL	Null	DLE	Data link escape
SOH	Start of heading	DC1	Device control 1
STX	Start of text	DC2	Device control 2
ETX	End of text	DC3	Device control 3
EOT	End of transmission	DC4	Device control 4
ENQ	Enquiry	NAK	Negative acknowledge
ACK	Acknowledge	SYN	Synchronous idle
BEL	Bell	ETB	End of transmission block
BS	Backspace	CAN	Cancel
HT	Horizontal tabulation (skip)	EM	End of medium
LF	Line feed	SUB	Substitute
VT	Vertical tabulation (skip)	ESC	Escape
FF	Form feed	FS	File separator
CR	Carriage return	GS	Group separator
SO	Shift out	RS	Record separator
SI	Shift in	US	Unit separator
DEL	Delete	SP	Space

is a block diagram of this encoder. If the decimal input 3 on the encoder is activated, then the logic circuit inside the unit outputs the BCD number 0011 as shown.

Active low inputs

A more accurate description of a 10-line-to-4-line priority encoder is shown in Fig. 6-6(*b*). This is a connection diagram furnished by National Semiconductor and shows the 74147 10-line-to-4-line priority encoder. Note the bubbles at both the inputs (1 to 9) and the outputs (*A* to *D*). The bubbles mean that the 74147 priority encoder has both *active low inputs* and

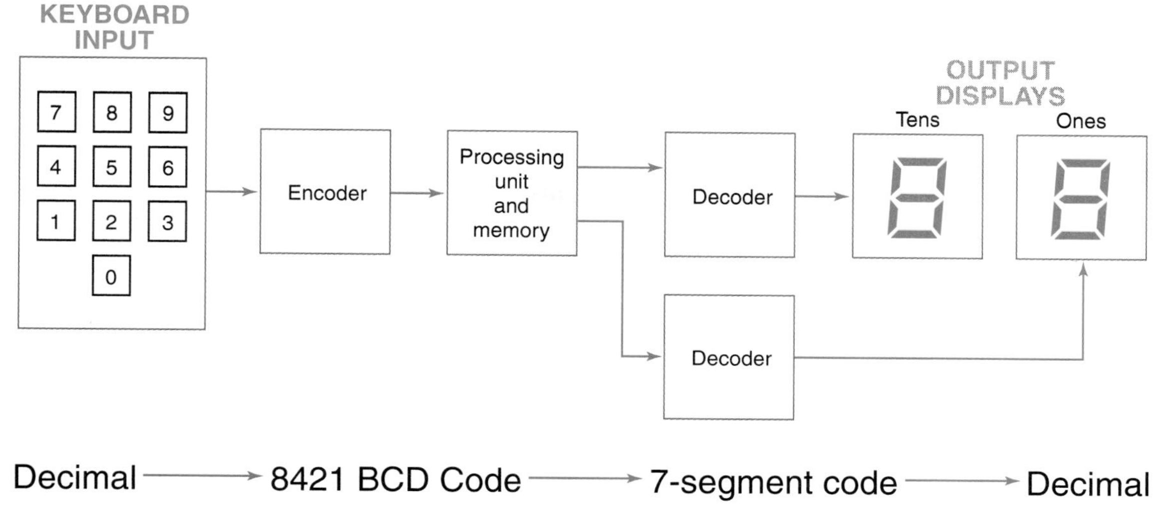

Fig. 6-5 A digital system.

active low outputs. A truth table is given for the 74147 priority encoder in Fig. 6-6(*c*). Note that only low logic levels (L on the truth table) activate the appropriate input. The active state for the outputs on this IC are also LOW. Notice that in the last line of the truth table in Fig. 6-6(*c*) the L (logical 0) at input 1 activates only the *A* output (the least significant bit of the four-bit group).

Active low outputs

The 74147 TTL IC in Fig. 6-6(*c*) is packaged in a 16-pin DIP. Internally, the IC consists of circuitry equivalent to about 30 logic gates.

The 74147 encoder in Fig. 6-6 has a *priority* feature. This means that if two inputs are activated at the same time, only the larger number will be encoded. For instance, if both the 9 and the 4 inputs were activated (LOW), then the output would be LHHL, representing decimal 9. Note that the outputs need to be complemented (inverted) to form the true binary number of 1001.

TEST

Answer the following questions.

14. Refer to Fig. 6-6. The 74147 encoder IC has active __________ (HIGH, LOW) inputs and active __________ (HIGH, LOW) outputs.
15. Refer to Fig. 6-6. If only input 7 of the 74147 encoder is LOW, what is the logic state at each of the four outputs?
16. Refer to Fig. 6-6(*b*). What is the meaning of a bubble on the logic symbol at input 4 (pin 1 on the 74147 IC)?
17. Refer to Fig. 6-6. If both inputs 2 and 8 go LOW, what is the logic state at each of the four outputs?

6-6 Seven-Segment LED Displays

The common task of decoding from machine language to decimal numbers is suggested in the system in Fig. 6-5. A very common output device used to display decimal numbers is the seven-segment display. The seven segments of the display are labeled *a* through *g* in Fig. 6-7(*a*). The displays representing decimal digits 0 through 9 are shown in Fig. 6-7(*b*). For instance, if segments *a*, *b*, and *c* are lit, a decimal 7 is displayed. If, however, all segments *a* through *g* are lit, a decimal 8 is displayed.

Seven-segment display

Several common seven-segment display packages are shown in Fig. 6-8 on page 128. The seven-segment LED display in Fig. 6-8(*a*) fits a regular 14-pin DIP IC socket. Another single-digit seven-segment LED display is shown in Fig. 6-8(*b*). This display fits crosswise into a wider DIP IC socket. Finally, the unit in Fig. 6-8(*c*) is a multidigit LED display widely used in digital clocks.

The seven-segment display may be constructed with each of the segments being a thin filament that glows. This type of unit is called an *incandescent* display and is similar to a regular lamp. Another type of display is the *gas-discharge tube,* which operates at high voltages. It gives off an orange glow. The modern

Incandescent display

Gas-discharge tube

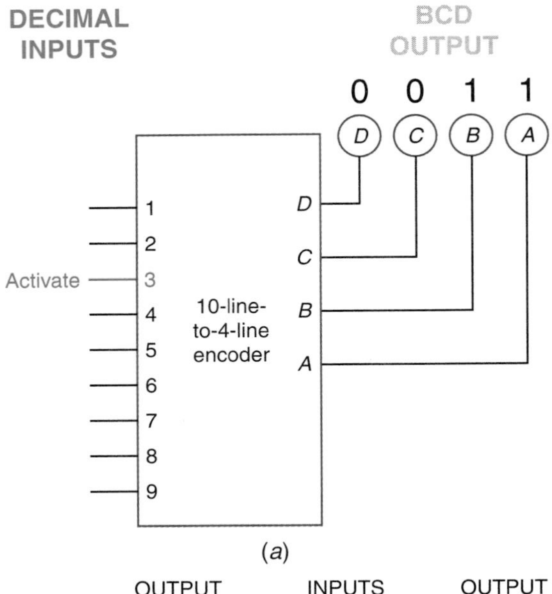

(*a*)

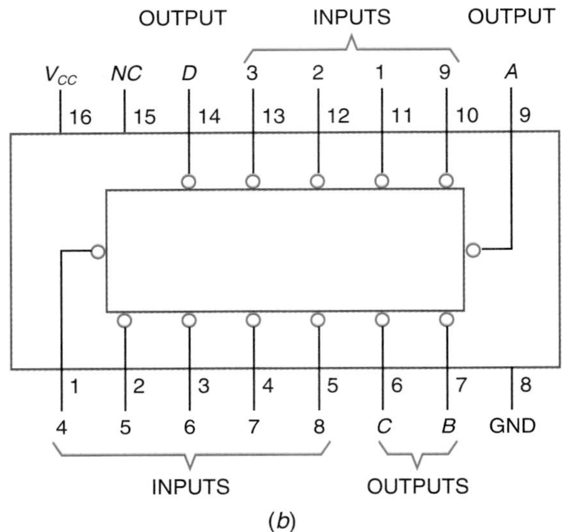

(*b*)

INPUTS									OUTPUTS			
1	2	3	4	5	6	7	8	9	*D*	*C*	*B*	*A*
H	H	H	H	H	H	H	H	H	H	H	H	H
X	X	X	X	X	X	X	X	L	L	H	H	L
X	X	X	X	X	X	X	L	H	L	H	H	H
X	X	X	X	X	X	L	H	H	H	L	L	L
X	X	X	X	X	L	H	H	H	H	L	L	H
X	X	X	X	L	H	H	H	H	H	L	H	L
X	X	X	L	H	H	H	H	H	H	L	H	H
X	X	L	H	H	H	H	H	H	H	H	L	L
X	L	H	H	H	H	H	H	H	H	H	L	H
L	H	H	H	H	H	H	H	H	H	H	H	L

H = HIGH logic level, L = LOW logic level, X = Don't care

(*c*)

Fig. 6-6 (*a*) 10-line-to-4-line encoder. (*b*) Pin diagram for 74147 encoder IC. (*c*) Truth table for 74147 encoder. *(Courtesy of National Semiconductor Corporation.)*

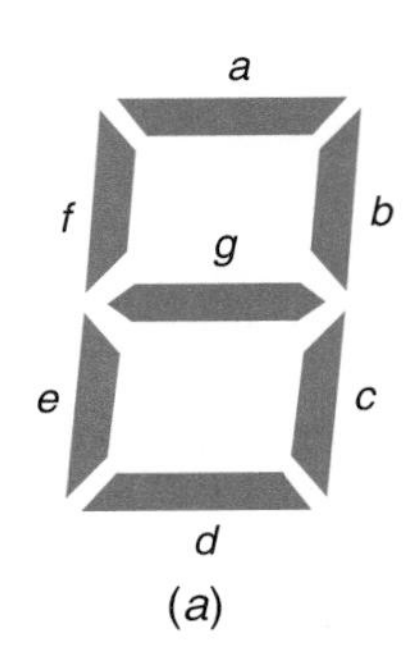

(*a*)

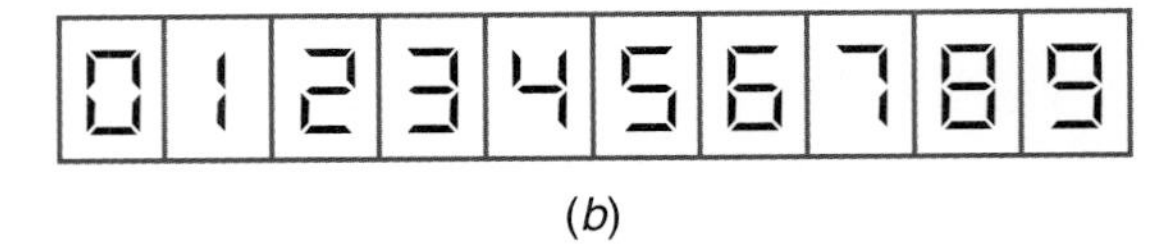

(*b*)

Fig. 6-7 (*a*) Segment identification. (*b*) Decimal numbers on typical seven-segment display.

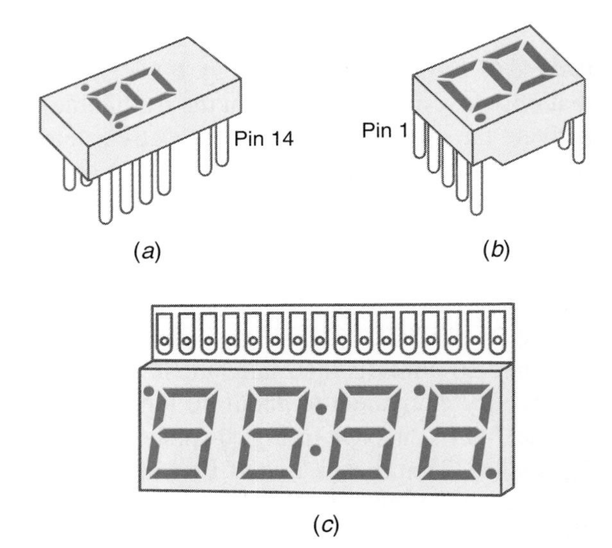

(*c*)

Fig. 6-8 (*a*) DIP seven-segment LED display. (*b*) A common 10-pin single-digit package. Note the location of pin 1. Pins are numbered counterclockwise from pin 1 when viewed from the top of the display. (*c*) Multidigit package.

Vacuum fluorescent (VF) display

Liquid crystal display (LCD)

74147 encoder IC

vacuum fluorescent (VF) display gives off a blue-green glow when lit and operates at low voltages. The newer *liquid-crystal display (LCD)* creates numbers in a black or silvery color. The common LED display gives off a characteristic reddish glow when lit.

A basic single *LED (light-emitting diode)* is illustrated in Fig. 6-9. The cutaway view of the LED in Fig. 6-9(*a*) shows the small exposed diode chip with a reflector to project the light upward toward the plastic lens. Of importance

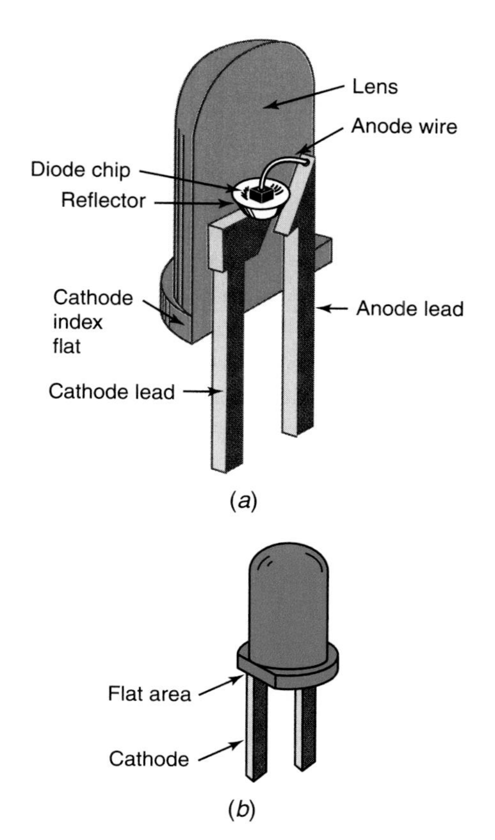

Fig. 6-9 (*a*) Cutaway view of standard light-emitting diode. (*b*) Identifying the cathode lead of the LED.

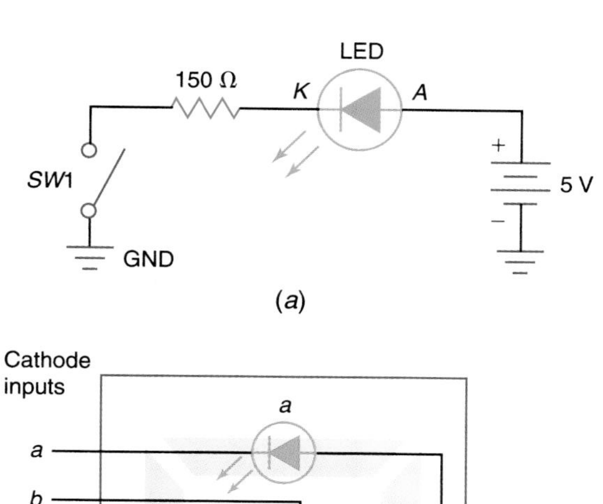

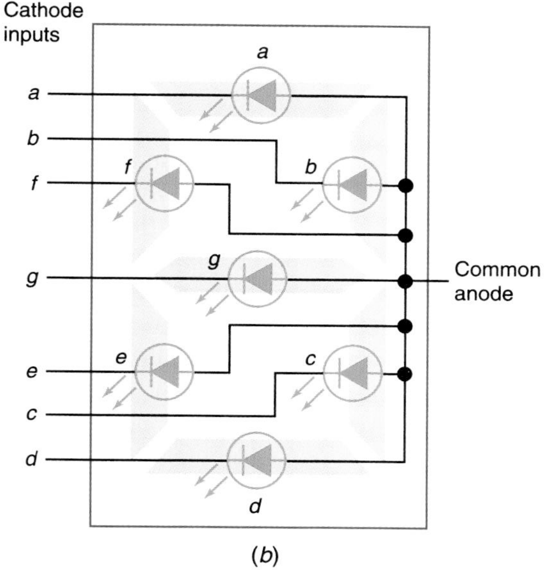

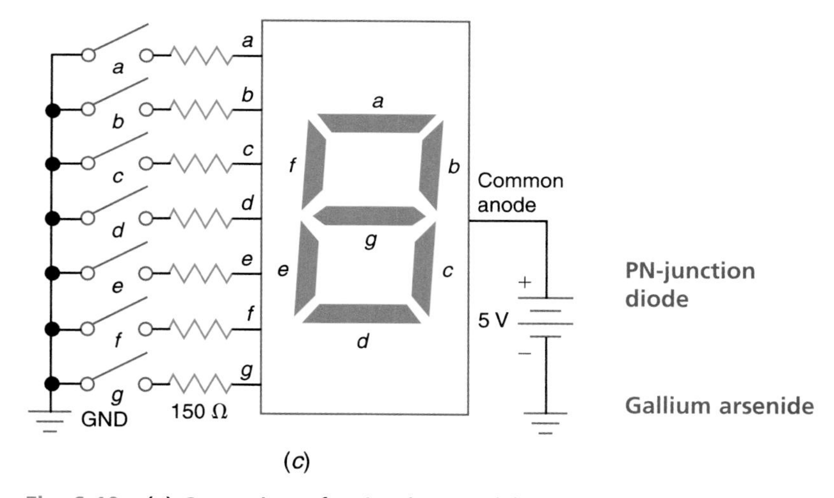

Fig. 6-10 (*a*) Operation of a simple LED. (*b*) Wiring a common-anode seven-segment LED display. (*c*) Driving a seven-segment LED display with switches.

during use is the index area on the rim of the LED shown in Fig. 6-9(*b*). The flat area shows the cathode side of the LED. The LED is basically a PN-junction diode. When the diode is forward-biased, current flows through the PN junction and the LED lights and is focused by the plastic lens. Many LEDs are fabricated from gallium arsenide (GaAs) and several related materials. LEDs come in several colors including red, green, orange, blue, and amber.

PN-junction diode

Gallium arsenide

A single LED is being tested in Fig. 6-10(*a*). When the switch (*SW*1) is closed, current flows from the 5-V power supply through the LED, causing it to light. The series resistor limits current to about 20 mA. Without the limiting resistor the LED would burn out. Typically, LEDs can accept only about 1.7 to 2.1 V across their terminals when lit. Being a diode, the LED is sensitive to polarity. Hence, the *cathode (K)* must be toward the negative (GND) terminal while the *anode (A)* must be toward the positive terminal of the power supply.

Seven-segment LED display

A seven-segment LED display is shown in Fig. 6-10(*b*). Each segment (*a* through *g*) contains an LED, as shown by the seven symbols. The display shown has all the anodes tied together and coming out the right side as a single connection (common anode). The inputs on the left go to the various segments of the display. The device in Fig. 6-10(*b*) is referred to as a *common-anode seven-segment LED display.* These units can also be purchased in common-cathode form.

Common anode

Common cathode

To understand how segments on the display are activated and lit, consider the circuit in Fig. 6-10(*c*). If switch *b* is closed, current flows from GND through the limiting resistor to the *b*-segment LED and out the common-anode connection to the power supply. Only segment *b* will light.

Suppose you wanted the decimal 7 to light on the display in Fig. 6-10(*c*). Switches *a*, *b*, and *c* would be closed, lighting the LED segments *a*, *b*, and *c*. The decimal 7 would light on the display. Likewise, if the decimal 5 were to be lit, switches *a*, *c*, *d*, *f*, and *g* would be closed. These five switches would ground the correct segments, and a decimal 5 would appear on the display. Note that it takes a GND voltage (LOW logic level) to activate the LED segments on this display.

Mechanical switches are used in Fig. 6-10(*c*) to drive the seven-segment display. Usually power for the LED segments is provided by an IC. The IC is called a *display driver.* In practice, the display driver is usually packaged in the same IC as the decoder. Therefore, it is common to speak of *seven-segment decoder/drivers.*

Decoder

Display driver

Seven-segment decoder/drivers

TEST

Supply the missing word or words in each statement.

18. Refer to Fig. 6-7(*a*). If segments *a*, *c*, *d*, *f*, and *g* are lit, the decimal number _________ will appear on the seven-segment display.
19. The seven-segment unit that gives off a blue-green glow is a(n) _________ (vacuum fluorescent, incandescent, LCD, LED) display.
20. The letters "LED" stand for _________.
21. Refer to Fig. 6-10(*c*). If switches *b* and *c* are closed, segments _________ and _________ will light. This _________ (LCD, LED) seven-segment unit will display the decimal number _________.
22. On a single LED, as in Fig. 6-9, the flat area on the rim of the plastic identifies the _________ lead.

6-7 Decoders

A *decoder,* like an encoder, is a code translator. Figure 6-5 shows two decoders being used in the system. The decoders are translating the 8421 BCD code to a seven-segment display code that lights the proper segments on the displays. The display will be a decimal number. Figure 6-11 shows the BCD number 0101 at

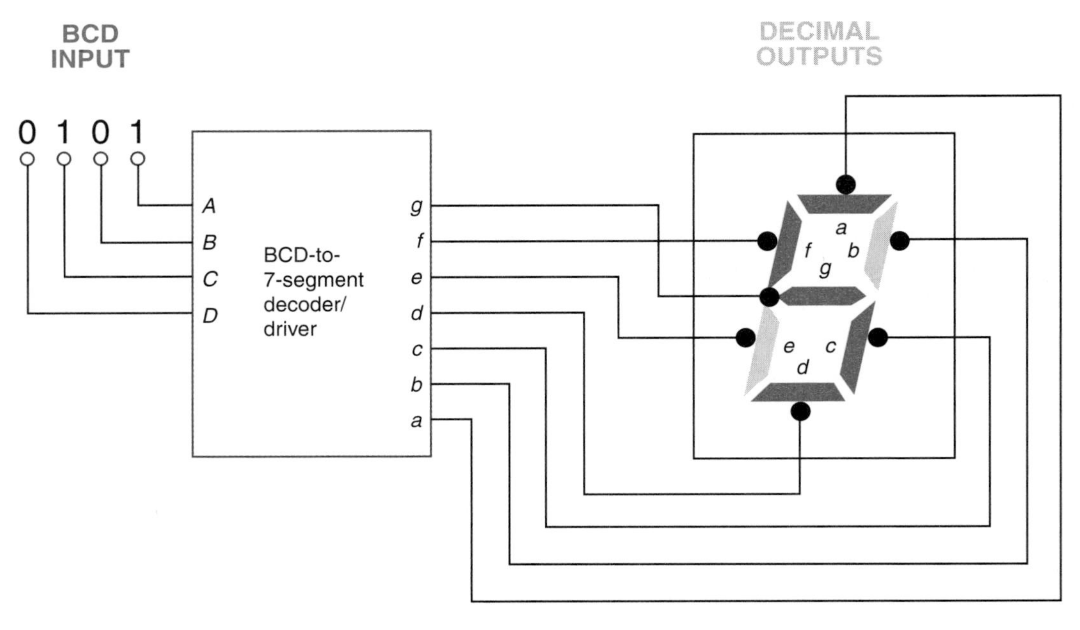

Fig. 6-11 A decoder driving a seven-segment display.

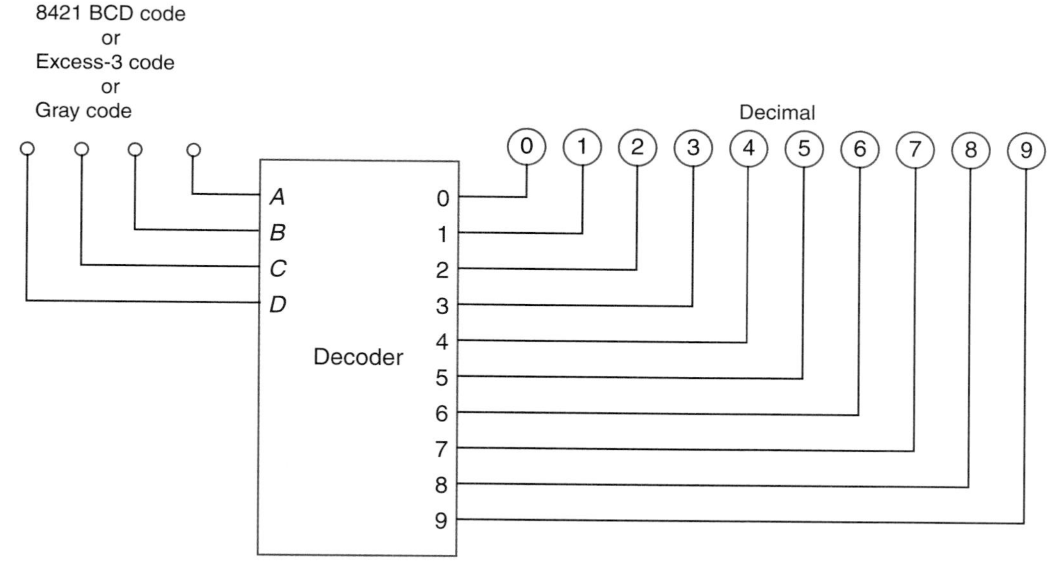

Fig. 6-12 A typical decoder block diagram. Note that inputs may be 8421 BCD, excess-3, or Gray code.

the input of the *BCD-to-seven-segment decoder/driver.* The decoder activates outputs *a*, *c*, *d*, *f*, and *g* to light the segments shown in Fig. 6-11. The decimal number 5 lights up on the display.

Decoders come in several varieties, such as the ones illustrated in Fig. 6-12. Notice in Fig. 6-12 that the same block diagram is used for the 8421 BCD, the excess-3, and the Gray decoder.

Other decoders are available such as BCD converters, BCD-to-binary converters, 4-to-16-line decoders, and 2-to-4-line decoders. Other encoders available are a decimal-to-octal and 8-to-3 line priority encoder.

Decoders, like encoders, are combinational logic circuits with several inputs and outputs. Most decoders contain from 20 to 50 gates. Most decoders and encoders are packaged in single IC packages.

TEST

Answer the following questions.

23. Refer to Fig. 6-11. If the BCD input to the decoder/driver is 1000, which segments on the display will light? The seven-segment LED display will read what decimal number?
24. List at least three types of decoders.

6-8 BCD-to-Seven-Segment Decoder/Drivers

7447A BCD-to-seven-segment decoder/driver

Blanking

Combinational logic circuits

A logic symbol for a commercial TTL 7447A BCD-to-seven-segment decoder/driver is shown in Fig. 6-13(*a*). The BCD number to be decoded is applied to the inputs labeled *D*, *C*, *B*, and *A*. When activated with a LOW, the lamp-test (LT) input activates all outputs (*a* to *g*). When activated with a LOW, the blanking input (BI) makes all outputs HIGH, turning all

About Electronics

Douglas Engelbart. What is commonly known today as the computer mouse was engineered in 1963 by Douglas Engelbart and was originally called an X-Y position indicator for a display system. Engelbart's early mouse was an electronic marble that was housed in a wooden box that had a red button and a copper wire tail. After retiring from Stanford Research Institute, Englebart developed a keyboard that has only five buttons.

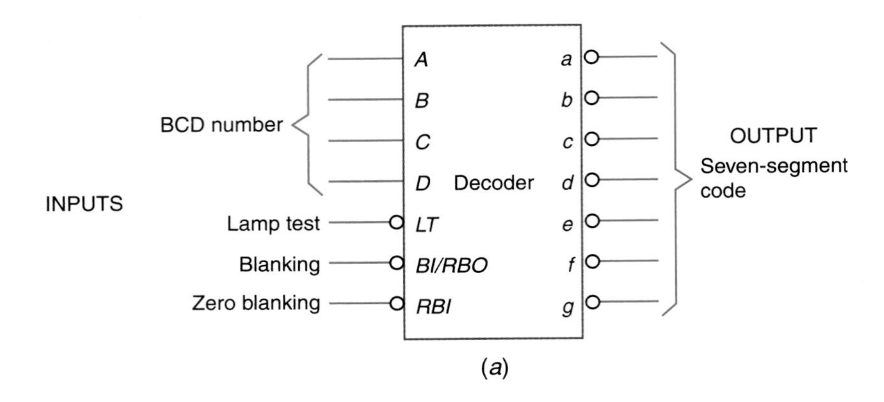

(*a*)

Decimal or function	INPUTS						BI/BRO	OUTPUTS							Note
	LT	*RBI*	*D*	*C*	*B*	*A*		*a*	*b*	*c*	*d*	*e*	*f*	*g*	
0	H	H	L	L	L	L	H	ON	ON	ON	ON	ON	ON	OFF	
1	H	X	L	L	L	H	H	OFF	ON	ON	OFF	OFF	OFF	OFF	
2	H	X	L	L	H	L	H	ON	ON	OFF	ON	ON	OFF	ON	
3	H	X	L	L	H	H	H	ON	ON	ON	ON	OFF	OFF	ON	
4	H	X	L	H	L	L	H	OFF	ON	ON	OFF	OFF	ON	ON	
5	H	X	L	H	L	H	H	ON	OFF	ON	ON	OFF	ON	ON	
6	H	X	L	H	H	L	H	OFF	OFF	ON	ON	ON	ON	ON	
7	H	X	L	H	H	H	H	ON	ON	ON	OFF	OFF	OFF	OFF	
8	H	X	H	L	L	L	H	ON	ON	ON	ON	ON	ON	ON	1
9	H	X	H	L	L	H	H	ON	ON	ON	OFF	OFF	ON	ON	
10	H	X	H	L	H	L	H	OFF	OFF	OFF	ON	ON	OFF	ON	
11	H	X	H	L	H	H	H	OFF	OFF	ON	ON	OFF	OFF	ON	
12	H	X	H	H	L	L	H	OFF	ON	OFF	OFF	OFF	ON	ON	
13	H	X	H	H	L	H	H	ON	OFF	OFF	ON	OFF	ON	ON	
14	H	X	H	H	H	L	H	OFF	OFF	OFF	ON	ON	ON	ON	
15	H	X	H	H	H	H	H	OFF	OFF	OFF	OFF	OFF	OFF	OFF	
BI	X	X	X	X	X	X	L	OFF	OFF	OFF	OFF	OFF	OFF	OFF	2
RBI	H	L	L	L	L	L	L	OFF	OFF	OFF	OFF	OFF	OFF	OFF	3
LT	L	X	X	X	X	X	H	ON	ON	ON	ON	ON	ON	ON	4

H = HIGH level, L = LOW level, X = Irrelevant

Notes:

1. The blanking input (*BI*) must be open or held at a HIGH logic level when output functions 0 through 15 are desired. The ripple-blanking input (*RBI*) must be open or HIGH if blanking of a decimal zero is not desired.
2. When a LOW logic level is applied directly to the blanking input (*BI*), all segment outputs are OFF regardless of the level of any other input.
3. When ripple-blanking input (*RBI*) and inputs *A*, *B*, *C*, and *D* are at a LOW level with the lamp test (*LT*) input HIGH, all segment outputs go OFF and the ripple-blanking output (*RBO*) goes to a LOW level (response condition).
4. When the blanking input/ripple-blanking output (*BI/RBO*) is open or held HIGH and a LOW is applied to the lamp test (*LT*) input, all segment outputs are ON.

(*b*)

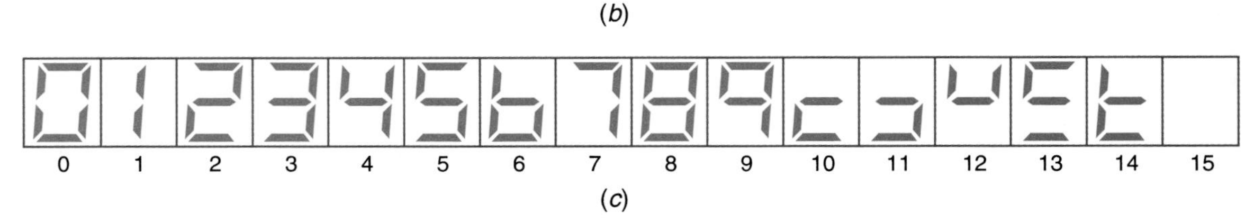

(*c*)

7447A TTL decoder IC

Fig. 6-13 (*a*) Logic symbol for 7447A TTL decoder IC. (*b*) Truth table for 7447A decoder. *(Courtesy of Texas Instruments, Inc.)* (*c*) Format of readouts on seven-segment display using the 7447A decoder IC.

attached displays OFF. When activated with a LOW, the ripple-blanking input (RBI) blanks the display *only if it contains a 0.* When the RBI input becomes active, the BI/RBO pin temporarily becomes the *ripple-blanking output (RBO)* and drops to a LOW. Remember that "blanking" means to cause no LEDs on the display to light.

The seven outputs on the 7447A IC are all active LOW outputs. In other words, the outputs are normally HIGH and drop to a LOW when activated.

The exact operation of the 7447A decoder/driver IC is detailed in the truth table furnished by Texas Instruments and reproduced in Fig. 6-13(*b*). The decimal displays generated by the 7447A decoder are shown in Fig. 6-13(*c*). Note that invalid BCD inputs (decimals 10, 11, 12, 13, 14, and 15) do generate a unique output on the 7447A decoder.

The 7447A decoder/driver IC is typically connected to a common-anode seven-segment LED display. Such a circuit is shown in Fig. 6-14. It is especially important that the seven 150-Ω limiting resistors be wired between the 7447A IC and the seven-segment display.

Assume that the BCD input to the 7447A decoder/driver in Fig. 6-14 is 0001 (LLLH). This is equal to line 2 of the truth table in Fig. 6-13(*b*). This input combination causes segments *b* and *c* on the seven-segment display to light (outputs *b* and *c* drop to LOW). Decimal 1 is displayed. The LT and two BIs are not shown in Fig. 6-14. When not connected, they are assumed to be "floating" HIGH and therefore disabled in this circuit. Good design practice suggests that these "floating" inputs should be connected to +5 V to make sure they stay HIGH.

"Floating" inputs

Good design practice

Many applications, such as a calculator or cash register, require that the *leading zeros be blanked.* The illustration in Fig. 6-15 (on page 134) shows the use of the 7447A decoder/drivers operating a group of displays as in a cash register. The six-digit display example details how the blanking of leading 0s would be accomplished using the 7447A IC driving LED displays.

Blanking leading zeros

The current inputs to the six decoders are shown across the bottom of the drawing in Fig. 6-15. The current BCD input is 0000 0000 0011 1000 0001 0000 (003810 in decimal). The two left 0s should be blanked, so the display reads 38.10. The blanking of the leading 0s is handled by wiring the RBI and RBO pins to each 7447A decoder IC together as shown in Fig. 6-15.

Invalid BCD inputs

Working from left to right in Fig. 6-15, notice that the RBI input of IC6 is grounded. From the 7447A decoder's truth table in Fig. 6-13(*b*) it can be determined that when RBI is LOW and when all BCD inputs are LOW, then all segments of the display are blanked or off. Also the RBO is forced LOW. This LOW is passed to the RBI of IC5.

Continuing in Fig. 6-15, with the BCD input to IC5 as 0000 and the RBI at LOW, the display is also blanked. The RBO of IC5 is forced LOW and is passed to the RBI input of IC4. Even with the RBI LOW, IC4 *does not* blank the display because the BCD is 0011. The RBO of IC4 remains HIGH, which is sent to IC3.

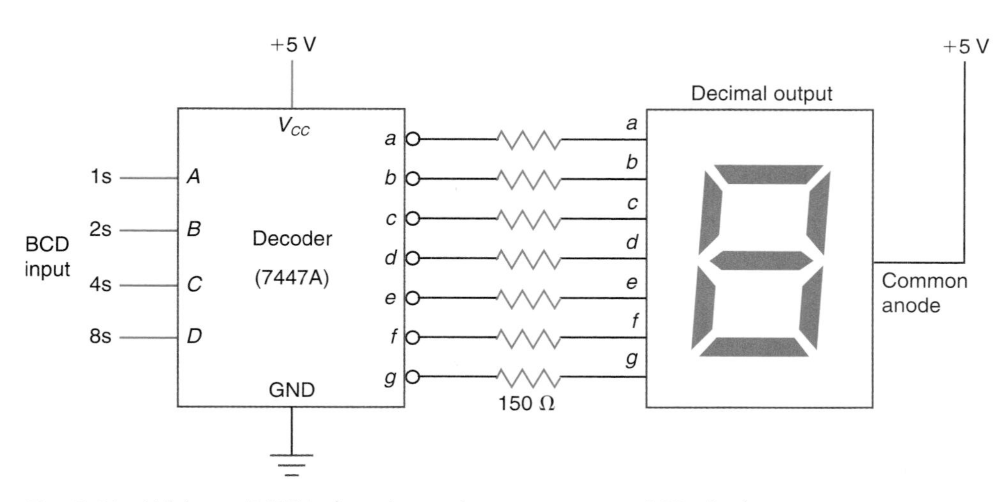

Fig. 6-14 Wiring a 7447A decoder and seven-segment LED display.

Driving a seven-segment LED display

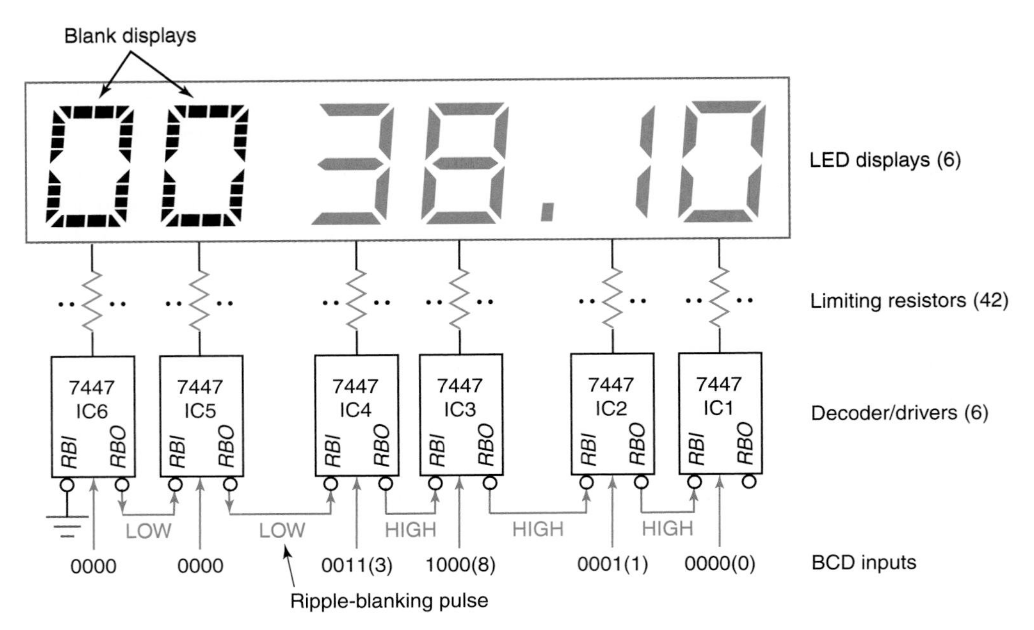

Fig. 6-15 Using the ripple-blanking input (RBI) of the 7447A decoder/driver to blank leading zeros in a multidigit display.

A question arises about the right LED display in Fig. 6-15. The BCD input to IC1 is 0000_{BCD} and a zero (0) appears on the display. The zero on the IC1 display is not blanked because the RBI input is not activated (RBI = HIGH). The first row of the truth table in Fig. 6-13(*b*) shows that the 7447A decoder/driver will display the zero when the RBI is HIGH.

TEST

Check your understanding by answering the following questions.

25. Refer to Fig. 6-13. The 7447A decoder/driver IC has active _________ (HIGH, LOW) BCD inputs and active _________ (HIGH, LOW) outputs.
26. Refer to Fig. 6-13. The lamp test, blanking, and zero-blanking inputs of the 7447A are active _________ (HIGH, LOW) inputs.
27. The RBI and RBO inputs of the 7447A are commonly used for blanking _________ on calculator and cash register multidigit displays.
28. What will the seven-segment display read during each time period (t_1 to t_7) for the circuit shown in Fig. 6-16?
29. List the segments on the seven-segment display that will light during each time period (t_1 to t_7) for the circuit in Fig. 6-16.

6-9 Liquid-Crystal Displays

Liquid-crystal displays

The LED actually *generates* light, where the LCD simply *controls* available light. The LCD has gained wide acceptance because of its very low power consumption. The LCD is also well suited for use in sunlight or in other brightly lit areas. The DMM (digital multimeter) in Fig. 6-17 uses a modern LCD.

The LCD is also suited for more complex displays than just seven-segment decimal. The LCD display in Fig. 6-17 contains an analog scale across the bottom as well as the larger digital readout. In practice you will find that the DMM LCD has several other symbols, which you can also observe on the DMM in Fig. 6-17.

Field-effect LCD

The construction of a common LCD unit is shown in Fig. 6-18(*a*) on page 136. This unit is called the *field-effect LCD*. When a segment is energized by a low frequency square-wave signal, the LCD segment appears black while the rest of the surface remains shiny. Segment *e* is energized in Fig. 6-18(*a*). The nonenergized segments are nearly invisible.

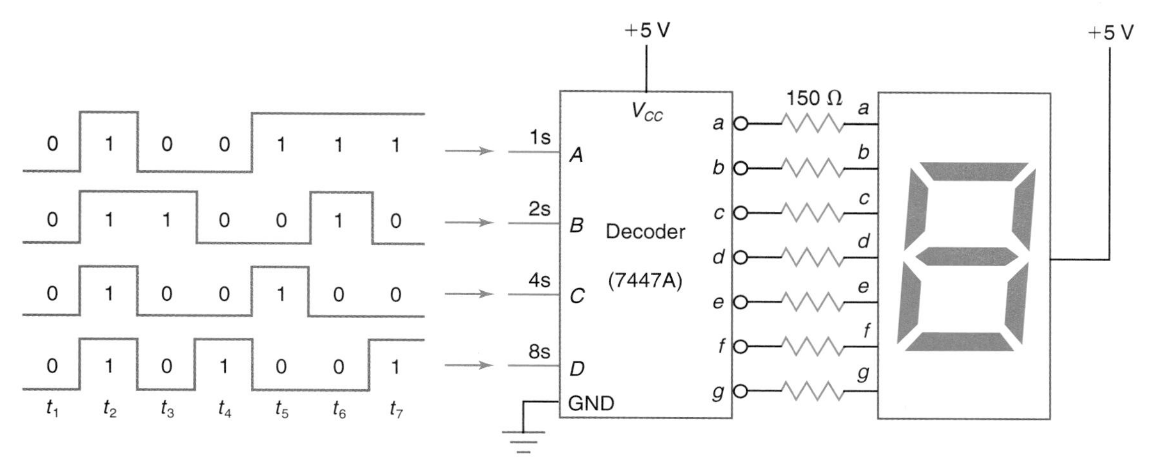

Fig. 6-16 Decoder-LED display circuit for problems 28 and 29.

The key to LCD operation is the liquid crystal, or *nematic fluid*. This nematic fluid is sandwiched between two glass plates. An ac voltage is applied across the nematic fluid, from the top metalized segments to the metalized backplane. When affected by the magnetic field of the ac voltage, the nematic fluid transmits light differently and the energized segment appears as black on a silvery background.

Nematic fluid

Field-effect LCD

The field-effect LCD uses a polarizing filter on the top and bottom of the display as shown in Fig. 6-18 (*a*). The backplane and segments are internally wired to contacts on the edge of the LCD. Only two of the many contacts are shown in Fig 6-18 (*a*).

Decimal 7 is displayed on the LCD shown in Fig 6-18 (*b*). The BCD-to-seven-segment decoder on the left is receiving a BCD input of 0111. This input activates the *a*, *b*, and *c* outputs of the decoder (*a*, *b*, and *c* are HIGH in this example). The remaining outputs of the decoder are LOW (*d*, *e*, *f*, and *g* = LOW). The 100-Hz square-wave input is always applied to the backplane of the display. This signal is also applied to each of the CMOS XOR gates used to drive the LCD. Note that the XOR gates produce an inverted waveform when activated (*a*, *b*, and *c* XOR gates are activated). The 180° out-of-phase signals to the backplane and segments *a*, *b*, and *c* cause these areas of the LCD to turn black. The in-phase signals from XOR gates *d*, *e*, *f*, and *g* do not cause these segments to be activated. Therefore, these segments remain nearly invisible.

JOB TIP

Interviewers may want to know about hobby interests such as computers, radio control model building, robotics, etc.

Digital multimeter

The XOR gates used as LCD drivers in Fig. 6-18(*b*) are CMOS units. TTL XOR gates are not used because they cause a small dc offset voltage to be developed across the LCD's ne-

Fig. 6-17 Digital multimeter (DMM) using a liquid-crystal display. *(Courtesy of Tektronix, Inc.)*

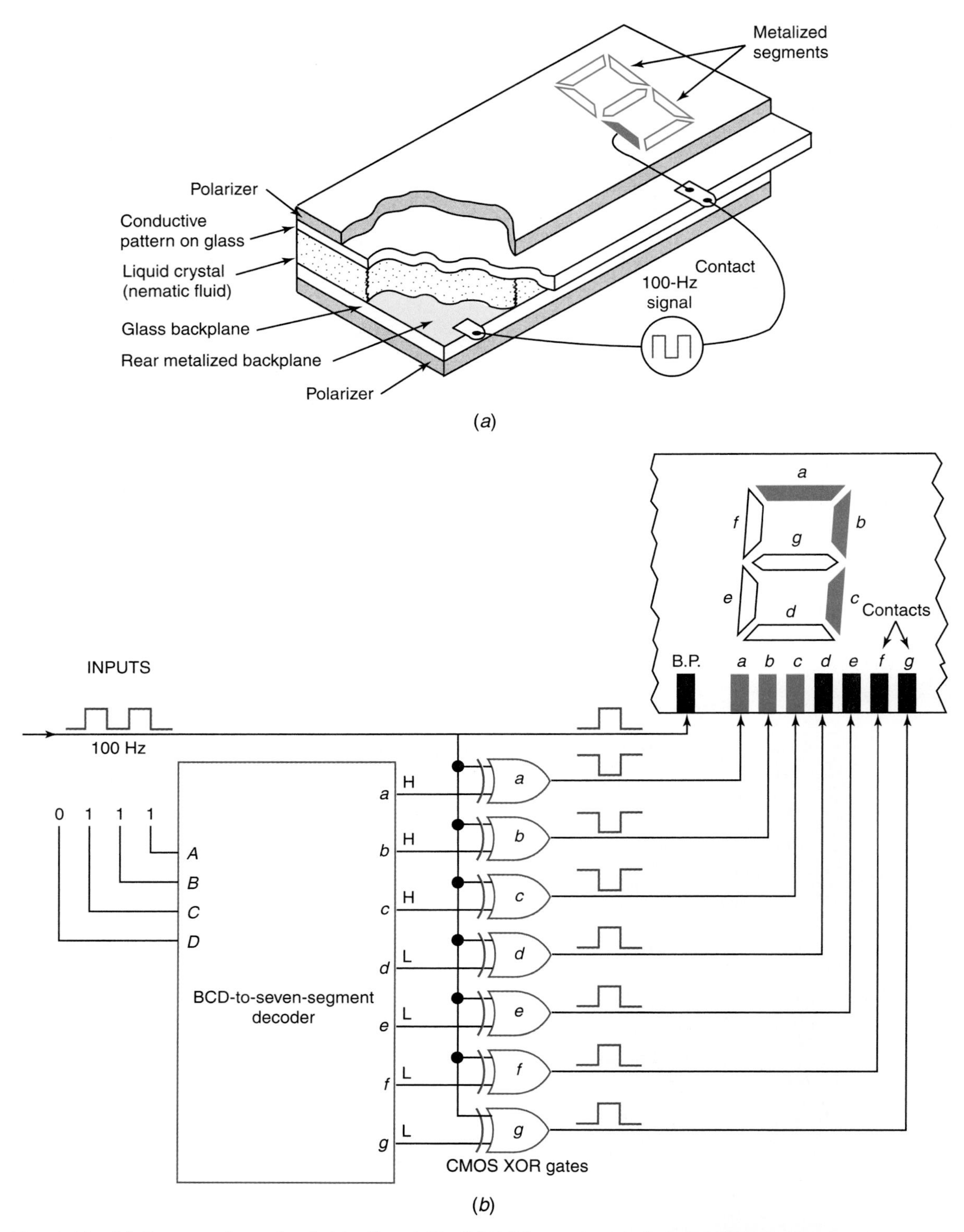

Construction of a field-effect LCD

Fig. 6-18 (*a*) Construction of a field-effect LCD. (*b*) Wiring a CMOS decoder/driver system to an LCD.

Driving an LCD

matic fluid. The *dc voltage* across the nematic fluid *will destroy the LCD* in a short time.

In actual practice, the decoder and XOR LCD display drivers pictured in Fig. 6-18(*b*) are usually packaged in a single CMOS IC. The 100-Hz square-wave signal frequency is not critical and may range from 30 to 200 Hz. Liquid-crystal displays are sensitive to low temperatures. At below-zero temperatures the LCD display's turn-on and turn-off times become very slow. However, the long lifetime and extremely low power consumption make them ideal for battery or solar cell operation.

Figure 6-19 illustrates two examples of typical commercial LCD devices. Note that both have pins which can be soldered into a printed circuit board. In the lab, these LCD displays may be plugged into solderless mounting boards. However, this must be done with great care because there are many fragile pins. Most labs will have the LCDs mounted on a printed circuit board with the appropriate connectors.

A simple two-digit seven-segment LCD is sketched in Fig. 6-19(*a*). Notice the use of two glass plates. Because of the thin glass used in LCDs, care must be taken not to drop or bend the display. Notice in Fig. 6-19(*a*) that two plastic headers with pins are fastened on each side of the glass backplane. The LCD illustrated in Fig. 6-19(*a*) has 18 pins. Only the common or backplane pin is labeled. Each seg-

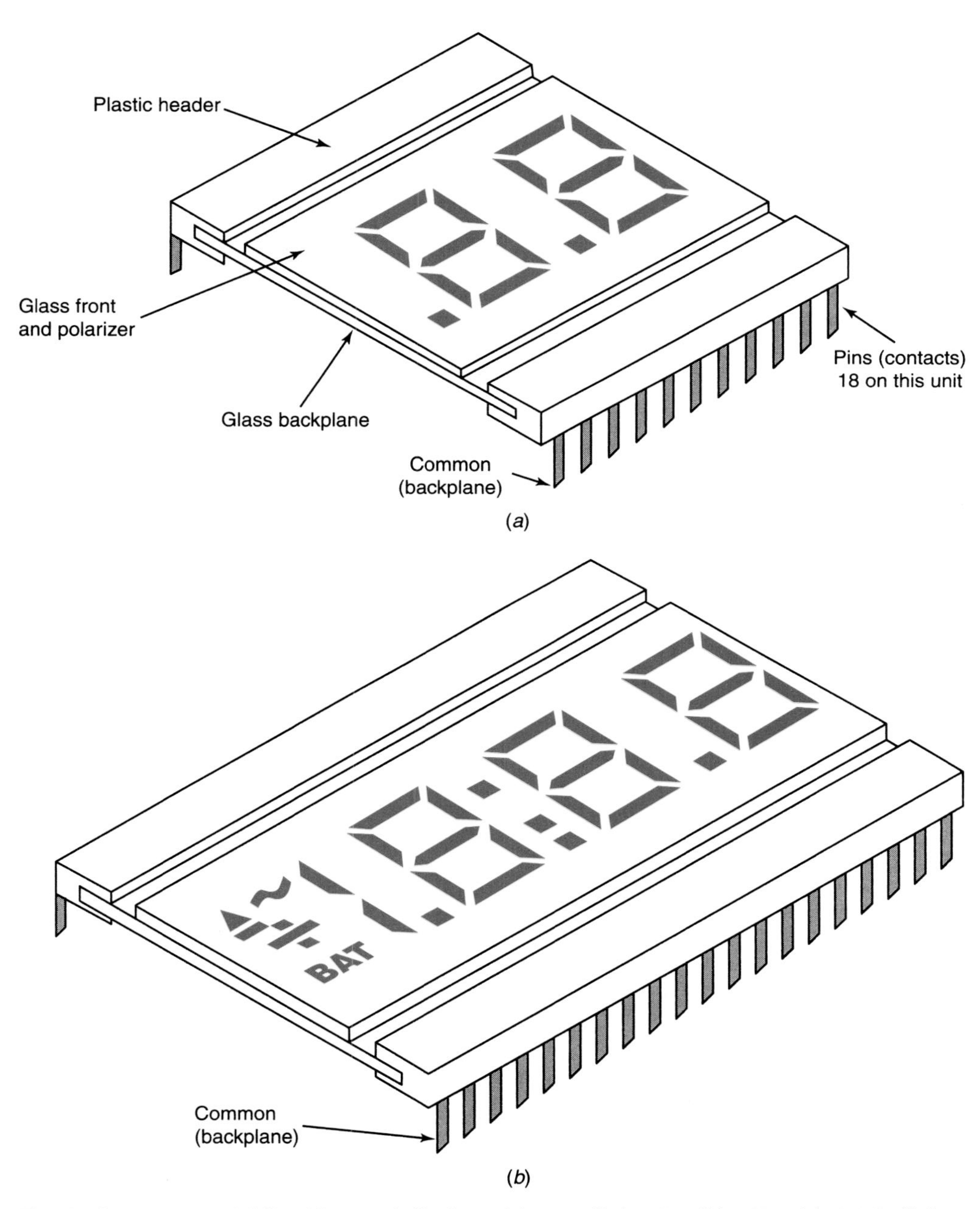

Fig. 6-19 Commercial liquid-crystal displays. (*a*) Two-digit LCD. (*b*) LCD with 3 1/2 digits and symbols.

Liquid-crystal displays

ment and decimal point has a pin connection on this LCD package.

Another commercial LCD is illustrated in Fig. 6-19(*b*). This LCD has a more complex display, including symbols. This unit comes in a 40-pin package. All segments, decimal points, and symbols are assigned a pin number. Only the backplane or common pin is noted on the drawing. Manufacturers' data sheets must be consulted for actual pin numbers.

BCD-to-seven-segment latch/decoder/driver for LCD

Liquid-crystal displays

Liquid-crystal displays that produce frosty white characters on a dark background are also available. This type is usually the *dynamic-scattering LCD.* The dynamic-scattering LCD uses a different nematic fluid and no polarizers. The dynamic-scattering LCD consumes more power than field-effect displays. Currently the field-effect LCD is the most popular. The black letters on a silvery background, as on the DMM in Fig. 6-17, tell you this is a field-effect LCD.

Dynamic-scattering LCD

Field-effect LCD

74HC4543 BCD-to-seven-segment latch/decoder/driver CMOS IC

TEST

Supply the missing word or words in each statement.

30. Digits appear ________ (black, silver) on a ________ (black, silver) background on the field-effect LCD.
31. The LCD uses a liquid crystal, or ________ fluid, which transmits light differently when affected by a magnetic field from an ac voltage.
32. A(n) ________ (ac, dc) voltage applied to an LCD will destroy the unit.
33. The LCD unit consumes a ________ (large amount, moderate amount, very small amount) of power.

6-10 Using CMOS to Drive an LCD Display

A block diagram of an LCD decoder-driver system is sketched in Fig. 6-20(*a*). The input is 8421 BCD. The latch is a temporary memory to hold the BCD data. The BCD-to-seven-segment decoder operates somewhat like the 7447A decoder that was studied earlier. Note that the output from the decoder in Fig. 6-20(*a*) is in seven-segment code. The last block before the display is the LCD driver. This consists of XOR gates as in Fig. 6-18(*b*). The drivers and backplane (common) of the display must be driven with a 100-Hz square-wave signal. In actual practice the latch, decoder, and LCD driver are all available in a single CMOS package. The 74HC4543 and 4543 ICs described by the manufacturer as a *BCD-to-seven-segment latch/decoder/driver* for LCDs are such packages.

A wiring diagram for a single LCD driver circuit is shown in Fig. 6-20(*b*). The 74HC4543 decoder/driver CMOS IC is being employed. The 8421 BCD input is 0011 (decimal 3). The 0011_{BCD} is decoded into seven-segment code. A separate 100-Hz clock feeds the signal to both the LCD backplane (common) and the Ph (phase) input of the 74HC4543 IC. The driving signals in this example are sketched for each segment of the LCD. Note that *only out-of-phase signals will activate a segment.* In-phase signals (such as segments *e* and *f* in this example) do not activate LCD segments.

A pin diagram for the 74HC4543 BCD-to-seven-segment latch/decoder/driver CMOS IC is reproduced in Fig. 6-21(*a*) on page 140. Detailed information on the operation of the 74HC4543 IC is contained in the truth table in Fig. 6-21(*b*). On the output side of the truth table, an "H" means the segment is on while an "L" means the segment is off. The format of the decimal numbers generated by the decoder is shown in Fig. 6-21(*c*). Note especially the numbers 6 and 9. The 74HC4543 decoder forms the 6 and 9 differently from the 7447A TTL decoder studied earlier. Compare Fig. 6-21(*c*) with Fig. 6-13(*c*) to see the differences.

TEST

Answer the following questions.

34. Refer to Fig. 6-20(*a*). The job of the decoder block is to translate from ________ code to ________ code.
35. Refer to Fig. 6-20(*b*). All of the drive lines going from the driver to the LCD carry a square-wave signal (T or F).
36. Refer to Fig. 6-22 on page 141. What is the decimal reading on the LCD for each input pulse (t_1 to t_5)?
37. Refer to Fig. 6-22. For input pulse t_5 only, which drive lines have an *out-of-phase* signal appearing on them?

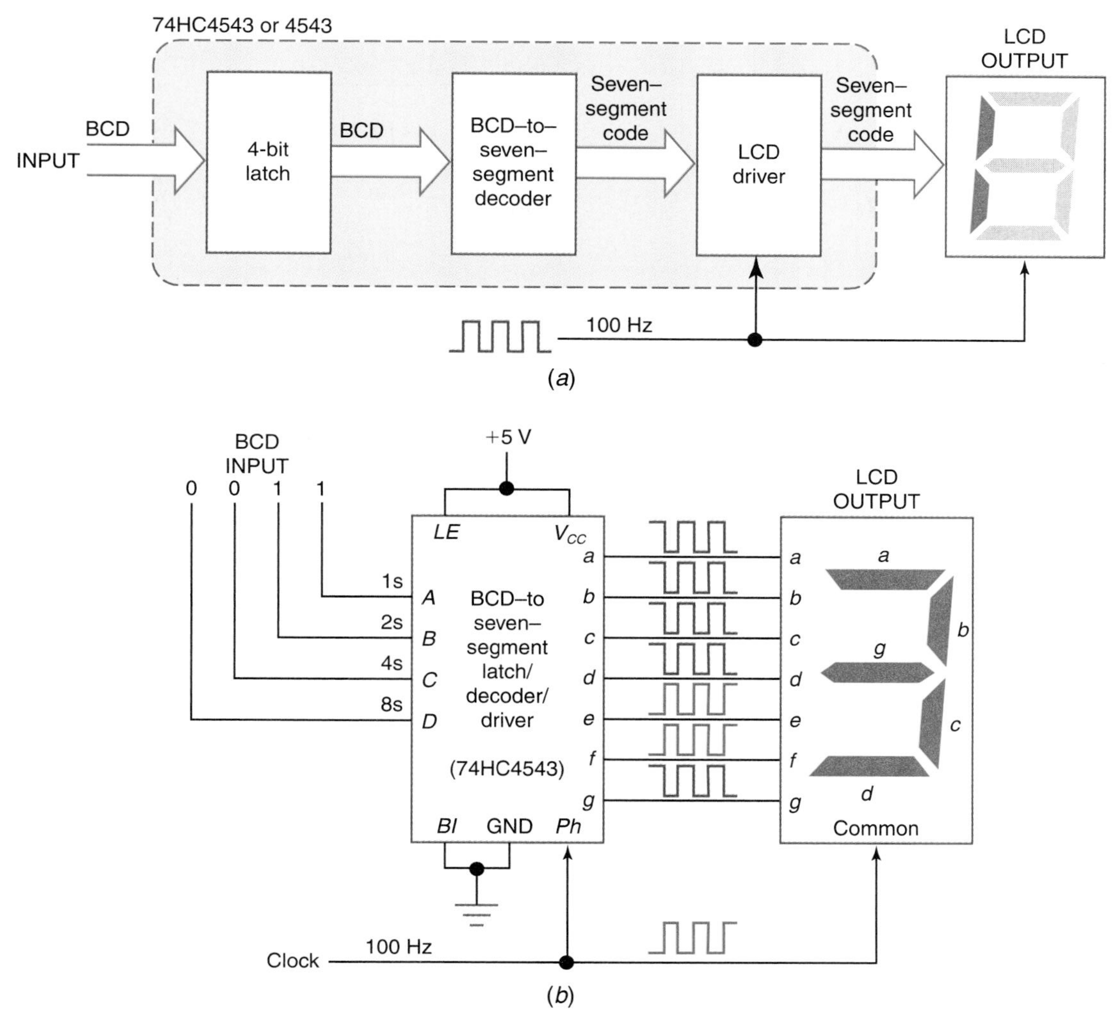

Fig. 6-20 Driving a seven-segment LCD. (*a*) Block diagram of system used to decode and drive LCD. (*b*) Using the 74HC4543 CMOS IC to decode and drive the LCD.

Driving seven-segment LCD

38. Refer to Fig. 6-20(*a*). The LCD driver block inside the 4543 IC will probably contain a group of ________ (AND, XOR) gates.
39. Refer to Fig. 6-20(*a*). What block in the 74HC4543 IC is a type of memory device?
40. Refer to Fig. 6-20(*b*). What activates segments *a*, *b*, *c*, *d*, and *g* on the LCD?

6-11 Vacuum Fluorescent Displays

The *vacuum fluorescent (VF) display* is a modern relative of the triode vacuum tube. A schematic symbol for a triode vacuum tube is sketched in Fig. 6-23(*a*) (page 142). The three parts of the triode tube are called the *plate (P),* *grid (G),* and *cathode (K).* The cathode is also called the *filament* or the *heater.* The plate is also called the *anode.*

Vacuum fluorescent (VF) display

Plate (P)

Grid (G)

Cathode (K)

Filament

Heater

The cathode/heater is a fine tungsten wire coated with a material such as barium oxide. The cathode gives off electrons when heated. The grid is a stainless steel screen. The plate can be thought of as the "collector of electrons" in the triode tube.

Assume that the cathode (K) of the triode tube in Fig. 6-23(*a*) is heated and has "boiled off" some electrons into the vacuum surrounding the cathode. Next, assume that the grid (G) becomes positive. The electrons are attracted to the grid. Next, assume that the plate (P) becomes positive. When the plate becomes positive, electrons will be attracted through the screenlike grid to the plate. Finally, the triode

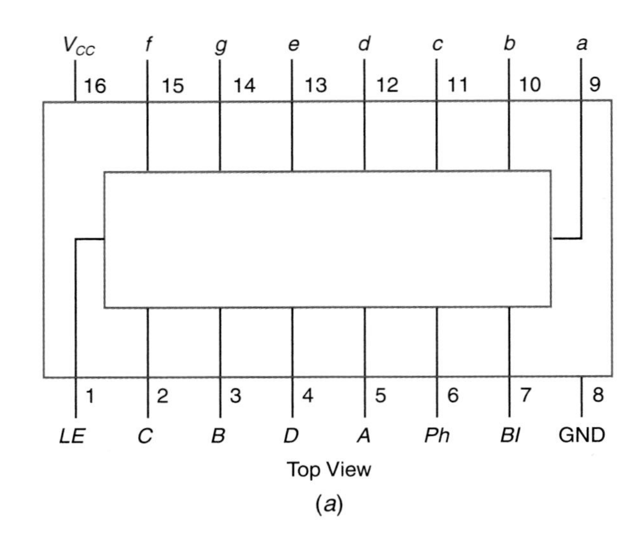

(*a*)

Truth Table

INPUTS							OUTPUTS							
LE	*BI*	*Ph**	*D*	*C*	*B*	*A*	*a*	*b*	*c*	*d*	*e*	*f*	*g*	Display
X	H	L	X	X	X	X	L	L	L	L	L	L	L	Blank
H	L	L	L	L	L	L	H	H	H	H	H	H	L	0
H	L	L	L	L	L	H	L	H	H	L	L	L	L	1
H	L	L	L	L	H	L	H	H	L	H	H	L	H	2
H	L	L	L	L	H	H	H	H	H	H	L	L	H	3
H	L	L	L	H	L	L	L	H	H	L	L	H	H	4
H	L	L	L	H	L	H	H	L	H	H	L	H	H	5
H	L	L	L	H	H	L	H	L	H	H	H	H	H	6
H	L	L	L	H	H	H	H	H	H	L	L	L	L	7
H	L	L	H	L	L	L	H	H	H	H	H	H	H	8
H	L	L	H	L	L	H	H	H	H	H	L	H	H	9
H	L	L	H	L	H	L	L	L	L	L	L	L	L	Blank
H	L	L	H	L	H	H	L	L	L	L	L	L	L	Blank
H	L	L	H	H	L	L	L	L	L	L	L	L	L	Blank
H	L	L	H	H	L	H	L	L	L	L	L	L	L	Blank
H	L	L	H	H	H	L	L	L	L	L	L	L	L	Blank
H	L	L	H	H	H	H	L	L	L	L	L	L	L	Blank
L	L	L	X	X	X	X	**							**
†	†	H	†				Inverse of Output Combinations Above							Display as above

X = don't care
† = same as above combinations
* = for liquid crystal readouts, apply a square wave to *Ph*
** = depends upon the BCD code previously applied when *LE* = H

(*b*)

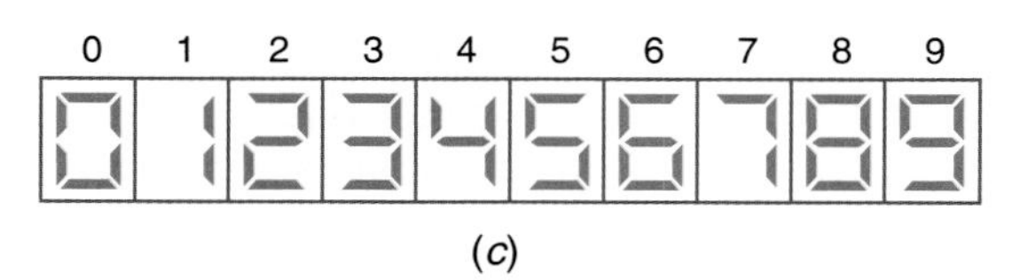

(*c*)

74HC4543 BCD-to-seven-segment latch/decoder/driver CMOS IC

Fig. 6-21 The 74HC4543 BCD-to-seven-segment latch/decoder/driver CMOS IC. (*a*) Pin diagram. (*b*) Truth table. (*c*) Format of digits formed by the 74HC4543 decoder IC.

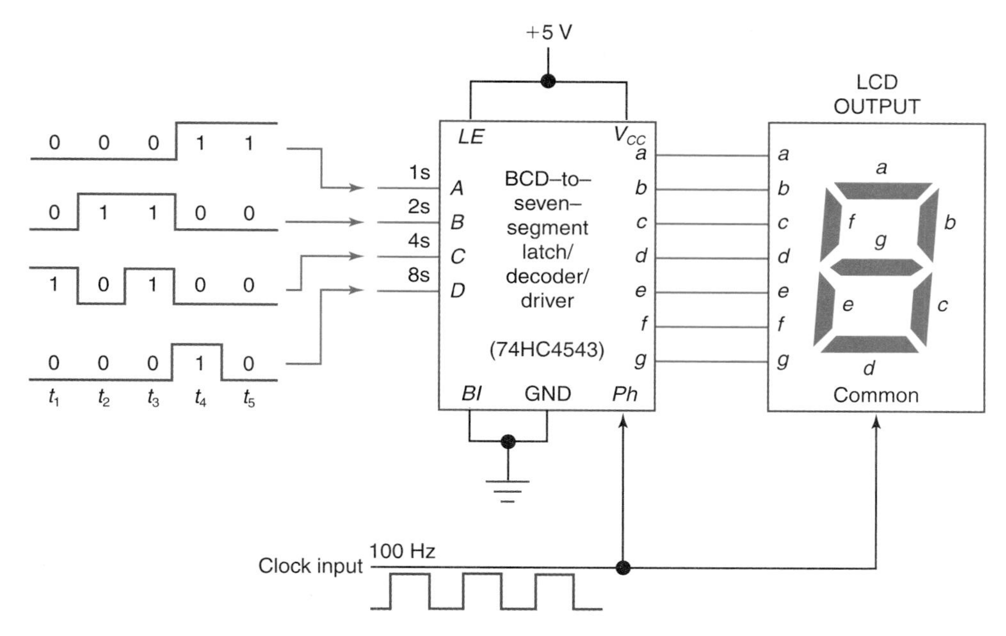

Fig. 6-22 Decoder-LCD circuit for problems 36 and 37.

is conducting electricity from cathode to anode.

You can stop the triode tube from conducting in one of two ways. First, make the grid slightly negative (leave the plate positive). This will repel the electrons and they will not pass through the grid to the plate. Second, leave the grid positive and drop the plate voltage to 0. The plate will not attract electrons and the triode tube does not conduct electricity from cathode to anode.

The schematic symbol in Fig. 6-23(*b*) represents a single digit of a VF display. Notice the single cathode (K), single grid (G), and seven plates (P_a to P_g). Each of the seven plates has been coated with a *zinc oxide fluorescent* material. Electrons striking the fluorescent material on the plate will cause it to glow a blue-green color. The seven plates in the schematic in Fig. 6-23(*b*) represent the seven segments of a normal numeric display. Note that the entire unit is held in a glass enclosure containing a vacuum.

A single-digit seven-segment display is being operated in Fig. 6-23(*c*). The cathode (heater) is being powered by direct current in this example, with +12 V applied to the grid (G). Two plates (P_c and P_f) are grounded. Each of the remaining five plates has +12 V applied. The high positive voltage on the five plates (P_a, P_b, P_d, P_e, and P_g) causes these plates to attract electrons. These five plates glow a blue-green color as electrons strike their surface.

In actual practice, the plates of the VF display tube are shaped like segments of a number or other shapes. Figure 6-24(*a*) on page 143 is a view of the physical arrangement of cathode, grid, and plates. Notice that the plates are arranged in seven-segment format in this display unit. The screen above the segments is the grid. Above the grid are the cathodes (filaments or heaters). Each segment, grid, or cathode lead comes out the side of the sealed glass vacuum tube. The VF display shown in Fig. 6-24(*a*) would be viewed from above the unit looking downward. The fine wire cathodes and grid would be almost invisible. Lighted segments (plates) show through the mesh (grid).

A commercial vacuum fluorescent display is sketched in Fig. 6-24(*b*). This VF display contains 4 seven-segment numeric displays, a colon, and 10 triangle-shaped symbols. The internal parts of most VF displays are visible through the sealed glass package. Visible in the display are the cathodes (filaments or heaters) stretched horizontally across the display. These are very fine wires and are barely visible on a commercial display. Next, the grids are shown in five sections. Each of the five grids can be activated individually. Finally, the fluorescent-coated plates form the numeric segments, colons, and other symbols.

interNET CONNECTION

For information on solid state displays, visit the website for Lumex, Inc.

Zinc-oxide fluorescent material

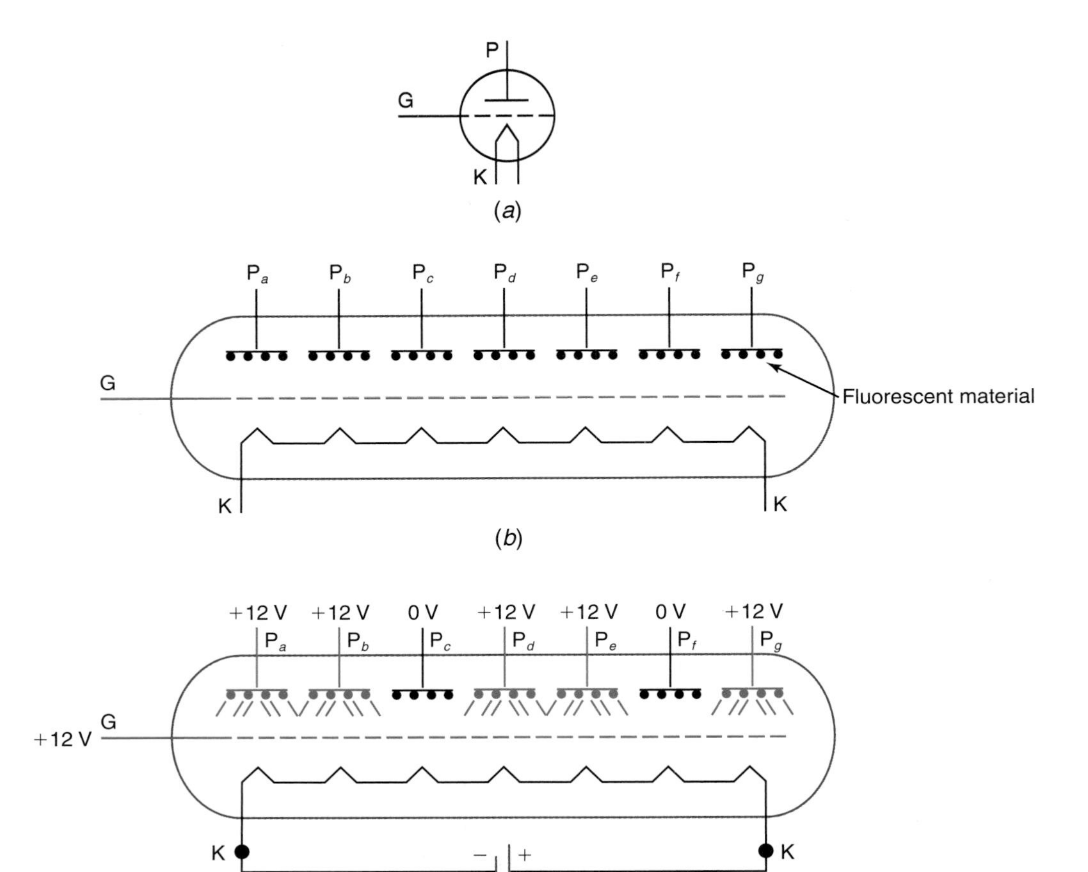

Triode vacuum tube

VF display

Fig. 6-23 (*a*) Schematic symbol for a triode vacuum tube. (*b*) Schematic symbol for a single digit of a VF display. (*c*) Lighting plates on the VF display.

Advantages of VF displays

Vacuum fluorescent displays are based on an older technology but they have gained great favor in recent years. This is because they can operate at relatively low voltages and power and have an extremely long life and fast response. They can display in various colors (with filters), have good reliability, and low cost. Vacuum fluorescent displays are compatible with the popular 4000 series CMOS family of ICs. They are widely used in readouts found in automobiles, VCRs, TVs, household appliances, and digital clocks.

Driving a VF display

Did You Know?

Sobering Sensors. A new "smart" car has an electronic steering wheel that interacts with the driver's palm perspiration and measures alcohol content. The car won't start if the driver is intoxicated or wearing gloves.

TEST

Answer the following questions.

41. A VF display glows with a(n) ________ color when activated.
42. Refer to Fig. 6-25. Which plates of this VF display will glow?
43. Name the parts of the vacuum fluorescent display labeled *A*, *B*, and *C* in Fig. 6-26.
44. Refer to Fig. 6-26. What segments of the VF display glow, and what decimal number is lit?

6-12 Driving a VF Display

The voltage requirements for operating VF displays are somewhat higher than for LED or LCD units. This requirement makes them com-

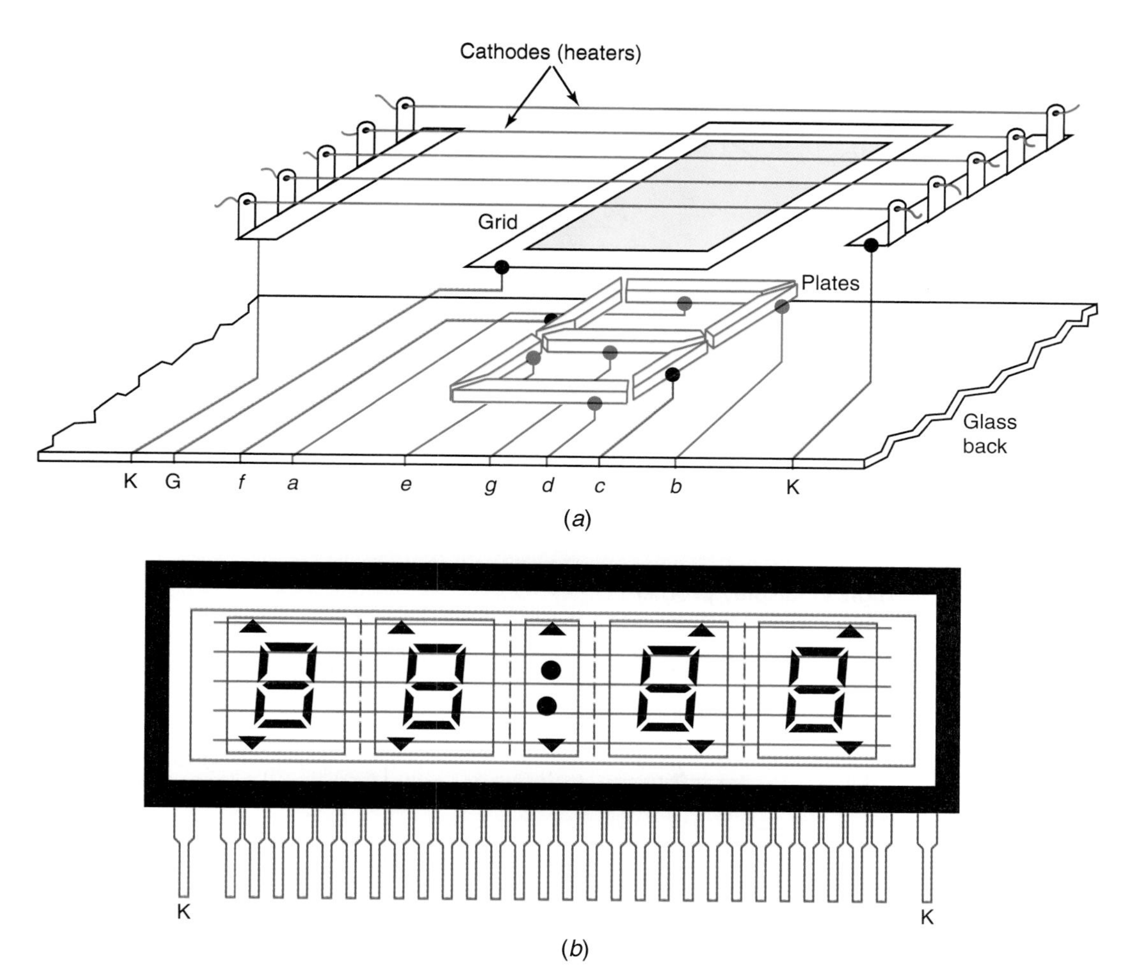

Fig. 6-24 (*a*) Typical construction of VF display. (*b*) Commercial four-digit VF display.

Construction of vacuum fluorescent display

patible with 4000 series CMOS ICs. Recall that 4000 series CMOS ICs can operate on voltages up to 18 V.

A wiring diagram of a simple BCD decoder/driver circuit is detailed in Fig. 6-27 on page 144. In this example, 1001_{BCD} is translated into the decimal 9 on the VF display. The circuit uses the 4511 BCD-to-seven-segment latch/decoder/driver CMOS IC. In this example, the *a*, *b*, *c*, *f*, and *g* output lines are HIGH (+12 V) with only the *d* and *e* lines LOW.

The +12-V supply is connected directly to the grid in Fig. 6-27. The cathode (filament or heater) circuit contains a resistor (R_1) to limit the current through the heaters to a safe level. The +12 V is also used to supply power for the 4511 decoder/driver CMOS IC. Note the labels on the power connections to the 4511

BCD-to-seven-segment latch/decoder/driver CMOS IC

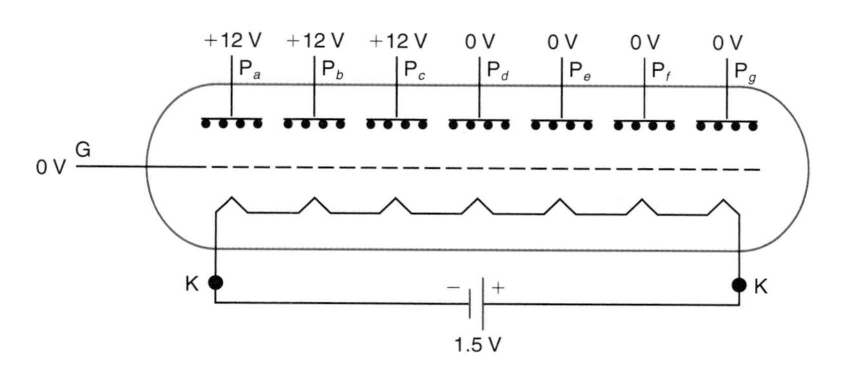

Fig. 6-25 VF display with no positive grid voltage.

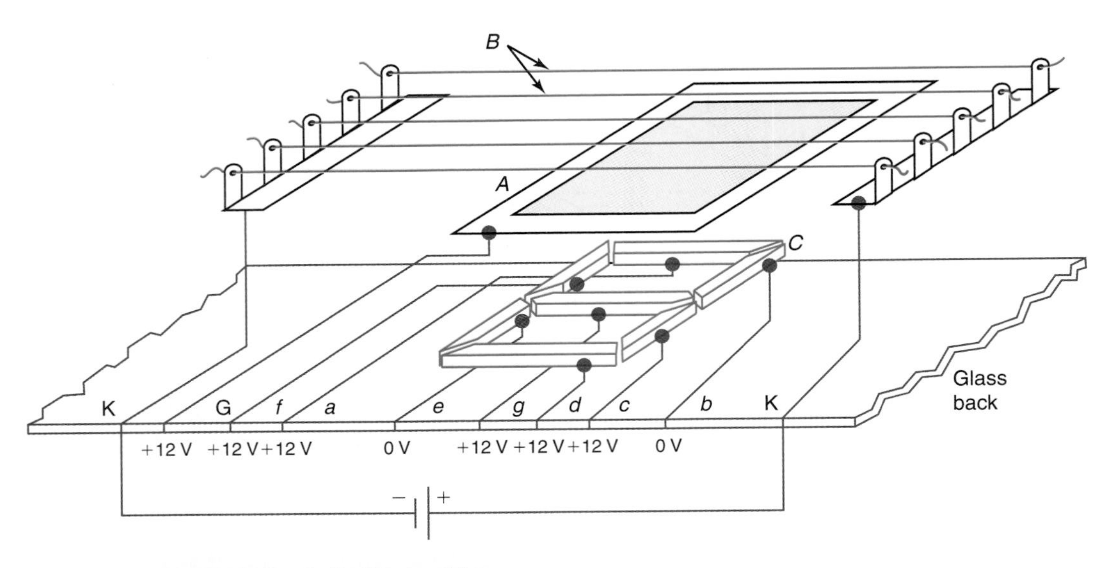

Fig. 6-26. Single-digit VF display problem.

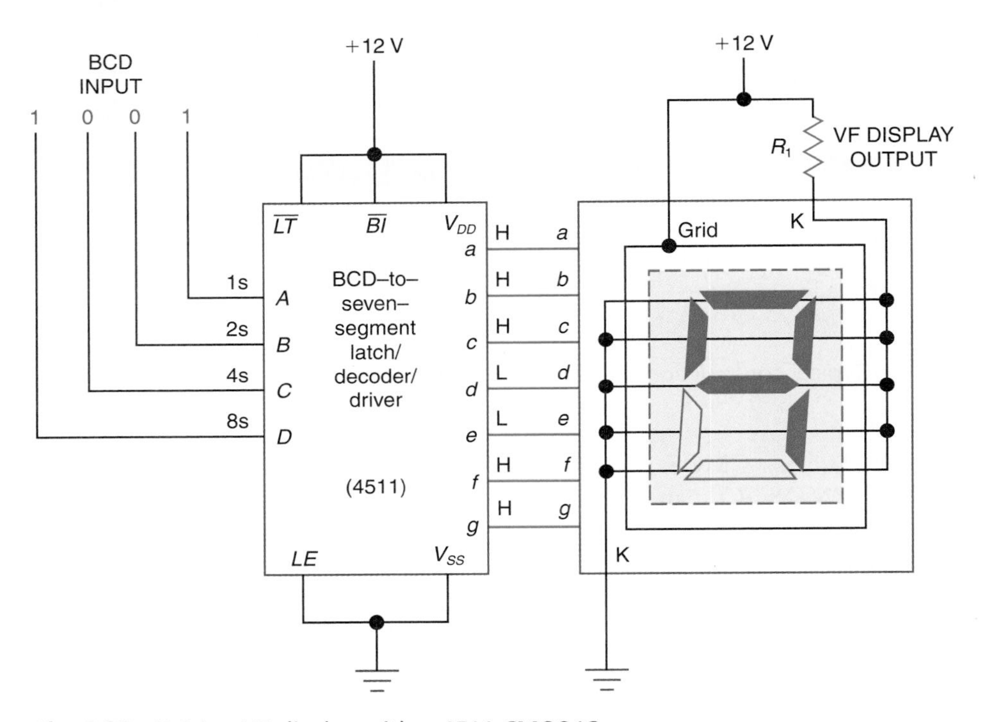

Driving VF display

Fig. 6-27 Driving VF display with a 4511 CMOS IC.

IC. The V_{DD} pin connects to +12 V while V_{ss} goes to ground (GND).

The pin diagram, truth table, and number formats for the 4511 CMOS IC are shown in Fig. 6-28. The 4511 BCD-to-seven-segment latch/decoder/driver IC's pin diagram is shown in Fig. 6-28(*a*). This is the top view of this 16-pin DIP CMOS IC. Internally, the 4511 IC is organized like the 74HC4543 unit. The latch, decoder, and driver sections are illustrated in the shaded section of the block diagram in Fig. 6-20(*a*).

The truth table in Fig. 6-28(*b*) shows seven inputs to the 4511 decoder/driver IC. The BCD data inputs are labeled *D*, *C*, *B*, and *A*. The $\overline{LT}$ input stands for lamp test. When activated by a LOW (row 1 on the truth table), all outputs go HIGH and light all segments of

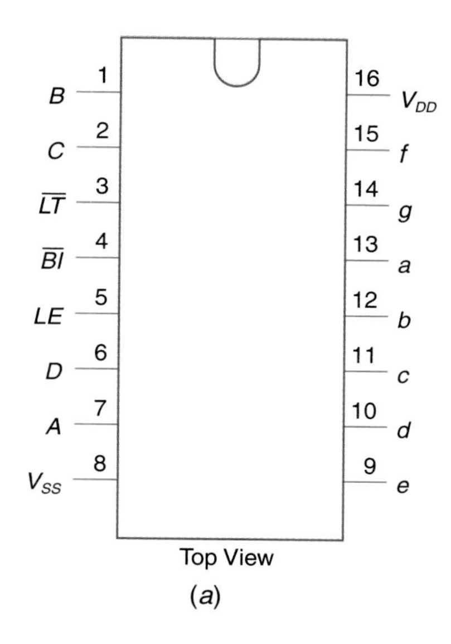

(*a*)

Truth Table

INPUTS							OUTPUTS							
LE	$\overline{BI}$	$\overline{LT}$	*D*	*C*	*B*	*A*	*a*	*b*	*c*	*d*	*e*	*f*	*g*	Display
X	X	0	X	X	X	X	1	1	1	1	1	1	1	B
X	0	1	X	X	X	X	0	0	0	0	0	0	0	
0	1	1	0	0	0	0	1	1	1	1	1	1	0	0
0	1	1	0	0	0	1	0	1	1	0	0	0	0	1
0	1	1	0	0	1	0	1	1	0	1	1	0	1	2
0	1	1	0	0	1	1	1	1	1	1	0	0	1	3
0	1	1	0	1	0	0	0	1	1	0	0	1	1	4
0	1	1	0	1	0	1	1	0	1	1	0	1	1	5
0	1	1	0	1	1	0	0	0	1	1	1	1	1	6
0	1	1	0	1	1	1	1	1	1	0	0	0	0	7
0	1	1	0	0	0	0	1	1	1	1	1	1	1	8
0	1	1	1	0	0	1	1	1	1	0	0	1	1	9
0	1	1	1	0	1	0	0	0	0	0	0	0	0	
0	1	1	1	0	1	1	0	0	0	0	0	0	0	
0	1	1	1	1	0	0	0	0	0	0	0	0	0	
0	1	1	1	1	0	1	0	0	0	0	0	0	0	
0	1	1	1	1	1	0	0	0	0	0	0	0	0	
0	1	1	1	1	1	1	0	0	0	0	0	0	0	
1	1	1	X	X	X	X				*				*

X = Don't care

* Depends upon the BCD code appiled during the 0 to 1 transition of *LE*.

(*b*)

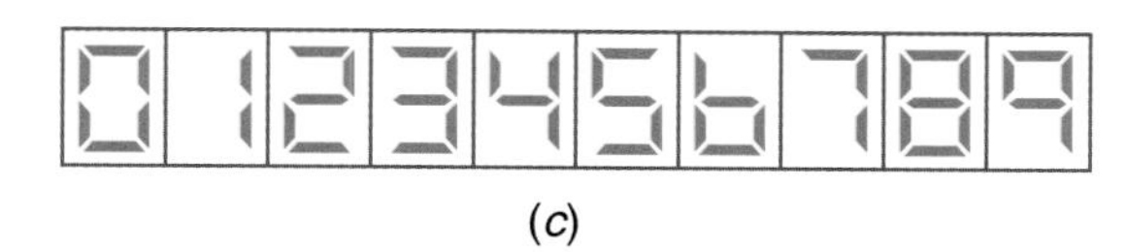

(*c*)

Fig. 6-28 The 4511 BCD-to-seven-segment latch/decoder/driver CMOS IC. (*a*) Pin diagram. (*b*) Truth table. (*c*) Format of digits using the 4511 decoder IC. *(Courtesy of National Semiconductor Corporation.)*

4511 BCD-to-seven-segment latch/decoder/driver CMOS IC

an attached display. The $\overline{BI}$ input stands for blanking input. When $\overline{BI}$ is activated with a LOW, all outputs go LOW and all segments of an attached display are blanked. The *LE* (latch enable) input can be used like a memory to hold data on display while the BCD input data changes. If $LE = 0$, then data passes through the 4511 IC. However, if $LE = 1$ then the last data present at the data inputs (*D*, *C*, *B*, *A*) is latched and held on the display. The *LE*, $\overline{BI}$, and $\overline{LT}$ inputs are all disabled in the circuit in Fig. 6-27.

Next, refer to the output side of the truth table in Fig. 6-28. On the 4511 IC, a HIGH or 1 is an active output. In other words, an output of 1 turns on a segment on the attached display. Therefore, a 0 output means the display's segment is turned off.

The format of the digits generated by the 4511 BCD-to-seven-segment decoder IC are illustrated in Fig. 6-28(*c*). Especially note the formation of the decimal numbers 6 and 9.

TEST

Answer the following questions.

45. Refer to Fig. 6-27. The +12-V power supply is being used because the ________ (CMOS, TTL) 4511 decoder/driver IC and the ________ (LCD, VF) display operate properly at this voltage.
46. Refer to Fig. 6-27. What is the purpose of resistor R_1 in this circuit?
47. Refer to Fig. 6-29. What is the decimal reading on the vacuum fluorescent display for each input pulse (t_1 to t_4)?
48. Refer to Fig. 6-29. During pulse (t_4) what voltages are applied to the seven plates (segments) of the VF display?

6-13 Troubleshooting a Decoding Circuit

Consider the BCD-to-seven-segment decoder circuit in Fig. 6-30. The problem is that segment *a* of the display does not light. The technician first checks the circuit visually. Then the IC is checked for signs of excessive heat. The V_{CC} and GND voltages are checked with a DMM or logic probe. In this example, the results of these tests did not locate the problem. Next, the temporary jumper wire from GND to the LT input of the 7447A IC should cause all segments on the display to light, giving a decimal 8 indication. Segment *a* on the display still does not light. The logic probe is used to check the logic levels at the outputs (*a*

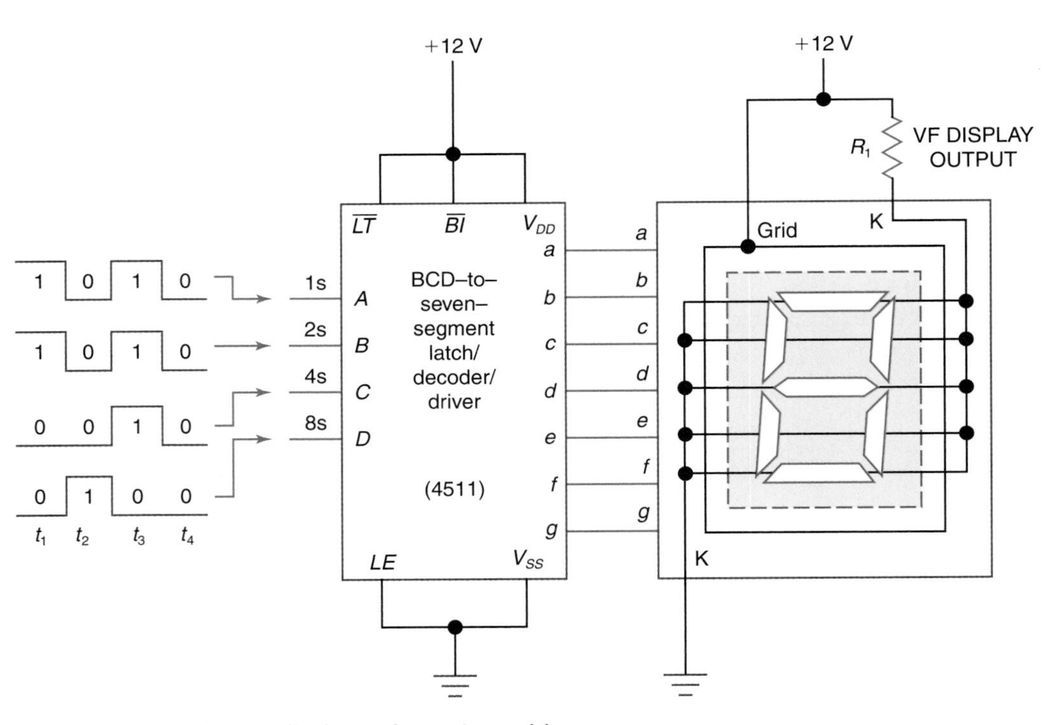

Fig. 6-29 Decoder–VF display pulse-train problem.

to *g*) of the 7447A decoder. They are all L (LOW) in Fig. 6-30, as required. Next, the logic levels are checked on the display side of the resistors. They are all H (HIGH) except the faulty line, which is LOW. The LOW and HIGH pattern in Fig. 6-30 indicates a voltage drop across each of the bottom six resistors. The LOW indications on both ends of the top resistor in Fig. 6-30 indicate an open circuit in the segment *a* section of the seven-segment display. Segment *a* of the display must be faulty. The entire seven-segment LED display is replaced. The replacement must have the same pin diagram and be a common-anode LED display. After replacement, the circuit is checked for proper operation.

The circuit shown in Fig. 6-31 produces no display. The hurried technician checks the V_{CC} and GND pins with a logic probe. The readings as shown in Fig. 6-31 seem all right. The test jumper wire from the LT to the GND should light all segments of the LED display. No segments on the display light. The logic probe shows faulty HIGH readings at all the outputs (*a* to *g*) of the 7447A IC. The technician checks the voltage at V_{CC} with a DMM. The reading is 4.65 V. This is quite low. The technician now touches the top of the 7447A IC. It is very hot. The chip (7447A) has an internal short circuit and must be replaced. The 7447A IC is replaced, and the circuit is checked for proper operation.

Internal short circuit

In this example, the technician forgot to use his or her own senses first. A simple touch of the top of the DIP IC circuit would have suggested a bad 7447A chip. Note that the HIGH

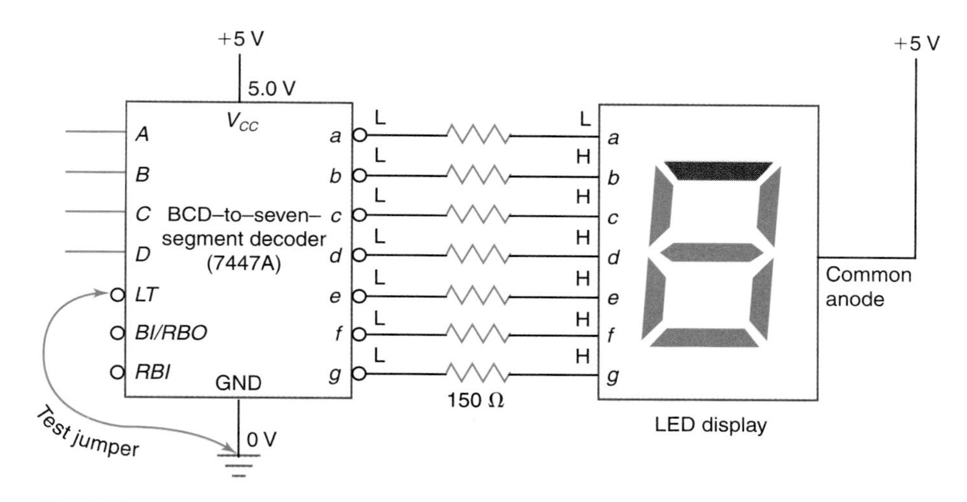

Fig. 6-30 Troubleshooting a faulty decoder/LED display circuit.

Troubleshooting faulty decoder/LED display circuit

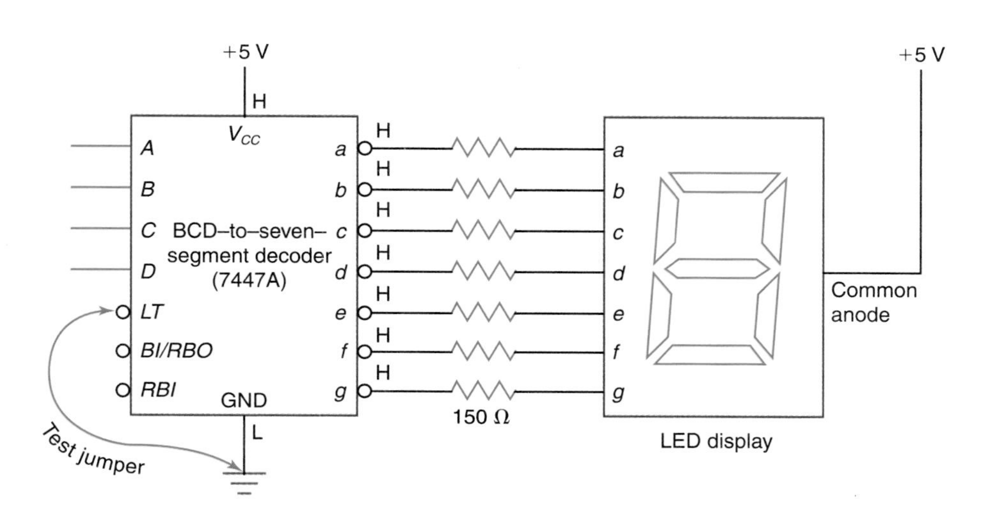

Fig. 6-31 Troubleshooting a faulty decoder circuit with blank LED display.

reading on the V_{CC} pin did not give the technician an accurate picture. The voltage was actually 4.65 V instead of the normal 5.0 V. In this case the voltmeter reading gave the technician a clue as to the difficulty in the circuit. The short circuit was dropping the power supply voltage to 4.65 V.

TEST

Answer the following questions.

49. What is the first step in troubleshooting a digital logic circuit?
50. An internal ________ (open, short) circuit in a TTL IC will many times cause the IC to become excessively hot.

Chapter 6 Review

Summary

1. Many codes are used in digital equipment. You should be familiar with decimal, binary, octal, hexadecimal, 8421 BCD, excess-3, Gray, and ASCII codes.
2. Converting from code to code is essential for your work in digital electronics. Table 6-4 will aid you in converting from several of the codes.
3. The most popular alphanumeric code is the 7-bit ASCII code. The ASCII code is widely used in microcomputer keyboard and display interfacing. Extended ASCII uses 8-bits.
4. Electronic translators are called encoders and decoders. The complicated logic circuits are manufactured in single IC packages.
5. Seven-segment displays are very popular devices for reading out numbers. Light-emitting diode (LED), liquid-crystal display (LCD), and vacuum fluorescent (VF) types are popular displays.
6. The BCD-to-seven-segment decoder/driver is a common decoding device. It translates from BCD machine language to decimal numbers. The decimal numbers appear on seven-segment LED, LCD, or VF displays.

TABLE 6-4

Decimal Number	Binary Number	BCD Codes				Gray code
		8421		Excess-3		
0	0000		0000		0011	0000
1	0001		0001		0100	0001
2	0010		0010		0101	0011
3	0011		0011		0110	0010
4	0100		0100		0111	0110
5	0101		0101		1000	0111
6	0110		0110		1001	0101
7	0111		0111		1010	0100
8	1000		1000		1011	1100
9	1001		1001		1100	1101
10	1010	0001	0000	0100	0011	1111
11	1011	0001	0001	0100	0100	1110
12	1100	0001	0010	0100	0101	1010
13	1101	0001	0011	0100	0110	1011
14	1110	0001	0100	0100	0111	1001
15	1111	0001	0101	0100	1000	1000
16	10000	0001	0110	0100	1001	11000
17	10001	0001	0111	0100	1010	11010
18	10010	0001	1000	0100	1011	11011
19	10011	0001	1001	0100	1100	11010
20	10100	0010	0000	0101	0011	11110

Chapter Review Questions

Answer the following questions.

6-1. Write the binary numbers for the decimal numbers in *a* to *f*:
a. 17 d. 75
b. 31 e. 150
c. 42 f. 300

6-2. Write the 8421 BCD numbers for the decimal numbers in *a* to *f*:
a. 17 d. 1632
b. 31 e. 47,899
c. 150 f. 103,926

6-3. Write the decimal numbers for the 8421 BCD numbers in *a* to *f*:
a. 0010 d. 0111 0001 0110 0000
b. 1111 e. 0001 0001 0000 0000 0000
c. 0011 0000 f. 0101 1001 1000 1000 0101

6-4. Write the excess-3 code numbers for the decimal numbers in *a* to *d*:
a. 7
b. 27
c. 59
d. 318

6-5. Why is the excess-3 code used in some arithmetic circuits?

6-6. List two codes you learned about that are classified as BCD codes.

6-7. Write the Gray code numbers for the decimal numbers in *a* to *f*.
   a. 1   d. 4
   b. 2   e. 5
   c. 3   f. 6

6-8. The ________ (Gray, XS3) code is commonly associated with optical encoding of a shaft's angular position.

6-9. The important characteristic of the Gray code is that only one digit changes as you decrement or increment the count (T or F).

6-10. Refer to Table 6-3. The 7-bit ASCII code for the capital letter S is ________.

6-11. The letters "ASCII" stand for ________.

6-12. Standard ASCII is a(n) ________ -bit ________ (alphanumeric, BCD) code that is used to represent number, letters, punctuation marks, and control characters.

6-13. List two general names for code translators, or electronic code converters.

6-14. A(n) ________ (decoder, encoder) is the electronic device used to convert the decimal input of a calculator keypad to the BCD code used by the central processing unit.

6-15. A(n) ________ (decoder, encoder) is the electronic device used to convert the BCD of the

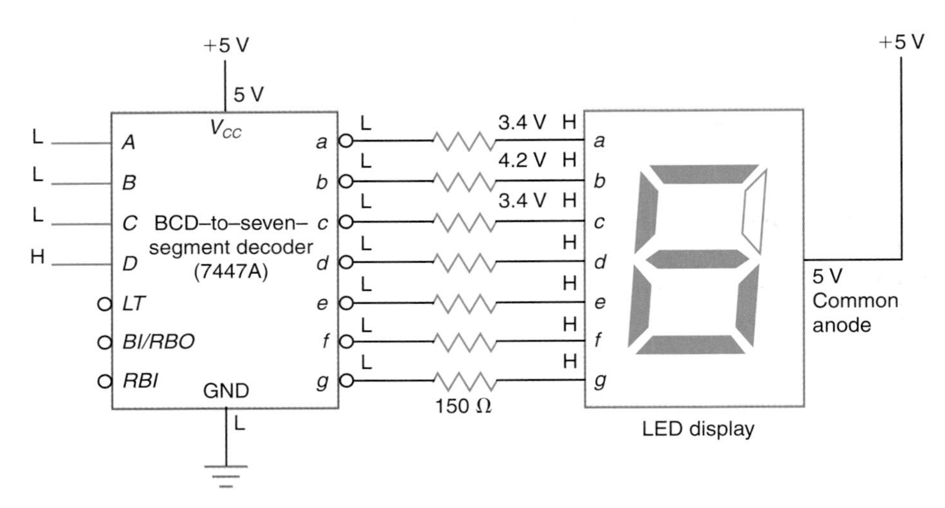

Fig. 6-32 Troubleshooting problems. Logic levels and voltages given on faulty decoder/LED display circuit.

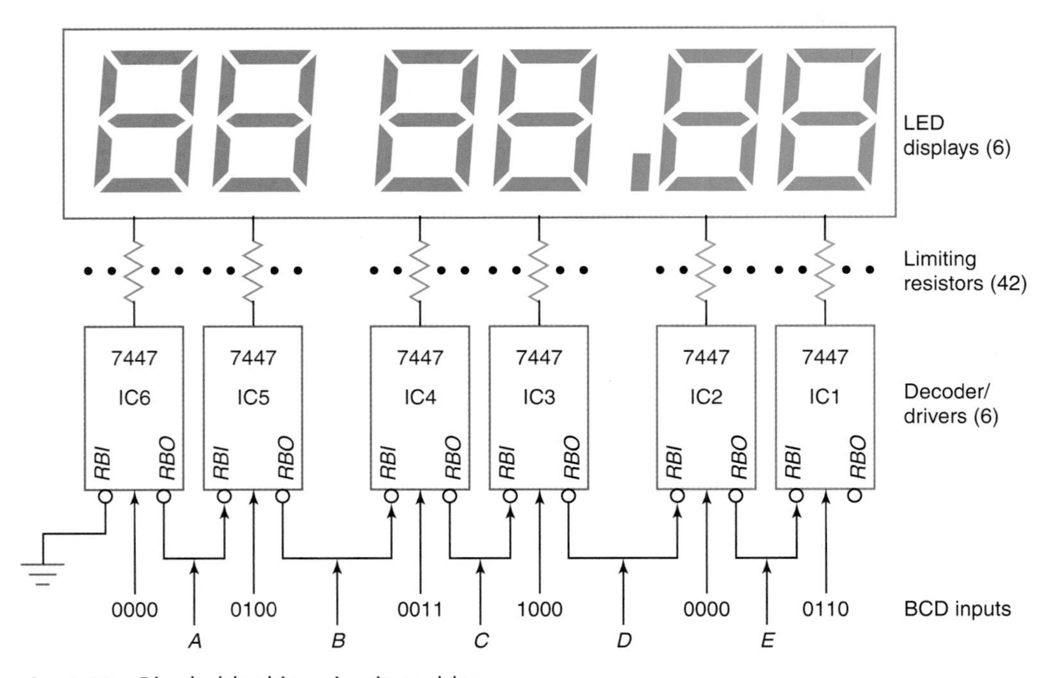

Fig. 6-33 Ripple-blanking circuit problem.

central processing unit of a calculator to the decimal display output.

6-16. Which segments of the seven-segment display *will light* when the following decimal numbers appear? Use the letters *a, b, c, d, e, f,* and *g* as answers.

a. 0	d. 3	g. 6	j. 9
b. 1	e. 4	h. 7	
c. 2	f. 5	i. 8	

6-17. The seven-segment displays that you will use emit light (usually red) and are of the ________ (LCD, LED) type.

6-18. The ________ (LCD, LED) seven-segment display is used where battery operation demands low power consumption.

6-19. The ________ (LCD, LED) display is used where the unit must be read in bright light.

6-20. The ________ display generates light while the LCD controls available light.

6-21. The ________ (LED, VF) seven-segment display operates on a slightly higher voltage power supply.

6-22. Refer to Fig. 6-32. With the 1000 (*HLLL*) input, the display should read decimal ________.

6-23. Refer to Fig. 6-32. All the outputs from the 7447A decoder are ________ (HIGH, LOW). This is ________ (correct, not correct) for this circuit.

6-24. Both a voltmeter and a ________ ________ are used to troubleshoot the circuit in Fig. 6-32.

6-25. Refer to Fig. 6-32. Segment *b* of the LED display appears to be ________ (open, partially short circuited). The display should be replaced with a common- ________ LED display having the same pin diagram as the one in the circuit.

6-26. Refer to Fig. 6-33. With the BCD input shown, the six-digit display reads ________.

6-27. The front and back panels of a ________ (LCD, LED) seven-segment display are made of glass and can be broken by rough handling.

6-28. Refer to Fig. 6-34. With the driving signals shown, the LCD will display the decimal number ________. The input must be the BCD number ________.

6-29. Vacuum fluorescent displays can operate on 12 V, which makes them very compatible with (CMOS, TTL) ICs and automotive applications.

6-30. Refer to Fig. 6-35 on page 152. What will the VF seven-segment display read for each input pulse?

6-31. Refer to Fig. 6-35. List the approximate voltages at each of the seven plates and the grid of the VF display during pulse t_4.

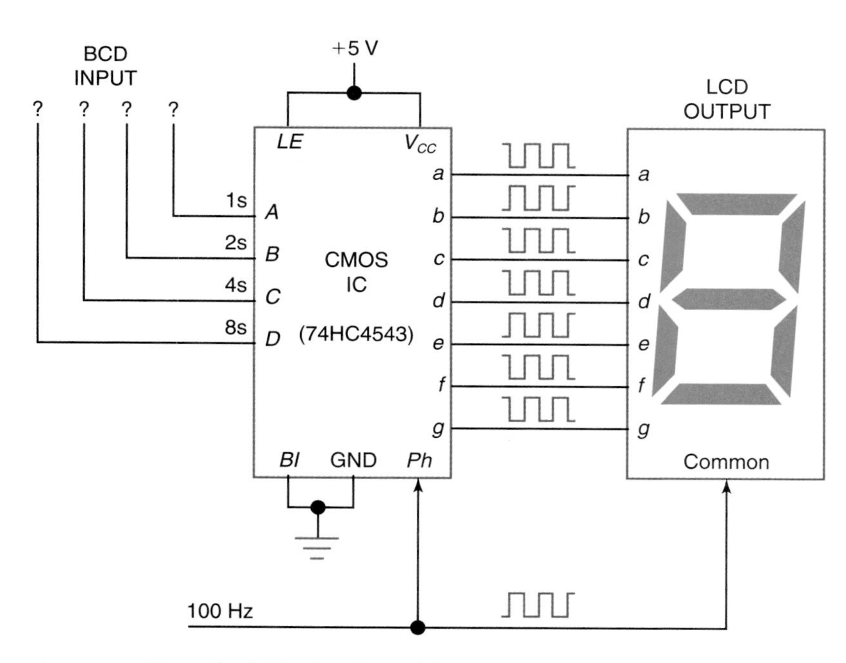

Fig. 6-34 Decoder/LCD circuit problem.

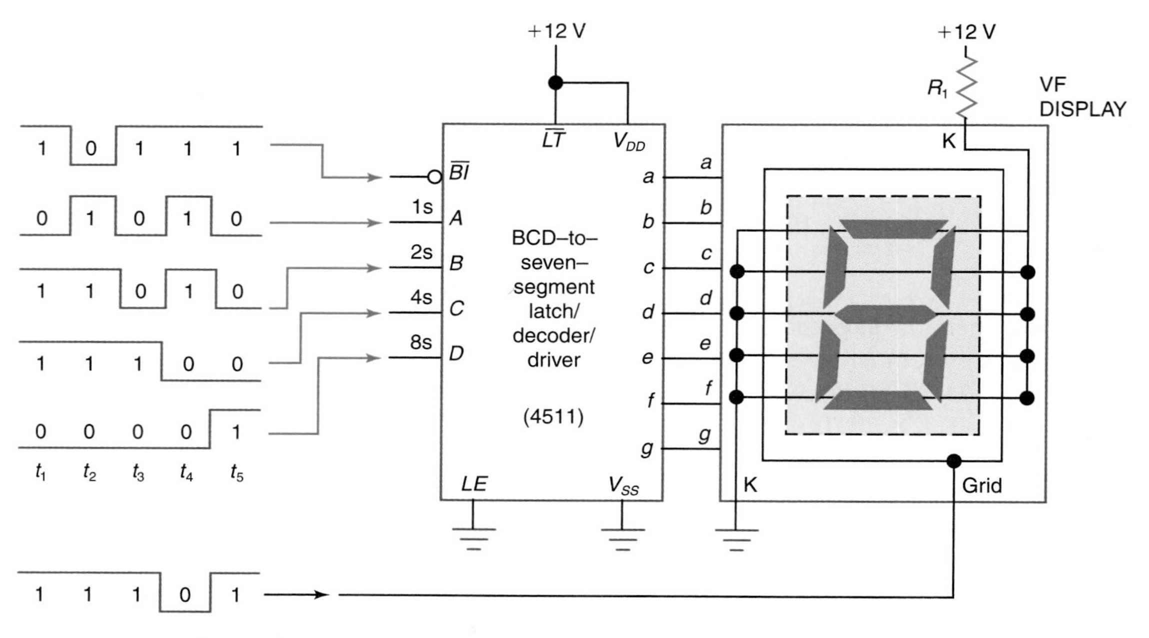

Fig. 6-35 Decoder/VF display pulse-train problem.

Critical Thinking Questions

6-1. Convert the following 8421 BCD numbers to binary.
a. 0011 0101
b. 1001 0110
c. 0111 0100

6-2. As you count in the Gray code, what is the most important characteristic?

6-3. Refer to Fig. 6-5. If the decoder chip is a 4511 and the circuit operates on a 12-V power supply, then the output displays are probably ________ (LED, VF) units.

6-4. Refer to Fig. 6-6. Why is the output of the 74147 10-line to four-line encoder 0111 when *both* inputs 2 and 7 are activated at the same time?

6-5. What is the purpose of the 7447A TTL IC, and with which type of seven-segment display is it compatible?

6-6. The 7447A decoder TTL IC contains 44 gates and is considered a ________ (combinational, sequential) logic circuit. The 7447A decoder has ________ (number) active-LOW inputs, ________ (number) active-HIGH inputs, and ________ (number) active-LOW outputs.

6-7. List the condition (HIGH or LOW) of each of the ripple blanking lines *A* to *E* in Fig. 6-33.

6-8. Refer to Fig. 6-34. List the three functions of the 74HC4543 CMOS IC.

6-9. What are some reasons a designer might select a VF display for an automotive application?

6-10. At the option of your instructor, use circuit simulation software to (1) draw the logic circuit sketched in Fig. 6-36, (2) generate a truth table for the logic circuit, and (3) determine if it is a Gray-to-binary decoder or a binary-to-Gray code decoder.

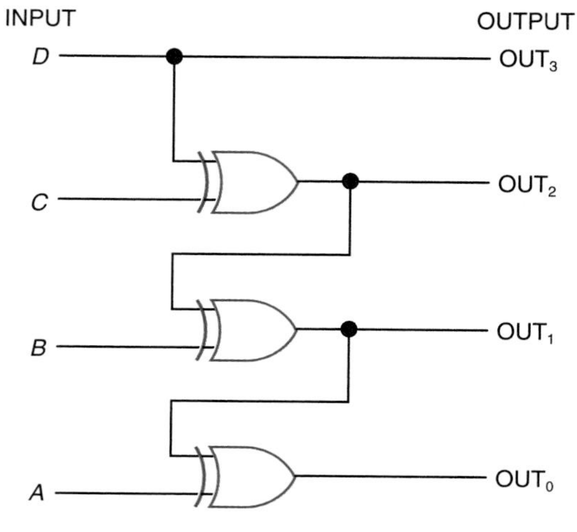

Fig. 6-36 Logic circuit.

Answers to Tests

1. 11101
2. 0010 1001
3. 8765
4. 0100 1011
5. 60
6. is not
7. Only one digit changes as you count in the Gray code.
8. Frank Gray
9. optical encoding
10. alphanumeric
11. American Standard Code for Information Interchange
12. 101 0010
13. $
14. LOW, LOW
15. output D = HIGH
    output C = LOW
    output B = LOW
    output A = LOW
16. The invert bubble means that input 4 is an active LOW input; it is activated by a logical 0.
17. output D = LOW
    output C = HIGH
    output B = HIGH
    output A = HIGH
18. 5
19. vacuum fluorescent
20. light-emitting diode
21. *b*, *c*, LED, 1
22. cathode
23. all segments, 8
24. 1. BCD-to-seven-segment
    2. 8421-BCD-to-decimal
    3. Excess-3-to-decimal
    4. Gray-code-to-decimal
    5. BCD-to-binary
    6. Binary-to-BCD
25. HIGH, LOW
26. LOW
27. leading zeros
28. pulse t_1 = 0
    pulse t_2 = blank display (not a BCD number)
    pulse t_3 = 2
    pulse t_4 = 8
    pulse t_5 = 5
    pulse t_6 = 3
    pulse t_7 = 9
29. pulse t_1 = *a*, *b*, *c*, *d*, *e*, *f*
    pulse t_2 = blank display
    pulse t_3 = *a*, *b*, *d*, *e*, *g*
    pulse t_4 = *a*, *b*, *c*, *d*, *e*, *f*, *g*
    pulse t_5 = *a*, *c*, *d*, *f*, *g*
    pulse t_6 = *a*, *b*, *c*, *d*, *g*
    pulse t_7 = *a*, *b*, *c*, *f*, *g*
30. black, silver
31. nematic
32. dc
33. very small amount
34. BCD, seven-segment
35. T
36. pulse t_1 = 4
    pulse t_2 = 2
    pulse t_3 = 6
    pulse t_4 = 9
    pulse t_5 = 1
37. *b* and *c*
38. XOR
39. 4-bit latch
40. out-of-phase signals between inputs and common (backplane)
41. blue-green (also blue or green)
42. none
43. part A = grid
    part B = cathode (heaters)
    part C = plates
44. *a*, *c*, *d*, *f*, *g*; *5*
45. CMOS, VF
46. limit current through cathodes to a safe level
47. pulse t_1 = 3
    pulse t_2 = 8
    pulse t_3 = 7
    pulse t_4 = 0
48. segment *a*–*f* = +12 V, segment *g* = GND
49. Use your senses to locate open or short circuits or ICs that are too hot.
50. short

Flip-Flops

Chapter Objectives

This chapter will help you to:

1. *Memorize* the block diagram and explain the function of each input and output on several types of flip-flops.
2. *Use* truth tables to determine the mode of operation and outputs of a flip-flop.
3. *Interpret* flip-flop waveform diagrams to determine the mode of operation, outputs, and the type of triggering.
4. *Discuss* the organization and use of a 4-bit latch and *predict* the operation of the IC.
5. *Classify* flip-flops as synchronous or asynchronous and *compare* the triggering of the synchronous units.
6. *Describe* the operation of Schmitt trigger devices and *cite* their applications.
7. *Compare* traditional with newer IEEE/ANSI flip-flop symbols.

Combination logic circuits

Sequential logic circuits

Counters

Shift registers

Memory devices

Engineers classify logic circuits into two groups. We have already worked with *combination logic circuits* using AND, OR, and NOT gates. The other group of circuits is classified as *sequential logic circuits.* Sequential circuits involve timing and memory devices. The basic building block for combinational logic circuits is the logic gate. The basic building block for sequential logic circuits is the *flip-flop (FF).* This chapter covers several types of flip-flop circuits. In later chapters you will wire flip-flops together. Flip-flops are wired to form *counters, shift registers,* and various *memory devices.*

7-1 The R-S Flip-Flop

R-S flip-flop

Complementary

Set and reset inputs

Set condition

Reset condition

Hold condition

The logic symbol for the *R-S flip-flop* is drawn in Fig. 7-1. Notice that the R-S flip-flop has two inputs, labeled S and R. The two outputs are labeled Q and $\overline{Q}$ (say "not Q" or "Q not"). In flip-flops the outputs are always opposite, or *complementary.* In other words, if output $Q = 1$, then output $\overline{Q} = 0$, and so on. The letters "S" and "R" at the inputs of the R-S flip-flop are often referred to as the *set* and *reset* inputs.

Set — S — FF — Q — Normal
INPUTS / OUTPUTS
Reset — R — $\overline{Q}$ — Complementary

Fig. 7-1 Logic symbol for an R-S flip-flop.

The truth table in Table 7-1 details the operation of the R-S flip-flop. When the S and R inputs are both 0, both outputs go to a logical 1. This is called a *prohibited state* for the flip-flop and is not used. The second line of the truth table shows that when input S is 0 and R is 1, the Q output is set to logical 1. This is called the *set condition.* The third line shows that when input R is 0 and S is 1, output Q is reset (cleared) to 0. This is called the *reset condition.* Line 4 in the truth table shows both inputs (R and S) at 1. This is the idle or at rest condition and leaves Q and $\overline{Q}$ in their previous complementary states. This is called the *hold condition.*

TABLE 7-1 Truth table for R-S flip-flop

Mode of operation	Inputs S	R	Q	$\overline{Q}$	Outputs Effect on output Q
Prohibited	0	0	1	1	Prohibited-Do not use
Set	0	1	1	0	For setting Q to 1
Reset	1	0	0	1	For resetting Q to 0
Hold	1	1	Q	$\overline{Q}$	Depends on previous state

From Table 7-1, it may be observed that it takes a logical 0 to activate the set (set Q to 1). It also takes a logical 0 to activate the reset, or clear (clear Q to 0). Because it takes a logical 0 to enable, or activate, the flip-flop, the logic symbol in Fig. 7-1 has invert bubbles at the R and S inputs. These invert bubbles indicate that the set and reset inputs are activated by a logical 0.

R-S flip-flops can be purchased in an IC package, or they can be wired from logic gates, as shown in Fig. 7-2. The NAND gates in Fig. 7-2 form an R-S flip-flop. This NAND-gate R-S flip-flop operates according to the truth table in Table 7-1.

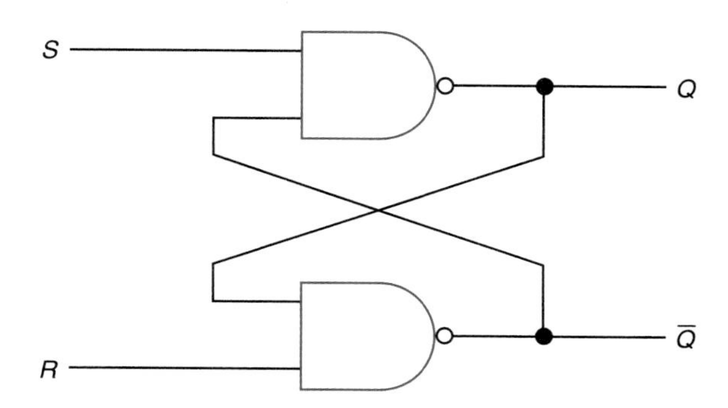

Fig. 7-2 Wiring an R-S flip-flop using NAND gates.

Timing diagrams

Waveform diagram

Many times *timing diagrams*, or *waveforms*, are given for sequential logic circuits. These diagrams show the voltage level and timing between inputs and outputs and are similar to what you would observe on an oscilloscope. The horizontal distance is *time*, and the vertical distance is *voltage*. Figure 7-3 shows the input waveforms (R, S) and the output wave forms (Q, $\overline{Q}$) for the R-S flip-flop. The bottom of the diagram lists the lines of the truth table from Table 7-1. The Q waveform shows the set and reset conditions of the output; the logic levels (0, 1) are on the right side of the waveforms. Waveform diagrams of the type shown in Fig. 7-3 are very common when dealing with sequential logic circuits. Study this diagram to see what it tells you. The waveform diagram is really a type of truth table.

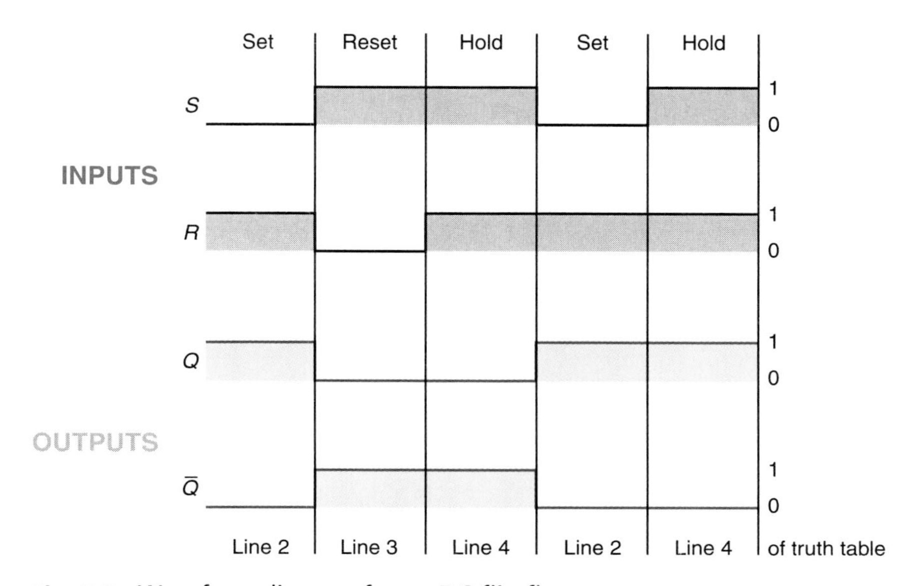

Fig. 7-3 Waveform diagram for an R-S flip-flop.

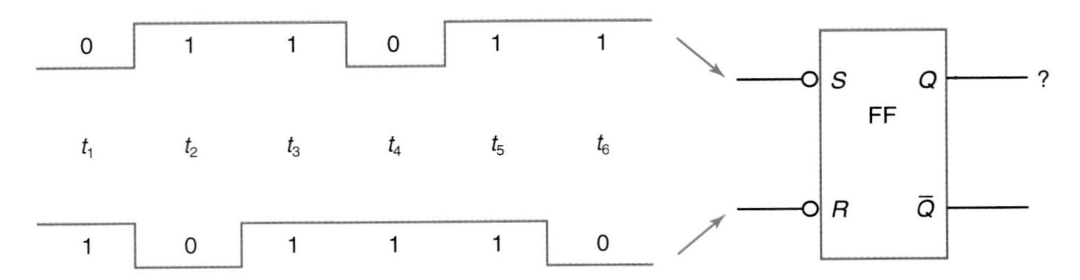

Fig. 7-4 R-S flip-flop pulse-train problem.

R-S latch

Set-reset flip-flop

The R-S flip-flop is also called an *R-S latch* or a *set-reset flip-flop.* Do you know the logic symbol and truth table for the R-S flip-flop? Do you know the four modes of operation for the R-S flip-flop?

TEST

Answer the following questions.

CLK input

1. The R-S flip-flop in Fig. 7-1 has active _________ (HIGH, LOW) inputs.
2. List the mode of operation of the R-S flip-flop for each input pulse shown in Fig. 7-4. Answer with the terms "set" and "reset," "hold," and "prohibited."
3. List the binary output at the normal output (Q) of the R-S flip-flop for each of the pulses shown in Fig. 7-4.

Clocked R-S flip-flop

Synchronous operation

Memory characteristic

7-2 The Clocked R-S Flip-Flop

The logic symbol for a *clocked R-S flip-flop* is shown in Fig. 7-5. Observe that it looks almost like an R-S flip-flop except that it has one ex-

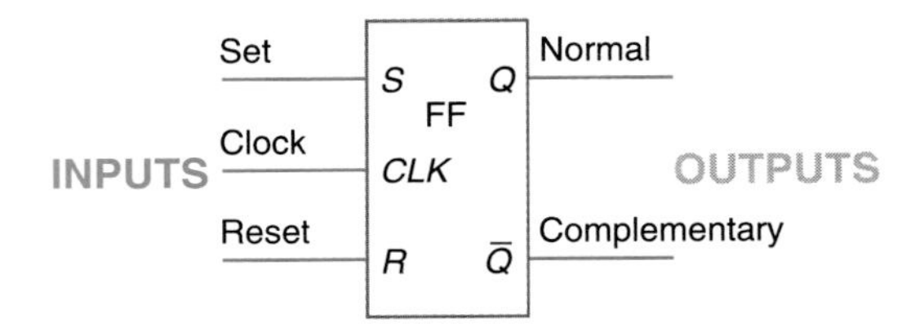

Fig. 7-5 Logic symbol for a clocked R-S flip-flop.

tra input labeled *CLK* (for clock). Figure 7-6 diagrams the operation of the clocked R-S flip-flop. The *CLK* input is at the top of the diagram. Notice that the clock pulse (1) has no effect on output Q with inputs S and R in the 0 position. The flip-flop is in the *idle,* or *hold,* mode during clock pulse 1. At the preset S position, the S (set) input is moved to 1, but output Q is not yet set to 1. The rising edge of clock pulse 2 permits Q to go to 1. Pulses 3 and 4 have no effect on output Q. During pulse 3 the flip-flop is in its set mode, while during pulse 4 it is in its hold mode. Next, input R is preset to 1. On the rising edge of clock pulse 5 the Q output is reset (or cleared) to 0. The flip-flop is in the reset mode during both clock pulses 5 and 6. The flip-flop is in its hold mode during clock pulse 7; therefore, the normal output (Q) remains at 0.

Notice that the outputs of the clocked R-S flip-flop *change only on a clock pulse.* We say that this flip-flop operates *synchronously;* it operates *in step with* the clock. Synchronous operation is very important in calculators and computers, where each step must happen in an exact order.

Another characteristic of the clocked R-S flip-flop is that once it is set or reset it stays that way even if you change some inputs. This is a *memory characteristic,* which is extremely valuable in many digital circuits. This characteristic is evident during the hold mode of operation. In the waveform diagram in Fig. 7-6,

About Electronics

Heart of Glass. *Preforms* are the pieces made to begin the construction of light guides used in fiber optics. In a preform, you can see the concentric rings of glass, each bonding with the next. At the start, they are about ½ inch in diameter. Then the layered core and its surrounding rings are stretched into a fiber no thicker than a hair and many kilometers long.

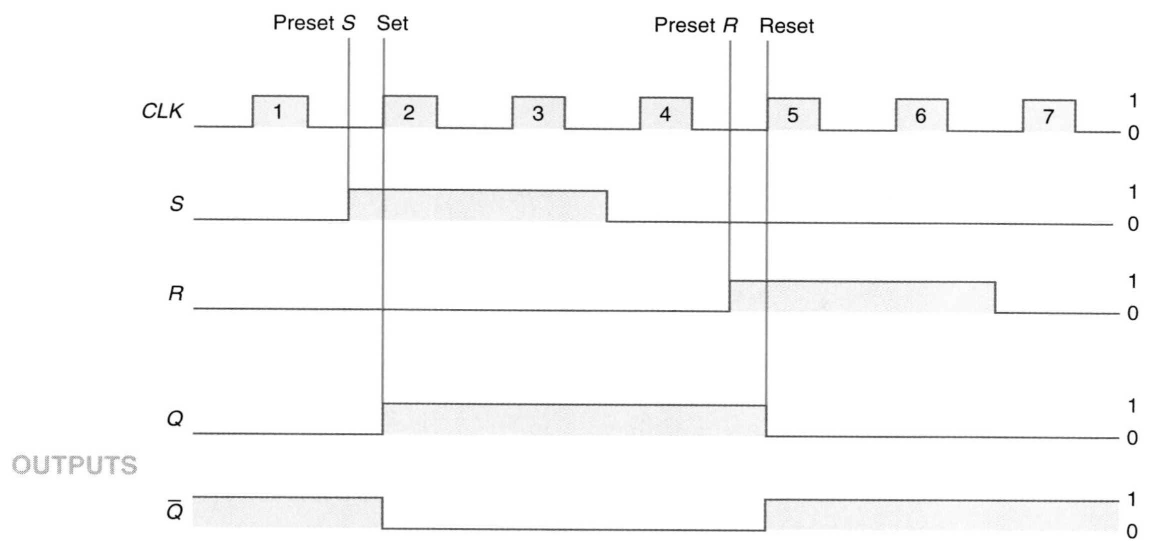

Fig. 7-6 Waveform diagram for a clocked R-S flip-flop.

Waveform diagram for a clocked R-S flip-flop

this flip-flop is in the hold mode during clock pulses 1, 4, and 7.

Figure 7-7(*a*) shows a truth table for the clocked R-S flip-flop. Notice that only the top three lines of the truth table are usable; the bottom line is prohibited and not used. Observe that the *R* and *S* inputs to the clocked R-S flip-flop are active HIGH inputs. That is, it takes a HIGH on input *S* while $R = 0$ to cause output *Q* to be set to 1.

Figure 7-7(*b*) shows a wiring diagram of a clocked R-S flip-flop. Notice that two NAND

Mode of operation	INPUTS			OUTPUTS		
	CLK	*S*	*R*	*Q*	$\overline{Q}$	Effect on output *Q*
Hold	⎍	0	0	No change		No change
Reset	⎍	0	1	0	1	Reset or cleared to 0
Set	⎍	1	0	1	0	Set to 1
Prohibited	⎍	1	1	1	1	Prohibited—do not use

(*a*)

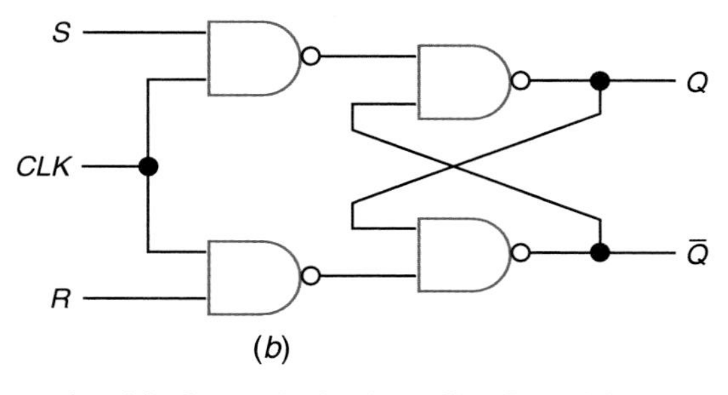

(*b*)

Fig. 7-7 (*a*) Truth table for a clocked R-S flip-flop. (*b*) Wiring a clocked R-S flip-flop using NAND gates.

Truth table for a clocked R-S flip-flop

Wiring a clocked R-S flip-flop using NAND gates

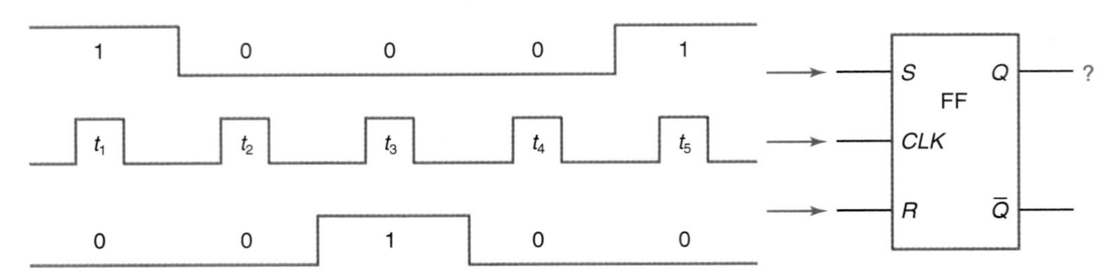

Fig. 7-8 Clocked R-S flip-flop pulse-train problem.

gates have been added to the inputs of the R-S flip-flop to add the clocked feature.

JOB TIP

All of the writing you do must be legible, accurate, clear, and concise.

It is important to remember that the memory characteristics exhibited by flip-flops is one of the fundamental reasons why digital technology has become so widely used in modern electronics products. It is strongly suggested that you actually experiment with R-S and clocked R-S flip-flops either on a circuit simulator or with actual ICs on a solderless breadboard. Operating flip-flops in the lab will help you better understand their operation.

TEST

Answer the following questions.

4. The set and reset inputs (*S*, *R*) of the clocked R-S flip-flop in Fig. 7-5 are active _________ (HIGH, LOW) inputs.
5. List the mode of operation of the clocked R-S flip-flop for each input pulse shown in Fig. 7-8. Answer with the terms "set," "reset," "hold," and "prohibited."
6. List the binary output at the normal output (*Q*) of the clocked R-S flip-flop for each of the pulses shown in Fig. 7-8.

7-3 The D Flip-Flop

D flip-flop

Delay flip-flop

The logic symbol for the *D flip-flop* is shown in Fig. 7-9(*a*). It has only one *data input* (*D*) and a clock input (*CLK*). The outputs are labeled *Q* and $\overline{Q}$. The D flip-flop is often called a *delay flip-flop.* The word "delay" describes what happens to the data, or information, at input *D*. The data (a 0 or 1) at input *D* is *delayed one clock pulse* from getting to output *Q*. A simplified truth table for the D flip-flop is shown in Fig. 7-9(*b*). Notice that output *Q* follows input *D* *after one clock pulse* (see Q^{n+1} column).

A D flip-flop may be formed from a clocked R-S flip-flop by adding an inverter, as shown in Fig. 7-10. More commonly you will use a D flip-flop contained in an IC. Figure 7-11(a) shows a typical commercial D flip-flop. Two extra inputs (PS (preset) and CLR (clear)) have been added to the D flip-flop in Fig. 7-11(*a*). The *PS* input sets output *Q* to 1 when enabled by a logical 0. The *CLR* input clears output *Q* to 0 when enabled by a logical 0. The PS and *CLR* inputs will override the *D* and *CLK* inputs. The *D* and *CLK* inputs operate as they did in the D flip-flops in Fig. 7-9.

7474 TTL D flip-flop

A more detailed truth table for the commercial *7474 TTL D flip-flop* is shown in Fig.

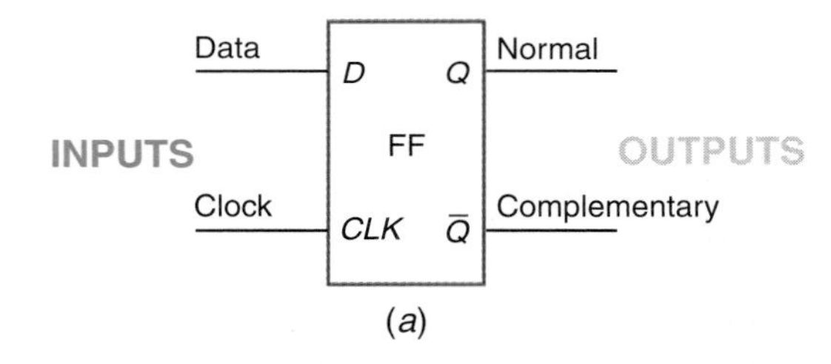

(*a*)

Input	Output
D	Q^{n+1}
0	0
1	1

(*b*)

Fig. 7-9 D flip-flop. (*a*) Logic symbol. (*b*) Simplified truth table.

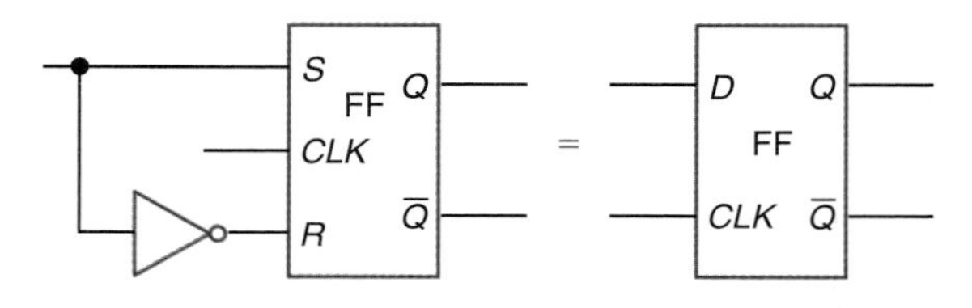

Fig. 7-10 Wiring a D flip-flop.

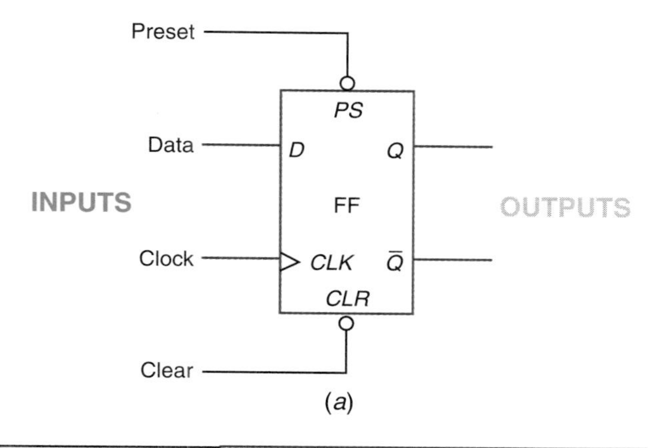

(a)

Mode of operation	INPUTS				OUTPUTS	
	Asynchronous		Synchronous			
	PS	CLR	CLK	D	Q	$\bar{Q}$
Asynchronous set	0	1	X	X	1	0
Asynchronous reset	1	0	X	X	0	1
Prohibited	0	0	X	X	1	1
Set	1	1	↑	1	1	0
Reset	1	1	↑	0	0	1

0 = LOW
1 = HIGH
X = Irrelevant
↑ = LOW-to-HIGH transition of clock pulse

(b)

Fig. 7-11 (a) Logic symbol for commercial D flip-flop.
(b) Truth table for 7474 D flip-flop.

7-11(*b*). Remember that the asynchronous (not synchronous) inputs (*PS* and *CLR*) override the synchronous inputs. The asynchronous inputs are in control of the D flip-flop in the first three lines of the truth table in Fig. 7-11(*b*). The synchronous inputs (*D* and *CLK*) are irrelevant as shown by the "X"s on the truth table. The prohibited condition, line 3 on the truth table, should be avoided. With both asynchronous inputs disabled ($PS = 1$ and $CLR = 1$), the D flip-flop can be set and reset using the *D* and *CLK* inputs. The last two lines of truth table use a clock pulse to transfer data from input *D* to output *Q* of the flip-flop. Being in step with the clock, this is called *synchronous operation.* Note that this flip-flop uses the LOW-to-HIGH transition of the clock pulse to transfer data from input *D* to output *Q*.

D flip-flops are sequential logic devices which are widely used temporary memory devices. D flip-flops are wired together to form *shift registers* and *storage registers.* These registers are commonly used in digital systems. Remember that the D flip-flop *delays* data from reaching output *Q* one clock pulse and is called a *delay flip-flop.* D flip-flops are sometimes also called *data flip-flops* or *D-type latches.* D flip-flops are available in both TTL and CMOS IC form. A few typical CMOS D flip-flops might be the 74HC74, 74AC74, 74FCT374, 74HC273, 74AC273, 4013, and 40174. D flip-flops are so popular with designers that more than 50 different ICs are available in just the FACT CMOS logic series.

Shift registers

Storage registers

TEST

Answer the following questions.

7. List the mode of operation of the 7474 D flip-flop for each input pulse shown in Fig. 7-12 on page 160. Answer with the terms "asynchronous set," "asynchronous reset," "prohibited," "set," and "reset."

Synchronous operation

7474 D flip-flop

8. List the binary output at the normal output (Q) of the D flip-flop for each of the pulses shown in Fig. 7-12.

7-4 The J-K Flip-Flop

J-K flip-flop

Universal flip-flop

Toggling

7476 TTL J-K flip-flop

The *J-K flip-flop* is considered the *universal flip-flop,* having the features of all the other types of flip-flops. The logic symbol for the J-K flip-flop is illustrated in Fig. 7-13(*a*). The inputs labeled J and K are the data inputs. The input labeled CLK is the clock input. Outputs Q and $\overline{Q}$ are the usual normal and complementary outputs on a flip-flop. A truth table for the J-K flip-flop is shown in Fig. 7-13(*b*). When the J and K inputs are both 0, the flip-flop is in the *hold* mode. In the hold mode the data inputs have no effect on the outputs. The outputs "hold" the last data present.

Lines 2 and 3 of the truth table show the reset and set conditions for the Q output. Line 4 illustrates the useful *toggle* position of the J-K flip-flop. When both data inputs J and K are at 1, repeated clock pulses cause the output to turn off-on-off-on-off-on, and so on. This off-on action is like a toggle switch and is called *toggling.*

The logic symbol for the commercial *7476*

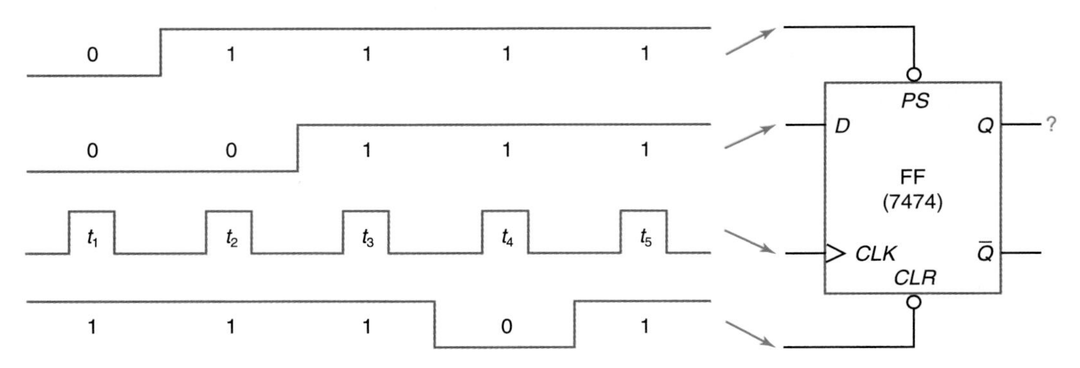

Fig. 7-12 D flip-flop problem for test items 7 and 8.

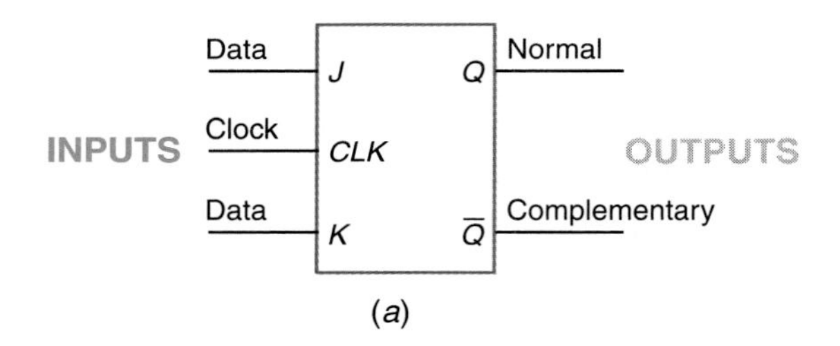

(*a*)

Mode of operation	INPUTS			OUTPUTS		
	CLK	*J*	*K*	*Q*	$\overline{Q}$	Effect on output *Q*
Hold		0	0	No change		No change—disable
Reset		0	1	0	1	Reset or cleared to 0
Set		1	0	1	0	Set to 1
Toggle		1	1	Toggle		Changes to opposite state

(*b*)

Fig. 7-13 J-K flip-flop. (*a*) Logic symbol. (*b*) Truth table.

TTL J-K flip-flop is shown in Fig. 7-14(*a*). Added to the symbol are two asynchronous inputs (preset and clear). The synchronous inputs are the *J* and *K* data and clock inputs. The customary normal (*Q*) and complementary ($\overline{Q}$) outputs are also shown. A detailed truth table for the commercial 7476 J-K flip-flop is drawn in Fig. 7-14(*b*). Recall that asynchronous inputs (such as *PS* and *CLR*) override synchronous inputs. The asynchronous inputs are activated in the first three lines of the truth table. The synchronous inputs are irrelevant (overridden) in the first three lines in Fig. 7-14(*b*); therefore, an "X" is placed under the *J*, *K*, and *CLK* inputs for these rows. The prohibited state occurs when both asynchronous inputs are activated at the same time. The prohibited state is not useful and should be avoided.

When both asynchronous inputs (*PS* and *CLR*) are disabled with a 1, the synchronous inputs can be activated. The bottom four lines of the truth table in Fig. 7-14(*b*) detail the *hold, reset, set,* and *toggle* modes of operation for the 7476 J-K flip-flop. Note that the 7476 J-K flip-flop uses the entire pulse to transfer data from the *J* and *K* data inputs to the *Q* and $\overline{Q}$ outputs.

7476 J-K flip-flop

JOB TIP

Read your employee handbook and all other materials supplied by your employer.

J-K flip-flops are widely used in many digital circuits. You will use the J-K flip-flop especially in *counters.* Counters are found in almost every digital system.

Counters

In summary, the J-K flip-flop is considered the "universal" flip-flop. Its unique feature is the toggle mode of operation so useful in designing counters. When the J-K flip-flop is wired for use only in the toggle mode, it is commonly called a *T flip-flop.* J-K flip-flops are available in both TTL and CMOS IC form. Typical CMOS J-K flip-flops are the 74HC76, 74AC109, and 4027 ICs.

T flip-flop

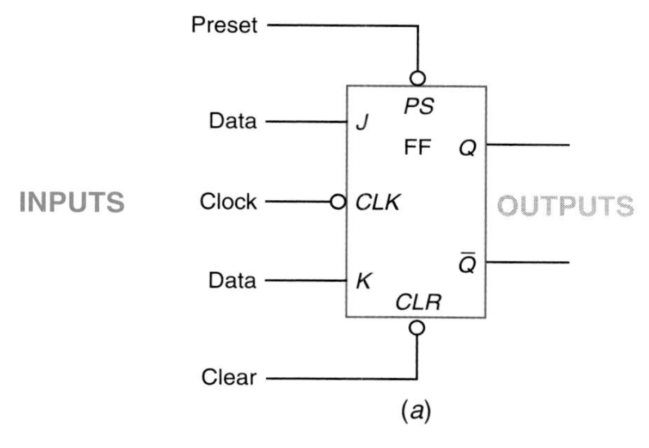

(*a*)

Mode of operation	INPUTS					OUTPUTS	
	Asynchronous		Synchronous				
	PS	*CLR*	*CLK*	*J*	*K*	*Q*	$\overline{Q}$
Asynchronous set	0	1	X	X	X	1	0
Asynchronous reset	1	0	X	X	X	0	1
Prohibited	0	0	X	X	X	1	1
Hold	1	1	⎍	0	0	No change	
Reset	1	1	⎍	0	1	0	1
Set	1	1	⎍	1	0	1	0
Toggle	1	1	⎍	1	1	Opposite state	

0 = LOW
1 = HIGH
X = Irrelevant
⎍ = Positive clock pulse

(*b*)

Fig. 7-14 (*a*) Logic symbol for commercial J-K flip-flop. (*b*) Truth table for 7476 J-K flip-flop.

Memory device

Latch

D flip-flop

7475 TTL 4-bit transparent latch

Electronic encoder/decoder system

TEST

Answer the following questions.

9. List the mode of operation of the 7476 J-K flip-flop for each input pulse shown in Fig. 7-15. Answer with the terms "asynchronous set," "asynchronous reset," "prohibited," "hold," "reset," "set," and "toggle."
10. List the binary output at the normal output (Q) of the J-K flip-flop after each of the pulses shown in Fig. 7-15.

7-5 IC Latches

Consider the block diagram of the digital system in Fig. 7-16(*a*). Press and hold the decimal number 7 on the keyboard. A 7 will be observed on the seven-segment display. Release the 7 on the keyboard and the 7 disappears from the display. It is obvious that a *memory device* is needed to hold the BCD code for 7 at the inputs to the decoder. A device that serves as a temporary buffer memory is called a *latch.* A 4-bit latch has been added to the system in Fig. 7-16(*b*). Now when the decimal number 7 on the keyboard is *pressed and released,* the seven-segment display continues to show a 7.

The term "latch" refers to a digital storage device. The D flip-flop is a good example of a device used to latch data. However, other types of flip-flops are also used for the latching function.

Manufacturers have developed many latches in IC form. The logic diagram for the *7475*

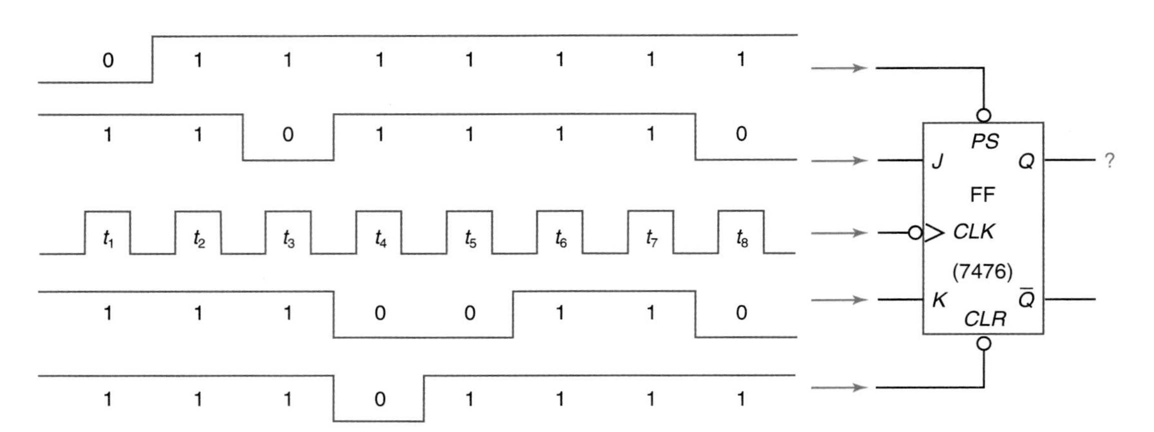

Fig. 7-15 J-K flip-flop problem for test items 9 and 10.

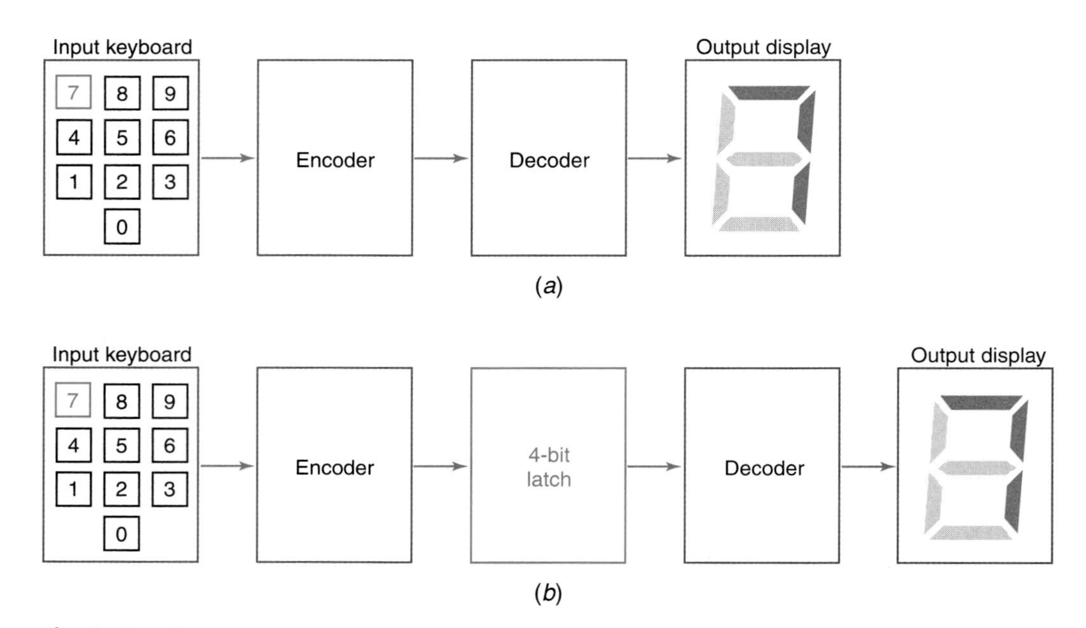

Fig. 7-16 Electronic encoder/decoder system. (*a*) Without buffer memory. (*b*) With buffer memory (latch) added.

Did You Know?

On-Board Car Electronics. This device can: make hotel reservations; get you into your locked-out car; give you sports scores, video directions, and traffic reports; and notify emergency service if you are having car trouble. Alpine Electronics has developed this system, which uses global positioning systems, to help drivers navigate. Using a dashboard CD-ROM computer, the system gives directions by map and voice, monitors the security system, and programs music. You can use this system to play music, find the shortest route, find the route using the fewest toll roads, or even find the nearest restroom or ATM.

TTL 4-bit transparent latch is shown in Fig. 7-17(*a*). This unit has four D flip-flops enclosed in a single IC package. The D_0 data input and the normal Q_0 and complementary $\overline{Q}_0$ outputs form the first D flip-flop. The enable input (E_{0-1}) is similar to the clock input on the D flip-flop. When E_{0-1} is enabled, both D_0 and D_1 are transferred to their outputs.

A simplified truth table for the 7475 latch IC is shown in Fig. 7-17(*b*). If the enable input is at a logical 1, data is transferred, without a separate clock pulse, from the *D* input to the *Q*

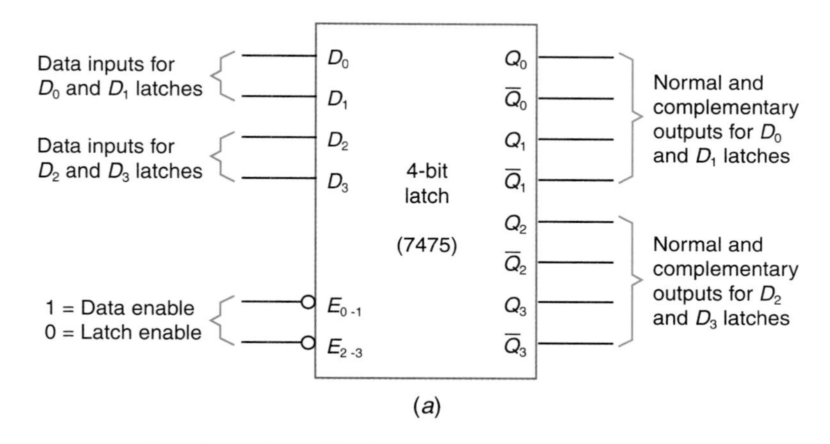

(*a*)

Mode of operation	INPUTS		OUTPUTS	
	E	*D*	*Q*	$\overline{Q}$
Data enabled	1	0	0	1
	1	1	1	0
Data latched	0	X	No change	

0 = LOW
1 = HIGH
X = Irrelevant

(*b*)

Fig. 7-17 (*a*) Logic symbol for commercial 7475 4-bit transparent latch. (*b*) Truth table for 7475 D latch.

About Electronics

Holographic Memory. A holographic data storage system is made by using laser beams to etch crystals containing holograms. Through this method, thousands of pages of material can be contained in a space the size of a dime. This information can be retrieved ten times faster than with other storage methods.

12. A _________ (HIGH, LOW) at the enable inputs places the 7475 latch IC in the data-latched mode of operation.
13. In the data-latched mode, a change at any of the *D* inputs to the 7475 latch IC has _________ (an immediate effect on their respective outputs; no effect on the outputs).
14. When a flip-flop is used to temporarily hold data, it is sometimes called _________.

Triggering flip-flops

Data-enabled mode

Synchronous flip-flops

Data-latched mode

7475 4-bit transparent latch

Positive-edge-triggered flip-flop

Counters

Shift registers

Delay unit

Frequency dividers

Negative-edge-triggered flip-flop

Flip-flop triggering

and $\overline{Q}$ outputs. As an example, if $E_{0-1} = 1$ and $D_1 = 1$, then without a clock pulse output Q_1 would be set to 1 while $\overline{Q}_1$ would be reset to 0. In the *data-enabled mode* of operation the *Q* outputs follow their respective *D* inputs on the 7475 latch.

Consider the last line of the truth table in Fig. 7-17(*b*). When the enable input drops to 0, the 7475 IC enters the *data-latched mode.* The data that was at *Q* remains the same even if the *D* inputs change. The data is said to be latched. The 7475 IC is called a *transparent* latch because when the enable input is HIGH, the normal outputs follow the data at the *D* inputs. Note that the D_0 and D_1 flip-flops in the 7475 IC are controlled by the E_{0-1} enable input whereas the E_{2-3} input controls the D_2 and D_3 pair of flip-flops.

One use of a flip-flop is to hold, or latch, data. When used for this purpose, the flip-flop is called a latch. Flip-flops have many other uses, including *counters, shift registers, delay units,* and *frequency dividers.*

Latches are available in all logic families. Several typical CMOS latches are the 4042, 4099, 74HC75, and 74HC373 ICs. Latches are sometimes built into other ICs such as the 4511 and 4543 BCD-to-seven segment latch/decoder/driver chips illustrated in Chap. 6.

One of the primary advantages of digital over analog circuitry is the availability of easy-to-use memory devices. The latch is the most fundamental memory device used in digital electronics. Almost all digital equipment contains simple memory devices called latches.

TEST

Supply the missing word in each statement.

11. When the 7475 latch IC is in its data-enabled mode of operation, the _________ outputs follow their respective *D* inputs.

7-6 Triggering Flip-Flops

We have classified flip-flops as synchronous or asynchronous in their operation. *Synchronous flip-flops* are all those that have a clock input. We found that the clocked R-S, the D, and the J-K flip-flop operate in step with the clock.

When using manufacturers' data manuals you will notice that many synchronous flip-flops are also classified as either *edge-triggered* or *master/slave.* Figure 7-18 on the next page shows two edge-triggered flip-flops in the toggle position. On clock pulse 1 the positive edge (positive-going edge) of the pulse is identified. The second waveform shows how the *positive-edge-triggered flip-flop* toggles each time a positive-going pulse comes along (see pulses 1 to 4). On pulse 1 in Fig. 7-18 the negative edge (negative-going edge) of the pulse is also labeled. The bottom waveform shows how the *negative-edge-triggered flip-flop* toggles. Notice that it changes state, or toggles, each time a negative-going pulse comes along (see pulses 1 to 4). Especially notice the difference in timing between the positive- and negative-edge-triggered flip-flops. This triggering time difference is quite important for some applications.

It is common to show the type of triggering on the flip-flop. The logic symbol for a D flip-flop with positive-edge triggering is shown in Fig. 7-19(*a*). Note the use of the small > inside the flip-flop near the clock input. This > symbol says data is transferred to the output on the edge of the pulse. A logic symbol for a D flip-flop using negative-edge triggering is shown in Fig. 7-19(*b*). The added invert bubble at the clock input shows that triggering occurs on the negative-going edge of the clock pulse. Finally, a typical D latch symbol is

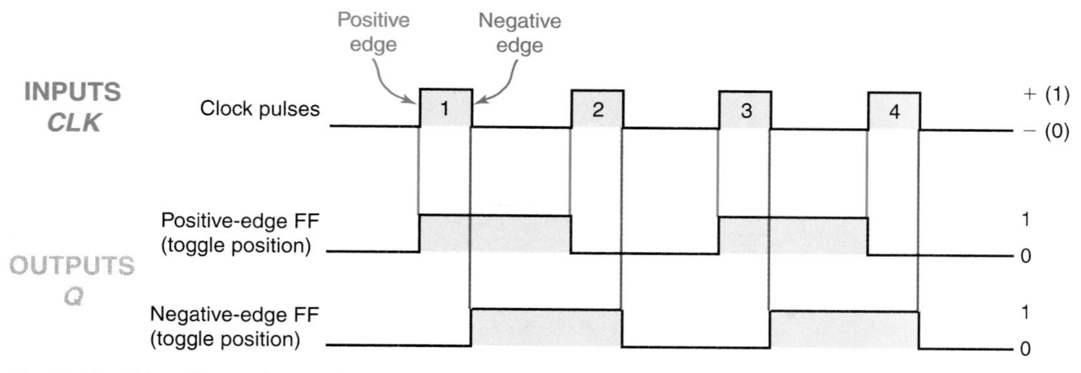

Fig. 7-18 Waveforms for positive- and negative-edge-triggered flip-flops.

shown in Fig. 7-19(*c*). Note the lack of a > symbol next to the enable (similar to a clock) input. This means that this unit is not considered an edge-triggered unit. Like the R-S flip-flop, the D latch is considered asynchronous. Recall that the D latch normal output (Q) follows its input (D) when the enable (E) input is HIGH. The data is latched when the enable input drops to LOW. Several manufacturers label the enable input with a "G" on the D latch.

Another class of flip-flop triggering is the master/slave type. The *J-K master/slave flip-flop* uses the entire pulse (positive edge and negative edge) to trigger the flip-flop. Figure 7-20 shows the triggering of a master/slave flip-flop. Pulse 1 shows four positions (*a* to *d*) on the waveform. The following sequences of operation takes place in the master/slave flip-flop at each point on the clock pulse:

Read the operator's manuals for the test equipment used on the job.

- Point *a*: leading edge—isolate input from output
- Point *b*: leading edge—enter information from J and K inputs
- Point *c*: trailing edge—disable J and K inputs
- Point *d*: trailing edge—transfer information from input to output

J-K master/slave flip-flop

A very interesting characteristic of the master/slave flip-flop is shown on pulse 2, Fig. 7-20. Notice that at the beginning of pulse 2 the outputs are disabled. For a very brief moment the J and K inputs are moved to the toggle positions (see point *e*) and then disabled. The J-K master/slave flip-flop "remembers" that the J and K inputs were in toggle positions, and it toggles at point *f* on the waveform diagram. This memory characteristic happens only while the clock pulse is high (at logical 1).

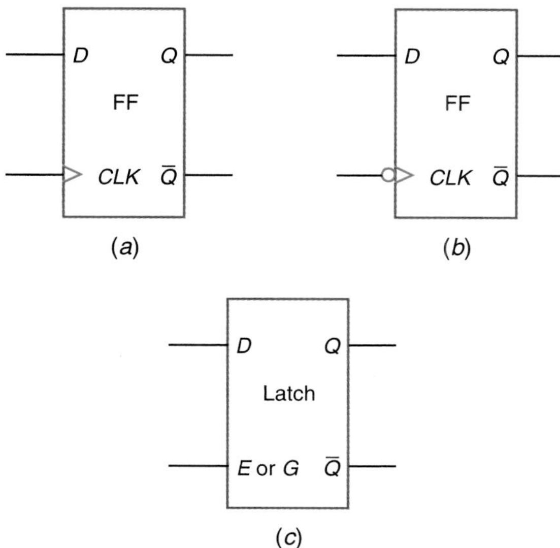

Fig. 7-19 (*a*) Logic symbol for positive-edge-triggered D flip-flop. (*b*) Logic symbol for negative-edge-triggered D flip-flop. (*c*) Logic symbol for D latch.

TEST

Supply the missing word in each statement.

15. A positive-edge-triggered flip-flop changes state on the _________ transition of the clock pulse.
16. A negative-edge-triggered flip-flop changes state on the _________ transition of the clock pulse.
17. The ">" near the clock input inside a flip-flop logic symbol means _________.
18. A(n) _________ J-K flip-flop uses both the positive and the negative edge of the clock pulse for data transfer.

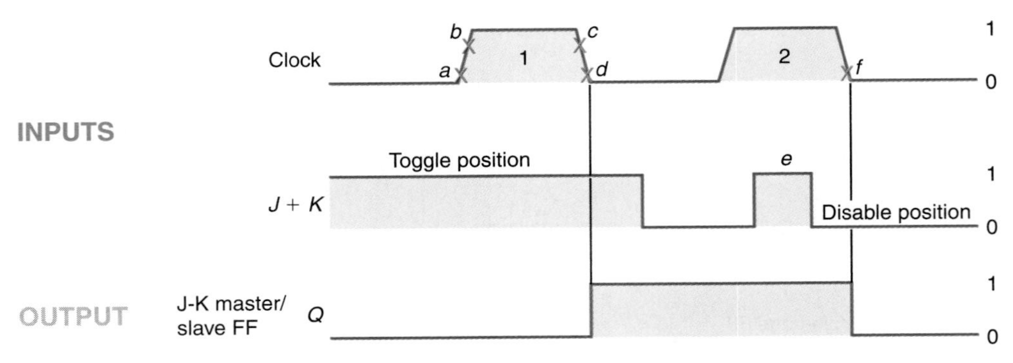

Fig. 7-20 Triggering the J-K master/slave flip-flop.

Triggering the J-K master/slave flip-flop

7-7 Schmitt Trigger

Rise and fall times

7414 Schmitt trigger inverter TTL IC

Switching threshold

Schmitt trigger inverter

Schmitt trigger used for wave shaping

Signal conditioning

Hysteresis

Digital circuits prefer waveforms with fast *rise and fall times.* The waveform on the right side of the inverter symbol in Fig. 7-21 is an example of a good digital signal. The square wave's L-to-H and H-to-L edges are vertical. This means that the rise and fall times are very fast (almost instantaneous).

The waveform to the left of the inverter symbol in Fig. 7-21 has very slow rise and fall times. The poor waveform on the left in Fig. 7-21 might lead to unreliable operation if fed directly into counters, gates, or other digital circuitry. In this example, a *Schmitt trigger inverter* is being used to "square up" the input signal and make it more useful. The Schmitt trigger in Fig. 7-21 is reshaping the waveform. This is called *signal conditioning.* Schmitt triggers are widely used in signal conditioning.

A voltage profile of a typical TTL inverter (7404 IC) is reproduced in Fig. 7-22(*a*). Of special interest is the *switching threshold* of the 7404 IC. The switching threshold may vary from chip to chip, but it is always in the undefined region. Figure 7-22(*a*) shows that a typical 7404 IC has a switching threshold of +1.2 V. In other words, when the voltage rises to +1.2 V, the output changes from HIGH to LOW. However, if the voltage drops below +1.2 V, the output switches from LOW to HIGH. Most regular gates have a single switching threshold voltage whether the input voltage is rising (L to H) or falling (H to L).

A voltage profile for a 7414 Schmitt trigger inverter TTL IC is sketched in Fig. 7-22(*b*). Note that the *switching threshold is different* for positive-going (V +) and negative-going (V −) voltages. The voltage profile for the 7414 IC shows that the switching threshold is 1.7 V for a positive-going (V +) input voltage. However, the switching threshold is 0.9 V for a negative-going (V −) input voltage. The difference between these switching thresholds (1.7V and 0.9V) is called *hysteresis.* Hysteresis provides for excellent noise immunity and helps the Schmitt trigger square up waveforms with slow rise and fall times.

Schmitt triggers are also available in CMOS. These include the 40106, 4093, 74HC14, and 74AC14 ICs.

One of the characteristics of a bistable multivibrator (or flip-flop) is that its outputs are either HIGH or LOW. When changing states (H to L or L to H), they do so rapidly without the outputs being in the undefined region. This "snap action" of the output is also characteristic of Schmitt triggers.

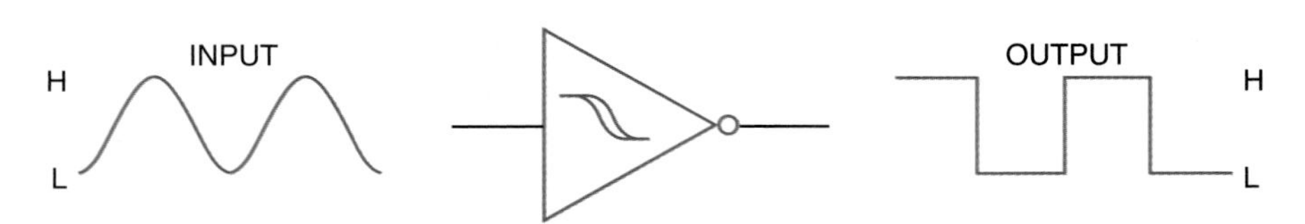

Fig. 7-21 Schmitt trigger used for wave shaping.

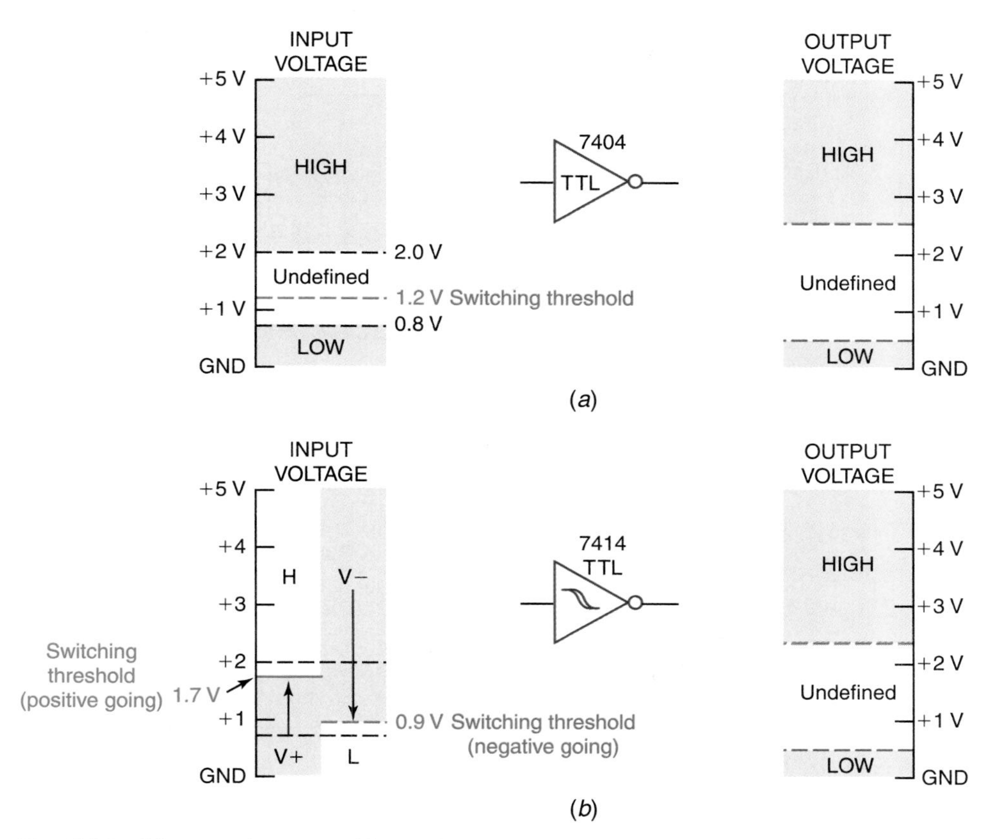

Fig. 7-22 (*a*) TTL voltage profiles with switching threshold. (*b*) Voltage profiles for 7414 TTL Schmitt trigger IC showing switching thresholds.

TEST

Answer the following questions.

19. The ________ is a good device for squaring up a waveform with slow rise and fall times.
20. Draw the schematic symbol for a Schmitt trigger inverter.
21. A Schmitt trigger is said to have ________ because its switching thresholds are different for positive-going and negative-going inputs.
22. Schmitt triggers are commonly used for ________ (memory, signal conditioning).

7-8 IEEE Logic Symbols

The flip-flop logic symbols you have learned are the traditional ones recognized by most workers in the electronics industry. Manufacturers' data manuals usually include traditional symbols and are recently including new IEEE standard logic symbols.

The table in Fig. 7-23 shows the traditional flip-flop and latch logic symbols learned in this chapter along with their IEEE counterparts. All IEEE logic symbols are rectangular and include the number of the IC directly above the symbol. Smaller rectangles show the number of duplicate devices in the package. Notice that all inputs are on the left of the IEEE symbol while outputs are on the right.

The IEEE 7474 D flip-flop symbol shows four inputs labeled "*S*" (for "set"), ">*C*1" (for "positive-edge-triggered clock"), "1*D*" (for "data"), and "*R*" (for "reset"). The triangles at the *S* and *R* inputs on the IEEE 7474 symbol identify them as active-LOW inputs. The 7474 outputs are on the right of the IEEE symbol with no internal identifying markings. The $\overline{Q}$ outputs have triangles suggesting active-LOW outputs. The markings inside the IEEE logic symbols are standard while the outside markings vary from manufacturer to manufacturer.

IEEE logic symbols

Consider the IEEE logic symbol for the 7476 dual master/slave J-K flip-flop in Fig. 7-23.

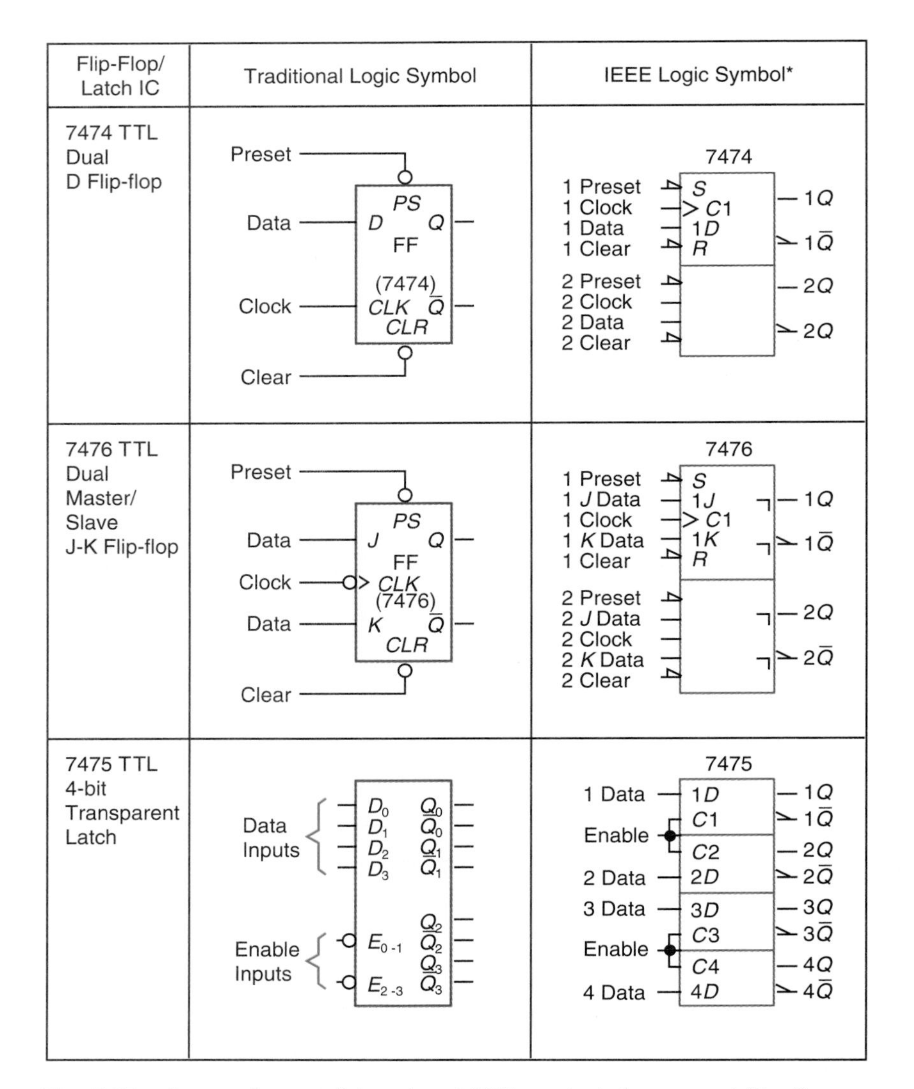

Fig. 7-23 Comparing traditional and IEEE symbols for several flip-flops.

The internal inputs are marked "*S*" ("set"), "1*J*" ("*J* data"), "*C*1" ("Clock"), "1*K*" ("*K* data"), and "*R*" ("reset"). The "7476" above the symbol identifies the specific TTL IC. The ⌉ markings near the Q and $\overline{Q}$ outputs are the unique IEEE symbols for pulse-triggering. The IEEE logic symbol for the 7476 IC shows that there are two active-LOW inputs (*S* and *R*) and a single active-LOW output ($\overline{Q}$) on each J-K flip-flop. The active-LOW inputs and outputs are marked with a small right triangle. The symbol is repeated below to indicate that the 7476 IC package contains two identical J-K flip-flops.

The IEEE standard logic symbol for the 7475 4-bit transparent latch is reproduced in Fig. 7-23. Note the four rectangles to represent the four D-type latches in the 7475 IC package. The four $\overline{Q}$ output leads are marked with small triangles.

TEST

Answer the following questions.

23. The "*C*" marking inside the IEEE symbol stands for the control input or the __________ inputs on the flip-flops.
24. The complementary ($\overline{Q}$) outputs of the flip-flops and latches on an IEEE symbol are designated with the __________ symbol.
25. The asynchronous clear on the 7474 and 7476 flip-flop are active-__________ inputs and are marked with the letter "*R*" which stands for __________.

Chapter 7 Review

Summary

1. Logic circuits are classified as combinational or sequential. Combinational logic circuits use AND, OR, and NOT gates and do not have a memory characteristic. Sequential logic circuits use flip-flops and involve a memory characteristic.
2. Flip-flops are wired together to form counters, registers, and memory devices.
3. Flip-flop outputs are always opposite, or complementary.
4. The table in Fig. 7-24 summarizes some basic flip-flops.
5. Waveform (timing) diagrams are used to describe the operation of sequential devices.
6. Flip-flops can be edge-triggered or master/slave types. Flip-flops can be pulse-or edge-triggered.
7. Special flip-flops called latches are widely used in most digital circuits as temporary buffer memories.
8. Schmitt triggers are special gates that are used for signal conditioning.
9. Figure 7-23 compares traditional flip-flop/latch symbols with the newer IEEE logic symbols.

Circuit	Logic Symbol	Truth Table	Remarks
R-S flip-flop	$\bar{S}$ (inverted input) S, Q FF $\bar{R}$ (inverted input) R, $\bar{Q}$	S R | Q 0 0 | prohibited 0 1 | 1 set 1 0 | 0 reset 1 1 | hold	R-S latch Set-reset flip-flop (asynchronous)
Clocked R-S flip-flop	S, Q FF CLK R, $\bar{Q}$	CLK S R | Q ⎍ 0 0 | hold ⎍ 0 1 | 0 reset ⎍ 1 0 | 1 set ⎍ 1 1 | prohibited	(synchronous)
D flip-flop	D, Q FF CLK, $\bar{Q}$	CLK D | Q ⎍ 0 | 0 ⎍ 1 | 1	Delay flip-flop Data flip-flop (synchronous)
J-K flip-flop	J, Q FF CLK K, $\bar{Q}$	CLK J K | Q ⎍ 0 0 | hold ⎍ 0 1 | 0 ⎍ 1 0 | 1 ⎍ 1 1 | toggle	Most universal FF (synchronous)

Fig. 7-24 Summary of basic flip-flops.

Answer the following questions.

7-1. Logic ________ are the basic building blocks of combinational logic circuits; the basic building blocks of sequential circuits are devices called ________.

7-2. List one type of asynchronous and three types of synchronous flip-flops (mark the synchronous types).

7-3. Draw a traditional logic symbol for the following flip-flops:
a. J-K
b. D
c. Clocked R-S
d. R-S

7-4. Draw a truth table for the following flip-flops:
a. J-K
b. D
c. Clocked R-S
d. R-S

7-5. If both synchronous and the asynchronous inputs on a J-K flip-flop are activated, which input will control the output?

7-6. When we say the flip-flop is in the set condition, we mean output ________ is at a logical ________.

7-7. When we say the flip-flop is in the reset, or clear, condition, we mean output ________ is at a logical ________.

7-8. On a timing, or waveform, diagram the horizontal distance stands for ________ and the vertical distance stands for ________.

7-9. Refer to Fig. 7-6. This waveform diagram is for a(n) ________ flip-flop. This flip-flop is ________-edge triggered.

7-10. List two types of edge-triggered flip-flops.

7-11. The "D" in "flip-flop" stands for ________, or data.

7-12. D flip-flops are widely used as temporary memories called ________.

7-13. If a flip-flop is in its toggle mode of operation, what will the output act like upon repeated clock pulses?

7-14. Identify these acronyms used on traditional flip-flop logic symbols:
a. *CLK*
b. *CLR*
c. *D*
d. FF
e. *PS*
f. *R*
g. *S*

7-15. Give a descriptive name for the following TTL ICs:
a. 7474
b. 7475
c. 7476

7-16. The 7474 IC is a(n) ________-edge-triggered unit.

7-17. List the modes of operation of the 7474 IC.

7-18. List the mode of operation of the 7476 J-K flip-flop for each input pulse shown in Fig. 7-25.

7-19. List the binary outputs at the normal output (Q) of the J-K flip-flop for each time period (t_1-t_7) shown in Fig. 7-25.

7-20. List the mode of operation of the 7475 4-bit latch for each time period (t_1 through t_7) shown in Fig. 7-26.

7-21. List the binary output (4-bit) at the output indicators of the 7475 4-bit latch for each time period (t_1 through t_7) shown in Fig. 7-26.

7-22. Refer to Fig. 7-27. The output waveform on the right of the logic symbol will be a ________ (sine, square) wave.

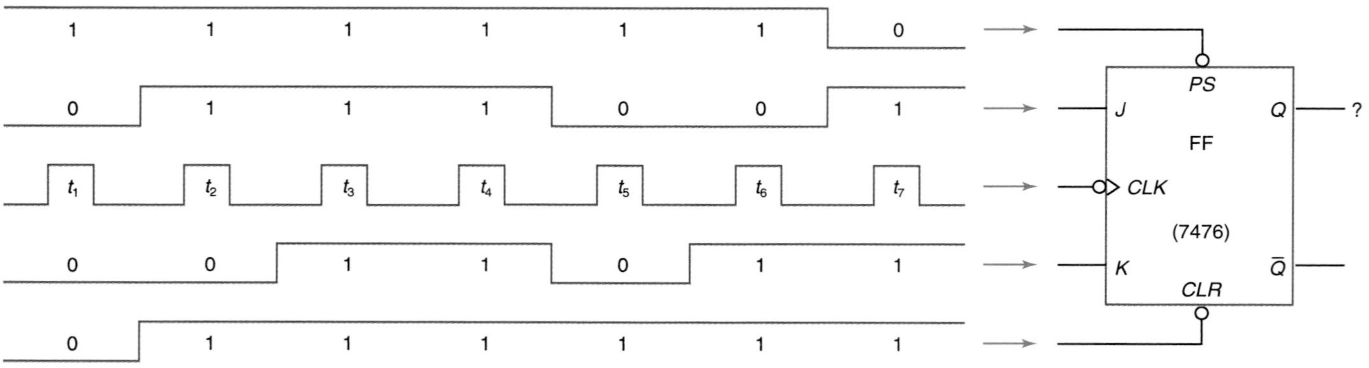

Fig. 7-25 Pulse-train problem.

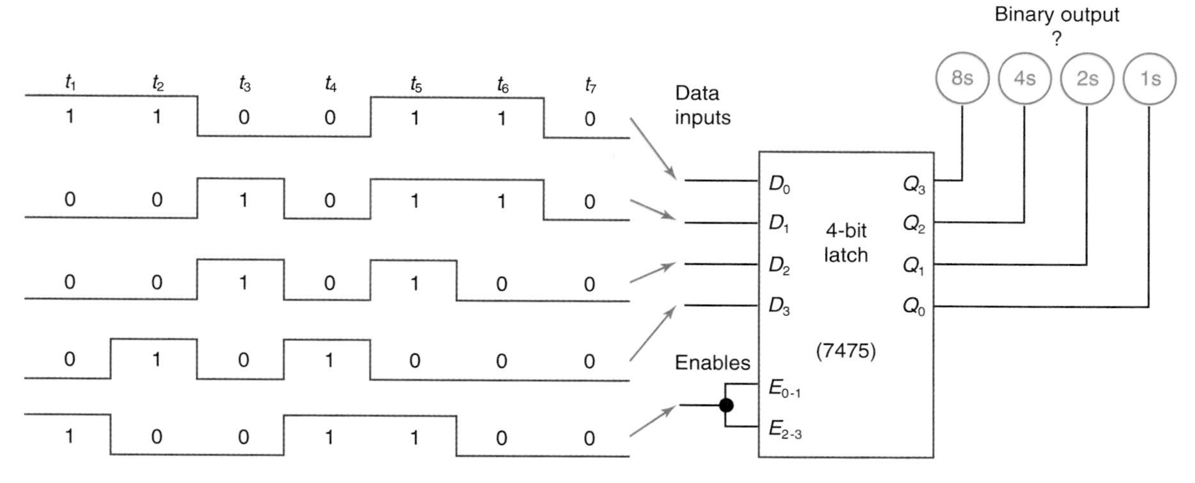

Fig. 7-26 Pulse-train problem.

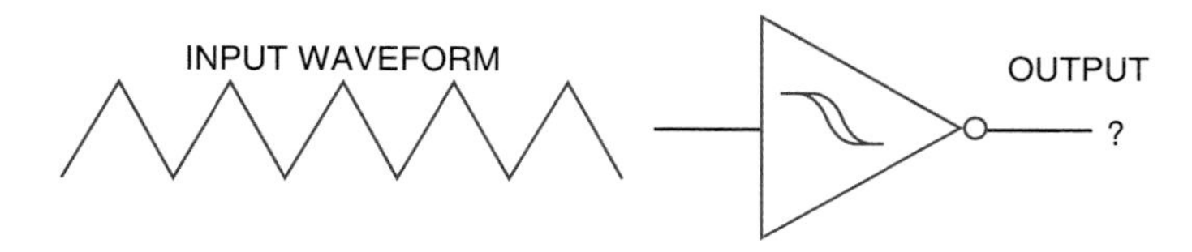

Fig. 7-27 Sample problem.

7-23. The inverter in Fig. 7-27 is being used as a signal ________ (conditioner, multiplexer) in this circuit.

7-24. The logic symbol in Fig. 7-27 is for a symbol ________ (two words) inverter IC.

7-25. Identify these markings found inside and on the leads of the IEEE flip-flop/latch logic symbols.

a. *C*
b. *S*
c. *R*
d. *D*
e. *J*
f. *K*
g. ⅂
h. >*C*

Critical Thinking Questions

7-1. List two other names sometimes given for an R-S flip-flop.

7-2. Explain the difference between asynchronous and synchronous devices.

7-3. Draw the traditional and IEEE logic symbols for a D flip-flop (7474 IC) and a J-K flip-flop (7476 IC).

7-4. Refer to Fig. 7-3. Notice that line 4 is listed two times across the bottom. Why does output $Q = 0$ in the first case and then 1 in the second case when inputs *R* and *S* are both 1 in each case?

7-5. Explain how a J-K master/slave flip-flop (such as the 7476 IC) is triggered.

7-6. Explain how a J-K master/slave flip-flop (such as the 7476 IC) can still toggle even when the device is in the hold mode on the H-to-L transition of the clock pulse.

7-7. What is the fundamental difference between a combinational logic and a sequential logic circuit?

7-8. List several devices that are built using J-K flip-flops.

7-9. Explain why Schmitt trigger devices tend to "square up" inputs with slow rise times.

7-10. At the option of your instructor, use circuit simulation software to (1) draw the flip-flop circuit sketched in Fig. 7-28, (2) test the operation of the flip-flop circuit, (3) make a truth table for the flip-flop (something like Table 7-1) listing the modes of operation as "set," "reset," "hold," and "prohibited," and (4) determine if it acts more like an R-S or a J-K flip-flop.

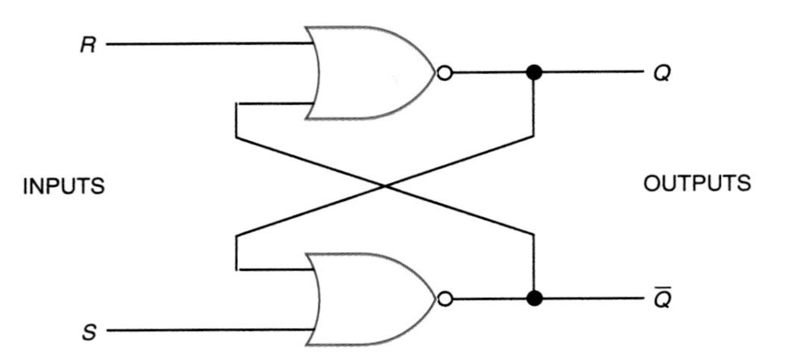

Fig. 7-28 Flip-flop circuit.

Answers to Tests

1. LOW
2. pulse t_1 = set
   pulse t_2 = reset
   pulse t_3 = hold
   pulse t_4 = set
   pulse t_5 = hold
   pulse t_6 = reset
3. pulse $t_1 = 1$
   pulse $t_2 = 0$
   pulse $t_3 = 0$
   pulse $t_4 = 1$
   pulse $t_5 = 1$
   pulse $t_6 = 0$
4. HIGH
5. pulse t_1 = set
   pulse t_2 = hold
   pulse t_3 = reset
   pulse t_4 = hold
   pulse t_5 = set
6. pulse $t_1 = 1$
   pulse $t_2 = 1$
   pulse $t_3 = 0$
   pulse $t_4 = 0$
   pulse $t_5 = 1$
7. pulse t_1 = asynchronous set (or preset)
   pulse t_2 = reset
   pulse t_3 = set
   pulse t_4 = asynchronous reset (or clear)
   pulse t_5 = set
8. pulse $t_1 = 1$
   pulse $t_2 = 0$
   pulse $t_3 = 1$
   pulse $t_4 = 0$
   pulse $t_5 = 1$
9. pulse t_1 = asynchronous set (or preset)
   pulse t_2 = toggle
   pulse t_3 = reset
   pulse t_4 = asynchronous reset (or clear)
   pulse t_5 = set
   pulse t_6 = toggle
   pulse t_7 = toggle
   pulse t_8 = hold
10. pulse $t_1 = 1$
    pulse $t_2 = 0$
    pulse $t_3 = 0$
    pulse $t_4 = 0$
    pulse $t_5 = 1$
    pulse $t_6 = 0$
    pulse $t_7 = 1$
    pulse $t_8 = 1$
11. Q (normal)
12. LOW
13. no effect on the outputs
14. latch
15. LOW-to-HIGH
16. HIGH-to-LOW
17. edge-triggering
18. master/slave
19. Schmitt trigger
20. See figure below

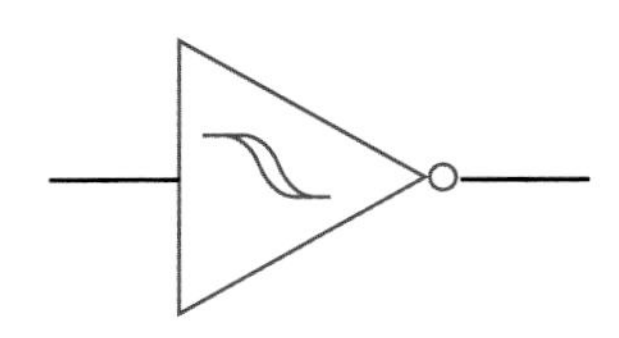

21. hysteresis
22. signal conditioning
23. clock
24. triangle
25. LOW, reset

Counters

Chapter Objectives

This chapter will help you to:

1. *Draw* a circuit diagram of a ripple counter using J-K flip-flops.
2. *Analyze* the circuit action of any mod-3 through mod-8 synchronous counter.
3. *Understand* the operation of and *draw* a block diagram of a frequency divider circuit.
4. *Interpret* data sheets for several commercial TTL and CMOS counter ICs.
5. *Predict* the operation of a 4-bit magnitude comparator IC from its truth table.
6. *Analyze* the operation of an electronic "guess the number" game containing a clock, a counter, and a 4-bit magnitude comparator.
7. *Determine* the output for a variety of counters based on a series of inputs.
8. *Understand* and *explain* the details of a counting system driven by an optical sensor.
9. *List* and *summarize* the use of several pieces of test equipment used in troubleshooting logic circuits.
10. *Troubleshoot* a faulty ripple counter circuit.

Almost any complex digital system contains several *counters.* A counter's job is the obvious one of counting events or periods of time or putting events into sequence. Counters also do some not so obvious jobs: dividing frequency, addressing, and serving as memory units. This chapter discusses several types of counters and their uses. Flip-flops are wired together to form circuits that count. Because of the wide use of counters, manufacturers also make self-contained counters in IC form. Many counters are available in all TTL and CMOS families.

8-1 Ripple Counters

Counting in binary and decimal is illustrated in Fig. 8-1 on page 174. With four binary places (*D, C, B,* and *A*) we can count from 0000 to 1111 (0 to 15 in decimal). Notice that column *A* is the 1s binary place, or least significant digit (LSD). The term "least significant bit" (LSB) is usually used. Column *D* is the 8s binary place, or most significant digit (MSD). The term "most significant bit" (MSB) is usually used. Notice that the 1s column changes state the most often. If we design a counter to count from binary 0000 to 1111, we need a device that has 16 different output states: a *modulo* (mod)-*16 counter.* The *modulus of a counter* is the number of different states the counter must go through to complete its counting cycle.

Modulus of a counter

A mod-16 counter using four J-K flip-flops is diagramed in Fig. 8-2(*a*) on page 174. Each J-K flip-flop is in its toggle position (*J* and *K* both at 1). Assume the outputs are cleared to 0000. As clock pulse 1 arrives at the clock (*CLK*) input of flip-flop 1 (FF 1), it toggles (on the negative edge) and the display shows 0001. Clock pulse 2 causes FF 1 to toggle again, returning output *Q* to 0, which causes FF 2 to toggle to 1. The count on the display now reads 0010. The counting continues, with each

Least significant bit

Most significant bit

Modulo-16 counter

BINARY COUNTING				DECIMAL COUNTING
D	*C*	*B*	*A*	
8s	4s	2s	1s	
0	0	0	0	0
0	0	0	1	1
0	0	1	0	2
0	0	1	1	3
0	1	0	0	4
0	1	0	1	5
0	1	1	0	6
0	1	1	1	7
1	0	0	0	8
1	0	0	1	9
1	0	1	0	10
1	0	1	1	11
1	1	0	0	12
1	1	0	1	13
1	1	1	0	14
1	1	1	1	15

Fig. 8-1 Counting sequence for a 4-bit electronic counter.

Binary counting

flip-flop output triggering the next flip-flop on its negative-going pulse. Look back at Fig. 8-1 and see that column *A* (1s column) must change state on every count. This means that FF 1 in Fig. 8-2(*a*) must toggle for each pulse. FF 2 must toggle only half as often as FF 1, as seen from column *B* in Fig. 8-1. Each more significant bit in Fig. 8-1 toggles less often.

The counting of the mod-16 counter is shown up to a count of decimal 10 (binary 1010) by waveforms in Fig. 8-2(*b*). The *CLK* input is shown on the top line. The state of each flip-flop (FF 1, FF 2, FF 3, FF 4) is shown on the waveforms below. The binary count is shown across the bottom of the diagram. Especially note the vertical lines on Fig. 8-2(*b*); these lines show that the clock triggers only FF 1. FF 1 triggers FF 2, FF 2 triggers FF 3, and so on. Because one flip-flop affects the next one, it takes some time to toggle all the flip-flops. For instance, at point *a* on pulse 8, Fig. 8-2(*b*), notice that the clock triggers FF 1, causing it to go to 0. This in turn causes FF 2 to toggle from 1 to 0. This in turn causes FF 3 to toggle from 1

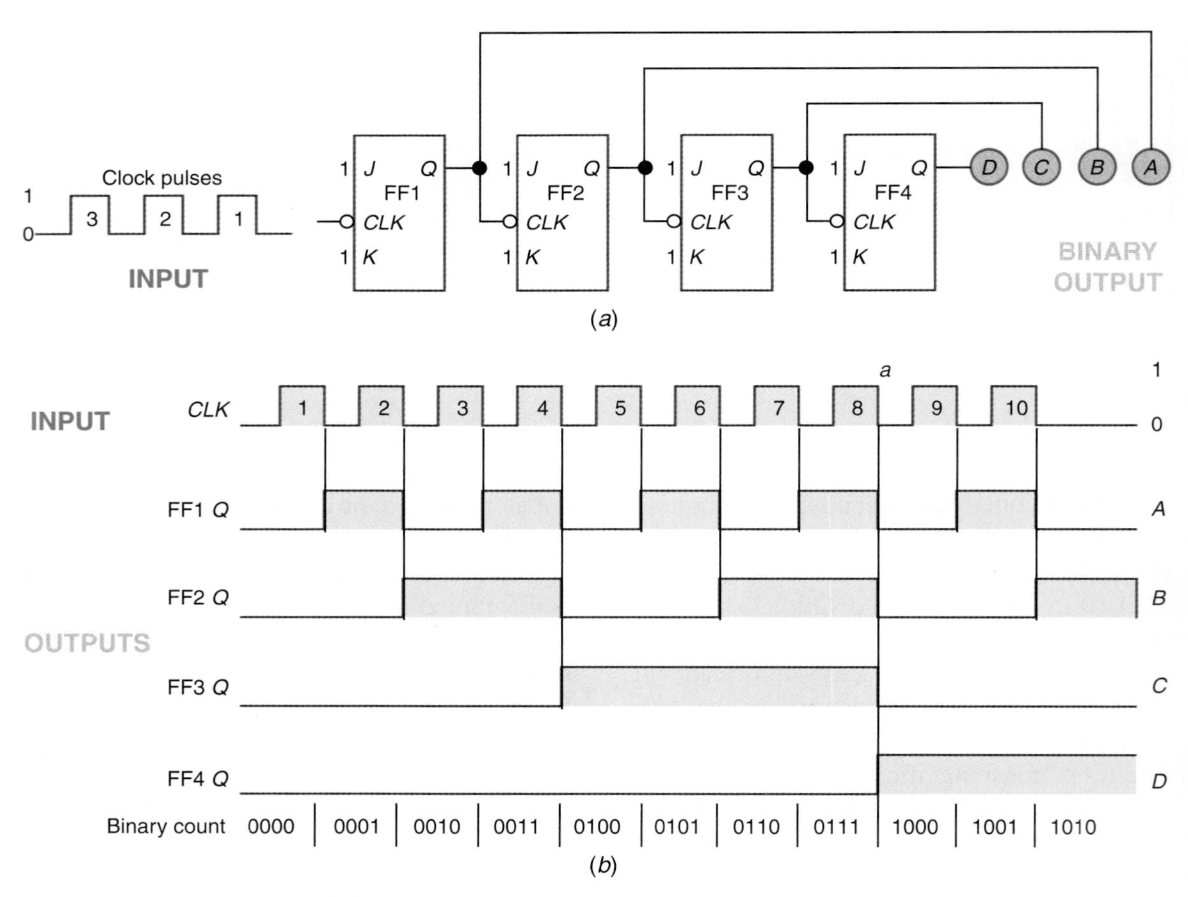

Fig. 8-2 Mod-16 counter. (*a*) Logic diagram. (*b*) Waveform diagram.

to 0. As output Q of FF 3 reaches 0 it triggers FF 4, which toggles from 0 to 1. We see that the changing of states is a chain reaction that *ripples* through the counter. For this reason this counter is called a *ripple counter.*

The counter we studied in Fig. 8-2 could be described as a ripple counter, a mod-16 counter, a 4-bit counter, or an asynchronous counter. All these names describe something about the counter. The ripple and *asynchronous* labels mean that all the flip-flops do not trigger at one time. The mod-16 description comes from the number of states the counter goes through. The *4-bit* label tells how many binary places there are at the output of the counter.

TEST

Answer the following questions.

1. The unit in Fig. 8-3 is a(n) ________ -bit ripple counter.
2. The unit in Fig. 8-3 is a mod- ________ counter.
3. Each J-K flip-flop in Fig. 8-3 is in the ________ (hold, reset, set, toggle) mode because inputs J and K are both HIGH.
4. List the binary output after each of the six input pulses shown in Fig. 8-3.

8-2 Mod-10 Ripple Counters

The counting sequence for a mod-10 counter is from 0000 to 1001 (0 to 9 in decimal). This is down to the heavy line in Fig. 8-1. This *mod-10 counter,* then, has four place values: 8s, 4s, 2s, and 1s. This takes four flip-flops connected as a ripple counter in Fig. 8-4. We must *add* a NAND gate to the ripple counter to clear all the flip-flops back to zero *immediately after* the 1001 (9) count. The *trick* is to look at Fig. 8-1 and determine what the *next count will be after 1001.* You will find it is 1010 (decimal 10). You must feed the two 1s in the 1010 into a NAND gate as shown in Fig. 8-4. The NAND gate then clears the flip-flop back to 0000. The counter then starts its count from 0000 up to 1001 again. We say we are using the NAND gate to reset the counter to 0000. By using a NAND gate in this manner we can make several other modulous counters. Fig. 8-4 illustrates a mod-10 ripple counter. This type of counter might also be called a *decade* (meaning 10) *counter.*

JOB TIP

Companies are very serious about safety. Use safety equipment properly.

Ripple counter

Asynchronous counter

4-bit counter

Decade counter

Ripple counters can be constructed from individual flip-flops. Manufacturers also produce ICs with all four flip-flops inside a single package. Some IC counters even contain the reset NAND gate, such as the one you used in Fig. 8-4.

TEST

Answer the following questions.

5. Refer to Fig. 8-4. This is the logic diagram for a mod-10 ________ (ripple, synchronous) counter. Because it has 10 states (counts from 0 through 9), it is also called a(n) ________ counter.
6. The circuit in Fig. 8-5 is a ________ (ripple, synchronous) mod- ________ counter.
7. List the binary output after each of the six input pulses shown in Fig. 8-5.

Mod-10 ripple counter

8-3 Synchronous Counters

The ripple counters we have studied are asynchronous counters. Each flip-flop does not trigger exactly in step with the clock pulse. For some high-frequency operations it is necessary to have all stages of the counter trigger together. There is such a counter: *a synchronous counter.*

Synchronous counter

A synchronous counter is shown in Fig. 8-6(*a*) on page 177. This logic diagram is for

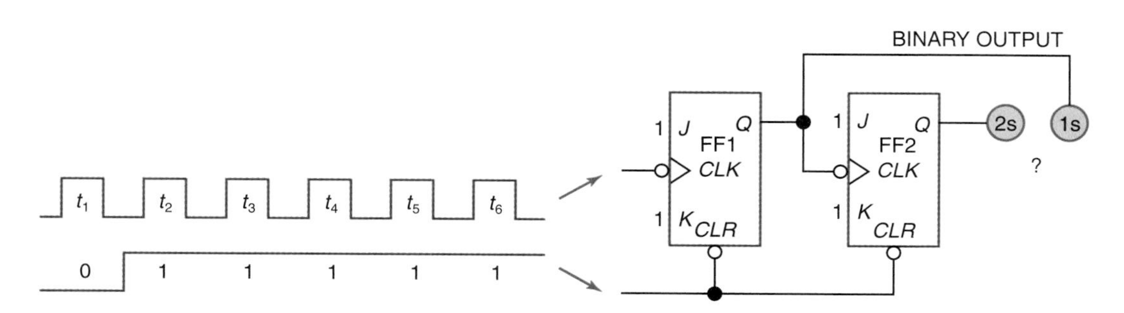

Fig. 8-3 Counter problem for test items 1 through 4.

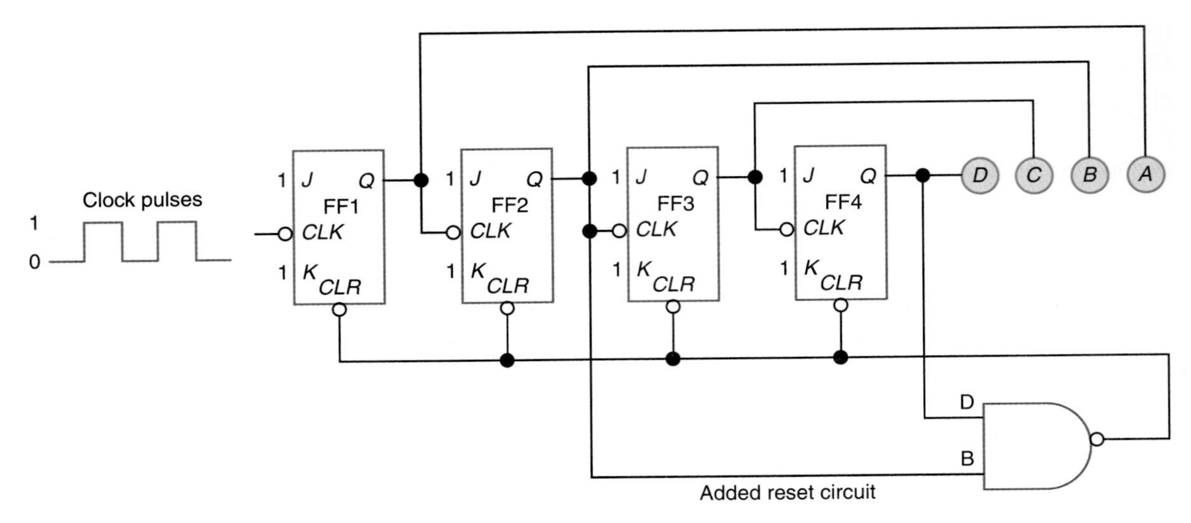

Fig. 8-4 Logic diagram for a mod-10 ripple counter.

a 3-bit (mod-8) counter. First notice the *CLK* connections. The clock is connected directly to the *CLK* input of each flip-flop. We say that the *CLK* inputs are connected in *parallel.* Figure 8-6(*b*) gives the counting sequence of this counter. Column *A* is the binary 1s column, and FF 1 does the counting for this column. Column *B* is the binary 2s column, and FF 2 counts this column. Column *C* is the binary 4s column, and FF 3 counts this column.

Let us study the counting sequence of this mod-8 counter by referring to Fig. 8-6(*a*) and (*b*):

Pulse 1—row 2
Circuit action: Each flip-flop is pulsed by the clock.
Only FF 1 can toggle because it is the only one with 1s applied to both *J* and *K* inputs.
FF 1 goes from 0 to 1.
Output result: 001 (decimal 1).

Pulse 2—row 3
Circuit action: Each flip-flop is pulsed.
Two flip-flops toggle because they have 1s applied to both *J* and *K* inputs.
FF 1 and FF 2 both toggle.
FF 1 goes from 1 to 0.
FF 2 goes from 0 to 1.
Output result: 010 (decimal 2).

Pulse 3—row 4
Circuit action: Each flip-flop is pulsed.
Only one flip-flop toggles.
FF 1 toggles from 0 to 1.
Output result: 011 (decimal 3).

Pulse 4—row 5
Circuit action: Each flip-flop is pulsed.
All flip-flops toggle to opposite state.
FF 1 goes from 1 to 0.
FF 2 goes from 1 to 0.

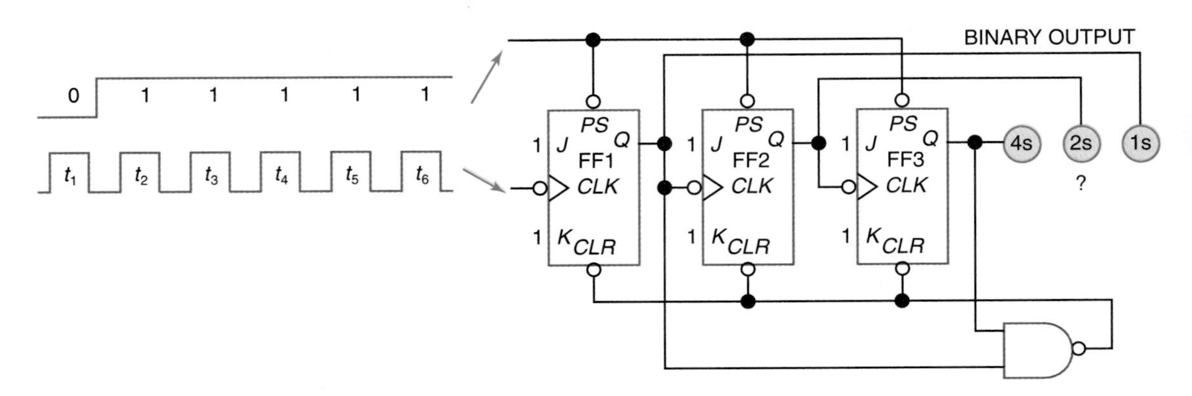

Fig. 8-5 Counter problem for test items 6 and 7.

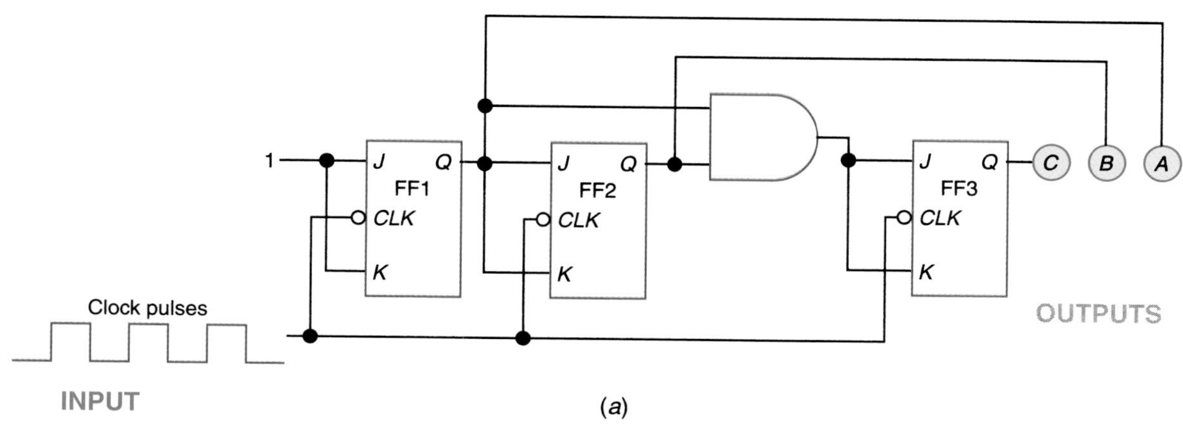

(a)

ROW	NUMBER OF CLOCK PULSES	BINARY COUNTING SEQUENCE			DECIMAL COUNT
		C	B	A	
1	0	0	0	0	0
2	1	0	0	1	1
3	2	0	1	0	2
4	3	0	1	1	3
5	4	1	0	0	4
6	5	1	0	1	5
7	6	1	1	0	6
8	7	1	1	1	7
9	8	0	0	0	0

(b)

Fig. 8-6 A 3-bit synchronous counter. (a) Logic diagram. (b) Counting sequence.

FF 3 goes from 0 to 1.
Output result: 100 (decimal 4).

Pulse 5—row 6
Circuit action: Each flip-flop is pulsed.
Only one flip-flop toggles.
FF 1 goes from 0 to 1.
Output result: 101 (decimal 5).

Pulse 6—row 7
Circuit action: Each flip-flop is pulsed.
Two flip-flops toggle.
FF 1 goes from 1 to 0.
FF 2 goes from 0 to 1.
Output result: 110 (decimal 6).

Pulse 7—row 8
Circuit action: Each flip-flop is pulsed.
Only one flip-flop toggles.
FF 1 goes from 0 to 1.
Output result: 111 (decimal 7).

Pulse 8—row 9
Circuit action: Each flip-flop is pulsed.
All three flip-flops toggle.
All flip-flops change from 1 to 0.
Output result: 000 (decimal 0).

We now have completed the explanation of how the *3-bit synchronous counter* works. Notice that the J-K flip-flops are used in their *toggle* mode (*J* and *K* at 1) or hold mode (*J* and *K* at 0).

3-bit synchronous counter

J-K flip-flops in either toggle or hold modes

Synchronous counters are most often purchased in IC form. Synchronous counters are available in both TTL and CMOS.

TEST

Supply the missing word in each statement.

8. A counter that triggers all the flip-flops at the same instant is called a _________ (ripple, synchronous) counter.
9. Clock inputs are connected in _________ (parallel, series) on a synchronous counter.

10. Refer to Fig. 8-6(a). FF 1 is always in the _________ (hold, reset, set, toggle) mode in this circuit.
11. Refer to Fig. 8-6. On clock pulse 4, _________ (only FF 1 toggles; both FF 1 and FF 2 toggle; only FF 3 toggles; all the flip-flops toggle), producing a binary count of 100 at the outputs of the counter.

8-4 Down Counters

Up to now we have used counters that count upward (0, 1, 2, 3, 4, . . .). Sometimes, however, we must count downward (9, 8, 7, 6, . . .) in digital systems. A counter that counts from higher to lower numbers is called a *down counter.*

Down counter

A logic diagram of a mod-8 asynchronous down counter is shown in Fig. 8-7(a); the counting sequence for this counter is listed in Fig. 8-7(b). Note how much the down counter in Fig. 8-7(a) looks like the up counter in Fig. 8-2(a). The only difference is in the "carry" from FF 1 to FF 2 and the carry from FF 2 to FF 3. The up counter carries from Q to the CLK input of the next flip-flop. The down counter carries from $\overline{Q}$ (not Q) to the CLK input of the next flip-flop. Notice that the down counter has a preset (PS) control to preset the counter to 111 (decimal 7) to start the downward count. FF 1 is the binary 1s place (column A) counter. FF 2 is the 2s place (column B) counter. FF 3 is the 4s place (column C) counter.

Mod-8 asynchronous down counter

TEST

Answer the following questions.

12. Refer to Fig. 8-7(a). All flip-flops are in the _________ (hold, reset, set, toggle) mode in this counter.
13. Refer to Fig. 8-7(a). It takes a (HIGH-to-LOW, LOW-to-HIGH) transition of the clock pulse to trigger these J-K flip-flops.
14. Refer to Fig. 8-7. On clock pulse 1, _________ (only FF 1 toggles; both FF 1 and FF 2 toggle; only FF 3 toggles; all the flip-flops toggle) producing a binary count of 110 at the outputs of the counter.
15. List the counter's binary output for each of the six input pulses shown in Fig. 8-8.

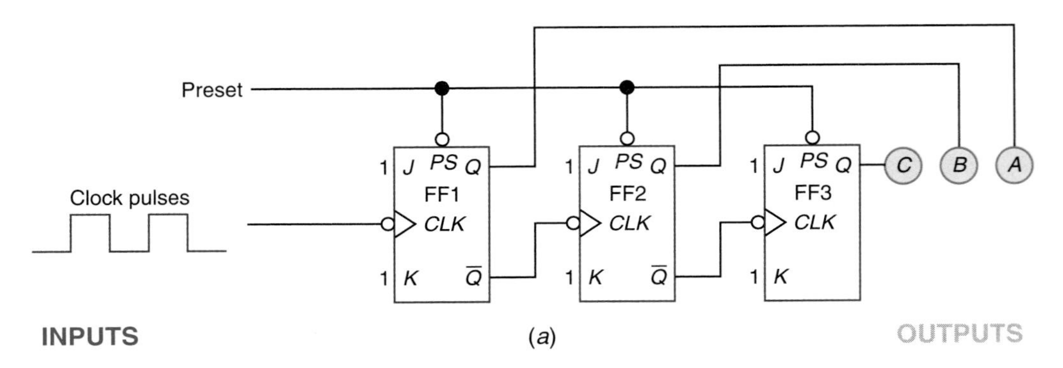

(a)

NUMBER OF CLOCK PULSES	BINARY COUNTING SEQUENCE			DECIMAL COUNT
	C	B	A	
0	1	1	1	7
1	1	1	0	6
2	1	0	1	5
3	1	0	0	4
4	0	1	1	3
5	0	1	0	2
6	0	0	1	1
7	0	0	0	0
8	1	1	1	7
9	1	0	0	6

(b)

Fig. 8-7 A 3-bit ripple down counter. (a) Logic diagram. (b) Counting sequence.

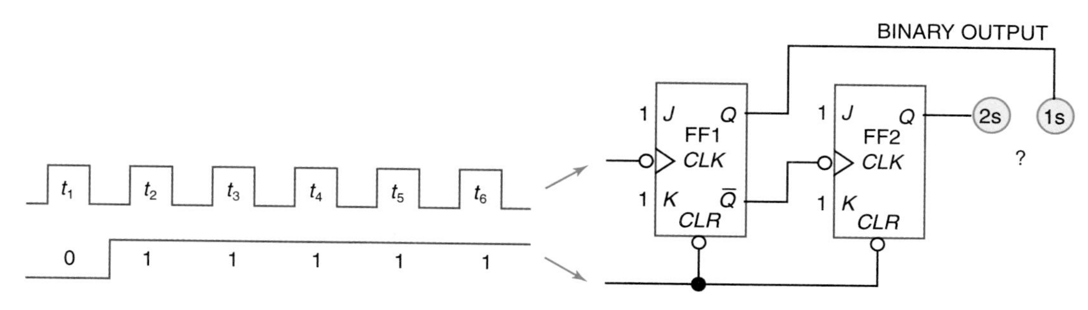

Fig. 8-8 Counter problem for test item 15.

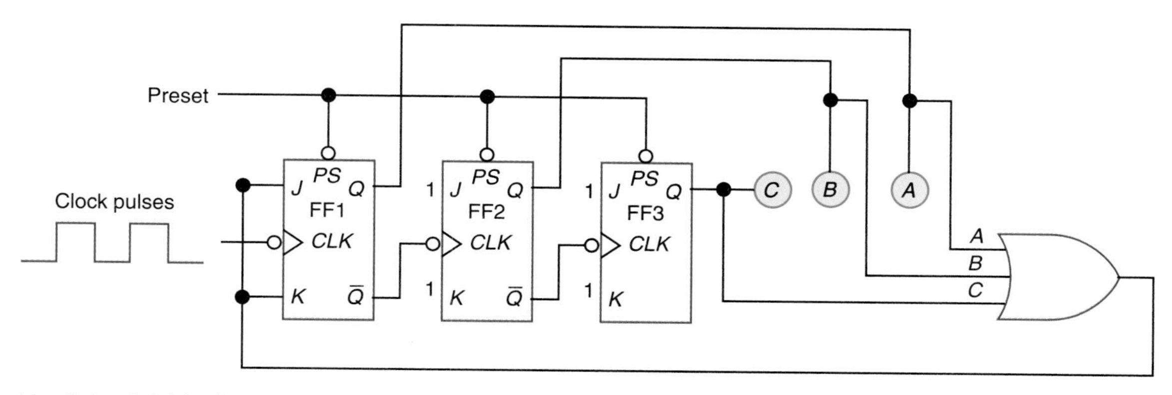

Fig. 8-9 A 3-bit down counter with self-stopping feature.

8-5 Self-Stopping Counters

3-bit down counter with self-stopping feature

Recirculating counter

The down counter shown in Fig. 8-7(*a*) *recirculates.* That is, when it gets to 000 it starts at 111, then 110, and so forth. However, sometimes you want a counter to *stop* when a sequence is finished. Figure 8-9 illustrates how you could stop the down counter in Fig. 8-7 at the 000 count. The counting sequence is shown in Fig. 8-7(*b*). In Fig. 8-9 we add an *OR* gate to place a logical 0 on the *J* and *K* inputs of FF 1 when the count at outputs *C, B,* and *A* reaches 000. The preset must be enabled (*PS* to 0) again to start the sequence at 111 (decimal 7).

Up or down counters can be stopped after any sequence of counts by using a logic gate or combination of gates. The output of the gate is fed back to the *J* and *K* inputs of the first flip-flop in a ripple counter. The logical 0s fed back to the *J* and *K* inputs of FF 1 in Fig. 8-9 place it in the hold mode. This stops FF 1 from toggling, thereby stopping the count at 000.

TEST

Supply the missing word or words in each statement.

16. Refer to Fig. 8-9. This is the logic diagram for a self-stopping 3-bit ________ (down, up) counter.
17. Refer to Fig. 8-9. With an output count of 000, the OR gate outputs a ________ (HIGH, LOW). This places FF 1 in the ________ (hold, toggle) mode.
18. Refer to Fig. 8-9. With an output count of 111, the OR gate outputs a ________ (HIGH, LOW). This places FF 1 in the ________ (hold, toggle) mode.

8-6 Counters as Frequency Dividers

Frequency division

An interesting and common use of counters is for *frequency division.* An example of a simple system using a frequency divider is shown in Fig. 8-10. This system is the basis for an electric clock. The 60-Hz input frequency is from the power line (formed into a square wave). The circuit must divide the frequency by 60,

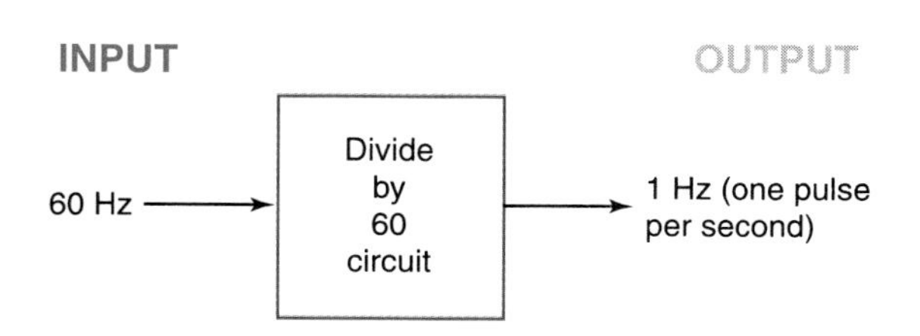

Fig. 8-10 A 1-second timer system.

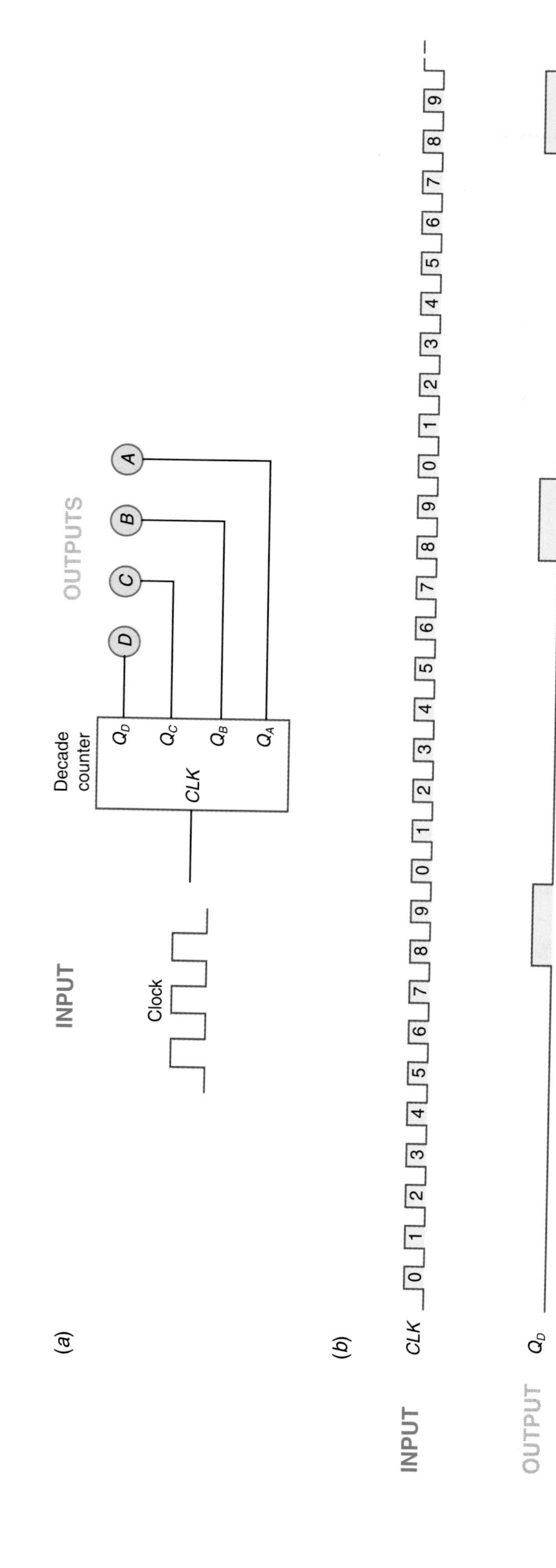

Fig. 8-11 Decade counter used as a divide-by-10 counter. (*a*) Logic diagram. (*b*) Waveform diagram.

and the output will be one pulse per second (1 Hz). This is a seconds timer.

A block diagram of a decade counter is drawn in Fig. 8-11(*a*). In Fig. 8-11(*b*) the waveforms at the *CLK* input and the binary 8s place (output Q_D) are shown. Notice that it takes 30 input pulses to produce 3 output pulses. Using division, we find that $30 \div 3 = 10$. Output Q_D of the decade counter in Fig. 8-11(*a*) is a *divide-by-10* counter. In other words, the output frequency at Q_D is only one-tenth the frequency at the input of the counter.

If we use the decade counter (divide-by-10 counter) from Fig. 8-10 and a mod-6 counter (divide-by-6 counter) in series, we get the divide-by-60 circuit we need in Fig. 8-10. A diagram of such a system is illustrated in Fig. 8-12. The 60-Hz square wave enters the divide-by-6 counter and comes out at 10 Hz. The 10 Hz then enters the divide-by-10 counter and comes out at 1 Hz.

You are already aware that counters are used as frequency dividers in *digital* timepieces, such as electronic digital clocks, automobile digital clocks, and digital wristwatches. Frequency division is also used in *frequency counters, oscilloscopes,* and *television receivers.*

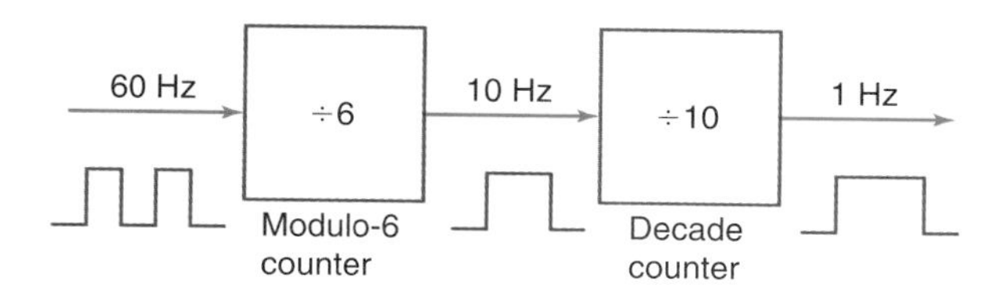

Fig. 8-12 Practical divide-by-60 circuit used as a 1-second timer.

Divide-by-60 circuit

TEST

Supply the missing word in each statement.

19. Refer to Fig. 8-12. If the input frequency on the left is 60,000 Hz, the output frequency from the decade counter is ________ Hz.
20. Refer to Fig. 8-11(*a*). Output *A* divides the input clock frequency by ________ (number).

8-7 TTL IC Counters

Manufacturers' IC data manuals contain long lists of counters. This section covers only two representative types of TTL IC counters.

The *7493 TTL 4-bit binary counter* is detailed in Fig. 8-13 on page 182. The block diagram in Fig. 8-13(*a*) shows that the 7493 IC houses four J-K flip-flops wired as a ripple counter. If you look carefully at Fig. 8-13(*a*), you will notice that the bottom three J-K flip-flops are prewired internally as a 3-bit ripple counter with output Q_B connected to the clock input of the next lower J-K flip-flop and output Q_C connected internally to the clock input of the bottom J-K flip-flop. Importantly, the top J-K flip-flop *does not* have its Q_A output internally connected to the next lower flip-flop. To use the 7493 IC as a 4-bit ripple counter (mod-16), you have to *externally connect output Q_A to input B,* which is the *CLK* input of the second flip-flop. A counting sequence for the 7493 IC wired as a 4-bit ripple counter is reproduced in Fig. 8-13(*c*). Consider the *J* and *K* inputs to each flip-flop in Fig. 8-13(*a*): it is understood that these inputs are permanently held HIGH so the flip-flops are in the toggle mode. Notice that the clock inputs suggest that the 7493 uses negative-edge triggering.

Digital clock

JOB TIP

Technicians with a combination of hardware and software skills are in great demand.

Frequency counter

Oscilloscopes

7493 TTL 4-bit binary counter

About Electronics

Star Gazing A powerful computerized telescope uses a laser beam that serves as a guide "star," reflecting off the upper atmosphere, from which the telescope can locate real stars. The telescope uses the reflection as a focal point. A very thin mirror aimed at the focal point has 127 actuators glued on its back. These actuators drive the computer to adjust the minute portions of the mirror 50 to 100 times per second. The result is that air turbulence does not affect clarity.

(a) BLOCK DIAGRAM

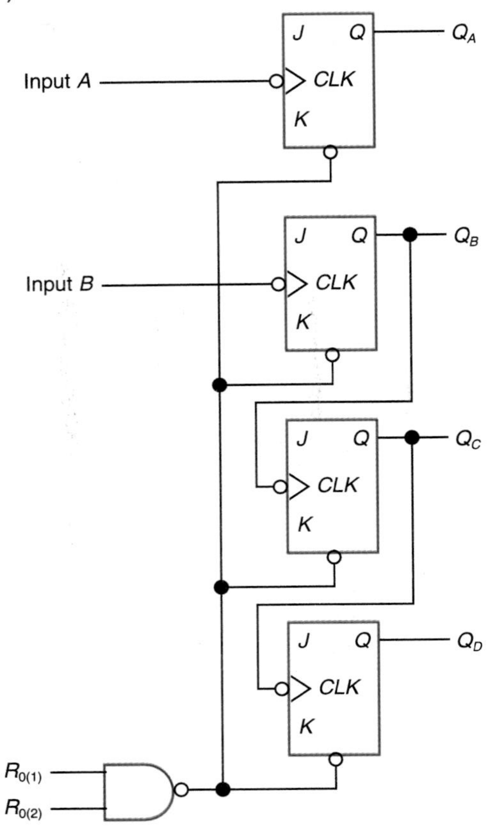

The J and K inputs shown without connection for reference only and are functionally at a high level.

(b) PIN CONFIGURATION

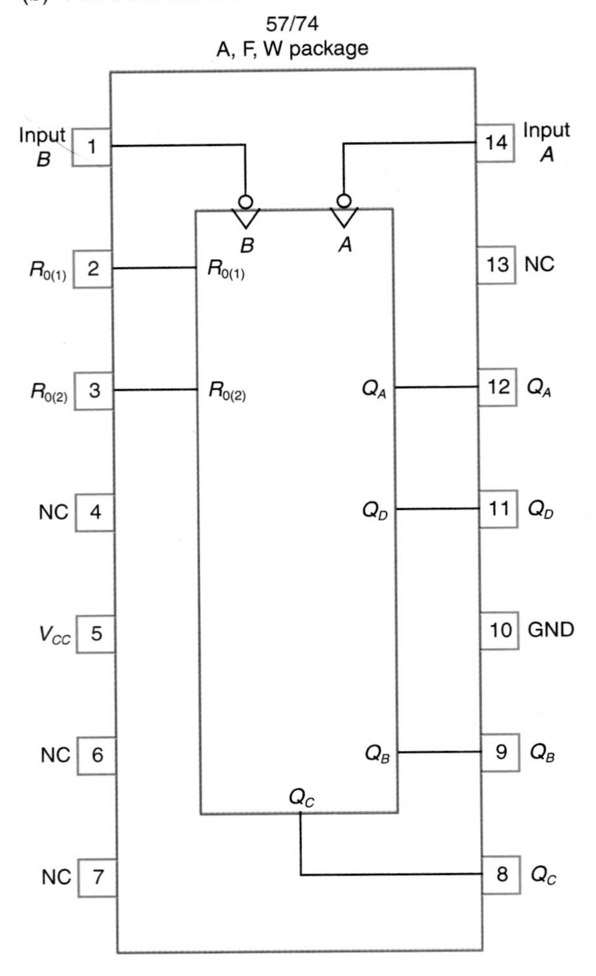

(c) COUNT SEQUENCE

COUNT	OUTPUT			
	Q_D	Q_C	Q_B	Q_A
0	L	L	L	L
1	L	L	L	H
2	L	L	H	L
3	L	L	H	H
4	L	H	L	L
5	L	H	L	H
6	L	H	H	L
7	L	H	H	H
8	H	L	L	L
9	H	L	L	H
10	H	L	H	L
11	H	L	H	H
12	H	H	L	L
13	H	H	L	H
14	H	H	H	L
15	H	H	H	H

Output Q_A is connected to input B.

(d) RESET/COUNT FUNCTION TABLE

RESET INPUTS		OUTPUT			
$R_0(1)$	$R_0(2)$	Q_D	Q_C	Q_B	Q_A
H	H	L	L	L	L
L	X	Count			
X	L	Count			

NOTES:
A. Output Q_A is connected to input B for BCD count (or binary count).
B. Output Q_D is connected to input A for biquinary count.
C. H = high level, L = low level, X = irrelevant.

4-bit binary counter IC (7493)

Fig. 8-13 A 4-bit binary counter IC (7493). (a) Block diagram. (b) Pin configuration. (c) Count sequence. (d) Reset/count function table. *(Courtesy of National Semiconductor Corp.)*

Recall the use of a two-input NAND gate to change the mod-16 ripple counter to a decade counter in Fig. 8-4. Figure 8-13(*a*) shows that such a two-input NAND gate is built into the 7493 counter IC. Inputs $R_{0(1)}$ and $R_{0(2)}$ are the inputs to the internal NAND gate. The reset/count function table in Fig. 8-13(*d*) shows that the 7493 counter will be reset (0000) when both $R_{0(1)}$ and $R_{0(2)}$ are HIGH. When either or both reset inputs are LOW, the 7493 IC will count. *Caution:* If the reset inputs ($R_{0(1)}$ and $R_{0(2)}$) are left disconnected, they will float HIGH and the 7493 IC will be in the reset mode and will not count. Note B in Fig. 8-13(*d*) suggests that you can use the 7493 IC as a *biquinary counter* by connecting output Q_D to Q_A with output Q_A becoming the most significant bit. The biquinary number system is used in the hand-manipulated abacus and soroban.

The 7493 4-bit ripple counter is packaged in a 14-pin DIP as shown in Fig. 8-13(*b*). Note especially the unusual location of the GND (pin 10) and V_{CC} (pin 5) connections to the 7493 counter, which are commonly on the corners of the many ICs. Counterparts of the 7493 in other families include the 74LS93 and 74C93 4-bit ripple counters.

A second TTL IC counter is detailed in Fig. 8-14 on page 184. It is the *74192 up/down decade counter IC.* Read the manufacturer's description of the IC counter in Fig. 8-14(*a*). Because the 74192 counter is a synchronous counter and has many features, it is quite complex, as shown in the logic diagram reproduced in Fig. 8-14(*b*). The 74192 IC is packaged in either a 16-pin dual in-line package or a 20-pin surface mount package. The pin configurations of both the DIP and surface mount packages are drawn in Fig. 8-14(*c*) on page 185. Both IC packages shown in Fig. 8-14(*c*) are viewed from the top. Note especially the unusual location of pin 1 on the surface mount package.

The waveform diagram in Fig. 8-14(*d*) details several sequences used on the 74192 counter IC. Useful sequences detailed in the waveform diagram are clear, preset (load), count up, and count down. The clear (*CLR*) input to the 74192 is an active-HIGH input while the load is an active-LOW input. Counterparts to the 74192 synchronous up/down counter are the 74LS192, 74F192, 74HC192, and 74C192 ICs.

You probably have already figured out that some of the features are not used on these IC counters for some applications. Figure 8-15(*a*) on page 185 shows the 7493 IC counter being used as a mod-8 counter. Look back at Fig. 8-13 and notice that several inputs and an output are not being used. Figure 8-15(*b*) shows the 74192 counter being used as a decade down counter. Six inputs and two outputs are not being used in this circuit. Simplified logic diagrams similar to those in Fig. 8-15 are more common than the complicated diagrams in Figs. 8-13(*a*) and 8-14(*b*).

Biquinary counter

74192 synchronous decade up/down counter

TEST

Answer the following questions.

21. Refer to Fig. 8-13. If both inputs to the NAND gate (pins 2 and 3 on the 7493 IC) are HIGH, the output from the 7493 counter will be ________ (4 bits).
22. Refer to Fig. 8-13. The 7493 IC is a(n) ________ -bit ________ (down, up) counter.
23. Refer to Fig. 8-14. The 74192 IC is a ________ (decade, mod-16) up/down ________ (ripple, synchronous) counter.
24. Refer to Fig. 8-14. The clock input to the 74192 for counting upward is pin ________ (number) on the DIP IC.
25. Refer to Fig. 8-14. The 74192 IC has an active ________ (HIGH, LOW) clear input.
26. List the output frequency at points *B, C,* and *D* in Fig. 8-16.
27. The 7493 IC is a ripple divide-by-2, divide-by-4, and divide-by- ________ unit in Fig. 8-16.

8-8 CMOS IC Counters

Manufacturers of CMOS chips offer a variety of counters in IC form. This section covers only two types of CMOS counters.

A manufacturer of CMOS ICs, provided the diagrams and tables in Fig. 8-17 on page 186. The diagrams are for a *74HC393 dual 4-bit binary ripple counter.* A *function diagram* (something like a logic diagram) of the 74HC393 counter IC is shown in Fig. 8-17(*a*). Note that the IC contains two 4-bit binary ripple counters. The table in Fig. 8-17(*b*) gives the names and functions of each input and

74192 decade up/down counter IC

74HC393 dual 4-bit binary ripple counter

(*a*) DESCRIPTION

This monolithic circuit is a synchronous reversible (up/down) counter having a complexity of 55 equivalent gates. Synchronous operation is provided by having all flip-flops clocked simultaneously so that the outputs change coincidentally with each other when so instructed by the steering logic. This mode of operation eliminates the output counting spikes which are normally associated with asynchronous (ripple-clock) counters.

The outputs of the four master-slave flip-flops are triggered by a low-to-high-level transition of either count (clock) input. The direction of counting is determined by which count input is pulsed while the other count input is high.

All four counters are fully programmable; that is, each output may be preset to either level by entering the desired data at the data inputs while the load input is low. The output will change to agree with the data inputs independently of the count pulses. This feature allows the counters to be used as modulo-N dividers by simply modifying the count length with the preset inputs.

A clear input has been provided which forces all outputs to the low level when a high level is applied. The clear function is independent of the count and load inputs. The clear, count, and load inputs are buffered to lower the drive requirements. This reduces the number of clock drivers, etc., required for long words.

These counters were designed to be cascaded without the need for external circuitry. Both borrow and carry outputs are available to cascade both the up- and down-counting functions. The borrow output produces a pulse equal to the count-down input when the counter underflows. Similarly, the carry output produces a pulse equal in width to the count-up input when an overflow condition exists. The counters can then be easily cascaded by feeding the borrow and carry outputs to the count-down and count-up inputs respectively of the succeeding counter.

(*b*) LOGIC DIAGRAM

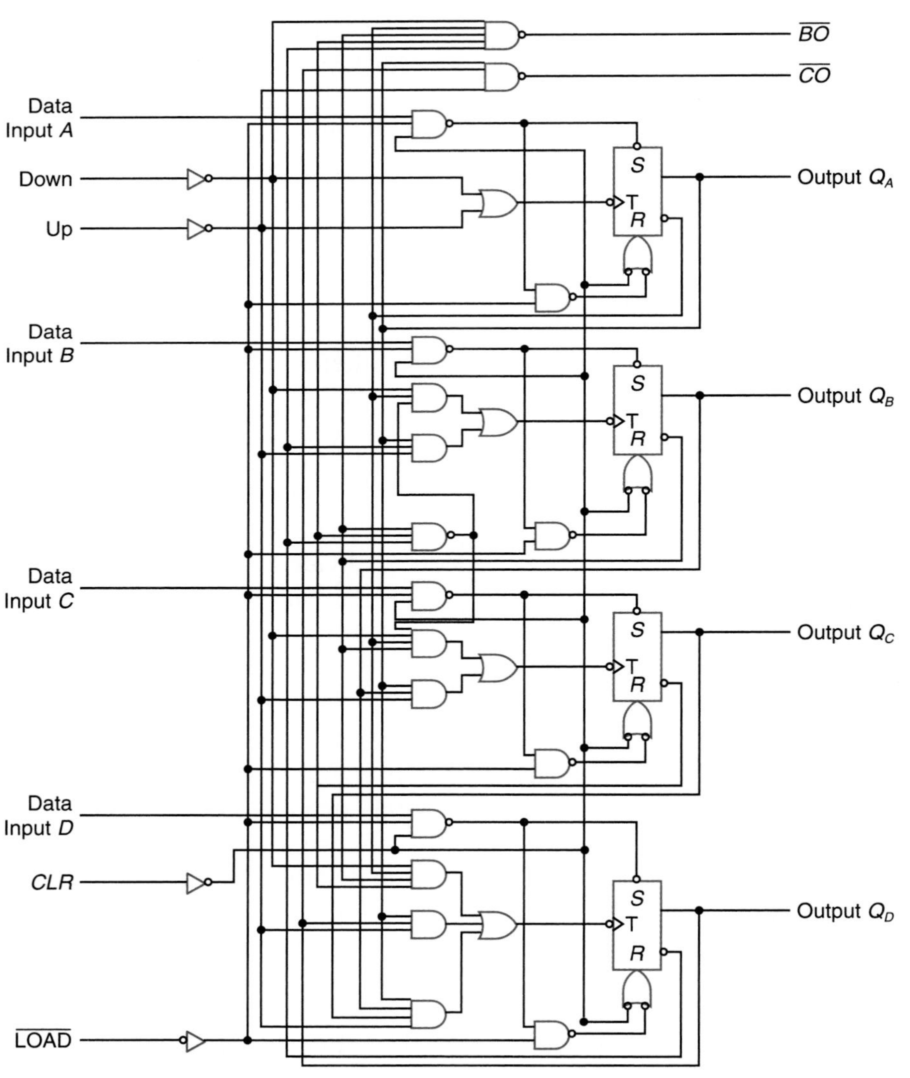

Synchronous decade up/down counter IC (74192)

Fig. 8-14 Synchronous decade up/down counter IC (74192). (*a*) Description. (*b*) Logic diagram.

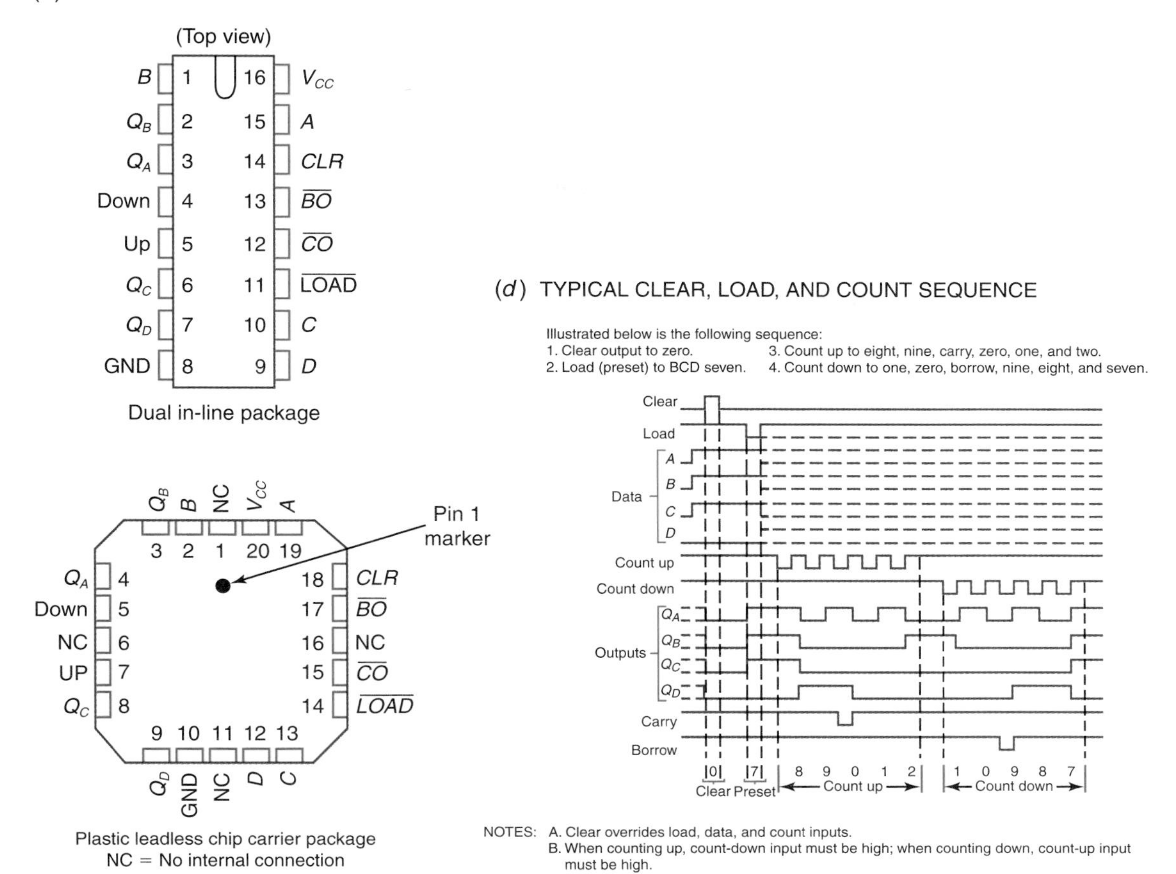

Fig. 8-14 (*Continued*) (*c*) Pin configurations. (*d*) Waveforms. *(Courtesy of Texas Instruments, Inc.)*

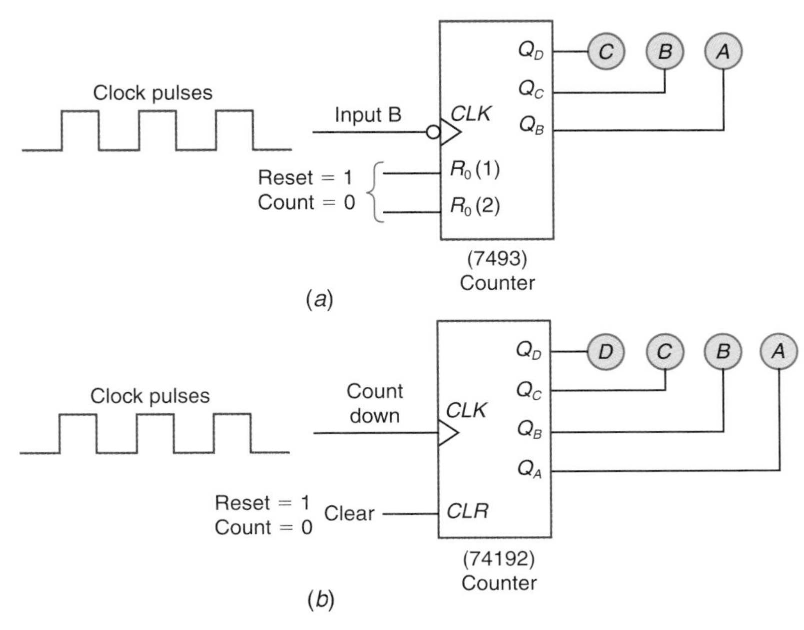

Fig. 8-15 (*a*) 7493 IC wired as a mod-8 up counter. (*b*) 74192 IC wired as a decade down counter.

Mod-8 counter

Decade down counter

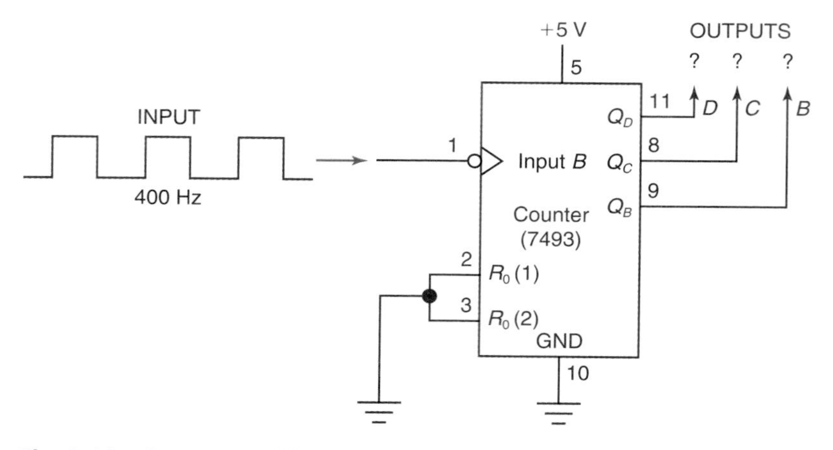

Fig. 8-16 Counter problem for test items 26 and 27.

output pin on the 74HC393 IC. Note that the clock inputs are labeled with the letters $\overline{CP}$ instead of *CLK*, as used earlier. Pin labels vary from manufacturer to manufacturer. For this reason, you must learn to use manufacturer's data manuals for exact information.

JOB TIP

Knowledge of normal operation of an electronic system is the key to successful troubleshooting.

T flip-flop

Each 4-bit counter in the 74HC393 IC package consists of four T flip-flops. A *T flip-flop* is any flip-flop that is in the toggle mode. This is shown in the detailed logic diagram drawn in Fig. 8-17(*c*).

Note that the *MR* input is an asynchronous master reset pin. The *MR* pins are active HIGH inputs. In other words, a HIGH at the *MR* input will override the clock and reset the individual counter to 0000.

A pin diagram for the 74HC393 IC is reproduced in Fig. 8-17(*d*). This dual in-line package IC is being viewed from the top. The counting sequence for the 74HC393 counter is binary 0000 through 1111 (0 to 15 in decimal).

The functional diagram in Fig. 8-17(*a*) and logic diagram in Fig. 8-17(*c*) both suggest that the counters are triggered on the HIGH-to-

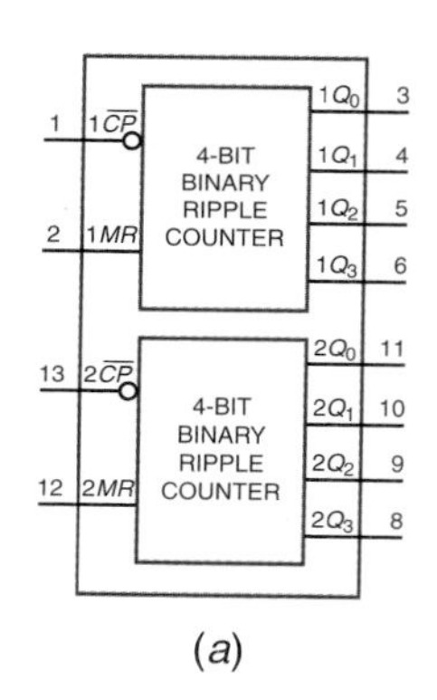

(*a*)

PIN DESCRIPTION

PIN NO.	SYMBOL	NAME AND FUNCTION
1, 13	$1\overline{CP}$, $2\overline{CP}$	clock inputs (HIGH-to-LOW, edge-triggered)
2, 12	1*MR*, 2*MR*	asynchronous master reset inputs (active HIGH)
3, 4, 5, 6, 11, 10, 9, 8	$1Q_0$ to $1Q_3$, $2Q_0$ to $2Q_3$	flip-flop outputs
7	GND	ground (0 V)
14	V_{CC}	positive supply voltage

(*b*)

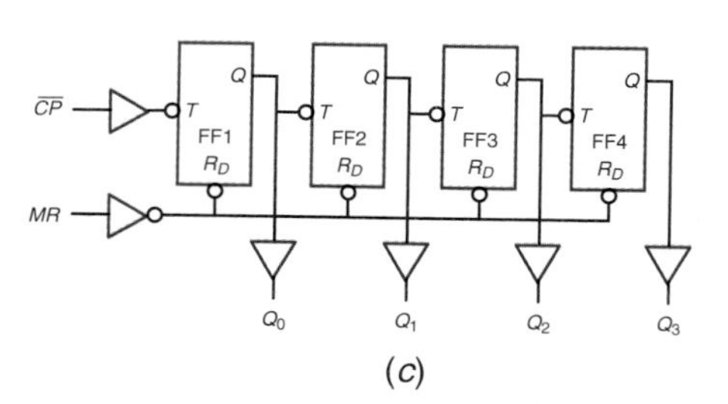

(*c*)

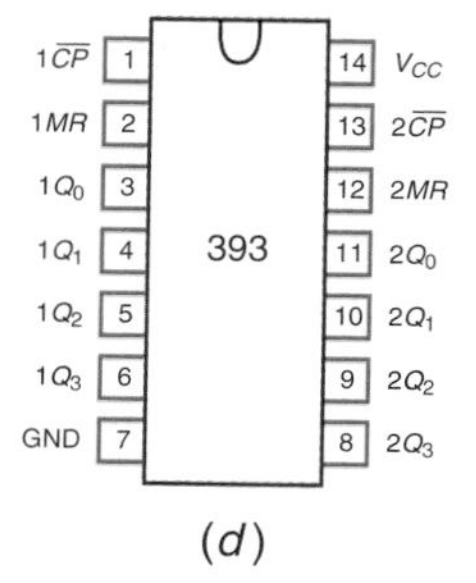

(*d*)

74HC393 4 bit-counter

Fig. 8-17 CMOS dual 4-bit binary counter IC (74HC393). (*a*) Function diagram. (*b*) Pin descriptions. (*c*) Detailed logic diagram. (*d*) Pin diagram. *(Courtesy of Signetics Corporation.)*

LOW transition of the clock pulse. The outputs (Q_0, Q_1, Q_2, Q_3) of the ripple counter are asynchronous (not exactly in step with the clock). As with all ripple counters, there is a slight delay in outputs because the first flip-flop triggers the second, the second the third, and so forth. Note that the > symbol at the clock ($\overline{CP}$ inputs) has been omitted by this manufacturer. Again, many variations occur in both labels and logic diagrams from manufacturer to manufacturer.

The second CMOS IC counter we shall discuss is the *74HC193 presettable synchronous 4-bit binary up/down counter IC*. The 74HC193 counter has more features than the 74HC393 IC. Manufacturer's information on the 74HC193 counter IC is detailed in Fig. 8-18.

A function diagram of the 74HC193 IC is drawn in Fig. 8-18(*a*) with pin descriptions following in Fig. 8-18(*b*). The 74HC193 has two clock inputs (CP_U and CP_D). One clock input is used for counting up (CP_U) and the other when counting down (CP_D). Figure 8-18(*b*) notes that the clock inputs are edge-triggered on the LOW-to-HIGH transition of the clock pulse.

A truth table for the 74HC193 counter is drawn in Fig. 8-18(*d*). The operating modes for the counter on the left give an overview of the many functions of the 74HC193 counter. Its modes of operation are *reset, parallel load, count up,* and *count down.* The truth table in Fig. 8-18(*d*) also makes it clear which pins are inputs and which are outputs.

Typical clear (reset), preset (parallel load), count up, and count down sequences are shown in Fig. 8-18(*e*). Waveforms are useful when investigating an IC's typical operations or timing.

Figures 8-19 and 8-20 (on page 189) show two possible applications for the CMOS counter ICs studied in this section. Figure 8-19 shows a logic diagram for a 74HC393 IC wired as a simple 4-bit binary counter. The *MR* (master reset) pin must be tied to either 0 or 1. The *MR* input is an active HIGH input so a 1 clears the binary outputs to 0000. With a logical 0 at the reset pin (*MR*), the IC is allowed to count upward from binary 0000 to 1111.

The 74HC193 CMOS IC is a more sophisticated counter. Figure 8-20 diagrams a mod-6 counter which starts at binary 001 and counts up to 110 (1 to 6 in decimal). This might be useful in a game circuit where the rolling of dice is simulated. The NAND gate in the mod-6 counter activates the asynchronous parallel load ($\overline{PL}$) input with a LOW just after the highest required count of binary 0110. The counter is then loaded with 0001 which is permanently connected to the data inputs (D_0 to D_3). Clock pulses enter the count-up clock input (CP_U). The count-down clock input (CP_D) must be tied to +5 V and the master reset (*MR*) pin must be grounded to disable these inputs and allow the counter to operate. The mod-6 counter circuit in Fig. 8-20 shows the flexibility of the CMOS 74HC193 presettable 4-bit up/down counter IC.

74HC193 presettable synchronous 4-bit binary up/down counter IC

TEST

Answer the following questions.

28. Refer to Fig. 8-17. The 74HC393 IC contains two ________ (4-bit binary, decade) counters.
29. Refer to Fig. 8-17. The reset pin (*MR*) on the 74HC393 counter is an active ________ (HIGH, LOW) input.
30. Refer to Fig. 8-17. The 74HC393 counter's clock inputs are triggered by the ________ (H-to-L, L-to-H) transition of the clock pulse.
31. The circuit drawn in Fig. 8-19 is a mod-________ (number) ________ (ripple, synchronous) counter.
32. Refer to Fig. 8-18. The 74HC193 is a presettable ________ (ripple, synchronous) 4-bit up/down counter IC.

Modes of operation for 74HC193 counter

Did You Know?

Devices in the Medical Field

- In the past, blood tests required several vials of blood because testing machines couldn't handle small quantities. A new method encapsulates blood within a dime-sized "vial" and moves blood electrically within a computer chip with channels that carry liquid instead of wires. For this procedure, less than a billionth of a liter needs to be sampled.
- Deeply embedded bodily ailments, such as kidney disorders, that cause scar tissue can be located when medical personnel use ultrasound together with touch. When organs do not move freely, the area is scarred.

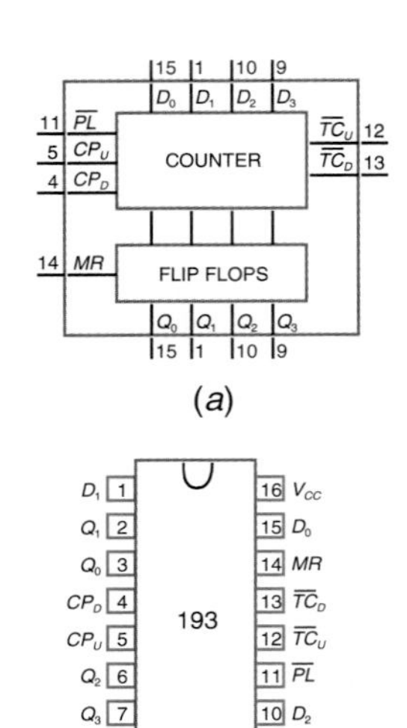

(*a*)

PIN DESCRIPTION

PIN NO.	SYMBOL	NAME AND FUNCTION
3, 2, 6, 7	Q_0 to Q_3	flip-flop outputs
4	CP_D	count down clock input*
5	CP_U	count up clock input*
8	GND	ground (0 V)
11	$\overline{PL}$	asynchronous parallel load input (active LOW)
12	$\overline{TC}_U$	terminal count up (carry) output (active LOW)
13	$\overline{TC}_D$	terminal count down (borrow) output (active LOW)
14	MR	asynchronous master reset input (active HIGH)
15, 1, 10, 9	D_0 to D_3	data inputs
16	V_{CC}	positive supply voltage

* LOW-to-HIGH, edge triggered

(*b*)

D_1 1 | 16 V_{CC}
Q_1 2 | 15 D_0
Q_0 3 | 14 MR
CP_D 4 | 13 $\overline{TC}_D$
CP_U 5 | 12 $\overline{TC}_U$
Q_2 6 | 11 $\overline{PL}$
Q_3 7 | 10 D_2
GND 8 | 9 D_3

193

7Z93712

(*c*)

OPERATING MODE	INPUTS								OUTPUTS					
	MR	$\overline{PL}$	CP_U	CP_D	D_0	D_1	D_2	D_3	Q_0	Q_1	Q_2	Q_3	$\overline{TC}_U$	$\overline{TC}_D$
reset (clear)	H	X	X	L	X	X	X	X	L	L	L	L	H	L
	H	X	X	H	X	X	X	X	L	L	L	L	H	H
parallel load	L	L	X	L	L	L	L	L	L	L	L	L	H	L
	L	L	X	H	L	L	L	L	L	L	L	L	H	H
	L	L	L	X	H	H	H	H	H	H	H	H	L	H
	L	L	H	X	H	H	H	H	H	H	H	H	H	H
count up	L	H	↑	H	X	X	X	X	count up				H*	H
count down	L	H	H	↑	X	X	X	X	count down				H	H**

* $\overline{TC}_U = CP_U$ at terminal count up (HHHH)
** $\overline{TC}_D = CP_D$ at terminal count down (LLLL)

H = HIGH voltage level
L = LOW voltage level
X = don't care
↑ = LOW-to-HIGH clock transition

(*d*)

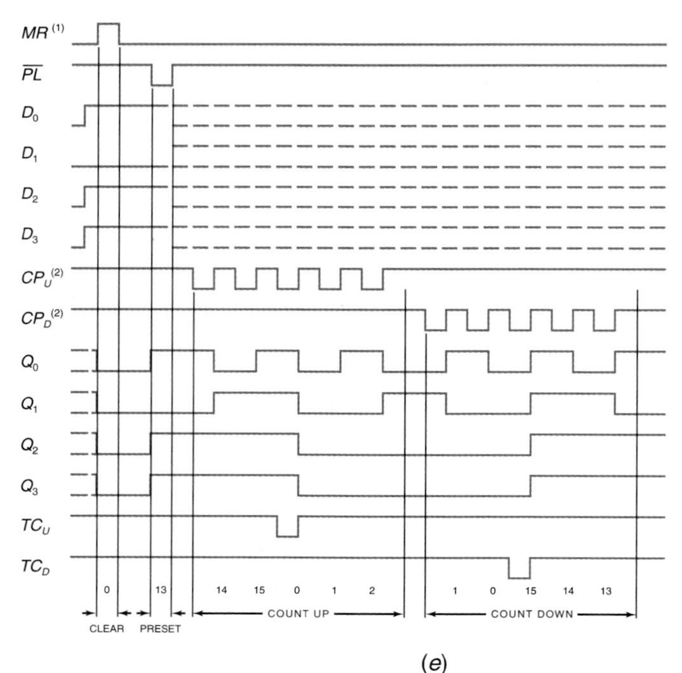

(1) Clear overrides load, data and count inputs.
(2) When counting up the count down clock input (CP_D) must be HIGH, when counting down the count up clock input (CP_U) must be HIGH.

Sequence
Clear (reset outputs to zero);
load (preset) to binary thirteen;
count up to fourteen, fifteen, terminal count up, zero, one and two;
count down to one, zero, terminal count down, fifteen, fourteen and thirteen.

(*e*)

CMOS presettable 4-bit synchronous up/down counter IC (74HC193)

Fig. 8-18 CMOS presettable 4-bit synchronous up/down counter IC (74HC193). (*a*) Function diagram. (*b*) Pin descriptions. (*c*) Pin diagram. (*d*) Truth table. (*e*) Typical clear, preset, and count sequence. *(Courtesy of Signetics Corporation.)*

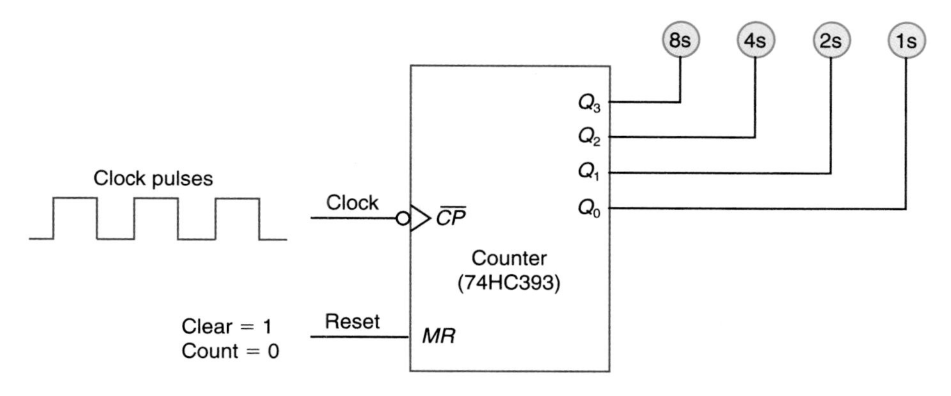

Fig. 8-19 A 74HC393 IC wired as a 4-bit binary counter.

4-bit binary counter

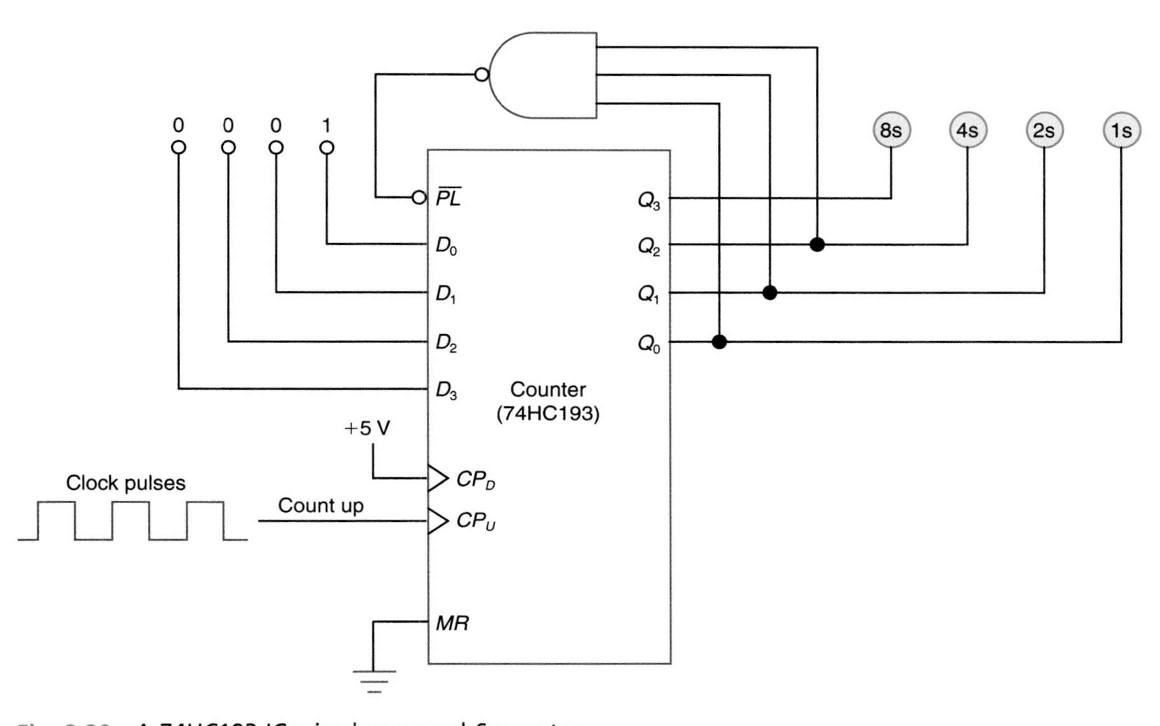

Fig. 8-20 A 74HC193 IC wired as a mod-6 counter.

Mod-6 counter

33. Refer to Fig. 8-18. The reset pin (*MR*) is _________ (asynchronous, synchronous) and overrides all other inputs on the 74HC193 IC.
34. Refer to Fig. 8-18. The outputs of the 74HC193 are labeled _________ (D_0–D_3, Q_0–Q_3).
35. Refer to Fig. 8-20. List the binary counting sequence for this counter circuit.
36. Refer to Fig. 8-20. What is the purpose of the three-input NAND gate in this counter circuit?
37. Refer to Fig. 8-17(*a*). How do you explain the lack of the > symbol near the clock inputs since the 74HC393 counters are edge-triggered?

8-9 Counting Real-World Events

As we have mentioned before, the processing power of digital circuits is not very useful if we cannot input data and output results. The block diagram in Fig. 8-21(*a*) is a summary of the systems that we have studied to this point. In the digital processing area, we have studied some combinational and some sequential logic. We have studied several encoders and decoders that handle interfacing. We have studied many output devices such as LEDs, seven-segment LED, LCD, and VF displays, incandescent bulbs, buzzers, relays, dc motors, and stepper motors. We have worked with only a few input

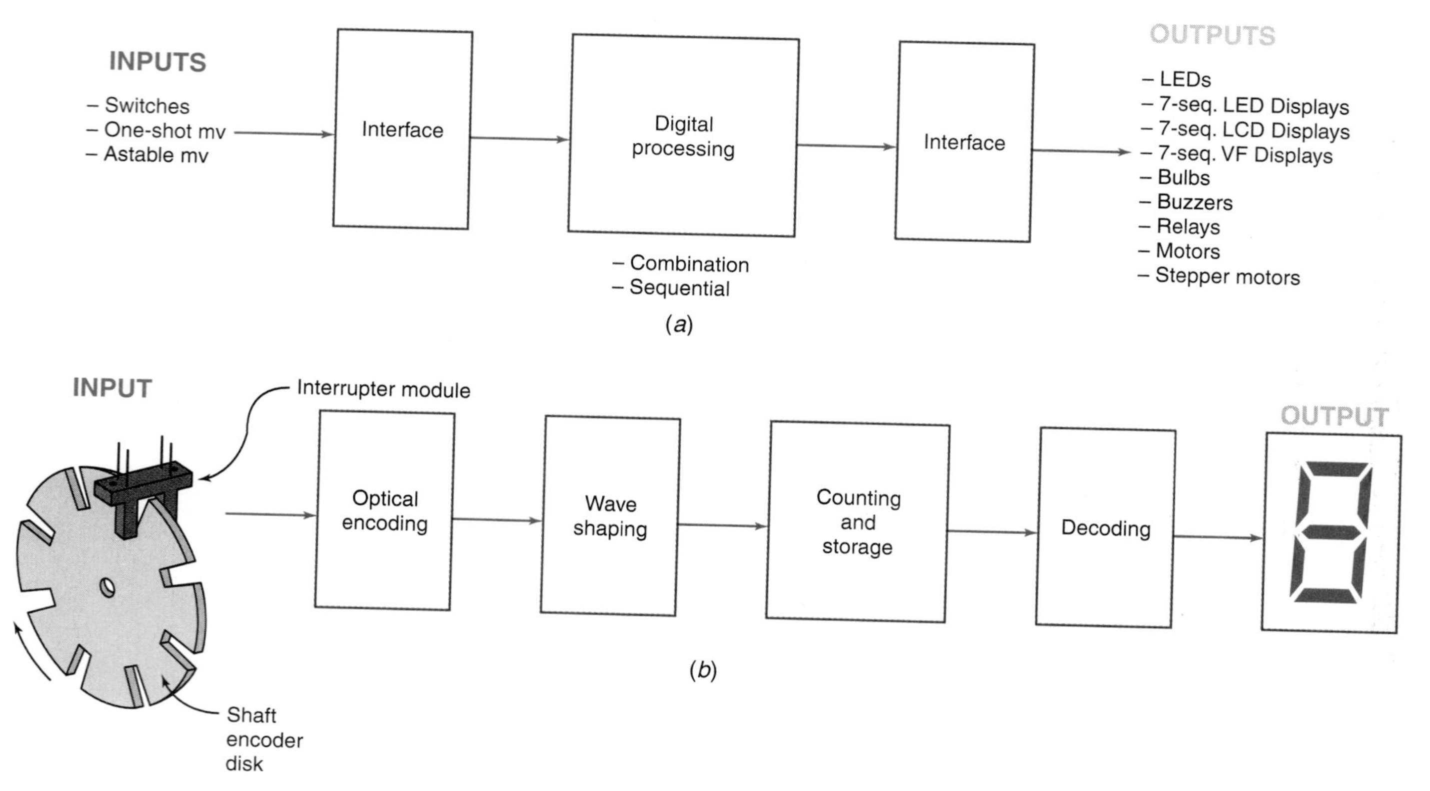

Fig. 8-21 (*a*) Typical inputs, processing, and outputs from a digital system. (*b*) Interrupter module optically encoding disk that drives counter.

devices such as clocks (both astable and monostable) and switches. In this section we will add a new input device.

A block diagram for the system we are studying in this section is drawn in Fig. 8-21(*b*). We will use *optical encoding* for input, the counter will accumulate the count, and the seven-segment display will form the output. The opto-coupled interrupter module will sense each time the infrared light beam is interrupted and send this as a signal to the wave shaper and then to the decade counter/accumulator. Finally, the BCD count will be decoded and the number of slots in the shaft encoder disk that have moved by the stationary opto-coupled interrupter module will be displayed.

The *opto-coupled interrupter module* or *optical sensor* is constructed with an infrared light-emitting diode aimed at a phototransistor across the slot. A schematic symbol for the opto-coupled interrupter module is reproduced in Fig. 8-22(*a*). If current flows through the infrared diode on the emitter (*E*) side, the NPN phototransistor is activated on the detector (*D*) side of the module. If the light from the LED is blocked, the phototransistor on the detector side of the module is deactivated (turned off). The H21A1 (ECG3100) opto-coupled interrupter module is shown in Fig. 8-22(*b*). Notice that pins 1 and 2 of the H21A1 interrupter module are for the emitter side or infrared light-emitting diode. Typical wiring of the emitter side of the interrupter module is also detailed in Fig. 8-22(*a*). Pins 3 and 4 of the H21A1 interrupter module are for the detector side or NPN phototransistor. Typical wiring of the detector side of the interrupter module using a 10-kΩ pull-up resistor is also shown in Fig. 8-22(*b*). The signal from the detector side of the interrupter module is then sent on to the wave shaping circuit.

A wiring diagram for a simple system that counts the number of pulses coming from the opto-coupled interrupter module is detailed in Fig. 8-23 on page 192. When an opaque object interrupts the light beam in the module, the phototransistor is deactivated (turned off) and the input to the 7414 Schmitt trigger inverter is pulled HIGH by the 10-k pull-up resistor. The output of the inverter goes from

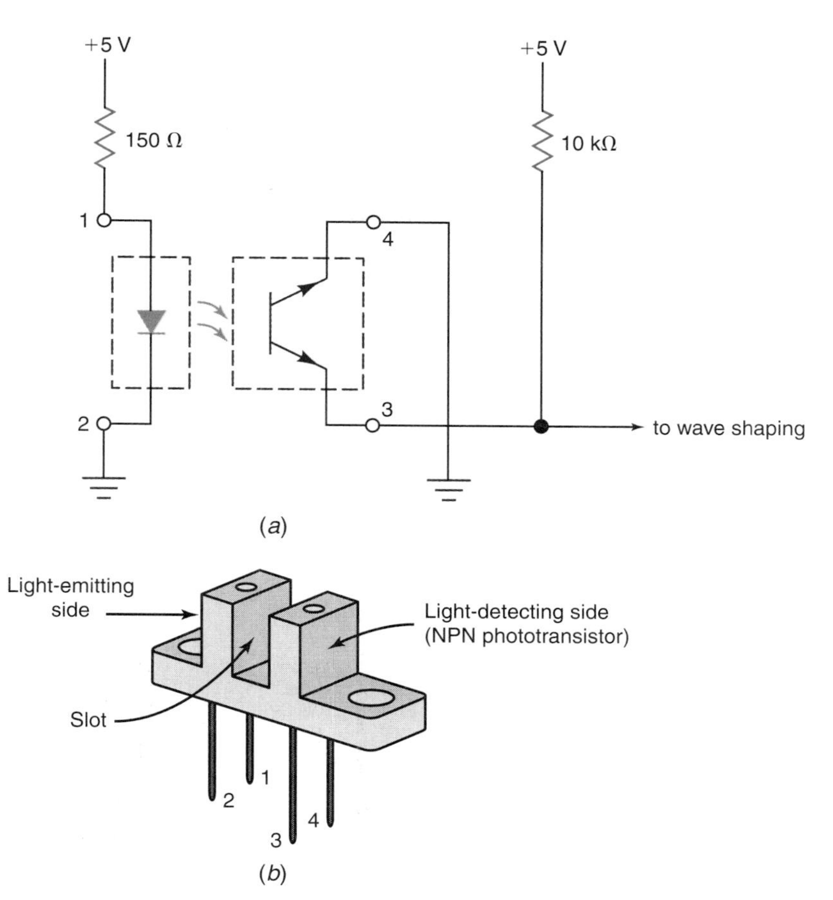

Fig. 8-22 (*a*) Schematic diagram of opto-coupled interrupter module's emitter and light detecting sides wired. (*b*) Drawing of H21A1 opto-coupled interrupted module (slot type).

HIGH to LOW. When the opaque object is removed from the slot of the opto-coupled interrupter module, the infrared light crosses the slot striking the base of the phototransistor. The phototransistor is activated (turned on) and the voltage at pin 3 of the inverter goes from HIGH to LOW. Then the output of the wave shaper goes from LOW to HIGH, which will trigger the 74192 to count upward by 1. The 7447 IC decodes the BCD input to a seven-segment code and lights the appropriate segments on the LED display.

In summary, the optical encoder/counter system in Fig. 8-21(*b*) *increments* (one count upward) the count each time an opening in the encoder disk passes through the slot in the interrupter module. The opto-coupled interrupted module uses infrared light so the ambient light does not cause false triggering. Remember that the infrared diode emits the correct wavelength of light for the phototransistor to detect.

Two common types of optical sensors are the slot-type module used in the last optical encoder/counter system and the *reflective-type sensor.* A sketch of a common reflective-type optical sensor is shown in Fig. 8-24(*a*) on page 193. Notice that it has two holes in the front. One is an infrared emitting diode while the other is the receiver part of the optical sensor, which is a phototransistor. This reflective-type optical sensor is carefully aimed at a target such as the disk shown in Fig. 8-24(*b*). The white areas reflect light and turn on the output phototransistor while the dark stripes absorb light and turn off the phototransistor.

TEST

Supply the missing word or words in each statement.

38. Refer to Fig. 8-21(*b*). The ________ (decoder, interrupter module) is the device that does the job of optical encoding in this circuit.

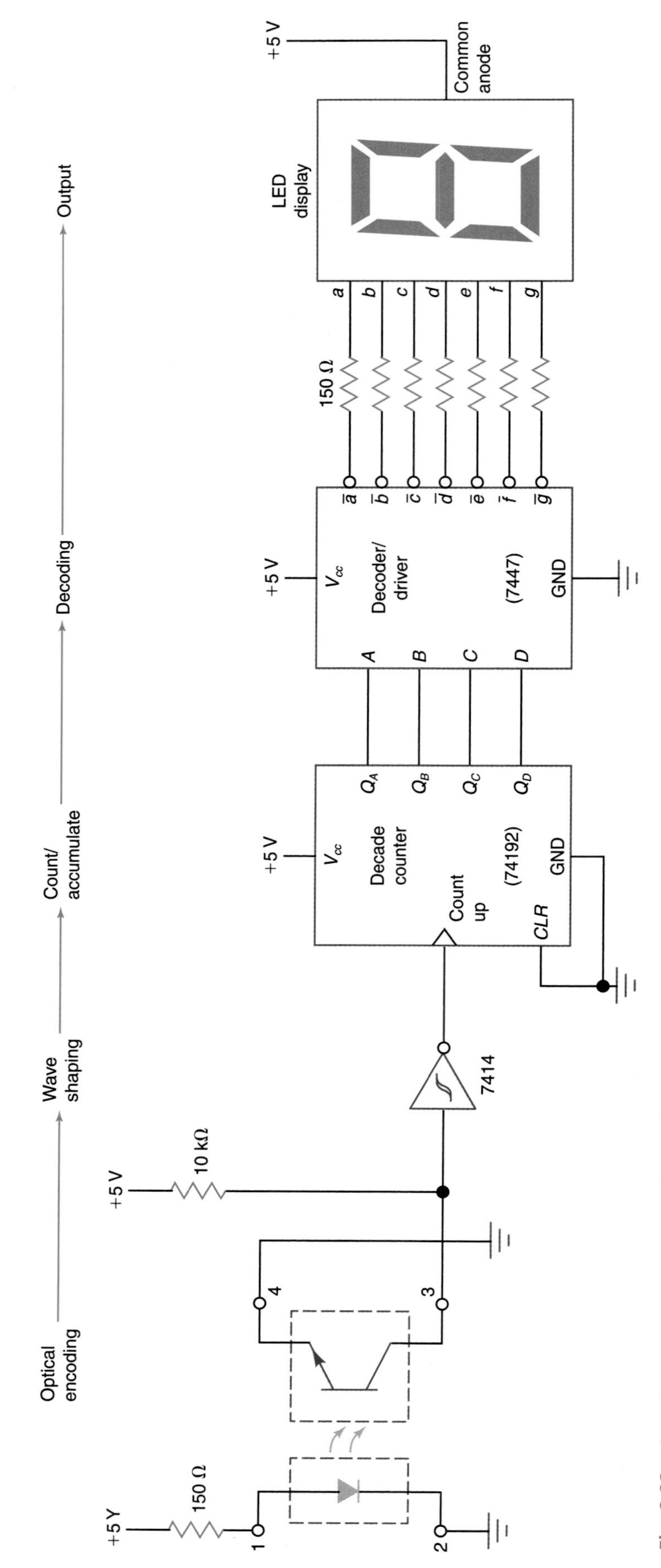

Fig. 8-23 Counter system using optical encoding.

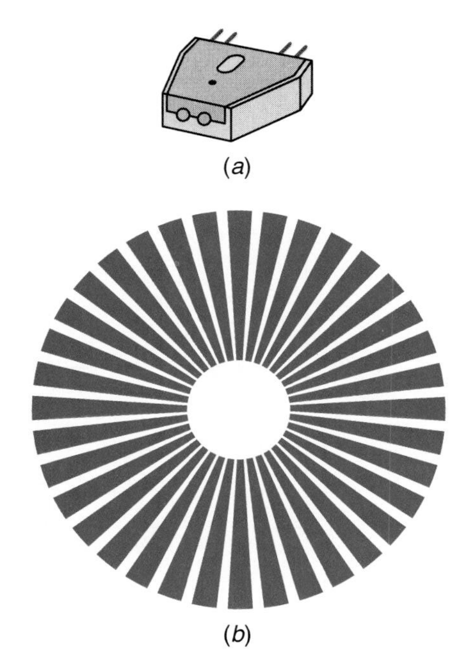

Fig. 8-24 (*a*) Reflective-type optical encoder. (*b*) Shaft encoder disk used with reflective-type optical encoder.

39. Refer to Fig. 8-22(*a*). The diode on the emitter side of the interrupter module gives off _________ (infrared, ultraviolet) light.
40. Refer to Fig. 8-22(*a*). The _________ (phototransistor, germanium transistor) on the detector side of the interrupter module is sensitive to infrared light and does not trigger because of white room light.
41. Refer to Fig. 8-21(*b*). The optical sensor being used in this system is classified as a _________ (reflective-type, slot-type) module.
42. Refer to Fig. 8-23. The count on the display increments when the output of the Schmitt trigger inverter goes from _________ (H to L, L to H).
43. Refer to Fig. 8-21(*b*). The count on the display increments when the encoder disk opening just _________ (enters, leaves) the interrupter module.
44. Refer to Fig. 8-23. The 74192 operates as a _________ (decade, mod-16) counter in this circuit and it also performs the task of _________ (decoding the count, temporarily storing the count).

8-10 Using a CMOS Counter in an Electronic Game

This section will feature a CMOS counter being used in an electronic game. The game is the classic computer game of "guess the number." In the computer version, a random number is generated and the player tries to guess the unknown number. The computer responds with one of three responses: correct, too high, or too low. The player can then guess again until he or she zeros in on the unknown number. The player who uses the fewest guesses wins the game.

The schematic for a simple electronic version of this game is drawn in Fig. 8-25. To operate the game, first press the push-button switch (SW_1). This allows the approximately 1-kHz signal into the clock input of the binary counter. When the push button is released, a random binary number (from 0000 to 1111) is held by the counter at the *B* inputs to the *74HC85 4-bit magnitude comparator.* The player's guess is entered at the *A* inputs to the comparator IC. If the random number (*B* inputs) and the guess (*A* inputs) are equal, then the $A = B_{OUT}$ output will be activated (HIGH) and the green LED will light. This means the guess was correct. After a correct guess a new random number should be generated by pressing SW_1.

74HC85 4-bit magnitude comparator

If a player's guess (*A* inputs) is lower than the random number (*B* inputs), the comparator will activate the $A < B_{OUT}$ output. The yellow LED will light. This means that the guess was too low and the player should try again by entering a somewhat higher number.

Finally, if a player's guess (*A* inputs) is higher than the random number (*B* inputs), the comparator will activate the $A > B_{OUT}$ output. The red LED will light. This means that the guess was too high and the player should try again.

Figure 8-26 on page 195 gives greater detail on the operation of the *74HC85 4-bit magnitude comparator IC.* The pin diagram is shown in Fig. 8-26(*a*). This is the top view of the DIP 74HC85 CMOS IC. The truth table for the 74HC85 comparator is reproduced in Fig. 8-26 (*b*).

The 74HC85 comparator has three "extra" inputs used for *cascading comparators.* Typical cascading of 74HC85 magnitude comparators is shown in Fig. 8-27 on page 196. This circuit compares the magnitude of the two 8-bit binary

Cascading comparators

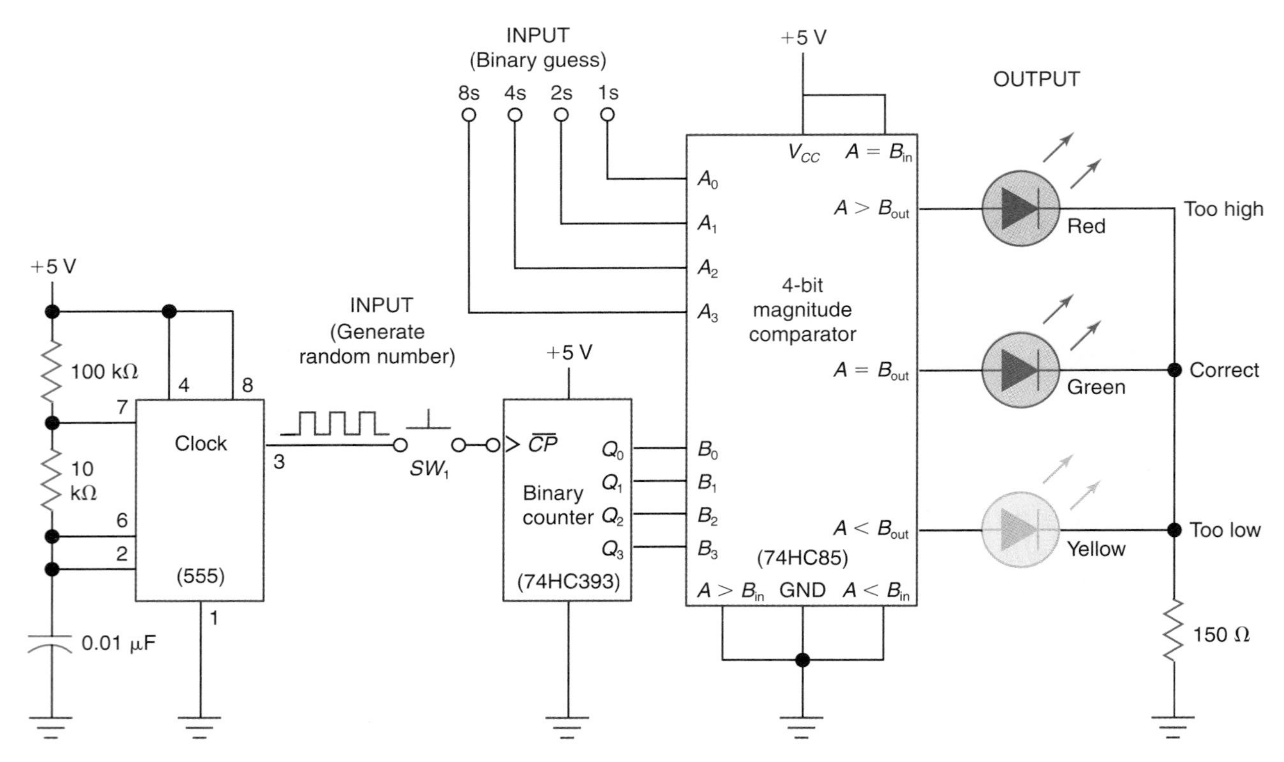

Electronic "guess the number" game

Fig. 8-25 Electronic "guess the number" game.

words $A_7 A_6 A_5 A_4 A_3 A_2 A_1 A_0$ and $B_7 B_6 B_5 B_4 B_3 B_2 B_1 B_0$. The output from IC_2 is one of three responses ($A > B$, $A = B$, or $A < B$).

TEST

Answer the following questions.

45. Refer to Fig. 8-25. If the binary counter holds the number 1001 and your guess is 1011, the ________ (color) LED will light indicating your guess is ________ (correct, too high, too low).
46. Refer to Fig. 8-25. How do you generate a random number before guessing?
47. Refer to Fig. 8-25. The 555 timer is wired as a(n) ________ (astable, monostable) multivibrator.
48. Refer to Fig. 8-28 on page 197. List the *color* of the output LED that is lit for each time period (t_1 to t_6).

8-11 Sequential Logic Troubleshooting Equipment

Troubleshooting equipment

Until now, troubleshooting has been performed on combinational logic circuits (gating circuits). Sequential logic circuits are composed of flip-flops and as such are somewhat more difficult to troubleshoot. Your favorite pieces of test equipment to this time have been the logic probe and the voltmeter. Besides these, several other pieces of test equipment are handy when troubleshooting sequential logic circuits. It is common to have available a logic pulser, logic monitors, and an oscilloscope. You may also have available a logic analyzer and an IC tester.

Commercial logic probe

A *commercial logic probe* is pictured in Fig. 8-29(*a*) on page 198. Notice that this logic probe has a diode-transistor/transistor-transistor logic (DTL/TTL) or CMOS selection switch. This switch selects the family of the IC being tested. If the MEM/PULSE switch is in the pulse position, any pulse as short as about 50 ns will be displayed on the pulse LED for about 0.3 s. If the MEM/PULSE switch is in the MEM (memory) position, any signal change at the tip (1 to 0 or 0 to 1) will activate the pulse LED. In the memory mode, the pulse LED lights continuously on any single pulse.

Logic monitor

Another logic level measuring device used with digital ICs is the *logic monitor.* One logic monitor is shown in Fig. 8-29(*b*). The ribbon cable connects to an IC test clip something like the ones shown in Fig. 8-29(*c*). The logic levels of all 16 pins on an IC are then visible on

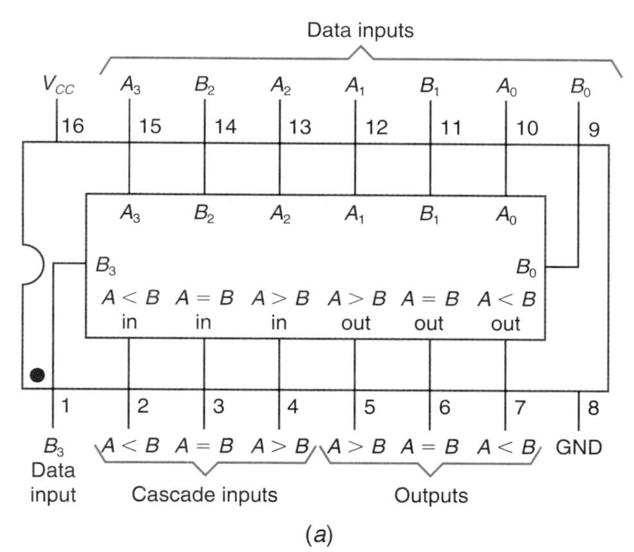

(*a*)

Truth Table—74HC85 Magnitude Comparator IC

COMPARING INPUTS				CASCADING INPUTS			OUTPUTS		
A_3, B_3	A_2, B_2	A_1, B_1	A_0, B_0	$A > B$	$A < B$	$A = B$	$A > B$	$A < B$	$A = B$
$A_3 > B_3$	X	X	X	X	X	X	H	L	L
$A_3 < B_3$	X	X	X	X	X	X	L	H	L
$A_3 = B_3$	$A_2 > B_2$	X	X	X	X	X	H	L	L
$A_3 = B_3$	$A_2 < B_2$	X	X	X	X	X	L	H	L
$A_3 = B_3$	$A_2 = B_2$	$A_1 > B_1$	X	X	X	X	H	L	L
$A_3 = B_3$	$A_2 = B_2$	$A_1 < B_1$	X	X	X	X	L	H	L
$A_3 = B_3$	$A_2 = B_2$	$A_1 = B_1$	$A_0 > B_0$	X	X	X	H	L	L
$A_3 = B_3$	$A_2 = B_2$	$A_1 = B_1$	$A_0 < B_0$	X	X	X	L	H	L
$A_3 = B_3$	$A_2 = B_2$	$A_1 = B_1$	$A_0 = B_0$	H	L	L	H	L	L
$A_3 = B_3$	$A_2 = B_2$	$A_1 = B_1$	$A_0 = B_0$	L	H	L	L	H	L
$A_3 = B_3$	$A_2 = B_2$	$A_1 = B_1$	$A_0 = B_0$	X	X	H	L	L	H
$A_3 = B_3$	$A_2 = B_2$	$A_1 = B_1$	$A_0 = B_0$	H	H	L	L	L	L
$A_3 = B_3$	$A_2 = B_2$	$A_1 = B_1$	$A_0 = B_0$	L	L	L	H	H	L

(*b*)

Fig. 8-26 CMOS magnitude comparator IC (74HC85). (*a*) Pin diagram. (*b*) Truth table. *(Courtesy of National Semiconductor Corporation.)*

74HC85 4-bit magnitude comparator IC

the LEDs. The LEDs on the logic monitor light when a HIGH logic level is present. They do not light when the pin is LOW or in the undefined region between HIGH and LOW. The threshold voltage for indicating a HIGH on the unit in Fig. 8-29(*b*) is about 2.3 V for TTL and 70 percent of V_{cc} for CMOS. Notice in Fig. 8-29(*b*) that the logic monitor can check either CMOS or TTL/DTL circuits. Lower (or higher) voltage thresholds can be used to indicate HIGH levels on the logic monitor by using the variable mode and adjusting the variable threshold control shown at the lower left in Fig. 8-29(*b*).

Logic monitors are many times constructed in an enlarged IC test clip like those in Fig. 8-29(*c*). One such logic monitor is pictured in Fig. 8-29(*d*). This logic monitor is referred to as a *logic clip* by some manufacturers and technicians. The logic monitor pictured in Fig. 8-29(*d*) clips over the 16 pins of a DIP IC. This unit is designed for TTL or DTL and has a threshold voltage of about 2.0 V.

Logic clip

A commercial *digital logic pulser* is pictured in Fig. 8-30 on page 199. This unit outputs a single pulse when the push button is pressed. According to the manufacturer, the pulse width is 1.5 μs with TTL selected. The

Digital logic pulser

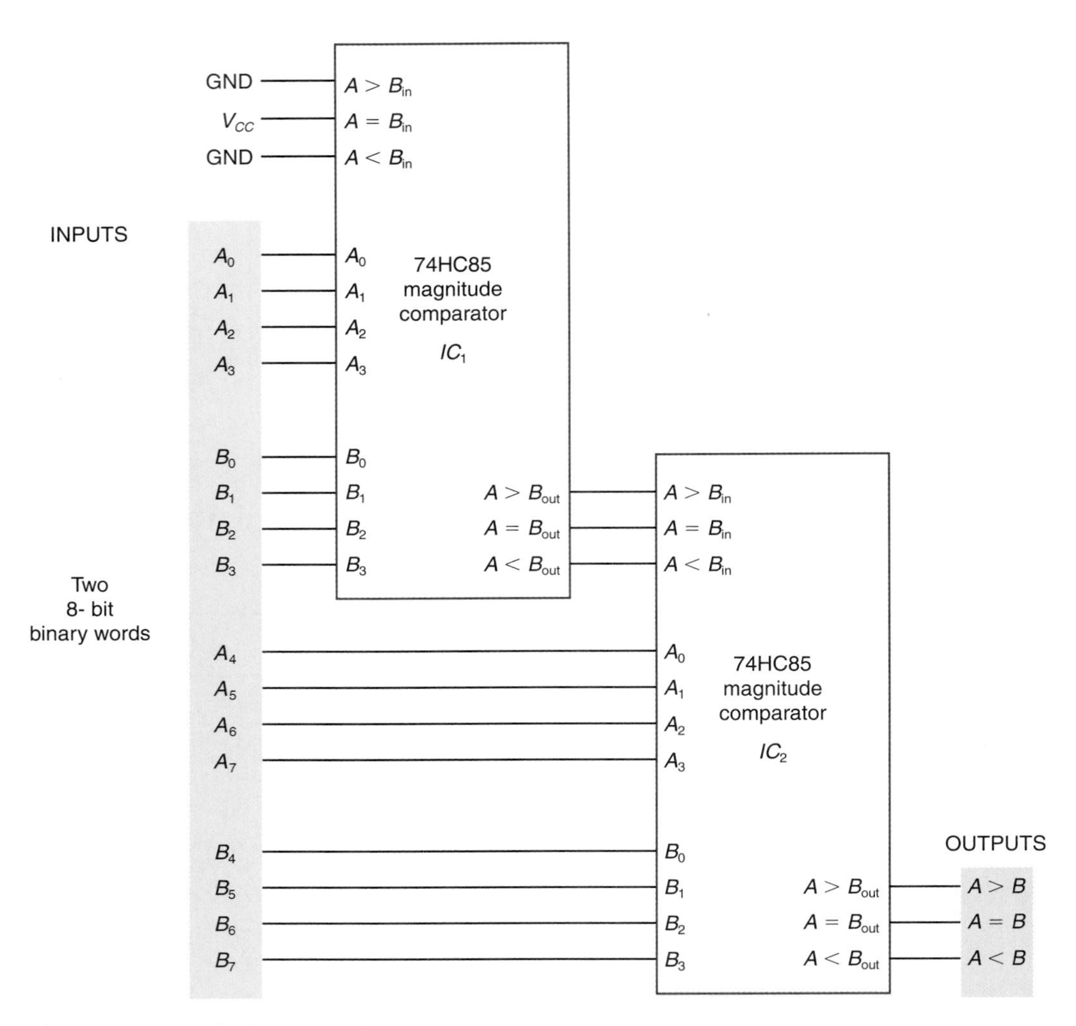

Cascaded magnitude comparators

Fig. 8-27 Cascaded magnitude comparators.

Tristating

DMM/logic probe

pulse width is 10 μs with CMOS selected. If the push button on this digital pulser is pressed and held down for a time, a pulse train of 100 pulses per second is generated at the tip. The tip is isolated by high impedance when the push button is not pressed. This is called *tristating*. The output can have three levels: LOW, HIGH, or high impedance.

Using the digital pulser in Fig. 8-30, a single press of the pulse push button emits a flash on the pulse indicator LED. When the push button is held down for a time, the pulse LED indicator lights continuously. This indicates a pulse train.

One of the fundamental test instruments used in digital electronics is the logic probe, which we have seen and used earlier. Sometimes, however, it is valuable to know a more exact voltage at a point in the circuit instead of simply HIGH, LOW, or undefined. Some of the undefined voltages can be high enough to be considered a HIGH by the next digital IC. In such cases, the probe-type combination *DMM/logic probe,* such as the one pictured in Fig. 8-31 on page 199, can be extremely handy. The logic-probe section can test either TTL or CMOS circuitry. When in the logic-probe mode, the dc voltmeter also displays the exact dc voltage. The DMM section has a full 3½ digit LCD, and can test ac or dc voltage, ac or dc current, and resistance. The DMM/logic probe features an audible continuity tester and a diode test. It also provides a *data hold function* and has high input impedance. A small handheld package, this DMM/logic probe is a favorite for simple digital troubleshooting.

Beginning students, busy instructors, and efficient technicians find IC testers useful in

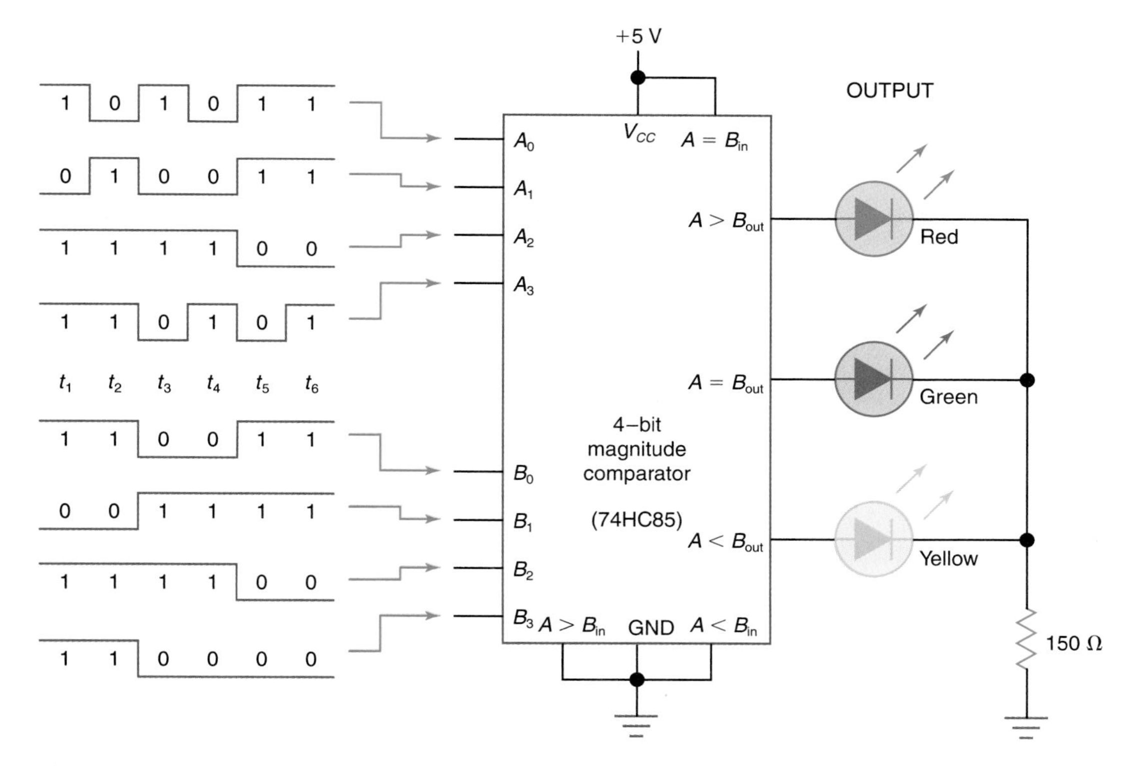

Fig. 8-28 Comparator problem for test item 48.

ensuring the ICs used in a circuit work according to specifications. The *portable IC tester* pictured in Fig. 8-32 on page 199 provides a quick method of verifying that a chip is satisfactory. This portable IC tester tests 14 to 20 pin devices plugged into a *zero-insertion force (ZIF) socket.* It includes test, search, type, and reset keys. It has a 16-character LCD to display results. The portable IC tester can test nearly all of the TTL series, CMOS series, *dynamic RAM* (DRAM) 41 and 44 series ICs. It can even identify unknown ICs. The portable IC tester is for out-of-circuit testing of ICs; more sophisticated programmable IC testers are available to test IC both in and out of the circuit.

As in most areas of electronics, the oscilloscope is probably the most important research, development, and troubleshooting instrument. A *dual-trace oscilloscope,* such as the one pictured in Fig. 8-33 on page 200, is particularly valuable in comparing waveforms and observing important timing relationships in digital circuits. Oscilloscopes are very good for observing repetitive or synchronous events in a digital system. However, they have difficulties when observing glitches or asynchronous events in digital systems. A *glitch* is a random signal that causes problems in a digital system. It may be caused by noise in the system or a design error. A glitch usually takes the form of a very short L-H-L or H-L-H pulse that appears when the waveform should be at a steady LOW or HIGH.

If you look very closely at the Tektronix TDS 220 scope in Fig. 8-33 you will observe that it is labeled a *Digital Real Time™ oscilloscope.* Scopes are classified as either *analog* or *digital.* The less expensive analog scopes display a signal in "real time" by tracing it (voltage versus time) on a cathode ray tube (CRT). The 100-mHz model pictured in Fig. 8-33 looks a lot like an analog scope you may have used, but your clue that it is digital in nature is the alphanumeric information and menus on the screen. The digital scope takes samples of the input voltages many times per cycle, stores this as digital information, and then reconstructs or graphs an output waveform from the digital information. Because the information is digitized, it is possible to store the information and then output it to the screen or to an external device like a printer or computer.

The Digital Real Time™ oscilloscope shown in Fig. 8-33 is lightweight, small in

Glitch

Portable IC tester

Zero-insertion force (ZIF) IC socket

Digital Real-Time™ oscilloscope

Dynamic RAM (DRAM)

Dual-trace oscilloscope

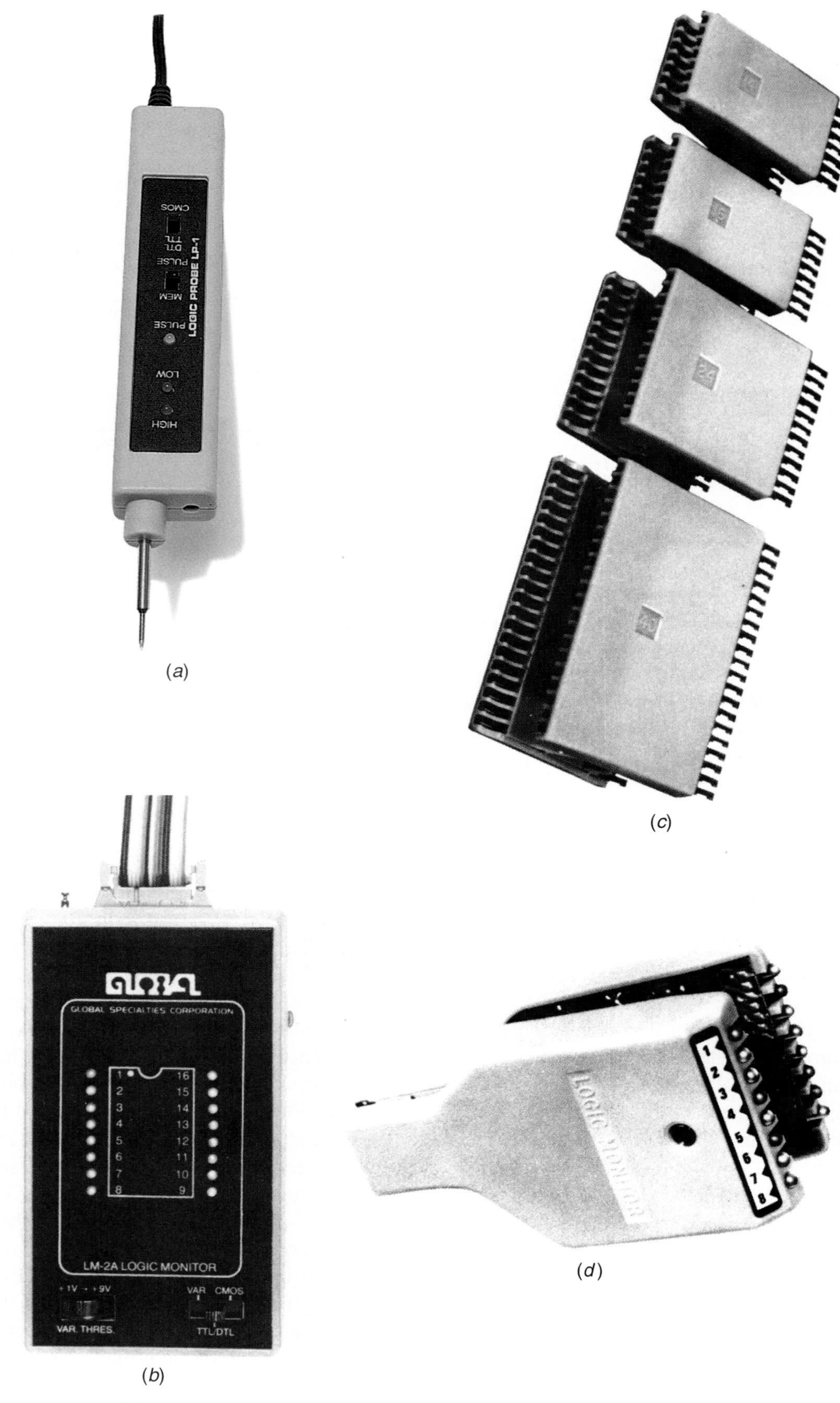

Fig. 8-29 (*a*) Logic probe. (*b*) Logic monitor. (*c*) IC clips. (*d*) Logic monitor (logic clip). *(Courtesy of Interplex Electronics, Inc.)*

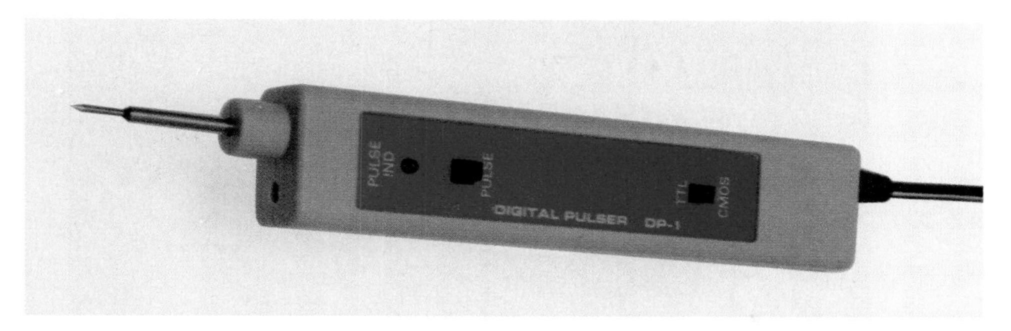

Fig. 8-30 Commercial digital logic pulser for TTL or CMOS circuits. *(Courtesy of Interplex Electronics, Inc.)*

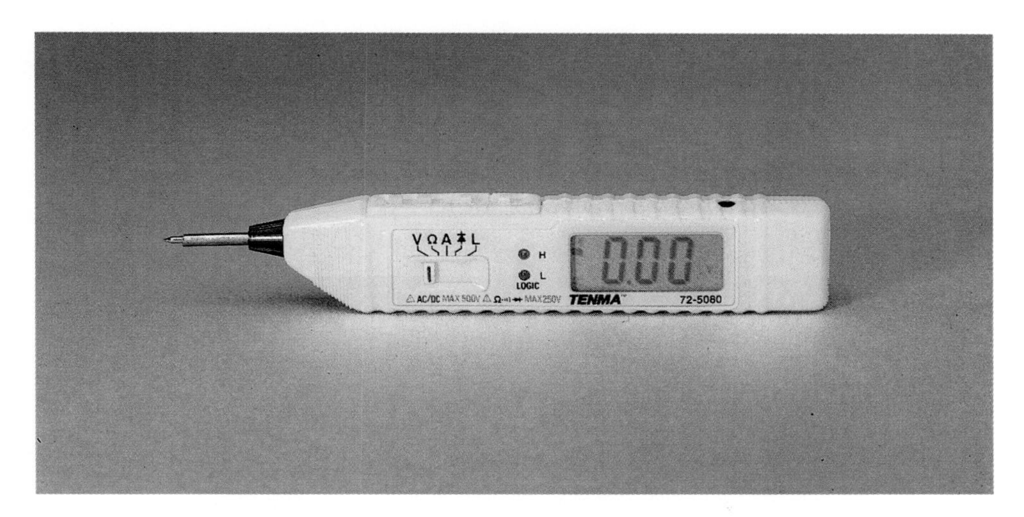

Fig. 8-31 DMM/logic probe instrument. *(Courtesy of MCM Electronics.)*

size, and easy to learn by those who have already used an analog scope. The Tektronix TDS 220 dual-trace digital scope features a sampling rate of 1 GS/s (a billion samples per second) and a bandwidth of 100 MHz. Automatic measurements are generated and displayed for time period and frequency. The storage capacity is 2500 sample points per channel. While most analog oscilloscopes use CRT, an older technology, for the visual display, the TDS 220 scope uses a backlit LCD. Because of the LCD, the front-to-back depth of the scope is less than 5 in. The overall weight of the scope is about 4 lb. The screen also has voltage and time cursors. The Tektronix TDS 220 is a step above the normal analog oscilloscope and is available for a modest price.

Fig. 8-32 Portable IC tester. *(Courtesy of Jameco.)*

Logic analyzer

The *logic analyzer* is a specialized instrument that typically has many input channels (16 to more than 100). It samples and stores multiple input signals and then displays them on a CRT. A typical logic analyzer is pictured in Fig. 8-34. The advantages of the logic analyzer over the oscilloscope are its many input channels, signal sampling methods, relatively large memory or storage, sophisticated output (CRT), computer compatability, and ability to detect asynchronous events or glitches. These

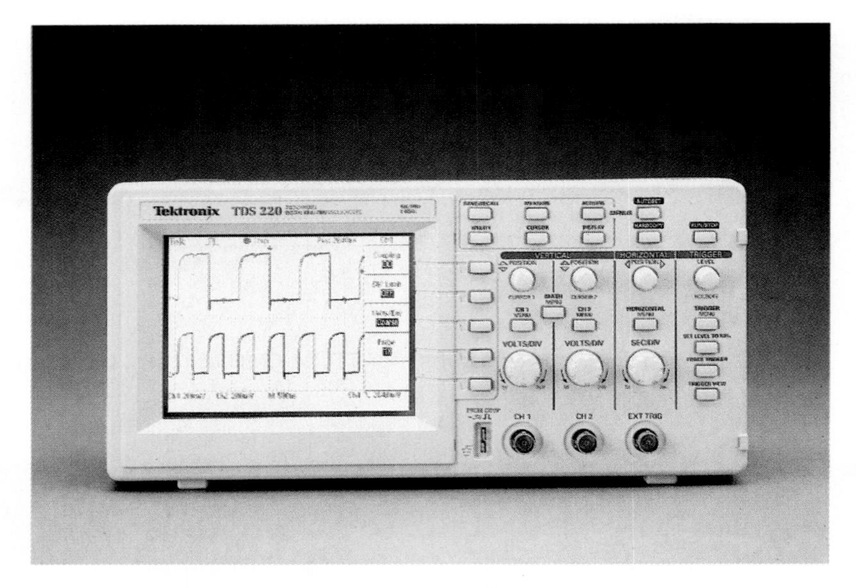

Dual-trace oscilloscope

Fig. 8-33 A dual-trace Digital Real Time™ oscilloscope. *(Courtesy of Tektronix, Inc.)*

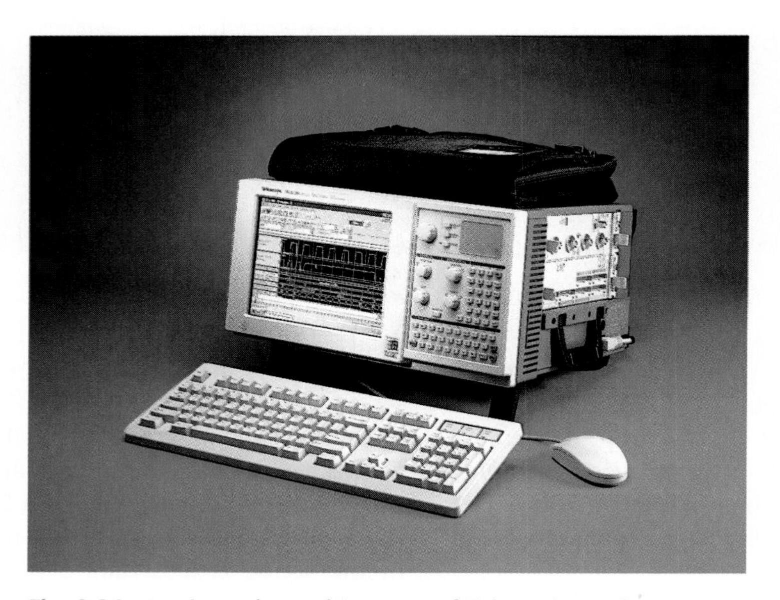

Fig. 8-34 Logic analyzer. *(Courtesy of Tektronix, Inc.)*

glitches can be stored for later analysis on the CRT. Logic analyzer are often expensive.

TEST

Answer the following questions.

49. Digital circuits that use flip-flops (such as counters) are classified as _________ (combinational, sequential) logic circuits.
50. A logic _________ (monitor, probe) can check the logic level of all the pins on an IC at one time.
51. Refer to Fig. 8-30. A single press of the pulse push button on the logic pulser sends out a _________ (stream of pulses at about 100 Hz; single pulse) to the tip.
52. An analog oscilloscope is a very good instrument for observing _________ (asynchronous, repetitive) digital signals.
53. Refer to Fig. 8-31. This instrument looks like a logic probe but is also a(n) _________ (ADC, DMM).
54. Refer to Fig. 8-32. This portable IC tester can test only the 7400 series of TTL ICs (T or F).

55. Refer to Fig. 8-33. This is a(n) _________ (analog, digital) oscilloscope.
56. The advantage of a digital over an analog scope is that input information that has been digitized can be stored, sent to a printer, used for calculations, or displayed on the screen (T or F).
57. Typically, _________ (analog, digital) oscilloscopes are more expensive.
58. An expensive test instrument called a logic _________ (analyzer, probe) features multiple input channels/displays, higher rates of signal sampling, greater signal storage, and computer compatibility.

8-12 Troubleshooting a Counter

Troubleshooting a counter

2-bit ripple counter

Consider the job of troubleshooting the faulty 2-bit ripple counter shown in Fig. 8-35(*a*). For your convenience, a pin diagram for the 74HC76 IC used in this circuit is shown in Fig. 8-35(*b*). Note that all of the input and output markings in Fig. 8-35(*a*) and (*b*) are *not* the same. For instance, the asynchronous preset

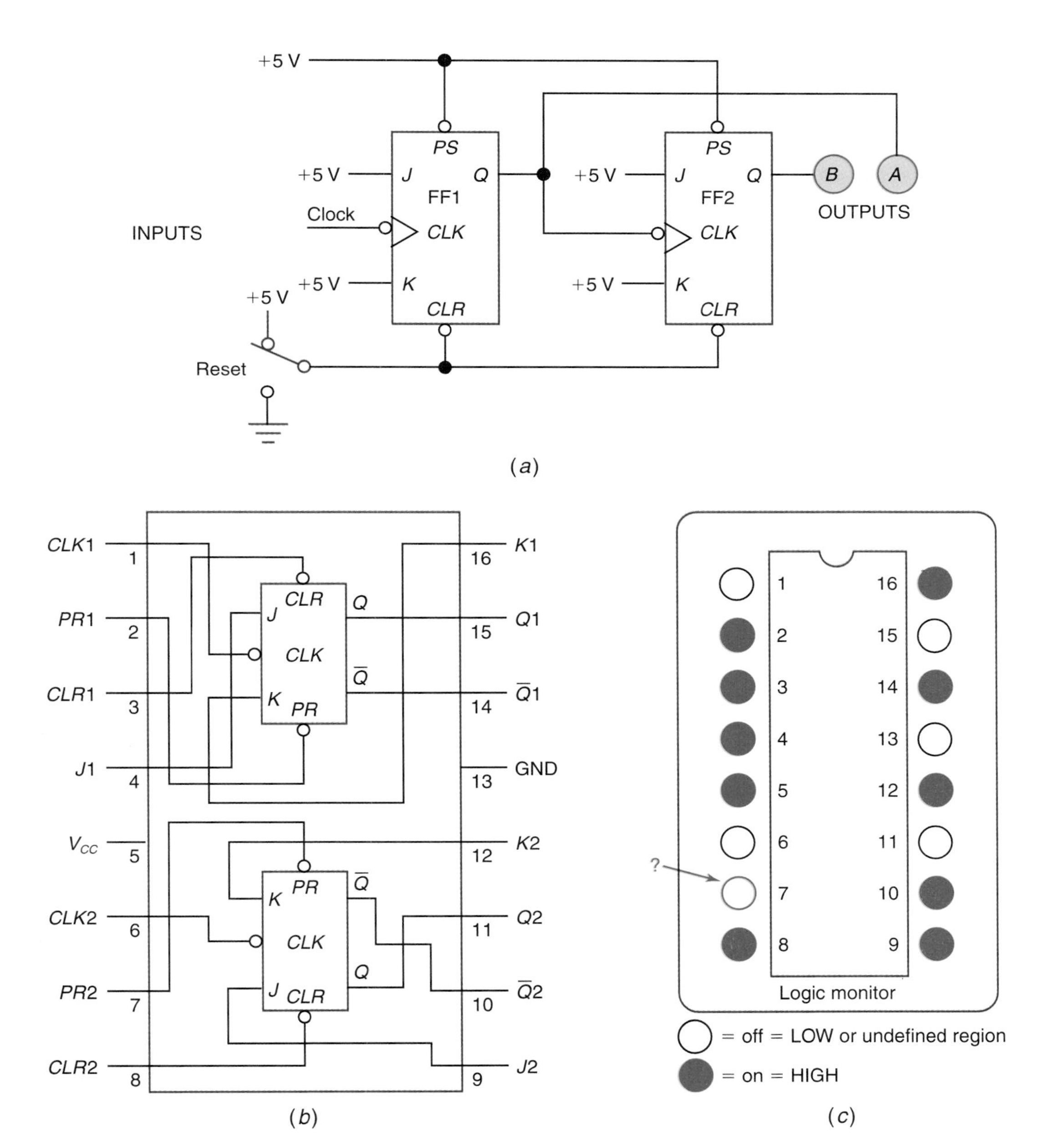

Troubleshooting a faulty counter

"Floating" input problem

Fig. 8-35 (*a*) Faulty 2-bit ripple counter circuit used in troubleshooting example. (*b*) Pin diagram for 7476 J-K flip-flop IC. *(Courtesy of National Semiconductor Corporation.)* (*c*) Logic monitor reading after momentarily resetting the faulty 2-bit counter.

inputs on the logic diagram are labeled *PS*. The same inputs are labeled *PR* (for preset) by National Semiconductor Corporation. The labels on the pins may be different from manufacturer to manufacturer. However, the pins on an IC labeled as a 74HC76 serve the same *function* even if the labeling is different.

It is found that the faulty 2-bit counter circuit can be cleared to 00 by the reset switch at the left in Fig. 8-35(*a*). The IC seems to be operating at the correct temperature, and the technician can see no signs of trouble.

A digital logic pulser (like the one in Fig. 8-30) is used to pulse the *CLK* input on FF 1. According to the pin diagram, the tip of the digital pulser must touch pin 1 of the 74HC76 IC. Upon repeated single pulses, the counting sequence is 00 (reset), 01, 10, 11, 10, 11, 10, 11, and so on. The *Q* output of FF 2 seems to be "stuck HIGH"; however, the asynchronous clear (*CLR*), or reset, switch can drive it LOW.

Power is then turned off in the circuit in Fig. 8-35(*a*). A TTL logic monitor like the one in Fig. 8-29(*b*) is clipped over the pins on the 74HC76 IC. Power is turned on again. The reset switch is activated. The results displayed on the logic monitor after the reset are shown in Fig. 8-35(*c*). Compare the logic levels shown on the logic monitor with your expectations. You must use the manufacturer's pin diagram furnished in Fig. 8-35(*b*). In looking over the logic levels pin by pin, the LOW or undefined logic level at pin 7 should cause concern. This is the asynchronous preset (*PS* or *PR*) input and should be HIGH according to the logic diagram in Fig. 8-35(*a*). If it is LOW or in the undefined region it can cause the *Q* output of FF 2 to be in the "stuck HIGH" condition.

Most important troubleshooting tools

A logic probe such as the one in Fig. 8-29(*a*) is used to check pin 7 of the 74HC76 IC. Both LEDs on the logic probe remain off. This means neither a LOW nor a HIGH logic level is present. Pin 7 appears to be floating in the undefined region between LOW and HIGH. The IC is interpreting this as LOW at some times and as HIGH at other times.

For related information, visit the website for Texas Instruments.

The IC is removed from the 16-pin DIP IC socket. It is found that pin 7 of the IC is bent under and not making contact with the IC socket. This caused it to float. The fault is illustrated in Fig. 8-36. This common fault is very hard to see when the IC is seated in the IC socket.

In this example, several tools were used in troubleshooting. First, the logic diagram and your knowledge of how it works are most important. Second, a manufacturer's pin diagram was used. Third, a digital logic pulser was used to inject single pulses. Fourth, a logic monitor checked the logic level at all pins of the 74HC76 IC. Fifth, a logic probe was used to check the suspected pin of the IC. Finally, your knowledge of the circuit and visual observation solved the problem. *Your knowledge of the circuit's normal operation and your powers of observation are probably the most important troubleshooting tools.* Logic pulses, logic probes, logic monitors, DMMs, logic analyzers, IC testers, and oscilloscopes are only aids to your knowledge and powers of observation.

The example of a floating input caused by a bent-under pin is a very common problem in student-constructed circuits. It is good practice to make sure all inputs go to the proper logic level. This is true for TTL and especially true for CMOS circuits.

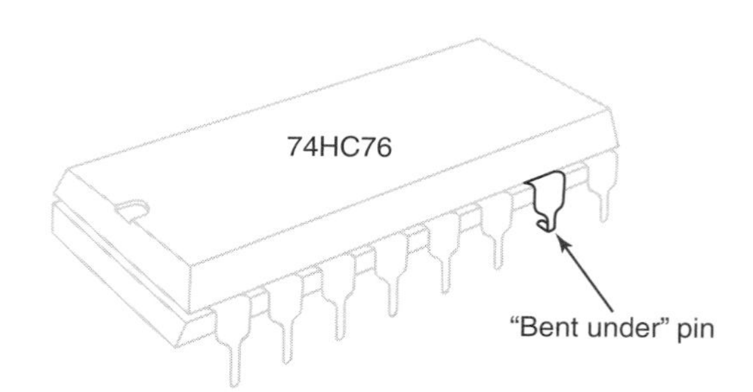

Fig. 8-36 Bent pin caused input to float.

TEST

Supply the missing word or words in each statement.

59. Refer to Fig. 8-35. Pins 4, 9, 12, and 16 to the *J* and *K* inputs of the flip-flops should all be ________ (HIGH, LOW) in this circuit.
60. Refer to Fig. 8-35. Pins 3 and 8 to the ________ inputs of the flip-flops should follow the logic state of the reset switch.
61. Refer to Fig. 8-35. Pins 2 and 7 to the ________ inputs of the flip-flops should be ________ (HIGH, LOW) in this circuit.
62. Refer to Fig. 8-35. The fault in this circuit was located at pin ________ (number). It was ________ rather than being at a HIGH logic level.

Chapter 8 Review

Summary

1. Flip-flops are wired together to form binary counters.
2. Counters can operate asynchronously or synchronously. Asynchronous counters are called ripple counters and are simpler to construct than synchronous counters.
3. The modulus of a counter is how many different states it goes through in its counting cycle. A mod-5 counter counts 000, 001, 010, 011, 100 (0, 1, 2, 3, 4 in decimal).
4. A 4-bit binary counter has four binary place values and counts from 0000 to 1111 (decimal 0 to 15).
5. Gates can be added to the basic flip-flops in counters to add features. Counters can be made to stop at a certain number. The modulus of a counter can be changed.
6. Counters are designed to count either up or down. Some counters have both features built into their circuitry.
7. Counters are used as frequency dividers. Counters are also widely used to count or sequence events and temporarily store data.
8. Manufacturers produce a wide variety of self-contained IC counters. They produce detailed data sheets for each counter IC. Several TTL and CMOS counter ICs were studied in this chapter.
9. Many variations in pin labeling and logic symbols occur from manufacturer to manufacturer.
10. A transducer such as an optical sensor can be used to count real-world events such as in shaft encoding. Optical encoders come in slot types and reflective types, and both are based on an infrared diode shining on an output phototransistor.
11. A magnitude comparator will compare two binary numbers and decide if $A = B$, $A > B$, or $A < B$. Magnitude comparator ICs can be cascaded to compare larger binary numbers.
12. The technician's knowledge of the circuit and powers of observation are the most important tools in troubleshooting. The logic probe, voltmeter, DMM, logic monitor, digital pulser, logic analyzer, IC tester, and oscilloscope aid the technician's observations when troubleshooting sequential logic circuits.

Chapter Review Questions

Answer the following questions.

8-1. Draw a logic symbol diagram of a mod-8 ripple up counter. Use three J-K flip-flops. Show input *CLK* pulses and three output indicators labeled *C*, *B*, and *A* (*C* indicator is MSB).

8-2. Draw a table (similar to Fig. 8-1) showing the binary and decimal counting sequence of the mod-8 counter in question 8-1.

8-3. Draw a waveform diagram (similar to Fig. 8-2(*b*)) showing the eight *CLK* pulses and the outputs (*Q*) of FF 1, FF 2, and FF 3 of the mod-8 counter from question 8-1. Assume you are using negative-edge-triggered flip-flops.

8-4. A(n) ________ (asynchronous, synchronous) counter is the more complex circuit.

8-5. Synchronous counters have the *CLK* inputs connected in ________ (parallel, series).

8-6. Draw a logic symbol diagram for a 4-bit ripple down counter. Use four J-K flip-flops in this mod-16 counter. Show the input *CLK* pulses, *PS* input, and four output indicators labeled *D, C, B,* and *A*.

8-7. If the ripple down counter in question 8-6 is a recirculating type, what are the next three counts after 0011, 0010, and 0001?

8-8. Redesign the 4-bit counter in question 8-6 to count from binary 1111 to 0000 and then *stop.* Add a four-input OR gate to your existing circuit to add this self-stopping feature.

8-9. Draw a block diagram (similar to Fig. 8-12)

showing how you would use two counters to get an output of 1 Hz with an input of 100 Hz. Label your diagram.

8-10. Refer to Fig. 8-13 for questions **a** to **f** on the 7493 IC counter:

a. What is the maximum count length of this counter?

b. This is a ________ (ripple, synchronous) counter.

c. What must be the conditions of the reset inputs for the 7493 to count?

d. This is a(n) ________ (down, up) counter.

e. The 7493 IC contains ________ (number) flip-flops.

f. What is the purpose of the NAND gate in the 7493 counter?

8-11. Refer to Fig. 8-14 for questions **a** to **f** on the 74192 counter:

a. What is the maximum count length of this counter?

b. This is a ________ (ripple, synchronous) counter.

c. A logical ________ (0, 1) is needed to clear the counter to 0000.

d. This is a(n) ________ (down, up, both up and down) counter.

e. How could we preset the outputs of the 74192 IC to 1001?

f. How do we get the counter to count downward?

8-12. Draw a diagram (similar to Fig. 8-15(*a*)) showing how you would wire the 7493 counter as a 4-bit (mod-16) ripple counter. Refer to Fig. 8-13.

8-13. Refer to Fig. 8-37. The 74192 counter is in the ________ (clear, count up, load) mode during pulse t_1.

8-14. List the binary output from the 74192 counter IC after each of the eight input pulses shown in Fig. 8-37. Start with t_1 and end with t_8.

8-15. Refer to Fig. 8-17 for questions **a** to **e** on the 74HC393 IC counter:

a. This is a ________ (ripple, synchronous) counter.

b. This is a(n) ________ (down, up, either up or down) counter.

c. The MR pins are ________ (asynchronous, synchronous) active ________ (HIGH, LOW) inputs that clear the outputs.

d. Each counter contains four ________ (R-S, T) flip-flops.

e. This is a ________ (CMOS, TTL) counter.

8-16. Refer to Fig. 8-18 for questions **a** to **e** on the 74HC193 IC counter:

a. When the MR pin is activated with a ________ (HIGH, LOW), all outputs are reset to ________ (0, 1).

b. This is a ________ (ripple, synchronous) counter.

c. Parallel data from the data inputs (D_0 to D_3) flow through to the outputs (Q_0 to Q_3) when the ________ input is activated with a LOW.

d. When a clock signal enters pin CP_U, the CP_D pin must be tied to ________ (+5 V, GND).

8-17. Refer to Fig. 8-38. List the mode of operation for the 74HC193 counter during each

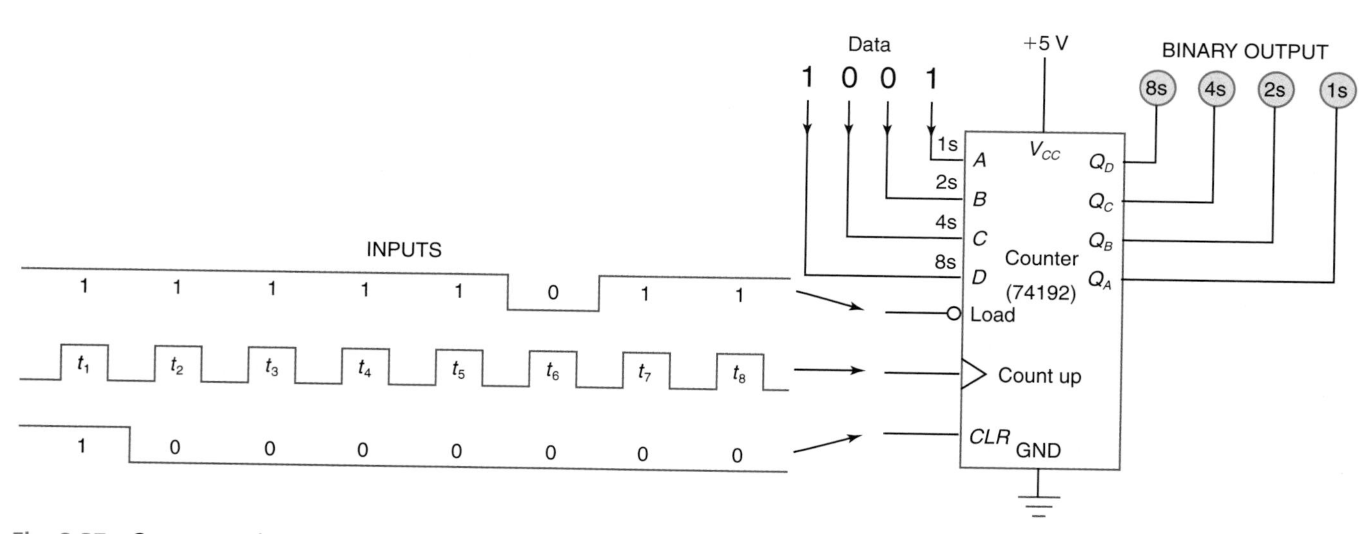

Fig. 8-37 Counter pulse-train problem.

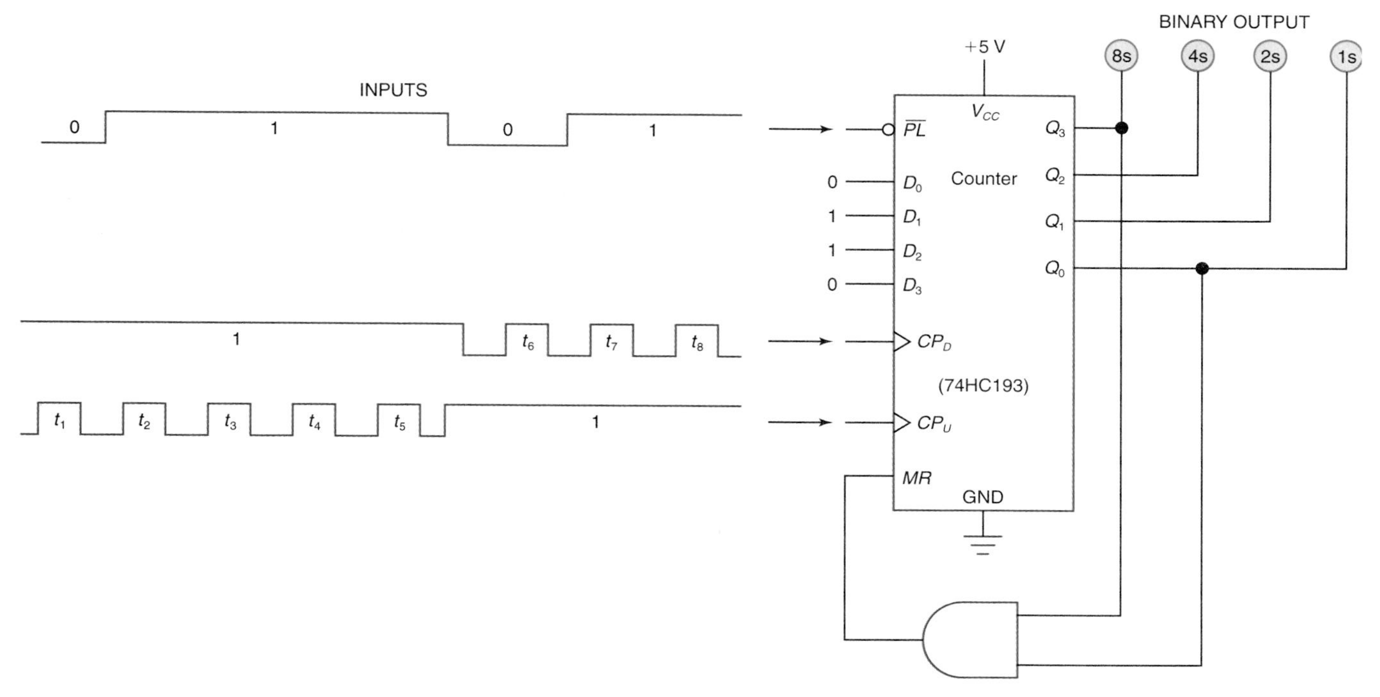

Fig. 8-38 Counter pulse-train problem.

pulse t_1 to t_8 (use answers *parallel load, count up, count down*).

8-18. Refer to Fig. 8-38. List the binary output for the 74HC193 counter IC after each pulse t_1 to t_8.

8-19. Refer to Fig. 8-21(*b*). The device optically sensing the opening in the shaft encoder disk and sending a signal to the wave-shaping circuit is a(n) ________.

8-20. Refer to Fig. 8-21(*b*). The optical encoder at the top of the shaft encoder disk is of the ________ (reflective type, slot type).

8-21. Refer to Fig. 8-22(*b*). The H21A1 interrupter module contains a(n) ________ on the emitter side and a phototransistor on the detector side of the optical sensor.

8-22. Refer to Fig. 8-23. The device that performs wave shaping in this circuit is the ________ (7414, 74192) IC.

8-23. Refer to Fig. 8-23. The device that performs as a decade counter in this circuit is the ________ (7447, 74192) IC.

8-24. Refer to Fig. 8-23. The 7447 IC is a decoder/driver that translates BCD data to ________ code and drives the display.

8-25. Refer to Fig. 8-24(*b*). The disk shown (black and white strips) would be an encoder disk used by a ________ (reflective-type, slot-type) optical sensor.

8-26. Refer to Fig. 8-27. The two 74HC85 magnitude comparator ICs are said to be ________ (cascaded, subdivided) so they can compare two ________ (number)-bit binary numbers.

8-27. Refer to Fig. 8-39 on page 206. List the *color* of the output LED that is lit for each time period (t_1 to t_6).

8-28. While an oscilloscope is very good for observing repetitive digital signals, a logic ________ (analyzer, monitor) is better for looking at asynchronous signals.

8-29. The equipment pictured in Fig. ________ (8-30, 8-32) is an easy-to-use instrument for testing if an IC is good or bad.

8-30. A digital ________ (IC tester, pulser) is an instrument for injecting a signal into a circuit.

8-31. List five pieces of electronic test equipment used in testing and troubleshooting sequential digital logic circuits.

8-32. The handy unit pictured in Fig. 8-31 contains what two electronic test instruments in one small package?

8-33. Refer to Fig. 8-33. This is a moderately priced, high-quality ________ (analog, digital) oscilloscope from Tektronix.

8-34. Refer to Fig. 8-33. Most oscilloscopes use a CRT display while the Tektronix TDS 220 dual-trace scope uses a(n) ________ (LCD, VF) display.

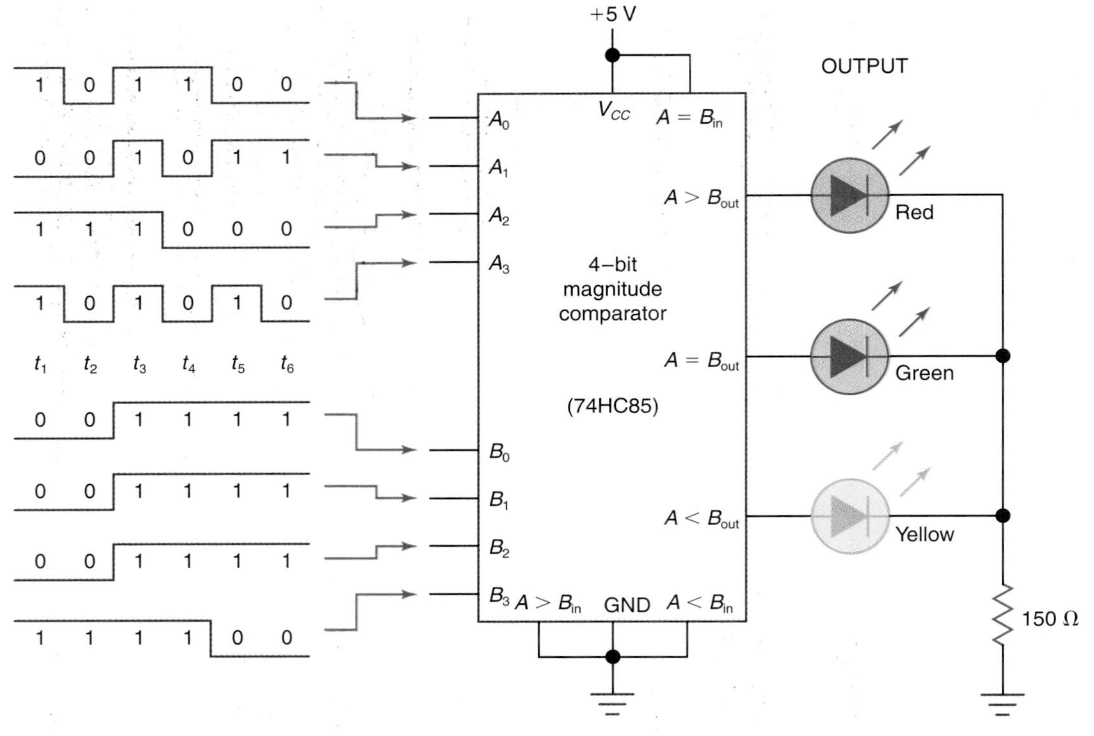

Fig. 8-39 Magnitude comparator pulse-train problem.

Critical Thinking Questions

8-1. What types of flip-flops are useful in wiring counters because they have a toggle mode?

8-2. Draw a logic symbol diagram of a mod-5 ripple up counter. Use three J-K flip-flops and a two-input NAND gate. Show input *CLK* pulses and three output indicators labeled *C, B,* and *A* (*C* indicator is MSB).

8-3. Draw a logic diagram for a mod-10 counter using a 7493 IC.

8-4. Draw a logic diagram of a divide-by-8 counter using a 7493 IC. Show which output of the 7493 IC is the divide-by-8 output.

8-5. Refer to Fig. 8-37. List the mode of operation during each of the input pulses t_1 to t_8.

8-6. Refer to Fig. 8-17. Why are the master reset inputs (1MR and 2MR) referred to as asynchronous?

8-7. Refer to Fig. 8-18. The 74HC193 IC counter is called presettable because of what operating mode?

8-8. Refer to Fig. 8-38. Give the modulus and list the counting sequence for this counter.

8-9. Design a decade up counter (0 to 9 in decimal) using a 74HC193 IC and a two-input AND gate.

8-10. On a(n) ________ (asynchronous, synchronous) counter, all outputs change to their new states at the same instant.

8-11. What is another name for an asynchronous counter?

8-12. If we refer to a divide-by-6 counter, the circuit will probably be used for what purpose?

8-13. Refer to Fig. 8-23. The counter and display will increment when the output of the inverter goes from ________ (H-to-L, L-to-H).

8-14. Refer to Fig. 8-23. The counter and display will increment when the beam of infrared light across the slot ________.
a. is broken (from light to no light)
b. begins again (from no light to light)

8-15. Compare the shaft encoder disks in Fig. 8-21(*b*) and 8-24(*b*). Which disk will provide the greater resolution?

Answers to Tests

1. two
2. 4
3. toggle
4. pulse t_1 = 00
   pulse t_2 = 01
   pulse t_3 = 10
   pulse t_4 = 11
   pulse t_5 = 00
   pulse t_6 = 01
5. ripple, decade
6. ripple, 5
7. pulse t_1 = 111, then cleared to 000 just before pulse t_2
   pulse t_2 = 001
   pulse t_3 = 010
   pulse t_4 = 011
   pulse t_5 = 100
   pulse t_6 = 000
8. synchronous
9. parallel
10. toggle
11. all the flip-flops toggle
12. toggle
13. HIGH-to-LOW
14. only FF 1 toggles
15. pulse t_1 = 00
    pulse t_2 = 11
    pulse t_3 = 10
    pulse t_4 = 01
    pulse t_5 = 00
    pulse t_6 = 11
16. down
17. LOW, hold
18. HIGH, toggle
19. 1000
20. 2
21. 0000 (reset)
22. four, up
23. decade, synchronous
24. 5
25. HIGH
26. point B = 200 Hz
    point C = 100 Hz
    point D = 50 Hz
27. 8
28. 4-bit binary
29. HIGH
30. H-to-L
31. 16, ripple
32. synchronous
33. asynchronous
34. Q_0–Q_3
35. 0001, 0010, 0011, 0100, 0101, 0110 (1 to 6 in decimal)
36. to preset the counter to 0001 after the highest count of 0110
37. manufacturers use different standards when drawing logic symbols and labeling
38. interrupter module
39. infrared
40. phototransistor
41. slot-type
42. L-to-H (LOW-to-HIGH)
43. enters
44. decade, temporarily storing the count
45. red, too high
46. press and release switch SW_1
47. astable
48. t_1 = green
    t_2 = red
    t_3 = yellow
    t_4 = red
    t_5 = green
    t_6 = red
49. sequential
50. monitor
51. single pulse
52. repetitive
53. DMM
54. F
55. digital
56. T
57. digital
58. analyzer
59. HIGH
60. clear
61. preset (asynchronous), HIGH
62. 7, floating

Chapter 9

Shift Registers

Chapter Objectives

This chapter will help you to:

1. *Draw* a circuit diagram of a serial-load shift register using D flip-flops.
2. *Define* terms such as *shift right, shift left, parallel load,* and *serial load* and *describe* procedures for performing these operations on various shift registers.
3. *Interpret* data sheets for several commercial TTL and CMOS shift register ICs.
4. *Predict* the operation of several TTL and CMOS shift register ICs based on a series of inputs.
5. *Analyze* the operation of a digital roulette game containing a voltage-controlled oscillator (VCO), a ring counter, a power-up initializing circuit, and an audio amplifier.
6. *Troubleshoot* a faulty shift register circuit.

Memory and shifting characteristics

A typical example of a *shift register* at work is found within a calculator. As you enter each digit on the keyboard, the numbers shift to the left on the display. In other words, to enter the number 268 you must do the following. First, you press and release the 2 on the keyboard; a 2 appears at the extreme right on the display. Next, you press and release the 6 on the keyboard causing the 2 to shift one place to the left allowing 6 to appear on the extreme right; 26 appears on the display. Finally, you press and release the 8 on the keyboard; 268 appears on the display. This example shows two important characteristics of a shift register: (1) It is a *temporary memory* and thus holds the numbers on the display (even if you release the keyboard number) and (2) it shifts the numbers to the left on the display each time you press a new digit on the keyboard. These *memory* and *shifting characteristics* make the shift register extremely valuable in most digital electronic systems. This chapter introduces you to shift registers and explains their operations.

Shift registers

Shift registers are constructed by wiring flip-flops together. We mentioned in Chaps. 7 and 8 that flip-flops have a memory characteristic. This memory characteristic is put to good use in a shift register. Instead of wiring shift registers by using individual gates or flip-flops, you can buy shift registers in IC form.

Shift registers often are used to store data momentarily. Figure 9-1 shows a typical example of where shift registers might be used in a digital system. This system could be that of a calculator. Notice the use of shift registers to hold information from the encoder for the processing unit. A shift register is also being employed for temporary storage between the processing unit and the decoder. Shift registers are also used at other locations within a digital system.

One method of describing shift register characteristics is by how data is *loaded into* and *read from* the storage units. Four categories of shift registers are illustrated in Fig. 9-2. Each storage device in Fig. 9-2 is an 8-bit register. The registers are classified as:

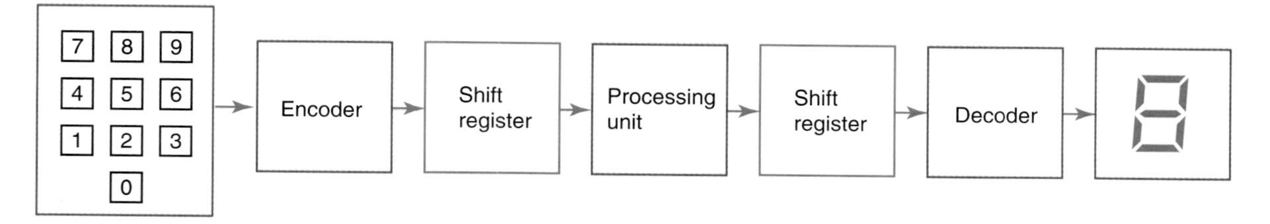

Fig. 9-1 A digital system using shift registers.

1. Serial in–serial out (Fig. 9-2(*a*))
2. Serial in–parallel out (Fig. 9-2(*b*))
3. Parallel in–serial out (Fig. 9-2(*c*))
4. Parallel in–parallel out (Fig. 9-2(*d*))

The diagrams in Fig. 9-2 illustrate the fundamental idea of each type of register. These classifications are often used in a manufacturer's literature.

9-1 Serial Load Shift Registers

Serial in–serial out

Serial in–parallel out

Parallel in–serial out

Parallel in–parallel out

Shift register characteristics

4-bit shift register

A basic shift register is shown in Fig. 9-3 on page 210. This shift register is constructed from four D flip-flops. This register is called a *4-bit shift register* because it has four places to store data: *A*, *B*, *C*, *D*.

With the aid of Table 9-1 and Fig. 9-3, let us operate this shift register. First, clear (*CLR*

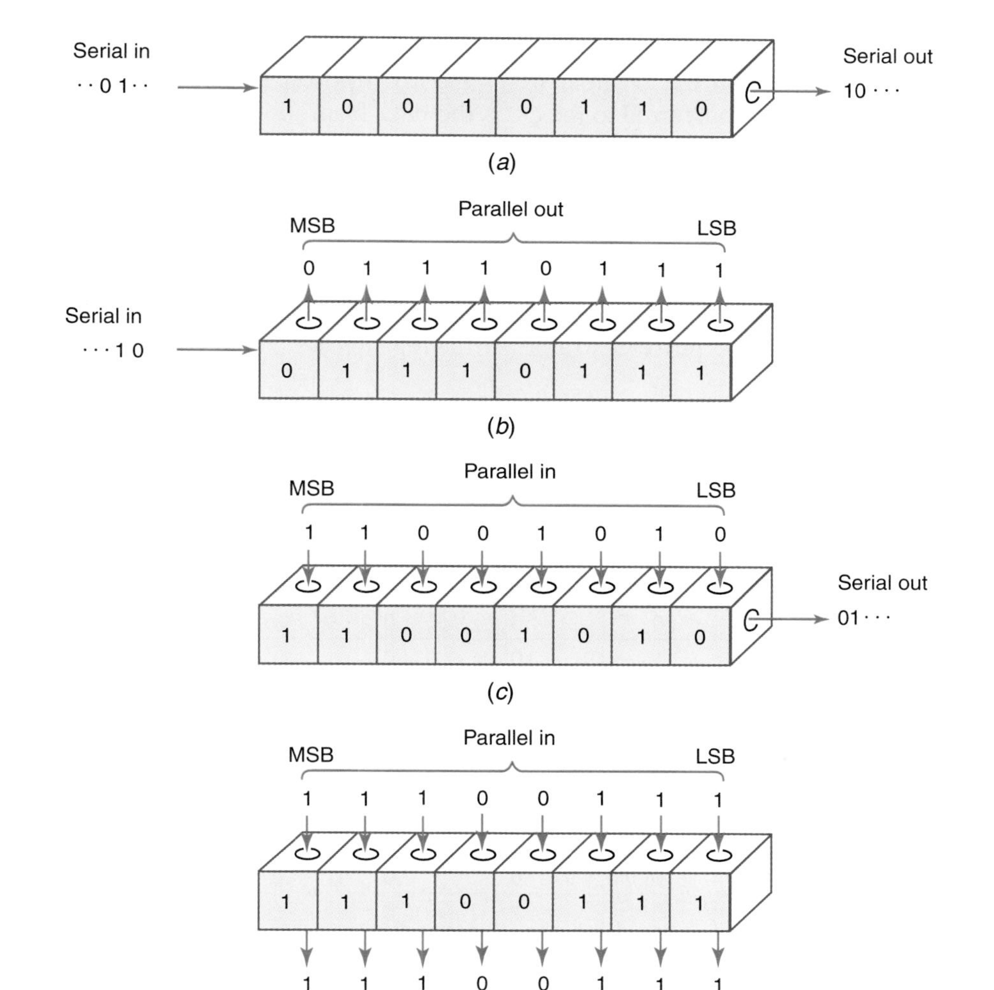

Fig. 9-2 Shift register characteristics. (*a*) Serial in–serial out. (*b*) Serial in–parallel out. (*c*) Parallel in–serial out. (*d*) Parallel in–parallel out.

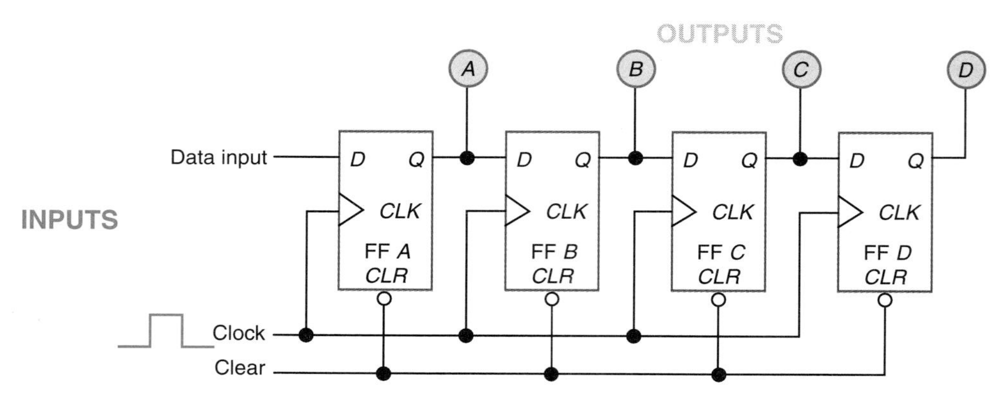

Fig. 9-3 A 4-bit serial load shift register using D flip-flops.

input to 0) all the outputs (*A*, *B*, *C*, *D*) to 0000. (This situation is shown in line 1, Table 9-1.) The outputs remain 0000 while they await a clock pulse. Pulse the *CLK* input once; the output now shows 1000 (line 3, Table 9-1) because the 1 from the *D* input of FF *A* has been transferred to the *Q* output on the clock pulse. Now enter 1s on the data input (clock pulses 2 and 3, Table 9-1); these 1s shift across the display to the right. Next, enter 0s on the data input (clock pulses 4 to 8, Table 9-1); you can see the 0s being shifted across the display (lines 6 to 10, Table 9-1). On clock pulse 9 (Table 9-1) enter a 1 at the data input. On pulse 10 the data input is returned to 0. Pulses 9 to 13 show the single 1 on display being shifted to the right. Line 15 shows the 1 being shifted out the right end of the shift register and being lost.

JOB TIP

Continuing education is essential in any technical occupation.

Serial load shift registers

Remember that the D flip-flop is also called a *delay* flip-flop. Recall that it simply transfers the data from input *D* to output *Q after a delay of one clock pulse.*

The circuit diagrammed in Fig. 9-3 is referred to as a *serial load shift register.* The term "serial load" comes from the fact that only one bit of data at a time can be entered in the register. For instance, to enter 0111 in the register, we had to go through the sequence

Table 9-1 Operation of a 4-bit Serial Shift Register

Line Number	Inputs: Clear	Inputs: Data	Inputs: Clock Pulse Number	Output: FF *A* (*A*)	Output: FF *B* (*B*)	Output: FF *C* (*C*)	Output: FF *D* (*D*)
1	0	0	0	0	0	0	0
2	1	1	0	0	0	0	0
3	1	1	1	1	0	0	0
4	1	1	2	1	1	0	0
5	1	1	3	1	1	1	0
6	1	0	4	0	1	1	1
7	1	0	5	0	0	1	1
8	1	0	6	0	0	0	1
9	1	0	7	0	0	0	0
10	1	0	8	0	0	0	0
11	1	1	9	1	0	0	0
12	1	0	10	0	1	0	0
13	1	0	11	0	0	1	0
14	1	0	12	0	0	0	1
15	1	0	13	0	0	0	0

from lines 1 through 6 in Table 9-1. It took five steps (line 2 was not needed) to serially load 0111 into the serial load shift register. To enter 0001 in this serial load shift register we need five steps, as shown in Table 9-1, lines 10 to 14. According to the classifications in Fig. 9-2, this would be a serial in–parallel out register. However, if data were taken from only FF *D*, it becomes a serial in–serial out register.

The shift register in Fig. 9-3 could become a 5-bit shift register just by adding one more D flip-flop. Shift registers typically come in 4-, 5-, and 8-bit sizes. Shift registers also can be wired using other flip-flops. J-K flip-flops and clocked R-S flip-flops are also used to wire shift registers.

TEST

Answer the following questions.

1. The unit shown in Fig. 9-4 is a shift-right ________ (parallel, serial) load shift register.
2. List the contents of the register in Fig. 9-4 after each of the six clock pulses (*A* = left bit, *C* = right bit).
3. A(n) ________ (entire 3-bit group, single bit) is loaded on each clock pulse in the serial load shift register in Fig. 9-4.
4. The clear (*CLR*) input to the shift register in Fig. 9-4 is active- ________ (HIGH, LOW).
5. The clear input in Fig. 9-4 must be ________ (HIGH, LOW) and a ________ (H to L, L to H) clock pulse at the *CLK* input will trigger a right shift in this register.

9-2 Parallel Load Shift Registers

Parallel load shift registers

The serial load shift register we studied in the last section has two disadvantages: it permits only one bit of information to be entered at a time, and it loses all its data out the right side when it shifts right. Figure 9-5(*a*) on page 212 illustrates a system that permits *parallel loading* of four bits at once. These inputs are the data inputs *A*, *B*, *C*, and *D* in Fig. 9-5. This system could also incorporate a *recirculating* feature that would put the output data back into the input so that it is not lost.

Recirculating feature

4-bit parallel load recirculating shift register

A wiring diagram of the *4-bit parallel load recirculating shift register* is drawn in Fig. 9-5(*b*). This shift register uses four J-K flip-flops. Notice the recirculating lines leading from the *Q* and $\overline{Q}$ outputs of FF *D* back to the *J* and *K* inputs of FF *A*. These feedback lines cause the data that would normally be lost out of FF *D* to recirculate through the shift register. The *CLR* input clears the outputs to 0000 when enabled by a logical 0. The parallel load data inputs *A*, *B*, *C*, and *D* are connected to the preset (*PS*) inputs of the flip-flops to set 1s at any output position (*A*, *B*, *C*, *D*). If the switches attached to the parallel load data inputs are even temporarily switched to a 0, that output will be preset to a logical 1. The clock pulsing the *CLK* inputs of the J-K flip-flops will cause data to be shifted to the right. The data from FF *D* will be recirculated back to FF *A*.

Table 9-2 on page 213 will help you understand the operation of the parallel load shift register. As you turn on the power, the outputs may assume any combination, such as the one in line 1. Line 2 shows the register being cleared with the *CLR* input. Line 3 shows 0100

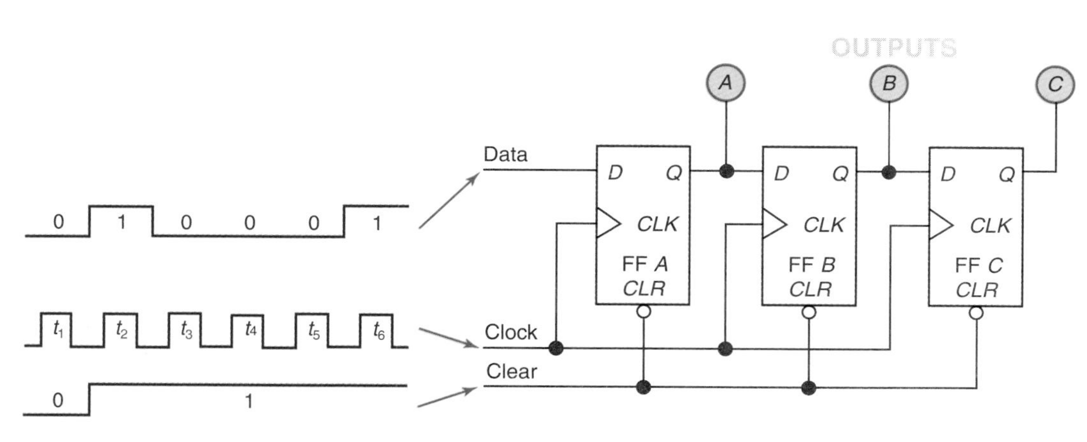

Fig. 9-4 A shift register problem for test items 1 through 5.

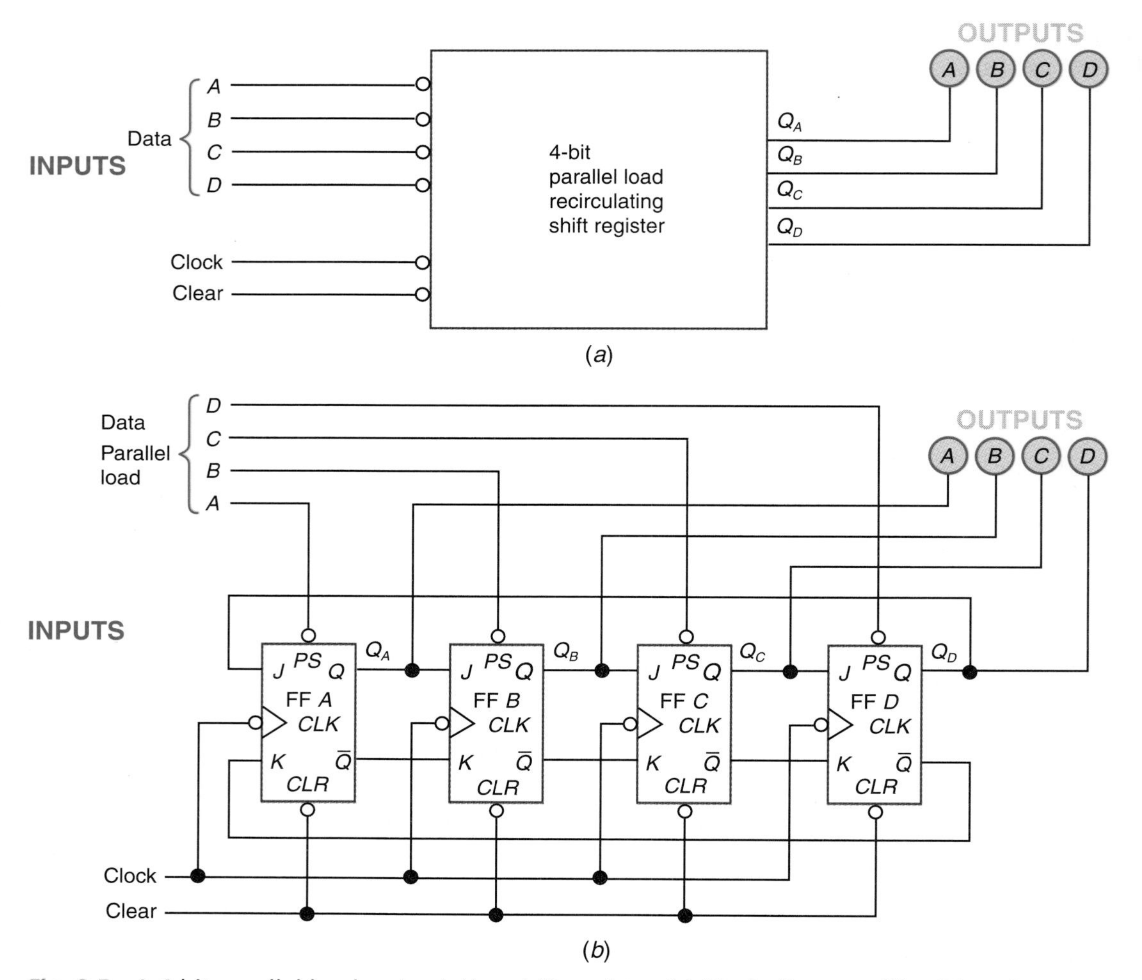

Fig. 9-5 A 4-bit parallel load recirculating shift register. (*a*) Block diagram. (*b*) Wiring diagram.

4-bit parallel load recirculating shift register

being loaded into the register using the parallel load data switches. Lines 4 to 8 show five clock pulses and the shifting of the data to the right. Look at lines 5 and 6: the 1 is being recirculated from the right end (FF *D*) of the register back to the left end (FF *A*). We say the 1 is being recirculated.

Line 9 shows the register being cleared again by the *CLR* input. New information (0110) is being loaded in the data inputs in line 10. Lines 11 to 15 illustrate the register being shifted five times by clock pulses. Note that it takes four clock pulses to come back to the original data in the register (compare lines 11 and 15 or lines 4 and 8 in Table 9-2). The register in Fig. 9-5 could be classified as a parallel in–parallel out storage device.

The recirculating feature of the shift register in Fig. 9-5(*b*) can be disabled by disconnecting the two recirculating lines. The register is then a parallel in–parallel out register. However, if only the output from FF *D* is considered, this register is a parallel in–serial out storage device.

TEST

Answer the following questions.

6. The unit in Fig. 9-6 is a shift-right _________ (serial, parallel) load recirculating shift register.
7. Refer to Fig. 9-6. List the mode of operation of the shift register during each of the eight clock pulses. Use as answers the terms "clear," "parallel load," and "shift-right."
8. List the contents of the register in Fig. 9-6 immediately after each of the eight clock pulses (A = left bit, C = right bit).
9. Refer to Fig. 9-6. This is a _________ (nonrecirculating, recirculating) 3-bit shift register.

TABLE 9-2 Operation of a 4-bit Parallel Load Recirculating Shift Register

Line Number	Inputs					Clock Pulse Number	Output			
	Clear	Parallel Load Data					FF *A*	FF *B*	FF *C*	FF *D*
		A	*B*	*C*	*D*		*A*	*B*	*C*	*D*
1	1	1	1	1	1	0	1	1	1	0
2	0	1	1	1	1	0	0	0	0	0
3	1	1	0	1	1	0	0	1	0	0
4	1	1	1	1	1	1	0	0	1	0
5	1	1	1	1	1	2	0	0	0	1
6	1	1	1	1	1	3	1	0	0	0
7	1	1	1	1	1	4	0	1	0	0
8	1	1	1	1	1	5	0	0	1	0
9	0	1	1	1	1		0	0	0	0
10	1	1	0	0	1		0	1	1	0
11	1	1	1	1	1	6	0	0	1	1
12	1	1	1	1	1	7	1	0	0	1
13	1	1	1	1	1	8	1	1	0	0
14	1	1	1	1	1	9	0	1	1	0
15	1	1	1	1	1	10	0	0	1	1

10. Refer to Fig. 9-6. The parallel-load inputs are ________ (asynchronous, synchronous) on this shift register.
11. Refer to Table 9-2. The shift register is in the clear mode during what two lines on the table?
12. Refer to Table 9-2. The shift register is in the parallel-load mode during what two lines on the table?

9-3 A Universal Shift Register

When reviewing data manuals you will see that manufacturers produce many shift registers in IC form. In this section one such IC shift register will be studied: the *74194 4-bit bidirectional universal shift register.*

74194 4-bit bidirectional universal shift register

The 74194 IC is a very adaptable shift register and has most of the features we have seen so far in one IC package. A 74194 IC register

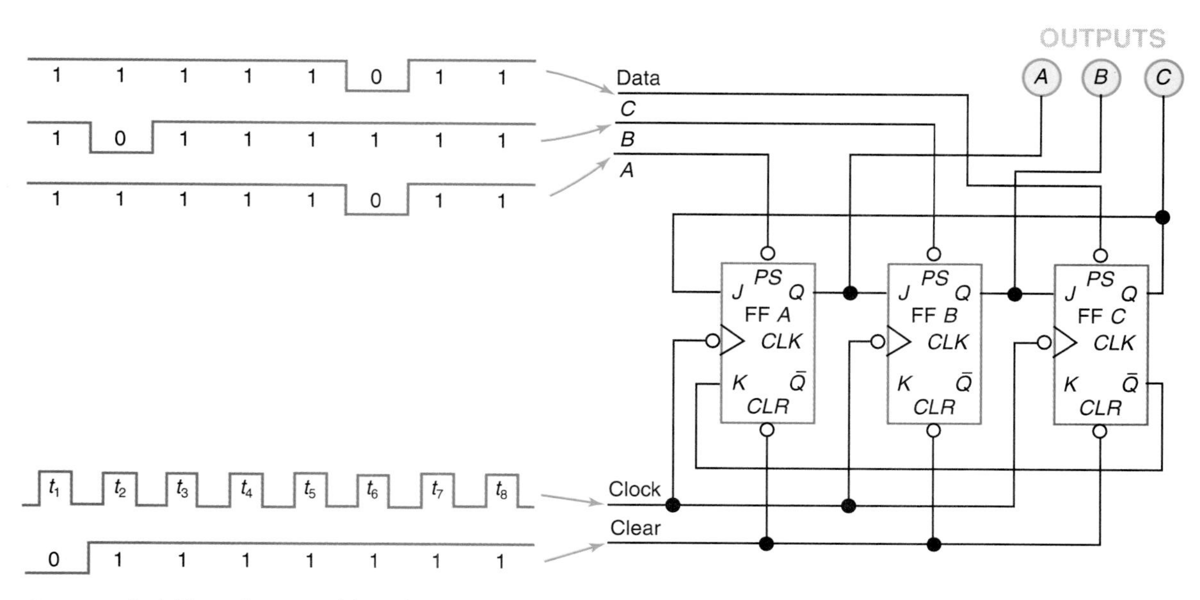

Fig. 9-6 A shift register problem for test items 6 through 10.

This bidirectional shift register is designed to incorporate virtually all of the features a system designer may want in a shift register. The circuit contains 45 equivalent gates and features parallel inputs, parallel outputs, right-shift serial inputs, operating-mode-control inputs, and a direct overriding clear line. The register has distinct modes of operating, namely:

- Parallel (broadside) load
- Shift right (in the direction Q_A toward Q_D)
- Shift left (in the direction Q_D toward Q_A)
- Inhibit clock (do nothing)

Synchronous parallel loading is accomplished by applying the four bits of data and taking both mode control inputs, $S0$ and $S1$, high. The data are loaded into the associated flip-flops and appear at the outputs after the positive transistion of the clock input. During loading, serial data flow is inhibited.

Shift right is accomplished synchronously with the rising edge of the clock pulse when $S0$ is high and $S1$ is low. Serial data for this mode is entered at the shift-right data input. When $S0$ is low and $S1$ is high, data shifts left synchronously and new data is entered at the shift-left serial input.

Clocking of the flip-flop is inhibited when both mode control inputs are low. The mode of the S54194/N74194 should be changed only while the clock input is high.

(*a*) Description

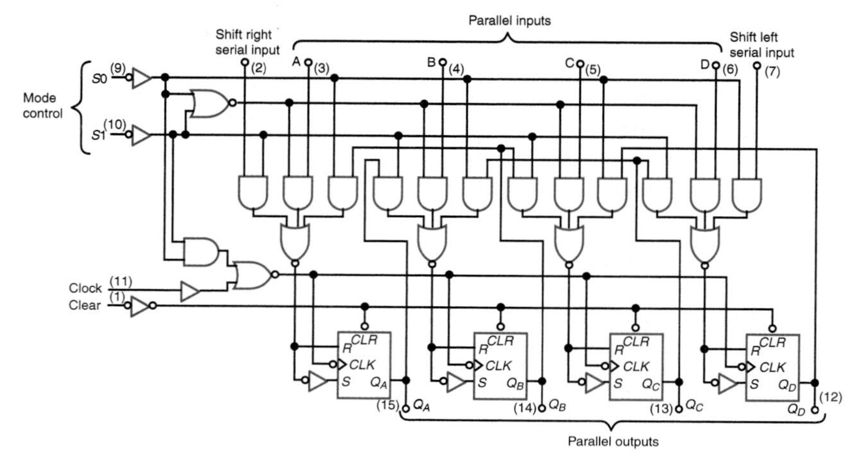

(*b*) Logic diagram

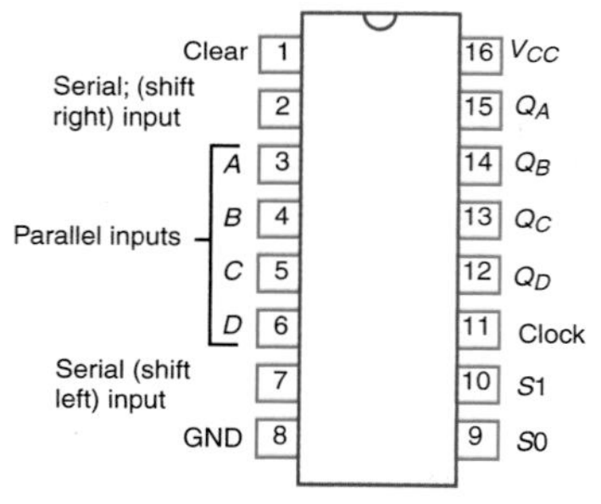

(*c*) Pin configuration

INPUTS										OUTPUTS			
CLEAR	MODE		CLOCK	SERIAL		PARALLEL				Q_A	Q_B	Q_C	Q_D
	$S1$	$S0$		LEFT	RIGHT	A	B	C	D				
L	X	X	X	X	X	X	X	X	X	L	L	L	L
H	X	X	L	X	X	X	X	X	X	Q_{A0}	Q_{B0}	Q_{C0}	Q_{D0}
H	H	H	↑	X	X	a	b	c	d	a	b	c	d
H	L	H	↑	X	H	X	X	X	X	H	Q_{An}	Q_{Bn}	Q_{Cn}
H	L	H	↑	X	L	X	X	X	X	L	Q_{An}	Q_{Bn}	Q_{Cn}
H	H	L	↑	H	X	X	X	X	X	Q_{Bn}	Q_{Cn}	Q_{Dn}	H
H	H	L	↑	L	X	X	X	X	X	Q_{Bn}	Q_{Cn}	Q_{Dn}	L
H	L	L	X	X	X	X	X	X	X	Q_{A0}	Q_{B0}	Q_{C0}	Q_{D0}

(*d*) Function table

H = high level (steady state)
L = low level (steady state)
X = irrelevant (any input, including transitions)
↑ = transition from low to high level
a,b,c,d, = the level of steady state input at inputs *A,B,C*, or *D*, respectively
Q_{A0}, Q_{B0}, Q_{C0}, Q_{D0} = the level of Q_A, Q_B, Q_C, Q_D, respectively, before the indicated steady state input conditions were established
Q_{An}, Q_{Bn}, Q_{Cn}, Q_{Dn} = the level of Q_A, Q_B, Q_C, Q_D, respectively before the most recent ↑ transition of the clock

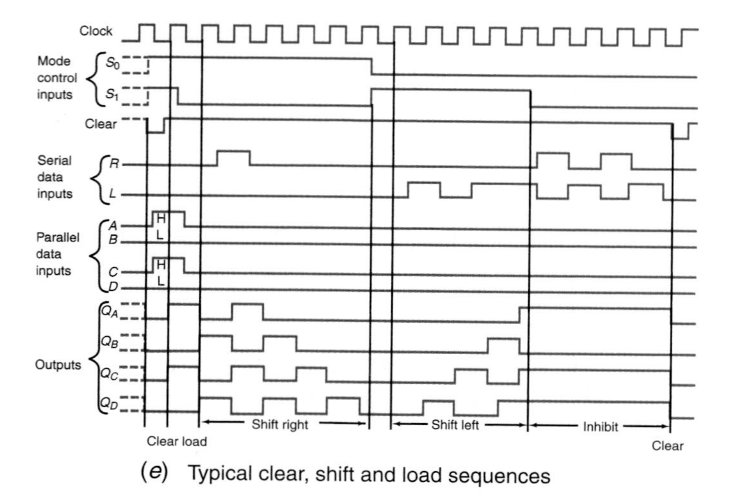

(*e*) Typical clear, shift and load sequences

74194 shift register

Fig. 9-7 A 4-bit TTL universal shift register (74194). (*a*) Description. (*b*) Logic diagram. (*c*) Pin configuration. (*d*) Function (truth) table. (*e*) Waveforms. (*Courtesy of Texas Instruments.*)

can shift right or left. It can be loaded serially or in parallel. Several 4-bit 74194 IC registers can be cascaded to make an 8-bit or longer shift register. And this register can be made to recirculate data.

The Texas Instruments data manual contains the descriptions, diagrams, and tables shown in Fig. 9-7. Read the description of the 74194 shift register in Fig. 9-7(*a*) for a good overview of what this shift register can do.

A logic diagram of the 74194 register is reproduced in Fig. 9-7(*b*). Because it is a 4-bit register, the circuit contains four flip-flops. Extra gating circuitry is needed for the many features of this universal shift register. The pin configuration in Fig. 9-7(*c*) will help you determine the labeling of each input and output. Of course, the pin diagram is also a must when actually wiring a 74194 IC.

The truth table and waveform diagrams in Fig. 9-7(*d*) and (*e*) are very helpful in determining exactly how the 74194 IC register works because they illustrate the clear, load, shift-right, shift-left, and inhibit modes of operation. As you use the 74194 universal shift register you will have occasion to look quite carefully at the truth table and waveform diagrams.

TEST

Answer the following questions.

13. List the five modes of operation for the 74194 universal shift register IC.
14. Refer to Fig. 9-7. If both mode control inputs (*S*0, *S*1) to the 74194 are HIGH, the unit is in the _________ mode.
15. Refer to Fig. 9-7. If both mode control inputs (*S*0, *S*1) to the 74194 IC are LOW, the unit is in the _________ mode.
16. Refer to Fig. 9-7. Shift right on the 74194 IC is accomplished when *S*0 is _________ (HIGH, LOW) and *S*1 is _________ (HIGH, LOW) and when the clock pulse goes from _________ to _________.

9-4 Using the 74194 IC Shift Register

In this section we shall use the 74194 universal shift register in several ways. Figure 9-8(*a*) and (*b*) on page 216 shows the 74194 IC being used as serial load registers. A *serial load shift-right register* is shown in Fig. 9-8(*a*). This register operates exactly like the serial shift register in Fig. 9-3. Table 9-1 could also be used to chart the performance of this new shift register. Notice that the *mode control inputs* (*S*0, *S*1) must be in the positions shown for the 74194 IC to operate in its shift-right mode. Shifting to the right is defined by the manufacturer as shifting from Q_A to Q_D. The register in Fig. 9-8(*a*) shifts data to the right, and as it leaves Q_D the data is lost.

The best interview policy is to be open and honest. Never exaggerate, but do not sell yourself short.

Serial load shift-right register

Mode control inputs

The 74194 IC has been revised slightly in Fig. 9-8(*b*). The shift-left serial input is used, and the mode control inputs *have been changed.* This register enters data at *D* (Q_D) and shifts it toward *A* (Q_A) with each pulse of the clock. This register is a *serial load shift-left register.*

Serial load shift-left register

Did You Know?

Guiding Fiber-Optic Construction. Light guide fibers used in telecommunications (fiber-optic cables) must be joined carefully so that minimal light escapes at the junction. Here a light guide fiber is readied to be spliced to another fiber. The fiber is held in place by the grooves in a pair of silicon chips. Once it is in this precisely fixed position, it can be merged with another fiber into the near-perfect alignment needed.

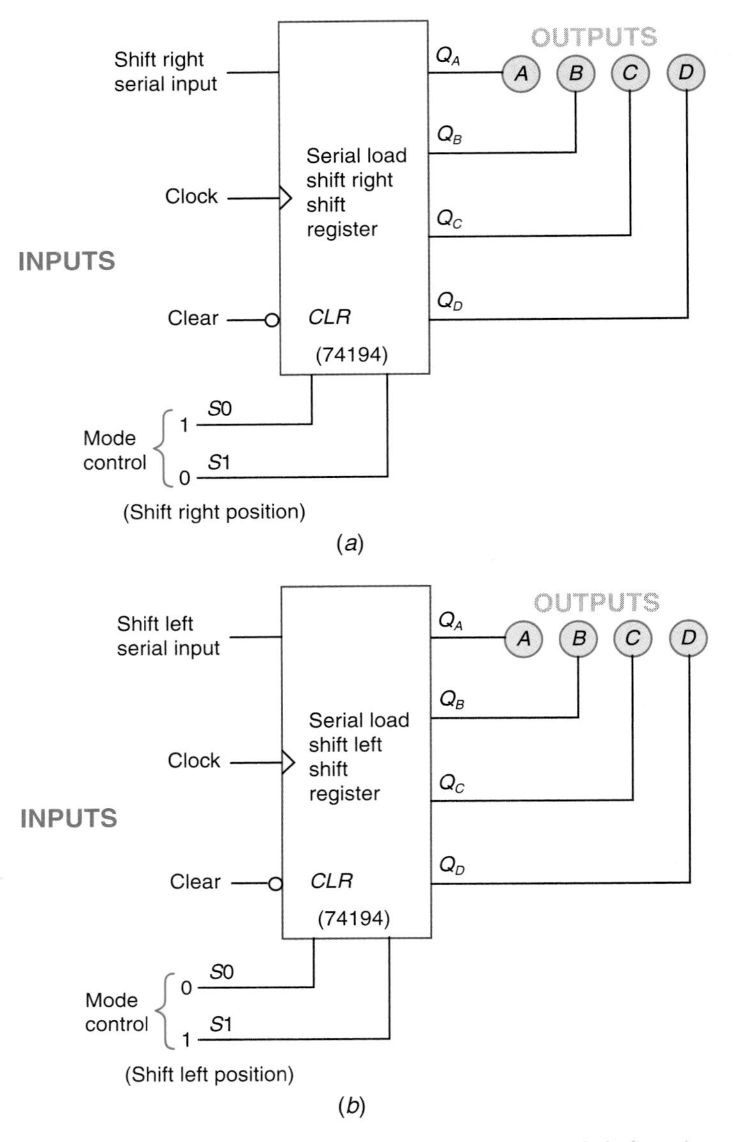

Using the 74194 shift register IC

Fig. 9-8 (*a*) A 74194 IC wired as a 4-bit serial load shift-right register. (*b*) A 74194 IC wired as a 4-bit serial load shift-left register.

Parallel load shift-right/left register

8-bit parallel load shift-right register

In Fig. 9-9 the 74194 IC is wired as a *parallel load shift-right/left register.* With a single clock pulse the data from the parallel load inputs *A*, *B*, *C*, and *D* appears on the display. The loading happens only when the mode controls (*S*0, *S*1) are set at 1, as shown. The mode control can then be changed to one of the three types of operations: shift right, shift left, or inhibit. The shift-right and shift-left serial inputs both are connected to 0 to feed in 0s to the register in the shift-right or shift-left mode of operation. With the mode control in the inhibit position ($S0 = 0$, $S1 = 0$), the data does not shift right or left but stays in position in the register. When using the 74194 IC you must remember the mode control inputs because they control the operation of the entire register. The $\overline{CLR}$ input clears the register to 0000 when enabled by a 0. The $\overline{CLR}$ input overrides all other inputs.

Two 74194 IC shift registers are connected in Fig. 9-10 to form an *8-bit parallel load shift-right register.* The *CLR* input clears the outputs to 0000 0000. The parallel load inputs *A* to *H* allow entry of all eight bits of data on a single clock pulse (mode control: $S0 = 1$, $S1 = 1$). With the mode control in the shift-right position ($S0 = 1$, $S1 = 0$), the register shifts right

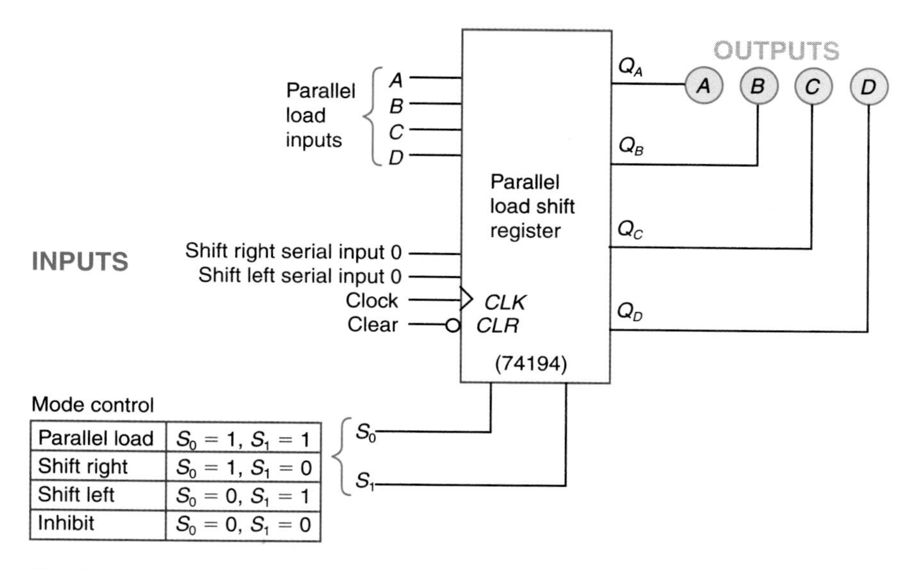

Parallel load	$S_0 = 1, S_1 = 1$
Shift right	$S_0 = 1, S_1 = 0$
Shift left	$S_0 = 0, S_1 = 1$
Inhibit	$S_0 = 0, S_1 = 0$

Fig. 9-9 A 74194 IC wired as a parallel load shift-right/left register.

Using the 74194 shift register IC

for each clock pulse. Notice that a recirculating line has been placed from output H (output Q_D of register 2) back to the shift-right serial input of shift register 1. Data that normally would be lost out of output H is recirculated back to position A in the register. Both inputs $S0$ and $S1$ at 0 will inhibit data shifting in the shift register.

As you have just seen, the 74194 IC 4-bit bidirectional universal shift register is very

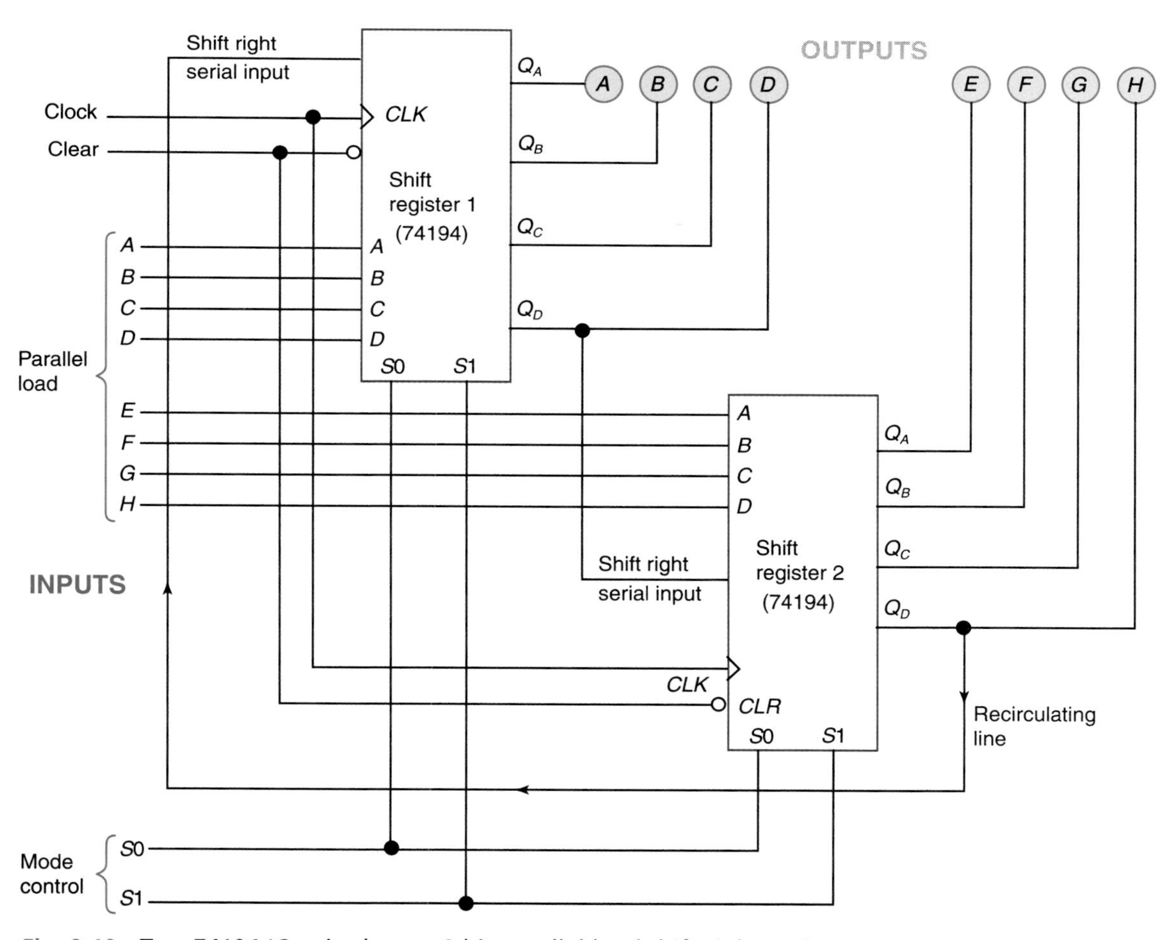

Fig. 9-10 Two 74194 ICs wired as an 8-bit parallel load shift-right register.

About Electronics

Devices for the Blind

- NOMAD, an electronic touch-sensitive screen with a speech synthesizer makes traveling to unfamiliar places easier for the blind. A graphic with raised lines is placed on the screen. The user identifies the graphic to the computer and touches a spot on the screen. As the user continues to touch the spot, the computer voice gives information about the location in increasing detail.
- Two high school students, Steven Daniels and Cuong Lai, designed an electronic vest for the blind that tells what is up ahead. The vest is studded with infrared sensors connected to a unit worn on the back of the hand. The hand feels data from the sensors in a patterened "picture" of obstacles ahead.

74HC164 8-bit serial in–parallel out shift register

useful. The circuits in this section are some examples of how the 74194 IC can be used. Remember that all shift registers use as their basis the memory characteristic of a flip-flop. Shift registers often are used as temporary memories. Shift registers also can be used to convert serial data to parallel data or parallel data to serial data. And shift registers can be used to delay information (delay lines). Shift registers are also used in some arithmetic circuits. Microprocessors and microprocessor-based systems make extensive use of registers similar to the ones used in this chapter. Counterparts to the 74194 are the 74S194, 74LS194A, 74F194, and 74HC194 ICs.

TEST

Supply the missing word or words in each statement.

17. The 74194 IC is in the parallel load mode when both mode control inputs ($S0$, $S1$) are ________ (HIGH, LOW). The four bits of data at the parallel load inputs are loaded into the registers by applying ________ (number) clock pulse(s) to the *CLK* input.
18. If the mode control inputs ($S0$, $S1$) to the 74194 IC are both LOW, the shift register is in the ________ mode.
19. For the 74194 IC to shift right, the mode controls are $S0 =$ ________ and $S1 =$ ________ and the serial data enters the ________ input.
20. Refer to Fig. 9-9. If $S0 = 1$, $S1 = 1$, shift left serial input = 1, and clear input = 0, then the outputs are ________.
21. Refer to Fig. 9-7. The 74194 IC is triggered on the ________ (H to L, L to H) transition of the clock pulse.
22. Refer to Fig. 9-7. An active ________ (clear, shift-left serial) input overrides all other inputs and resets the register outputs to 0000 on the 74194 IC.
23. Refer to Fig. 9-7. To ________ (shift left, shift right) means to shift data from Q_D toward the Q_A output on the 74194 IC.

9-5 An 8-Bit CMOS Shift Register

This section will detail the operation of one of many CMOS shift registers available from manufacturers. A manufacturer furnished the technical information in Fig. 9-11 on the *74HC164 8-bit serial in–parallel out shift register.*

The 74HC164 CMOS IC is an 8-bit edge-triggered register with serial data entry. Parallel outputs are available from each internal D flip-flop. The detailed logic diagram in Fig. 9-11(*a*) shows the use of eight D flip-flops with parallel data outputs (Q_0 to Q_7).

The 74HC164 IC featured in Fig. 9-11 is described as having a serial input. Data is entered serially through one of two inputs (D_{sa} and D_{sb}). Observe on Fig. 9-11(*a*) that the data inputs (D_{sa} and D_{sb}) are ANDed together. The data inputs may be tied together as a single input or one may be tied HIGH using the other for data entry.

The master reset input ($\overline{MR}$) to the 74HC164 IC is shown at the lower left in Fig. 9-11(*a*). It is an active LOW input. The truth table in Fig. 9-11(*b*) shows that the $\overline{MR}$ input overrides all other inputs and clears all flip-flops to 0 when activated.

The 74HC164 IC shifts data one place to the right on each LOW-to-HIGH transition of the clock (*CP*) input. The clock pulse also enters data from the ANDed data inputs (D_{sa} and D_{sb}) into output Q_0 of FF 1 (see Fig. 9-11(*a*)).

For your reference, a pin diagram for the 74HC164 shift register IC is reproduced in Fig. 9-11(*c*). The helpful table in Fig. 9-11(*d*) describes the function of each pin on this CMOS IC.

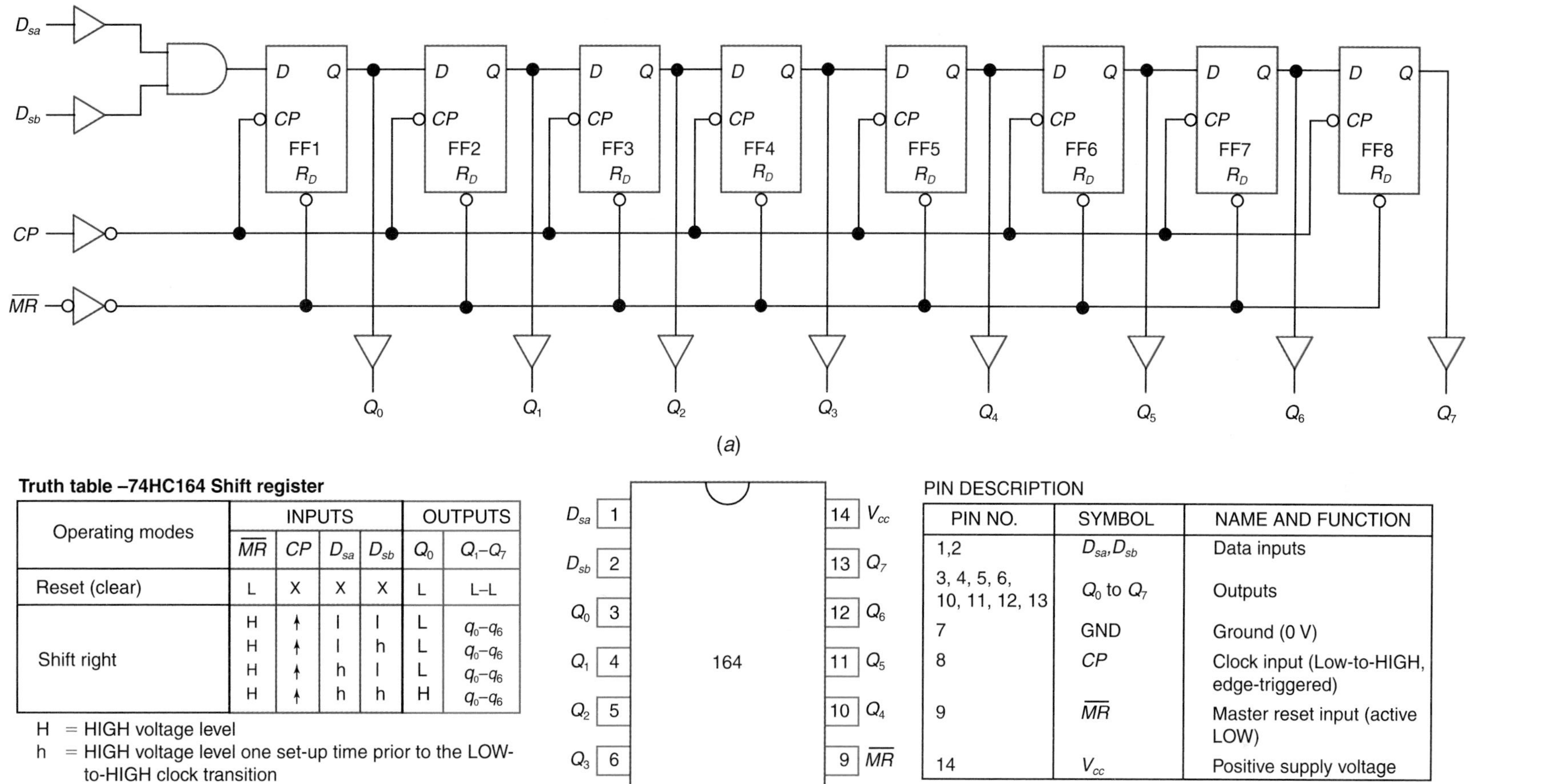

Truth table –74HC164 Shift register

Operating modes	INPUTS				OUTPUTS	
	$\overline{MR}$	CP	D_{sa}	D_{sb}	Q_0	Q_1–Q_7
Reset (clear)	L	X	X	X	L	L–L
Shift right	H	↑	l	l	L	q_0–q_6
	H	↑	l	h	L	q_0–q_6
	H	↑	h	l	L	q_0–q_6
	H	↑	h	h	H	q_0–q_6

H = HIGH voltage level
h = HIGH voltage level one set-up time prior to the LOW-to-HIGH clock transition
L = LOW voltage level
l = LOW voltage level one set-up time prior to the LOW-to-HIGH clock transition
q = lowercase letters indicate the state of the referenced input one set-up time prior to the LOW-to-HIGH clock transition
↑ = LOW-to-HIGH clock transition

(b)

PIN DESCRIPTION

PIN NO.	SYMBOL	NAME AND FUNCTION
1,2	D_{sa}, D_{sb}	Data inputs
3, 4, 5, 6, 10, 11, 12, 13	Q_0 to Q_7	Outputs
7	GND	Ground (0 V)
8	CP	Clock input (Low-to-HIGH, edge-triggered)
9	$\overline{MR}$	Master reset input (active LOW)
14	V_{cc}	Positive supply voltage

(d)

Fig. 9-11 An 8-bit CMOS serial in–parallel out shift register (74HC164). (*a*) Detailed logic diagram. (*b*) Truth table. (*c*) Pin diagram. (*d*) Pin descriptions. (*Courtesy of Signetics Corporation.*)

74HC164 8-bit shift register IC

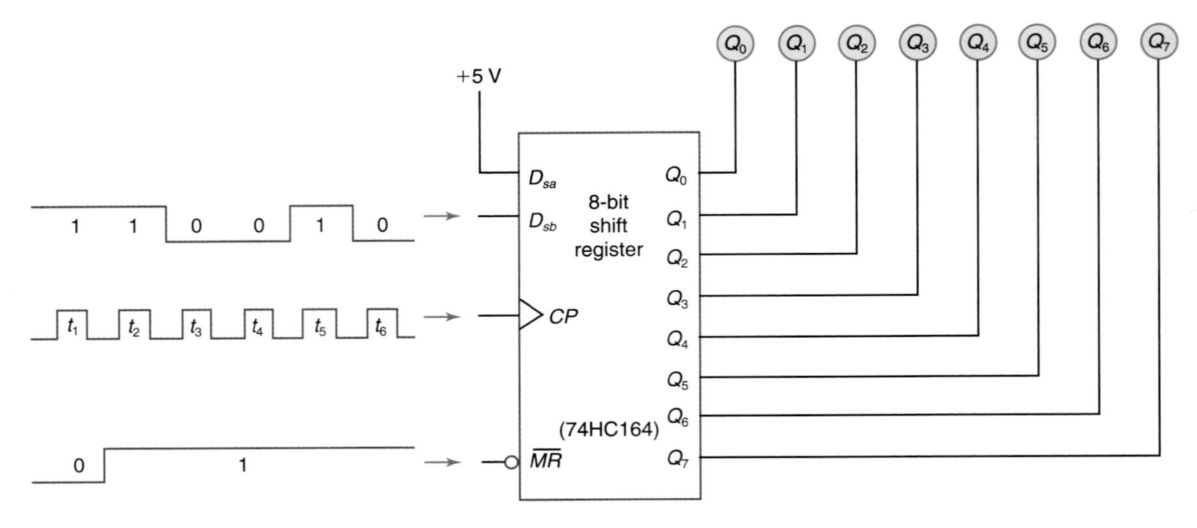

Fig. 9-12 Shift-register problem for test items 26 and 27.

TEST

Answer the following questions.

24. The 74HC164 IC's master reset pin is an active ________ (HIGH, LOW) input.
25. The clock input to the 74HC164 IC responds to a(n) ________ (H to L, L to H) transition of the clock pulse.
26. Refer to Fig. 9-12 above. List the shift register's mode of operation for each clock pulse (t_1 through t_6).
27. Refer to Fig. 9-12. List the 8-bit output (Q_0 bit on left, Q_7 bit on right) after each of the six clock pulses.
28. The 74HC164 is a ________ (CMOS, TTL) shift register IC.
29. The 74HC164 is a(n) ________ (4-bit, 8-bit) ________ (parallel-load, serial-load) shift register IC.
30. On the 74HC164 IC, the serial data inputs (D_{sa} and D_{sb}) are ________ (ANDed, ORed) together inside the chip to form the serial data input.

9-6 Using Shift Registers—Digital Roulette

The roulette wheel holds great fascination for people of all ages. Variations are used in game shows and of course in gambling casinos. This section explores an electronic version of the mechanical roulette wheel. Digital roulette is a favorite project for many students.

Digital roulette wheel

A block diagram of a *digital roulette wheel* is sketched in Fig. 9-13. This simple roulette wheel design uses only eight number markers. The number markers are LEDs in this electronic version of roulette. Only a single LED (number marker) must light at a time. A *ring counter* is a circuit that will cause the LEDs to light, one at a time, in sequence. A ring counter is simply a shift register with some added circuitry.

Ring counter

When turning on the power, the shift register in Fig. 9-13 must first be cleared to all zeros. Note that the system on-off switch is not represented in the block diagram. Second, when the "spin wheel" switch is pressed a *single* HIGH must be loaded into position 0 on the display lighting LED 0. The *voltage-controlled oscillator (VCO)* puts out a string of clock pulses which gradually decrease in frequency and stop. The clock pulses are directed to the ring counter (shift register) and the *audio amplifier* sections of the digital roulette game. Each clock pulse entering the ring counter will shift the single light around the roulette wheel. The lighting sequence should be 0, 1, 2, 3, 4, 5, 6, 7, 0, 1, and so forth, until the VCO stops emitting clock pulses. When clock pulses stop, a single LED should remain lit on the roulette wheel in some random position.

Voltage-controlled oscillator (VCO)

Audio amplifier

The VCO in Fig. 9-13 also sends clock pulses to the audio amplifier section. Each clock pulse is amplified to sound like the click of a roulette wheel. The frequency gradually decreases and stops, simulating a mechanical wheel coasting to a stop.

The ring counter block of the digital roulette

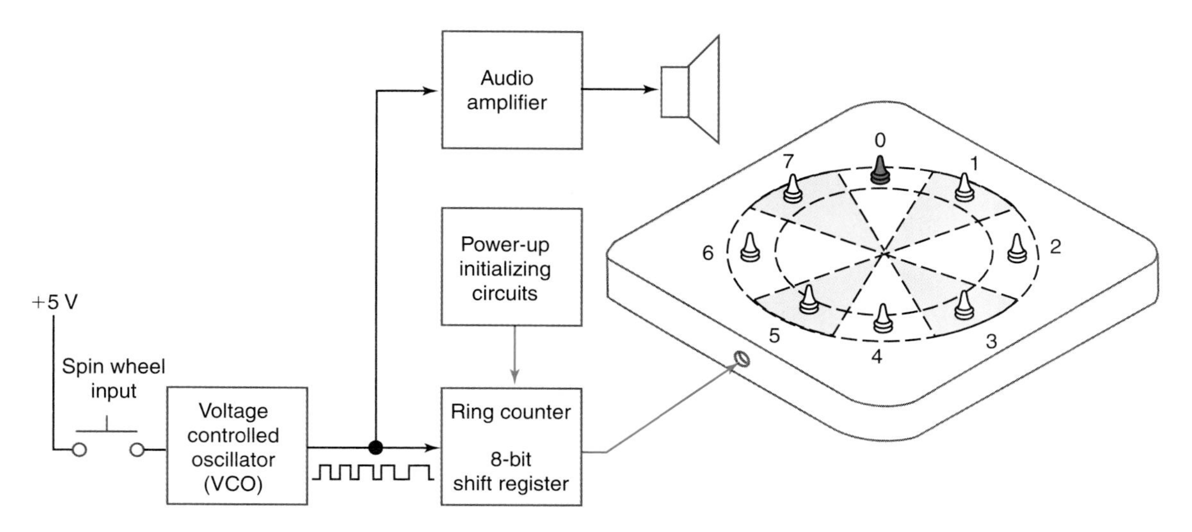

Fig. 9-13 Block diagram for simplified electronic digital roulette wheel.

Digital roulette circuit

game is detailed in Fig. 9-14(*a*) on page 222. Notice that the ring counter makes use of the 74HC164 8-bit serial in–parallel out shift register IC studied earlier. When power is turned on, the circuits in the power-up initializing block clear all outputs to zero (all LEDs are off). Upon pressing the "spin wheel" input switch, the first pulse loads a single HIGH into the shift register. This situation is illustrated in Fig. 9-14(*a*). The clock pulses that follow move the single light across the display. This is illustrated in Fig. 9-14(*b*). Notice that on each L-to-H transition of the clock the single HIGH in the 74HC164 8-bit register shifts one position to the right. When the HIGH reaches output Q_7 (after clock pulse eight in Fig. 9-14(*b*)), a *recirculating line (feedback)* is run back to the data inputs to transfer the HIGH back to the left LED (output Q_0). In the example in Fig. 9-14(*b*), the switch is opened after the twelfth pulse. This stops the light at Q_3. This is the "winning number" on the roulette wheel for this spin.

The 74HC164 8-bit shift register IC is wired as a ring counter in Fig. 9-14(*a*). The circuit has the two characteristics that make it a ring counter. First, it has feedback from the last flip-flop (Q_7) to the first FF (Q_0). Second, it is loaded with a given pattern of 1s and 0s and these recirculate as long as clock pulses reach the *CP* input of the shift register. In this case, a single 1 is loaded into the shift register and is recirculated.

In summary, the circuit in Fig. 9-14(*a*) is a very simple *electronic roulette wheel.* Pressing the spin wheel input causes the single light to be circulated through the LEDs. When the switch opens the shifting stops.

For added appeal, the simple digital roulette circuit in Fig. 9-14 can be changed by adding a clock that will continue to run for a time after the push button is released. Sound could also be added for a more realistic simulation. Figure 9-15 on page 223 adds both features to the digital roulette wheel.

The versatile 555 timer IC is wired as a VCO in Fig. 9-15. Pressing the spin wheel input switch turns on transistor Q_1. The 555 timer operates as a free-running MV. This square-wave output from the *VCO* drives both the clock input (*CP*) of the *ring counter* and the *audio amplifier.* Pulses from the VCO alternately turn transistor Q_2 on and off, clicking the speaker.

When the spin wheel input switch is opened the 47-μF capacitor holds a positive charge for a time which is applied to the base (*B*) of transistor Q_1. This keeps the transistor turned on for several seconds before the capacitor becomes discharged. As the 47-μF capacitor discharges, the voltage at the base of Q_1 becomes less and the resistance of the transistor (from emitter to collector) increases. This decreases the frequency of the oscillator. This causes the shifting light to slow down. The clicking from the speaker also decreases in frequency. This simulates the slowing of a mechanical roulette wheel.

To review, the *power-up initializing circuitry* block in Fig. 9-15 must first clear the shift reg-

Recirculating line (feedback)

74HC164 8-bit shift register IC

Digital roulette circuit

VCO

Ring counter

Audio amplifier

Ring counter circuit

Power-up initializing circuitry

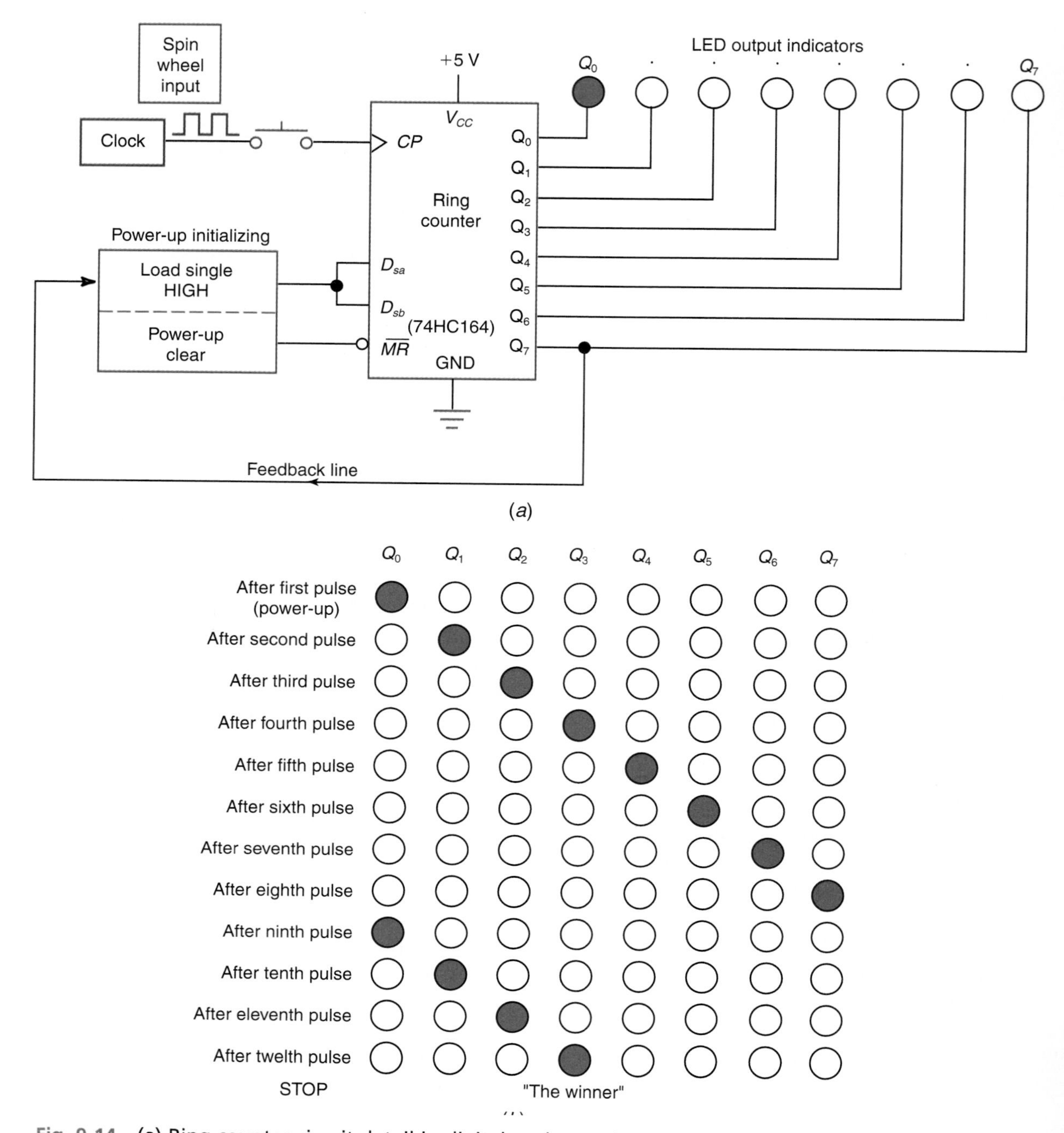

Fig. 9-14 (*a*) Ring counter circuit detail in digital roulette wheel. (*b*) Output from ring counter for first 12 clock pulses.

ister and then set only the first output HIGH. These two circuits have been added to the digital roulette wheel in Fig. 9-16 on page 224.

Automatic clear circuit

An *automatic clear circuit* has been added to the roulette wheel in Fig. 9-16. It consists of the resistor-capacitor combination (R_7 and C_4). When power is turned on, the voltage at the top of the 0.01-μF capacitor starts LOW and increases quickly to a HIGH as it charges through resistor R_7. The master reset ($\overline{MR}$) input to the 74HC164 register is held LOW just long enough for the output of the shift register to be cleared to 00000000. At this point all the LEDs are off.

The circuit that loads a single 1 into the ring counter consists of the four NAND gates and two resistors (R_5 and R_6). The NAND gates are wired as an R-S latch. The two resistors (R_5 and R_6) force the output of the NAND gate (IC*a*) HIGH when the power is first turned on.

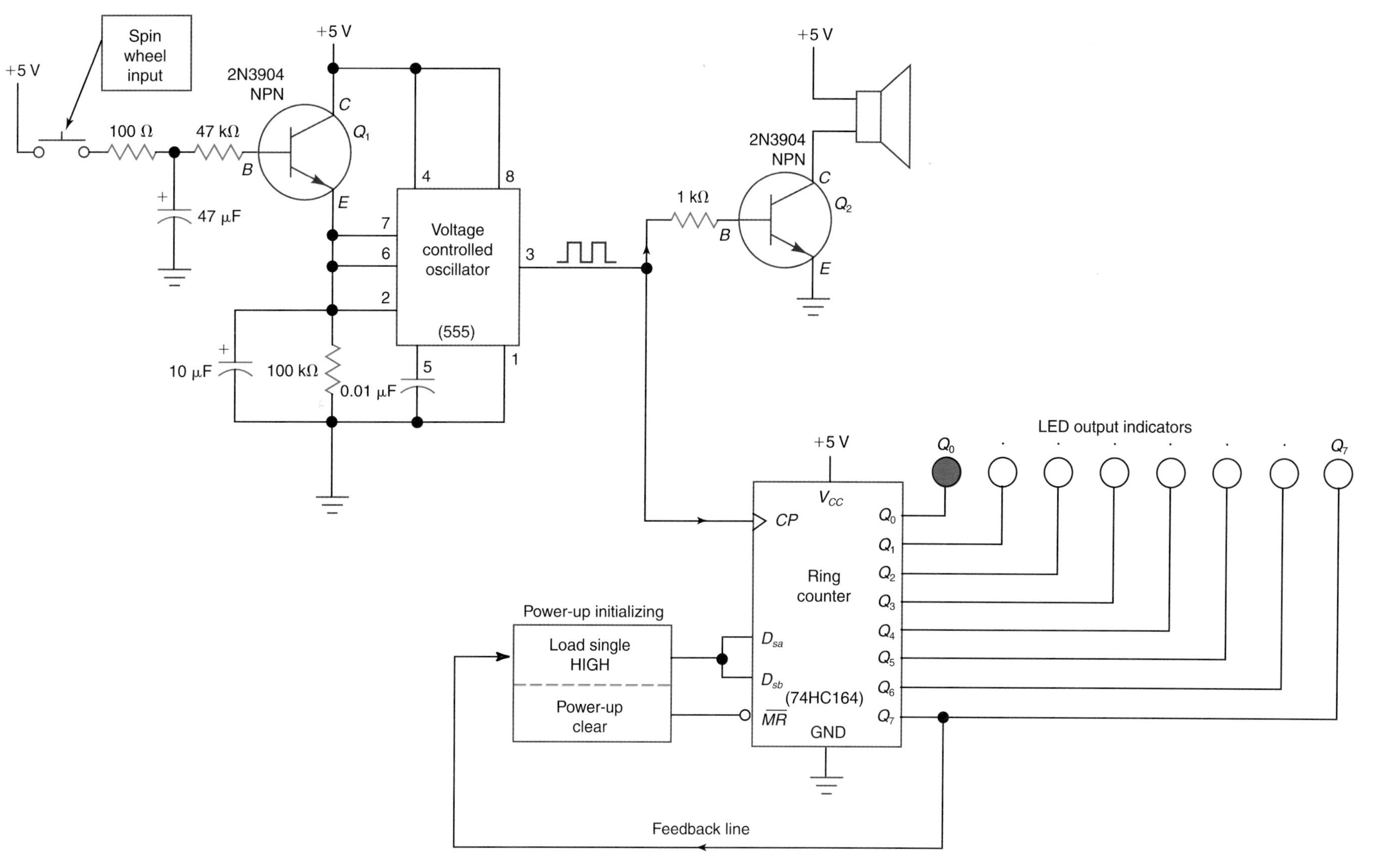

Fig. 9-15 Voltage-controlled oscillator circuit detail in digital roulette wheel.

Digital roulette wheel circuit

Digital roulette game

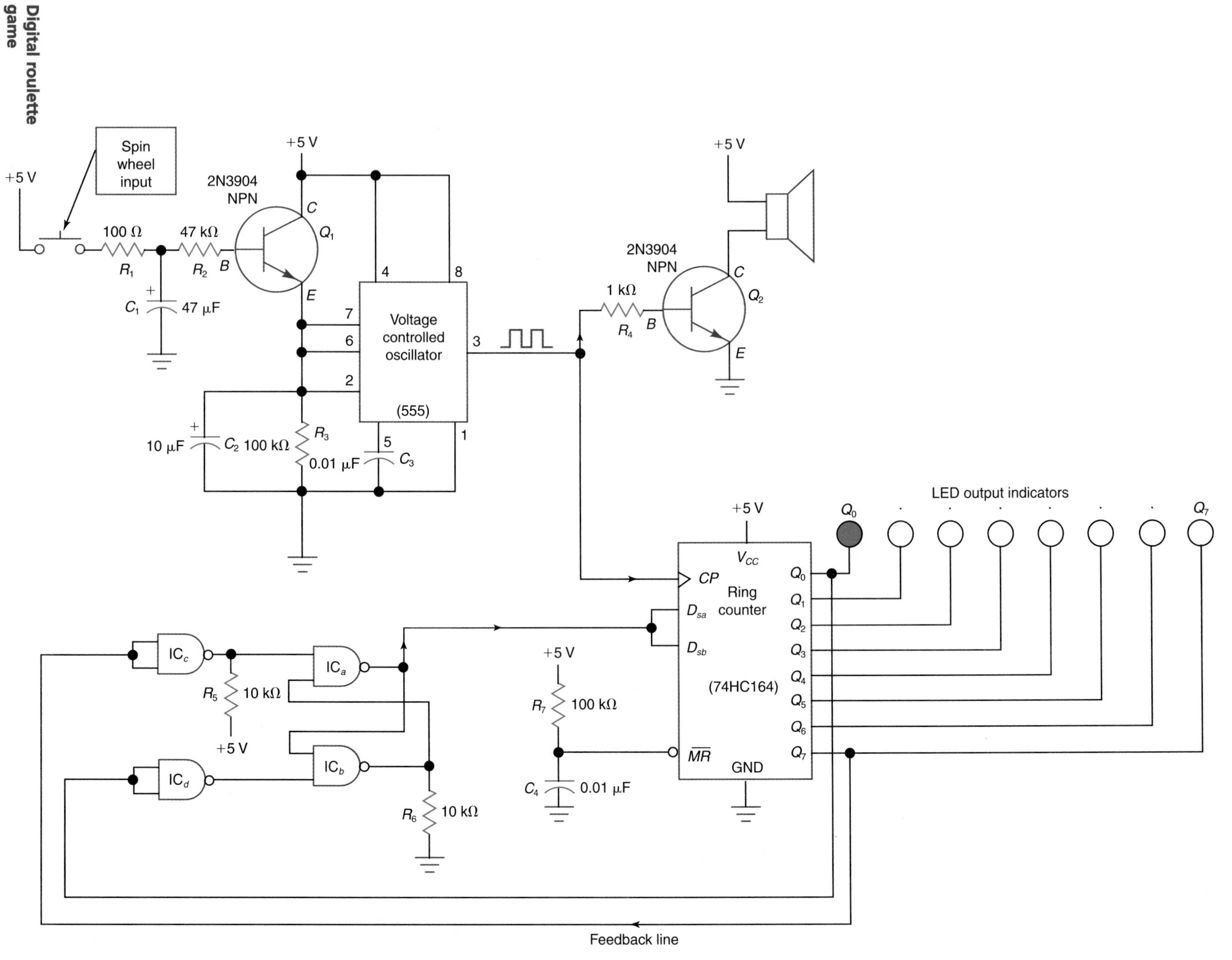

Fig. 9-16 Completed digital roulette wheel. Power-up initializing circuits have been added.

This HIGH is applied to the data inputs (D_{sa} and D_{sb}) of the *ring counter.* On the very first L-to-H transition of the clock the HIGH at the data inputs is transferred to output Q_0 of the 74HC164 IC. Immediately this HIGH is fed back to the input of IC*d* and resets the latch so that a LOW now appears at the data inputs (D_{sa} and D_{sb}). Only a single HIGH was loaded into the ring counter. Repeated clock pulses move the HIGH (light) across the display until Q_7 of the ring counter goes HIGH. This HIGH is fed back to the input of IC*c* setting the latch so that a 1 appears at the data inputs of the ring counter. The single HIGH has been recirculated back to Q_0.

TEST

Answer the following questions.

31. Refer to Fig. 9-16. Components R_4 and Q_2 form the ________ block of the digital roulette wheel circuit.
32. Refer to Fig. 9-16. The 74HC164 8-bit shift register is wired as a(n) ________ in this circuit.
33. Refer to Fig. 9-16. What components cause the 74HC164 IC to reset all outputs to 0 when the power is first turned on?
34. Refer to Fig. 9-16. Clock pulses are fed to the ring counter by the output of the 555 timer IC wired as a(n) ________.
35. Refer to Fig. 9-16. The four NAND gates used for loading a single 1 in the ring counter are wired as a(n) ________ circuit.

9-7 Troubleshooting a Simple Shift Register

Troubleshooting

Ring counter

Consider the faulty serial load shift-right register drawn in Fig. 9-17. Four D flip-flops (two 7474 ICs) have been wired together to form this 4-bit register.

After checking for obvious mechanical and temperature problems, the student or technician runs the following sequence of tests to observe the problem:

1. *Action:* Clear input to 0 and back to 1.
   Result: Output indicators = 0000 (not lit).
   Conclusion: Clear function operating correctly.
2. *Action:* Data input = 1.
   Single pulse to *CLK* of flip-flops from logic pulser.
   Result: Output indicators = 1000.
   Conclusion: FF *A* loading 1s properly.
3. *Action:* Data input = 1.
   Single pulse to *CLK* of flip-flops from logic pulser.
   Result: Output indicators = 1100.
   Conclusion: FF *A* and FF *B* loading 1s correctly.
4. *Action:* Data input = 1.
   Single pulse to *CLK* of flip-flops from logic pulser.
   Result: Output indicators = 1110.
   Conclusion: FF *A*, FF *B*, and FF *C* loading 1s correctly.

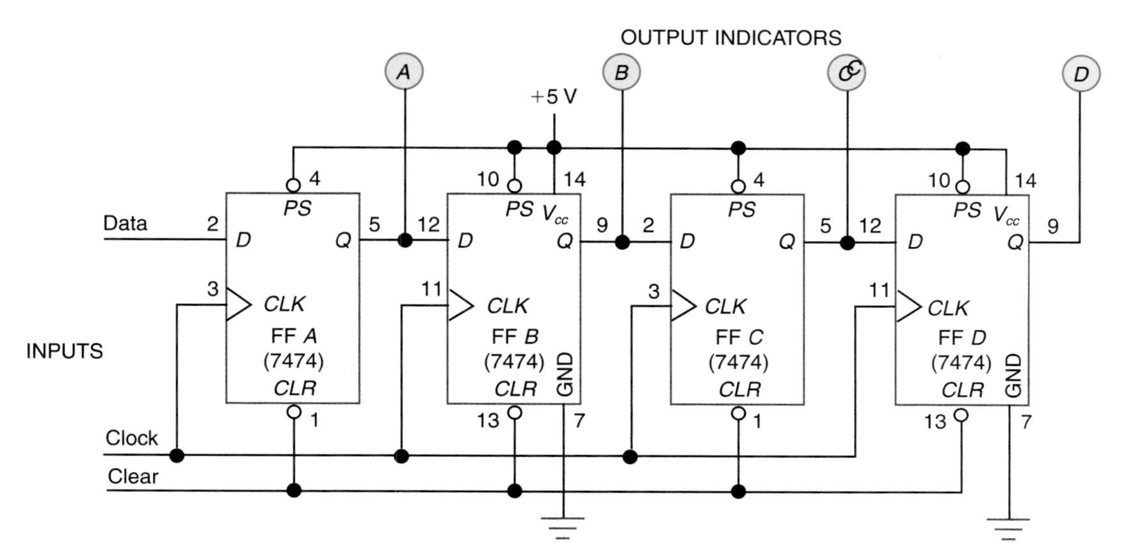

Fig. 9-17 Faulty 4-bit serial load shift-right register used as a troubleshooting example.

Faulty 4-bit serial load shift-right register

5. *Action:* Data input = 1.
   Single pulse to *CLK* of flip-flops from logic pulser.
   Result: Output indicators = 1110.
   Conclusion: Suspect problem near or in FF *D* since it did not load a HIGH properly.
6. *Action:* Logic probe at *D* input to FF *D* to see if *D* = 1.
   Result: *D* = 1 on FF *D*.
   Conclusion: HIGH data at *D* of FF *D* is correct.
7. *Action:* One pulse to *CLK* (pin 11) of FF *D* from logic pulser.
   Result: Output indicator remains at 1110.
   Conclusion: Data not being transferred from input *D* of FF *D* to output *Q* on a clock pulse.
8. *Action:* Logic probe to output *Q* of FF *D* (pin 9).
   Result: Neither HIGH nor LOW indicator lights on logic probe.
   Conclusion: Output *Q* (pin 9) of FF *D* floating between HIGH and LOW. Probably a faulty FF *D* in second 7474 IC.
9. *Action:* Remove and replace second 7474 IC (FF *C* and FF *D*) with exact replacement.
10. *Action:* Retest circuit, starting at step 1.
    Result: All flip-flops load 1s and 0s.
    Conclusion: Shift register circuit is now operating properly.

According to the sequence of tests, the *Q* output of FF *D* seemed to be stuck LOW while it was actually floating between LOW and HIGH. This fact made our conclusion in step 1 incorrect. This fault was caused by an open circuit within the second 7474 IC itself. Again, the technician's knowledge of how the circuit operates along with observations helped locate the fault. The logic probe and digital logic pulser aided the technician in making observations.

Redundant circuitry

Sometimes the technician is not exactly sure of the appropriate logic level. In a circuit with *redundant circuitry* (circuits repeated over and over), the technician could go back to FF *A* and FF *B* and compare these readings with those on FF *C* and FF *D*. Digital circuits have much redundant circuitry, and at times this technique is helpful in troubleshooting.

TEST

Answer the following questions.

36. Refer to Fig. 9-17. Describe the observed problem in this circuit.
37. Refer to Fig. 9-17. What is wrong with this circuit?
38. Refer to Fig. 9-17. How can the fault in this circuit be repaired?
39. What test equipment can be used to troubleshoot this shift register circuit?

Chapter 9 Review

Summary

1. Flip-flops are wired together to form shift registers.
2. A shift register has both a memory and a shift characteristic.
3. A serial load shift register is one that permits only one bit of data to be entered per clock pulse.
4. A parallel load shift register is one that permits all data bits to be entered at one time (one clock pulse).
5. A recirculating register feeds output data back into the input.
6. Shift registers can be designed to shift either left or right.
7. Manufacturers produce many adaptable universal shift registers.
8. Shift registers are widely used as temporary memories and for shifting data. They also have other uses in digital electronic systems.
9. A ring counter is a shift register that (1) has a recirculating line, and (2) is loaded with a pattern of 0s and 1s, which is repeated over and over as the unit is clocked.

Chapter Review Questions

Answer the following questions.

9-1. Draw a logic symbol diagram of a 5-bit serial load shift-right register. Use five D flip-flops. Label inputs data, *CLK*, and *CLR*. Label outputs *A*, *B*, *C*, *D*, and *E*. The circuit will be similar to the one in Fig. 9-3.

9-2. Explain how you would clear to 00000 the 5-bit register you drew in question 9-1.

9-3. After clearing the 5-bit register, explain how you would enter (load) 10000 into the register you drew in question 9-1.

9-4. After clearing the 5-bit register, explain how you would enter (load) 00111 into the register you drew in question 9-1.

9-5. Refer to the register you drew in question 9-1. List the contents of the register after each clock pulse shown in **b** to **e** (assume data input = 0).
   a. Original output = 01001 ($A = 0$, $B = 1$, $C = 0$, $D = 0$, $E = 1$)
   b. After one clock pulse =
   c. After two clock pulses =
   d. After three clock pulses =
   e. After four clock pulses =

9-6. Refer to Fig. 9-9. The parallel load register using the 74194 IC needs ________ (no, one, three, four) clock pulse(s) to load data from the parallel load inputs.

9-7. A ________ (serial, parallel) load shift register is the simplest circuit to wire.

9-8. A ________ (serial, parallel) load shift register is the easiest to load.

9-9. Refer to Fig. 9-7 for questions **a** to **i** on the 74194 IC shift register:
   a. How many bits of information can this register hold?
   b. List the four modes of operation for this register.
   c. What is the purpose of the mode control inputs (*S*0, *S*1)?
   d. The ________ input overrides all other inputs on this register.
   e. What type are the flip-flops and how many are used in this shift register?
   f. The register shifts on the ________ (negative-, positive-) going edge of the clock pulse.
   g. What does the inhibit mode of operation mean?
   h. By definition, to shift left means to shift data from ________ to ________ (use letters).
   i. This register can be loaded ________ (serially, in parallel, either serially or in parallel).

9-10. Refer to Fig. 9-18. List the 74194 shift register's mode of operation during each of the eight clock pulses. Use as answers "clear," "in-

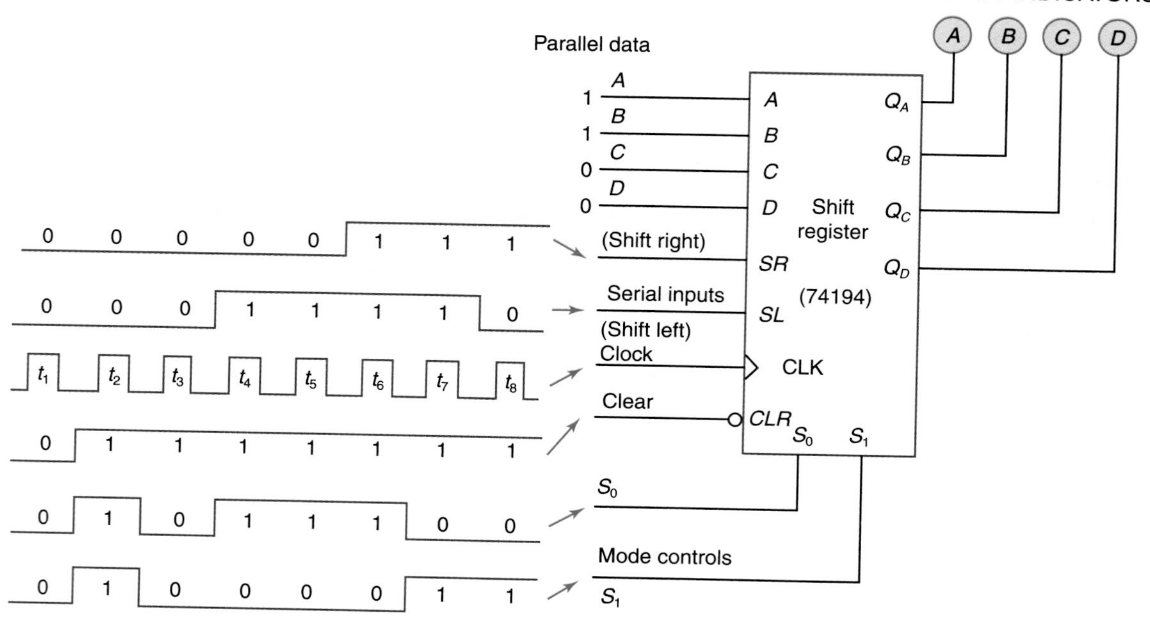

Fig. 9-18 Shift register problem for Critical Thinking Question 4.

hibit," "shift right," "shift left," and "parallel load."

9-11. Refer to Fig. 9-11 for questions **a** to **f** on the 74HC164 shift register.
   a. How many bits of information can this register store?
   b. This shift register is a ________ (CMOS, TTL) IC.
   c. This is a ________ (parallel, serial)-load shift register.
   d. The master reset is an active ________ (HIGH, LOW) input.
   e. The register shifts data on the ________ (H to L, L to H) transition of the clock pulse.
   f. The register has two data inputs which are ________ (ANDed, ORed) together for loading data into FF 1.

9-12. Refer to Fig. 9-19. List the contents of the register during each of the eight clock pulses (Q_0 = left bit, Q_7 = right bit).

9-13. Refer to Fig. 9-13. The device that generates clock pulses in the digital roulette circuit is called a(n) ________.

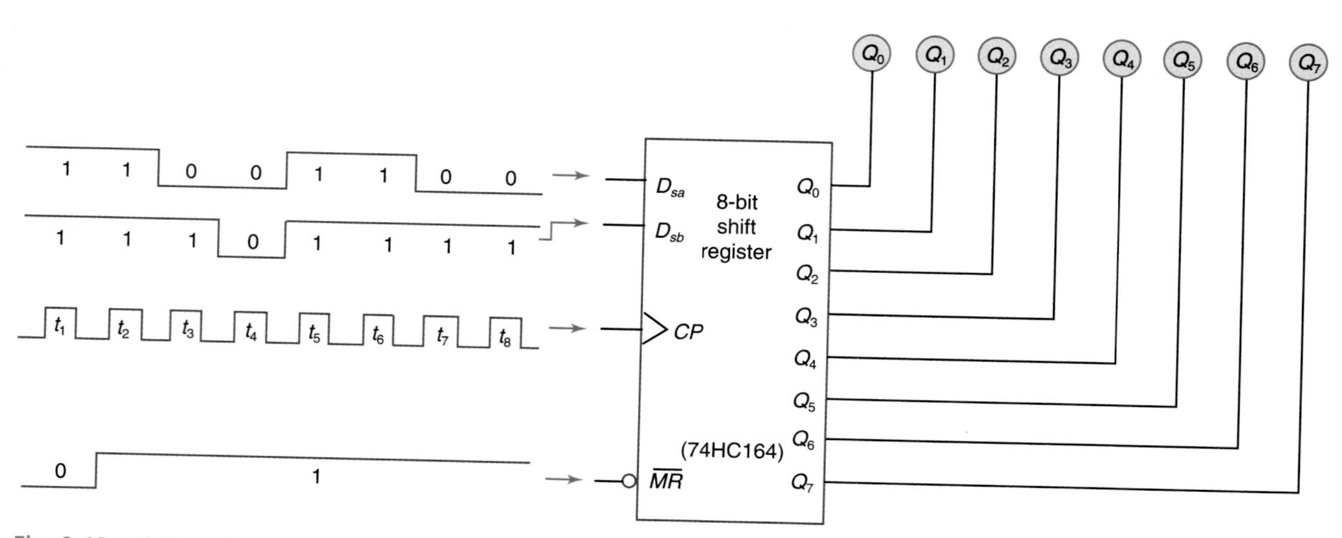

Fig. 9-19 Shift register problem for Review Question 12.

9-14. Refer to Fig. 9-14(*a*). The 74HC164 shift register is wired as a(n) __________ in this circuit.

9-15. Refer to Fig. 9-16. The frequency of the VCO decreases as the voltage at the top of capacitor __________ (C_1, C_2, C_4) decreases.

9-16. Refer to Fig. 9-16. What is the purpose of resistor R_7 and capacitor C_4?

9-17. Refer to Fig. 9-16. Resistors R_5 and R_6 force the output of IC*a* __________ (HIGH, LOW) when the power is first turned on.

9-18. Refer to Fig. 9-16. If only Q_0 of the ring counter is HIGH (as shown), the R-S latch forces the output of IC*a* __________ (HIGH, LOW).

Critical Thinking Questions

9-1. The shift register in Fig. 9-5(*b*) needs __________ (no, one, four) clock pulse(s) to load data from the parallel data inputs.

9-2. The shift register in Fig. 9-5(*b*) can only load __________ (0, 1s) using the parallel load inputs.

9-3. List several uses of shift registers in digital systems.

9-4. List the contents of the register in Fig. 9-18 after each of the eight clock pulses (A = left bit, D = right bit).

9-5. Describe in general terms the nature of the output from the VCO in Fig. 9-13.

9-6. Refer to Fig. 9-5. Describe the procedure you would follow when loading the data 1101 into this 4-bit parallel-load shift register.

9-7. Refer to Fig. 9-9. Parallel loading of data is a(n) __________ (asynchronous, synchronous) operation when using the 74194 shift register IC.

9-8. A ring counter is classified as a type of __________ (shift register, VCO).

9-9. Draw a block diagram of a 16-bit electronic roulette wheel with VCO, audio amplifier, power-up initializing circuits, and ring counter blocks. It should look something like the 8-bit electronic roulette wheel in Fig. 9-13.

9-10. At the option of your instructor, use Electronics Workbench® circuit simulation software to (1) draw the 8-bit serial-load shift register drawn in Fig. 9-20, (2) test the opera-

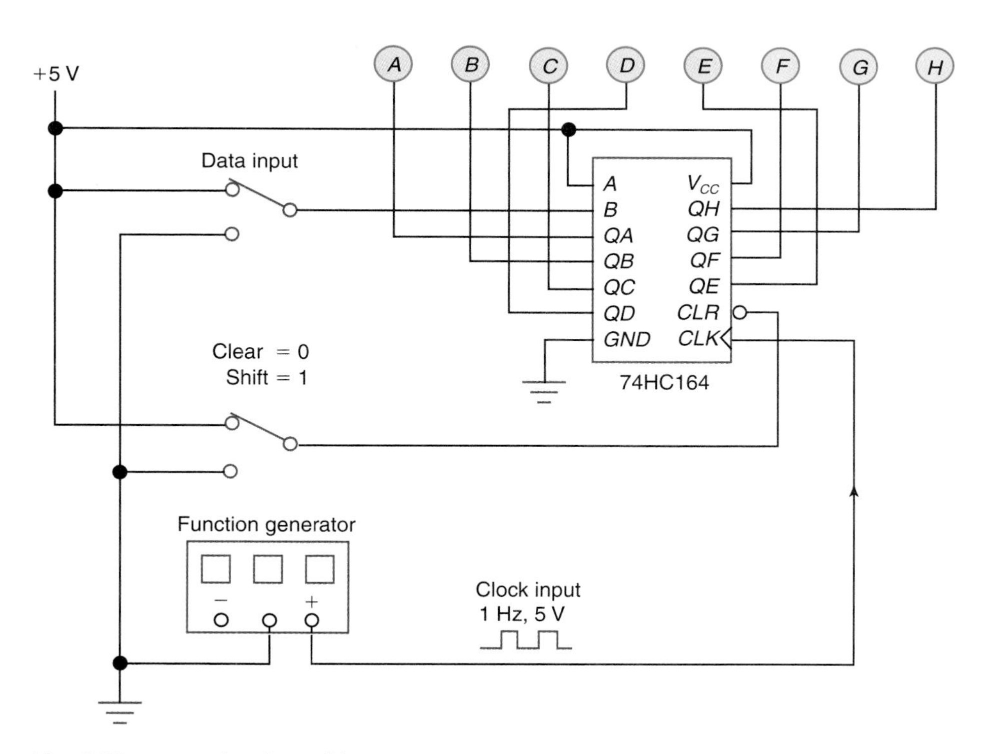

Fig. 9-20 EWB circuit problem.

tion of the shift register, and (3) save the circuit and show your instructor your design.

9-11. At the option of your instructor, use Electronics Workbench® circuit simulation software to (1) add a recirculating line to your 8-bit shift register you designed in question 9-10 (Hint: OR recirculating line and data input), (2) test the operation of the shift register with recirculating feature, and (3) save the circuit and show your instructor your design.

Answers to Tests

1. serial
2. after pulse t_1 = 000
   after pulse t_2 = 100
   after pulse t_3 = 010
   after pulse t_4 = 001
   after pulse t_5 = 000
   after pulse t_6 = 100
3. single bit
4. LOW
5. HIGH, L-to-H
6. parallel
7. pulse t_1 = clear
   pulse t_2 = parallel load
   pulse t_3 = shift-right
   pulse t_4 = shift-right
   pulse t_5 = shift-right
   pulse t_6 = parallel load
   pulse t_7 = shift-right
   pulse t_8 = shift-right
8. after pulse t_1 = 000
   after pulse t_2 = 010
   after pulse t_3 = 001
   after pulse t_4 = 100
   after pulse t_5 = 010
   after pulse t_6 = 101
   after pulse t_7 = 110
   after pulse t_8 = 011
9. recirculating
10. asynchronous
11. lines 2 and 9
12. lines 3 and 10
13. 1. clear
    2. parallel load
    3. shift-right
    4. shift-left
    5. inhibit (do nothing)
14. parallel load
15. inhibit
16. HIGH, LOW, LOW, HIGH
17. HIGH, one
18. inhibit
19. 1, 0, shift-right serial
20. 0000 (cleared)
21. L-to-H
22. clear
23. shift left
24. LOW
25. L-to-H or LOW-to-HIGH
26. during pulse t_1 = reset
    during pulse t_2 = shift-right
    during pulse t_3 = shift-right
    during pulse t_4 = shift-right
    during pulse t_5 = shift-right
    during pulse t_6 = shift-right
27. during pulse t_1 = 00000000
    during pulse t_2 = 10000000
    during pulse t_3 = 01000000
    during pulse t_4 = 00100000
    during pulse t_5 = 10010000
    during pulse t_6 = 01001000
28. CMOS
29. 8-bit, serial-load
30. ANDed
31. audio amplifier
32. ring counter
33. R_7 and C_4
34. voltage-controlled oscillator or VCO
35. R-S latch or latch
36. will not shift a HIGH into the *D* position
37. Output *Q* (pin 9) of FF *D* floating; 7474 IC that contains FF *C* and FF *D* faulty
38. A new 7474 IC should be inserted, replacing FF *C* and FF *D*.
39. logic pulser, logic probe

Arithmetic Circuits

Chapter Objectives

This chapter will help you to:

1. *Memorize* and *draw* the block diagrams for a half adder, a full adder, a half subtractor, and a full subtractor.
2. *Solve* binary addition and subtraction problems by hand and from a truth table.
3. *Solve* binary subtraction problems using the 1s complement and end-around carry methods.
4. *Design* and *draw* block-style logic diagrams for several parallel adder and sub tractor circuits using half adders, full adders, and gates.
5. *Solve* several binary multiplication problems.
6. *Convert* decimal numbers to 2s complement notation and 2s complement to decimal numbers.
7. *Add* and *subtract* signed numbers using 2s complement addition and subtraction.
8. *Troubleshoot* a faulty full-adder circuit.

The public's imagination has been captured by computers and modern-day calculators, probably because these machines perform arithmetic tasks with such fantastic speed and accuracy. This chapter deals with some logic circuits that can add and subtract. (Of course, the adding and subtracting is done in binary.) Regular logic gates will be wired together to form *adders* and *subtractors.* Basic adder and subtractor circuits are combinational logic circuits, but they are commonly used with various latches and registers to hold data.

In the *central processing unit (CPU)* of a computer, arithmetic is handled in a section commonly called the *arithmetic-logic unit (ALU).* This section within the CPU can usually add and subtract, multiply and divide, complement, compare, shift and rotate, increment and decrement, and perform logic operations such as AND, OR, and XOR. Many older *microprocessors* and several modern *microcontrollers* (a miniature microprocessor used mainly for control purposes) do not have multiply and divide commands in their instruction set.

Central processing unit (CPU)

Arithmetic-logic unit (ALU)

Microcontroller

10-1 Binary Addition

Remember that in a binary number, such as 101011, the leftmost digit is the MSB and the rightmost digit is the LSB. Also remember the place values given to the binary number is: 1s, 2s, 4s, 8s, 16s, and 32s.

You probably still recall learning your addition and subtraction tables when you were in elementary school. This is a difficult task in the decimal number system because there are so many combinations. This section deals with the simple task of adding numbers in binary. Because they have only two digits (0 and 1), the binary addition tables are simple. Figure 10-1(*a*) on page 232 shows the binary addition tables. Just as in the case of adding with decimals, the first three problems are easy. The next problem is 1 + 1. In decimal that would be 2. In binary a 2 is written as 10. Therefore, in binary 1 + 1 = 0, with a carry of 1 to the next most significant place value.

Binary addition

MSB

LSB

Figure 10-1(*b*) shows some examples of adding numbers in binary. The problems are also shown in decimal so that you can check your understanding of binary addition. The first

0	1	0	1
+0	+0	+1	+1
0	1	1	0 carry 1

(*a*)

101	5	1010	10	11010	26
+ 10	+2	+ 11	+ 3	+ 1100	+12
111	7	1101	13	100110	38

(carries: 1010 + 11 has one carry; 11010 + 1100 has two carries)

(*b*)

Fig. 10-1 (*a*) Binary addition tables. (*b*) Sample binary addition problems.

Binary addition tables

problem is adding binary 101 to 10, which equals 111 (decimal 7). This problem is simple using the addition tables in Fig. 10-1(*a*). The second problem in Fig. 10-1(*b*) is adding binary 1010 to 11. Here you must notice that a 1 + 1 = 0 plus a carry from the 2s place to the 4s place, as shown in the diagram. The answer to this problem is 1101 (decimal 13). In the third problem in Fig. 10-1(*b*) the binary number 11010 is added to 1100. In the figure, note two carries with the solution as 100110 (decimal 38).

Another sample addition problem is shown in Fig. 10-2(*a*). The solution looks simple until we get to the 2s column and find 1 + 1 + 1 in binary. This equals 3 in decimal, which is 11 in binary. This one situation we left out of the first group of binary addition tables. Looking carefully at Fig. 10-2, you see that the 1 + 1 + 1 situation can arise in any column except the 1s column. So the binary addition table in Fig. 10-1(*a*) is complete for the *1s column only.* The new short-form addition table in Fig. 10-2(*b*) adds the other possible combination of 1 + 1 + 1. The addition table in Fig. 10-2(*b*), then, is for all the place values (2s, 4s, 8s, 16s, and so on) except the 1s column.

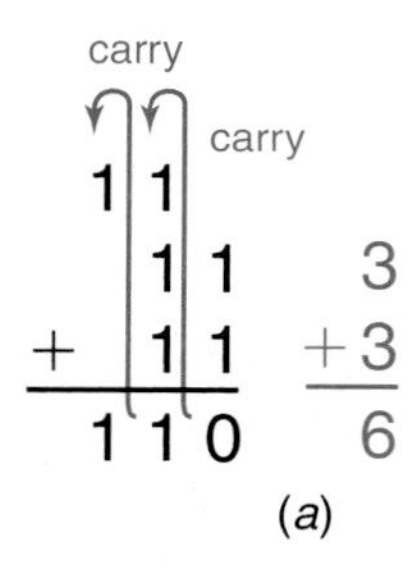

(*a*)

0	1	1	1
			1
+0	+0	+1	+1
0	1	0 carry 1	1 carry 1

(*b*)

Fig. 10-2 (*a*) Sample binary addition problem. (*b*) Short-form addition table.

To be an intelligent worker on digital equipment you must master binary addition. Several practice problems are provided in the first test.

TEST

Answer the following questions.

1. What is the sum of binary 1010 + 0100? (Check your answer using decimal addition.)
2. What is the sum of binary 1010 + 0111?
3. What is the sum of binary 1111 + 1001?
4. What is the sum of binary 10011 + 0111?

Half adder

10-2 Half Adders

The addition table in Fig. 10-1(*a*) can be thought of as a truth table. The numbers being added are on the input side of the table. In Fig. 10-3(*a*) these are the *A* and *B* input columns. The truth table needs *two* output columns, one column for the sum and one column for the carry. The sum column is labeled with the summation symbol Σ. The carry column is labeled with a C_O. The C_O stands for carry output or *carry out.* A convenient block symbol for the adder that performs the job of the truth table is shown in Fig. 10-3(*b*). This circuit is called a

INPUTS		OUTPUTS	
B	*A*	Σ	C_O
0	0	0	0
0	1	1	0
1	0	1	0
1	1	0	1
Binary digits to be added		Sum	Carry out
		XOR	AND

(*a*)

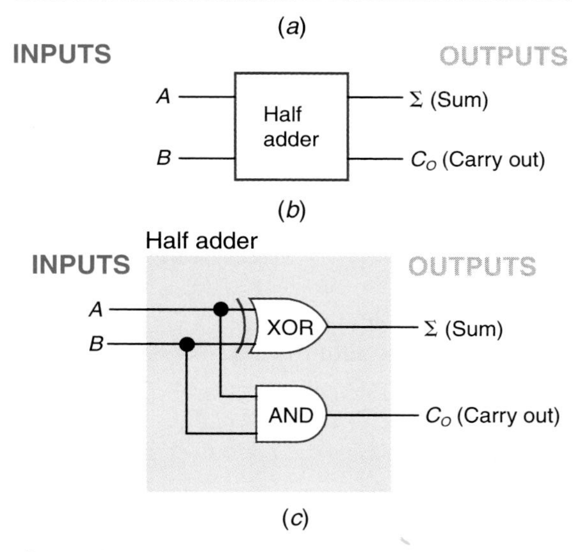

Fig. 10-3 Half adder. (*a*) Truth table. (*b*) Block symbol. (*c*) Logic diagram.

half-adder circuit. The half-adder circuit has two inputs (*A, B*) and two outputs, (Σ, C_O).

Take a careful look at the half-adder truth table in Fig. 10-3(*a*). What is the Boolean expression needed for the C_O output? The Boolean expression is $A \cdot B = C_O$. You need a two-input AND gate to take care of output C_O.

Now what is the Boolean expression for the sum (Σ) output of the half adder in Fig. 10-3(*a*)? The Boolean expression is $\overline{A} \cdot B + A \cdot \overline{B} = \Sigma$. We could use two AND gates and one OR gate to do the job. If you look closely you will notice that this pattern is also that of an XOR gate. The simplified Boolean expression is then $A \oplus B = \Sigma$. In other words, we find that only one 2-input XOR gate is needed to produce the sum output.

Using a two-input AND gate and a two-input XOR gate, a logic symbol diagram for a half adder is drawn in Fig. 10-3(*c*). The half-adder circuit adds only the LSB column (1s column) in a binary addition problem. A circuit called a *full adder* must be used for the 2s, 4s, 8s, and 16s, and higher places in binary addition.

Full adder

Half-adder circuit

TEST

Answer the following questions.

5. Draw a block diagram of a half adder. Label inputs *A* and *B*; label outputs Σ and C_O.
6. Draw a truth table for a half adder.
7. A half-adder circuit is used for adding only the ________ (1s, 2s, 4s, 8s) column of a binary addition problem.
8. Refer to Fig. 10-4. List the outputs from both the sum (Σ) and carry out (C_O) terminals of the half-adder circuit for each input pulse (t_1 to t_4).

10-3 Full Adders

Figure 10-2(*b*) is the short form of the binary addition table, with the 1 + 1 + 1 situation shown. The truth table in Fig. 10-5(*a*) on page 234 shows all the possible combinations, of *A*, *B*, and C_{in} (carry in). This truth table is for a full adder. Full adders are used for all binary place values except the 1s place. The full adder must be used when it is possible to have an extra *carry input.* A block diagram of a full adder is shown in Fig. 10-5(*b*). The full adder has three inputs: C_{in}, *A*, and *B*. These three inputs must be added to get the Σ and C_O outputs.

One of the easiest methods of forming the combinational logic for a full adder is diagramed in Fig. 10-5(*c*); two half-adder circuits and an OR gate are used. The expression for this arrangement is $A \oplus B \oplus C = \Sigma$. The expression for the carry out is $A \cdot B + C_{in} \cdot (A \oplus B) = C_O$. The logic circuit in Fig. 10-6(*a*) on page 235 is a full adder. This circuit is based upon the block diagram using two half adders shown in Fig. 10-5(*c*). Directly below this logic diagram is a logic circuit that is somewhat eas-

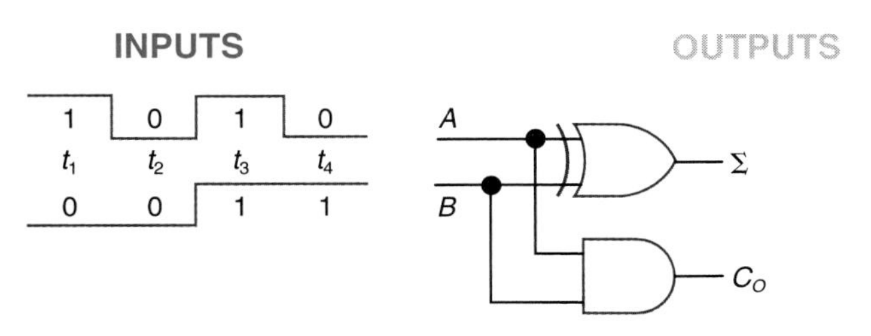

Fig. 10-4 Half adder pulse-train problem for test question 8.

INPUTS			OUTPUTS	
C_{in}	B	A	Σ	C_O
0	0	0	0	0
0	0	1	1	0
0	1	0	1	0
0	1	1	0	1
1	0	0	1	0
1	0	1	0	1
1	1	0	0	1
1	1	1	1	1
Carry + B + A			Sum	Carry out

(*a*)

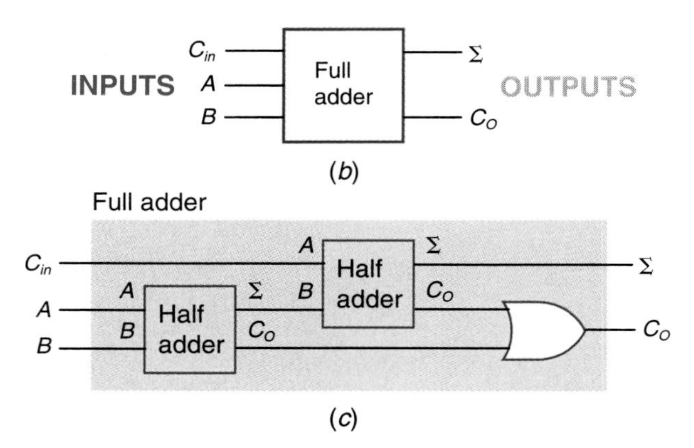

Fig. 10-5 Full adder. (*a*) Truth table. (*b*) Block symbol. (*c*) Constructed from half adders and an OR gate.

ier to wire. Figure 10-6(*b*) contains two XOR gates and three NAND gates, which makes the circuit fairly easy to wire. Notice that the circuit in Fig 10-6(*b*) is exactly the same as the one in Fig. 10-6(*a*), except that NAND gates have been substituted for AND and OR gates.

Half and full adders are used together. For the problem in Fig. 10-2(*a*) we need one half adder for the 1s place and two full adders for the 2s place and the 4s place value. Half and full adders are rather simple circuits. However, many of these circuits are needed to add longer problems (more binary digits).

Arithmetic-logic unit (ALU)

Many circuits similar to half and full adders are part of a microprocessor's *arithmetic-logic unit (ALU)*. These circuits are then used for adding 8-bit or 16- or 32-bit binary numbers in a microcomputer system. The microprocessor's ALU can also subtract using the same half- and full-adder circuits. Later in this chapter, you will use adders to perform binary subtraction.

TEST

Answer the following questions.

9. Draw a block diagram of a full adder. Label inputs *A*, *B*, and C_{in}; label outputs Σ and C_O.
10. Draw a truth table for a full adder.
11. Adder circuits are widely used in the __________ section of a microprocessor.
12. A __________ (half-adder, full-adder) circuit must be used for the 2s, 4s, 8s, and more significant bits in a binary addition problem.
13. Refer to Fig. 10-7 on page 236. List the outputs from both the sum (Σ) and carry out (C_O) terminals of the full-adder circuit for each of the input pulses (t_1 to t_8).

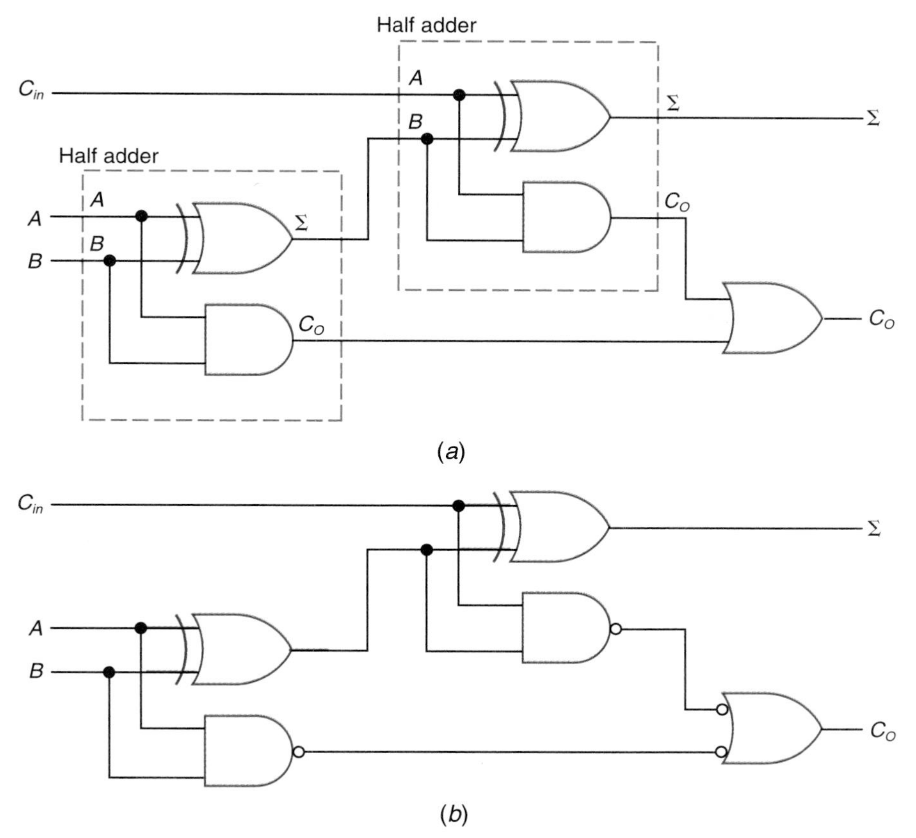

Fig. 10-6 Full adder. (*a*) Logic diagram. (*b*) Logic diagram using XOR and NAND gates.

10-4 Three-Bit Adders

Half and full adders are connected to form adders that add several binary digits (bits) at one time. The system in Fig. 10-8, on page 236 adds two 3-bit numbers. The numbers being added are written as $A_2A_1A_0$ and $B_2B_1B_0$. Numbers from the 1s place value column are entered into the 1s adder which is a half adder. The inputs to the 2s adder are the carry from the half adder (C_{in}) and the new bits A_1 and B_1 from the problem. The 4s adder adds A_2 and B_2 and the carry in from the 2s adder. The total sum is shown in binary at the lower right. The output also has an 8s place value to take care of any binary number over 111 in the sum. Notice that the 4s adder's output (C_O) is connected to the 8s sum indicator.

The 3-bit binary adder is organized as you would *add and carry* by hand. The electronic adder in Fig. 10-8 on page 236, is very much faster than doing the same problem by hand. Notice that multibit adders use a half adder for the 1s column only; all other bits use a full adder. This type of adder is called a *parallel adder*.

In a parallel adder, all bits are applied to the inputs at the same time. The sum appears at the output almost immediately. The parallel adder shown in Fig. 10-8 is a combinational logic circuit and typically needs various registers to latch data at the inputs and outputs.

3-bit adders

Full adder

Parallel adder

Combinational logic circuit

TEST

Supply the missing word (or words) in each statement.

14. The unit in Fig. 10-8 uses a(n) ________ for adding the 1s column and a ________ for the more significant columns.
15. Parallel adders are ________ (combinational, sequential) logic circuits.
16. If the inputs to the 3-bit binary adder in Fig. 10-8 are 110_2 and 111_2, the output indicators will show a sum of ________ in binary.

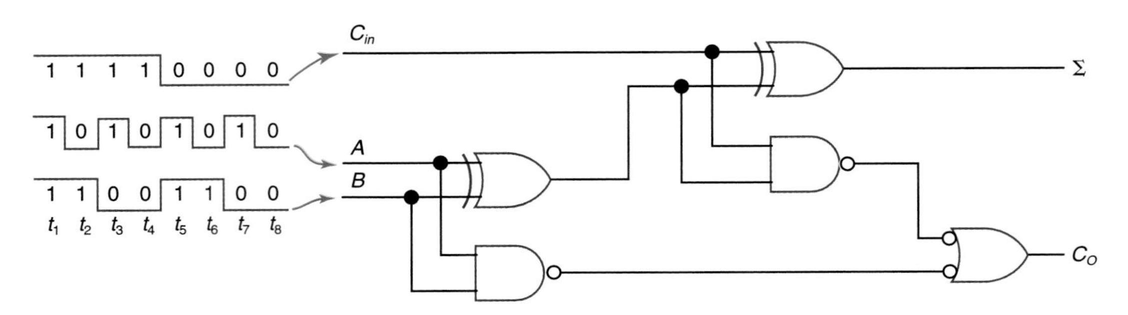

Fig. 10-7 Full adder pulse-train problem for test question 13.

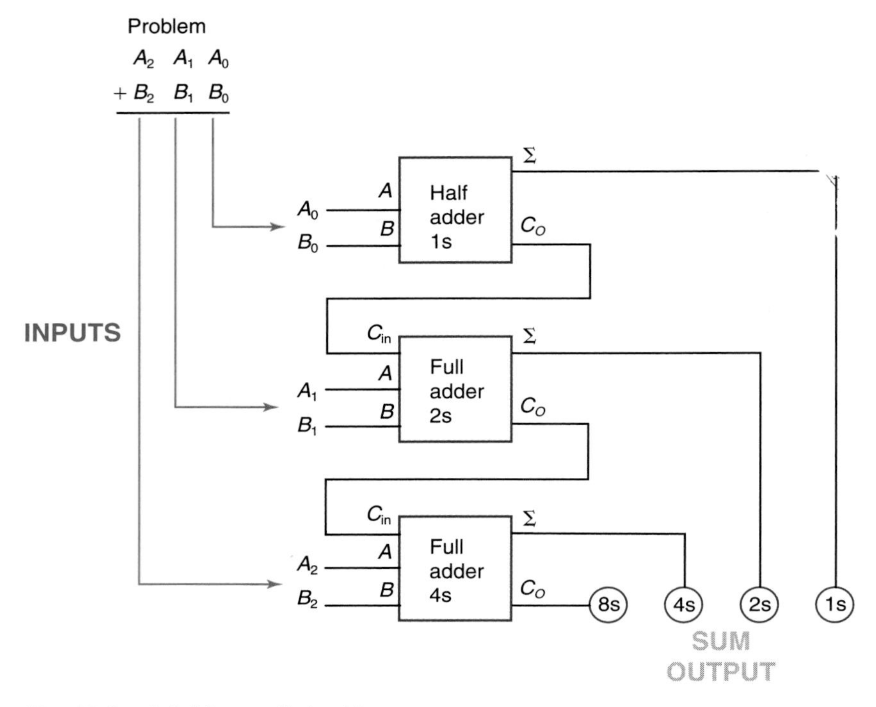

Fig. 10-8 A 3-bit parallel adder.

17. If the inputs to the 3-bit binary adder in Fig. 10-8 are 010_2 and 110_2, the output indicators will show a sum of ________ in binary.
18. If the inputs to the 3-bit binary adder in Fig. 10-8 are 111_2 and 111_2, the output indicators will show a sum of ________ in binary.

10-5 Binary Subtraction

Binary subtraction

Half subtractors

Full subtractors

You will find that *adders* and *subtractors* are very similar. You use *half subtractors* and *full subtractors* just as you use half and full adders. Binary subtraction tables are shown in Fig. 10-9(*a*). Converting these rules to truth-table form gives the table in Fig. 10-9(*b*). On the input side, *B* is subtracted from *A* to give output *Di* (difference). If *B* is larger than *A*, such as in line 2, we need a *borrow,* which is shown in the column labeled B_O (borrow out).

A block diagram of a half subtractor is shown in Fig. 10-9(*c*). Inputs *A* and *B* are on the left. Outputs *Di* and B_O are on the right side of the diagram. Looking at the truth table in Fig. 10-9(*b*), we can determine the Boolean expressions for the half subtractor. The expression for the *Di* column is $A \oplus B = Di$. This is the same as for the half adder (see Fig. 10-3(*a*)). The Boolean expression for the B_O column is $\overline{A} \cdot B = B_O$. Combining these two expressions in a logic diagram gives the logic circuit in Fig. 10-9(*d*). This is the logic circuit for a half subtractor; notice how much it looks like the half-adder circuit in Fig. 10-4.

When you subtract several columns of bi-

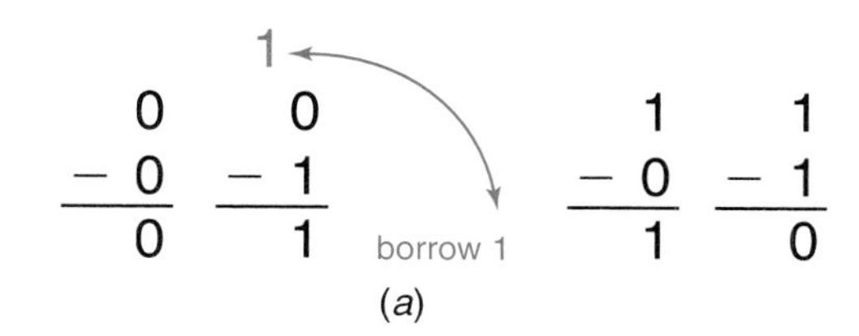

(*a*)

INPUTS		OUTPUTS	
A	*B*	*Di*	B_O
0	0	0	0
0	1	1	1
1	0	1	0
1	1	0	0
A − *B*		Difference	Borrow out

(*b*)

INPUTS

OUTPUTS

A, *B* → Half subtractor → *Di* (Difference), B_O (Borrow)

(*c*)

A, *B* → *Di* (Difference), B_O (Borrow)

(*d*)

Fig. 10-9 (*a*) Binary subtraction tables. (*b*) Truth table for the half subtractor. (*c*) Block symbol of half subtractor. (*d*) Logic diagram for half subtractor.

nary digits, you must take into account the borrowing. Suppose you are subtracting the numbers in Fig. 10-10(*a*) on page 238. You might keep track of the differences and borrows as shown in the figure. Look over the subtraction problem carefully, and check if you can do binary subtraction by this longhand method. (You can check yourself on the next test.)

A truth table that considers all the possible combinations in binary subtraction is shown in Fig. 10-10(*b*). For instance, line 5 of the table is the situation in the 1s column in Fig. 10-10(*a*). The 2s column equals line 3, the 4s column line 6, the 8s column line 3, the 16s column line 2, and the 32s column line 6 of the truth table.

A block diagram of a full subtractor is drawn in Fig. 10-11(*a*), on page 239. The inputs *A*, *B*, and B_{in} are on the left; the outputs *Di* and B_O are on the right. Like the full adder, the full subtractor can be wired using two half subtractors and an OR gate. Figure 10-11(*b*) is a full subtractor showing how half subtractors are used. A logic diagram for a full subtractor is shown in Fig. 10-11(*c*). This circuit performs as a full subtractor as specified in the truth table in Fig. 10-10(*b*). The AND-OR circuit on the B_O output can be converted to three NAND gates if you want. The circuit would then be similar to the full-adder circuit in Fig. 10-6(*b*).

Humor is useful, but crude and rude remarks are not appropriate.

TEST

Answer the following questions.

19. Do the binary subtraction problems in **a** to **f**. (Check yourself using decimal subtraction.)
    a. 11 − 10
    b. 100 − 10
    c. 111 − 111
    d. 1010 − 101
    e. 10010 − 11
    f. 1000 − 01
20. Draw a block diagram of a half subtractor. Label inputs *A* and *B*; outputs *Di* and B_O.
21. Draw a truth table for a half subtractor.

Binary subtraction tables

Half subtractors

About Electronics

Wilderness Locator One of the most dangerous aspects of a hike in the wilderness has always been the problem of finding help in the event of an accident—at least until recently. New handheld devices, such as this one from Magellan, utilize satellite technology to send messages such as distress signals and to identify the exact coordinates of the device to rescuers.

32s 16s 8s 4s 2s 1s

```
      1
   ↗10 10  0 ↗10
   1̸  0̸  0̸  1̸  0̸  1      A
−           1  0  1  0    − B
-------------------------------
      1  1  0  1  1      Di
```

(*a*)

INPUTS			OUTPUTS	
A	***B***	B_{in}	***Di***	B_O
0	0	0	0	0
0	0	1	1	1
0	1	0	1	1
0	1	1	0	1
1	0	0	1	0
1	0	1	0	0
1	1	0	0	0
1	1	1	1	1
A − *B* − B_{in}			Difference	Borrow out

(*b*)

Truth table for a full subtractor

Fig. 10-10 (*a*) Sample binary subtraction. problem (*b*) Truth table for a full subtractor.

22. Draw a block diagram for a full subtractor. Label inputs *A*, *B*, and B_{in}; label outputs *Di* and B_O.
23. Draw a truth table for a full subtractor.

10-6 Parallel Subtractors

Parallel subtractor

Half and full subtractors are wired together to perform as a *parallel subtractor.* You have already seen adders connected as parallel adders. An example of a parallel adder is the 3-bit adder in Fig. 10-8. A parallel subtractor is wired in a similar manner. The adder in Fig. 10-8 is considered a parallel adder because all the digits from the problem flow into the adder at the same time.

Figure 10-12 on page 240 diagrams the wiring of a single half subtractor and three full subtractors. This forms a 4-bit parallel subtractor that can subtract binary number $B_3B_2B_1B_0$ from binary number $A_3A_2A_1A_0$. Notice that the top subtractor (half subtractor) subtracts the LSBs (1s place). The B_O output of the 1s subcontractor is tied to the B_{in} input of the 2s subtractor. Each subtractor's B_O output is connected to the next more significant bit's borrow input. These borrow lines keep track of the borrows we discussed earlier.

Did You Know?

Compact Data Storage Information storage is getting smaller and faster. Today, researchers continue to go beyond the limits of silicon to find ways to transmit, store, and retrieve data with molecules. The first step in meshing molecular systems with electronics is to magnetize single molecules.

TEST

Answer the following questions.

24. Refer to Fig. 10-12. This is a block diagram of a 4-bit ________ (parallel adder, parallel subtractor, serial adder, serial subtractor) circuit.
25. Refer to Fig. 10-12. The lines between subtractors (B_O to B_{in}) serve what purpose in this circuit?

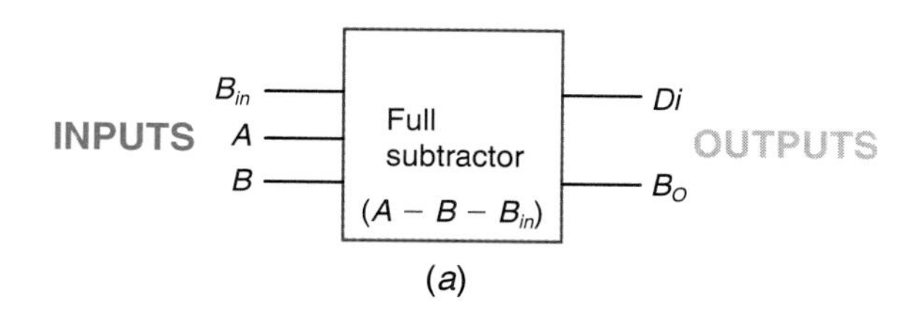

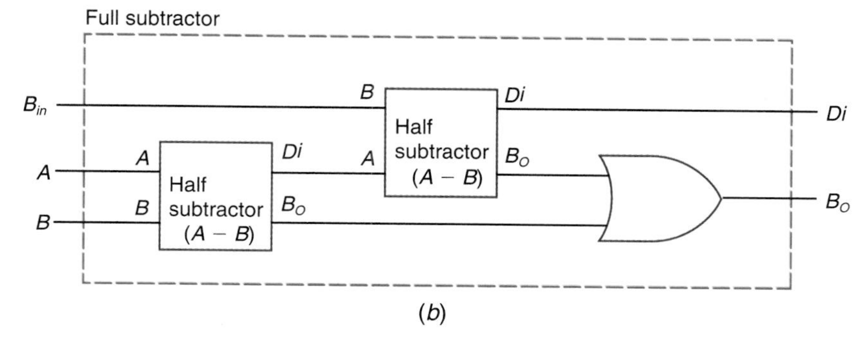

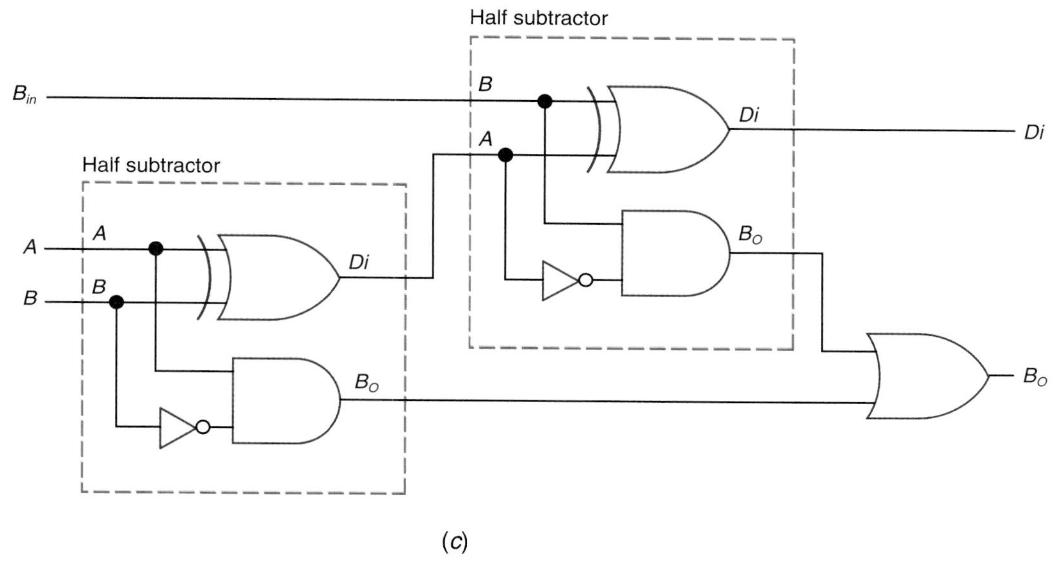

Fig. 10-11 Full subtractor. (*a*) Block symbol. (*b*) Constructed with half subtractors and an OR gate. (*c*) Logic diagram.

Full subtractor

10-7 Using Adders for Subtraction

In earlier sections we found that there are circuits for adding or subtracting binary numbers. To simplify circuitry in a calculating machine, it is convenient to have a more universal device for calculations. With a few little tricks we can use an adder to also do subtraction.

There is a mathematical technique that helps us use an adder to do binary subtraction. The technique is outlined in Fig. 10-13 on page 240. The problem is to subtract decimal 6 from decimal 10 (binary: 1010 − 0110). The subtracting is shown being done first in decimal, then in binary, and finally by the special technique. In the special technique the steps are first to write the *1s complement* of the number being subtracted (change all 1s to 0s and all 0s to 1s) and then add. The 1s complement of 0110 is 1001, as shown. The temporary answer to this addition is shown as 10011. Next, the last carry on the left is carried around to the 1s place (see the arrow on the diagram). This is called an *end-around carry*. When the end-around carry is added to the rest of the number, the result is the *difference* between the original binary numbers, 1010 and 0110. The answer to this problem is shown as binary 100 (decimal 4).

1s complement

Using the 1s complement and end-around carry method is somewhat difficult in long-

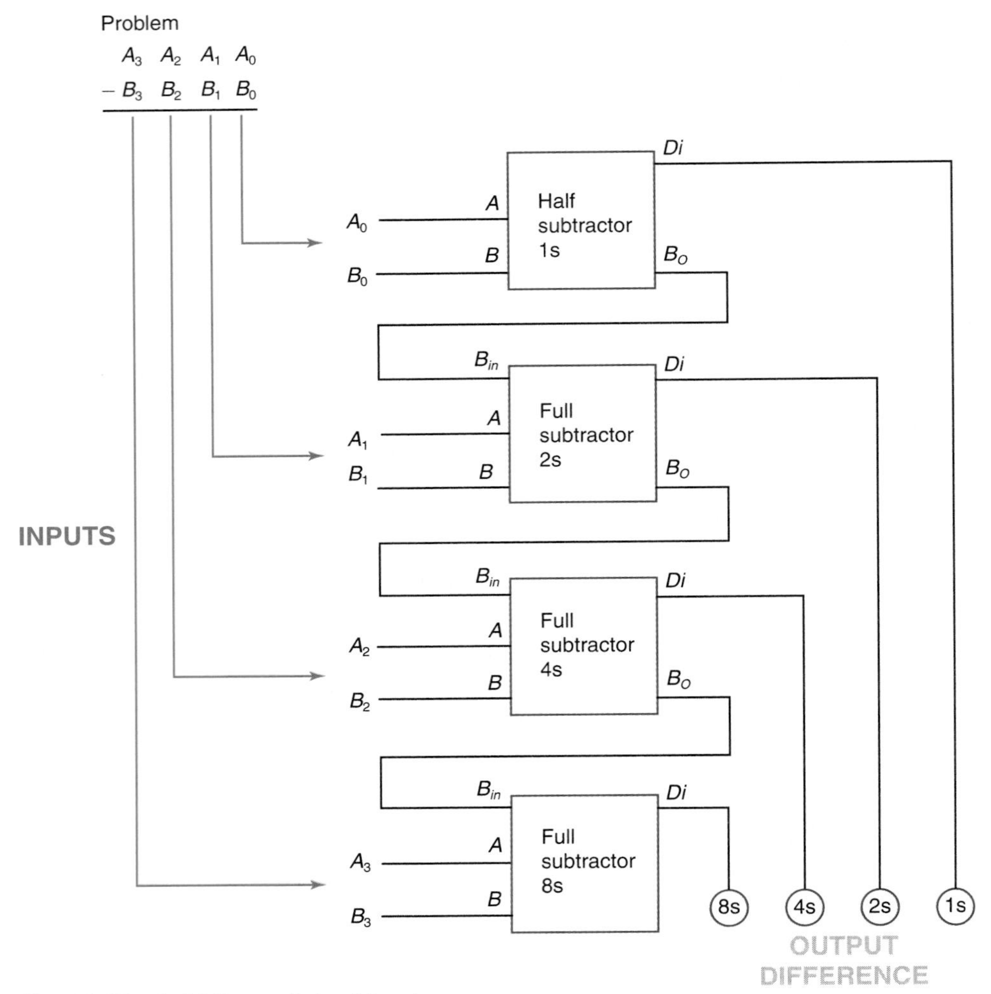

Fig. 10-12 A 4-bit parallel subtractor.

hand. However, this method is quite easy to do with logic circuits. You can see that this method uses an adder to do subtraction. You should know how to do 1s complement and end-around carry subtraction. (There are a few practice problems in the next test.)

Now let us use adders to do binary subtraction. Figure 10-14 on page 241 shows four *full adders* (labeled *FA*) being used to perform binary subtraction. Pay special attention to the four inverters that complement the binary number represented by $B_3B_2B_1B_0$. The inverters give the 1s complement input at the B inputs to each full adder. The adder adds the binary numbers represented by $A_3A_2A_1A_0$ and $\overline{B_3}\,\overline{B_2}\,\overline{B_1}\,\overline{B_0}$. The extra carry at C_O of the 8s full adder is carried back to the 1s adder by the end-around carry line shown. The indicators at

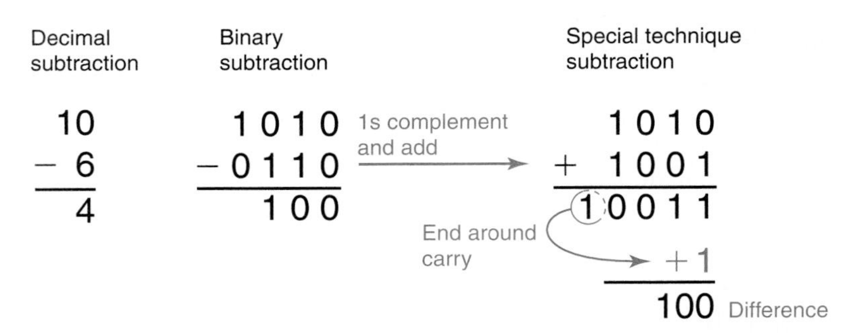

Fig. 10-13 An example of 1s complement and end-around carry subtraction.

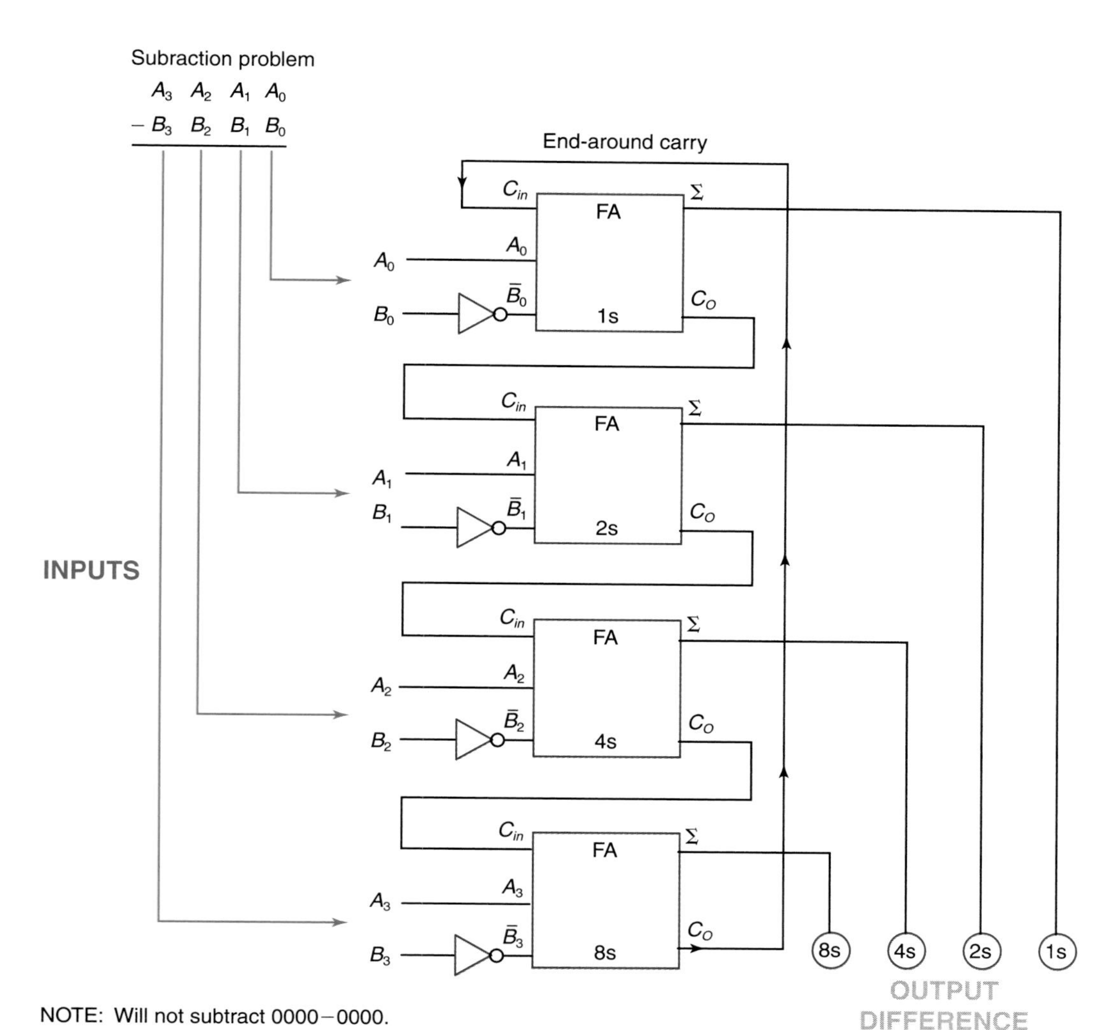

Fig. 10-14 An experimental 4-bit binary subtractor.

the lower right show the *difference* between the binary numbers $A_3A_2A_1A_0$ and $B_3B_2B_1B_0$.

The 4-bit binary subtraction circuit in Fig. 10-14 works fine except it has a problem when subtracting binary 0000 – 0000: It generates the incorrect difference of 1111. Because of this problem, computing devices use 2s complement numbers. With 2s complement numbers, we can add or subtract signed numbers. Later sections will cover 2s complement notation, subtraction, and subtractors. Adders and subtractors using 2s complement numbers are accurate and quite versatile.

TEST

Answer the following questions.

26. Do the binary subtraction problems in **a** to **f** using the 1s complement and end-around carry method.

a. 11 – 10 =
b. 111 – 010 =
c. 1000 – 0111 =
d. 110 – 100 =
e. 1001 – 0111 =
f. 1011 – 0110 =

27. The 4-bit subtractor based on 1s complement end-around-carry is an impractical circuit because of an inaccuracy (T or F).
28. Refer to Fig. 10-14. This subtractor circuit uses inverters to form the 1s complement of the subtrahend, four ________, and an end-around carry line.

10-8 IC Adders

IC manufacturers produce several adders. One elementary arithmetic IC is the TTL *7483 4-bit binary full adder.* A block symbol for the 7483 IC adder is drawn in Fig. 10-15 on page 242. The problem of addition of the two 4-bit binary numbers ($A_3A_2A_1A_0$ and $B_3B_2B_1B_0$) is

TTL 7483 4-bit binary full adder

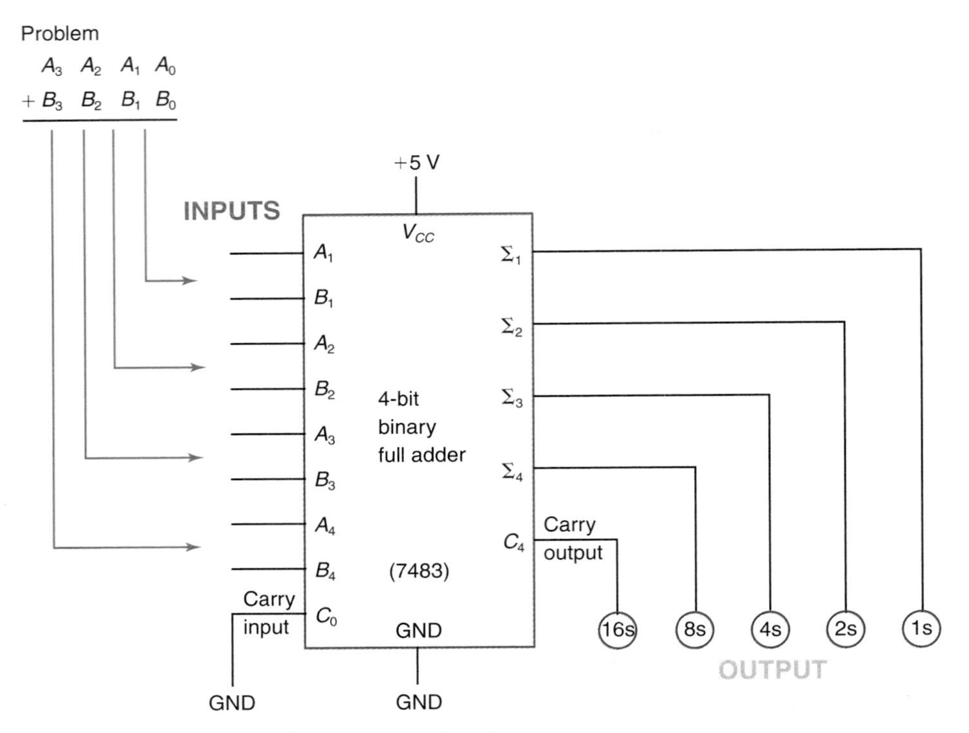

Fig. 10-15 The 7483 4-bit binary adder IC.

shown being entered into the eight inputs of the 7483 IC. Notice a difference in numbering systems on the problem and the IC (subscripts don't match). For adding just two 4-bit numbers, the C_O input is held at 0. The C_O input is marked as the C_{in} input by some manufacturers. The sum outputs are shown attached to output indicators. The C_4 output is attached to the 16s output indicator. The C_4 output is marked as the C_O output by some manufacturers. This binary adder can indicate a sum as high as 11110 (decimal 30) when adding binary 1111 to 1111.

Internally the 7483 IC adder is organized very much like the unit in Fig. 10-14 (without the four inverters). The C_4 (carry out) on the 7483 IC is the same as the C_O on the 8s full adder in Fig. 10-14. The carry input labeled C_O on the 7483 IC is the same as the C_{in} on the 1s full adder in Fig. 10-14.

Cascading adders

The 7483 IC adder can be *cascaded* by connecting the C_4 output of the first IC to the C_O carry input of the next 7483 IC. With two 7483 IC adders cascaded, an *8-bit binary adder* is produced. The 7483 IC can also be a 4-bit subtractor using the arrangement shown in Fig. 10-14. The B inputs are inverted, or complemented, and the C_4 carry out is the end-around carry line to the C_O carry input of the 7483 IC.

8-bit binary adder

Counterparts to the 7483 4-bit adder are the 74LS83 and 74C83 ICs. Other 4-bit adders that function the same as the 7483 IC but have a different pin configuration are the 74283, 74LS283, 74S283, 74F283, and 74HC283.

A more complex arithmetic chip is the 74LS181 IC. The 74LS181 and its relatives, the 74LS381, are described as *arithmetic-logic units/function generators.* These units perform many of the tasks of the ALUs in simple microprocessors and microcontrollers. These functions include add, subtract, shift, magnitude comparison, XOR, AND, NAND, OR, NOR, and other logic operations. The 74LS181 has CMOS relatives including the 74HC181 and MC14581.

TEST

Supply the missing word in each statement.

29. The 7483 IC contains a 4-bit binary __________.
30. Two 7483 ICs can be __________ to form an 8-bit parallel binary adder.
31. An adder such as the 7483 IC does not have a memory device, such a latch, built into the chip and is classified as a __________ (combinational, sequential) logic device.

32. The ________ (74LS32, 74LS181) is a more complex IC that performs many of the same operations (such as add, subtract, shift, compare, AND, OR, etc.) as the ALU of a microprocessor or microcontroller.

10-9 Binary Multiplication

In elementary school you learned how to multiply. You learned to lay out your multiplication problem similar to that in Fig. 10-16(*a*). You learned that the top number is called the *multiplicand* and the bottom number is the *multiplier.* The solution to the problem is called the *product.* The product of 7 × 4, then, is 28, as shown in Fig. 10-16(*a*).

Figure 10-16(*b*) shows that multiplication really is just *repeated addition.* The problem 7 × 4 = 28 is represented by the multiplicand (7) being added four times, because 4 is the multiplier. The product is 28.

If you want to multiply 54 × 14, the repeated addition system is complicated and takes a long time. The multiplicand (54) must be added 14 times to get a product of 756. Most of us were taught to multiply 54 × 14 in the manner shown in Fig. 10-17(*a*). To solve the multiplication problem 54 × 14, we first multiply the multiplicand, 54, by 4. This results in the first partial product (216) shown in Fig. 10-17(*b*). Next we multiply the multiplicand by 1. Actually the multiplicand is multiplied by a multiplier of 10, as shown in Fig. 10-17(*c*). The second partial product is 540. The first and second partial products (216 and 540) are then added for a final product of 756.

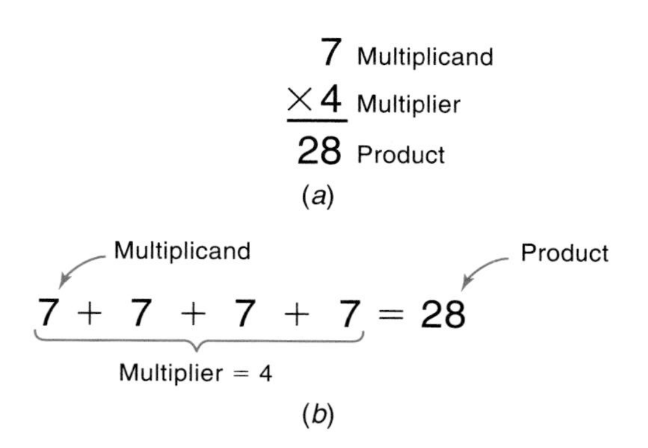

Fig. 10-16 (*a*) Decimal multiplication problem. (*b*) Multiplying using the repeated addition method.

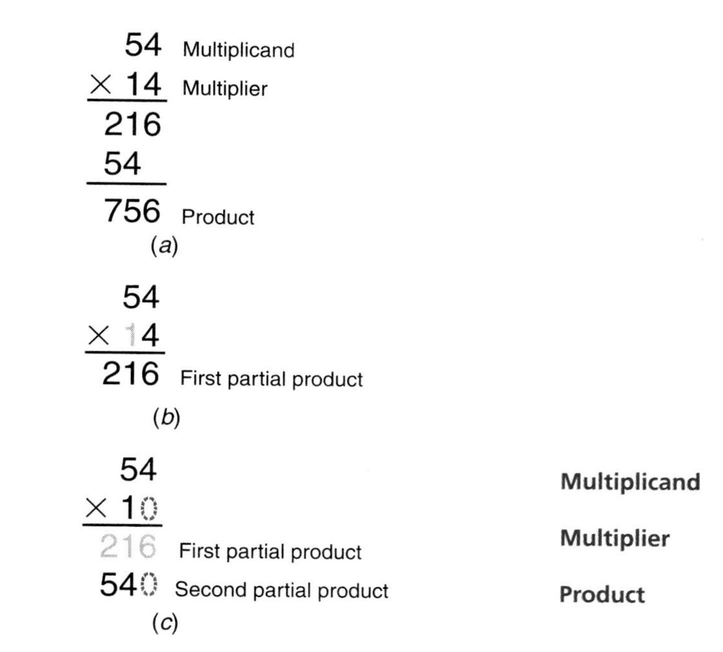

Fig. 10-17 (*a*) Decimal multiplication problem. (*b*) Calculating the first partial product. (*c*) Calculating the second partial product.

It is normal to omit the 0 in the second partial product, as in Fig. 10-17(*a*).

It is important to notice the *process* in the problem in Fig. 10-17. The multiplicand is first multiplied by the LSD of the multiplier. This gives the first partial product. The second partial product is then calculated by multiplying the multiplicand by the MSD of the multiplier. The two partial products are then added, producing the final product. This same process is used in *binary multiplication.*

Binary multiplication is much simpler than decimal multiplication. The binary system has only two digits (0 and 1), which makes the rules for multiplying simple. Figure 10-18(*a*) on page 244 shows the rules for binary multiplication.

Multiplication with binary numbers is done just as with decimal numbers. Figure 10-18(*b*) details a problem where binary 111 is multiplied by binary 101. First, the multiplicand (111) is multiplied by the 1s bit of the multiplier. The result is the first *partial product,* shown as 111 in Fig. 10-18(*b*). Next, the multiplicand is multiplied by the 2s bit of the multiplier. The result is the second partial product (0000). Notice that the LSB of the second partial product, 0000, is left off in Fig. 10-18(*b*).

Multiplicand

Multiplier

Product

Binary multiplication

Partial product

Rules for binary multiplication

0	0	1	1
×0	×1	×0	×1
0	0	0	1

(*a*)

Decimal	Binary	
7	111	Multiplicand
×5	×101	Multiplier
35	111	First partial product
	000	Second partial product
	111	Third partial product
	100011	Product

(*b*)

Repeated addition-type multiplier system

Rules for binary multiplication

Fig. 10-18 (*a*) Rules for binary multiplication. (*b*) Sample multiplication problem.

Third, the multiplicand is multiplied by the 4s bit of the multiplier. The result is the third partial product of 11100, shown in Fig. 10-18(*b*) as 111, with the two blank spaces in the 1s and 2s places. Finally, the first, second, and third partial products are added, resulting in a product of binary 100011. Notice that the same multiplication problem in decimal is shown at the left of Fig. 10-18(*b*) for your convenience. The binary product 100011 equals the decimal product 35.

Another binary multiplication problem is shown in Fig. 10-19. At the left the problem is in the familiar decimal form; the same problem is repeated in binary form at the right, where binary 11011 is multiplied by 1100. As in decimal multiplication, the 0s in the multiplier can simply be brought down to hold the 1s and 2s places in the binary number. The binary product is shown as 101000100, which equals decimal 324.

You will gain experience in solving binary multiplication problems by answering the questions that follow.

TEST

Answer the following questions.

33. Find the product for binary 111 × 10.
34. Find the product for binary 1101 × 101.
35. Find the product for binary 1100 × 1110.

10-10 Binary Multipliers

We can multiply numbers by repeated addition, as illustrated in Fig. 10-16(*b*). The multiplicand (7) could be added four times to obtain the product of 28. A block diagram of a circuit that performs repeated addition is shown in Fig. 10-20. The multiplicand is held in the top register. In our example the multiplicand is a decimal 7, or a binary 111. The multiplier is held in the down counter shown on the left in Fig. 10-20. The multiplier in our example is a decimal 4, or a binary 100. The lower product register holds the product.

The repeated addition technique is shown in operation in Fig. 10-21. This chart shows how the multiplicand (binary 111) is multiplied by the multiplier (binary 100). The product register is cleared to 00000. After one count downward, a partial product of 00111 (decimal 7) appears in the products register. After the second count downward, a partial product of 01110 (decimal 14) appears in the product register. After the third count downward, a partial product of 10101 (decimal 21) appears in the product register. After the fourth downward count, the *final product* of 11100 (decimal 28) appears in the product register. The multiplication problem (7 × 4 = 28) is complete. The circuit of Fig. 10-20 has added 7 four times for a total of 28.

This type of circuit is not widely used because of the long time it takes to do the re-

Decimal	Binary	
27	11011	Multiplicand
×12	× 1100	Multiplier
54	1101100	Third partial product
27	11011	Fourth partial product
324	101000100	Product

Fig. 10-19 Sample multiplication problem.

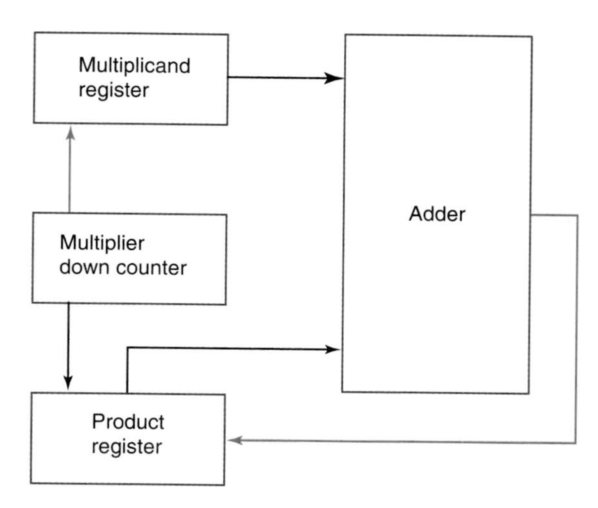

Fig. 10-20 Block diagram of a repeated addition-type multiplier system.

peated addition when large numbers are multiplied. A more practical method of multiplying in digital electronic circuits is the *add and shift method* (also called the shift and add method). Figure 10-22 shows a binary multiplication problem. In this problem binary 111 is multiplied by 101 (7 × 5 in decimal). This hand-done procedure is standard except for the temporary product in line 5. Line 5 has been added to help you understand how multiplication might be done by digital circuits. Close observation of binary multiplication shows the following three important facts:

1. Partial products are always 000 if the multiplier is 0 and equal to the multiplicand if the multiplier is 1.
2. The product register needs twice as many bits as the multiplicand register.
3. The first partial product is shifted one place to the *right* (*relative to* the second partial product) when adding.

You can observe each characteristic by looking at the sample problem in Fig. 10-22.

Binary multiplication

The important characteristics of longhand multiplication have been given. A binary multiplication circuit can be designed by using these characteristics. Figure 10-23(*a*) on page 246 shows a circuit that does binary multiplication. Notice that the multiplicand (111) is loaded into the register at the upper left. The accumulator register is cleared to 0000. The multiplier (101) is loaded into the register at the lower right. Notice, too, that the accumulator and the multiplier are considered

JOB TIP

Learn to disagree with fellow workers without offending them.

Add and shift method

	Load with binary	After 1 down count	After 2 down counts	After 3 down counts	After 4 down counts
Multiplicand register	111	111	111	111	111
Multiplier counter	100	011	010	001	000
Product register	00000	00111	01110	10101	11100
	Load				Stop

Fig. 10-21 Multiplying binary 111 and 100 using the repeated addition circuit.

Line 1	111	Multiplicand
Line 2	× 101	Multiplier
Line 3	111	First partial product
Line 4	000	Second partial product
Line 5	0111	Temporary product (line 3 + line 4)
Line 6	111	Third partial product
Line 7	100011	Product

Fig. 10-22 Binary multiplication problem.

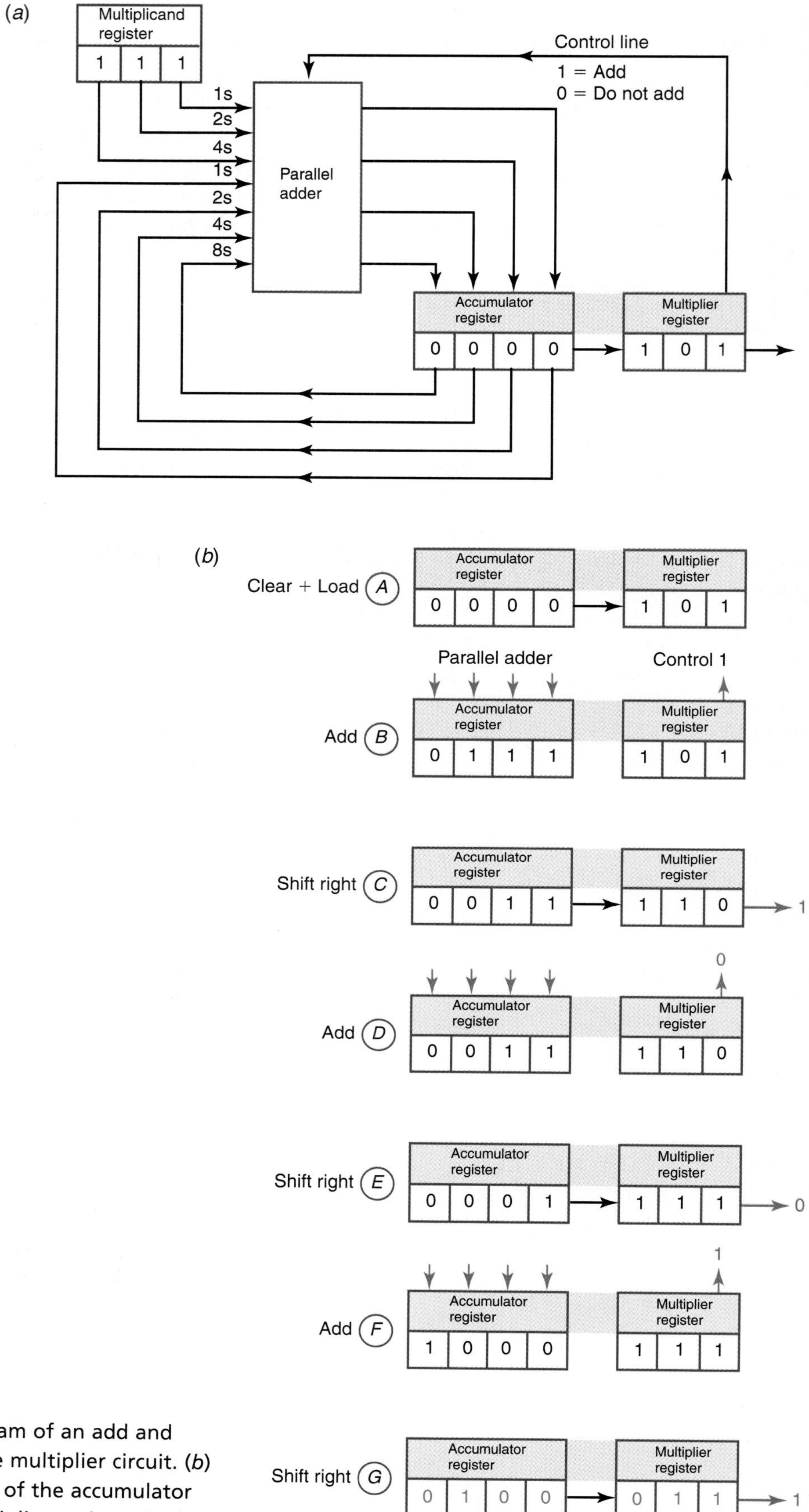

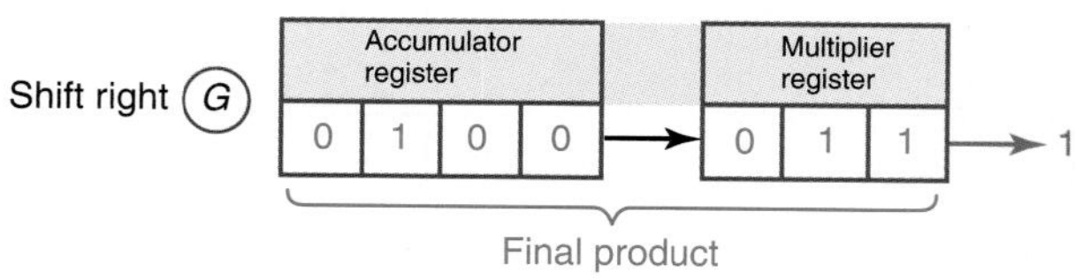

Add and shift-type multiplier circuit

Fig. 10-23 (*a*) Diagram of an add and shift-type multiplier circuit. (*b*) Contents of the accumulator and multiplier registers in the add and shift multiplier circuit.

together. This is shown by the shading connecting the two registers.

Let us use the circuit in Fig. 10-23(*a*) to go through the detailed procedure for multiplying. The diagram in Fig. 10-23(*b*) is a step-by-step review of how binary 111 is multiplied by 101 using the add and shift method. The binary 111 is loaded into the multiplicand register. The accumulator and multiplier registers are loaded in step *A* in Fig. 10-23(*b*). Step *B* shows the 0000 and the 111 from the accumulator and multiplicand registers being added when the 1 is applied to the control line. This is comparable to line 3 of the multiplication problem in Fig. 10-22. Step *C* shifts both the accumulator and the multiplier register one place to the right. The LSB of the multiplier (1) is shifted out the right end and lost. Step *D* represents another *add* step. This time a 0 is applied to the control line. A 0 on the control line means *no* addition. The register contents remain the same. Step *D* is comparable to lines 4 and 5, Fig. 10-22. Step *E* shows the registers being shifted one place to the right. This time the 2s bit of the multiplier is lost as it is shifted out the right end of the register. Step *F* shows the 4s bit of the multiplier (1) commanding the adder to add. The accumulator contents (0001) and the multiplicand (111) are added. The result of that addition is deposited in the accumulator register (1000). This step is comparable to the left section of lines 5 to 7, Fig. 10-22. Step *G* is the final step in the *add and shift multiplication*; it shows a single shift to the right for both registers. The MSB (4s bit) of the multiplier is lost out of the right end of the register. The final product appears across both registers as 100011. Binary 111 multiplied by 101 resulted in a product of 100011 ($7 \times 5 = 35$ in decimal). The final product calculated by the multiplier circuit is the same result we got in line 7, Fig. 10-22, when we multiplied by hand.

Two types of multiplier circuits have just been illustrated. The first uses repeated addition to arrive at the product. That system is shown in Fig. 10-20. The second circuit uses the add and shift method of multiplying. The add and shift system is shown in Fig. 10-23.

In many computers *the procedure,* such as the add and shift method, can be programmed into the machine. Instead of permanently wiring the circuit, we simply *program,* or instruct, the computer to follow the procedure shown in Fig. 10-23(*b*). We are thus using *software* (a program) to do multiplication. This use of software cuts down on the amount of electronic circuits needed in the CPU of a computer.

Simpler 8-bit microprocessors, such as the Intel 8080/8085, Motorola 6800, and the 6502/65C02 do not have circuitry in their ALUs to do multiplication. To perform binary multiplication on these processors, the programmer must write a program (a list of instructions) that multiplies numbers. Either the add and shift or the repeated addition method can be used for programming these microprocessor-based machines to do multiplication. Most advanced microprocessors do have multiply instructions. Some more expensive microcontrollers do have a multiply instruction but many do not have this instruction.

Add and shift method of multiplying

Microprocessors

TEST

Answer the following questions.

36. Refer to Fig. 10-20. This circuit uses what method of binary multiplication?
37. A widely used technique for multiplying using digital circuits is the ________ method.
38. Refer to Fig. 10-23. This circuit uses what method of binary multiplication?
39. Most elementary 8-bit microprocessors ________ (do, do not) have a multiply instruction.
40. All microcontrollers have a multiply instruction (T or F).

10-11 2s Complement Notation, Addition, and Subtraction

The 2s complement method of representing numbers is widely used in microprocessors. To now, we have assumed that all numbers are positive. However, microprocessors must process both positive and negative numbers. Using 2s complement representations, the *sign as well as the magnitude* of a number can be determined.

2s complement representations

2s Complement 4-bit

For simplicity, assume we are using a 4-bit processor. This means that all data is transferred and processed in groups of four. The MSB is the sign bit of the number. This is

Sign bit

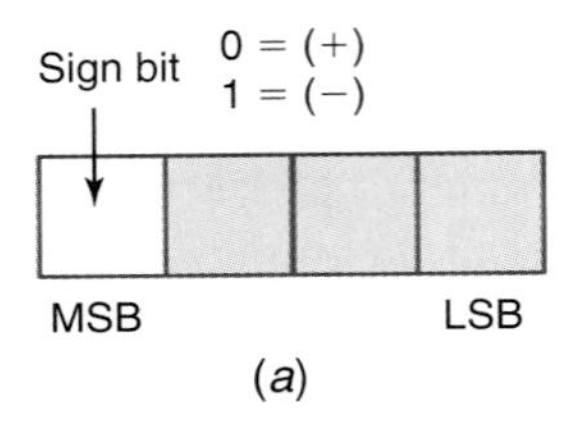

2s complement numbers

Signed decimal	4-bit 2s complement representation
+7	0111
+6	0110
+5	0101
+4	0100
+3	0011
+2	0010
+1	0001
0	0000
−1	1111
−2	1110
−3	1101
−4	1100
−5	1011
−6	1010
−7	1001
−8	1000

Same as binary numbers (+7 through 0)

(b)

Sign bit

Fig. 10-24 (*a*) MSB of 4-bit register is a sign bit. (*b*) 2s complement representation of positive and negative numbers.

shown in Fig. 10-24(*a*). A 0 sign bit means a positive number, while a 1 sign bit means a negative number.

The table in Fig. 10-24(*b*) shows the *2s complement representation* for all the 4-bit positive and negative numbers from +7 to −8. The MSBs in Fig. 10-24(*b*) of the positive 2s complement numbers are 0s. All negative numbers (−1 to −8) start with a 1. Note that 2s complement representations of positive numbers are the same as binary. Therefore, +7 (decimal) = 0111 (2s complement) = 0111 (binary).

Converting signed decimal numbers to 2s complement form

The 2s complement representation of a negative number is found by first taking the 1s complement of the number and then adding 1. An example of this process is shown in Fig. 10-25(*a*). The negative decimal number −4 is to be converted to its 2s complement form:

1. Convert the decimal number to its binary equivalent. In this example, convert -4_{10} to 0100_2.
2. Convert the binary number to its 1s complement by changing all 1s to 0s and all 0s to 1s. In this example, convert 0100_2 to 1011 (1s complement).
3. Add 1 to the 1s complement number, using regular binary addition. In this example, 1011 + 1 = 1100. The answer (1100 in this example) is the 2s complement representation. Therefore, -4_{10} = 1100 (2s complement).

This answer can be verified by referring back to the table in Fig. 10-24(*b*).

To convert from 2s complement form to binary, follow the procedure shown in Fig. 10-25(*b*). In this example, the 2s complement number (1100) is being converted to its binary equivalent. Its equivalent decimal number can then be found from the binary.

1. Form the 1s complement of the 2s complement number by changing all 1s to 0s and all 0s to 1s. In this example, convert 1100 to 0011.
2. Add 1 to the 1s complement number, using regular binary addition. In this example, 0011 + 1 = 0100. The answer (0100 in this example) is in binary. Therefore, $0100_2 = 4_{10}$.

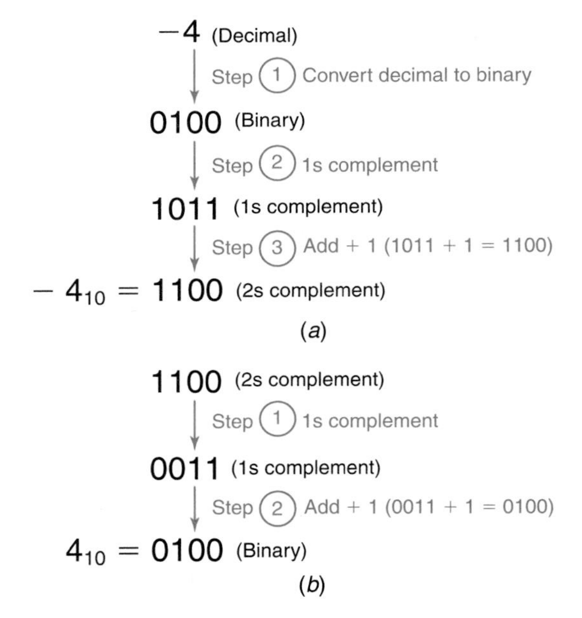

Fig. 10-25 (*a*) Converting signed decimal numbers to 2s complement form. (*b*) Converting from 2s complement form to binary numbers.

Because the MSB of the 2s complement number (1100) is a 1, the number is negative. Therefore, 2s complement 1100 equals -4_{10}.

2s Complement Addition

2s complement notation is widely employed because it makes it easy to add and subtract signed numbers. Four examples of adding 2s complement numbers are shown in Fig. 10-26. Two positive numbers are added in Fig. 10-26(*a*). 2s complement addition looks just like adding in binary in this example. Two negative numbers (-1_{10} and -2_{10}) are added in Fig. 10-26(*b*). The 2s complement numbers representing −1 and −2 are given as 1111 and 1110. The MSB (overflow from 4-bit register) is discarded, leaving the 2s complement sum of 1101, or −3 in decimal. Look over examples (*c*) and (*d*) in Fig. 10-26 to see if you understand the procedure for adding signed numbers using 2s complement notation.

(+4)	0100	
+(+3)	+ 0011	
$+7_{10}$	0111	(2s complement SUM)
	(*a*)	
(−1)	1111	
+(−2)	+ 1110	
-3_{10}	1 1101 (1 → Discard)	(2s complement SUM)
	(*b*)	
(+1)	0001	
+(−3)	+ 1101	
-2_{10}	1110	(2s complement SUM)
	(*c*)	
(+5)	0101	
+(−4)	+ 1100	
$+1_{10}$	1 0001 (1 → Discard)	(2s complement SUM)
	(*d*)	

Fig. 10-26 Four sample signed addition problems using 4-bit 2s complement numbers.

2s Complement Subtraction

2s complement notation is also useful in subtracting signed numbers. Four subtraction problems are shown in Fig. 10-27 on page 250. The first problem is $(+7) - (+3) = +4_{10}$. The subtrahend (+3 in this case) is converted to its binary form. Next, the 2s complement of this is formed, yielding 1101. Then 0111 is *added* to 1101, yielding 1 0100. The MSB (overflow from 4-bit register) is discarded, leaving the *difference* of 0100, or $+4_{10}$. Note that an adder is used for subtraction. This is done by converting the subtrahend to its 2s complement and adding. Any carry or overflow into the fifth binary place is discarded.

Add or subtract signed numbers

Look over the sample 2s complement subtraction problems using an adder in Fig. 10-27(*b*), (*c*), and (*d*). See if you can follow the procedure in these remaining subtraction problems.

For information on new product developments, visit the webiste for the Consumer Electronics Manufacturers Association (CEMA).

2s Complement 8-bit

Only 4-bit 2s complement representations have been used in previous examples. Most microprocessors and microcontrollers use 8-, 16-, or 32-bit groupings. The procedures used with 4-bit 2s complement descriptions of binary numbers also apply to 8-bit, 16-bit, or 32-bit representations.

In an 8-bit 2s complement of a number, the MSB is the sign bit as illustrated in Fig. 10-28(*a*) on page 251. This allows both the sign and magnitude of the number to be represented. A sampling of some 8-bit 2s complement representations of positive and negative numbers are shown in Fig. 10-28(*b*). Notice that the range of numbers for an 8-bit 2s complement is from −128 to +127. Notice from the top half of the chart in Fig. 10-28(*b*) that *decimal numbers from 0 through +127 (positive numbers) have 2s complements that are the same as binary numbers.* As an example, +125 is represented by 0111 1101 in either binary or 2s complement.

Converting a negative decimal number (from −1 to −128) to its 8-bit 2s complement is accomplished by the same process shown earlier to Fig. 10-25(*a*). Follow the three-step process in the example below:

1. Convert the decimal number −126 to its binary equivalent.

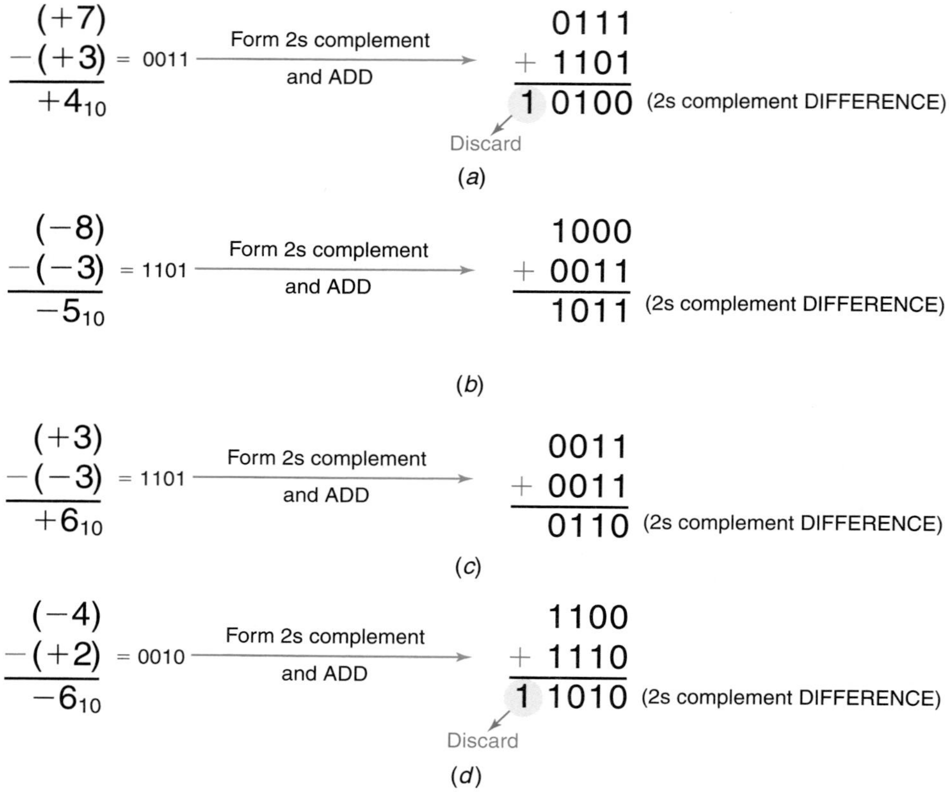

$$\begin{array}{r} (+7) \\ -(+3) \\ \hline +4_{10} \end{array} \quad -(+3) = 0011 \xrightarrow[\text{and ADD}]{\text{Form 2s complement}} \quad \begin{array}{r} 0111 \\ +\ 1101 \\ \hline 1\ 0100 \end{array} \text{ (2s complement DIFFERENCE)}$$

Discard (leading 1)

(*a*)

$$\begin{array}{r} (-8) \\ -(-3) \\ \hline -5_{10} \end{array} \quad -(-3) = 1101 \xrightarrow[\text{and ADD}]{\text{Form 2s complement}} \quad \begin{array}{r} 1000 \\ +\ 0011 \\ \hline 1011 \end{array} \text{ (2s complement DIFFERENCE)}$$

(*b*)

$$\begin{array}{r} (+3) \\ -(-3) \\ \hline +6_{10} \end{array} \quad -(-3) = 1101 \xrightarrow[\text{and ADD}]{\text{Form 2s complement}} \quad \begin{array}{r} 0011 \\ +\ 0011 \\ \hline 0110 \end{array} \text{ (2s complement DIFFERENCE)}$$

(*c*)

$$\begin{array}{r} (-4) \\ -(+2) \\ \hline -6_{10} \end{array} \quad -(+2) = 0010 \xrightarrow[\text{and ADD}]{\text{Form 2s complement}} \quad \begin{array}{r} 1100 \\ +\ 1110 \\ \hline 1\ 1010 \end{array} \text{ (2s complement DIFFERENCE)}$$

Discard (leading 1)

(*d*)

Fig. 10-27 Four sample signed subtraction problems using 4-bit 2s complement numbers.

Example: $126_{10} = 0111\ 1110_2$

2. Convert the binary number to its 1s complement. Example:
 0111 1110_2 = 1000 0001 (1s c)
3. Add 1 to the 1s complement forming the 2s complement. Example:
 1000 0001 (1s c) + 1 = 1000 0010 (2s c)

Result: -126_{10} = 1000 0010 in 2s complement

Next convert a 2s complement representation of a negative number to its decimal equivalent. Follow the three-step process in the example below:

2s complement addition

1. Convert the 2s complement to its 1s complement form. Example:
 1001 1100 (2s c) = 0110 0011 (1s c)
2. Add +1 to the 1s complement to form the binary number. Example:
 0110 0011 (1s c) + 1 = 0110 0100_2
3. Convert the binary number to its decimal equivalent. Example:

2s complement subtraction

0110 0100_2 = (64 + 32 + 4 = 100) = 100_{10}
Result: 1001 1100 (2s c) = -100_{10}

In the previous examples, you converted a negative decimal number to its 2s complement. Later, you reversed the process and converted a 2s complement to a negative decimal number. Because these conversions are time-consuming and prone to errors, Appendix A includes a *2s complement number conversion chart.* Appendix A contains 2s complements of decimal numbers −1 through −128.

Several 8-bit *2s complement addition* problems are solved in Fig. 10-29(*a*) on page 252. Remember when overflows (more than 8 bits) occur, they are discarded. The sums are in 2s complement notation, but remember that for positive numbers the 2s complement and binary number are the same. Review these addition problems to see if you understand the procedure. You will have practice problems later.

Several 8-bit *2s complement subtraction* problems are solved in Fig. 10-29(*b*). Remember

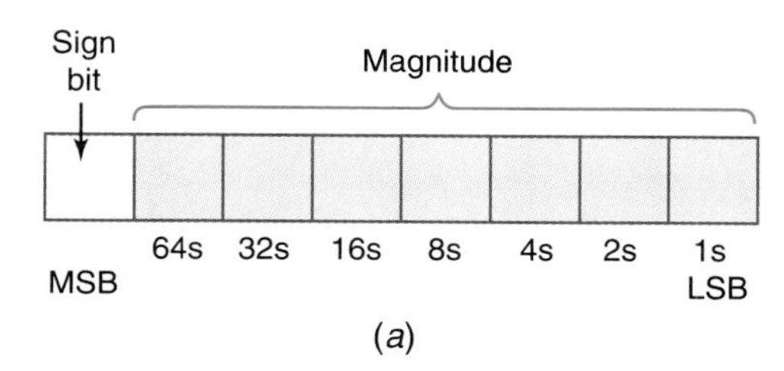

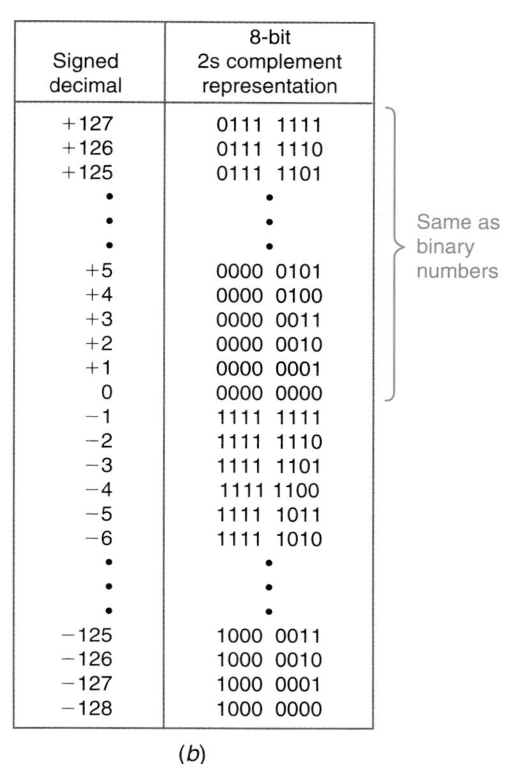

Signed decimal	8-bit 2s complement representation
+127	0111 1111
+126	0111 1110
+125	0111 1101
•	•
•	•
•	•
+5	0000 0101
+4	0000 0100
+3	0000 0011
+2	0000 0010
+1	0000 0001
0	0000 0000
−1	1111 1111
−2	1111 1110
−3	1111 1101
−4	1111 1100
−5	1111 1011
−6	1111 1010
•	•
•	•
•	•
−125	1000 0011
−126	1000 0010
−127	1000 0001
−128	1000 0000

Fig. 10-28 (*a*) MSB of 8-bit register is a sign bit. (*b*) 2s complement representation of selected positive and negative numbers.

when overflows (more than 8 bits) occur, they are discarded. Notice that only the subtrahends are 2s complemented before they are added to the minuend. The differences are in 2s complement notation but remember that for positive numbers, the 2s complement and binary number are the same. Review these subtraction problems to see if you understand the procedure. You will have practice problems later.

In summary, 2s complement notation is used because they show both the sign and magnitude of a number. Remember that 2s complement and binary numbers are identical for positive numbers. Twos complement numbers can be used with adders to either *add* or *subtract signed numbers.* The next section in the textbook will diagram an adder/subtractor system that makes use of 2s complement notation.

TEST

Answer the following questions.

41. When microprocessors process both positive and negative numbers, _________ representations are used.
42. The 2s complement number 0111 represents _________ in binary and _________ in decimal.
43. The 2s complement number 1111 represents _________ in decimal.
44. In 2s complement representation, the MSB is the _________ bit. If the MSB is 0 the number is _________ (negative, positive), whereas if the MSB is 1, the number is _________ (negative, positive).
45. The decimal number −6 equals _________ in 2s complement 4-bit representation.
46. The decimal number +5 equals _________ in 2s complement 4-bit representation.
47. Calculate the sum of the 2s complement numbers 1110 and 1101. Give the answer in 2s complement and in decimal.
48. Calculate the sum of the 2s complement numbers 0110 and 1100. Give the answer in 2s complement and in decimal.
49. Decimal 90 equals _________ in binary and _________ in 2s complement.
50. Decimal −90 equals _________ in 2s complement.
51. Adding 0111 1111 (2s c) and 1111 0000 (2s c) yields _________ in 2s complement or _________ in decimal.
52. Adding 1000 0000 (2s c) and 0000 1111 (2s c) yields _________ in 2s complement or _________ in decimal.
53. Subtracting 0001 0000 (2s c) from 1110 0000 (2s c) yields _________ in 2s complement or _________ in decimal.
54. Subtracting 1111 1111 (2s c) from 0011 0000 (2s c) yields _________ in 2s complement or _________ in decimal.

$$\begin{array}{r} (+60) \\ +(+20) \\ \hline +80_{10} \end{array} \qquad \begin{array}{r} 0011\ 1100 \\ +\ 0001\ 0100 \\ \hline 0101\ 0000 \end{array} \text{ (2s complement SUM)}$$

$$\begin{array}{r} (-50) \\ +(-30) \\ \hline -80_{10} \end{array} \qquad \begin{array}{r} 1100\ 1110 \\ +\ 1110\ 0010 \\ \hline 1\ 1011\ 0000 \end{array} \text{ (2s complement SUM)}$$

Discard

$$\begin{array}{r} (+30) \\ +(-90) \\ \hline -60_{10} \end{array} \qquad \begin{array}{r} 0001\ 1110 \\ +\ 1010\ 0110 \\ \hline 1100\ 0100 \end{array} \text{ (2s complement SUM)}$$

$$\begin{array}{r} (+90) \\ +(-80) \\ \hline +10_{10} \end{array} \qquad \begin{array}{r} 0101\ 1010 \\ +\ 1011\ 0000 \\ \hline 1\ 0000\ 1010 \end{array} \text{ (2s complement SUM)}$$

Discard

(*a*)

$$\begin{array}{r} (+65) \\ -(+35) \\ \hline +30_{10} \end{array} = 0010\ 0011 \xrightarrow[\text{and ADD}]{\text{Form 2s complement}} \begin{array}{r} 0100\ 0001 \\ +\ 1101\ 1101 \\ \hline 1\ 0001\ 1110 \end{array} \text{ (2s complement DIFFERENCE)}$$

Discard

$$\begin{array}{r} (-78) \\ -(-35) \\ \hline -43_{10} \end{array} = 1101\ 1101 \xrightarrow[\text{and ADD}]{\text{Form 2s complement}} \begin{array}{r} 1011\ 0010 \\ +\ 0010\ 0011 \\ \hline 1101\ 0101 \end{array} \text{ (2s complement DIFFERENCE)}$$

$$\begin{array}{r} (+40) \\ -(-21) \\ \hline +61_{10} \end{array} = 1110\ 1011 \xrightarrow[\text{and ADD}]{\text{Form 2s complement}} \begin{array}{r} 0010\ 1000 \\ +\ 0001\ 0101 \\ \hline 0011\ 1101 \end{array} \text{ (2s complement DIFFERENCE)}$$

$$\begin{array}{r} (-45) \\ -(+22) \\ \hline -67_{10} \end{array} = 0001\ 0110 \xrightarrow[\text{and ADD}]{\text{Form 2s complement}} \begin{array}{r} 1101\ 0011 \\ +\ 1110\ 1010 \\ \hline 1\ 1011\ 1101 \end{array} \text{ (2s complement DIFFERENCE)}$$

Discard

(*b*)

Fig. 10-29 (*a*) Four sample signed addition problems using 8-bit 2s complement numbers. (*b*) Four sample signed subtraction problems using 8-bit 2s complement numbers.

10-12 2s Complement Adders/Subtractors

A 2s complement *4-bit adder/subtractor system* is drawn in Fig. 10-30. Note the use of four full adders to handle the two 4-bit numbers. XOR gates have been added to the *B* inputs of each full adder to control the mode of operation of the unit. With the mode control at 0, the system *adds* the 2s complement numbers $A_3A_2A_1A_0$ and $B_3B_2B_1B_0$. The sum appears in 2s complement notation at the output indicators at the lower right. The LOW at the *A* inputs of the XOR gates permit the *B* data to flow through the gate with *no inversion.* If a HIGH enters B_0 input to the XOR gate, then a HIGH exits the gate at *Y*. The C_{in} input to the top 1s full adder is held at a 0 during the time the mode control is in the add position. In the add mode, the 2s complement adder operates just like a binary adder except that the carry out (C_O) from the 8s full adder is discarded. In Fig. 10-30, the C_O output from the 8s full adder is left disconnected.

The mode control input is placed at logical 1 for the unit to subtract 2s complement numbers. This causes the XOR gates *to invert the data at the B inputs.* The C_{in} input to the 1s full adder also receives a HIGH. The combination of the XOR gate's inversion plus adding the 1 at the C_{in} input of the 1s full adder is the same as complementing and adding 1. This is comparable to forming the 2s complement of the subtrahend (*B* number in Fig. 10-30).

2s complement adder/subtractor system

Remember that the system in Fig. 10-30 only uses 2s complement numbers. The 4-bit adder/subtractor system in Fig. 10-30 could be extended to 8 bits or 16 bits to handle larger 2s complement numbers.

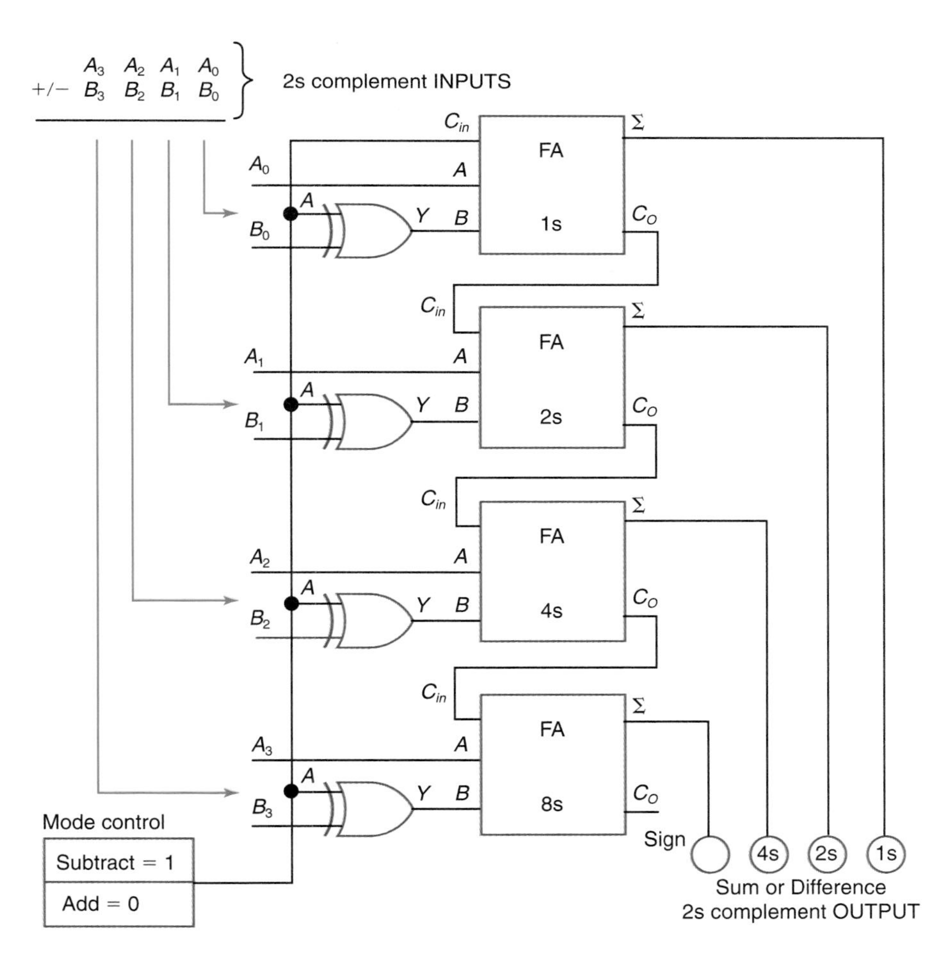

Fig. 10-30 Adder/subtractor system using 2s complement numbers.

Troubleshooting a full adder

Combinational logic circuit

■ TEST

Supply the missing word or words in each statement.

55. Refer to Fig. 10-30. The numbers to be added or subtracted in this system must be in _________ (binary, BCD, 1s complement, 2s complement) form.
56. Refer to Fig. 10-30. The sum or difference output from this system will be in _________ (binary, BCD, 1s complement, 2s complement) form.
57. Refer to Fig. 10-30. This system can add or subtract _________ (signed, only unsigned) numbers.
58. Refer to Fig. 10-30. If the system is adding 0011 to 1100, the output will read _________. This is the 2s complement representation for decimal _________.
59. Refer to Fig. 10-30. If the system is subtracting 0010 from 0101, the output will read _________. This is the 2s complement representation of decimal _________.

10-13 Troubleshooting a Full Adder

A faulty full adder circuit is sketched in Fig. 10-31(*a*). The student or technician first checks the circuit visually and for signs of excessive heat. No problems are found.

The full-adder is a combinational logic circuit. For your convenience, its truth table with normal outputs is shown at the left in Fig. 10-31(*b*). The student or technician manipulates the full-adder inputs and using a logic probe checks the outputs (Σ and C_O). The actual logic probe outputs are shown in the right-hand columns of the truth table in Fig. 10-31(*b*). H stands for a HIGH logic level, while L stands for a LOW logic level. Two errors seem to appear in the C_O column in lines 6 and 7 of the truth table. These are noted in Fig. 10-31(*b*). A look at the truth-table results of the faulty full adder indicates no trouble in the Σ column. The Σ circuitry involves the two XOR gates labeled 1 and 2 in Fig. 10-31(*a*). It appears that these gates are operating properly.

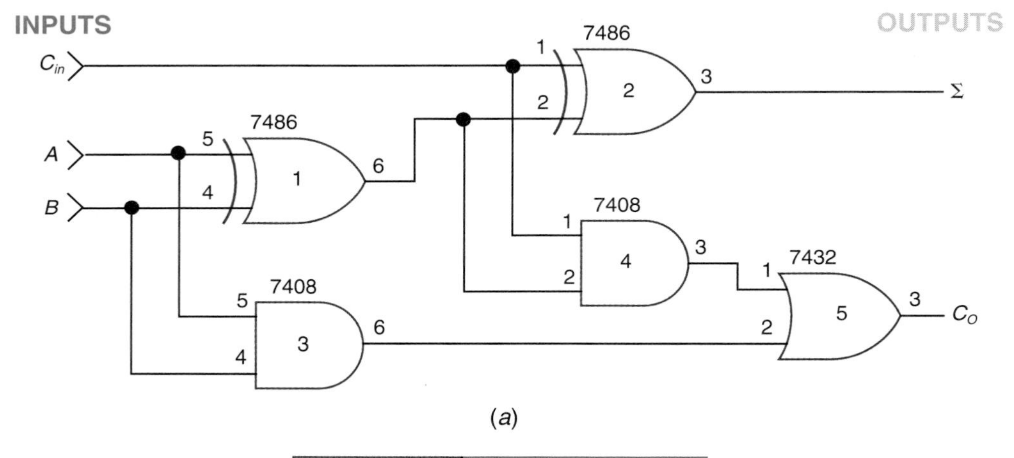

(*a*)

INPUTS			NORMAL OUTPUTS		ACTUAL OUTPUTS	
C_{in}	B	A	Σ	C_O	Σ	C_O
0	0	0	0	0	L	L
0	0	1	1	0	H	L
0	1	0	1	0	H	L
0	1	1	0	1	L	H
1	0	0	1	0	H	L
1	0	1	0	1	L	L ?
1	1	0	0	1	L	L ?
1	1	1	1	1	H	H

(*b*)

Fig. 10-31 (*a*) Faulty full-adder circuit used for troubleshooting problem. (*b*) Full-adder truth table with normal and actual outputs.

The troubleshooter expects the problem to be in the OR gate or two AND gates. The bottom line of the truth table suggests that the bottom AND gate and OR gate work. The upper AND gate (labeled 4) is suspect. The technician manipulates the inputs to line 6 on the truth table ($C_{in} = 1, B = 0, A = 1$). Pins 1 and 2 of the AND gate labeled 4 should both be 1. Both inputs to gate 4 indicate a HIGH logic level when a logic probe is touched to pins 1 and 2. Output 3 of AND gate 4 is checked and remains LOW. This indicates a stuck LOW output at gate 4.

The technician carefully checks the 7408 IC and surrounding circuit board for possible short circuits to GND. None are found. Gate 4 is assumed to have a stuck LOW output, and the 7408 IC is replaced with an exact duplicate.

After replacement of the 7408 IC, the troubleshooter checks the full-adder circuit for proper operation. The circuit works according to its normal truth table. Truth tables help both technicians and students with troubleshooting. Such tables define how a *normal circuit should respond.* The truth table becomes part of the technician's knowledge of the circuit. Knowledge of normal circuit operation is critical to good troubleshooting.

TEST

Supply the missing word or words in each statement.

60. Refer to Fig. 10-31. The fault in the ________ (combinational, sequential) logic circuit seems to be in the ________ (carry out, sum) part of the circuit.
61. Refer to Fig. 10-31. The fault in the circuit is in gate ________ (number); the output is stuck ________ (HIGH, LOW).

Chapter 10 Review

Summary

1. Arithmetic circuits, such as adders and subtractors, are combinational logic circuits constructed with logic gates.
2. The basic addition circuit is called a half adder. Two half adders and an OR gate can be wired to form a full adder.
3. The basic subtraction circuit is called a half subtractor. Two half subtractors and an OR gate can be wired to form a full subcontractor.
4. Adders (or subtractors) can be wired together to form parallel adders.
5. A 4-bit parallel adder adds two 4-bit binary numbers at one time. This adder contains a single half adder (1s place) and three full adders.
6. By using the 1s complement and end-around carry method of subtraction, an adder can be used for binary subtraction.
7. Manufacturers produce several arithmetic ICs, such as the 7483 4-bit binary adder.
8. Adder/subtractor units are often part of the CPU of calculating machines.
9. Binary multiplication performed by digital circuits may use repeated additions or the add and shift method.
10. Microprocessors use 2s complement notation when dealing with signed numbers. Adders can be used to perform both addition and subtraction using 2s complement numbers.
11. Truth tables are a great aid in troubleshooting combinational logic circuits since they define the normal operation of the circuits.

Chapter Review Questions

Answer the following questions.

10-1. Do binary addition problems **a** to **h** (show your work):

a. 101 + 011 =
b. 110 + 101 =
c. 111 + 111 =
d. 1000 + 0011 =
e. 1000 + 1000 =
f. 1001 + 0111 =
g. 1010 + 0101 =
h. 1100 + 0101 =

10-2. Draw a block diagram for a half adder (label two inputs and two outputs).

10-3. Draw a block diagram for a full adder (label three inputs and two outputs).

10-4. Do binary subtraction problems **a** to **h** (show your work):

a. 1100 − 0010 =
b. 1101 − 1010 =
c. 1110 − 0011 =
d. 1111 − 0110 =
e. 10000 − 0011 =
f. 1000 − 0101 =
g. 10010 − 1011 =
h. 1001 − 0010 =

10-5. Draw a block diagram of a half subtractor (label two inputs and two outputs).

10-6. Draw a block diagram of a full subtractor (label three inputs and two outputs).

10-7. Draw a block diagram of a 2-bit parallel adder (use a half and a full adder).

10-8. Do binary subtraction problems **a** to **h** using the 1s complement and end-around carry method (show your work):

a. 111 − 101 =
b. 1000 − 0011 =
c. 1001 − 0010 =
d. 1010 − 0100 =
e. 1011 − 1010 =
f. 1100 − 0110 =
g. 1110 − 0100 =
h. 1111 − 0111 =

10-9. Do binary multiplication problems **a** to **h** (show your work). Check your answers using decimal multiplication.

a. 101 × 011 =
b. 111 × 011 =
c. 1000 × 101 =
d. 1001 × 010 =
e. 1010 × 011 =
f. 110 × 111 =
g. 1100 × 1000 =
h. 1010 × 1001 =

10-10. List two methods of doing binary multiplication with digital electronic circuits.

10-11. If the CPU of your computer has only an adder and shift registers, how can you still do binary multiplication with this machine?

10-12. Convert the following signed decimal numbers to their 4-bit 2s complement form:

a. +1 =
b. +7 =
c. −1 =
d. −7 =

10-13. Convert the following 4-bit 2s complement numbers to their signed decimal form:
a. 0101 = c. 1110 =
b. 0011 = d. 1000 =

10-14. Convert the following 8-bit 2s complement numbers to their signed decimal form:
a. 0111 0000 c. 1000 0001
b. 1111 1111 d. 1100 0001

10-15. Convert the following signed decimal numbers to their 8-bit 2s complement form:
a. +50 c. −50
b. −32 d. −96

10-16. Add the following 4-bit 2s complement numbers. Give each sum as a *4-bit 2s complement number.* Also give each sum as a signed decimal number.
a. 0110 + 0001 =
b. 1101 + 1011 =
c. 0001 + 1100 =
d. 0100 + 1110 =

10-17. Subtract the following 4-bit 2s complement numbers. Give each difference as a *4-bit 2s complement number.* Also give each difference as a signed decimal number.
a. 0110 − 0010 =
b. 1001 − 1110 =
c. 0010 − 1101 =
d. 1101 − 0001 =

10-18. Add the following 8-bit 2s complement numbers. Give the sum in 8-bit 2s complement notation. Also give each sum as a signed decimal number.
a. 0001 0101 + 0000 1111 =
b. 1111 0000 + 1111 1000 =
c. 0000 1111 + 1111 1100 =
d. 1101 1111 + 0000 0011 =

10-19. Subtract the following 8-bit 2s complement numbers. Give the difference in 2s complement notation. Also give each difference as a signed decimal number.
a. 0111 0000 − 0001 1111 =
b. 1100 1111 − 1111 0000 =
c. 0001 1100 − 1110 1111 =
d. 1111 1100 − 0000 0010 =

10-20. See Table 10-1. The problem with the faulty half-adder circuit appears to be in the ________ (C_O, sum) output, which seems to be ________ (stuck HIGH, stuck LOW).

10-21. See Table 10-1. In an attempt to repair the faulty half-adder circuit, you might start by substituting a good ________ (AND gate IC, XOR gate IC) and then testing the circuit for correct operation.

TABLE 10-1 Logic Probe Results on Faulty Half-Adder Circuit

Inputs		Outputs	
B	*A*	Sum	C_O
L	L	L	H
L	H	H	H
H	L	H	H
H	H	L	H

Critical Thinking Questions

10-1. Draw a block diagram of a 3-bit parallel subtractor (use three full adders and three inverters).

10-2. Draw a logic symbol diagram of a 2-bit parallel adder using XOR, AND, and OR gates.

10-3. Draw a logic symbol diagram of a full subtractor circuit using XOR, NOT, and NAND gates. Use Fig. 10-11 as a guide.

10-4. Draw a logic diagram of an 8-bit binary adder using two 7483 4-bit adder ICs.

10-5. Convert the signed number +127 to its 8-bit 2s complement form. Remember that the leftmost bit will be 0, which means the number is positive.

10-6. Convert the signed number −25 to its 8-bit 2s complement form. Remember that the leftmost bit will be 1, which means the number is negative.

10-7. Twos-complement numbers are widely used in digital systems (such as microprocessors) because they can be used to represent ________ numbers.

10-8. Describe how you would form a 2s complement of a binary number.

10-9. The negative of a binary number is its ________ (2s complement, 9s complement).

10-10. Why might we say that decimal 0 would be represented as a positive number in 2s complement notation?

10-11. At the option of your instructor, use circuit simulation software to (1) construct an adder/subtractor system using 2s complement numbers (see Fig. 10-30), (2) test the circuit by adding and subtracting 2s complement numbers (see Figs. 10-26 and 10-27 for samples), and (3) show your instructor your circuit and results.

Answers to Tests

1. 1110
2. 10001
3. 11000
4. 11010
5. 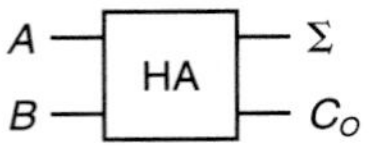 A, B → HA → Σ, C_O
6.

B	A	Σ	C_O
0	0	0	0
0	1	1	0
1	0	1	0
1	1	0	1

7. 1s
8. t_1: sum = 1, $C_O = 0$
   t_2: sum = 0, $C_O = 0$
   t_3: sum = 0, $C_O = 1$
   t_4: sum = 1, $C_O = 0$
9. 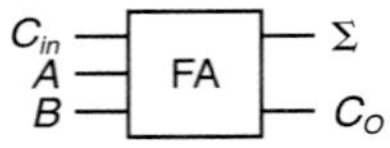 C_{in}, A, B → FA → Σ, C_O
10.

C_{in}	B	A	Σ	C_O
0	0	0	0	0
0	0	1	1	0
0	1	0	1	0
0	1	1	0	1
1	0	0	1	0
1	0	1	0	1
1	1	0	0	1
1	1	1	1	1

11. arithmetic-logic unit (ALU)
12. full-adder
13. t_1: sum = 1, $C_O = 1$
    t_2: sum = 0, $C_O = 1$
    t_3: sum = 0, $C_O = 1$
    t_4: sum = 1, $C_O = 0$
    t_5: sum = 0, $C_O = 1$
    t_6: sum = 1, $C_O = 0$
    t_7: sum = 1, $C_O = 0$
    t_8: sum = 0, $C_O = 0$
14. half adder, full adders
15. combinational
16. 1101
17. 1000
18. 1110
19. a. 01
    b. 10
    c. 000
    d. 101
    e. 1111
    f. 111
20. 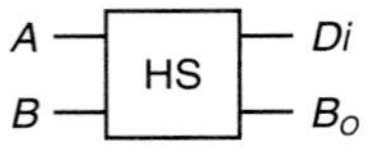 A, B → HS → Di, B_O
21.

A	B	Di	B_O
0	0	0	0
0	1	1	1
1	0	1	0
1	1	0	0

22. 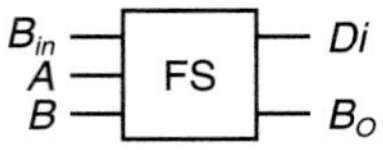 B_{in}, A, B → FS → Di, B_O
23.

A	B	B_{in}	Di	B_O
0	0	0	0	0
0	0	1	1	1
0	1	0	1	1
0	1	1	0	1
1	0	0	1	0
1	0	1	0	0
1	1	0	0	0
1	1	1	1	1

24. parallel subtractor
25. borrow lines
26. a. 0001
    b. 0101
    c. 0001
    d. 0010
    e. 0010
    f. 0101
27. T
28. full adders
29. adder
30. cascaded
31. combinational
32. 74LS181
33. 1110
34. 1000001
35. 10101000
36. repeated addition
37. add and shift
38. add and shift
39. do not
40. F
41. 2s complement
42. 0111, +7
43. −1
44. sign, positive, negative
45. 1010
46. 0101
47. 1011, −5
48. 0010, +2
49. 0101 1010, 0101 1010
50. 1010 0110
51. 0110 1111, +111
52. 1000 1111, −113
53. 1101 0000, −48
54. 0011 0001, +49
55. 2s complement
56. 2s complement
57. signed
58. 1111, −1
59. 0011, +3
60. combinational, carry out
61. 4, LOW

Memories

Chapter Objectives

This chapter will help you to:

1. *List* and *characterize* common memory and storage devices used in a microcomputer system.
2. *Sketch* the general organization of a computer, including the CPU, control bus, address bus, data bus, internal RAM, ROM, NVRAM, and bulk storage memory devices.
3. *Match* certain semiconductor memory cell types with specific characteristics and common uses.
4. *Associate* specific storage devices with their fundamental technology, such as magnetic, mechanical, optical, or semiconductor.
5. Given small semiconductor memory organization, *draw* the memory in table form, *sketch* a logic symbol for the memory, and *explain* the programming of the memory.
6. *Define* the read and write processes in a memory and *describe* how these processes might be carried out using a specific device.
7. *Identify* several specifications often associated with semiconductor memories.
8. *Identify* several common memory packages.
9. *List* and *describe* several bulk-storage methods.
10. *List* several emerging memory technologies that hold great promise.

It has been said that the most important characteristic that a digital system has over an analog system is its *ability to store data* for short or long periods. The availability and use of memory and digital storage devices has fueled what writers have called the *information revolution.* The entire Internet system is dependent on the transfer of data from one storage/memory device to another. Of course, computers and telecommunication systems are dependent on large amounts of digital storage.

The *compact disk read-only memory* (CD-ROM) pictured in Fig. 11-1 on page 260 is a wonderful modern example of an optical storage device that features very high storage capacity at low cost. A single CD-ROM can store the equivalent of more than 200,000 typed pages of information, which is more data storage than is available on more than 400 standard floppy disks. The CD-ROM has a storage capacity of 650 Mbytes per single 4.75-in. diameter disk. The manufacturing cost of a high-quality CD-ROM is inexpensive.

The flip-flop, which we have already studied, forms a basic "memory cell" in some semiconductor memories. You have already used a simple shift register, latches, and counters, which use the flip-flop as a temporary memory. Several more types of semiconductor memory cells will be investigated in this chapter. Several types of bulk storage devices will also be surveyed. Bulk storage devices are commonly classified as either magnetic, mechanical, optical, or semiconductor in nature.

Semiconductor memory ICs

Fig. 11-1 Storage using CD-ROMs. *(Courtesy of Imation, Inc.)*

11-1 Overview of Memory

Memory Devices in Computer

The sketch in Fig. 11-2 is an overview of a typical microcomputer system featuring the many types of memory and storage devices used in an everyday machine. The *CPU* is the *central processing unit,* which is the section of a computer or microprocessor that contains the arithmetic, logic, and control sections. The CPU is the focus of most data transfers. Flowing from the CPU in Fig. 11-2 are the *address bus* and *control bus* lines. A *bus* is a group of parallel conductors whose job it is to transfer information to other parts of the computer or microprocessor. The address bus and control bus are one-way communication lines that tell memory, storage, and other peripheral devices *who does what and when.* The data bus is a two-way communication channel for sending information to and receiving information from memory, storage, and other peripheral devices. The simplified block diagram in Fig. 11-2 shows some of the common internal semiconductor memory devices used in computers such as the *RAM, ROM,* and *NVRAM.* Notice that data from the data bus can flow into *(to write in memory)* or out of *(to read from memory)* both *random-access memory (RAM)* and *nonvolatile RAM (NVRAM).* The *read-only memory (ROM)* is different because it is permanently programmed and data can flow out of this semiconductor device only as shown by the arrow in Fig. 11-2. A variety of semiconductor read-only memory devices such as PROMs or EPROMs could be substituted for the ROM in this computer system.

Other memory components commonly associated with a modern microcomputer are listed under bulk storage devices in Fig. 11-2. They

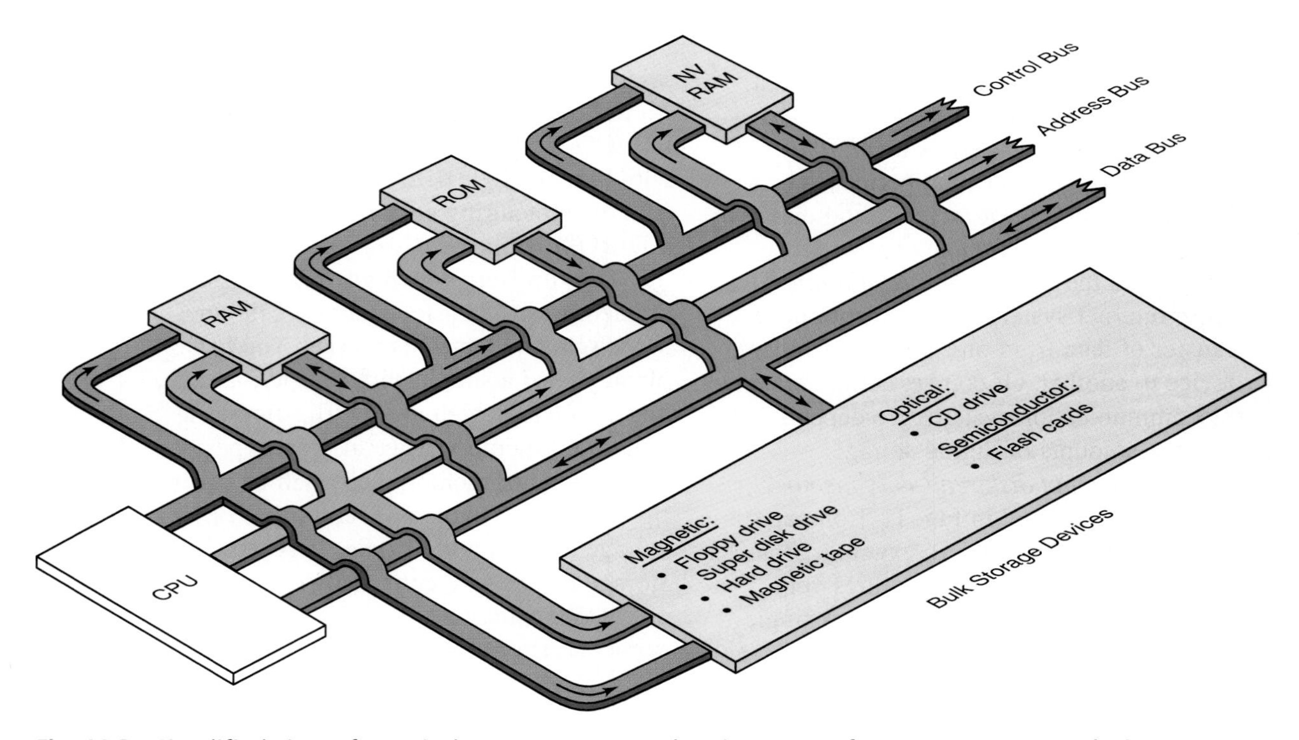

Fig. 11-2 Simplified view of a typical computer system showing types of memory or storage devices.

are divided according to the type of storage medium such as magnetic, optical, or semiconductor. *Floppy disk drives* are standard on almost all microcomputers, the 3.5-in. *double-sided high-density (DSHD)* are the most common. They store data in the thin coating of metal oxide coated on a flexible plastic disk. Alignment of magnetic domains in one direction or the other in the metal oxide surface represent digital 0s or 1s. A write head places 0s and 1s on the magnetic material while a read head detects or reads a 0 or 1 as the magnetic medium moves past it. Another almost universal magnetic bulk storage device is the *hard drive,* which operates on the same principles as the floppy disk except it holds much more data. On the hard drive, magnetic material is coated on rigid disks that typically rotate at much higher speeds than floppy disk drives. Floppy disks have the advantage that they can be removed and transported, whereas the hard drive is commonly a permanent part of the computer system. On larger systems, *magnetic tape* is commonly used to back up the data periodically on the hard drive.

The SuperDisk™ drive is a promising new entry into the bulk storage area and is expected to be included in many new microcomputers. A photograph of Imation's SuperDisk™ drive and diskette is reproduced in Fig. 11-3. It is a very high capacity "floppy disk drive" that will read from and write to a standard floppy disk or the newer SuperDisk™ diskette. The SuperDisk™ is removable and is same size as the 3.5-in. floppy disk but will hold up to 120 Mbytes of data, which is about 80 times that of a standard floppy disk. Imation's SuperDisk™ drive, the LS-120 pictured in Fig. 11-3, can also operate five times faster than a standard floppy disk drive for quicker data retrieval and less waiting time. SuperDisks™ make backing up a hard drive more feasible.

Fig. 11-3 The SuperDisk™ is a type of floppy that holds 120 Mbytes. The SuperDisk™ drive can read a regular floppy disk or a SuperDisk™. *(Courtesy of Imation, Inc.)*

Now many microcomputer systems come equipped with a CD drive that can read either music CDs or CD-ROMs. As mentioned previously, a CD-ROM has great capacity (650 Mbytes) and is commonly associated with the multimedia feature of many home, school, and business computers: CD-ROMs are commonly associated with text, still photos, video clips, and sound. In an industrial setting, CD-ROMs can be used in advertising, promotion, and graphic arts. CD-ROMs are also commonly used in storing and indexing data bases for reference. The CD is classified as an optical device because laser beams are aimed at a track of microscopic pits and lands (no pits) molded into the plastic of the CD-ROM. The reflected light from these tiny pits and lands is sensed and is interpreted as logical 0s and 1s. Higher capacity CD-ROMs are also becoming available.

A single semiconductor-type of bulk storage device is listed in the microcomputer system sketched in Fig. 11-2. Flash memories can appear in regular IC packages or in memory card form. A memory card looks something like a thick credit card. Digital cameras commonly use flash memory cards to store photos. A decade ago, flash memories were available only in small sizes, but recently 64 Mbit chips have become available. Ultimately, semiconductor flash memories may become *solid-state drives* as they replace the hard drive in some portable computers and other devices such as personal organizers.

Semiconductor Storage Cells

Semiconductor storage devices are commonly classified in about six categories: SRAM, DRAM, ROM, EPROM, EEPROM, and Flash memory (Flash EEPROM). Some of these technologies are better than others for certain jobs in a digital system. Following is a brief description of these technologies:

- *SRAM (static random-access memory)*—high access speed, read or write, requires continuous power (volatile memory), low density, high cost
- *DRAM (dynamic random-access memory)*—good access speed, read or write, volatile memory plus a need for refresh circuitry, high density, lower cost, RAM type used in most modern PCs
- *ROM (read-only memory)*—high density, nonvolatile (cannot be altered), reliable, low cost especially at high volumes
- *EPROM (electrically programmable read-only memory)*—high density, nonvolatile (can be updated although not easily), ultraviolet light erasable before reprogramming
- *EEPROM (electrically erasable programmable read-only memory)*—nonvolatile but electrically erasable by bytes for reprogramming, lower density, high cost
- *Flash Memory*—very high density, low power, nonvolatile but rewritable (bit-by-bit) within the digital system, fairly new and developing technology holding great promise as a solid-state hard drive, can be portable (like floppy disk) in memory card form

The diagram in Fig. 11-4 suggests three important characteristics of a semiconductor memory represented by the three large circles: nonvolatility, high density, and the capacity of being electrically updated. Notice in Fig. 11-4 that the newer flash memory has the best combination of nonvolatility, high density, and read/write capability (electrically updatable). Flash memory is a developing technology, and it can be expected that densities will go up and the price will fall, making the technology widely applied.

Consider the advantage of using flash memory in the system sketched in Fig. 11-2. In a common microcomputer system, the control unit of the computer would direct the hard drive or floppy disk drive to transfer a file or files to the RAM (probably DRAM in most systems). This takes quite a bit of time. If the RAM were replaced with flash memory this seek time (disk-to-DRAM loading) is eliminated making users

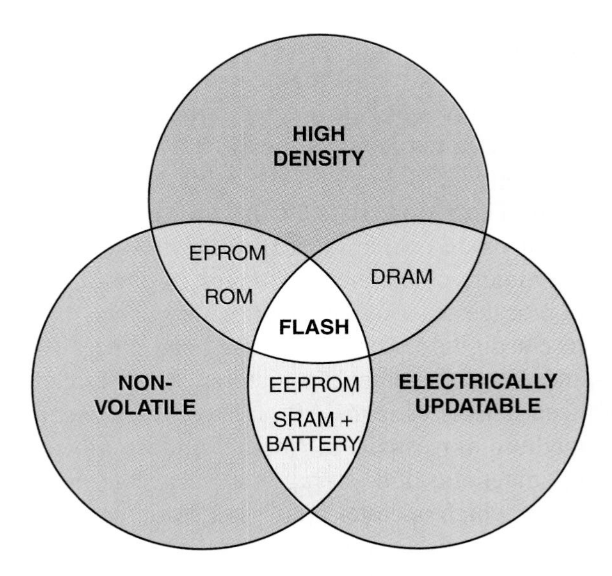

Fig. 11-4 Important semiconductor memory characteristics. *(Courtesy of Intel Corporation.)*

experience higher speed operation. Refer again to the system sketched in Fig. 11-2. The NVRAM is commonly implemented using SRAM with a battery backup. The system loses much of its intelligence when this battery fails. The NVRAM could be replaced by a flash memory, which is particularly suitable in portable systems that require very low power consumption and must be rugged and very compact.

TEST

Answer the following questions.

1. The section of a computer system that contains the arithmetic, logic, and control section and is the center of most data transfers is called the ________.
2. List two one-way buses in a microcomputer system that direct memory, storage, and peripheral devices.
3. List three general categories of bulk storage devices based on the technology each uses.
4. List at least two bulk storage devices commonly found in microcomputer systems.
5. Spell out the full term for each of the following abbreviations.
   a. RAM   d. EEPROM
   b. ROM   e. SRAM
   c. EPROM   f. DRAM

6. Based on the information in Fig. 11-4, which semiconductor memory type would be the best choice if you wanted a nonvolatile memory with read/write capabilities, and high density (memory cells are very small)?

11-2 Random-Access Memory (RAM)

One type of semiconductor memory device used in digital electronics is the *random-access memory.* The *RAM* is a memory that you can "teach." After the "teaching-learning" process (called *writing*), the RAM remembers the information for a while and the RAM's stored information can be recalled, or "remembered," at any time. We say that we can *write* information (0s and 1s) into the memory and *read out,* or recall, information. The RAM is also called a *read/write memory* or a *scratch-pad memory.*

A semiconductor memory with 64 cells in which to place 0s and 1s is illustrated in Fig. 11-5. The 64 squares (mostly blank) represent the 64 cells that can be filled with data. Notice that the 64 bits are organized into 16 groups called *words.* Each of the 16 words contains 4 *bits* of information. This memory is said to be organized as a 16 × 4 memory. That is, it contains 16 words, and each word is 4 bits long. A 64-bit memory could be organized as a 32 × 2 memory (32 words of 2 bits each), a 64 × 1 memory (64 words of 1 bit each), or an 8 × 8 memory (8 words of 8 bits each).

Address	Bit *D*	Bit *C*	Bit *B*	Bit *A*
Word 0				
Word 1				
Word 2				
Word 3	0	1	1	0
Word 4				
Word 5				
Word 6				
Word 7				
Word 8				
Word 9				
Word 10				
Word 11				
Word 12				
Word 13				
Word 14				
Word 15				

Fig. 11-5 Organization of a 64-bit memory.

The memory in Fig. 11-5 looks very much like a truth table on a scratch pad. On the table after word 3 we have written the contents of word 3 (0110). We say we have stored, or *written,* a word into the memory; this is the "write" operation. To see what is in the memory at word location 3, just read from the table in Fig. 11-5; this is the "read" operation. The write operation is the process of putting new information into the memory. The read operation is the process of copying information from memory. The read operation is also referred to as the *sense* operation because it senses, or reads, the contents of the memory.

Write operation

Random-access memory

Read operation

You could write any combination of 0s and 1s in the table in Fig. 11-5 rather like writing on a scratch pad. You could then read any word(s) from the memory, as from a scratch pad. Notice that the information in the memory remains even after it is read. Now it should be obvious why this memory is sometimes called a 64-bit scratch-pad memory. The memory has a place for 64 bits of information, and the memory can be written into or read from very much like a scratch pad.

Read/write memory

Memory organization

The memory in Fig. 11-5 is called a random-access memory because you can go directly to word 3 or word 15 and read its contents. In other words, you have access to any bit (or word) at any instant. You merely skip down to its word location and read that word. A location in the memory, such as word 3, is referred to as the storage location or *address.* In the case of Fig. 11-5, the address of word 3 is 0011_2 (3_{10}). However, the data stored at this address is 0110.

Address

The RAM cannot be used for permanent memory because it loses its data when the power to the IC is turned off. The RAM is considered a *volatile memory* because of this loss of data. Volatile memories are thus used for the *temporary* storage of data. However, some memories are permanent; they do not "forget" or lose their data when the power goes off. Such permanent memories are called *nonvolatile storage devices.*

Volatile memory

Nonvolatile storage devices

RAMs are used where only a temporary memory is needed. RAMs are used for calculator memories, buffer memories, cache memories, and microcomputer user memories.

TEST

Supply the missing word or words in each statement.

7. The letters "RAM" stand for ________.
8. Copying information into a storage location is called ________ into memory.
9. Copying information from a storage location is called ________ from memory.
10. A RAM is also called a(n) ________ or scratch-pad memory.
11. Refer to Fig. 11-5. This 64-bit unit is organized as a(n) ________ memory.
12. A disadvantage of the RAM is that it is ________; it loses its data when the power is turned ________ (off, on).

11-3 Static RAM ICs

7489 64-bit RAM TTL IC

The *7489 read/write TTL RAM* is a 64-bit data storage unit in IC form. Figure 11-6(*a*) is a logic symbol for the 7489 RAM. The memory cells are arranged like the layout of the table in Fig. 11-5. The memory can hold 16 words; each word in the 7489 IC is 4 bits wide. The 7489 RAM is said to be organized as a 16 × 4-bit memory. A pin diagram for the 7489 IC is given in Fig. 11-6(*b*).

A simplified truth table for the 7489 RAM is shown in Fig. 11-6(*c*). The *memory enable* ($\overline{ME}$) input is used to "turn on" or "select" the RAM for either reading or writing. The top line in the truth table shows both the $\overline{ME}$ and the *write enable* ($\overline{WE}$) inputs LOW. The 4 bits at the data inputs ($D1$ to $D4$) are stored in the memory location selected by the address inputs ($A3$ to $A0$). The RAM is in the *write mode.*

Let us write data into the 7489 memory chip. Suppose we want to write 0110 into word 3 location, as shown in Fig. 11-5. The address for word 3 is $A3 = 0$, $A2 = 0$, $A1 = 1$, and $A0 = 1$. Word 3 is located in the memory by placing a binary 0011 on the *address inputs* of the 7489 RAM (see Fig. 11-6(*a*)). Next, place the correct input data at the *data inputs.* To enter 0110, place a 0 at input A, a 1 at input B, a 1 at input C, and a 0 at input D. Next, place a LOW at the write enable ($\overline{WE}$) input. Last, place a LOW at the memory enable ($\overline{ME}$) input. Data is written into the memory in the storage location called word 3.

Now let us *read,* or *sense,* what is in the memory. If we want to read out the data stored at word 3, we first set the address inputs to binary 0011 (decimal 3). The write enable ($\overline{WE}$) input should be in the read position, or HIGH according to the truth table in Fig. 11-6(*c*). The memory enable ($\overline{ME}$) input should be LOW. The data outputs will indicate 1001. This output is the *complement* of the actual memory contents, which is 0110. Inverters could be attached to the outputs of the 7489 IC to make the output data the same as that in the memory. This illustrates the use of the *read mode* on the 7489 RAM.

The last two lines in the truth table in Fig. 11-6(*c*) inhibit both the read and write processes. When both $\overline{ME}$ and $\overline{WE}$ inputs are HIGH, all outputs go HIGH. When the $\overline{ME}$ input is HIGH and the $\overline{WE}$ input is LOW, the outputs are the complement of the inputs but no reading or writing is taking place.

74189 64-bit RAM

Tristate output

The 7489 RAM has open-collector outputs. This is suggested by the use of pull-up resistors on the outputs in the diagram in Fig. 11-6(*a*). A close relative of the 7489 is the *74189 64-bit RAM* with the same configuration and pins except its outputs are of the tristate type instead of the open-collector type. A *tristate output* has three levels: LOW, HIGH, or high impedance.

You will find that although different manufacturers use various labels for the inputs and outputs on this IC, all 7489 ICs have the inputs and outputs shown in Fig. 11-6. IC manufacturers usually include even very small memories like the 7489 RAM in separate data manuals that cover semiconductor memories.

Did You Know?

CD-Rewritable Drives CD-Rewritable drives are a versatile piece of computer hardware. They are like having four drives in one. (1) A CD-Rewritable disk can be used like a 500-megabyte floppy disk. (2) With special software, the CD-Rewritable disk drive can be used to record audio CDs. (3) Finally, a CD-Rewritable disk drive can double as a CD-ROM drive.

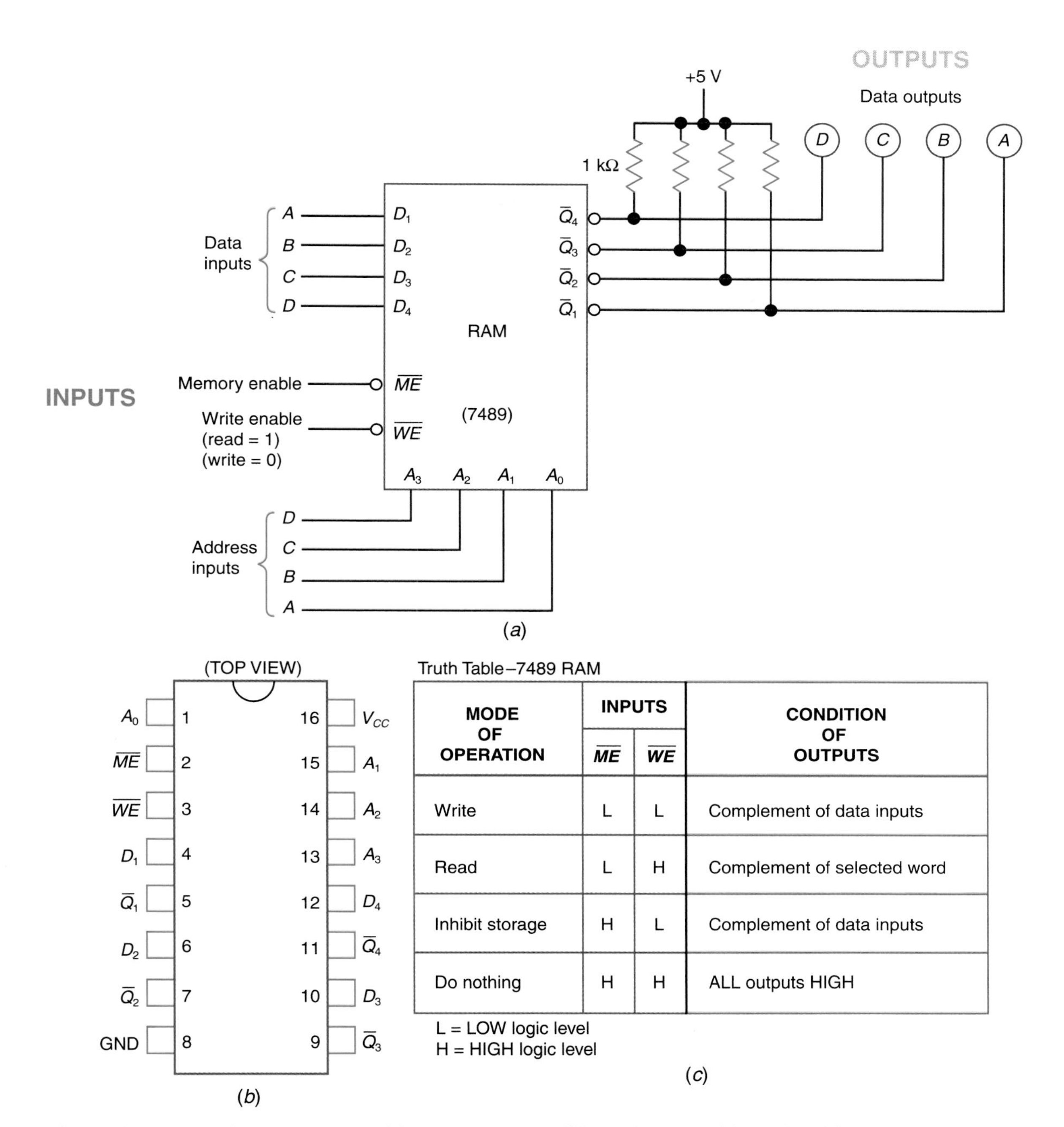

Truth Table–7489 RAM

MODE OF OPERATION	INPUTS		CONDITION OF OUTPUTS
	$\overline{ME}$	$\overline{WE}$	
Write	L	L	Complement of data inputs
Read	L	H	Complement of selected word
Inhibit storage	H	L	Complement of data inputs
Do nothing	H	H	ALL outputs HIGH

L = LOW logic level
H = HIGH logic level

(c)

Fig. 11-6 7489 64-bit RAM TTL IC. (*a*) Logic diagram. (*b*) Pin diagram. (*c*) Truth table.

The 7489 RAM is an obsolete IC that is used for experimental purposes in lab experiments to help show how many semiconductor memory chips are addressed, read from, and written to. Microprocessor-based equipment makes extensive use of semiconductor read/write RAMs in IC form.

Semiconductor RAM ICs are subdivided by manufacturers into static and dynamic types. The *static RAM* stores data in a flip-flop-like element. It is called a static RAM because it holds its 0 or 1 as long as the IC has power. The *dynamic RAM* IC stores its logic state as an electric charge in an MOS device. The stored charge leaks off after a very short time and must be refreshed many times per second. Refreshing the logic elements of a dynamic RAM requires rather extensive refresh circuitry. The dynamic RAM logic element is simpler and therefore takes up less space on the silicon chip. Dynamic RAMs come in larger sizes than static RAMs. The newer dynamic

Dynamic RAM

Static RAM

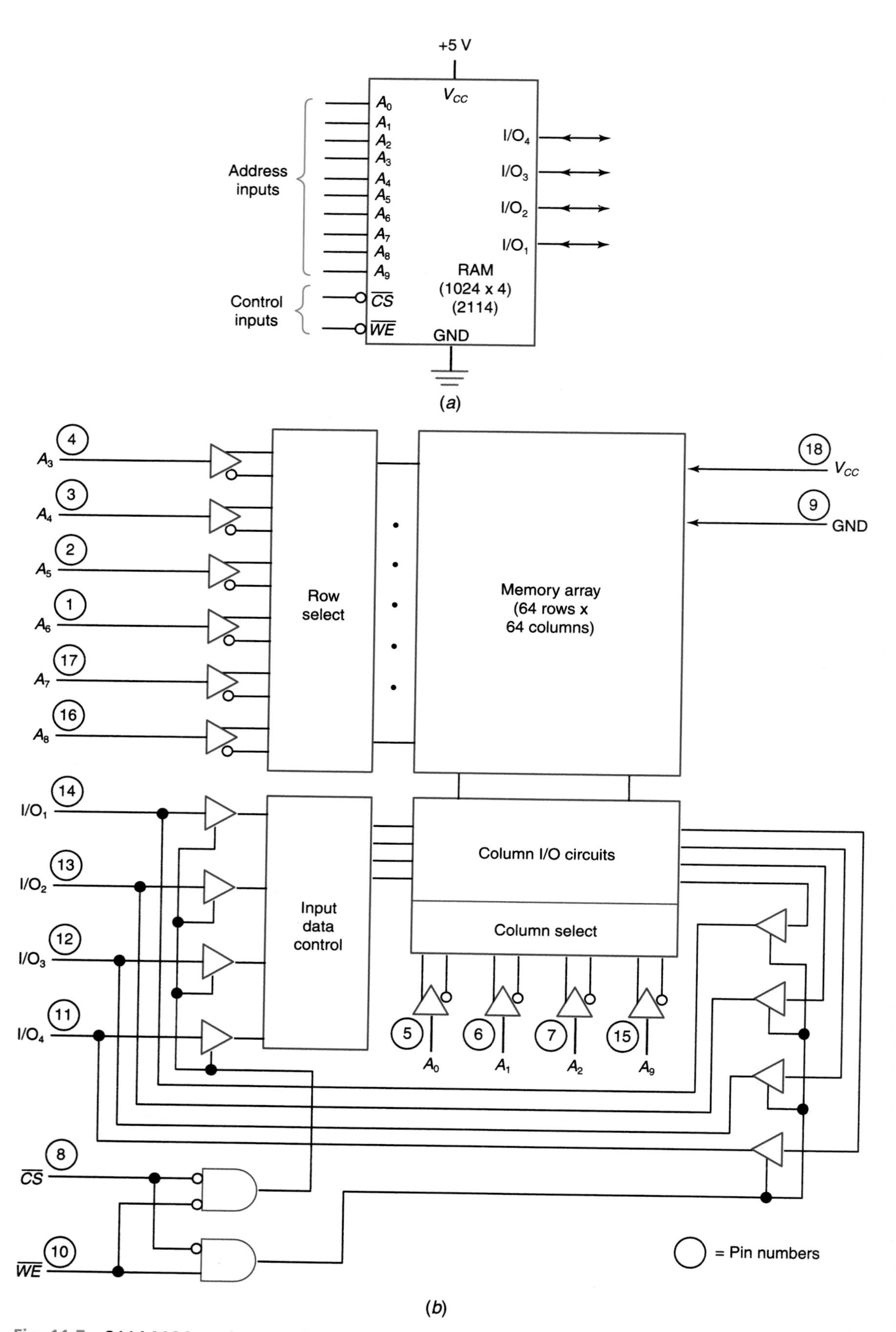

2114 MOS static RAM IC

Fig. 11-7 2114 MOS static RAM. (*a*) Logic diagram. (*b*) Block diagram of RAM chip. *(Courtesy of Intel Corporation.)*

RAMs have the refresh circuitry on the chip. Because of their ease of use, static RAMs will be used in this chapter.

One MOS memory IC is the *2114 static RAM.* The 2114 RAM will store 4096 bits which are organized into 1024 words of 4 bits each. A logic diagram of the 2114 RAM is sketched in Fig. 11-7(*a*). The 2114 RAM has 10 address lines which can access 1024 (2^{10}) words. It has *chip select* ($\overline{CS}$) and $\overline{WE}$ control inputs. The $\overline{CS}$ input is similar to the $\overline{ME}$ input on the 7489 RAM. The four input/output (I/O_1, I/O_2, I/O_3, I/O_4) pins serve as inputs when the RAM is in the write mode and outputs when the IC is in the read mode. The 2114 RAM is powered by a +5-V power supply.

A block diagram of the 2114 RAM is illustrated in Fig. 11-7(*b*). Especially note the *three-state buffers* used to isolate the input/output (I/O) pins from a computer data bus. Note that the address lines are also buffered. The 2114 RAM comes in 18-pin DIP IC form.

An important characteristic of a RAM is its access time. The *access time* is the time it takes to locate and output (or input) a piece of data. The access time of the 7489 TTL RAM is about 33 ns. The access time of the 2114 MOS RAM ranges between 100 and 250 ns depending on what version of the chip you purchase. The TTL RAM is said to be faster than the 2114 memory chip because of its shorter access time.

TEST

Supply the missing word in each statement.

13. The 7489 IC is a 64-bit ________.
14. The 7489 memory IC can hold ________ words, each word being ________ bits wide.
15. Refer to Fig. 11-6. If the address inputs = 1111, write enable = 0, memory enable = 0, and data inputs = 0011, then the 7489 IC is in the ________ (read, write) mode. Input data 0011 is being ________ (read from, written into) memory location ________ (decimal number).
16. A ________ (dynamic, static) RAM must be refreshed many times per second.
17. The 2114 RAM IC will store ________ bits of data, and each of the 1024 words is ________ bits wide.
18. Both the 7489 and 2114 memory chips are ________ (dynamic, static) RAMs.
19. Refer to Fig. 11-7. When both control inputs ($\overline{CS}$ and $\overline{WE}$) are LOW, the four I/O pins of the 2114 RAM serve as ________ (inputs, outputs).

2114 static RAM

11-4 Using a SRAM

We need some practice in using the 7489 read/write RAM. Let us *program* it with some usable information. To program the memory is to write in the information we want in each memory cell.

JOB TIP

Prepare your résumé and cover letters carefully.

Probably you cannot remember how to count from 0 to 15 in the *Gray code,* so let us take the Gray code and program it into the 7489 RAM. The RAM will remember the Gray code for us, and we can then use the RAM to convert from binary numbers to Gray code numbers.

Table 11-1 shows the Gray code numbers from 0 to 15. For convenience, binary numbers are also included in Table 11-1. The 64 logical 1s and 0s in the Gray code number column of the table must be written into the 64-bit RAM. The 7489 IC is perfect for this job because it contains 16 words; each word is 4 bits long. This is the same pattern we have in the Gray code column of Table 11-1. The decimal num-

Chip select

Gray code

Three-state buffers

Access time

Table 11-1 Gray Code

Decimal Number	Binary Number	Gray Code Number
0	0000	0000
1	0001	0001
2	0010	0011
3	0011	0010
4	0100	0110
5	0101	0111
6	0110	0101
7	0111	0100
8	1000	1100
9	1001	1101
10	1010	1111
11	1011	1110
12	1100	1010
13	1101	1011
14	1110	1001
15	1111	1000

ber in the table will be the word number (see Fig. 11-5). The binary number is the number applied to the address input of the 7489 RAM (see Fig. 11-6). The Gray code number is applied to the data inputs of the RAM (see Fig. 11-6(*a*)). When the $\overline{ME}$ and $\overline{WE}$ inputs are activated, the Gray code is written into the 7489 RAM. The RAM remembers this code as long as the power is not turned off.

Programming RAM

After the 7489 RAM is programmed with the Gray code, it is a *code converter.* Figure 11-8(*a*) shows the basic system. Notice that we input a binary number. The code converter reads out the equivalent Gray code number. The system is a *binary-to-Gray-code converter.*

Binary-to-Gray-code converter

How do you convert binary 0111 (decimal 7) to the Gray code? Figure 11-8(*b*) shows the binary number 0111 being applied to the address inputs of the 7489 RAM. The $\overline{ME}$ input is at 0. The $\overline{WE}$ input is in the read position (logical 1). The 7489 IC then reads out the stored word 7 in inverted form. The four inverters complement the output of the RAM. The result is the correct Gray code output. The Gray code output for binary 0111 is shown as 0100 in Fig. 11-8(*b*). You can input any binary number from 0000 to 1111 and get the correct Gray code output.

The binary-to-Gray-code converter in Fig. 11-8 works fine. It demonstrates how you can program and use the 7489 RAM. It is not very practical, however, because the RAM is a volatile memory. If the power is turned off for even an instant, the storage unit loses all its memory and "forgets" the Gray code. We say the memory has been *erased.* You then have to again program, or teach, the Gray code to the 7489 RAM.

Each time your home or school computer boots up when it is first started, it loads codes/programs into its RAM section of memory. This is much like loading the Gray code into the tiny 7489 RAM.

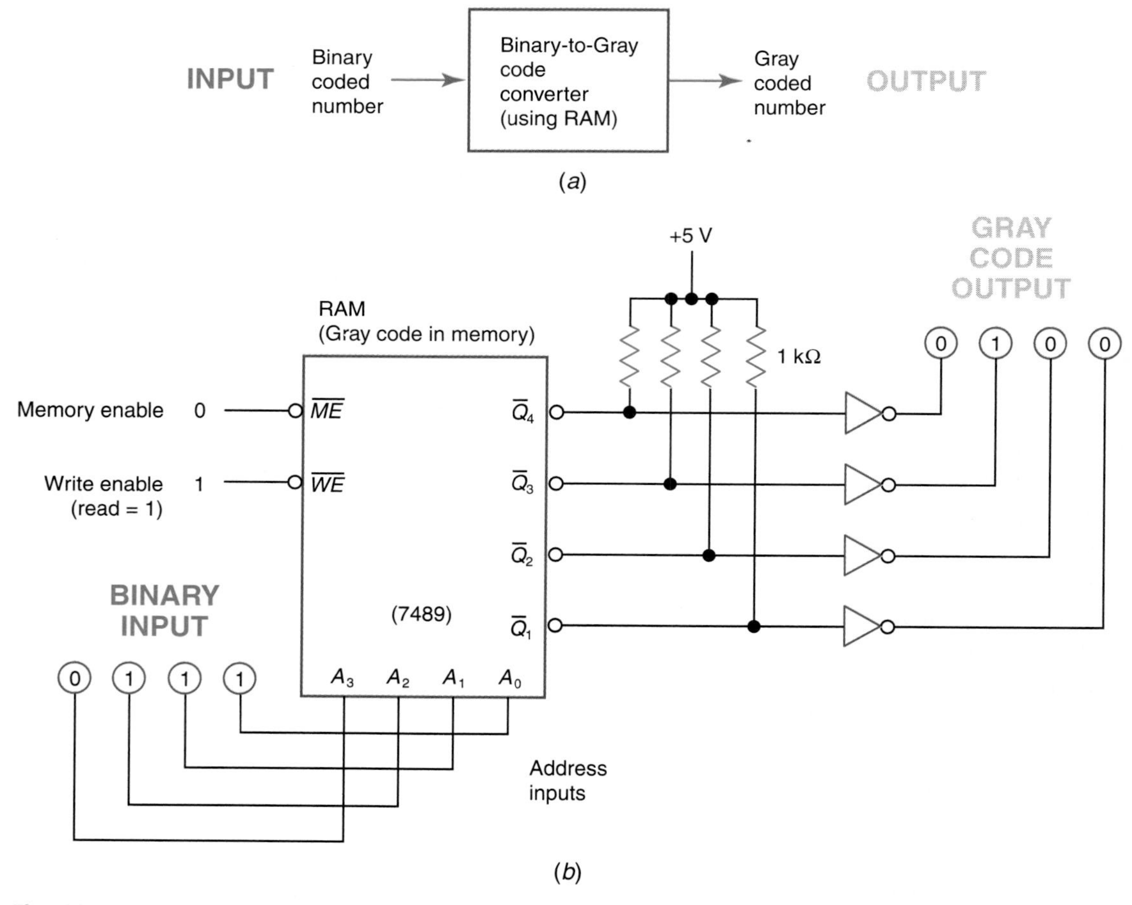

Fig. 11-8 Binary-to-Gray-code converter. (*a*) System diagram. (*b*) Wiring diagram using RAM.

TEST

Supply the missing word or words in each statement.

20. Refer to Fig. 11-8. The RAM is programmed as a(n) ________-code converter in this example.
21. Refer to Fig. 11-8. If the address inputs = 1000, $\overline{WE}$ = 1, and $\overline{ME}$ = 0, then the output at the displays on the right will be ________. This is the ________ code equivalent of binary ________.
22. If power to the 7489 IC in Fig. 11-8 is turned off for an instant, the RAM will ________ (lose its program and have to be reprogrammed, still hold the Gray code in its memory cells).

11-5 Read-Only Memory (ROM)

Many digital devices including microcomputers must store some information permanently. This is typically stored in a *read-only memory* or *ROM.* The ROM is programmed by the manufacturer to the user's specifications. Smaller ROMs can be used to solve combinational logic problems like decoding.

ROMs are classified as *nonvolatile memories* because they do not lose their data when power is turned off. The read-only memory is also referred to as the *mask-programmed ROM.* The ROM is used in only high-volume production applications because of the expensive initial setup costs. Programmable read-only memories (PROMs) are used for lower-volume applications where a permanent memory is required.

The primitive diode ROM circuit in Fig. 11-9 can perform the task of translating from binary to Gray code. The Gray code along with decimal and binary equivalents is listed in Table 11-1.

If the rotary switch in Fig. 11-9(*a*) has selected the decimal 6 position, what will the ROM output indicators display? The outputs (*D*, *C*, *B*, *A*) will indicate LHLH or 0101. The *D* and *B* outputs are connected directly to ground through the resistors and read LOW. The *C* and *A* outputs are connected to +5 V through two forward-biased diodes and the output voltage will read about +2 to +3 V, which is a logical HIGH. Notice that the pattern of diodes in the diode ROM matrix in Fig. 11-9(*a*) is similar to the pattern of 1s in the Gray code column in Table 11-1. Each new position of the rotary switch will give the correct Gray code output. In a memory, such as the ROM in Fig. 11-9, each position of the rotary switch is referred to as an *address.*

Address

A refinement in the diode ROM is shown in Fig. 11-9(*b*). The diode ROM circuit in Fig. 11-9(*b*) uses a *1-of-10 decoder* (7442 TTL IC) and inverters for row selection. This example shows a binary input of 0101 (decimal 5). This activates output 5 of the 7442 with a LOW. This drives the inverter, which outputs a HIGH. The HIGH forward biases the three diodes connected to the row 5 line. The outputs would be LHHH or 0111. This is the Gray code equivalent for binary 0101 according to Table 11-1.

1-of-10 decoder

The diode ROM suffers many disadvantages. Their logic levels are marginal. The diode ROM also suffers in that it has very limited drive capability. The diode ROMs do not have input and output buffering needed when working with systems that contain data and address buses.

Diode ROM disadvantages

Practical ROMs are available from many manufacturers. These can range from very small bipolar TTL units to quite large capacity CMOS or NMOS ROMs. Most commercial ROMs can be purchased in DIP form. As examples, a very small capacity unit might be the TTL 74S370 2048-bit ROM organized as a 512-word by 4-bit memory. A larger-capacity unit might be CMOS TMS47C512 524,288-bit ROM organized as a 65,536-word by 8-bit memory. The 65,536 × 8 unit has an access time of from 200 to 350 ns depending on the version you purchase. Personal computers have ROMs of this capacity and larger.

Read-only memory

ROM

Nonvolatile memories

Mask-programmed ROM

> **JOB TIP**
>
> Job application forms should be filled out legibly with no spelling errors.

As an example of a commercial product, the Texas Instruments TMS4764 ROM will be fea-

About Electronics

Protein-Based Memory Are protein-based 3D RAM memories in the future? A small cube of optically sensitive protein (such as Rhodopsin), suspended in a transparent plastic, might be the basis for a 20 gigabit RAM memory. Two laser beams might intersect at a point in the cube of protein to switch that "organic memory cell" from one logic state to another.

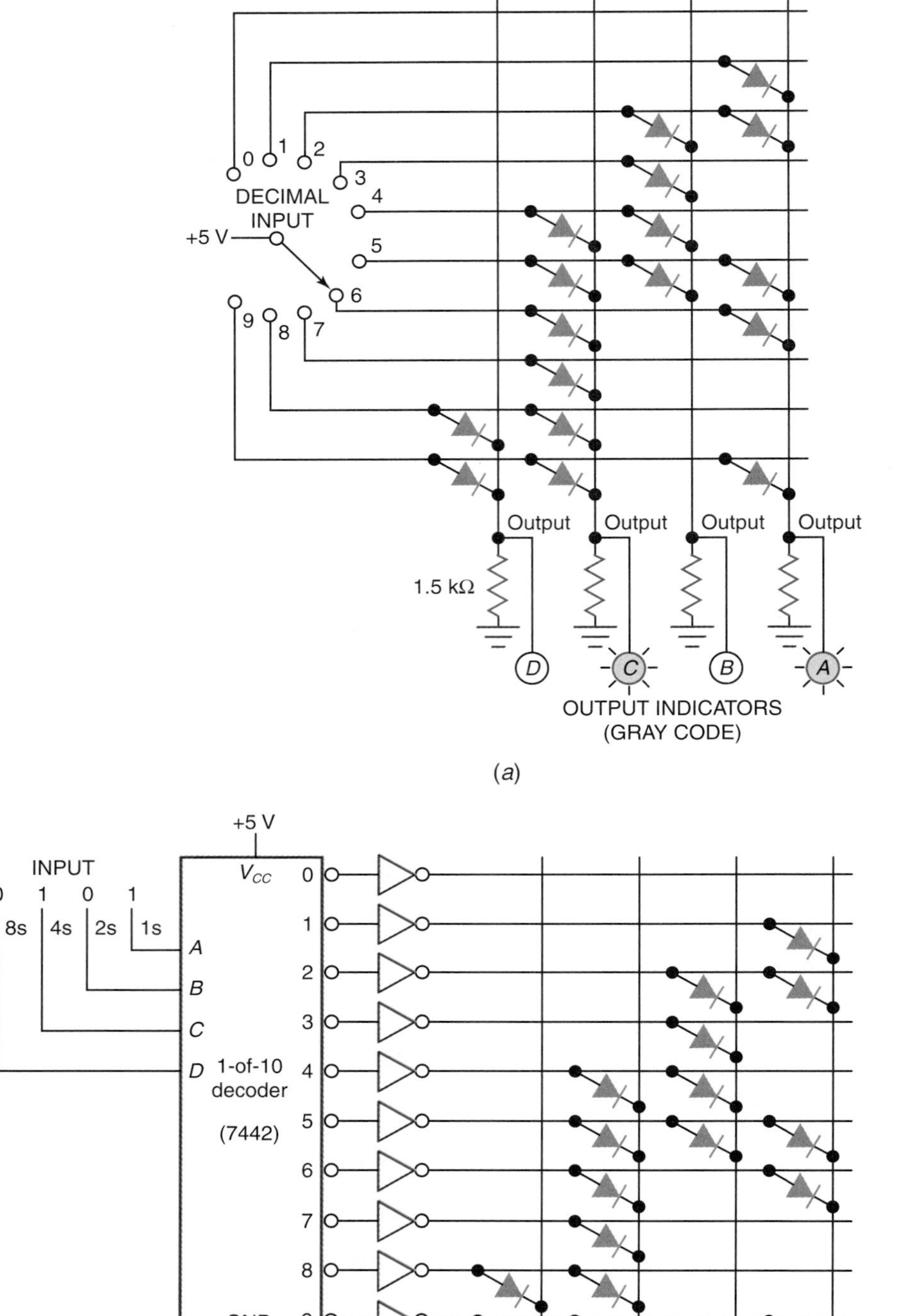

Primitive diode ROM

Fig. 11-9 Diode ROMs. (*a*) Primitive diode ROM programmed with Gray code. (*b*) Diode ROM with input decoding (programmed with Gray code).

tured. The *TMS4764* is an *8192-word by 8-bit ROM.* Its 8192 × 8 organization makes it useful in microprocessor-based systems which might store data in 8-bit groups, or bytes.

TMS4764 8192 × 8-bit ROM

A pin diagram for the TMS4764 ROM is reproduced in Fig. 11-10(*a*). The ROM is housed in a 24-pin DIP. The names and functions of the pins are given in the chart in Fig. 11-10(*b*). Notice that a total of 13 addresses lines (A_0 to A_{12}) are needed to address the 8192 (2^{13}) memory locations. A_0 is the LSB and A_{12} is the MSB of the word address. The access time of the TMS4764 ROM varies from 150 to 250 ns depending on the version of the chip you purchase. Permanently stored data is output via the pins labeled Q_1 through Q_8. Q_1 is considered the LSB while Q_8 is the MSB. The output pins (Q_1 to Q_8) are enabled by pin 20. Pin 20 may be programmed by the manufacturer to be an active HIGH or active LOW $\overline{CS}$ or $\overline{CE}$ input. When the three-state outputs are disabled, they are in a high-impedance state, which means they may be connected directly to a data bus in a microcomputer system.

Read-only memories are used to store permanent data and programs. Computer system programs, look-up tables, decoders, and character generators are but a few uses of the ROM. ROMs can also be used for solving combinational logic problems. General-purpose microcomputers allocate a larger proportion of their internal memory to RAM. However, dedicated computers allocate more addresses to ROM and usually contain only small amounts of RAM. About 500 different ROMs were available in one recent listing.

A computer program is typically referred to as *software.* However, when a computer program is stored in a ROM it is called *firmware* because of the difficulty of making changes.

Software

Firmware

For a summary, look back at Fig. 11-4. Notice that the ROM is a high-density memory device and is nonvolatile. The ROM is a permanent storage device that cannot be reprogrammed.

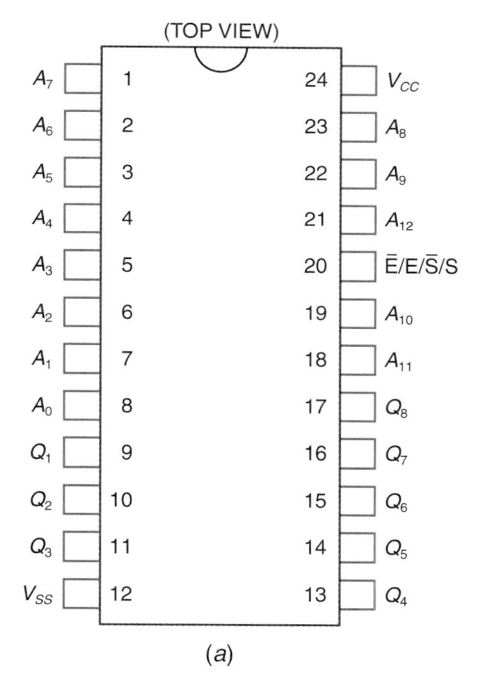

(*a*)

PIN NOMENCLATURE	
A_0–A_{12}	Address inputs
$\overline{E}$/E/$\overline{S}$/S	Chip Enable/Power Down or Chip Select
Q_1–Q_8	Data out
V_{CC}	5-V supply
V_{SS}	Ground

(*b*)

Fig. 11-10 TMS4764 ROM IC. (*a*) Pin diagram. (*b*) Pin nomenclature. *(Courtesy of Texas Instruments Incorporated.)*

TMS4764 ROM IC

TEST

Supply the missing word or words in each statement.

23. The letters "ROM" stand for ________.
24. Read-only memories never forget data and are called ________ memories.
25. The term ________ is used to describe microcomputer programs that are permanently held in ROM.
26. Read-only memories are programmed by the ________ (manufacturer, computer operator) to your specifications.
27. A back-up battery ________ (is, is not) needed to power the ROM when the computer is turned off so it can retain its programs and data.
28. Refer to Fig. 11-9(*a*). If the input switch is at 3 (binary 0011), the Gray code output will be ________.
29. Refer to Fig. 11-9(*b*). If the input is binary 1001, the Gray code output will be ________.
30. The typical ROM is a ________ (high-, low-) density memory device.

11-6 Using a ROM

Suppose you have to design a device that will give the decimal counting sequence shown in Table 11-2: 1, 117, 22, 6, 114, 44, 140, 17, 0, 14, 162, 146, 134, 64, 160, 177, and then back to 1. These numbers are to read out on seven-segment displays and must appear in the order shown.

Knowing you will use digital circuits, you convert the decimal numbers to BCD numbers. This is shown in Table 11-2. You find you have 16 rows and 7 columns of logical 0s and 1s. This section forms a truth table. As you look at the truth table, the problem seems quite complicated to solve with logic gates or data selectors. You decide to try a ROM. You think of the inside of a memory as a truth table. The BCD section of Table 11-2 reminds you that a memory organized as a 16 × 7 storage unit will do the job. This 16 × 7 ROM will have 16 words for the 16 rows on the truth table. Each word will contain seven bits of data for seven columns on the truth table. This will take a 112-bit ROM.

Table 11-2 Counting Sequence Problem

Decimal Readout			Binary-Coded Decimal Number		
100s	10s	1s	100s	10s	1s
		1	0	000	001
1	1	7	1	001	111
	2	2	0	010	010
		6	0	000	110
1	1	4	1	001	100
	4	4	0	100	100
1	4	0	1	100	000
	1	7	0	001	111
		0	0	000	000
	1	4	0	001	100
1	6	2	1	110	010
1	4	6	1	100	110
1	3	4	1	011	100
	6	4	0	110	100
1	6	0	1	110	000
1	7	7	1	111	111

A 112-bit ROM is shown in Fig. 11-11. Notice that it has four address inputs to select one of the 16 possible words stored in the ROM. The 16 different addresses are shown in the left columns of Table 11-3. Suppose the address inputs are binary 0000. Then the first line in Table 11-3 shows that the stored word is 0 000 001 (*a* to *g*). After decoding in Fig. 11-11, this stored word reads

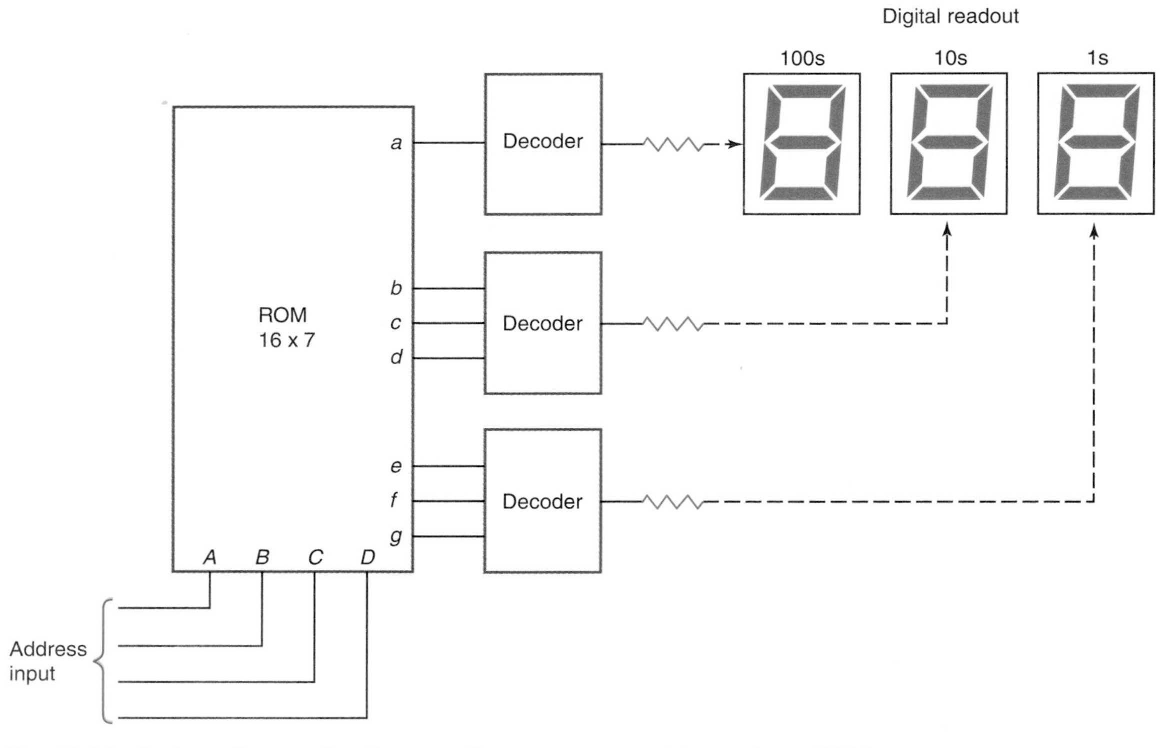

Fig. 11-11 System diagram for the counting sequence problem using a ROM.

Using a ROM

TABLE 11-3 Counting Sequence Problem

Inputs				ROM Outputs							Decimal Readout		
Address or Word Location				100s	10s			1s					
				1s	4s	2s	1s	4s	2s	1s			
D	*C*	*B*	*A*	*a*	*b*	*c*	*d*	*e*	*f*	*g*	100s	10s	1s
0	0	0	0	0	0	0	0	0	0	1			1
0	0	0	1	1	0	0	1	1	1	1	1	1	7
0	0	1	0	0	0	1	0	0	1	0		2	2
0	0	1	1	0	0	0	0	1	1	0			6
0	1	0	0	1	0	0	1	1	0	0	1	1	4
0	1	0	1	0	1	0	0	1	0	0		4	4
0	1	1	0	1	1	0	0	0	0	0	1	4	0
0	1	1	1	0	0	0	1	1	1	1		1	7
1	0	0	0	0	0	0	0	0	0	0			0
1	0	0	1	0	0	0	1	1	0	0		1	4
1	0	1	0	1	1	1	0	0	1	0	1	6	2
1	0	1	1	1	1	0	0	1	1	0	1	4	6
1	1	0	0	1	0	1	1	1	0	0	1	3	4
1	1	0	1	0	1	1	0	1	0	0		6	4
1	1	1	0	1	1	1	0	0	0	0	1	6	0
1	1	1	1	1	1	1	1	1	1	1	1	7	7

out on the digital displays as a decimal 1 (100s = 0, 10s = 0, 1s = 1).

Let us consider another example. Apply binary 0001 to the address inputs of the ROM in Fig. 11-11. The second row on Table 11-3 shows us that the stored word is 1 001 111 (*a* to *g*). When decoded, this word reads out on the digital display as decimal 117 (100s = 1, 10s = 1, 1s = 7). Remember that the 0s and 1s in the center section of Table 11-3 are *permanently* stored in the ROM. When the address at the left appears at the address input of the ROM, a row of 0s and 1s or word appears at the outputs.

You have solved the difficult counting sequence problem. Figure 11-11 diagrams the basic system to be used. The information in Table 11-3 shows the addressing and programming of the 112-bit ROM and the decoded BCD as a decimal readout. You would give the information in Table 11-3 to a manufacturer, who would custom-make as many ROMs as you need with the correct pattern of 0s and 1s.

It is quite expensive to have just a few ROMs custom-programmed by a manufacturer. You probably would not use the ROM if you did not have need for many of these memory units. Remember that this problem also could have been solved by a combinational logic circuit using logic gates.

Semiconductor memories usually come in 2^n sizes or 64-, 256-, 1024-, 4096-, 8192-bit and larger units. A 112-bit memory is an unusual size. The 112-bit memory was used in the example because its truth table in Table 11-3 is exactly the truth table of the 7447 IC. You used the 7447 IC as BCD-to-seven-segment decoder in Chap. 6. You will want to use the 7447 IC as a ROM in the laboratory.

Read-only memories are used for encoders, code converters, look-up tables, microprograms, character generators, function generators, microcomputer system firmware, and microcontroller firmware.

TEST

Supply the missing word or words in each statement.

31. Refer to Fig. 11-11. If power is turned off and then back on, the counting sequence programmed into the ROM will ________ (be lost from, remain in) memory.

32. Refer to Table 11-3 and Fig. 11-11. If the ROM address input = 1111, the digital readout will be _________.
33. Refer to Table 11-3 and Fig. 11-11. If the ROM address input = 1001, the digital readout will be _________.
34. A mask-ROM is programmed by the _________ (manufacturer, user).
35. A group of programs and data held permanently in a microcomputer's _________ (RAM, ROM) would be called firmware.

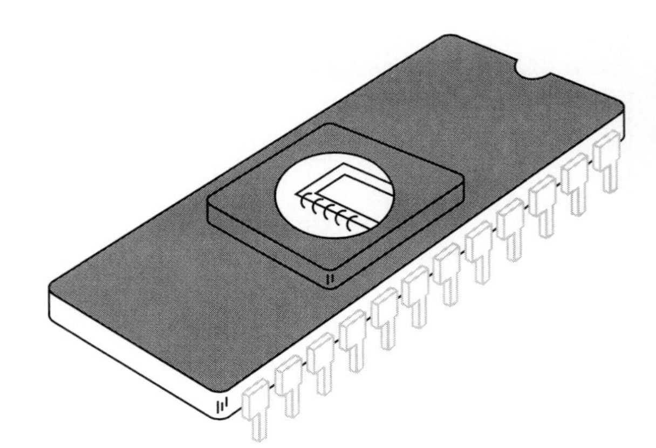

Fig. 11-12 EPROM. Note window in top used to erase EPROM with ultraviolet light.

11-7 Programmable Read-Only Memory (PROM)

Mask-programmable ROMs

Flash EEPROM

Field-programmable ROMs (PROMs)

Mask-programmable ROMs are programmed by the manufacturer using photographic masks to expose the silicon die. *Mask-programmable ROMs* have long development times and the initial costs are high. Mask-programmable ROMs are usually simply called ROMs.

Field-programmable ROMs (PROMs) are also available. They shorten development time and many times lower costs. It is also much easier to correct program errors and update products when PROMs can be programmed (burned) by the local developer. The regular PROM can only be programmed once like a ROM, but its advantage is that it can be made in limited quantities and can be programmed in the local lab or shop. The PROM is also called a *fusible-link PROM.*

EPROM (erasable programmable read-only memory)

PROM burner

"Burning" a PROM

The *EPROM (erasable programmable read-only memory)* is a variation of the PROM. The EPROM is programmed or burned in the local lab using a *PROM burner.* If an EPROM needs to be reprogrammed, a special window on the top of the IC is used. Ultraviolet (UV) light is directed at the chip under the window of the EPROM for about an hour. The UV light erases the EPROM by setting all the memory cells to a logical 1. The EPROM can then be reprogrammed. A 24-pin EPROM DIP IC is shown in Fig. 11-12. The actual EPROM chip is visible through the window on top of the IC. These units are sometimes called *UV erasable PROMs* or *UV EPROMs.*

EEPROM

Electrically erasable PROM

27XXX-series EPROM family

The *EEPROM* is a third variation of a programmable read-only memory. The EEPROM is an *electrically erasable PROM* also referred to as an E^2PROM. Because EEPROMs can be erased electrically, it is possible to erase and reprogram them without removing them from the circuit board. Parts of the code on the EEPROM can be reprogrammed one byte at a time.

The *flash EEPROM* is a fourth variation of a programmable read-only memory. The newer flash EEPROM is like an EEPROM in that it can be erased and reprogrammed while on the circuit board. Flash EEPROMs are gaining favor because they use a simpler storage cell, thereby allowing more memory cells on a single chip. We say they have greater density. Flash EEPROMs can be erased and reprogrammed faster than EEPROMs. While parts of the code can be erased and reprogrammed on an EEPROM, the entire flash EEPROM must be erased and reprogrammed.

The basic idea of a PROM is illustrated in Fig. 11-13. This simplified 16-bit (4 × 4) PROM is similar to the diode ROM studied in the previous section. In Fig. 11-13(*a*), each memory cell contains a diode and a good fuse. This indicates that all of the memory cells are storing a logical 1. This is how the PROM might look before programming.

The PROM in Fig. 11-13(*b*) has been programmed with seven 0s. To program or *burn* the PROM, tiny fuses must be blown as shown in Fig. 11-13(*b*). A blown fuse in this case disconnects the diode and means a logical 0 is permanently stored in this memory cell. Because of the permanent nature of burning a PROM, the unit cannot be reprogrammed. A PROM of the type illustrated in Fig. 11-13 can only be programmed once.

A popular EPROM family is the 27XXX series. These are available from many manufacturers. A short summary of some models in the

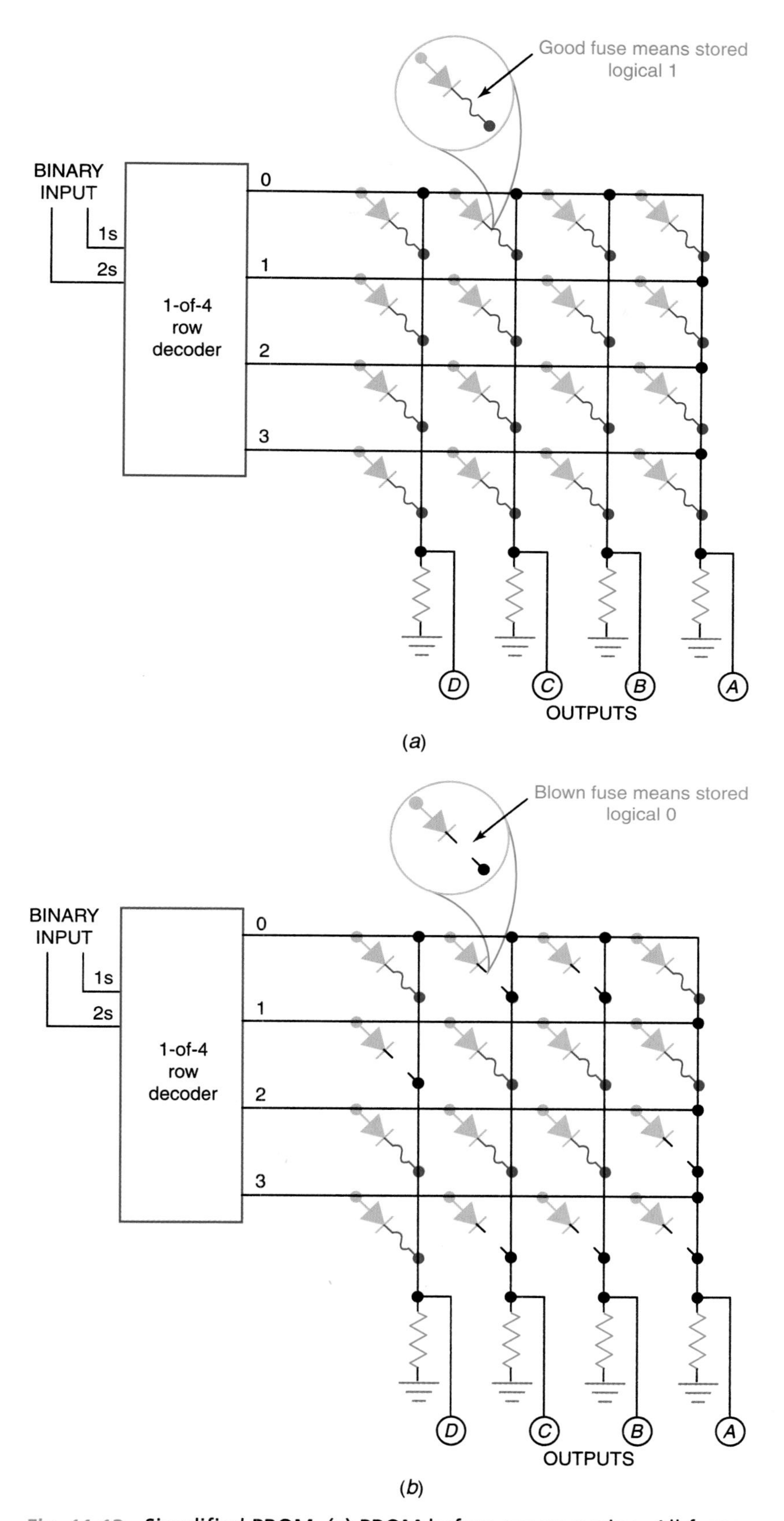

Fig. 11-13 Simplified PROM. (*a*) PROM before programming. All fuses good (all 1s). (*b*) PROM after programming. Seven fuses blown (seven 0s programmed).

Simplified PROM

TABLE 11-4 27XXX Series EPROMs

EPROM 27XXX	Organization	Number of Bits
2708	1024 × 8	8192
2716	2048 × 8	16384
2732	4096 × 8	32768
2764	8192 × 8	65536
27128	16384 × 8	131072
27256	32768 × 8	262144
27512	65536 × 8	524288

27XXX series is shown in Table 11-4. Notice that they are all organized with byte-wide (8-bit-wide) outputs, making them compatible with many microprocessor-based systems. Many versions of each of these basic numbers are available such as low-power CMOS units, EPROMs with different access times, and even pin-compatible PROMs, EEPROMs, and ROMs.

2732A 32K (4K × 8) ultraviolet-erasable PROM

A sample IC from the 27XXX series EPROM family is illustrated in Fig. 11-14. The pin diagram in Fig. 11-14(*a*) represents the *2732A 32K (4K × 8) ultraviolet-erasable PROM.* The 2732 EPROM has 12 address pins (A_0 to A_{11}) which can access 4096 (2^{12}) byte-wide words in the memory. The 2732 EPROM uses a 5-V power supply and can be erased using UV light. The $\overline{CE}$ input is like the chip select $\overline{CS}$ inputs on some other memory chips. The $\overline{CE}$ input is activated with a LOW. The $\overline{OE}/V_{PP}$ pin serves a dual purpose. It has one purpose during reading and another during writing. Under normal use the EPROM is being read. A LOW at the output enable ($\overline{OE}$) pin during a memory read activates the outputs driving the data bus of the computer system. The eight output pins are labeled O_0 to O_7 on the 2732 EPROM. A block diagram is drawn in Fig. 11-14(*c*) to show the organization of the 2732 EPROM chip.

Erasing the UV EPROM

When the 2732 EPROM is erased, all memory cells are returned to logical 1. Data is introduced by changing selected memory cells to 0s. The 2732 is in the *programming mode* (writing into the EPROM) when the dual-purpose $\overline{OE}/V_{PP}$ input is at 21 V. During programming (writing), the input data is applied to the data output pins (O_0 to O_7). The word to be programmed into the EPROM is addressed using the 12 address lines. A very short (less than 55 ms) TTL level LOW pulse is then applied to the $\overline{CE}$ input to complete the write process.

Erasing and programming an EPROM is handled by special equipment called PROM burners. After erasing and reprogramming it is common to protect the EPROM window (see Fig. 11-12) with an opaque sticker. The sticker over the EPROM window protects the chip from UV light from fluorescent lights and sunlight. The EPROM can be erased by direct sunlight in about one week or room-level fluorescent lighting in about three years.

TEST

Supply the missing word or words in each statement.

36. The letters "PROM" stand for ________.
37. The letters "EPROM" stand for ________.
38. The letters "EEPROM" stand for ________.
39. Erasing EPROMs can be done by shining ________ light through a special window in the top of the IC.
40. See Table 11-4. The 27512 EPROM can store a total of ________ bits of data organized as ________ words, each 8 bits wide.

11-8 Nonvolatile Read/Write Memory

Nonvolatile read/ write memories

Battery backup SRAM

NVSRAM

Flash EPROM

Both static and dynamic RAMs have the disadvantage of being volatile. When power is turned off, the data is lost. To solve this problem, *nonvolatile read/write memories* were developed. These are currently implemented by (1) using *battery backup* for a CMOS SRAM, (2) using a newer *nonvolatile static RAM (NVSRAM),* or (3) using a recently developed *flash EEPROM* or flash memory.

SRAM-Battery Backup

Battery backup of CMOS SRAM

Battery backup is a common method of solving the volatility problem of a SRAM. CMOS RAMs are used with battery backup systems because they consume little power. A long-life lithium battery is typically used to back up data on the CMOS SRAM. Backup batteries

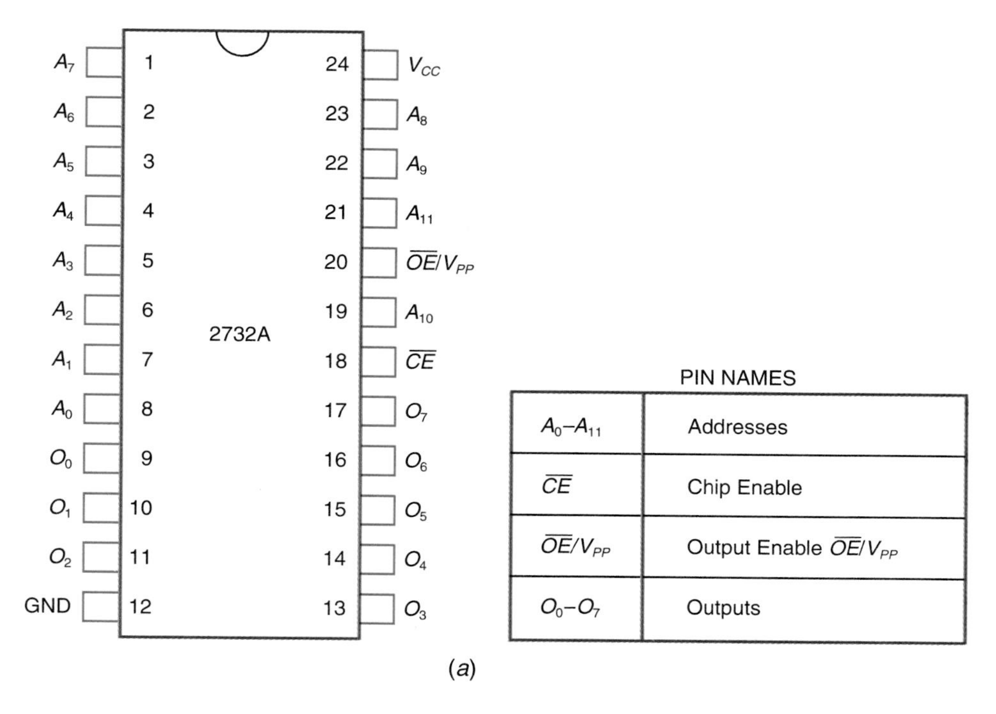

PIN NAMES	
A_0–A_{11}	Addresses
$\overline{CE}$	Chip Enable
$\overline{OE}/V_{PP}$	Output Enable $\overline{OE}/V_{PP}$
O_0–O_7	Outputs

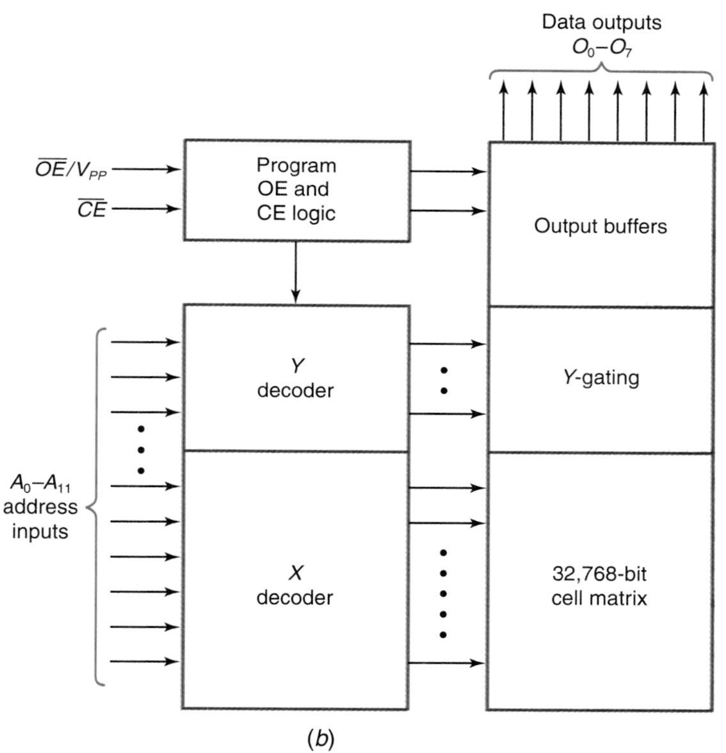

Fig. 11-14 2732 EPROM IC. (*a*) Pin diagram. (*b*) Pin names. (*c*) Block diagram. *(Courtesy of Intel Corporation.)*

2732 EPROM IC

have a life expectancy of about 10 years and may be imbedded in the memory package. Under normal operating conditions, the SRAM is powered by the equipment's power supply. When the power supply voltage drops to some predetermined lower level, voltage-sensing circuitry switches to backup battery power to maintain the contents of the SRAM until power is restored. Battery backup SRAMs are common in microcomputer systems.

NVSRAM

NVSRAM

Flash EPROM

Flash memory

Recently developed nonvolatile RAMs can solve the volatility problem. The nonvolatile RAM may be referred to as a *NVRAM (nonvolatile RAM), NOVRAM (nonvolatile RAM), NVSRAM (nonvolatile static RAM),* or *shadow RAM.* The NVRAM combines the read/write capabilities of a SRAM with the nonvolatility of an EEPROM. A block diagram of a small NVSRAM is detailed in Fig. 11-15. Note that the NVSRAM has two parallel memory arrays. The front array is a SRAM, while the back is a shadow EEPROM. During normal operation, the read/write SRAM is used. When the power supply voltage drops, a duplicate of all data in the SRAM is automatically stored in the nonvolatile EEPROM array. The *store* operation is represented in Fig. 11-15 with an arrow pointing toward the EEPROM array. On power up, the NVSRAM automatically executes the recall operation, which copies all data from the EEPROM to the SRAM. The *recall* operation is symbolized in Fig. 11-15 by the arrow pointing toward the front static RAM array.

NVSRAMs seem to have a slight advantage over battery backup SRAMs. NVSRAMs have better access speed and generally better overall life. NVSRAM ICs are smaller than the more bulky battery backup SRAM packages and therefore save PC board space. Currently, NVSRAMs are more expensive and are manufactured in limited sizes.

Flash Memory

Flash EPROMS may become a *low-cost* alternative to battery backup SRAMs and NVSRAMs. Flash memories are expected to be widely used in laptop computers and even may eventually replace the bulky power-hungry disk drives. The newer flash memories have the potential of becoming a "universal" storage device.

The commercial flash memory by Intel is featured in Fig. 11-16. Intel's *28F512 512K (64K × 8) CMOS Flash Memory* will store 524,288 (2^{19}) bits organized into 65,536 (2^{16}) words, each 8 bits wide. The block diagram and pin descriptions in Fig. 11-16 give an

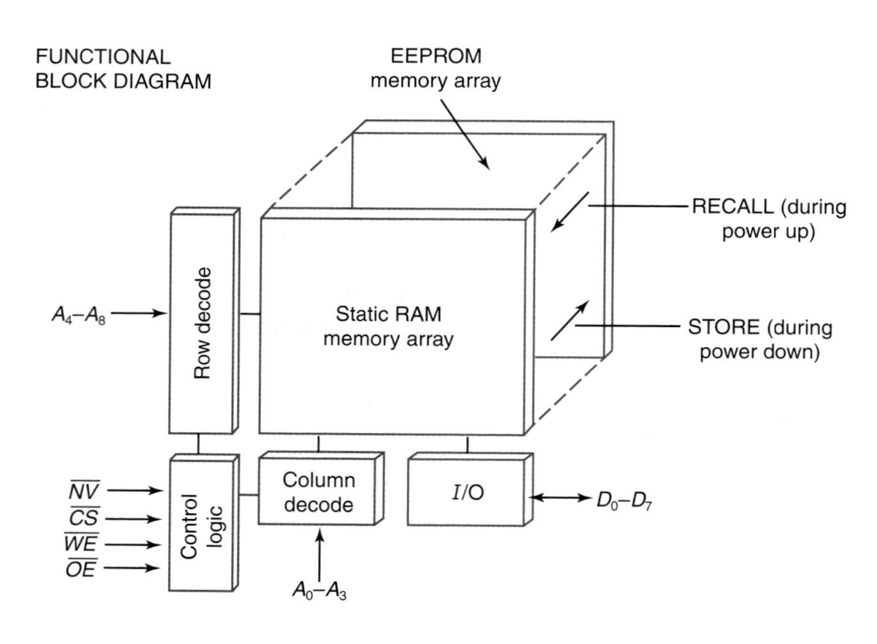

PIN NAMES

A_0–A_8	Address inputs
D_0–D_7	Data I/O
$\overline{CS}$	Chip Select
$\overline{NV}$	Non-Volatile Enable

$\overline{WE}$	Write Enable
$\overline{OE}$	Output Enable
V_{CC}	+ 5 volts ± 10%

Fig. 11-15 Block diagram and pin names on a typical NVSRAM.

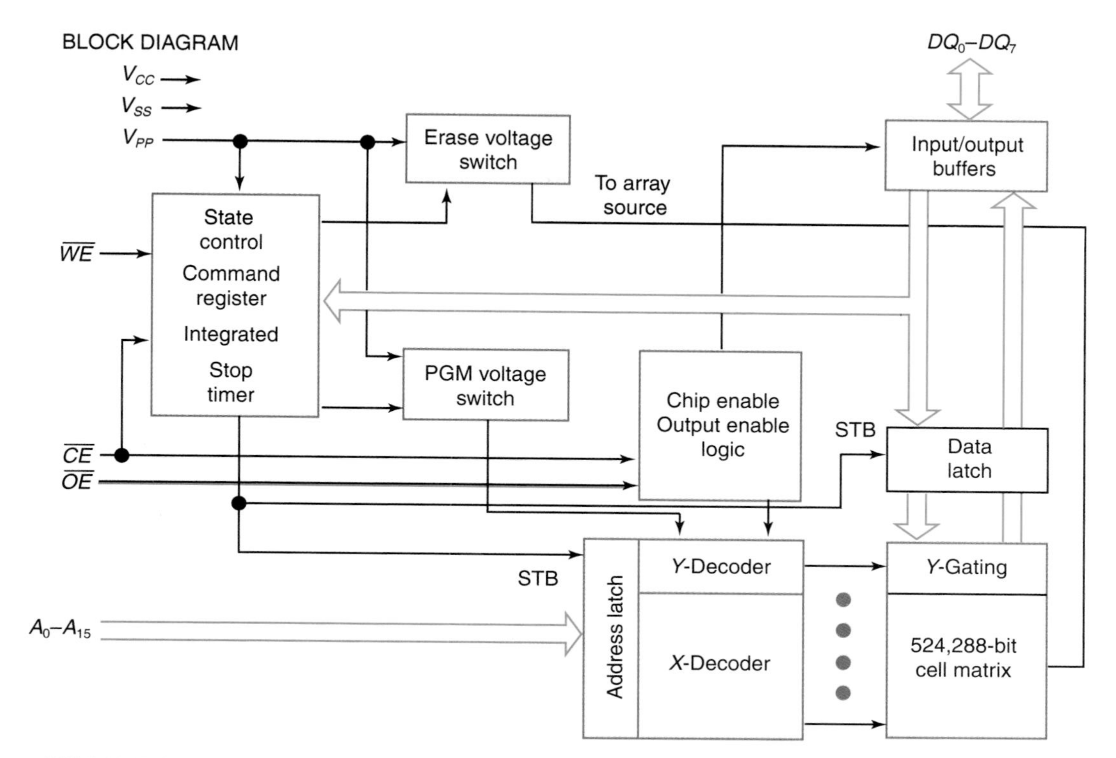

PIN DIRETIONS

Symbol	Type	Name and Function
A_0–A_{15}	INPUT	**ADDRESS INPUTS** for memory addresses. Addresses are internally latched during a write cycle.
DQ_0–DQ_7	INPUT/OUTPUT	**DATA INPUT/OUTPUT:** Inputs data during memory write cycles; outputs data during memory read cycles. The data pins are active high and float to tri-state OFF when the chip is deselected or the outputs are disabled. Data is internally latched during a write cycle.
$\overline{CE}$	INPUT	**CHIP ENABLE:** Activates the device's control logic, input buffers, decoders and sense amplifiers. $\overline{CE}$ is active low; $\overline{CE}$ high deselects the memory device and reduces power consumption to standby levels.
$\overline{OE}$	INPUT	**OUTPUT ENABLE:** Gates the device's output through the data buffers during a read cycle. $\overline{OE}$ is active low.
$\overline{WE}$	INPUT	**WRITE ENABLE:** Controls writes to the command register and the array. Write enable is active low. Addresses are latched on the falling edge and data is latched on the rising edge of the $\overline{WE}$ pulse. **Note:** With $V_{PP} \leq 6.5$ V, memory contents cannot be altered.
V_{PP}		**ERASE/PROGRAM POWER SUPPLY** for writing the command register, erasing the entire array, or programming bytes in the array.
V_{CC}		**DEVICE POWER SUPPLY** (5 V ± 10%)
V_{SS}		**GROUND**
NC		**NO INTERNAL CONNECTION** to device. Pin may be driven or left floating.

Fig. 11-16 Block diagram and pin descriptions for 28F512 512K CMOS Flash Memory. *(Courtesy of Intel Corporation.)*

28F512 flash memory IC

overview of the flash memory. The 28F512 flash memory reacts like a read-only memory when the V_{PP} erase/program power supply pin is LOW. When the V_{PP} pin goes HIGH (about +12 V), the memory can be quickly erased or programmed based on commands sent to the command register by the attached microprocessor or microcontroller. The 28F512 flash memory uses a 5-V supply to power the chip, but +12 V is required at the V_{PP} pin during erasing and programming.

In summary, flash EEPROMs or flash memories are an emerging memory technology which will become even more popular in the

Write process

future. Flash memories have many desirable characteristics including being nonvolatile, in-system rewritable (read/write), highly reliable and having low power consumption. Flash memories currently boast high densities (the single transistor storage cells are very tiny). Recent developments by Intel suggest that even higher densities will be available in flash memories. Intel has announced the StrataFlash™ memory which will store multiple bits of information in each cell. Intel is now producing both 32- and 64-Mbit StrataFlash™ memory chips. Figure 11-4 provides a good overview of the desirable characteristics of flash memory compared to other semiconductor memory devices.

Magnetic Core Memory

Magnetic-core memory

Read process

Historically, the tiny ferrite core used in *magnetic-core memories* was the first nonvolatile read/write memory. Magnetic-core memories were used before semiconductor memories were available as the central memories in computers. A highly magnified view of a single ferrite core is shown in Fig. 11-17(*a*). A typical core might measure 1/16 in. across.

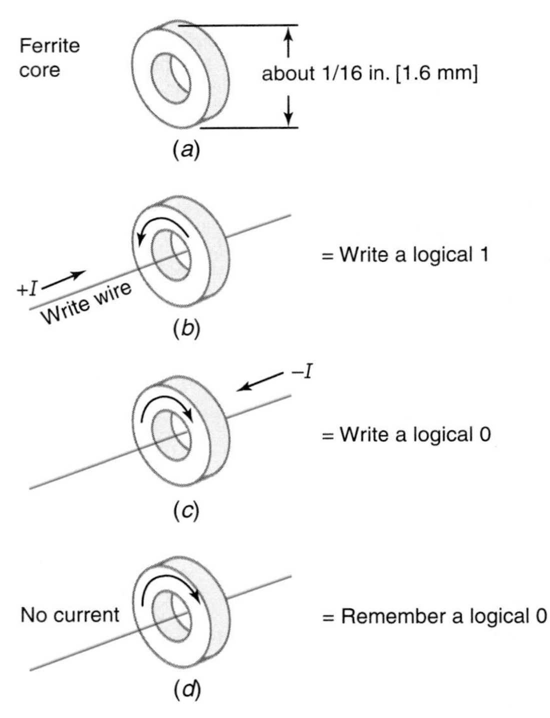

Fig. 11-17 Ferrite core. (*a*) Size. (*b*) Writing a 1. (*c*) Writing a 0. (*d*) Remembering a 0.

The ferrite core is used as a small magnet. Figure 11-17(*b*) shows a *write wire* threaded through the ferrite core. When current passes through the write wire in a given direction, the magnetic flux travels in a counterclockwise (ccw) direction. The magnetic flux direction is shown by an arrow on the core. We have defined the situation in Fig. 11-17(*b*) as a logical 1. In other words, ccw movement of the magnetic flux in the core means it is storing a logical 1.

Figure 11-17(*c*) shows the current being reversed. With the $-I$ pulse we find that the magnetic flux in the core reverses. The magnetic flux is now traveling in a clockwise (cw) direction in the core. The situation in Fig. 11-17(*c*) is defined as a logical 0. With no current flow in the write wire the ferrite core still is a magnet. Depending upon which direction the core was magnetized, it stores either a logical 0 or a logical 1. Figure 11-17(*d*) shows the ferrite core with no current flowing in the write wire. The core still has magnetic flux moving in a cw direction. We say the core is storing a logical 0.

The *read process* requires another wire threaded through the ferrite core. The added wire is called a *sense wire,* as shown in Fig. 11-18(*a*). To read the contents of the ferrite core, we apply a $-I$ pulse to the core, as shown in Fig. 11-18(*b*). Assuming the core is a logical 0, there will be *no change* in magnetic flux in the core. With no change in magnetic flux we have no current, or 0 A, *induced* in the sense wire. No induced pulse in the sense wire means that the core contains a 0.

In Fig. 11-18(*c*) we assume that the core contains a logical 1. This is shown with the ccw arrow on the core. To read the contents of the core, we apply a $-I$ pulse in Fig. 11-18(*d*). The magnetic flux changes direction from ccw to cw, as shown by the arrow. When the magnetic flux *changes direction,* a pulse is *induced* in the sense wire. The pulse in the sense wire tells us a logical 1 was stored in the core. Notice that the 1 which was stored in the ferrite core was destroyed by the read process. The core must be restored to the 1 state.

Magnetic-core memories have been replaced by much cheaper, more efficient, and lighter-weight semiconductor memories. Magnetic-core memories may still be used in applications where high temperatures and radiation are severe problems.

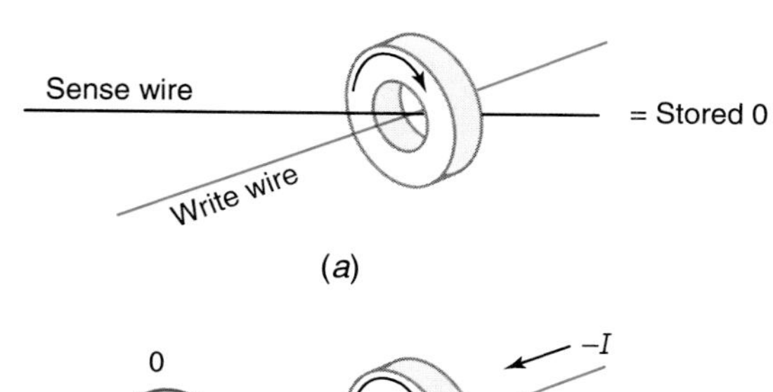

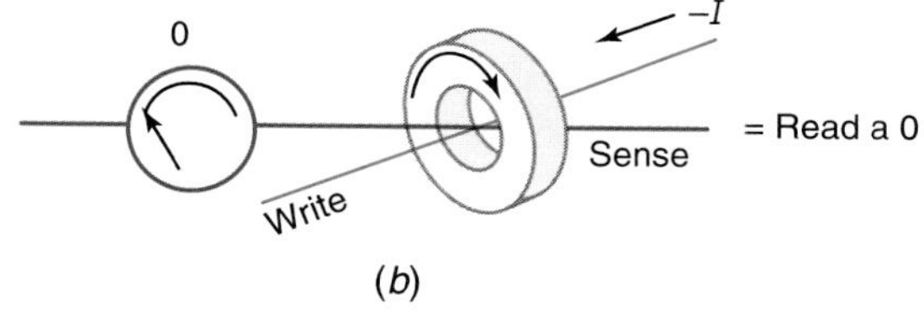

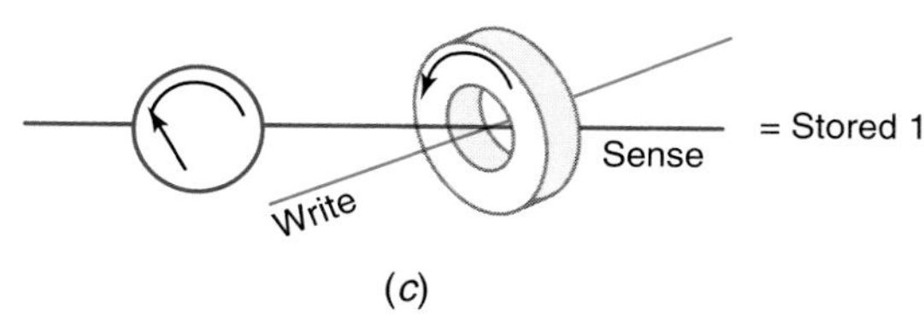

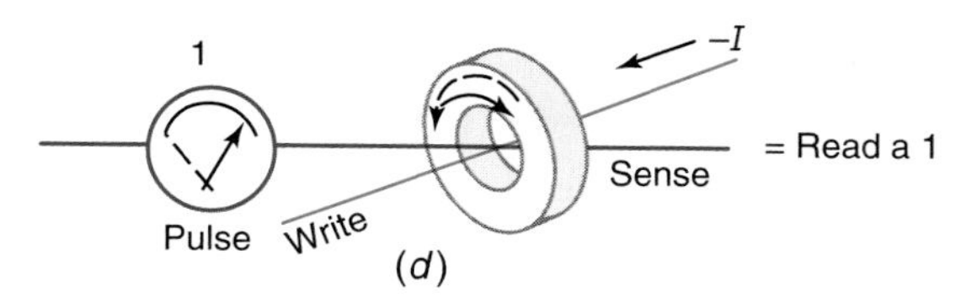

Fig. 11-18 Ferrite core. (*a*) Adding a sense wire. (*b*) Reading a 0. (*c*) Stored 1. (*d*) Reading a 1.

TEST

Supply the missing word or words in each statement.

41. The abbreviation "NVRAM" stands for ________.
42. Battery backup SRAMs commonly use a ________ (carbon-zinc, lithium) battery, which has a long life and maintains the data in the memory when power is lost.
43. A NVSRAM contains a static RAM array and a shadow ________ (EEPROM, ROM) memory array.
44. During power up using a NVSRAM, the ________ (recall, store) operation automatically occurs, duplicating all the data from the EEPROM into the SRAM memory array.
45. The newer ________ memory has the potential of becoming somewhat of a "universal" memory device that is finding wide application in laptop computers.
46. Refer to Fig. 11-16. The 28F512 flash memory can be erased/reprogrammed when +12 V is applied to the ________ pin of the IC.
47. Magnetic-core memory is based on the characteristics of the tiny ________.
48. The ________ (ROM, flash memory) is quite inexpensive, reliable, and is rewritable.
49. The ________ (SRAM, flash memory) is a read/write memory that is quite expensive and very fast.
50. Flash EEPROMs are a good substitute for a ROM but cannot replace DRAMs because they are not rewritable (T or F).

11-9 Memory Packaging

Four common methods of packaging semiconductor memory are shown in Fig. 11-19 on page 282. The familiar *dual in-line package (DIP)* is represented in Fig. 11-19(*a*). The DIP is the traditional IC package but it takes up a fair amount of surface area on a printed circuit board. Also, it is not a surface mount device, which is being used more often to save PC board space. The *single in-line package (SIP)*, illustrated in Fig. 11-19(*b*), sits the memory unit on edge to save PC board space. More memory density was possible with the *zig-zag in-line package (ZIP)*, sketched in Fig. 11-19(*c*). Finally, the *single in-line memory module (SIMM)*, shown in Fig. 11-19(*d*), has packed more memory in a smaller space. The SIMM is like a miniature PC board that holds an array of memory chips. Unlike the older DIPs, SIPs, and ZIPs, the SIMMs do not plug in but are held in contact with the socket by retainers. SIMMs have a vertical orientation to save PC board space and increase memory density. SIMMs are currently quite popular.

Another packaging method gaining popularity is the memory card. The *Personal Computer Memory Card International Association (PCMCIA)* defines standard physical and electrical characteristics of the PCMCIA card. This memory card is about the width and length of a standard credit card; its thickness varies (in four thicknesses) from about 3 to 19 mm. The memory card can house arrays of memory chips and other electronics using almost any type of memory device (PROM, DRAM, battery-backed SRAM, flash EEPROM, and so on). The flash memory card is very popular because of its very high density, low power consump-

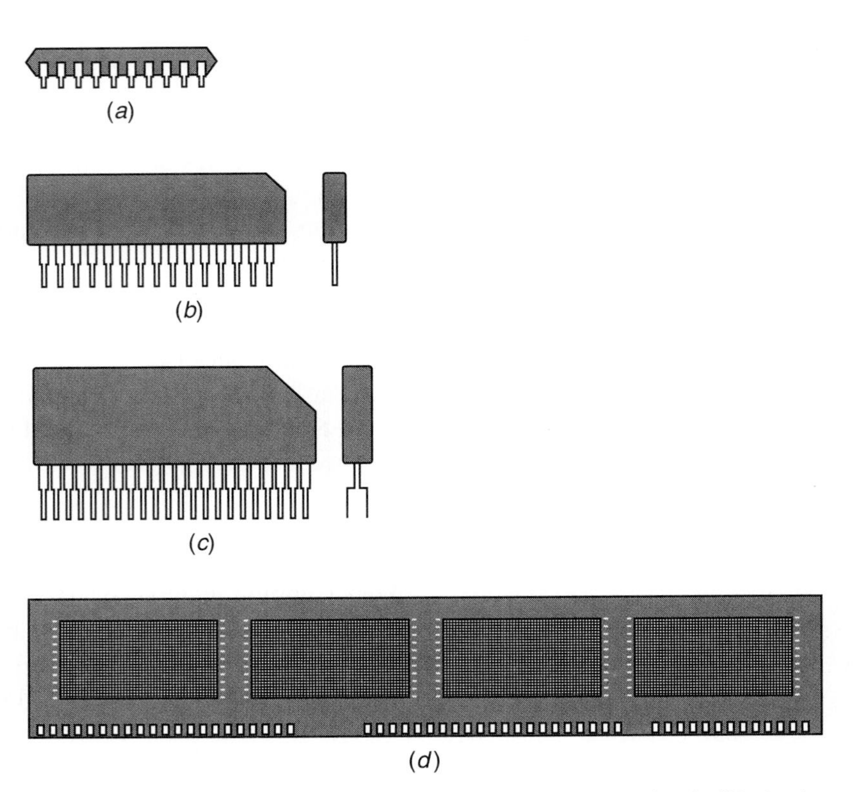

Fig. 11-19 Memory packages. (*a*) Dual in-line package (DIP). (*b*) Single in-line package (SIP). (*c*) Zig-zag in-line package (ZIP). (*d*) Single in-line memory module (SIMM).

tion, read/write capabilities, nonvolatility, and modest cost. A PCMCIA device containing flash memory would probably be referred to as a *flash memory card.* Flash memory cards are currently available with 20- or even 40-Mbyte storage capacities. The memory card enables a method of adding memory to laptop and palmtop computing devices or even a copier. Large-capacity flash memory cards can be used as *solid-state disk drives.* The standard PCMCIA memory card has an edge connector with 68 pins, which are assigned tasks (address lines, data lines, power supply, ground, and so forth). The 68-pin PCMCIA memory card allocates 26 pins for address lines, which allows addressing of a large memory (2^{26} = 64 Mbytes). You should show some caution when plugging in a memory card because other standards are used, such as the PCMCIA 88-pin interface, the Panasonic 34-pin interface, the Maxwell 36- or 38-pin interface, the Epson 40- or 50-pin interface, and others.

The disk drive (rigid or floppy disk) is an electromagnetic device that consumes much power and can mechanically wear, thereby causing problems. Disk drives are particularly vulnerable to shock, vibration, dust, and dirt. The *solid-state disk* using flash memory cards or similar devices that appear in portable computers and other equipment must be very small. They use very little power, and withstand shock and vibration. In a *solid-state computer,* the traditional DRAM and magnetic drive (floppy and/or rigid) combination would be replaced by some fast SRAM and flash memory. The SRAM/flash memory combination is very fast when transferring data from disk to RAM.

TEST

Answer the following questions.

51. The memory package in Fig. 11-19(*a*), which is the most traditional, is called the ________ (DIP, SIP, ZIP).
52. The memory package in Fig. 1-19(*d*) provides great density and is called the ________ (SIMM, SIP, ZIP).
53. The letters "SIMM" stand for what when referring to a memory package?
54. The letters "PCMCIA" stand for what when referring to a memory card?
55. A PCMCIA flash memory card is about

the size of a thick ________ (credit card, 5.25-in. floppy disk).

56. All memory cards follow the PCMCIA 68-pin standard for electrical connections and physical dimensions (T or F).

11-10 Computer Bulk Storage Devices

Generally, semiconductor memories are used for internal storage in most modern computers. Historically, magnetic-core memory serves the same purpose in old units. The computer's internal storage is also called *primary storage.* It is not possible to store all data inside the computer itself. For instance, it is neither necessary nor desirable to store last month's payroll information inside the computer after the checks are printed and cashed. Thus most data is stored outside the computer. External storage is also called *secondary storage.* Several methods are used to store information for immediate and future use by a computer. External storage devices are usually classified as either *mechanical, magnetic, optical,* or *semiconductor.*

Mechanical Devices

Mechanical bulk storage devices include the punched paper card and punched or perforated paper tape. The punched card was developed before 1900 by Herman Hollerith, who adapted them for use in the 1890 United States census. These cards commonly have holes punched in them to represent alphanumeric data. The code used is called the *Hollerith card code.* A typical Hollerith punched card is made of heavy paper and measures about 3.25 × 7.5 in. A common punched card can hold 80 characters. Punched paper cards are now obsolete.

Perforated paper tape is another method of mechanically storing data. The paper tape is a narrow strip of paper with holes punched across the tape at places selected according to a code. The paper tape can be stored on reels.

Magnetic Devices

Common magnetic bulk storage devices are the magnetic tape, the magnetic disk, and the magnetic drum. Each device operates much like a common tape recorder. Information is recorded (stored) on the magnetic material. Information can also be read from the magnetic material. Magnetic tape has been widely used for many years as a secondary storage medium. It is still very popular for backing up data because it is quite inexpensive. The main disadvantage of magnetic tapes is that they are sequential-access devices. That is, to find information on the tape you must search through the tape sequentially, which makes the access time long.

Magnetic disks have become particularly popular in recent years. Magnetic disks are random-access devices, which means that any data can be accessed easily and in a short time. Magnetic disks are manufactured in both rigid and floppy (flexible) disk form. The floppy disk is an extremely popular form of secondary storage used by most microcomputers. The rigid or hard disk is more expensive than the floppy disk.

Hard Disks

The hard disk or rigid disk drive is currently the most important bulk storage memory device used on modern computer systems. One of the original sealed dust-free hard-disk drives was developed by IBM and was referred to as a *Winchester drive* (after famous 30-30 rifle—30 Mbytes with 30 msec access time). The hard disk has proved to be reliable, fast, and today has a very large storage capacity. A picture of a modern hard-disk drive by Seagate Technology is reproduced in Fig. 11-20.

Fig. 11-20 A high-speed, high-capacity 4.5-Gbyte Cheetah hard-disk drive by Seagate. *(Courtesy of Seagate Technology.)*

Computer bulk storage devices

Magnetic disks

Primary storage

Floppy disk

Rigid or hard disk

Secondary storage

Mechanical bulk storage

Hollerith card code

Magnetic bulk storage

Did You Know?

PCs Increase Portability. Smaller seems to always be the goal in the ever-developing computer industry. With the soaring popularity of powerful laptop computers, manufacturers are developing even smaller computers. One of the first "palmtop" computers is the Phenom, produced by LG Electronics. This tiny computer weighs 0.71 pounds without batteries and measures 6½ × 3½ × 1 inches.

The cover has been removed from this normally sealed unit to expose four 3.5-in. rigid disks (called *platters*) made of aluminum, glass, or ceramic. The platters are probably coated with a *thin-film medium,* which is a microscopic layer of metal bonded to the disk. The hard drive featured in Fig. 11-20 has eight read/write heads (only one is visible), one on each side of the four platters. When the platters spin, the read/write heads float just above the surface of the disk. The read/write arm pivots to locate a specific circular track on the surface of the platters. The spindle speed on many hard drives is 3600 rpm. The Seagate Cheetah drive pictured in Fig. 11-20 has the extremely high spindle speed of about 10,000 rpm. The higher spindle speed allows the read/write heads to locate data more quickly. The high performance hard-disk drive shown in Fig. 11-20 measures about 1-in. thick, 4-in. wide, and 5.75-in. long. This small package has a storage capacity of 4.5 Gbytes (a Gbyte (*gigabyte*) is one billion bytes). A hard-disk drive with more platters would have an even greater storage capacity. Some larger hard-disk drives have capacities of more than 20 Gbytes.

The specification sheet for a hard-disk drive gives information such as the total storage capacity, number and size of platters, number or read/write heads, average seek time (read/write), average latency, spindle speed, physical dimensions, power requirements, and operating temperatures. The organization of the data on the disks might be given as the number of sectors (a sector commonly holds 512 bytes of data plus other information such as an address), number of tracks (concentric circles of data), or number of cylinders (like the number of tracks but three-dimensional, including both sides of all of the platters). Some of these specifications might look like the following for the Seagate Cheetah hard-disk drive pictured in Fig. 11-20.

Capacity	4.55 Gbytes
Average seek, read/write (msec)	7.5/8.5
Average latency (msec)	2.99
Spindle speed (rpm)	10,033
Discs/heads	4/8
Sectors per drive (512 bytes/sector)	8,890,080
Cylinders	6,526
Power requirements	
+12VDC +/−5%	
(amps typ)	0.82
(amps max)	2.0
+5VDC +/−5%	
(amps typ)	0.74
power (idle watts)	12.0
Operating temperature (°C)	5 to 50
Height (in./mm)	1.0/25.4
Width (in./mm)	4.0/101.6
Depth (in./mm)	5.75/146.1
Weight (lbs./kgs)	1.33/0.60

Floppy Disks

Currently the 3.5-in. floppy disk is one of the most common bulk storage devices used for long-term storage and transporting of data. Commonly used floppy disks or diskettes come in either 5.25- or 3.5-in. versions. The 3.5-in. floppy disk has become the standard. A diagram of a common 3.5-in. floppy disk is shown in Fig. 11-21(*a*) on page 286. The drawing shows the bottom view of the disk. It labels the rigid plastic case as well as the sliding metal shutter, both of which help protect the delicate floppy disk housed inside. The sliding metal shutter is shown open, exposing the floppy disk. When released, the cover snaps back to cover the floppy disk inside. The read/write heads of the disk drive can store or retrieve data from both sides of the floppy disk. The center has a metal hub attached to the bottom of the floppy for gripping the disk. The rectangular hole in the hub is an index hole used by some disk drives for timing purposes. The write-protect notch is located at the lower right in Fig. 11-21(*a*). If the write-protect hole is closed (as shown in the drawing), you can both write to or read from the disk. If the hole is open (move plastic slider down), the drive can only read the disk; we say that the disk is "write protected."

The most common 3.5-in. high-density (HD) floppy disk can store up to 2 Mbytes of data. A common format allows 1.44 Mbytes of data to be stored on a high-density, double-sided (HD DS) 3.5-in. diskette. Older 3.5-in. floppy disks can store much less data (400-, 720-, or 800 kbytes). Your floppy disk drive can tell which 3.5-in. floppy is inserted by the hole or no hole at the lower left of the disk. This is illustrated in Fig. 11-21(*a*). Many older 3.5-in. floppy disks are called double density (DD). As a practical matter, you will have trouble using a new HD disk in an older disk drive that was designed for older 3.5-in. DD floppy disks.

Data on a floppy disk is organized during the formatting process. The organization can be visualized by looking at the 3.5-in. floppy disk in Fig. 11-21(*b*). Notice that the disk is organized by *tracks* and *sectors*. The 3.5-in. HD disk is commonly organized in concentric circles or 80 tracks; each track then divides into 18 sectors. Remember that both sides of the disk are organized in tracks and sectors. Each sector can hold 512 bytes of data plus other information, such as an address, as illustrated in Fig. 11-21(*b*). An older DD 3.5-in. floppy disk is formatted with 80 tracks, both sides, but with only 9 sectors per track. The track and sector organization is also used by hard disks.

In summary, the 3.5-in floppy disk may be the most recognized storage device in the world. It is a cheap read/write bulk storage device that can be transported and read from or written to by almost any compatible computer. Its storage capacity is approximately 1.4 Mbytes. If more storage is needed, Imation's SuperDisk™, pictured in Fig. 11-3, with 120 Mbytes is an alternative. It is expected that the 3.5-in. floppy disk and its variations will be used for decades.

Optical Devices

Optical disks

An emerging bulk-storage technique is the *optical disk.* Optical disk technology is popular because it is reliable, high capacity, transportable, and very inexpensive. Optical disks are available in three types: (1) read-only or CD-ROM, (2) write-once, read-many (WORM), and (3) read/write. The *CD-ROM* is currently the standard read-only optical device. The CD-ROM is extremely popular in home, school, and business computer systems. Many computer applications (word processors, spreadsheets, data base, CAD, and so on) come in CD-ROM format instead of the older, lower-capacity, floppy disk format. The removable CD-ROM was introduced and pictured previously in Fig. 11-1.

The *WORM (write-once, read-many) optical disk* can be written to once and then read from many times. The WORM optical disk is sometimes referred to as a *CD-R* or *CD recordable disk.* The CD recordable disk looks like a 4.75-in. CD-ROM and can be read on most standard CD-ROM drives. A special drive (CD-R or CD-RW) is needed to record data on the CD recordable disk or WORM optical disk. Once information is recorded on the CD recordable disk, it cannot be changed. The CD recordable disk has a capacity of about 500 to 650 Mbytes. Using your own computer you can record text, audio, graphics, or video on the CD recordable disk. Like the CD-ROM, the CD recordable disk is permanent, reliable, and transportable, features high capacity, and is an inexpensive form of data storage.

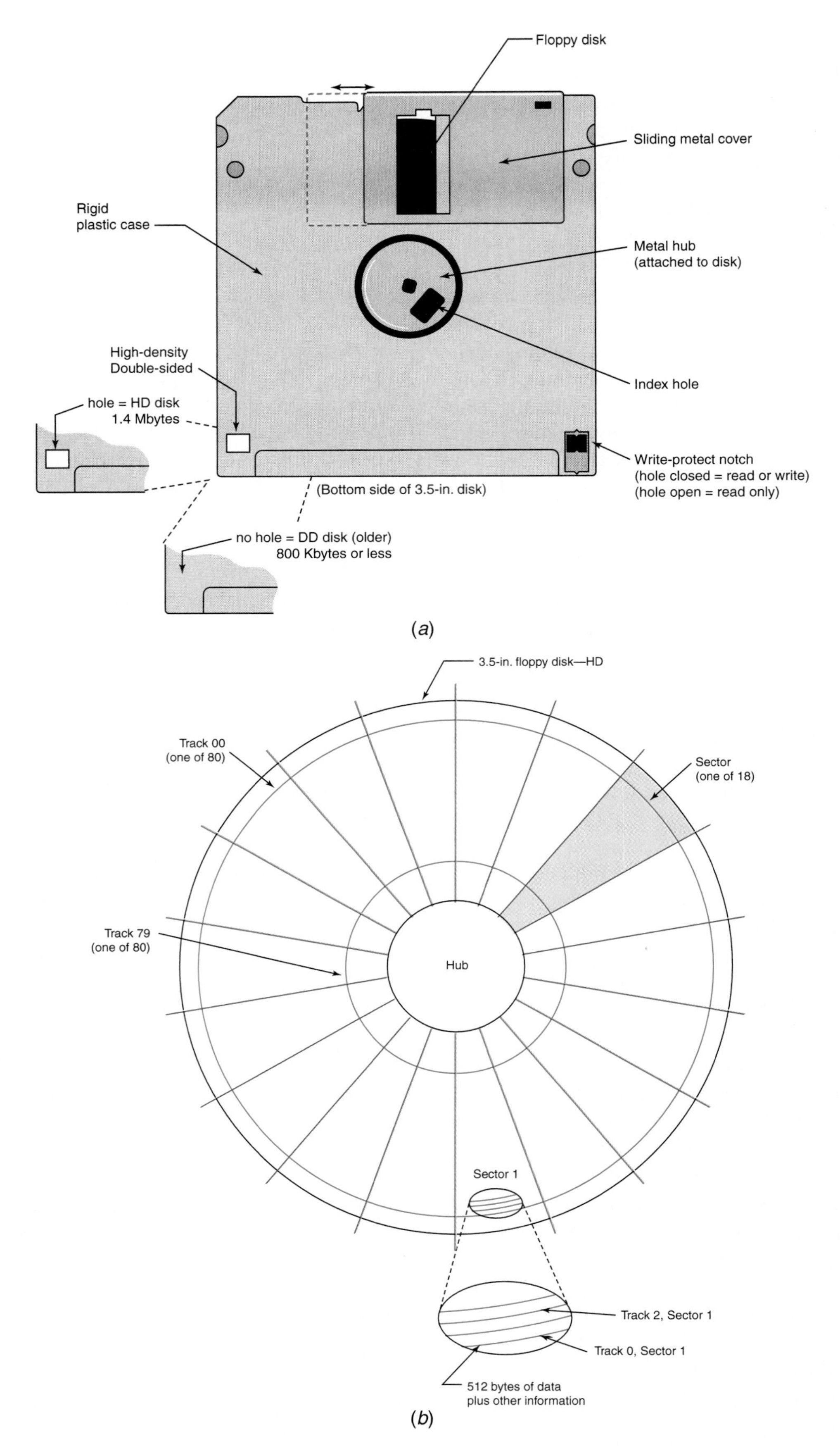

Fig. 11-21 A 3.5-in. floppy disk. (*a*) Physical characteristics. (*b*) Typical formatting into 80 tracks and 18 sectors.

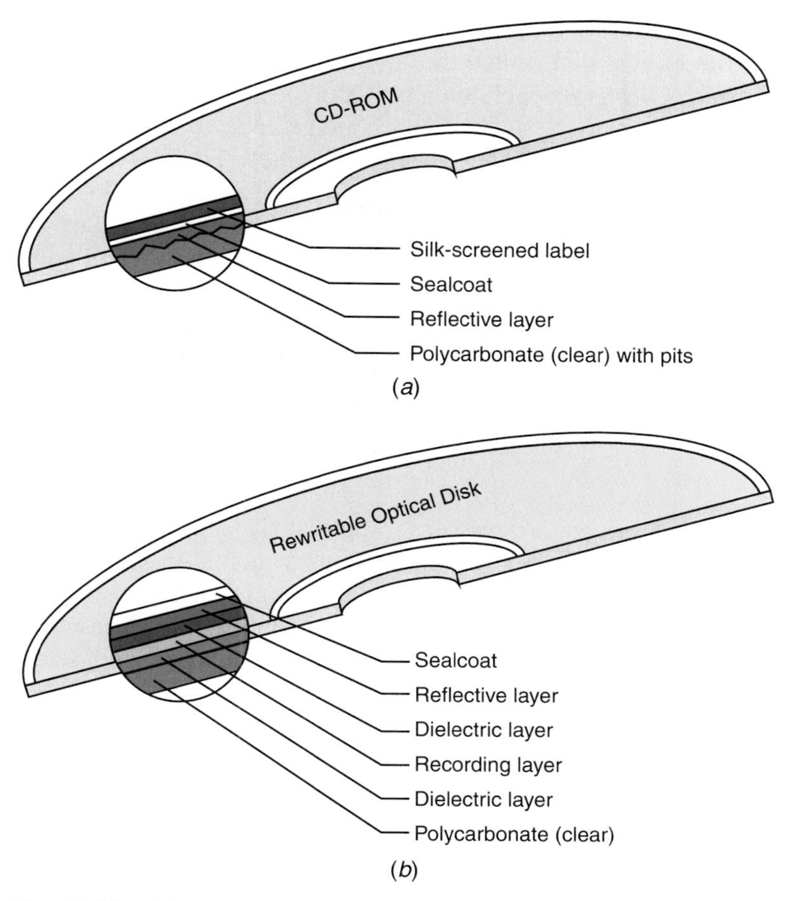

Fig. 11-22 (*a*) CD-ROM construction. (*b*) PD-rewritable optical disk construction.

Read/write or *rewritable optical disks* are manufactured in several diameters including CD size (4.75 in.) and 5.25 in. The construction of two popular optical disks is compared in Fig. 11-22. CD-ROM construction is sketched in Fig. 11-22(*a*). The 4.75-in. clear polycarbonate disk has pits (the data) molded into its upper surface. A reflective layer is coated over the pits with a protective sealcoat on top. The label is then silkscreened on the top of the CD-ROM. The light from a laser is focused from the bottom of the CD-ROM. The pits and lands (no pits) reflect light differently while a photodetector observes the reflections and interprets them as logical 0s and 1s.

A cross-section of a rewritable optical disk is sketched in Fig. 11-22(*b*). The base material is clear polycarbonate but several layers have been added unlike the CD-ROM. Sandwiched between the polycarbonate and reflective dielectric layers is the *recording layer.* During the *write process* on the rewritable optical disk, the laser changes the recording layer from its amorphous state to a crystalline state. Whereas the recording layer has low reflectivity in its amorphous state, it has high reflectivity in its crystalline state. During the *read process* a low-power laser focuses on the dull and shiny areas, and a photodetector observes the reflections and interprets them as logical 0s and 1s. The *second write process* is slightly different in that a higher power laser heats the crystalline areas; as they cool, they return to their amorphous state. The write process then proceeds as described previously. The read/write disk we described is sometimes called a *PD rewritable optical disk.* The PD part of the title refers to the "phase change" in the recording layer (from amorphous-to-crystalline or crystalline-to-amorphous). These optical read/write disks might be called *phase-change devices.* The PD rewritable optical disk is housed in a protective case something like that of the 3.5-in floppy disk. PD rewritable optical disks use a special disk drive to write and read the data. Rewritable disks in protective cases cannot be read by a standard CD-ROM drive. A newer PD read/write optical disk called the compact disk-erasable (CD-E) is readable on a standard CD-ROM drive.

Access Time

Various bulk-storage devices are compared on an access time/storage capacity basis in Fig. 11-23. Whereas access time is given in seconds, storage capacity is graphed in Mbytes. *Access time* is the time in seconds it takes to retrieve a piece of data from memory. The highest performance (shortest access time) device on the chart is the flash memory card. The disadvantage of the flash memory card, however, is the high cost. Mechanical methods of storing data (paper tape and punched cards) have the lowest performance and, as such, have been phased out for most applications. Magnetic tape and digital audiotape (DAT) have poor access time but have very large storage capacity at low cost. Hard disks are extremely popular because of their ease of use, large storage capacities, good access times, reasonable cost, and universal usage. Floppy disks continue to be popular because they are easy to use, are available for a very low cost, are portable, have medium access times, and are used universally. Optical storage shows promise because of high storage capacities, low cost, and moderate access times. The CD-ROM is becoming more universally used on home, school, and business PCs. Flash cards are an emerging technology, providing excellent access times and good storage capacities, but they are still a bit expensive.

Rewritable magneto-optical disk

Read/write optical disks have very large storage capacities. The read/write *magneto-optical* disk drive uses a laser and a coil of wire to write to, read from, and erase the metal-coated optical disk. One popular magneto-optical disk drive used for some microcomputer systems has a storage capacity of 120 Mbytes (120 million bytes) on a removable 3.5-in. optical disk. These *rewritable magneto-optical disks* look much like the 3.5-in. floppy disk except they are thicker and house an optical disk.

Flash memory modules

Flash EEPROM semiconductor memory devices may replace floppy and hard disks in laptop computers. Flash memory modules might be housed in a case that looks like the ROM cartridge used on some home video games.

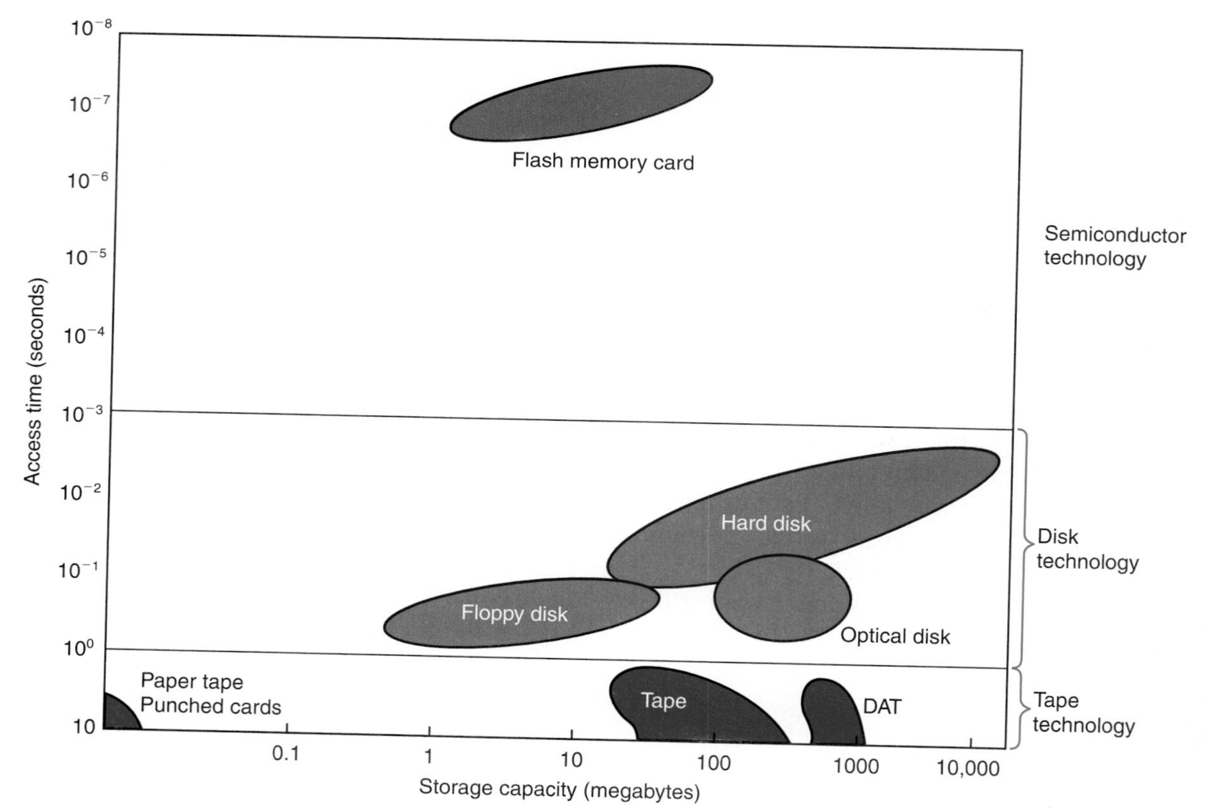

Fig. 11-23 Comparison of several bulk-storage devices.

They will act something like the ROM cartridges except flash memories are nonvolatile read/write devices.

TEST

Answer the following questions.

57. External computer bulk storage devices can be classified as mechanical, _________, _________, or _________.
58. List several computer bulk storage devices.
59. Magnetic disks are manufactured in both floppy (flexible) and _________ (amorphous, rigid) form.
60. The most important bulk storage device used in almost all computer systems is the _________ (magneto-optical disk drive, hard-disk drive).
61. The _________ (floppy disk, rigid disk) is faster and has much greater storage capacity.
62. Refer to Fig. 11-20. The rigid disks are commonly called _________ (platters, spindles) on this disk drive.
63. A gigabyte equals _________ bytes.
64. Refer to Fig. 11-20. This disk drive has four platters and _________ (1, 4, 8) read/write head(s).
65. Refer to Fig. 11-20. This modern hard-disk drive has a storage capacity of about _________ (30 Mbytes, 4.5 Gbytes).
66. A Winchester drive was an early name for a(n) _________ (optical-disk drive, hard-disk drive) developed by IBM.
67. The _________ (read-only optical disk, rigid disk) is reliable, high capacity, transportable, and inexpensive.
68. A CD-ROM is classified as a(n) _________ (magnetic, optical) bulk storage device.
69. The letters "WORM" stand for what in reference to an optical disk?
70. The PD rewritable optical disk is still in the research stages and is not yet available (T or F).
71. The letters "PD" in *PD rewritable optical disk* stand for phase-change device (T or F).
72. The rewritable magneto-optical disk is a removable disk with a _________ (small, large) storage capacity.

Chapter 11 Review

Summary

1. The availability of memory and data storage is why many electronic devices are designed using digital instead of analog circuitry.
2. Internal memory devices in a computer are usually in the form of RAM, ROM, and NVRAM. The CPU also contains other smaller memory devices like registers, counters, and latches.
3. External bulk storage devices are commonly classified as to their basic technology: magnetic, mechanical, optical, or semiconductor.
4. Bulk storage devices include floppy disks, hard disks, magnetic tape, CD-ROMs, and perhaps flash memory cards.
5. Semiconductor storage cells are classified as SRAM, DRAM, ROM, EPROM, EEPROM, and flash EEPROM. Some important characteristics of semiconductor memory devices are density, reliability, cost, power consumption, read-only or read/write, and nonvolatile/volatile.
6. A RAM is a semiconductor read/write random-access memory device. RAM comes in two forms: SRAM (static RAM) and DRAM (dynamic RAM).
7. A ROM is considered a permanent storage unit that has a read-only characteristic.
8. A PROM operates like a ROM. PROMs are one-time write devices. PROMs come in many varieties, generally known as EPROM, EEPROM, and NVSRAM. These "E" prefixed PROMs can be erased electrically or by shining ultraviolet light through a special transparent "window" on the top of the IC.
9. The write process stores information in the memory. The read, or sense, process detects the contents of the memory cell.
10. A NVRAM is a nonvolatile RAM that contains a RAM and an EEPROM. Nonvolatile RAM memory is also implemented in microcomputers using RAMs with battery backup.
11. A flash memory is a newer low-cost EEPROM that can be quickly erased and reprogrammed while in the circuit. Flash memory chips can be packaged as removable flash memory cards.
12. Magnetic core memory is an obsolete read/write memory based upon the magnetic characteristics of the ferrite core.
13. Computer external storage methods include magnetic tapes, floppy disks, rigid disks, optical disks (such as CD-ROMs, WORMs, and PD rewritable disks), and flash memory cards.
14. Microcomputers typically use both RAM, ROM, and NVRAM for internal main memory. Floppy disks, rigid disks, and CD-ROMs are the most common popular bulk storage devices used on smaller computer systems.
15. A byte is an 8-bit word. One Gbyte of memory means one billion bytes (actually 2^{30}). One Mbyte of memory means one million bytes (actually 2^{20}). One Kbyte of memory means one thousand (actually 2^{10} or 1024) bytes of memory.
16. DIP, SIP, ZIP, and SIMM are common memory packages. Memory cards are commonly packaged as a PCMCIA (Personal Computer Memory Card International Association) device.
17. The flash memory holds promise for use as a "solid-state drive" in tiny portable computer equipment.

Answer the following questions.

11-1. The most important characteristic of a digital system compared to an analog system is its ________ (ability to store data, ease of interfacing with real-world events).

11-2. The CD-ROM is an example of a bulk storage device using ________ (mechanical, optical) technology.

11-3. The ________ (CPU, RAM) is the section of the computer system that contains the arithmetic, logic, and control sections and is the focus of most data transfers.

11-4. Three common internal semiconductor memory devices used in most computer systems are the ________ (floppy, rigid disk and read/write, RAM, ROM, and NVRAM).

11-5. Semiconductor RAM is a ________ (read-only, read/write) type memory device.

11-6. Semiconductor ROM is a ________ (read-only, read/write) type memory device.

11-7. Semiconductor RAM is a ________ (nonvolatile, volatile) memory device.

11-8. Semiconductor ROM is a ________ (nonvolatile, volatile) memory device.

11-9. Semiconductor NVRAM is a ________ (read-only, read/write) memory device.

11-10. Name the three buses used in a typical personal computer system.

11-11. The ________ (address, data) bus in a typical PC system is a one-way bus used for selecting a specific memory location or peripheral.

11-12. The most common floppy used in today's PC systems is described as ________ (HD 3.5-in. disks, DD 5.25-in. disks).

11-13. Both floppy and rigid disks use ________ (magnetic, optical) principles for storing data.

11-14. Compared to the typical floppy disk, the hard-disk drive can store much ________ (less, more) data.

11-15. The SuperDisk™ is a ________ (3.5-in. magnetic disk, 4.75-in. optical disk) that can store 120 Mbytes of data.

11-16. The CD-ROM is an optical memory device that can store about ________ (30 Mbytes, 650 Mbytes) of data.

11-17. What semiconductor memory device may become a "solid-state drive" in small portable computers and personal assistants?

11-18. List at least five semiconductor memory devices.

11-19. Flash memory is an emerging technology and, as such, is very expensive; so far the flash memories have very small storage capacities (T or F).

11-20. Press the store key on a calculator. This activates the ________ (read, write) process in the memory section.

11-21. Press the recall key on a calculator. This activates the ________ (read, write) process in the memory section.

11-22. The following abbreviations stand for what?

a. RAM	d. EPROM
b. ROM	e. EEPROM
c. PROM	f. NVRAM

11-23. A ________ (RAM, ROM) has both the read and write capability.

11-24. A ________ (RAM, ROM) is a permanent memory.

11-25. A ________ (RAM, PROM) is a nonvolatile memory.

11-26. A ________ (RAM, ROM) has a read/write input control.

11-27. A ________ (RAM, ROM) has data inputs.

11-28. A 32×8 memory can hold ________ words. Each word is ________ bits long.

11-29. List at least three advantages of semiconductor memories.

11-30. A ________ (RAM, ROM) can be erased easily.

11-31. A ________ (flash memory, ROM) can be quickly erased and reprogrammed.

11-32. A ________ (PROM, NVSRAM) contains both a RAM and a shadow EEPROM.

11-33. A ________ (flash memory, UV EPROM) can be erased electrically in a very short time.

11-34. A ________ (flash memory, ROM) is a read/write nonvolatile memory device.

11-35. A(n) ________ (EEPROM, UV EPROM) can be erased and reprogrammed byte-by-byte without being removed from the equipment.

11-36. A ________ (DRAM, SRAM with battery backup) is a nonvolatile read/write memory device.

11-37. Memory or data storage is much easier to implement using ________ (analog, digital) electronic circuitry.

11-38. The 2114 IC is a ________ (dynamic, static) RAM.

11-39. The access time of the TTL 7489 RAM is ________ (faster, slower) than that of the MOS 2114 RAM.

11-40. Refer to Fig. 11-9(*b*). If the input to the decoder is binary 0010, the output from the ROM will be ________ in Gray code.

11-41. Computer programs that are permanently held in ROM are called ________.

11-42. Refer to Table 11-4. Which 27XXX series EPROM could be used to implement a 16K ROM in a microcomputer?

11-43. Refer to Fig. 11-16. The 16-address line inputs to the 28F512 flash memory IC can address ________ (number) words, each 8 bits wide.

11-44. Refer to Fig. 11-16. To erase and/or reprogram the 28F512 flash memory IC, the ________ (CE, V_{PP}) input must be pulled to a high of about ________ (+5, +12) volts.

11-45. The ferrite core is the memory cell in a ________ -type memory.

11-46. A ________ (magnetic core, RAM) unit has a nonvolatile memory.

11-47. List at least five common types of computer bulk (external) storage.

11-48. Magnetic disks have much ________ (faster, slower) access time than magnetic tapes.

11-49. The ________ (flash, hard) disk is one of the most popular bulk-storage methods used on current microcomputers.

11-50. Short access time in a memory device is a measure of ________ (good, poor) performance.

11-51. A microcomputer with 1 MB of memory has one million (2^{20}) ________ (bits, bytes) of storage.

Critical Thinking Questions

11-1. Draw a diagram of how a 32 × 8 memory might look in table form. The table will be similar to the one in Fig. 11-5.

11-2. List at least three uses of read-only memories.

11-3. List several common names for the NVRAM.

11-4. Why are many microcomputer systems equipped with both floppy- and hard-disk drives?

11-5. List several precautions you should take when handling floppy disks.

11-6. If a computer has 4 Mbytes of RAM, how many bytes of read/write memory does it contain?

11-7. Explain the difference between software and firmware.

11-8. Explain the difference between a mask-programmable ROM and a fusible-link PROM.

11-9. Explain the difference between a UV EPROM and an EEPROM.

11-10. Why have hard disks become almost standard bulk storage devices on most microcomputers?

11-11. List several types of nonvolatile read/write memory.

Answers to Tests

1. CPU or central processing unit
2. address bus, control bus
3. magnetic, optical, semiconductor, or mechanical
4. floppy-disk drive, hard-(rigid) disk drive, CD-ROM
5. a. random-access memory
   b. read-only memory
   c. electrically programmable read-only memory
   d. electrically erasable programmable read-only memory
   e. static RAM
   f. dynamic RAM
6. flash memory
7. random-access memory
8. writing
9. reading
10. read/write
11. 16 × 4 bit
12. volatile, off
13. RAM
14. 16, 4
15. write, written into, 15
16. dynamic
17. 4096, 4
18. static
19. inputs
20. binary-to-Gray
21. 1100, Gray, 1000
22. lose its program and have to be reprogrammed
23. read-only memory
24. nonvolatile
25. firmware
26. manufacturer
27. is not
28. 0010
29. 1101
30. high-
31. remain in

32. 177
33. 14
34. manufacturer
35. ROM
36. programmable read-only memory
37. erasable programmable read-only memory
38. electrically erasable programmable read-only memory
39. electrically, ultraviolet
40. 524,288, 65,536
41. nonvolatile RAM
42. lithium
43. EEPROM
44. recall
45. flash
46. V_{PP}
47. ferrite core
48. flash memory
49. SRAM
50. F
51. DIP
52. SIMM
53. single in-line memory module
54. Personal Computer Memory Card International Association
55. credit card
56. F
57. magnetic, optical, semiconductor
58. magnetic tapes, magnetic disks (floppy and rigid), magnetic drum, optical disks, flash memory or cards, paper tapes, paper punched cards
59. rigid
60. hard-disk drive
61. rigid disk
62. platters
63. one billion
64. 8
65. 4.5 Gbytes
66. hard-disk drive
67. read-only optical disk
68. optical
69. write-once, read-many
70. F
71. T
72. large

Digital Systems

Chapter Objectives

This chapter will help you to:

1. *Identify* six elements found in most systems.
2. *Describe* the internal organization of a typical calculator.
3. *Diagram* the general organization of a computer and a microcomputer and *detail* the execution of a program.
4. *Analyze* the operation of a simple microcomputer address decoding system.
5. *Discuss* several aspects of both serial and parallel data transmission.
6. *Answer* selected questions about error-detection and correction techniques.
7. *Describe* the operation of a simple adder/subtractor system.
8. *Analyze* the operation of a digital clock system including display multiplexing.
9. *Analyze* the operation of a digital frequency counter system.
10. *Analyze* the operation of an LCD timer system.
11. *Analyze* the operation of several digital dice game circuits.
12. *Characterize* and *answer* selected questions about a programmable logic controller (PLC).
13. *Convert* relay schematics to relay logic diagrams, and to logic gate diagrams and Boolean expressions.
14. *Summarize* the differences between microprocessors and microcontrollers.
15. *List* some applications of a microcontroller.
16. *Analyze* the operation of a microcontroller driving an LED display circuit.

Most digital devices we use every day are *digital systems,* such as handheld calculators, digital wristwatches, or even digital computers. Calculators, digital clocks, and computers are assemblies of *subsystems.* Typical subsystems might be adder/subtractors, counters, shift registers, RAMs, ROMs, encoders, decoders, data selectors, clocks, and display decoder/drivers. You have already used most of these subsystems. This chapter discusses various digital systems and how they transmit data. A digital system is formed by the proper assembly of digital subsystems.

Subsystems

Elements of a system

Input

Output

Processing

Control

Transmission

Storage

12-1 Elements of a System

Most mechanical, chemical, fluid, and electrical systems have certain features in common. Systems have an *input* and an *output* for their product, power, or information. Systems also act on the product, power, or information; this is called *processing.* The entire system is organized and its operation directed by a *control* function. The *transmission* function transmits products, power, or information. More complicated systems also contain a *storage* function. Figure 12-1 illustrates the overall organization of a system. Look carefully and you can see that this diagram is general enough to apply to nearly any system, whether it is transportation, fluid, school, or electronic. The transmission from device to device is shown by the colored

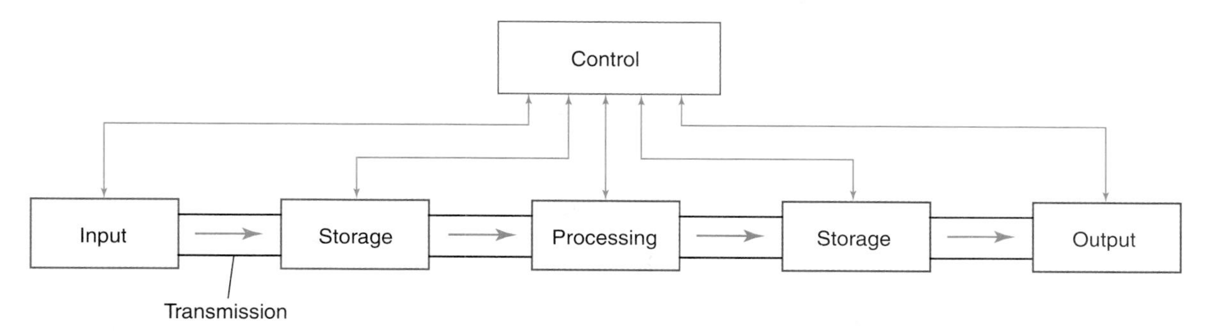

Fig. 12-1 The elements of a system.

lines and arrows. Notice that the data or whatever is being transferred always moves in one direction. It is common to use double arrows on the control lines to show that the control unit is directing the operation of the system as well as receiving feedback from the system.

The general system shown in Fig. 12-1 will help explain several digital systems in this chapter. In a digital system we shall be dealing with transmitting data (usually numbers/codes).

TEST

Answer the following questions.

1. There is a two-way path from the _________ section of a system to all other parts.
2. The keyboard of a microcomputer is classified as what part of the system?

12-2 A Digital System on an IC

We have learned that all digital systems can be wired from individual AND and OR gates and inverters. We have also learned that manufacturers produce subsystems on a single IC (counters, registers, and so on). We shall find that manufacturers have gone even a step further: some ICs contain nearly an entire digital system.

The least complex digital integrated circuits are classified as *small-scale integrations (SSI)*. An SSI contains circuit complexity of less than 12 gates or circuitry of similar complexity. Small-scale integrations include the logic gate and flip-flop ICs you have used.

A *medium-scale integration (MSI)* has a complexity of from 12 to 99 gates. ICs that are classified as MSIs belong to the small subsystem group. Typical examples include adders, registers, comparators, code converters, counters, data selectors/multiplexers, and small RAMs. Most of the ICs you have studied and used so far have been either SSIs or MSIs.

A *large-scale integration (LSI)* has the complexity of from 100 to 9999 gates. A major subsystem or a simple digital system is fabricated on a single chip. Examples of LSI chips are digital clocks, calculators, microprocessors, ROMs, RAMs, PROMs, EPROMs, and flash memories.

A *very large-scale integration (VLSI)* has the complexity of from 10,000 to 99,999 gates. VLSI ICs are usually digital systems on a chip. The term "chip" refers to the single silicon wafer (perhaps ¼-in. square) that contains all the electronic circuitry in an IC. Large memory chips and advanced microprocessors are examples of VLSI ICs.

An *ultra large-scale integration (ULSI)* is the next higher level of circuit complexity with more than 100,000 gates on a single chip. Various manufacturers tend to define SSI, MSI, LSI, VLSI, and ULSI differently.

In the 1960s, families of digital ICs were being developed using SSI and MSI technology. Late in the 1960s, large-scale integration technology developed many specialized ICs. Higher production LSIs included single chip calculators and memories. After the development of calculator chips, the architecture of a computer was designed into a single chip called the microprocessor. The *microprocessor* forms the CPU of a computer system. Improvements in CPU design and chip manufacturing have produced the latest generation of microprocessors that contain the equivalent of millions of transistors. In the 1980s, manufacturers combined some of the separate sections of a computer system (CPU, RAM, ROM, and input/output) into a single inexpensive IC.

Large-scale integration (LSI)

Very large-scale integration (VLSI)

Chip

Ultra large-scale integration (ULSI)

Calculator chip

Small-scale integration (SSI)

Microprocessor

Medium-scale integration (MSI)

These "tiny computers on a chip" were used mainly for control purposes and were not used in general-purpose computers. These inexpensive computers on a chip are generally referred to as *microcontrollers.*

Microcontroller

TEST

Supply the missing word in each statement.

3. A medium-scale integration is an IC that contains the equivalent of ________ gates.
4. A VLSI is an IC that contains the equivalent of more than ________ gates.
5. Using SSI and MSI technology, digital families of ICs (like TTL) were developed in the ________ (1940s, 1960s).
6. A(n) ________ (adder IC, microcontroller IC) could be described as a digital system on a chip.
7. Identify an ULSI device that forms the CPU of modern general-purpose computers.

12-3 The Calculator

The pocket calculator in nearly everyone's pocket or desk is a very complicated digital system. Knowing this, it is revealing to take apart a modern miniature calculator. You will find a battery, the tiny readout displays, a few wires from the keyboard, and a circuit board with an IC attached. That single IC is most of the digital system we call a calculator and contains an LSI chip that performs the task of thousands of logic gates. The single IC performs the storage, processing, and control functions of the calculating system. The keyboard is the input, and the displays are the output of the calculator system.

What happens inside the calculator chip when you press a number or add two numbers? The diagram in Fig. 12-2 will help us figure out how a calculator works. Figure 12-2 shows three components: the keyboard, the seven-segment displays, and the power supply. These parts are the only functional ones *not* contained in the single LSI IC in most small calculators. The keyboard is obviously the input device. The keyboard contains simple, normally open switches. The decimal display is the output. The readout unit in Fig. 12-2 contains only 6 seven-segment displays. The power supply is a battery in most inexpensive handheld calculators. Many modern calculators use solar cells as their power supply. CMOS ICs, and LCDs make this feasible.

Adder/subtractor subsystem

Clock subsystem

The calculator chip (the IC) is divided into several functional subsystems, as shown in Fig. 12-2. The organization shown is only one of several ways to get a calculator to operate. The heart of the system is the *adder/subtractor subsystem,* which operates very much like the adders you studied. The *clock subsystem* pulses all parts of the system at a constant frequency. The clock frequency is fairly high, ranging from 25 to 500 kHz. When the calculator is turned on, the clock runs constantly and the circuits "idle" until a command comes from the keyboard.

Display register

Suppose we add 2 + 3 with this calculator. As we press the 2 on the keyboard, the encoder translates the 2 to a BCD 0010. The 0010 is directed to the *display register* by the control circuitry and is stored in the display register. This information is also applied to the seven-segment decoder, and lines *a, b, d, e,* and *g* are activated. The first (1s display) seven-segment display shows a 2 when the scan line pulses that unit briefly. This scanning continues at a high frequency, and the display appears to be lit continuously even though it is being turned on and off many times per second. Next, we press + on the keyboard. This operation is transferred to and stored in code form in an extra register (X register). Now we press the 3 on the keyboard. The encoder translates the 3 to a BCD 0011. The 0011 is transferred to the display register by the controller and is passed to the *display decoder/driver,* which also places a 3 on the display. Meanwhile, the controller has moved the 0010 (decimal 2) to the operand register. Now we press the = key. The controller checks the X register to see what to do. The X register says to add the BCD numbers in the operand and display registers. The controller applies the contents of the display and *operand registers* to the adder inputs. The results of the addition collect in the *accumulator register.* The result of the addition is a BCD 0101. The controller routes the answer to the display register, shown on the readout as a 5.

Display decoder/driver

Operand register

Accumulator register

Instruction register

For longer and more complex numbers containing decimal points, the controller follows directions in the *instruction register.* For complicated problems the unit may cycle through hundreds of steps as programmed into the

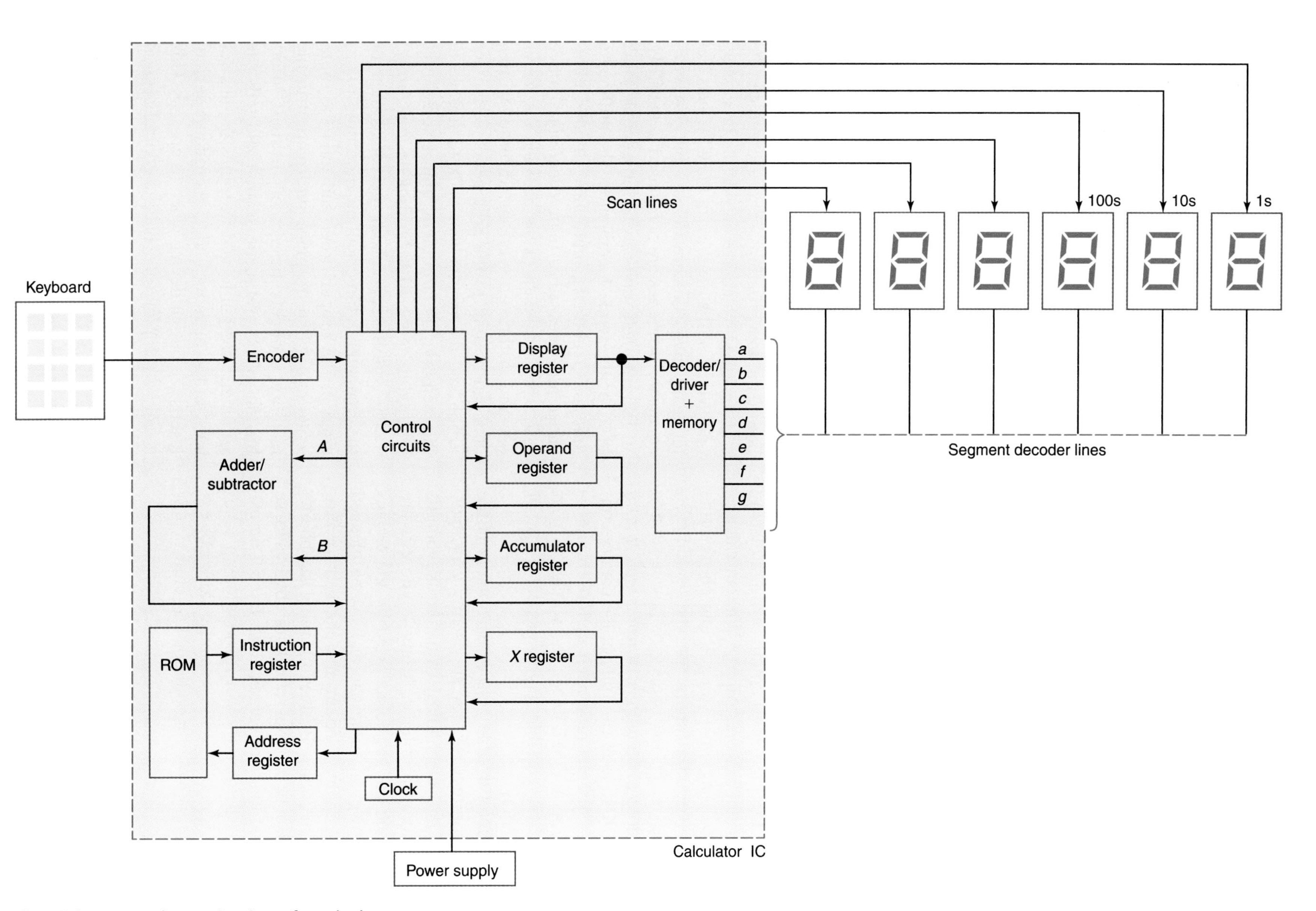

Fig. 12-2 Internal organization of a calculator.

ROM

ROM. Amazingly, however, even hundreds of operations take less than 1/10 s.

The registers in Fig. 12-2 are rather large units compared to the ones you have used in the laboratory. The ROM also has a large capacity (many thousands of bits). The calculator in Fig. 12-2 is but one example of how a calculator can operate. Each commercial design operates in its own unique way. This discussion serves only to point out that many of the subsystems you have already used are found in a complex digital system like the calculator.

Classic computer organization

Only the original IC designers need to know the organization of the subsystems in Fig. 12-2. This organization is sometimes referred to as the *architecture* of the calculator. Notice that all the elements of a system are present in this electronic calculator.

Architecture

You may be asked to troubleshoot and repair a calculator. Most inexpensive handheld calculators are designed as throwaway items.

Modern calculator designs use a keyboard, an LSI chip, perhaps a circuit board, a display module, and a power supply. Without calculator wiring diagrams, you still can check some obvious trouble spots in the calculator. Carefully check the battery or power supply. Either replace the *battery* or use a *load test* on the cell. Also look carefully at the battery connectors for signs of broken or loose connections. With a VOM, check the voltage as near to the IC as possible if you suspect a loss of power.

Battery load test

A second trouble spot in calculators is the keyboard. Because they are usually very inexpensive and mechanical in nature, keyboards cause many problems. Many keyboards are sealed and cannot be accessed by the technician. However, some can at least be cleaned out with compressed air and checked for broken connections. Besides cleaning keyboards and replacing batteries, check for obvious loose or broken wires or connections. Except on expensive calculators or in reconditioning laboratories, troubleshooting usually does not extend to the display module or LSI chip.

TEST

Answer the following questions.

8. A single ________ chip performs the storage, processing, and control functions in a modern calculator.
9. The display is the ________ device in a calculator system.
10. The keyboard is the ________ device in a calculator system.
11. List two typical problem areas to look at when troubleshooting an inexpensive calculator.

12-4 The Computer

The most complex digital systems include *computers*. Most digital computers can be divided into the five functional sections shown in Fig. 12-3. The input device may be a keyboard, mouse, joystick, graphics tablet, card reader, magnetic tape unit, scanner, network connection, or telephone line. This equipment lets us pass information from *person to machine* (or machine to machine). The input device often must *encode* human language into the binary language of the computer.

The memory section is the storage area for both data and programs. This storage can be supplemented by storage outside the processing unit. Much of the memory in the CPU was traditionally magnetic-core memory, but now semiconductor memories are being used in the CPU.

The arithmetic unit is what most people think of as being inside a computer. The arithmetic unit adds, subtracts, multiplies, divides, compares, and does other logic functions. Notice that a two-way path exists between the memory and arithmetic sections. In other words, data can be sent to the arithmetic section for action and the results sent back to storage in the memory. The arithmetic unit is sometimes referred to as the ALU (arithmetic-logic unit).

The control section is the nervous system of the computer. It directs all other sections to operate in the proper order and tells the input when and where to place information in the memory. It directs the memory to route information to the arithmetic section and tells the arithmetic section to add. It routes the answer back to the memory and to the output device. It tells the output device when to operate. This is only a sampling of what the control section can do.

The output section is the link between the *machine and a person* (or to a device or network). It can communicate to humans through a printer. It can output information on a CRT display. Output information can also be placed

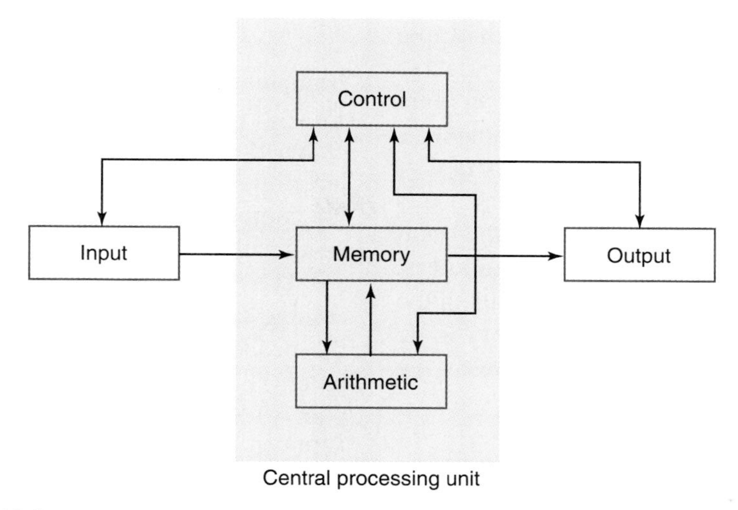

Fig. 12-3 Sections of a digital computer.

on bulk storage devices, such as magnetic tape, disk, or optical disks. The output section often must *decode* the language of the computer into human language.

The three middle blocks of Fig. 12-3 are often called the CPU. The arithmetic and memory sections and most of the control section are frequently found on a single circuit board. Devices located outside the CPU are often called *peripheral devices.*

The block diagram of the computer in Fig. 12-3 could well be the diagram for a calculator. Up to this point the basic systems operate the same. The basic difference between the calculator and computer is *size* and the use of a *stored program* in a computer. Computers are also faster and are multipurpose machines. Figure 12-4 shows that two types of information are put into a computer. One is the program (instructions) telling the control unit how to proceed in solving the problem. This program, which has to be carefully written by a programmer, is stored in the central memory while the problem is being solved. The second type of information fed to the computer is *data,* to be acted on by the computer. Data includes the facts and figures needed to solve the problem. Notice that the program information is placed in storage in the memory and used only by the control unit. The data information, however, is directed to various positions within the computer and is processed by the ALU. The data need never go to the control unit. The auxiliary memory is extra memory that may be needed to store partial results in some complex problems. It may not be located in the CPU. Data may be stored in peripheral devices such as a hard drive.

Central processing unit (CPU)

Peripheral devices

In summary, the computer is organized into five basic functional sections: input, memory,

Computer organization

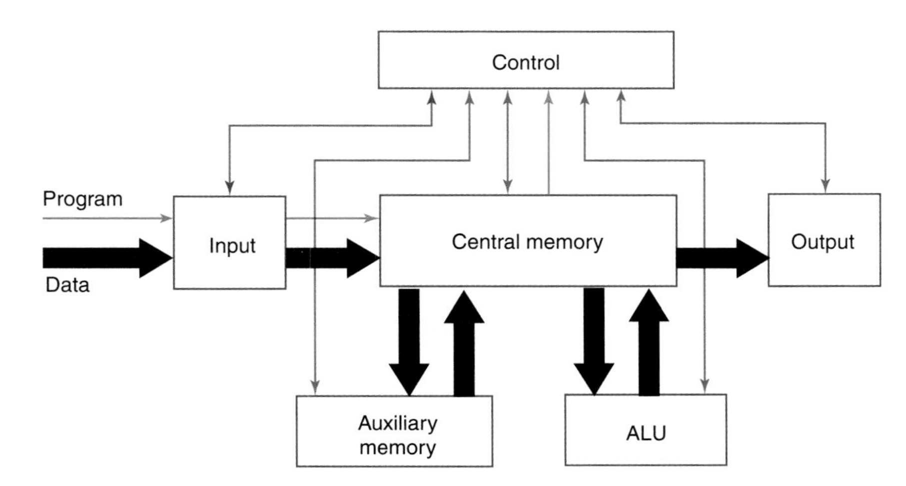

Fig. 12-4 Flow of program instructions and data in a computer system.

control, ALU, and output. Information fed into the CPU is either program instructions or data to be acted upon. The computer's stored program and size make it different from the calculator.

Computers, one of the most complex of digital systems, are not covered in depth in this section. There are entire books about the organization and architecture of computers. Remember, however, that all the circuits in the digital computer are constructed from logic gates, flip-flops, and subsystems such as the ones you have studied.

Microprocessor (μP)

Stored-program digital computer

Microcomputer organization

TEST

Answer the following questions.

12. Devices located outside the computer's CPU are often called ________ devices.
13. List several of the fundamental differences between a computer and a calculator.
14. List the two types of information fed into a digital computer.

12-5 The Microcomputer

Computers have been in general use since the 1950s. Formerly, digital computers were large, expensive machines used by governments or large businesses. The size and shape of the digital computer have changed in the past decades as a result of a device called the microprocessor. The *microprocessor* (*MPU,* for "microprocessing unit") is an IC that contains much of the processing capabilities of a larger computer. The MPU is a small but extremely complex VLSI device that is *programmable.* The MPU IC forms the heart of a microcomputer. The *microcomputer* is a *stored-program digital computer.*

The organization of a typical smaller microcomputer system is diagramed in Fig. 12-5. This microcomputer contains all the five basic sections of a computer: the *input* unit, the *control* and *arithmetic* units contained within the MPU, the *memory* units, and the *output* unit.

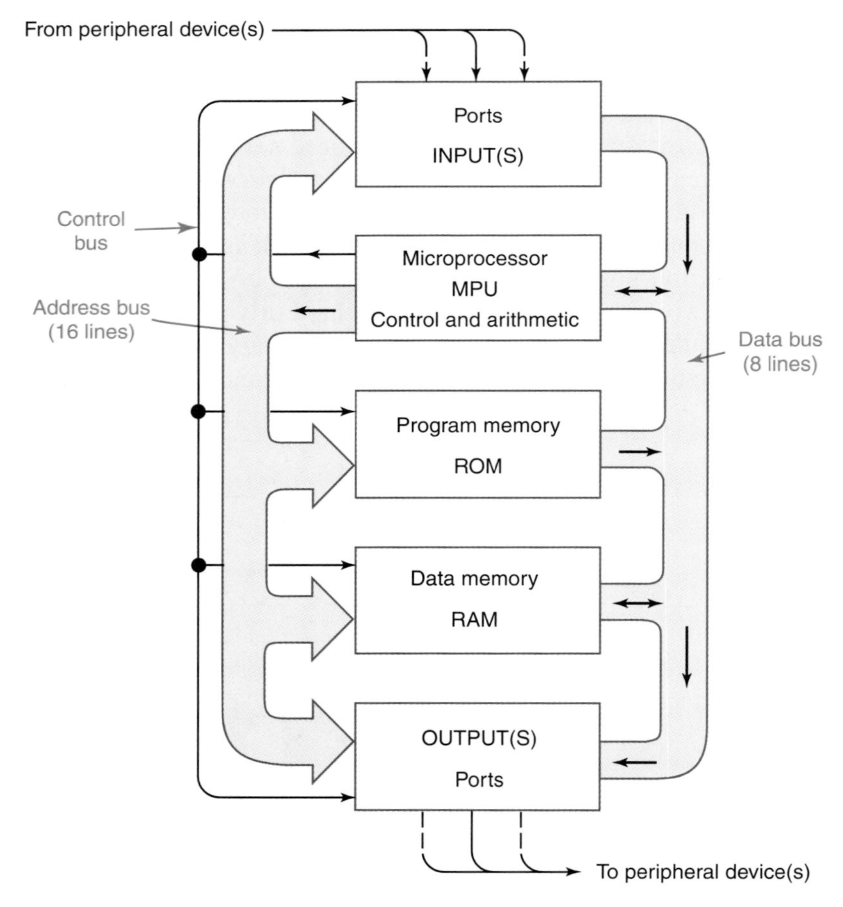

Microcomputer system

Fig. 12-5 Block diagram of a microcomputer system.

The MPU controls all the units of the system using the control lines shown at the left in Fig. 12-5. Besides the control lines, the *address bus* (16 parallel conductors) selects a certain memory location, input port, or output port. The *data bus* (eight parallel conductors) on the right in Fig. 12-5 is a *two-way path* for transferring data into and out of the MPU. It is important to note that the MPU can send data to memory or an output port or receive data from memory or an input port.

MPU

The microcomputer's ROM commonly contains a program. A *program* is a list of specially coded instructions that tell the MPU *exactly* what to do. The ROM in Fig. 12-5 is the place where the program resides in this example. In actual practice, the ROM contains a start-up or initializing program and perhaps other programs. Separate programs can also be loaded into RAM from auxiliary memory. These are user programs.

The RAM area in Fig. 12-5 is identified in this example as the data memory. Data used in the program resides in this memory.

The CPU and memory sections of the microcomputer are not very useful by themselves. The CPU must be interfaced with *peripheral devices* for input, output, and storage. Typical peripheral devices used for input, output, and storage on modern microcomputers are diagramed in Fig. 12-6. The keyboard, mouse, and joystick are probably the most common input devices connected to most microcomputers. Several other input devices connected to microcomputers are shown at the left in Fig. 12-6.

The floppy disk drive is a popular secondary storage device connected to most microcomputers. Other secondary storage devices interfaced with microcomputers are hard and optical disks and to a lesser extent tape drives. The CRT monitor and printer are the most common output peripheral devices used with typical microcomputers. Other output devices include TVs, speakers and sound systems, plotters, and laser printers.

Modem

A *modem* (modulator-demodulator) is the peripheral device that enables the microcomputer to transmit and receive data over telephone lines. Notice that the modem is classified as an *input/output* peripheral device in Fig. 12-6. It is an output device when transmitting data and an input device when receiving data.

TEST

Supply the missing word in each statement.

15. Refer to Fig. 12-5. The address bus is a one-way path, whereas the ________ bus is a two-way pathway for information.
16. Refer to Fig. 12-5. The ROM typically holds ________ (data, programs).
17. Refer to Fig. 12-5. The exact memory location, or input/output port, is selected by the MPU's output on the ________ bus and ________ bus.

Peripheral devices

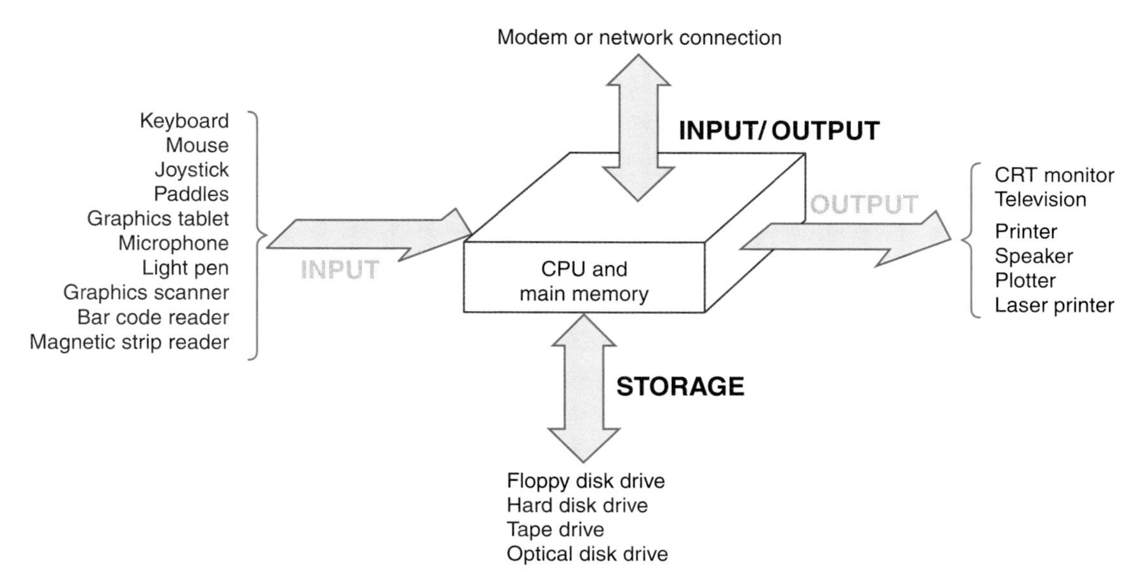

Fig. 12-6 Peripheral devices commonly attached to the CPU of a microcomputer.

Microcomputer peripheral devices

18. A(n) ________ is an input/output peripheral device that enables the microcomputer to send and receive data over telephone lines.
19. Refer to Fig. 12-6. The ________ is probably the most popular peripheral output device used with low-cost microcomputers.
20. Refer to Fig. 12-6. The ________ is a hand-operated input device that has a ball on the bottom and switch on the top. It is used to control the direct movement of the cursor on the CRT screen.

12-6 Microcomputer Operation

As an example of microcomputer operation, refer to Fig. 12-7. In this example, the following things are to happen:

1. Press the "A" key on the keyboard.
2. Store the letter "A" in memory.
3. Print the letter "A" on the screen of the CRT monitor.

The input-store-output procedure outlined in Fig. 12-7 is a typical microcomputer system operation. The electronic hardware used in a system like that in Fig. 12-7 is complicated. However, the transfer of data within the system will help explain the use of several different units within the microcomputer.

Program memory

The more detailed diagram in Fig. 12-8 will aid understanding of the typical microcomputer input-store-output procedure. First, look carefully at the *contents* section of the program memory in Fig. 12-8. Note that instructions have already been loaded into the first six memory locations. From Fig. 12-8, it is determined that the instructions currently listed in the program memory are:

1. Input data from input port 1.
2. Store data from port 1 in data memory location 200.
3. Output data to output port 10.

Parts of instruction: operation and operand

Note that there are only three instructions in the above program. It appears that there are six instructions in the program memory in Fig. 12-8. The reason for this is that instructions are sometimes broken into parts. The first part of instruction 1 above is to input data. The second part tells where the data comes from (from port 1). The first, *action* part of the instruction is called the *operation* and the second part the

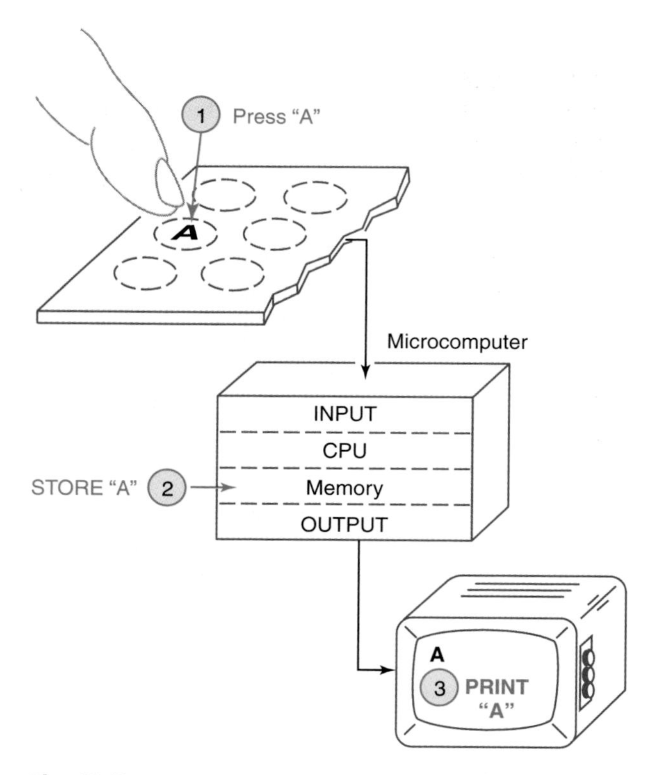

Fig. 12-7 An example of a common input-store-output microcomputer operation.

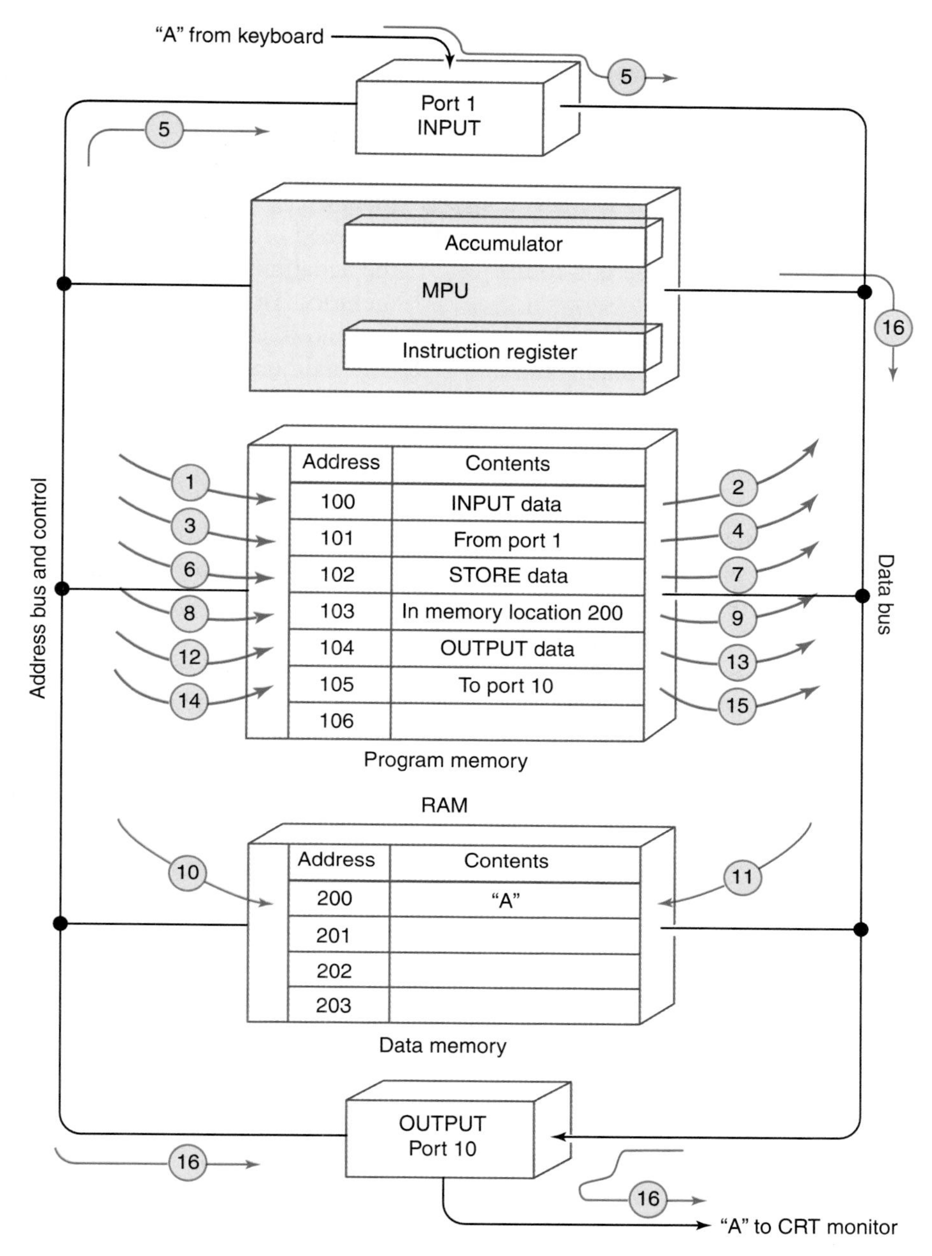

Fig. 12-8 Sequence of microcomputer operations in executing the input-store-output program.

operand. The operation and operand are located in separate memory locations in the program memory in Fig. 12-8. For the first instruction in Fig. 12-8, program memory location 100 holds the input operation while memory location 101 holds the operand (port 1) telling where information will be input from.

Two new sections are identified inside the *MPU* in Fig. 12-8. These two sections are called registers. These special registers are the *accumulator* and the *instruction register.*

The sequence of events happening within the microcomputer in the input-store-output "A" example is outlined in Fig. 12-8. The flow of instructions and data can be followed by keying on the circled numbers in the diagram. Remember that *the MPU is the center of all data transfers and operations.* Refer to Fig. 12-8 for all steps below.

1. The MPU sends out address 100 on the address bus. A control line *enables* the read input on the program memory IC. This step is symbolized in Fig. 12-8 by the encircled number 1.

MPU

Accumulator

Instruction register

2. The program memory sends the first instruction (INPUT data) on the data bus, and the MPU receives this coded message. The instruction is transferred to a special memory location within the MPU called the instruction register. The MPU *decodes,* or interprets, the instruction and determines that it needs the operand to the input data instruction.
3. The MPU sends out address 101 on the address bus. The control line enables the read input of the program memory.
4. The program memory places the operand (from port 1) on the data bus. The operand was located at address 101 in program memory. This coded message (the address for port 1) is received off the data bus and transferred to the instruction register. The MPU now decodes the entire instruction (input data from port 1).
5. The MPU uses the address bus and control lines to the input unit to cause port 1 to open. The coded form for "A" is transferred to and stored in the accumulator of the MPU.

Fetch-decode-execute sequence

It is important to note that the MPU always follows a *fetch-decode-execute sequence.* It first fetches the instruction from program memory. Second, the MPU decodes the instruction. Third, the MPU executes the instruction. Try to notice this fetch-decode-execute sequence in the next two instructions. Continue with the program listed in the program memory in Fig. 12-8.

6. The MPU addresses location 102 on the address bus. The MPU uses the control lines to enable the read input on the program memory.
7. The code for the store data instruction is sent on the data bus and received by the MPU, where it is transferred to the instruction register.
8. The MPU decodes the store data instruction and determines that it needs the operand. The MPU addresses the next memory location (103) and enables the program memory read input.
9. The code for "in memory location 200" is placed on the data bus by the program memory. The MPU accepts this operand and stores it in the instruction register. The entire "store data in memory location 200" has been fetched from memory and decoded.
10. The execute process now starts. The MPU sends out address 200 on the address bus and enables the *write* input of the data memory.
11. The MPU sends the information stored in the accumulator on the data bus to data memory. The "A" is received off the data bus and is written into location 200 in data memory. The second instruction has been executed. This store process does not destroy the contents of the accumulator. The accumulator still also contains the coded form of "A."
12. MPU must fetch the next instruction. It addresses location 104 and enables the read input of the program memory.
13. The code for the output data instruction is sent to the MPU on the data bus. The MPU receives the instruction and transfers it to the instruction register. The MPU decodes the instruction and determines that it needs an operand.
14. The MPU places address 105 on the address bus and enables the read input of the program memory.
15. The program memory sends the code for the operand (to port 10) to the MPU via the data bus. The MPU receives this code in the instruction register.
16. The MPU decodes the entire instruction "output data to port 10." The MPU activates port 10, using the address bus and control lines to the output unit. The MPU sends the code for "A" (still stored in the accumulator) on the data bus. The "A" is transmitted out of port 10 to the CRT monitor.

MPU-based systems

Most *MPU-based systems* transfer information in a fashion similar to the one detailed in Fig. 12-8. The greatest variations are probably in the input and output sections. Several more steps may be required to get the input and output sections to operate properly.

It is important to notice that the MPU is the center of and controls all operations. The MPU follows the fetch-decode-execute sequence. The actual operations of the MPU system, however, are dictated by the instructions listed in program memory. Instructions are usually performed in sequence (100, 101, 102, and so on).

All three instructions in the example would be fetched, decoded, and executed in a few microseconds or less by most small microcomputers. The advantage of MPU-based systems is their fast operation and flexibility. They are flexible because they can be reprogrammed to perform many tasks.

Microcomputers are complex digital systems containing an MPU IC (or set of ICs), some memory, and inputs and outputs. The MPU chip itself is a complex, highly integrated subsystem that can process instructions at a high rate of speed. It is expected that microcomputers will be a growth industry for many years to come. The last two sections gave only a brief overview of the basic operation and organization of a microcomputer.

TEST

Supply the missing word or words in each statement

21. The action part of a microcomputer instruction is called the ________. The second part of the instruction is called the ________.
22. Refer to Fig. 12-8. Program memory location ________ holds the operation part of the first instruction, whereas location ________ holds the operand part of the instruction.
23. Refer to Fig. 12-8. In this microcomputer, the ________ is the center of all data transfers and operations.
24. The microcomputer's MPU always follows a fetch-________-________ sequence when running.
25. Program instructions are usually performed in ________ (random, sequential) order in a microcomputer.

12-7 Microcomputer Address Decoding

Consider the simple 4-bit MPU-based system shown in Fig. 12-9(*a*) on page 306. This system uses only eight conductors in the address bus and four conductors in the data bus. The RAMs are tiny, 64-bit (16 × 4) units. These RAMs are like the 7489 RAMs you studied in the last chapter.

Two problems become apparent when working with a system like the one shown in Fig. 12-9(*a*). First, how does the MPU select which RAM to read data from when it sends the same 4-bit address to each? Second, how can several devices send data over a common data bus if, generally, outputs of logic devices cannot be tied together? The solutions to both these problems are shown in Fig. 12-9(*b*).

Address decoding

The *address decoder* shown in Fig. 12-9(*b*) decodes which RAM is to be used and sends the enabling signal over the chip select line. *Only one chip select line is activated at a time.* The address decoder block consists of familiar combinational logic gates. RAM 0 is selected when the address is 0 through 15. However, RAM 1 is selected when the address is 16 through 31.

Three-state buffer

The *three-state buffers* shown in Fig. 12-9(*b*) disconnect the RAM outputs from the data bus when the memory is not sending data. Only one device is allowed to send on the shared data bus at a given time. For this reason, the chip select line is also used to control, or turn on, the three-state buffers. When the three-state buffers are in the turned off mode, it is said that the buffer outputs are in their *high-impedance state* and are effectively disconnected from the four data lines at the inputs of the buffers.

High-impedance state

The logic circuits used in a simple address decoder are shown in Fig. 12-10 on page 307. In this example, only when the four address lines (A_7 to A_4) are all zero is the output of the bottom four-input OR gate LOW. When address lines A_7 to A_4 are 0000, then RAM 0 is enabled with a LOW at its memory enable ($\overline{ME}$) input.

When the four address lines going into the address decoder in Fig. 12-10 are 0001 ($A_7 = 0$, $A_6 = 0$, $A_5 = 0$, $A_4 = 1$), the top OR gate is activated. The 0001 causes the top OR gate in the address decoder to generate a LOW output, which activates the bottom device-select line. This enables the bottom RAM (RAM 1).

Microcomputer address decoding

The address decoder in Fig. 12-10 decodes only the four most significant address lines to generate the correct $\overline{ME}$ logic level. The RAMs internally decode the four least significant address lines (A_0 to A_3) to locate the exact 4-bit word in RAM.

The MPU-based system in Figs. 12-9 and 12-10 uses eight address lines. This means that the MPU can generate 256 (2^8) unique addresses. In the systems in Figs. 12-9 and 12-10, the first 16 addresses are used by RAM 0 while the next 16 addresses are used by RAM

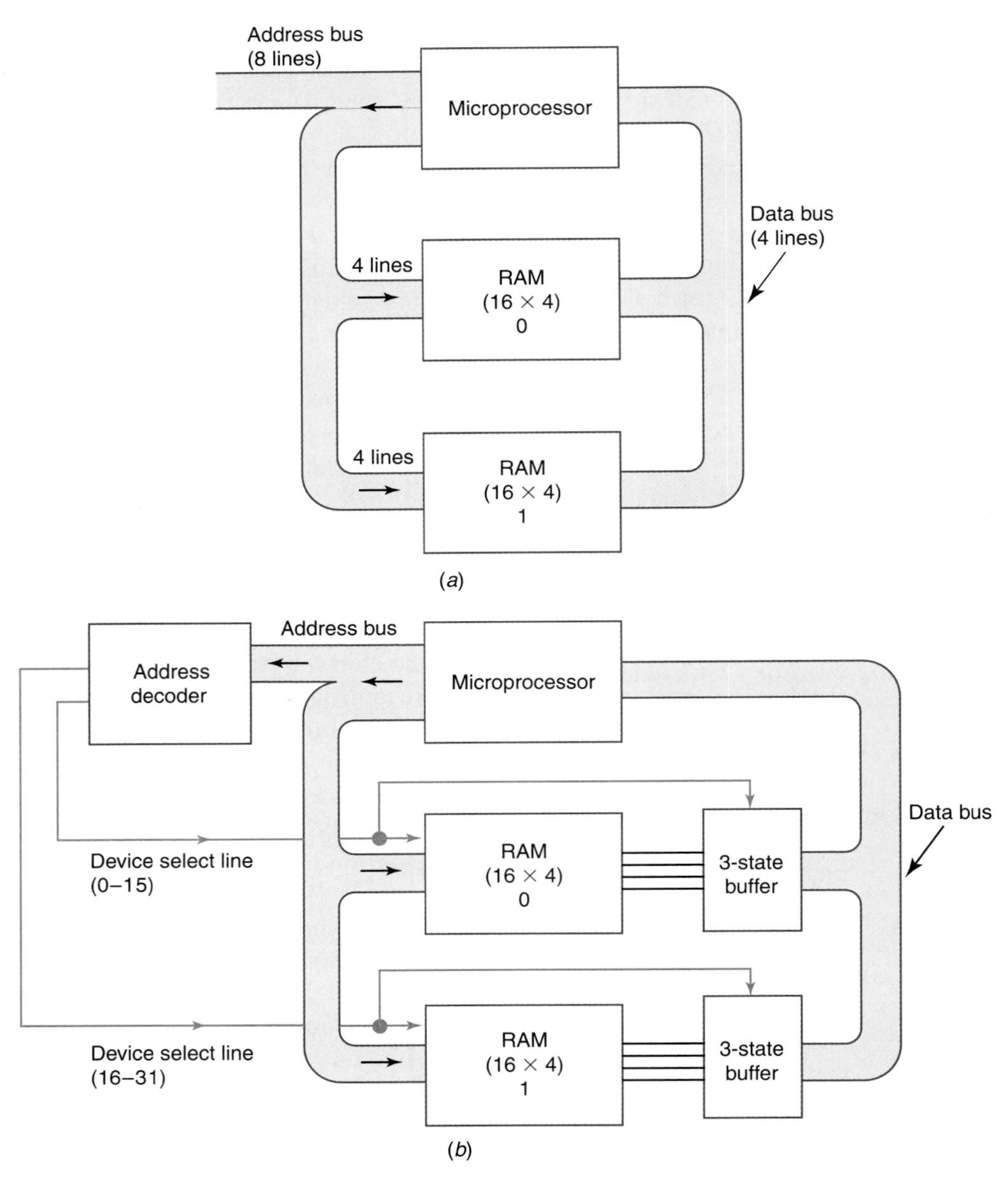

4-bit microprocessor-based system

Fig. 12-9 (*a*) Simplified 4-bit microprocessor interfaced with two 64-bit RAMs. (*b*) Address decoder and three-state buffers added to 4-bit microprocessor-based system.

RAM interfacing

Memory map

74125 quad three-state buffer IC

1. It is customary to draw a *memory map* of an MPU-based system. The memory map of our sample system is drawn in Fig. 12-11. This shows that the first 16 (0F in hexadecimal) addresses are used by RAM 0. These addresses range from 0 to 15 (00 to 0F in hexidecimal). The second 16 addresses are used by RAM 1. These addresses range from 16 to 31 (10 to 1F in hexadecimal). The third through sixteenth groups of addresses are not used in this very tiny system. It is customary to use hexadecimal notation in specifying addresses in an MPU-based system.

In Fig. 12-9(*b*), two blocks are labeled three-state buffers. The symbol for a buffer is drawn in Fig. 12-12(*a*) on page 308. It has a data input (*A*) and noninverted output (*Y*). When the control input (*C*) is deactivated with a 1, output *Y* goes to its high-impedance (high-Z) state and is effectively disconnected from the input.

A commercial version of the three-state buffer is shown in Fig. 12-12(*b*). This is the pin diagram for the *74125 quad three-state buffer TTL IC*. The truth table for the 74125 IC is shown in Fig. 12-12(*c*).

In summary, an address decoder is used to

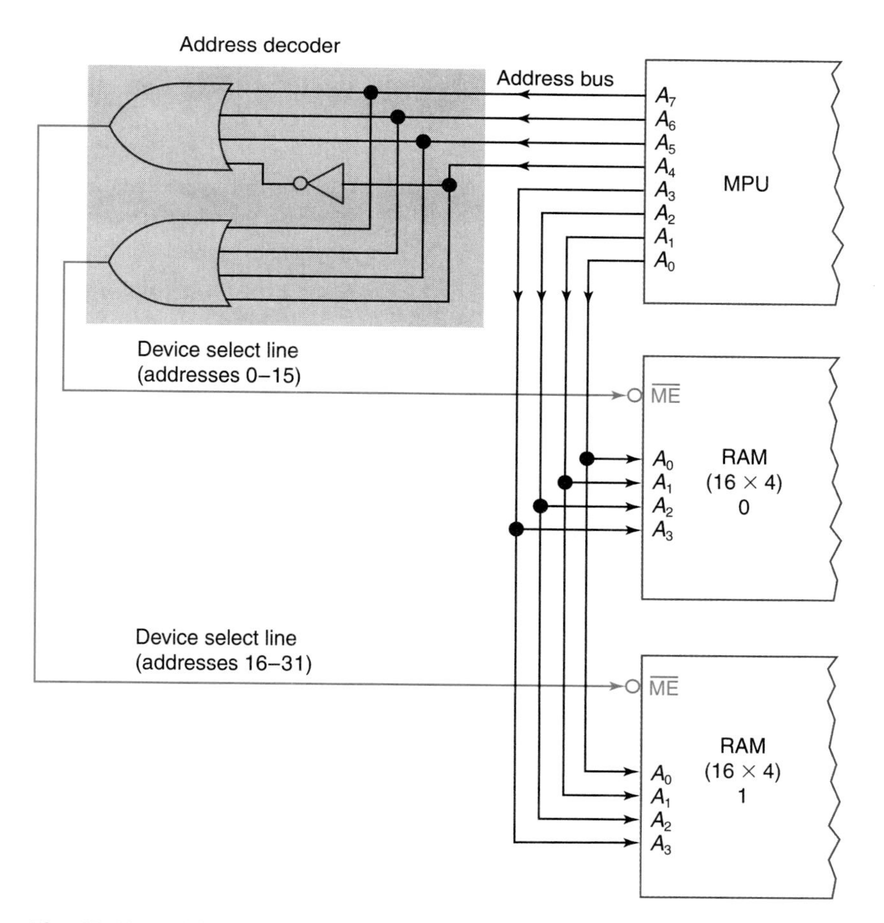

Fig. 12-10 Address decoder gating to generate correct device select signals.

Address decoder

select *which* device will be connected to the data bus in an MPU-based system. *Address decoders* are usually constructed of combinational logic circuits (simple gating circuits).

Address decoders

To permit many devices to use a common data bus, three-state buffers are used. A three-state buffer has a control input that, when disabled, places the output in the high-impedance (high-Z) state.

Both address decoders and three-state buffers are widely used in microcomputers. The three-state buffers are usually part of MPUs, larger RAMs, ROMs, and peripheral interface adapter ICs.

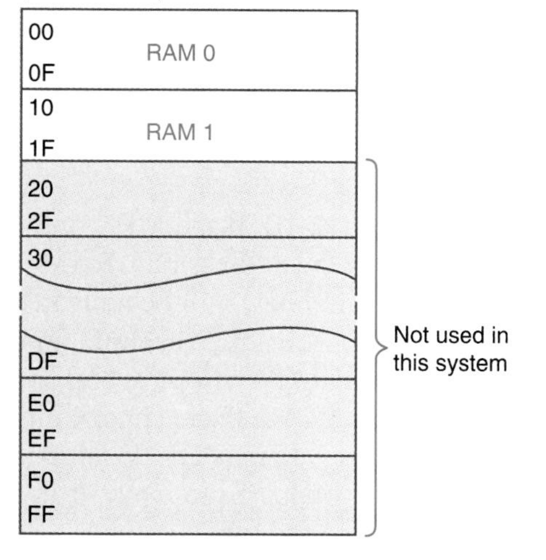

Fig. 12-11 Memory map of small microprocessor-based system using two 16 × 4 RAMs.

TEST

Supply the missing word or words in each statement.

26. Refer to Fig. 12-9. The ________ in this system selects which RAM will be used.
27. Refer to Fig. 12-9. When not in use, RAMs are isolated from the data bus by ________.

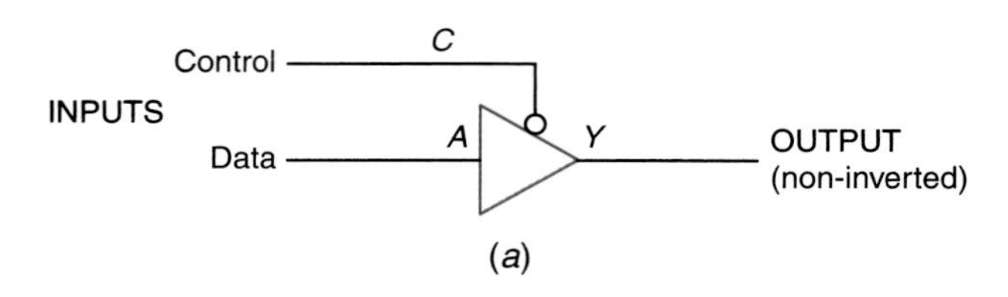

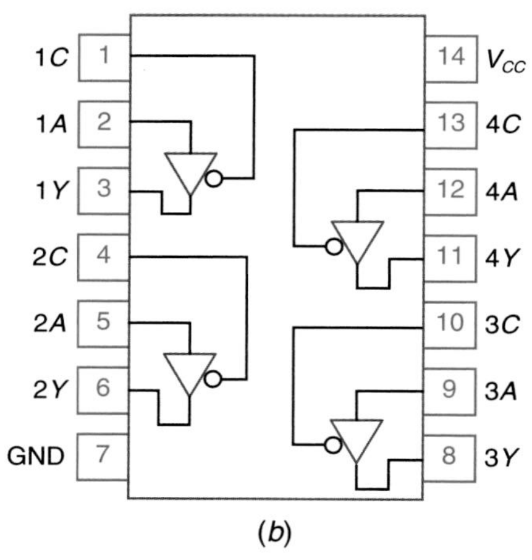

TRUTH TABLE

INPUTS		OUTPUT
C	A	Y
L	L	L
L	H	H
H	X	(Z)

L = LOW voltage level
H = HIGH voltage level
X = Don't care
(Z) = High impedance (off)

(c)

Fig. 12-12 (*a*) Logic symbol for a three-state buffer. (*b*) Pin diagram for commercial 74125 quad three-state buffer IC. (*c*) Truth table for 74125 three-state buffer IC.

28. Refer to Fig. 12-10. If the MPU outputs 00001000 on the address bus, RAM _________ (number) will be activated and storage area _________ (decimal number) located in the RAM will be accessed.
29. Refer to Fig. 12-12. If the control input on the three-state buffer is HIGH, output *Y* is _________ (connected to input *A*; in its high-impedance state).

12-8 Data Transmission

Data transmission

74150 MUX IC

Serial data

Multiplexers (MUX)

Demultiplexers (DEMUX)

Most data in digital systems is transmitted directly through wires and PC boards. Many times bits of data must be transmitted from one place to another. Sometimes the data must be transmitted over telephone lines or cables to points far away. If all the data were sent at one time over *parallel* wires, the cost and size of these cables would be too expensive and large. Instead, the data is sent over a single wire in *serial* form and reassembled into parallel data at the receiving end. The devices used for sending and receiving serial data are called *multiplexers (MUX)* and *demultiplexers (DEMUX).*

The basic idea of a MUX and DEMUX is shown in Fig. 12-13. Parallel data from one digital device is changed into *serial* data by the MUX. The serial data is transmitted by a single wire. The serial data is reassembled into parallel data at the output by the DEMUX. Notice the control lines that must also connect the MUX and DEMUX. These control lines keep the MUX and DEMUX synchronized. Notice that the 16 input lines are cut down to only a few transmission lines.

The system in Fig. 12-13 works in the following manner. The MUX first connects input 0 to the serial data transmission line. The bit is then transmitted to the DEMUX, which places this bit of data at output 0. The MUX and DEMUX proceed to transfer the data at input 1 to output 1, and so on. The bits are transmitted one bit at a time.

A MUX works much like a single-pole, many-position rotary switch, as shown in Fig. 12-14. Rotary switch 1 shows the action of a MUX. The DEMUX operates like rotary switch 2 in Fig. 12-14. The mechanical control in this diagram makes sure input 5 on SW 1 is delivered to output 5 on SW 2. Notice that the mechanical switches in Fig. 12-14 permit data to travel in either direction. Being made from logic gates, MUXs and DEMUXs permit data to travel only from input to output, as in Fig. 12-13.

You used a MUX before, in Chap. 4. The other name for MUX is *data selector.* DEMUXs are sometimes called *distributors* or *decoders.* The term "distributor" describes the action of SW 2 in Fig. 12-14, as it distributes the serial data first to output 1, then to output 2, then to output 3, and so forth.

Figure 12-15 on page 310 is a detailed wiring diagram of an experimental transmission system using the MUX/DEMUX arrangement. A word (16 bits long) is entered at the inputs (0 to 15) of the *74150 MUX IC.* The 7493 counter starts at binary 0000. This would be shown as 0 on the seven-segment display. With

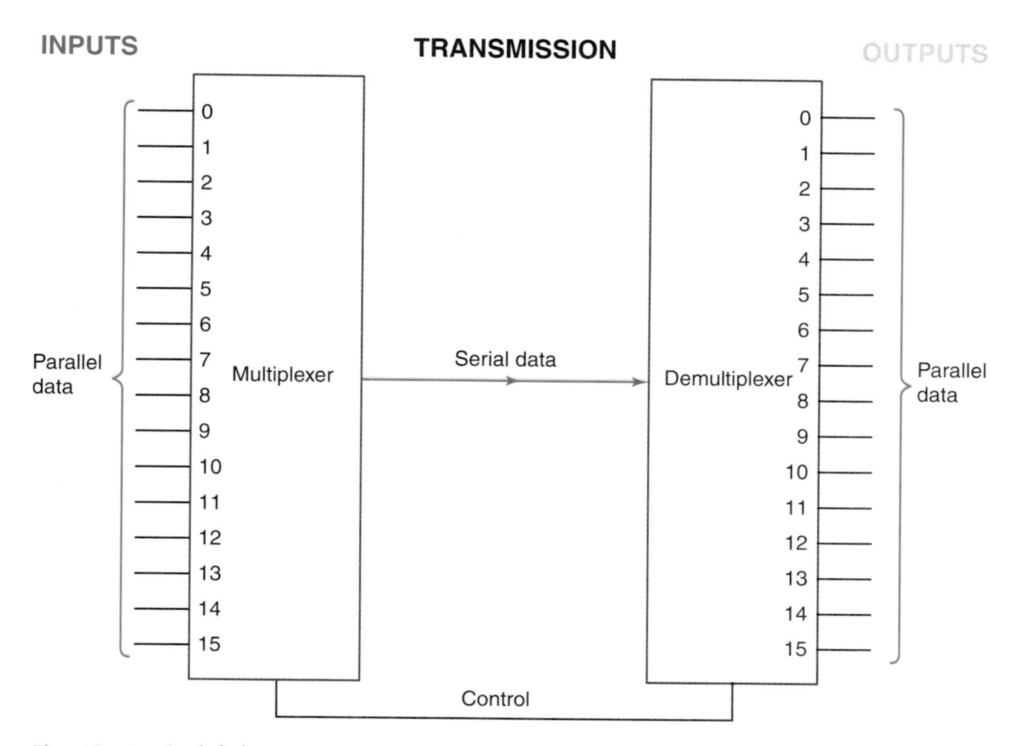

Fig. 12-13 Serial data transmission using a multiplexer and demultiplexer.

the data select inputs (*D, C, B, A*) of the 74150 MUX at 0000, the data is taken from input 0, which is shown as a logical 0. The logical 0 is transferred to the *74154 DEMUX IC,* where it is routed to output 0. Normally the output of the 74154 IC is inverted, as shown by the invert bubbles. A 7404 inverter complements the output back to the original logical 0.

The counter increases to binary 0001. This is shown as a 1 on the decimal readout. This binary 0001 is applied to the data select inputs of both ICs (74150 and 74154). The logical 1 at the input of the 74150 MUX is transferred to the transmission line. The 74154 DEMUX routes the data to output 1. The 7404 inverter complements the output, and the logical 1 appears as a lighted LED, as shown in the diagram. The counter continues to scan each input of the 74150 IC and transfer the contents to the output of the 74154. Notice that the counter must count from binary 0000 to 1111 (16 counts) to transfer just one parallel word from the input to the output of this system. The seven-segment LED readout provides a convenient way of keeping track of which input is being transmitted. If the clock is pulsed very fast, the parallel data can be transmitted quite quickly as serial data to the output.

74154 DEMUX IC

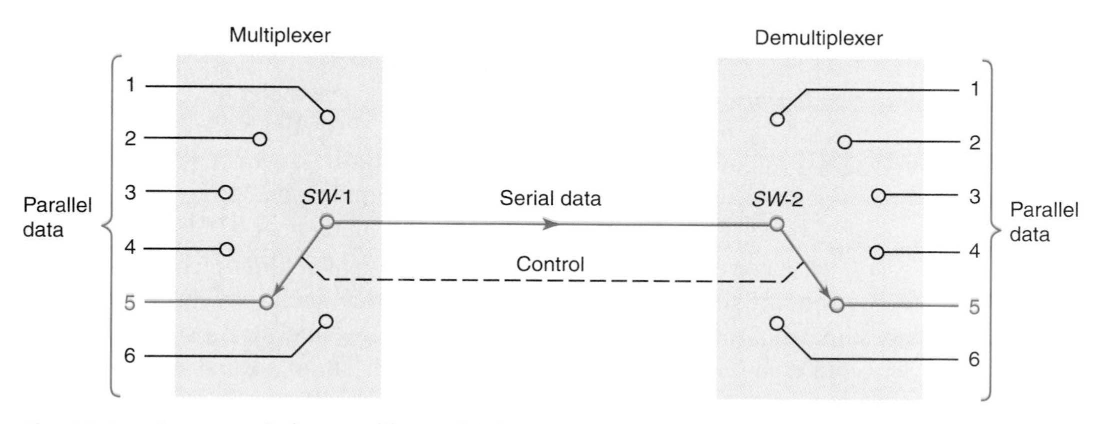

Fig. 12-14 Rotary switches act like multiplexers and demultiplexers.

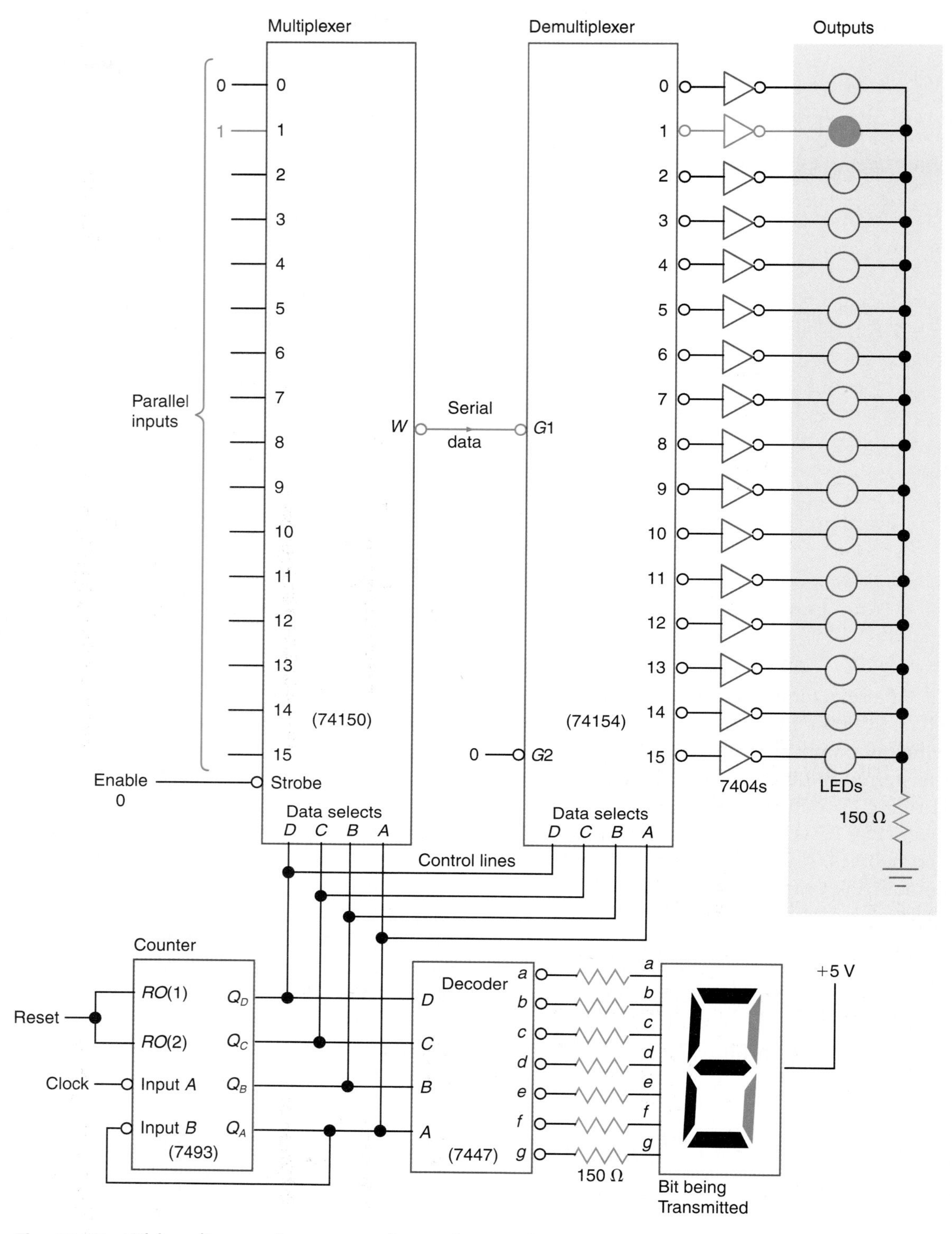

Fig. 12-15 Wiring diagram for an experimental transmission system.

Notice from Fig. 12-15 that we have saved many pieces of wire by sending the data in *serial* form. This takes somewhat more time, but the rate at which we send data over the transmission line can be very high.

Parallel interface

One common example of data transmission is the link between a microcomputer and a peripheral device such as a printer or modem. The computer's interface may send data either in parallel or serial format depending on the design of the printer.

A *parallel interface* transmits 8 bits (1 byte)

of data at one time. Figure 12-16 shows how the microcomputer's CPU controls a special IC called a *peripheral-interface adapter (PIA)*. The PIA IC communicates with the printer through the *handshaking* line to check if it is ready to receive data. If the printer signals the PIA that it is ready, bytes are transmitted from the CPU to the PIA and then on to the printer's *buffer memory*. The CPU can send data much faster than the printer can print the information. For this reason, the printer signals the PIA when its buffer memory is full. The PIA then signals the CPU to stop sending data temporarily until there is more room for data in the printer's buffer memory.

Peripheral interface adapters are not standardized. For instance, Motorola calls their unit a 6820 PIA while Intel's name for a similar input/output adapter unit is the 8255 *PPI (programmable peripheral interface)*. The PIAs are general-purpose ICs that can be programmed for either input or output. They have several parallel 8-bit I/O ports.

A serial interface transmits data 1 bit at a time. ICs called *UARTs (universal asynchronous receiver-transmitters)* are often used as the interface between the CPU and the data lines (also called *data links*). A UART consists of three sections as shown in Fig. 12-17. They are a *receiver*, a *transmitter*, and a *control block*. The receiver converts serial to parallel data. The transmitter section converts parallel data (as from the data bus of the CPU) to serial data. The control section manages the UART's functions and handles communications with the CPU and the peripheral device. The UART also encodes and decodes the serial signal including start, stop, and parity bits.

The speed at which serial data is transmitted is called the baud rate. The *baud rate* is the number of bits per second being transmitted through a data link. The baud rate is *not* the same as the number of characters or words transmitted per second. Common baud rates might be 110, 300, 1200, 2400, 9600, 19,200, and 38,400. The 16,550 UART can operate up to 115,200 bits per second.

The signal levels found in data lines are many times defined by standards. Two *serial interface standards* are the EIA RS-232C stan-

Peripheral-interface adapter (PIA)

Handshaking

Buffer memory

PPI (programmable peripheral interface)

UART

Data links

Baud rate

Serial interface standards

About Electronics

One Powerful Chip A memory chip is smaller than a fingernail, and yet it stores 262,000 bits of data in 600,000 electronic components. Pictured is the center of a 256-K dynamic random access memory (DRAM) chip, enlarged about 1000 times its original size.

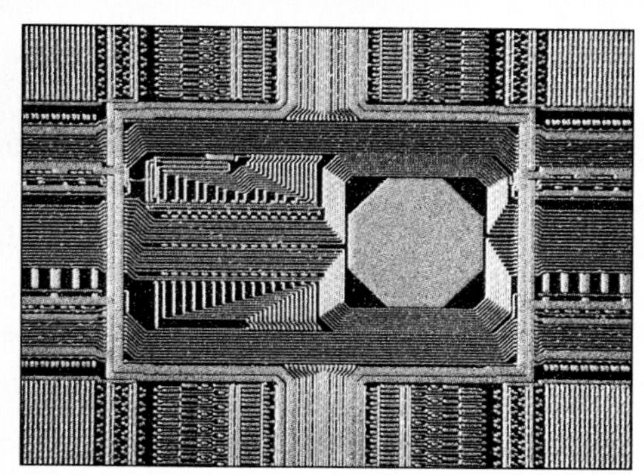

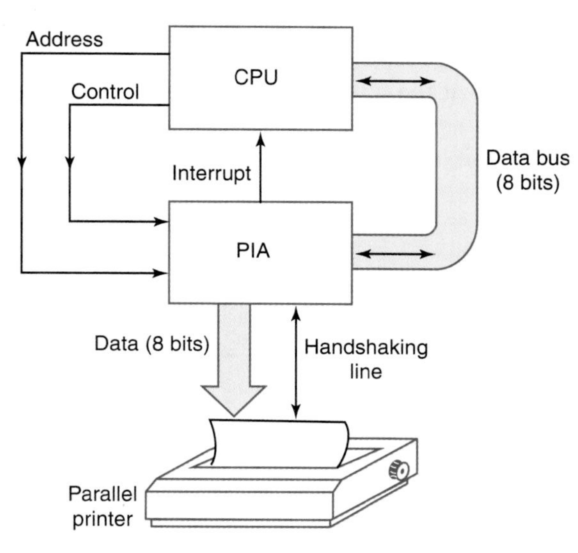

Fig. 12-16 Parallel data transmission from the CPU to printer using a peripheral-interface adapter (PIA) IC.

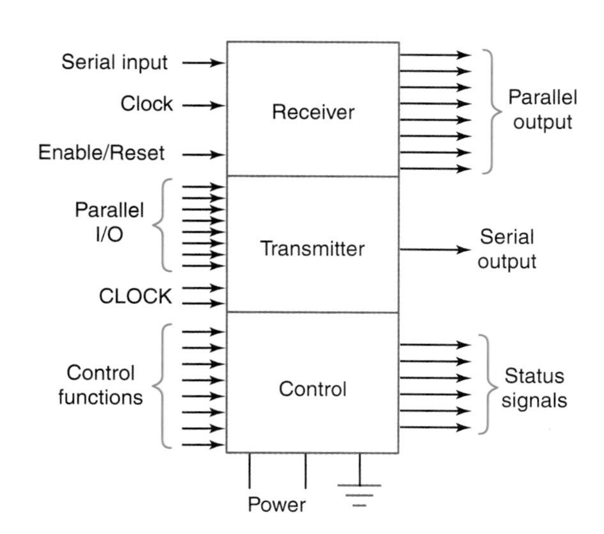

Fig. 12-17 Block diagram of a typical UART.

dard and the older 20-mA current loop teletype standard.

Parallel interface standards

Two common parallel interfaces are the Centronics standard and the IEEE-488 standard. The Centronics standard is used between many microcomputers and printers. The IEEE-488 interface is used between computers and scientific instrumentation.

Parity bit

TEST

Answer the following questions.

30. Refer to Fig. 12-13. A(n) ________ changes parallel data to serial data, whereas a(n) ________ changes serial data to parallel data for transmission.
31. Refer to Fig. 12-15. The 7493 IC is used to sequence the data selects from 0000 through ________ (binary number).
32. Refer to Fig. 12-16. A complex chip called a(n) ________ is used to output parallel data to the printer in many microcomputer systems.
33. An LSI IC used for asynchronous data transmission is called a(n) ________.
34. A measure of the speed of serial data transmission is called the ________ rate.
35. The EIA RS-232C standard might be used for ________ (parallel, serial) interfacing between a microcomputer and a peripheral device.

12-9 Detecting Errors in Data Transmissions

Digital equipment, such as a computer, is valuable to people because it is fast and *accurate.* To help make digital devices accurate, special *error detection* methods are used. You can well imagine an error creeping into a system when data is transferred from place to place.

Error detection

To detect errors, we must keep a constant check on the data being transmitted. To check accuracy we can generate and transmit an extra *parity bit.* Figure 12-18 shows such a system. In this system three parallel bits (*A, B,* and *C*) are being transmitted over a long distance. Near the input they are fed into a *parity bit generator* circuit. This circuit generates what is called a parity bit. The parity bit is transmitted with the data, and near the output the results are checked. If an error occurs during transmission, the *error detector* circuit sounds an alarm. If all the parallel data is the same at the output as it was at the input, no alarm sounds.

Table 12-1 will help you understand how the error-detection system works. This table is really a truth table for the parity bit generator in Fig. 12-18. Notice that the inputs are labeled *A, B,* and *C* for the three data transmission lines. The output is determined by looking across a horizontal row. We want an *even number of 1s* in each row (zero 1s, or two 1s, or four 1s). Notice that row 1 has no 1s. Row 2 has a single 1 plus the parity bit 1. Row 2 has two 1s. As you look down Table 12-1 you will notice that each horizontal row contains an even number of 1s. Next, the truth table is converted to a logic circuit. The logic circuit for the parity bit generator is drawn in Fig. 12-19(*a*). You can see that a three-input XOR gate will do the job for generating a parity bit. The three-input XOR gate in Fig. 12-19, then, is the logic circuit you would substitute for the parity bit generator block in Fig. 12-18.

Look at the entire truth table in Table 12-1. We can see that under normal circumstances each horizontal row contains an *even number*

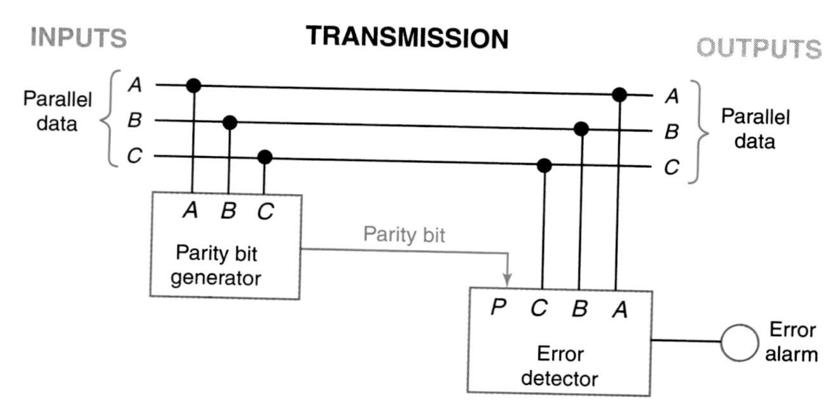

Fig. 12-18 Error-detection system using a parity bit.

TABLE 12-1 Truth Table for Parity Bit Generator			
Inputs			Output
Parallel Data			Parity Bit
C	*B*	*A*	*P*
0	0	0	0
0	0	1	1
0	1	0	1
0	1	1	0
1	0	0	1
1	0	1	0
1	1	0	0
1	1	1	1

of 1s. Were an error to occur, we might then have an *odd number* of 1s appear. A circuit that gives a logical 1 output any time an odd number of 1s appear is shown in Fig. 12-19(*b*). A four-input XOR gate would detect an odd number of 1s at the inputs and turn on the alarm light. Figure 12-19(*b*) diagrams the logic circuit that can substitute for the error-detector block in Fig. 12-18.

The parity bit can be generated for longer words such as a 7-bit ASCII character. For instance, the ASCII code for T is 1010100 (from Table 6-3). If transmitted using an even parity bit, an extra 1 would have to be added (to get an even number of 1s or four 1s). Another example, the ASCII code for S is 1010011 (from Table 6-3). If transmitted using an even parity bit, an extra 0 would have to be added (four 1s already). A seven-input XOR gate would generate the correct even parity bit for 7-bit ASCII characters. An 8-bit XOR gate at the receiver end would serve as an error detector circuit

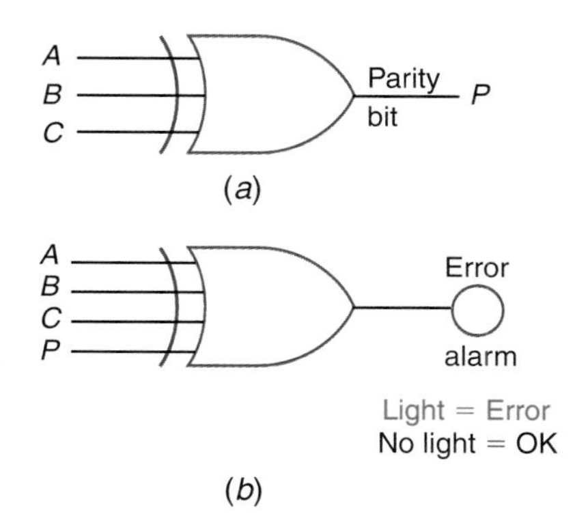

Fig. 12-19 (*a*) Parity bit generator circuit. (*b*) Error detector circuit.

(H = error, L = no error). Either an *even* or *odd parity bit* may be transmitted or received. An XNOR gate is used to generate an odd parity bit.

The use of the parity bit only warns you of an error. The system *does not* automatically correct the error. Some codes such as the *Hamming code* are *error correcting*. The Hamming code uses several extra parity bits when transmitting data. Other methods of assuring accuracy in data transmissions have also been developed.

Error correcting code

Hamming code

TEST

Supply the missing word or words in each statement.

36. Refer to Table 12-1. This is the truth table for an ________ (even, odd) parity bit generator.
37. Refer to Fig. 12-18. The parity bit generator block could be replaced with a three-input ________ gate, whereas the error detector block could be replaced with a four-input ________ gate.
38. Using even parity, what bit would be transmitted with the 7-bit ASCII code 1011000 as a parity bit?
39. Using odd parity, what bit would be transmitted with the 7-bit ASCII code 1011000 as a parity bit?
40. A seven-input ________ (AND, XOR) gate will generate an even parity bit for a 7-bit ASCII code.

12-10 Adder/Subtractor System

Adder/subtractor system

XOR gate used for parity bit generation and error detection

The adder/subtractor featured in this section combines subsystems you have already studied to form a simple system. A block diagram of the adder/subtractor is sketched in Fig. 12-20.

Did You Know?

Transistors as Far as the Eye Can See Sun Microsystems is developing a processor which may contain more than 30 million transistors. Some estimate that there may be 200 quadrillion transistors in use each day worldwide. This amounts to about 40 million transistors in use for each person on earth.

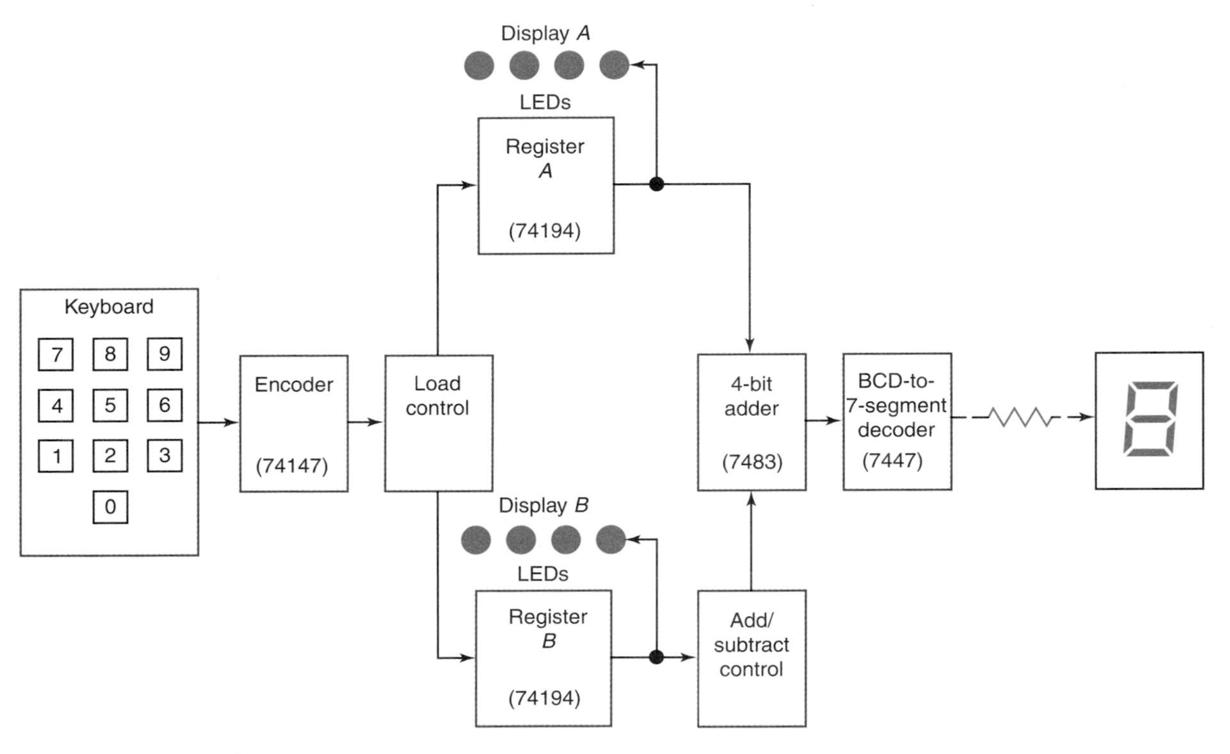

Fig. 12-20 Block diagram of an adder/subtractor system.

Some familiar digital functions are named in the blocks. For instance, the encoder (74147 decimal-to-BCD), the decoder (7447 BCD-to-7-segment), and the seven-segment LED display were studied in Chap. 6. The registers (74194 shift registers) were detailed in Chap. 9, and the adder/subtractor circuits were featured in Chap. 10. The purpose of this experimental adder/subtractor is to give you experience in using the digital functions you have already studied. If you wire this circuit in the lab, you will soon appreciate the complexity of even the cheapest calculators. This crude adder/subtractor system is much more difficult to operate than a calculator.

A detailed wiring diagram of the experimental adder/subtractor circuit is shown in Fig. 12-21. Four control inputs (clear, load A, load B, and add/subtract) are detailed on the wiring diagram. A clock input (single positive pulse) is also shown in Fig. 12-21. Register A (top display A) holds the top number in the addition/subtraction problem. Register B (bottom display B) holds the bottom number in the addition/subtraction problem.

The following steps might be used to operate the 4-bit adder/subtractor system detailed in Fig. 12-21:

1. Activate the *clear input.* All displays are cleared to zero.
2. Set the *add/subtract input* to the proper mode (start with add).
3. Load register A using the *load A input.* With load A input at 1 (load B = 0), press the number to be placed in register and at the same time apply a single + pulse to clock inputs.
4. Load register B using the *load B input.* With Load B input at 1 (Load A = 0), press the number to be placed in register and at the same time apply a single + pulse to clock inputs.
5. The sum (from 0 through 9) will immediately appear on the decimal display.
6. Assumptions:
   a. The sum will not exceed decimal 9.
   b. We are using unsigned decimal numbers.

To do a *subtraction problem* with the system in Fig. 12-21, the add/subtract mode control will be set for subtraction. The minuend is loaded into register A while the subtrahend is loaded into register B. The solution of the subtraction problem (difference) appears on the decimal display. Using the adder/subtractor in Fig. 12-21

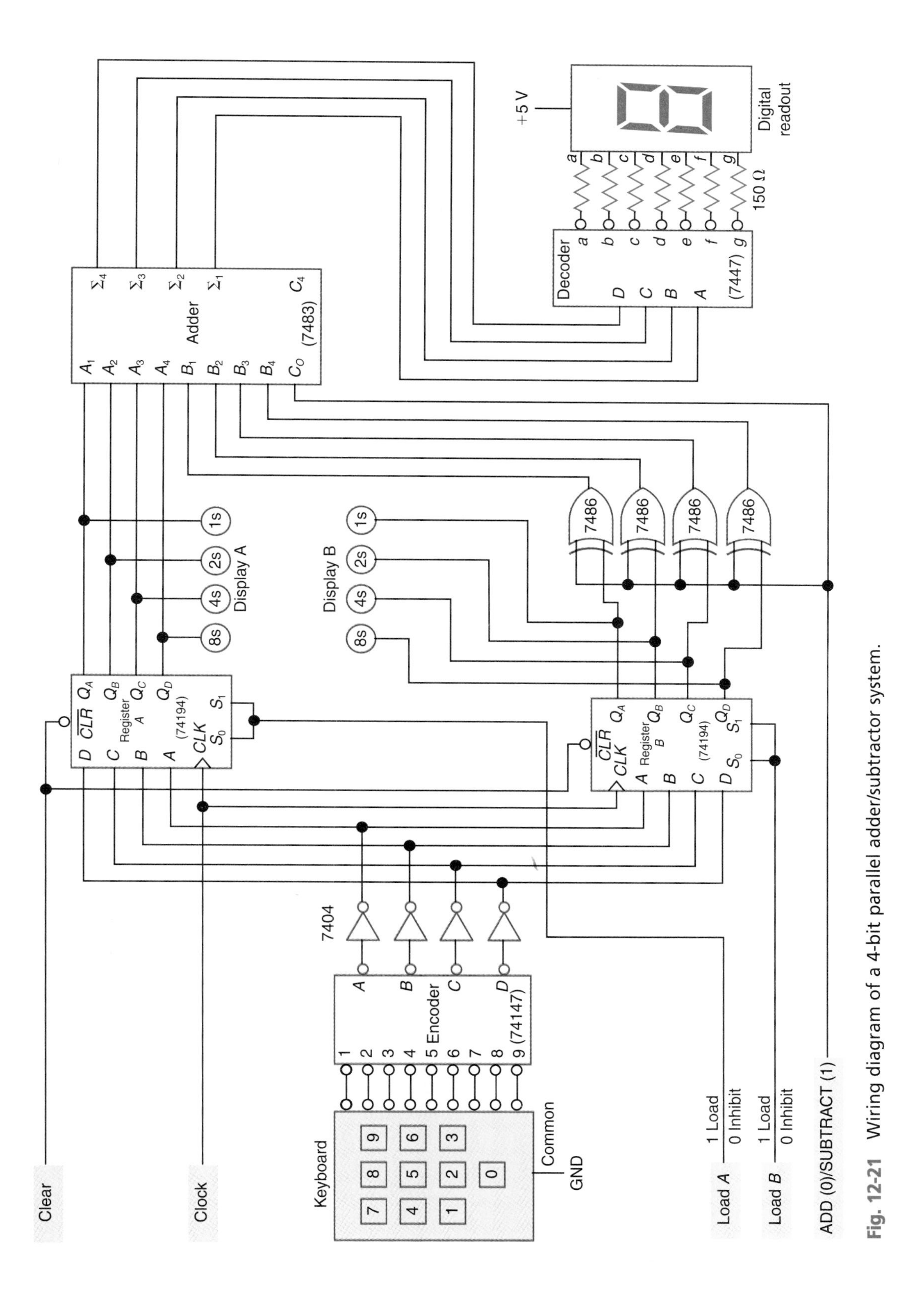

Fig. 12-21 Wiring diagram of a 4-bit parallel adder/subtractor system.

for subtraction, the assumptions are (1) the minuend is equal to or larger than the subtrahend, and (2) we are using unsigned numbers.

The 74147 encoder changes the decimal input from the keyboard to a binary number in Fig. 12-21. Note that the 74147 encoder has active LOW inputs and outputs. The 7404 IC inverts the outputs of the 74147 encoder back to true binary. From the inverters, the binary is fed into the parallel load data inputs of both registers A and B. With one clock pulse the parallel data at the input of the register is transferred and stored in the registers (only if S_0 and S_1 are both HIGH). Displays A and B indicate the binary numbers stored in registers A and B. These numbers are applied to the inputs of the 7483 4-bit adder. The 7483 IC adds binary numbers $A_4A_3A_2A_1$ and $B_4B_3B_2B_1$, with the sum appearing in binary at the outputs of the adder. The sum is decoded by the 7447 IC with the decimal sum appearing on the digital readout.

The 7486 XOR gates have no effect on the data flowing out of register B to the 7483 IC when the adder/subtractor system in Fig. 12-21 is in the *add mode*. The system is converted to the *subtract mode* when the add/subtract mode control input goes HIGH. The XOR gates now invert the subtrahend to its 1s complement form, which is applied to the $B_4B_3B_2B_1$ inputs of the adder. The C_0 input to the 7483 adder IC is held HIGH during subtraction and LOW during addition.

TEST

Supply the missing word in each statement.

41. Refer to Fig. 12-20. The ________ translates the decimal input from the keyboard to BCD.
42. Refer to Fig. 12-20. Registers *A* and *B* hold data at the inputs of the ________ while a calculation is performed.
43. Refer to Fig. 12-21. To perform addition, the add/subtract mode control must be ________ (HIGH, LOW).
44. Refer to Fig. 12-21. When activated, the clear input zeros both registers and the digital readout will display ________ (0, the previous number).
45. Refer to Fig. 12-21. Assume these inputs: clear = 1, load A = 1, load B = 1, add/subtract = 0. Pressing the number 4 on the keyboard and at the same time applying a positive pulse to the clock input will yield these results: display A = ________, display B = ________, and the digital display = ________.
46. Refer to Fig. 12-21. The 7486 XOR gates change the data flowing from register B to the adder IC to its 1s complement form during ________ (addition, subtraction).
47. Refer to Fig. 12-21. During subtraction, display A holds the ________ (minuend, subtrahend) while display B contains the ________ (minuend, subtrahend) and the result or difference appears in decimal form on the seven-segment display.
48. Refer to Fig. 12-21. Pressing 7 on the keyboard activates input 7 on the 74147 IC with a ________ (HIGH, LOW) causing the outputs of the encoder to give the 4-bit 1s complement of 7 or ________.
49. Refer to Fig. 12-21. The ________ (7404, 74194) ICs act as temporary memory devices in this adder/subtractor system.
50. Refer to Fig. 12-21. The ________ (7447, 7486) IC converts the binary output of the adder to decimal form in this adder/subtractor system.

12-11 The Digital Clock

We introduced a digital electronic clock in Chap. 8 and noted that various *counters* are the heart of a digital clock system. Figure 12-22(*a*) is a simple block diagram of a digital clock system. Many clocks use the power-line frequency of 60 Hz as their input or frequency standard. This frequency is divided into seconds, minutes, and hours by the *frequency divider* section of the clock. The one-per-second, one-per-minute, and one-per-hour pulses are then counted and stored in the *count accumulator* section of the clock. The stored contents of the count accumulators (seconds, minutes, hours) are then *decoded,* and the correct time is shown on the output *time displays.* The digital clock has the typical elements of a system. The input is the 60-Hz alternating current. The processing takes place in the frequency divider, count accumulator, and decoder sections. Storage takes place in the count accumulators. The control section is illustrated by the *time-set* control, as shown in Fig. 12-22(*a*). The output section is the digital time display.

Frequency divider

Count accumulator

It was mentioned that all systems consist of logic gates, flip-flops, and subsystems. The diagram in Fig. 12-22(*b*) shows how subsystems are organized to display time in hours, minutes, and seconds. This is a more detailed diagram of a digital clock. The input is still a 60-Hz signal. The 60 Hz may be from the low-voltage secondary coil of a transformer. The 60 Hz is divided by 60 by the first frequency di-

Digital clock features

Frequency divider circuits

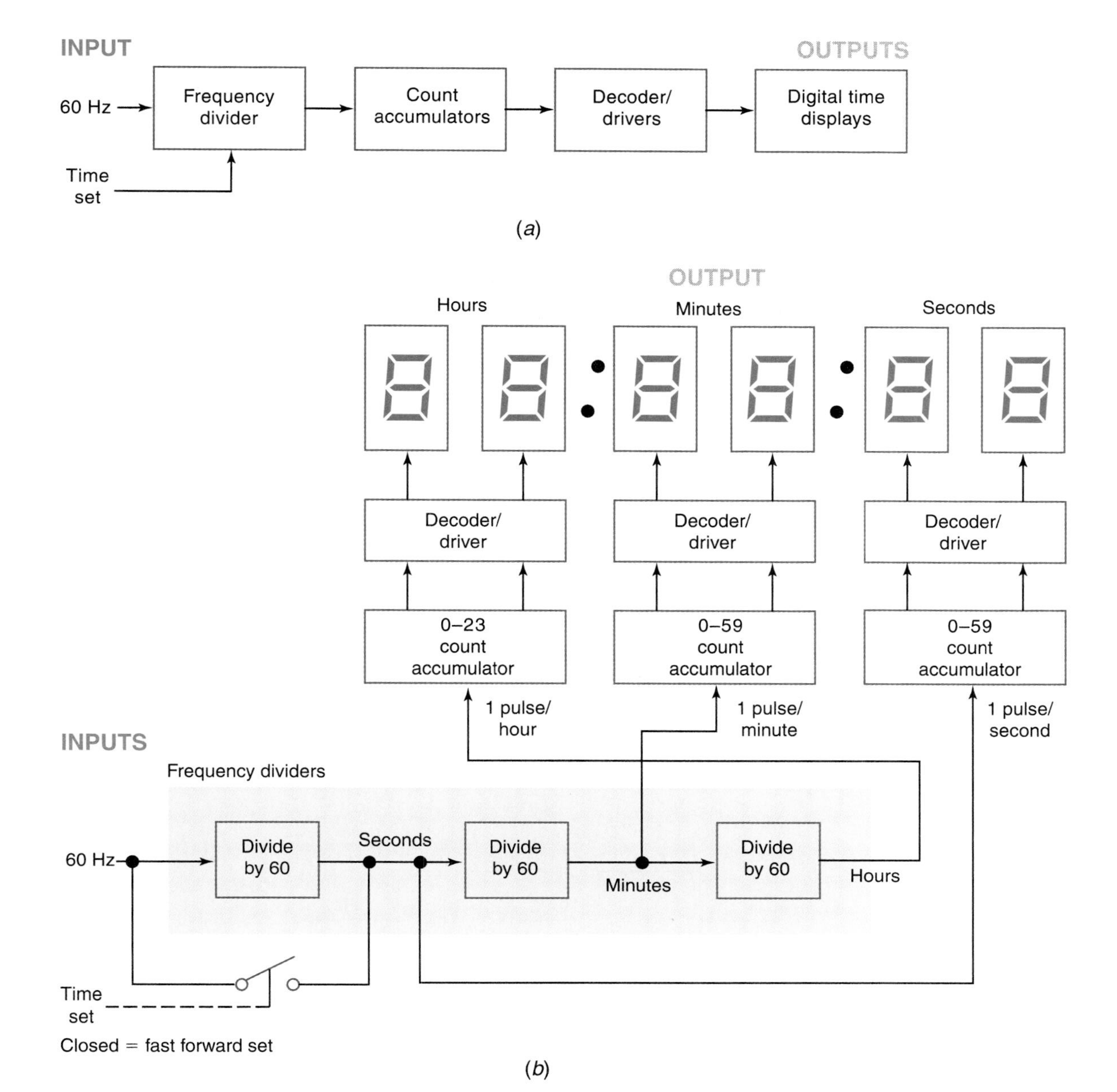

Fig. 12-22 (*a*) Simplified block diagram of a digital clock. (*b*) More detailed block diagram of a digital clock.

Digital clock as a system

vider. The output of the first divide-by-60 circuit is 1 pulse per second. The 1 pulse per second is fed into an *up counter* that counts upward from 00 through 59 and then resets to 00. The seconds counters are then decoded and displayed on the 2 seven-segment LED displays at the upper right, Fig. 12-22(*b*).

Consider the middle frequency-divider circuit in Fig. 12-22(*b*). The input to this divide-by-60 circuit is 1 pulse per second; the output is 1 pulse per minute. The 1-pulse-per-minute output is transferred into the 0 to 59 minutes counter. This up counter keeps track of the number of minutes from 00 through 59 and then resets to 00. The output of the minutes count accumulator is decoded and displayed on the two 7-segment LEDs at the top center, Fig. 12-22(*b*).

Now for the divide-by-60 circuit on the right in Fig. 12-22(*b*). The input to this frequency divider is 1 pulse per minute. The output of this circuit is 1 pulse per hour. The 1-pulse-per-hour output is transferred to the hours counter on the left. This hours count accumulator keeps track of the number of hours from 0 to 23. The output of the hours count accumulator is decoded and transferred to the 2 seven-segment LED displays at the upper left,

Fig. 12-22(*b*). You probably have noticed already that this is a 24-h digital clock. It easily could be converted to a 12-h clock by changing the 0 to 23 count accumulator to a 1 to 12 counter.

For setting the time a time-set control has been added to the digital clock in Fig. 12-22(*b*). When the switch is closed (a logic gate may be used), the display counts forward at a fast rate. This enables you to set the time quickly. The switch bypasses the first divide-by-60 frequency divider so that the clock moves forward at 60 times its normal rate. An even faster *fast-forward* set could be used by bypassing both the first and the second divide-by-60 circuits. The latter technique is common in digital clocks.

What is inside the divide-by-60 frequency dividers in Fig. 12-22(*b*)? In Chap 8. we spoke of a counter being used to divide frequency. Figure 12-23(*a*) is a block diagram of how a divide-by-60 frequency divider might be organized. Notice that a divide-by-6 counter is feeding a divide-by-10 counter. The entire unit divides the incoming frequency by 60. In this example, the 60-Hz input is reduced to 1 Hz at the output.

A detailed wiring diagram for a divide-by-60 counter circuit is drawn in Fig. 12-23(*b*). The three JK flip-flops and NAND gate form the divide-by-6 counter while the 74192 decade counter performs as a divide-by-10 unit. If 60 Hz enters at the left, the frequency will be reduced to 1 Hz at output Q_D of the 74192 counter.

The seconds and minutes count accumulators in Fig. 12-22(*b*) are also counters. The 0 to 59 is a decade counter cascaded with a 0 to 5 counter. The decade counter drives the 1s place of the displays. The mod-6 counter drives the 10s place of the displays. In a like manner, the hours count accumulator is a decade counter cascaded with a 0 to 2 counter. The decade counter drives the 1s place in the hours display. The mod-3 counter drives the 10s place of the hours display.

In many practical digital clocks the output may be in hours and minutes only. Most digital clocks are based upon one of many inexpensive ICs. Large-scale-integrated *clock chips*

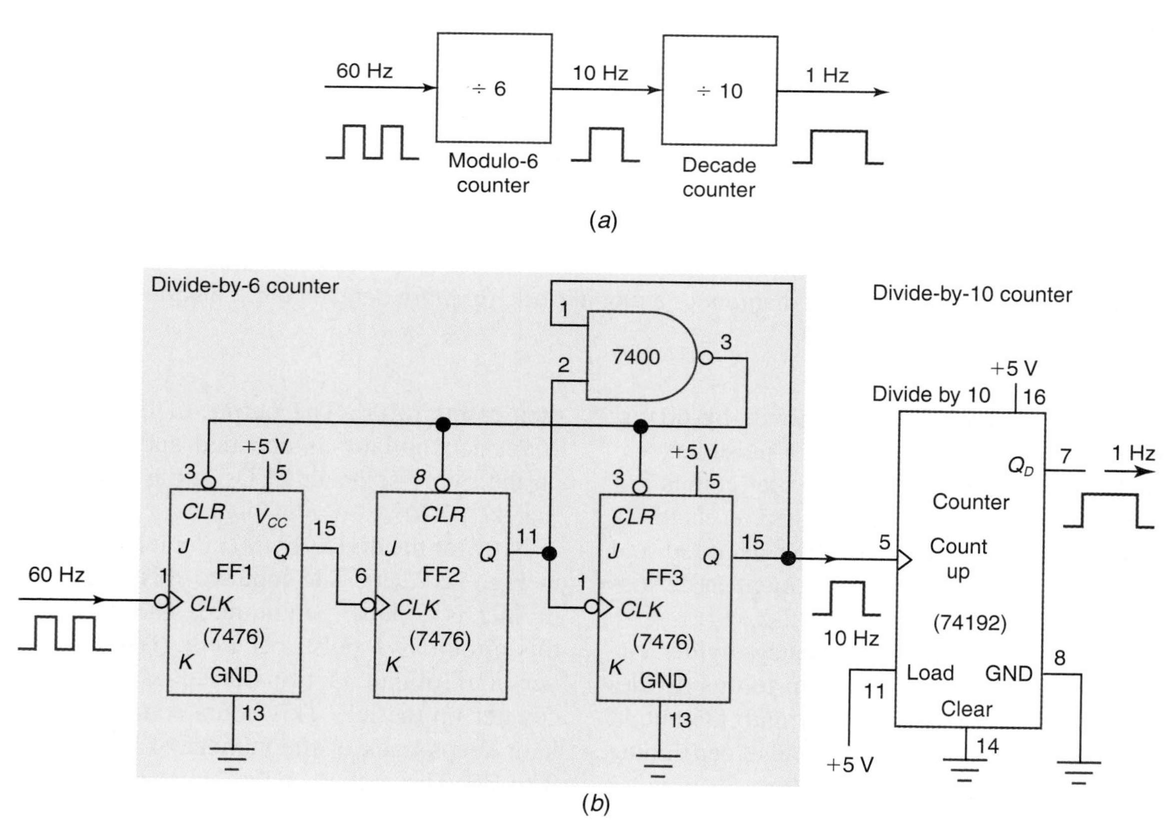

Fig. 12-23 Divide-by-60 counter. (*a*) Block diagram. (*b*) Wiring diagram using TTL ICs.

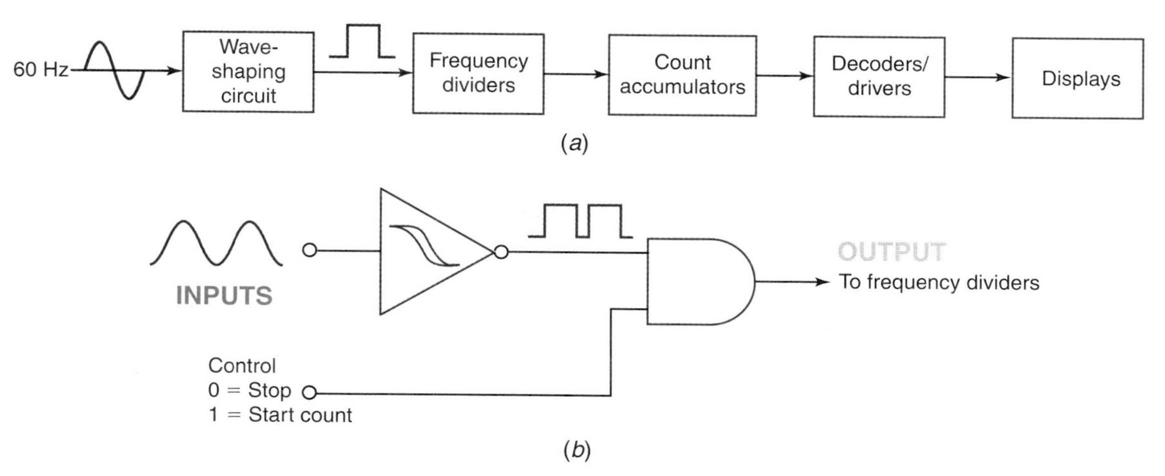

Fig. 12-24 Wave shaping. (*a*) Adding a wave-shaping circuit to the input of the digital clock system. (*b*) Schmitt trigger inverter used as a wave shaper.

Wave shaping using Schmitt trigger device

have all the frequency dividers, count accumulators, and decoders built into a single IC. For only a few extra dollars, clock chips have other features, such as 12- or 24-h outputs, calendar features, alarm controls, and radio controls.

An added feature you will use when you construct a digital timepiece is shown in Fig. 12-24(*a*). A *wave-shaping circuit* has been added to the block diagram of our digital clock. The IC counters that make up the frequency-divider circuit do not work well with a sine-wave input. The sine wave (shown at the left in Fig. 12-24(*a*)) has a *slow rise time* that does not trigger the counter properly. The sine-wave input must be converted to a square wave. The wave-shaping circuit changes the sine wave to a square wave. The square wave will now properly trigger the frequency-divider circuit.

Commercial LSI clock chips have wave-shaping circuitry built into the IC. In the laboratory you may use a *Schmitt trigger inverter IC* to square up the sine waves as you did in Chap. 7. A simple wave-shaping circuit is shown in Fig. 12-24(*b*). This circuit uses the TTL 7414 Schmitt trigger inverter IC to convert the sine wave to a square wave. The circuit in Fig. 12-24(*b*) also contains a *start/stop control.* When the control input is HIGH, the square wave from the Schmitt trigger inverter passes through the AND gate. When the control input goes LOW, the square-wave signal is inhibited and does not pass through the AND gate. The counter is stopped.

You will want to get some practical knowledge of how counters are used in dividing frequency. Remember that the counter subsystem is used for two jobs in the digital timepiece: first to *divide frequency* and second to *count* upward and *accumulate* or store the number of pulses at its input.

Frequency division

Count accumulation

Wave-shaping circuit

Slow rise time

Schmitt trigger inverter IC

Start/stop control

Clock module

TEST

Answer the following questions.

51. Refer to Fig. 12-22(*a*). Counters are used in the ________ and ________ sections of a digital clock.
52. Refer to Fig. 12-24. When operating a clock with a sine wave, a(n) ________ circuit is added to the clock.
53. Refer to Fig. 12-24(*b*). What is the purpose of the Schmitt trigger inverter in this circuit?
54. Refer to Fig. 12-24(*b*). What is the purpose of the AND gate in this circuit?

12-12 The LSI Digital Clock

The LSI clock chip forms the heart of modern digital timepieces. These digital clock chips are made as monolithic MOS ICs. Many times, the MOS LSI chip, or die, is mounted in an 18-, 24-, 28-, or 40-pin DIP IC. Other times, the MOS LSI chip is mounted directly on the PC board of a clock module. The tiny silicon die is sealed under an epoxy coating. Examples of the two packaging methods are shown in Fig. 12-25 on page 320. An MOS LSI clock IC packaged in a 24-pin DIP is illustrated in Fig. 12-25(*a*). Pin 1 of the DIP IC is identified in the normal manner (pin 1 is immediately ccw from the notch). A National Semiconductor clock module is sketched in Fig. 12-25(*b*). The

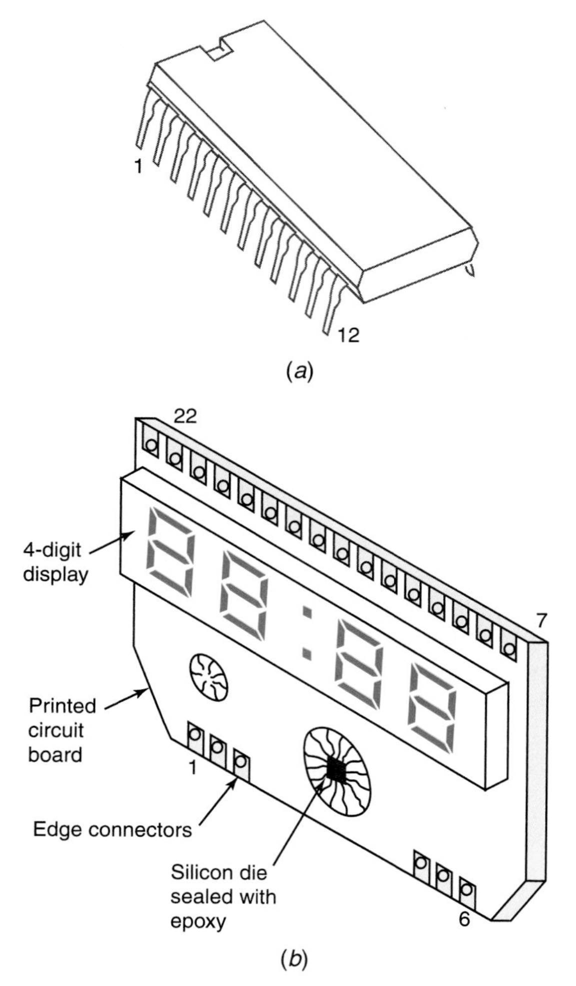

Fig. 12-25 (*a*) An LSI clock chip in a 24-pin dual in-line package. (*b*) A typical clock module containing a MOS/LSI die.

back is a PC board with 22 edge connectors. The numbering of the edge connectors is shown. A four-digit LED display is premounted on the board with all connections complete. Some clock modules have some discrete components and a DIP clock IC mounted on the board. The clock module in Fig. 12-25(*b*) has the tiny silicon chip, or die, mounted on the PC board. It is sealed with a protective epoxy coating.

MM5314 MOS LSI clock IC

A block diagram of National Semiconductor's *MM5314 MOS LSI clock IC* is shown in Fig. 12-26(*a*). The pin diagram is shown in Fig. 12-26(*b*). Refer to Fig. 12-26(*a*) and (*b*) for the following functional description of the MM5314 digital clock IC.

50- or 60-Hz Input (Pin 16)

Alternating current or rectified ac is applied to this input. The *wave-shaping circuit* squares up the waveform. The shaping circuit drives a chain of counters which perform the time-keeping job.

50- or 60-Hz Select Input (Pin 11)

This input programs the *prescale counter* to divide by either 50 or 60 to obtain a 1-Hz, or 1-pulse-per-second time base. The counter is programmed for 60-Hz operation by connecting this input to V_{DD} (GND). If the 50/60-Hz select input pin is left unconnected, the clock is programmed for 50-Hz operation.

Time-setting Inputs (Pins 13, 14, and 15)

Slow- and fast-setting inputs as well as a hold input are provided on this clock IC. These inputs are enabled when they are connected to V_{DD} (GND). Typically, a normally open push-button switch is connected from these pins to V_{DD}. The three gates in the counter chain are used for setting the time. For *slow set,* the prescale counter is bypassed. For *fast set,* the prescale counter and seconds counter are bypassed. The *hold* input inhibits any signal from passing through gate *A* to the prescale counter. This stops the counters, and time does not advance on the output display.

12- or 24-H Select Input (Pin 10)

This input is used to program the hours counter to divide by either 12 or 24. The 12-h display format is selected by connecting this input to V_{DD} (GND). Leaving pin 10 unconnected selects the 24-h format.

Output MUX Operation (Pins 3 to 9 and 17 to 22)

The seconds, minutes, and hours counters continuously reflect the time of day. Outputs from each counter are *multiplexed* to provide digit-by-digit sequential access to the time data. In other words, only one display digit is turned on for a very short time, then the second, then the third, and so forth. By multiplexing the dis-

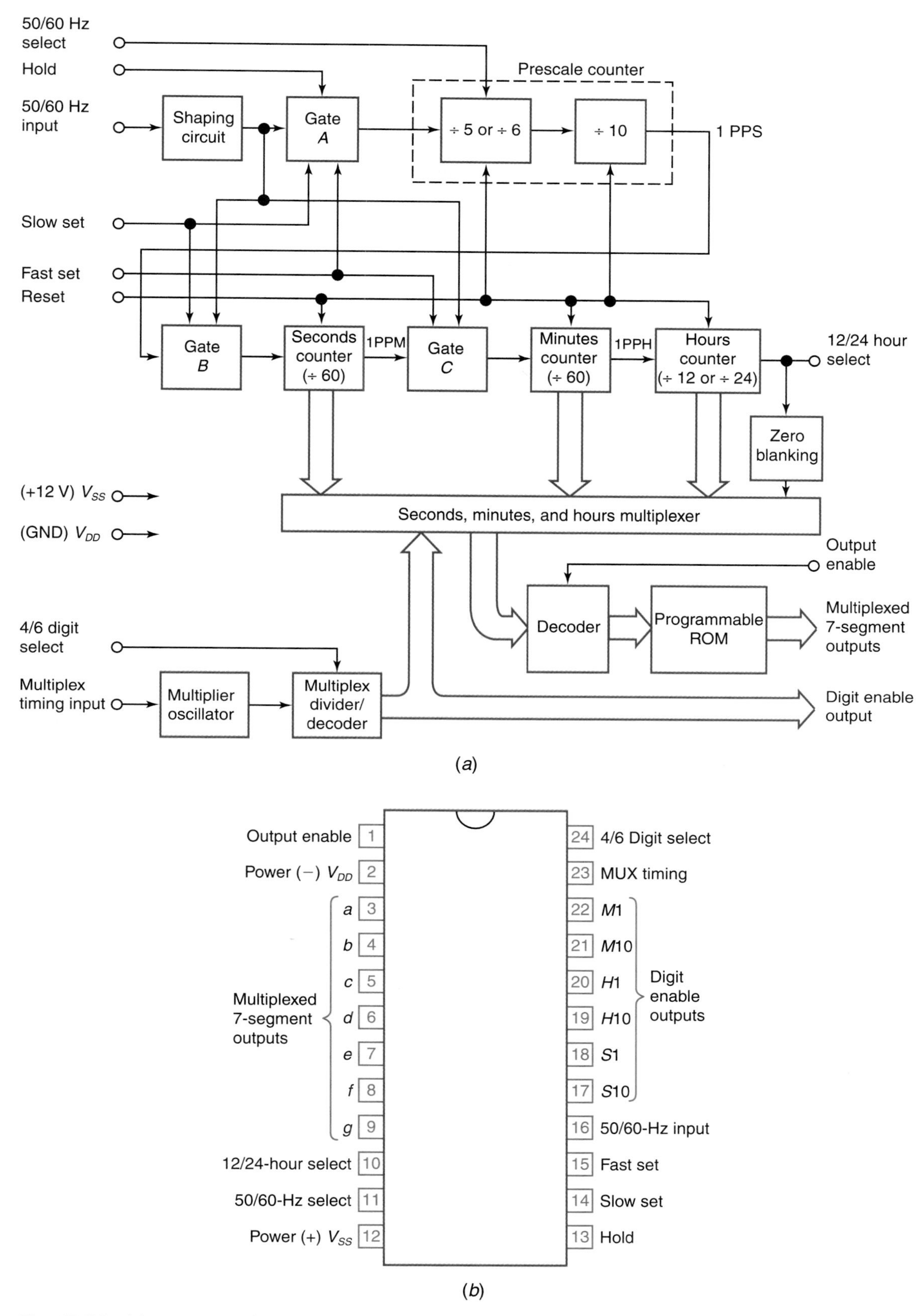

Fig. 12-26 (*a*) Functional block diagram of the MM5314 MOS/LSI clock chip. (*Courtesy of National Semiconductor Corporation.*) (*b*) Pin diagram for the MM5314 digital clock IC. (*Courtesy of National Semiconductor Corporation.*)

MM5314 MOS/LSI clock chip

plays instead of using 48 leads to the six displays (8 pins each × 6 = 48), only 13 output pins are required. These 13 outputs are the multiplexed seven-segment outputs (pins 3 through 9) and the digit enable outputs (pins 17 through 22).

Multiplexing and driving displays

The MUX is addressed by a *multiplex divider/decoder,* which is driven by a *multiplex oscillator.* The oscillator uses external timing components (resistor and capacitor) to set the frequency of the multiplexing function. The four/six-digit select input controls if the MUX turns on all six or just four displays in sequence. The *zero-blanking* circuit suppresses the 0 that would otherwise sometimes appear in the tens-of-hours display. The MUX addresses also become the display digit enable outputs (pins 17 to 22). The MUX outputs are applied to a decoder which is used to address a *PROM.* The PROM generates the final seven-segment output code. The displays are enabled in sequence from the unit seconds through the tens-of-hours display.

PROM

Multiplex Timing Input (Pin 23)

Relaxation oscillator (multiplex oscillator)

Adding a resistor and capacitor to the MM5314 clock IC forms a *relaxation oscillator.* The external resistor and capacitor are connected to the MUX timing input as shown in Fig. 12-27. Typical timing resistor and capacitor values might be 470 kΩ and 0.01 μF.

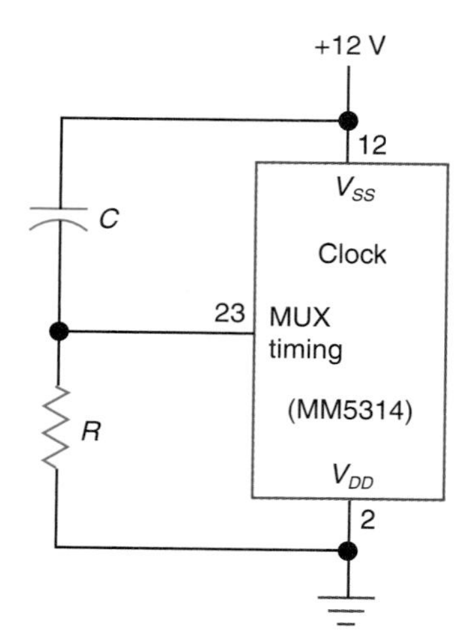

Fig. 12-27 Placement of the external resistor and capacitor used to set the frequency of the multiplex oscillator in the MM5314 clock IC.

Four/Six-Digit Select Input (Pin 24)

The four/six-digit select input controls the MUX. With no input connection, the clock outputs data for a four-digit display. Applying V_{DD} (GND) to this pin provides a six-digit display.

Output Enable Input (Pin 1)

With this pin unconnected, the seven-segment outputs are enabled. Switching V_{DD} (GND) to this input inhibits these outputs.

Power Inputs (Pins 2 and 12)

A dc 11- to 19-V nonregulated power supply operates the clock IC. The positive of the power supply connects to the V_{SS} (pin 12), while the negative connects to V_{DD} (pin 2).

TEST

Supply the missing word in each statement.

55. Digital clock LSI chips are made using ________ (bipolar, MOS) technology.
56. Refer to Fig. 12-26. If pin 16 of the MM5314 is at GND, the clock IC is programmed for ________-Hz operation.
57. Refer to Fig. 12-26. If the slow set input to the MM5314 IC is grounded, the ________ counter is bypassed.
58. Refer to Fig. 12-26. The MM5314 MOS/LSI clock chip requires a(n) ________-V nonregulated power supply.
59. The MM5314 clock chip ________ (directly drives, multiplexes) seven-segment displays.
60. The positive of the 12-V power supply is connected to the ________ (V_{DD}, V_{SS}) pin of the MM5314 clock chip.
61. The MM5314 clock chip ________ (has internal, needs external) wave-shaping circuitry to square up the incoming 60-Hz signal.
62. The MM5314 clock chip needs an external ________ (crystal, resistor and capacitor) connected to the MUX timing pin of the IC.

12-13 A Practical LSI Digital Clock System

Six-digit clock

MM5314 clock IC features

A *six-digit clock* using the MM5314 IC is sketched in Fig. 12-28(*a*). This student-built unit uses six common-anode seven-segment

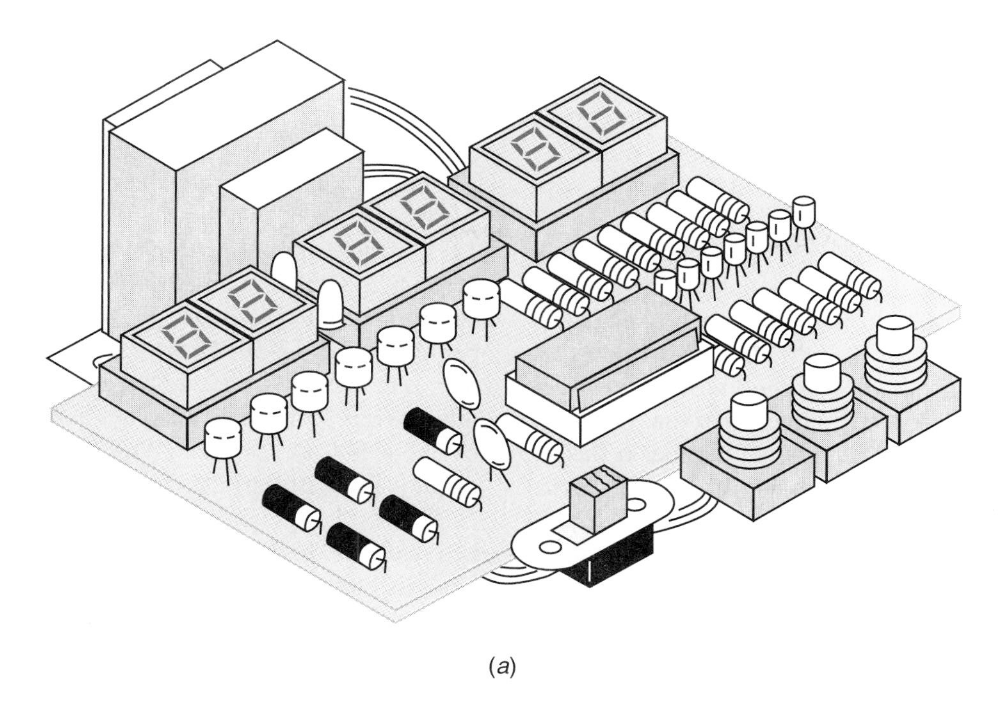

(a)

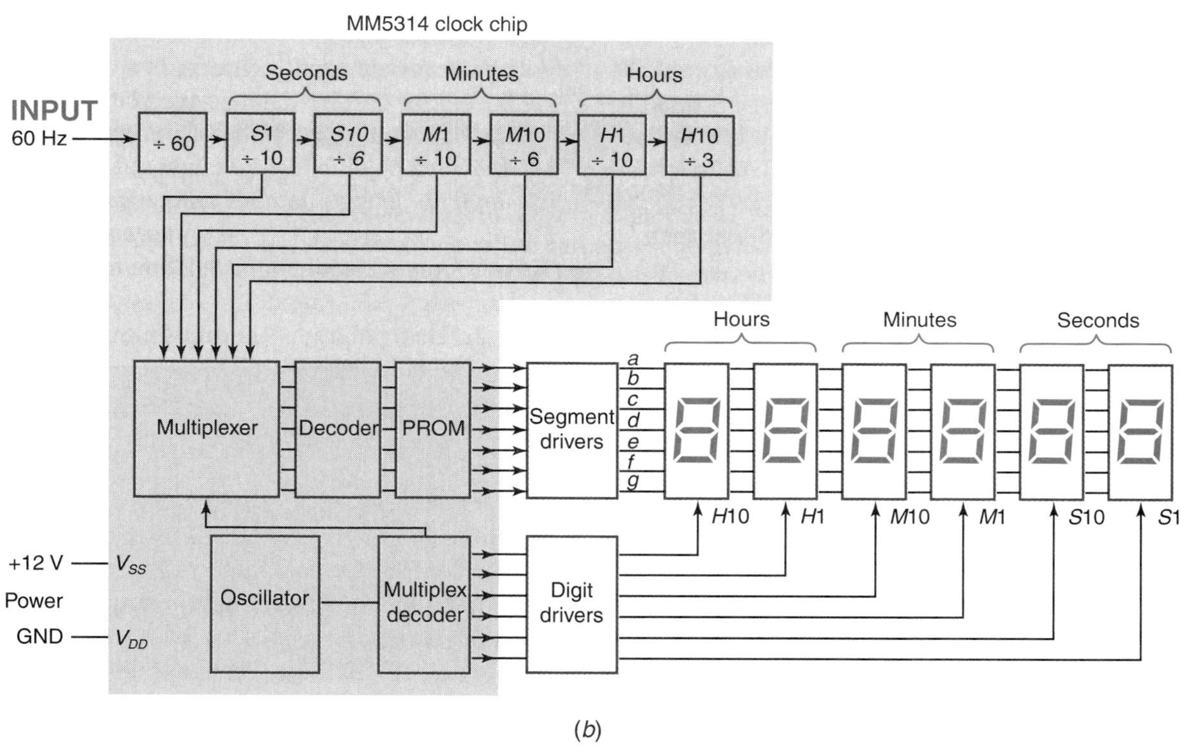

(b)

Fig. 12-28 (a) Sketch of a practical six-digit clock project. (b) Block diagram of the practical six-digit clock project using the MM5314 clock chip.

LED displays. Also notice the many extra parts used in the clock system. A block diagram of this system is shown in Fig. 12-28(*b*). The National Semiconductor MM5314 clock chip is used in this system. The 60 Hz is divided down to seconds, minutes, and hours by counters across the top in Fig. 12-28(*b*). These feed the MUX. The *oscillator* at the lower left produces a frequency of about 1 kHz.

Oscillator

Outside the MM5314 clock chip are 6 seven-segment common-anode LED displays. Because of the higher currents used by the LED

Segment drivers

Digit drivers

MUX

displays, *segment drivers* are used to sink the current from the cathodes of the displays. The *digit drivers* furnish an adequate amount of current to the anodes of the selected digit.

To help explain how the *MUX* works, suppose the time is 12:34:56. This information is held in the counters in the clock chip. The multiplex decoder first selects the *S*1 display. The MUX takes data from the *S*1 counter and places it on the decoder PROM. Segment lines *c, d, e, f,* and *g* are *activated to all the displays.* The multiplex decoder activates *only* the *S*1 line of the digit driver. The decimal number 6 flashes on for an instant, as shown in Fig. 12-29(*a*). The segments *c, d, e, f,* and *g* have been activated on all the displays, but only the right *S*1 unit has had its common-anode lead activated, or connected to V_{SS}. Therefore only the *S*1 display lights up.

RC filter network

Second, the clock IC's multiplex decoder selects the *S*10 display. The MUX finds that the *S*10 counter holds 5. The decoder-PROM-segment driver activates segments *a, c, d, f,* and *g*. Next, the common anode of the *S*10 display is activated, or connected to V_{SS}. A decimal 5 flashes on the *S*10 display. This is shown in Fig. 12-29(*b*).

One at a time, each display is activated by the multiplex decoder and digit driver. At the same time the MUX-decoder-PROM activates the proper segments. This is based on the present contents of the counters. Look over Fig. 12-29. This is one cycle through the six displays. This whole sequence (*a* through *f*) happens more than 100 times each second. This multiplexing, or scanning, occurs at a fast rate, and so the eye does not notice a flickering in the displays.

Visit the Intel website to learn about the latest developments in microprocessors.

A schematic diagram of the digital clock system using the MM5314 IC is shown in Fig. 12-30 on page 326. The step-down 12-V transformer (*T*1 with the bridge rectifier (*D*1–4) and filter capacitor (*C*1) form the dc power supply section of the clock. Alternating current voltage is taken off the transformer and coupled to the 50/60-Hz input (pin 16) of the clock chip through resistor *R*3. Capacitor *C*3 and resistor *R*4 determine the frequency of the multiplex oscillator. Placing a much larger value capacitor (perhaps 1 to 5 μF) across *C*3 slows down the multiplexing process to a point where you can see each display light in sequence.

The fast-set, slow-set, and hold normally open push-button switches (S_2, S_3, and S_4) are located at the lower left in Fig. 12-30. The action (fast set, slow set, or hold) is taken when these pins are connected through the switch to V_{DD}.

The *segment drivers* are the seven NPN transistors (Q_7 to Q_{13}) on the right of the IC in Fig. 12-30. These transistors sink the current from the displays when activated. The *digit drivers* are the six PNP transistors (Q_1 to Q_6) at the upper left in Fig. 12-30. These transistors connect *only one display anode at a time* to V_{SS}. The digit drivers scan the six displays at a frequency of about 500 to 1500 Hz. This activates each display about 100 to 200 times each second.

The two LEDs (D_6 and D_7 in Fig. 12-30) are activated 100 to 200 times each second and appear lit continuously. These LEDs form the colon between the hours and minutes displays on the completed clock. This colon may be seen in Fig. 12-28(*a*). Resistor *R*3, capacitor *C*2, and diode *D*5 form an *RC filter network.* This network is used to remove possible line voltage transients that could either cause the clock to gain time or damage the IC.

The 12/24-h select input (pin 10) on the MM5314 clock chip in Fig. 12-30 is connected to V_{DD}. This selects the 12-h format. The 50/60-Hz select input (pin 11) is connected to V_{DD}. This programs the IC for 60-Hz operation. The four/six-digit select input (pin 24) is connected to V_{DD}. This programs the multiplex decoder to provide a six-digit display.

TEST

Answer the following questions.

63. Refer to Fig. 12-29. This is an example of the MM5314 clock chip _________ (counting, decoding, multiplexing) six LED decimal displays.
64. Refer to Fig. 12-30. The six PNP transistors function as _________ in this digital clock circuit.
65. Refer to Fig. 12-30. The seven NPN transistors function as _________ in this digital clock circuit.
66. Refer to Fig. 12-30. With the 12/24-h select input grounded, the clock is programmed as a(n) _________-h clock.
67. Refer to Fig. 12-30. Which two components (outside the clock chip) are responsible for determining the frequency of the multiplexer?

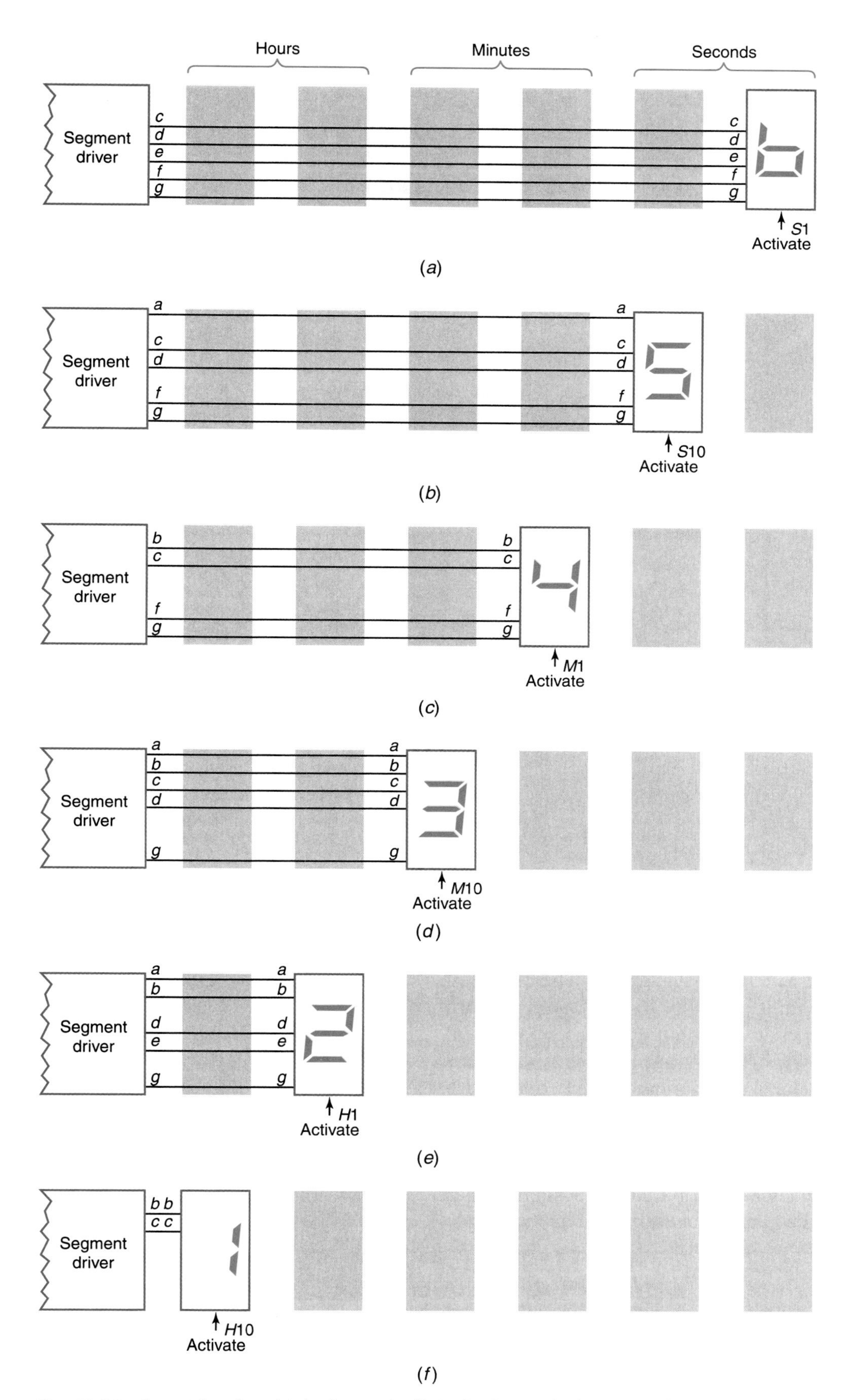

Fig. 12-29 Example of multiplexing a six-digit display with the time of day at 12:34:56. The entire sequence from (a) through (f) occurs in about 0.01 s.

Multiplexing a six-digit display

Digital clock circuit

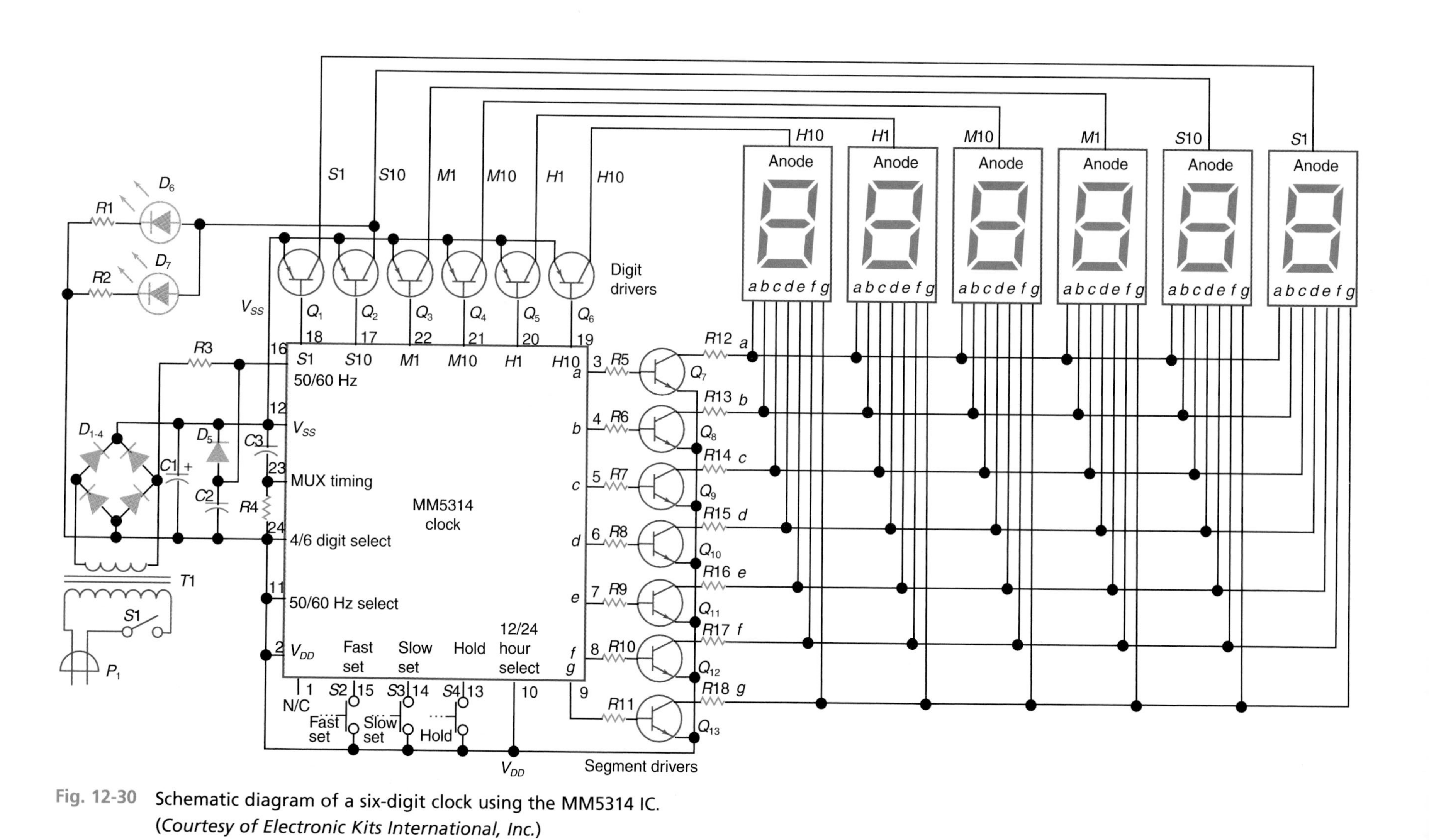

Fig. 12-30 Schematic diagram of a six-digit clock using the MM5314 IC. (*Courtesy of Electronic Kits International, Inc.*)

68. Refer to Fig. 12-30. The 4/6 digital select (pin 24) input to the MM5314 clock chip is connected to _________ (V_{DD}, V_{SS}), which selects the six-digit mode.
69. Refer to Fig. 12-29. The multiplexing of the seven-segment LED displays occurs at about _________ Hz in this digital clock.
70. Refer to Fig. 12-30. The dc power supply consists of the ac source, step-down transformer, bridge rectifier (D_1 to D_4), and filter capacitor _________ (C_1, C_3).
71. Refer to Fig. 12-30. The voltage that is applied to pin 16 of the IC through resistor R_3 is _________ (pure dc, 60 Hz pulsating dc or ac).
72. Refer to Fig. 12-28. The _________ (oscillator, PROM) in the clock IC selects which segments of the seven-segment LED display will light.

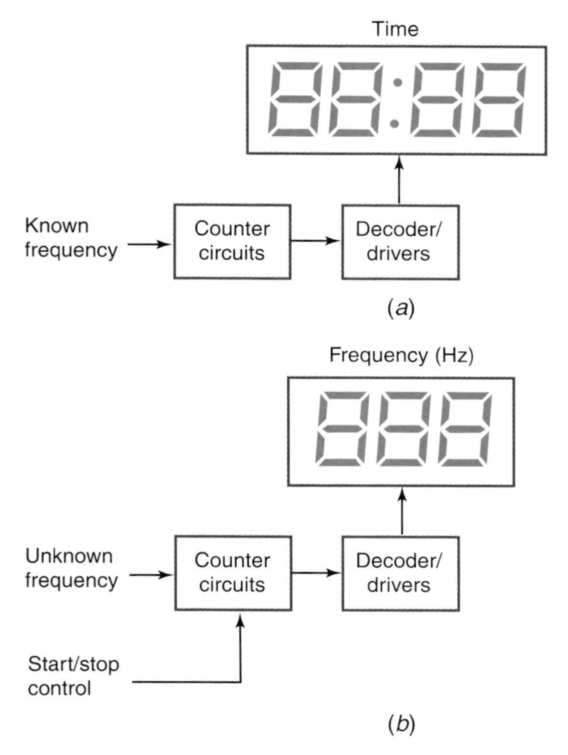

Fig. 12-31 (*a*) Simplified block diagram of a digital clock. (*b*) Simplified block diagram of a digital frequency counter.

12-14 The Frequency Counter

Frequency counter

An instrument used by technicians and engineers is the *frequency counter.* A digital frequency counter shows in decimal numbers the frequency in a circuit. Counters can measure from low frequencies of a few cycles per second (hertz, Hz) up to very high frequencies of thousands of megahertz (MHz). Like a digital clock, the frequency counter uses decade counters.

As a review, the block diagram for a digital clock is shown in Fig. 12-31(*a*). The known frequency is divided properly by the counters in the clock. The counter outputs are decoded and displayed in the time display. Figure 12-31(*b*) shows a block diagram of a frequency counter. Notice that the frequency counter circuit is fed an *unknown* frequency instead of the known frequency in a digital clock. The counter circuit in the frequency counter in Fig. 12-31(*b*) also contains a *start/stop control.*

The frequency counter has been redrawn in Fig. 12-32(*a*) on page 328. Notice that an AND gate has been added to the circuit. The AND gate controls the input to the decade counters. When the start/stop control is at logical 1, the unknown frequency pulses pass through the AND gate and on to the decade counters. The counters count upward until the start/stop control returns to logical 0. The 0 turns off the control gate and stops the pulses from getting to the counters.

Figure 12-32(*b*) is a more exact timing diagram of what happens in the frequency counter. Line *A* shows the start/stop control at logical 0 on the left and then going to 1 for *exactly 1 s.* The start/stop control then returns to logical 0. Line *B* diagrams a continuous string of pulses from the unknown frequency input. The unknown frequency and the start/stop control are ANDed together as we saw in Fig. 12-32(*a*). Line *C* in Fig. 12-32(*b*) shows only the pulses that are allowed through the AND gate. These pulses trigger the up counters. Line *D* shows the count observed on the displays. Notice that the displays start cleared to 00. The displays then count upward to 11 during the 1 s. The unknown frequency in line *B* in Fig. 12-32(*b*) is shown as 11 Hz (11 pulses/s).

Start/stop control

A somewhat higher frequency is fed into the frequency counter in Fig. 12-32(*c*). Again line *A* shows the start/stop control beginning at 0. It is then switched to logical 1 for *exactly 1 s.* It is then returned to logical 0. Line *B* in Fig. 12-32(*c*) shows a string of higher-frequency pulses. This is the unknown frequency being measured by this digital frequency counter. Line *C* shows the pulses that trigger the decade

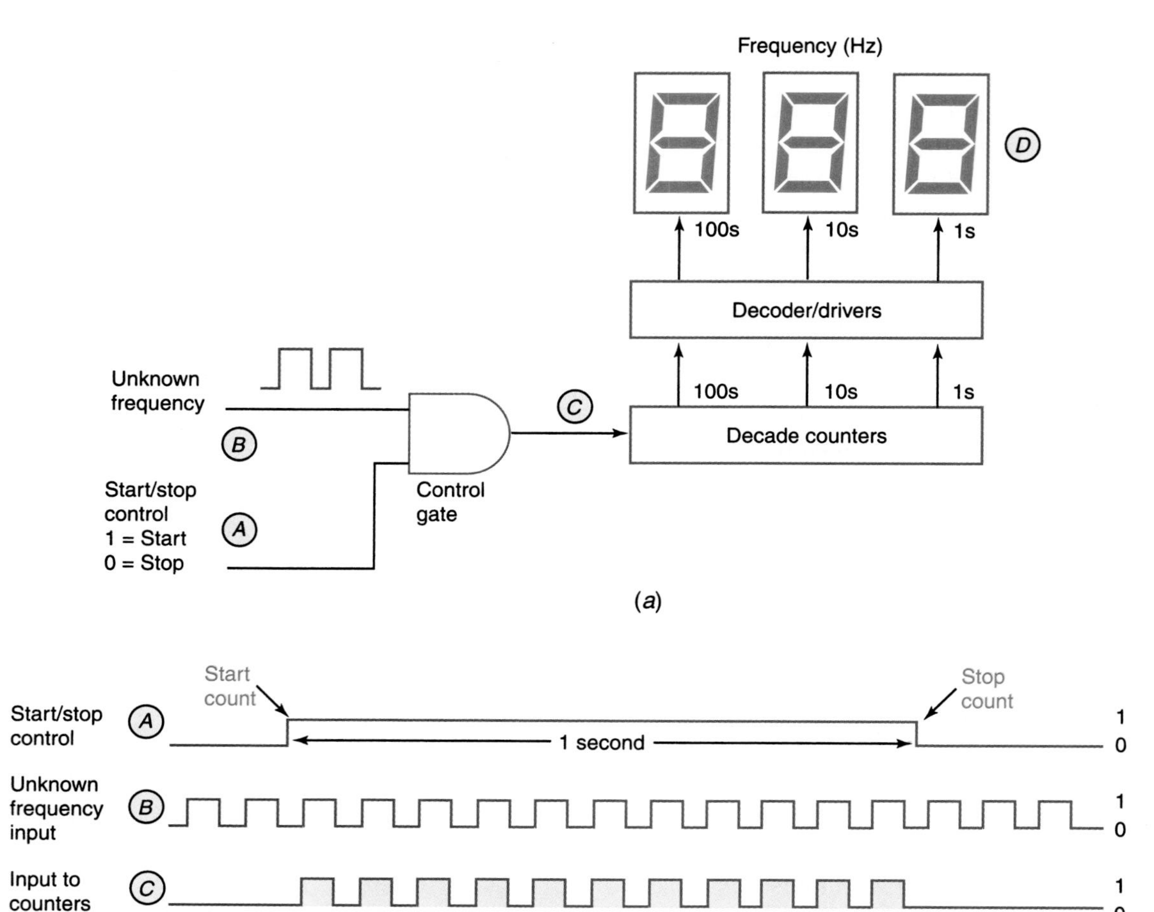

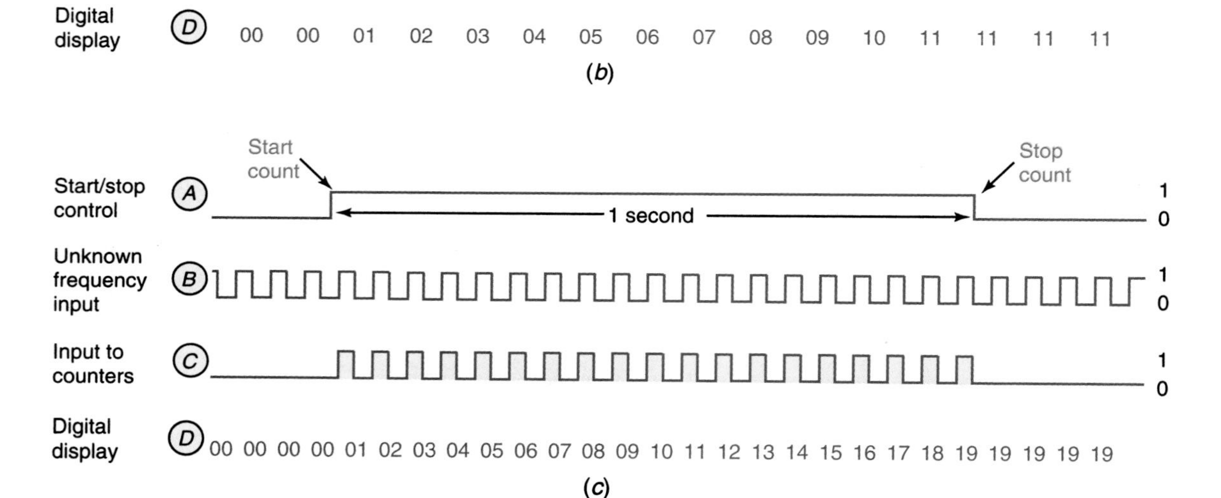

Digital frequency counter

Fig. 12-32 (*a*) Block diagram of a digital frequency counter showing the start/stop control. (*b*) Waveform diagram for an unknown frequency of 11 Hz. (*c*) Waveform diagram for an unknown frequency of 19 Hz.

Reset-count-display sequence

counters during the 1-s count-up period. The decade counters sequence upward to 19, as shown in line *D*. The unknown frequency in Fig. 12-32(*c*) is then 19 Hz.

If the unknown frequency were 870 Hz, the counter would count from 000 to 870 during the 1-s count period. The 870 would be displayed for a time, and then the counters would be reset to 000 and the frequency counted again. This *reset-count-display sequence* is repeated over and over.

Notice that the start/stop control pulse (count

pulse) must be *very accurate.* Figure 12-33 shows how a count pulse can be generated by using an accurate known frequency, such as the 60 Hz from the power line. The 60-Hz sine wave is converted to a square wave by the wave-shaping circuit. The 60-Hz square wave triggers a counter that divides the frequency by 60. The output is a pulse *1 s* in length. This *count pulse* turns on the control circuit when it goes high and permits the unknown frequency to trigger the counters. The unknown frequency is applied to the counters for 1 s.

Remember that the frequency counter goes through the reset-count-display sequence. So far we have shown only the count part of this sequence. The *counter reset circuit* is a group of gates that reset, or clear, the decade counters to 000 at the correct time—just before the count starts. Next, the 1-s count pulse permits the counters to count upward. The count pulse ends, and the unknown frequency is *displayed* on the seven-segment displays. In this circuit, the frequency is displayed in hertz. It is convenient to leave this display on the LEDs for a time. To do this, the divide-by-10 counter sends a pulse to the control circuit, which turns off the count sequence for 9 s. Events then happen like this. *Reset* the counters to 000. *Count* upward for 1 s. *Display* the unknown frequency for 9 s with no counts. Repeat the reset-count-display procedure every 10 s.

The frequency counter in Fig. 12-33 measures frequencies from 1 to 999 Hz. Notice the extensive use of counters in the divide-by-60, divide-by-10, and three decade counter circuits—hence the name frequency counter. The digital frequency counter actually counts the pulses in a given amount of time.

JOB TIP

Computer savvy is in high demand in most jobs.

One limitation of the counter diagramed in Fig. 12-33 is its top frequency; the top frequency that can be measured is 999 Hz. There are two ways to increase the top frequency of our counter. The first method is to add one or more counter-decoder-display units. We could extend the range of the frequency counter in Fig. 12-33 to a top limit of 9999 Hz by adding a single counter-decoder-display unit.

Counter reset circuit

The second method of increasing the frequency range is to count by 10s instead of 1s. This idea is illustrated in Fig. 12-34 on page 330. A divide-by-6 counter replaces the divide-by-60 unit in our former circuit. This makes the *count pulse* only 0.1 s long. The count pulse permits only one-tenth as many pulses through the control as with the 1-s pulse. This is the same as counting by 10s. Only three LED displays are used. The 1s display in Fig. 12-34 is only to show that a 0 must be added to the right of the three LED displays. The range of this frequency counter is from 10 to 9990 Hz.

In the circuit in Fig. 12-34, the decade counters count upward for 0.1 s. The display is held on the LEDs for 0.9 s. The counters are then reset to 000. The count-display-reset procedure is then repeated. The circuit in Fig. 12-34 has

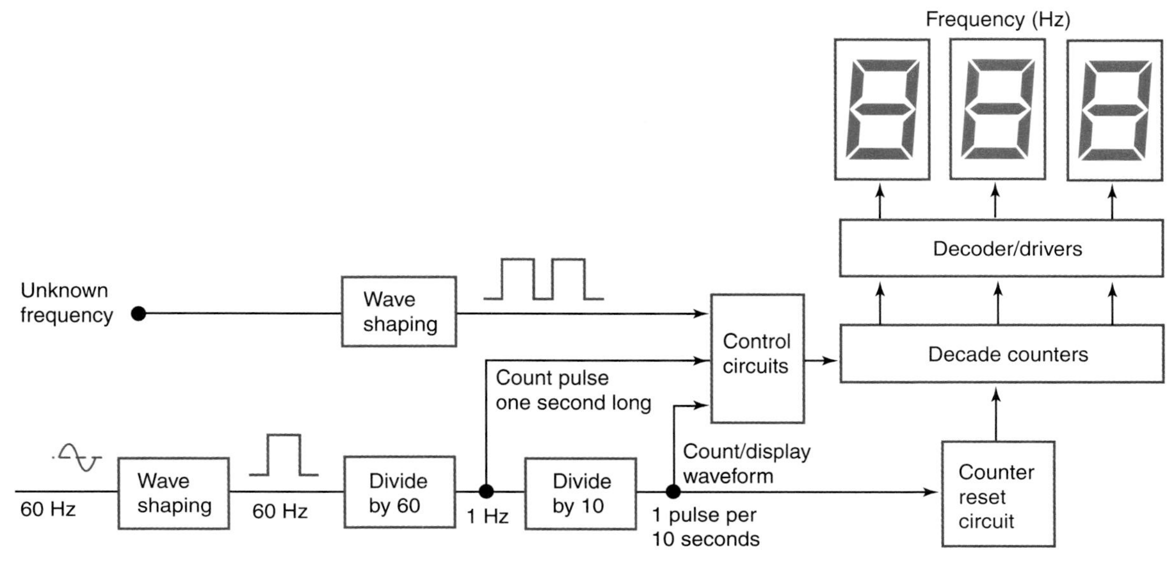

Fig. 12-33 More detailed block diagram of a digital frequency counter.

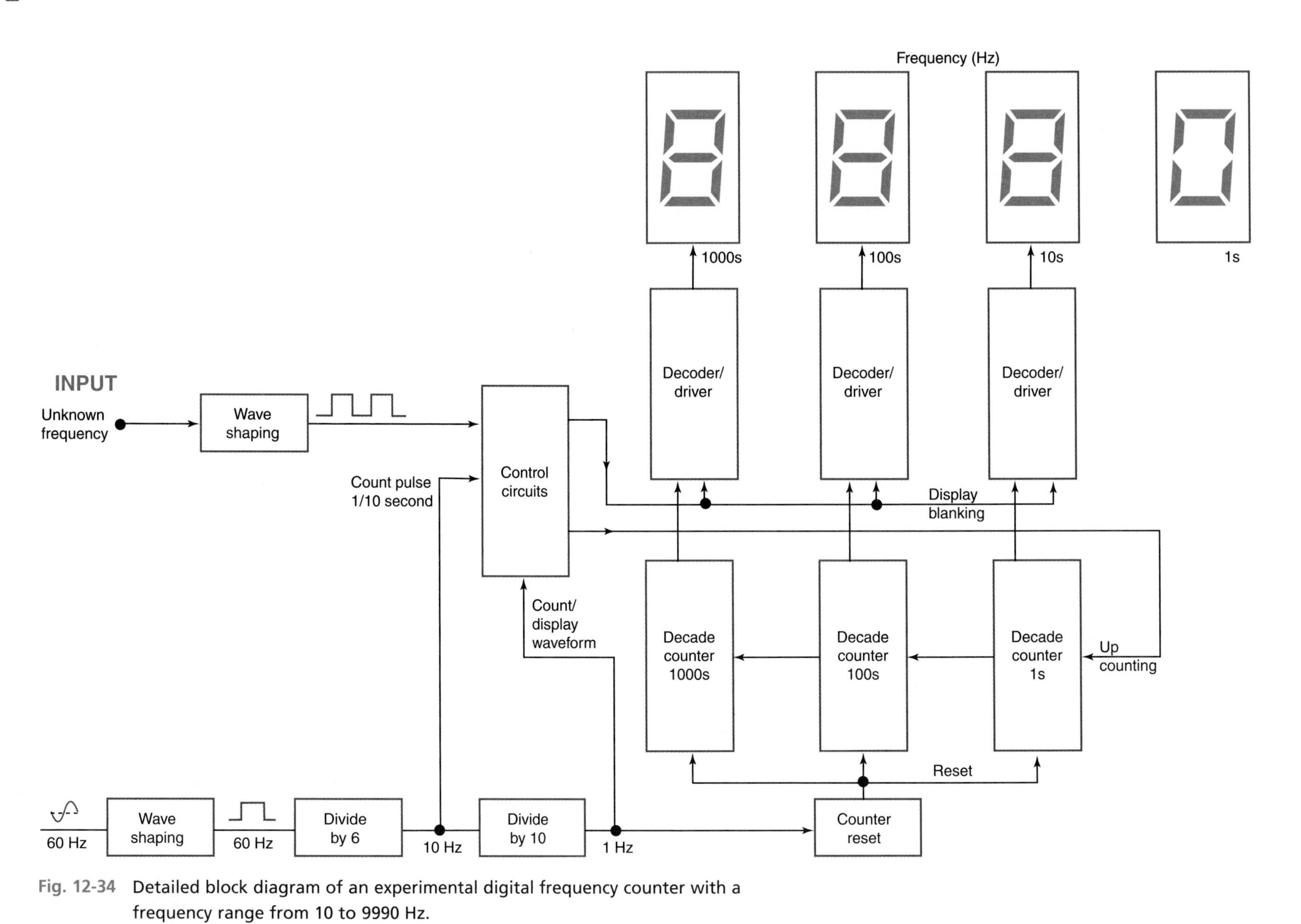

Fig. 12-34 Detailed block diagram of an experimental digital frequency counter with a frequency range from 10 to 9990 Hz.

Experimental frequency counter

one other new feature: during the count time the displays are blanked out. They are then turned on again when the unknown frequency is on the display. The sequence for this frequency counter is then reset, count (with displays blank), and, finally, the longer display period. This sequence repeats itself once every second while the instrument is being used.

The frequency counter diagramed in Fig. 12-34 is similar to one you can assemble in the laboratory from gates, flip-flops, and subsystems. It is suggested that you set up this complicated digital system because practical experience will teach you the details of the frequency counter system.

TEST

Supply the missing word in each statement.

73. Refer to Fig. 12-32. This digital frequency counter counts the number of pulses passing through the AND gate in ________ s.
74. Refer to Fig. 12-32. If the start/stop control to the AND gate is LOW, the signal at output *C* is a ________ (HIGH, LOW, square wave).
75. Refer to Fig. 12-33. The wave-shaping blocks might be implemented using TTL ________ (XOR gates, Schmitt trigger inverters).
76. Refer to Fig. 12-33. The divide-by-60 block might be implemented using ________ (counters, shift registers).
77. Refer to Fig. 12-34. The count pulse is ________ s long in this frequency counter.
78. Refer to Fig. 12-34. The unknown frequency input waveform is conditioned by a ________ circuit before entering the control circuitry of the counter.
79. Refer to Fig. 12-34. The decade counters serve the dual purpose of counting upward and ________ the count for display

12-15 An Experimental Frequency Counter

This section is based upon a frequency counter you can construct in the laboratory. Figure 12-35 on page 332 is a detailed wiring diagram of that frequency counter. This instrument was purposely designed using only components you have already used earlier in the book. This experimental frequency counter is not quite as accurate or stable as commercial units. Its maximum frequency is also limited to 9990 Hz, and its inputs are somewhat primitive.

The purposes for including the experimental frequency counter are as follows:

1. To show how SSI and MSI chips may be used to build digital subsystems and systems.
2. To demonstrate the concepts involved in the design and operation of a frequency counter.

Figure 12-34 is a block diagram of the frequency counter. Most components in the wiring diagram are in the same general position as in the block diagram.

At the lower left in Fig. 12-35, a 60-Hz sine wave is shaped into a square wave. The 60-Hz signal may come from the secondary of a low-voltage power transformer. The *wave shaping* is done by the 7414 Schmitt trigger inverter. This is the same unit we used earlier in Chap. 7 and in the digital clock to square up a sine wave. Remember that the divide-by-6 counter needs a square-wave input to operate properly.

Wave shaping

To the right of the lower 7414 inverter is a *divide-by-6 counter.* Three flip-flops (FF1, FF2, and FF3) and a NAND gate are wired to form the mod-6 counter. The frequency going into the divide-by-6 counter is 60 Hz; the frequency coming out of the counter (at *Q* of FF3) is 10 Hz. The 10 Hz is fed into the 7493 IC wired as a decade or divide-by-10 counter.

Divide-by-6 counter

Divide-by-10 counter

Figure 12-35 shows that the four outputs from the 7493 counter are NORed together (OR gate and inverter). The four-input NOR gate generates a 1-Hz signal. This 1-Hz signal

Experimental frequency counter

About Electronics

Electronic Vision

- You can't see your child through walls or metal objects— but a new wireless modem can. The miniature device, worn by both parent and child, locates a child who has wandered more than 15 feet away.
- Two roboticists took a 2800-mile vacation in a car steered by a notebook PC plugged into the lighter. The PC watched the road from a video camera. The vacationers operated only the brake and accelerator.

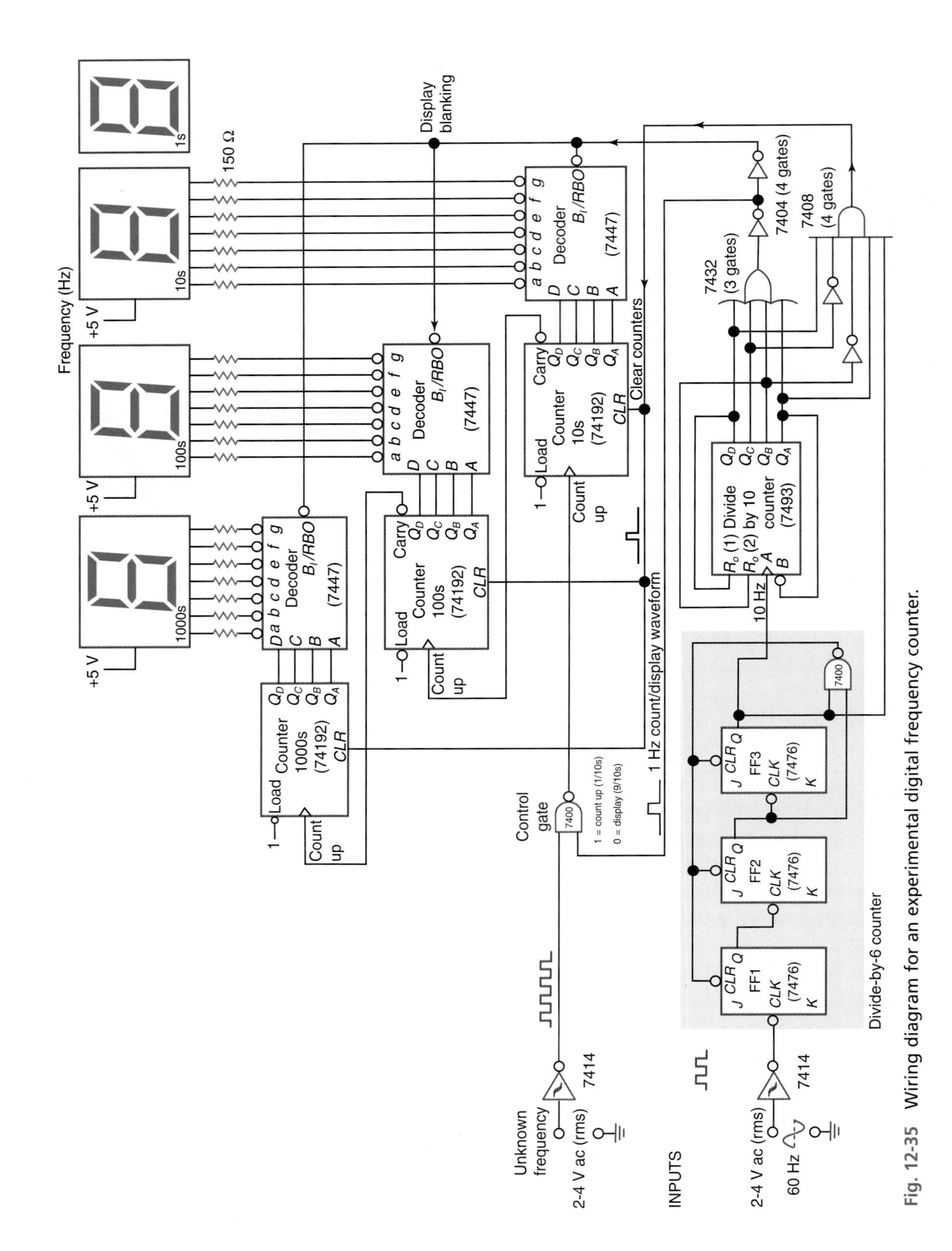

Fig. 12-35 Wiring diagram for an experimental digital frequency counter.

Experimental frequency counter

is called the *count/display waveform.* The count/display waveform is HIGH for exactly 0.1 s and low for 0.9 s. The count/display waveform is fed back into the 7400 control gate. When the counter/display waveform is HIGH for 0.1 s, the unknown frequency passes through the NAND gate on to the clock input of the 10s counter. When the count/display waveform is LOW for 0.9 s, the unknown frequency is blocked from passing through the

NAND control gate. It is during the 0.9 s that you may read the frequency off the seven-segment LED displays.

The frequency counter goes through a *reset-count-display* sequence. The *reset* pulse is generated by the five-input AND gate near the lower right in Fig. 12-35. It clears the 10s, 100s, and 1000s counters to zero. The reset (or counter clear pulse) is a very short positive pulse that occurs just before the counting occurs.

Next in the reset-count-display sequence is the *count* or *sampling time.* When the count/display waveform goes HIGH, the control gate is enabled and the unknown frequency passes through the NAND gate to the clock input of the 10s counter. Each pulse during this *sampling time* increments the 10s counter. When the 10s counter goes from 9 to 10, it carries the 1 to the 100s counter. After 0.1 s the count/display waveform goes LOW. This is the end of the sampling time. You will notice that the unknown frequency that was sampled causes the frequency to increase by 10s.

Last in the reset-count-display sequence is the *display time.* When the count/display waveform goes LOW the control gate is disabled. It is during this time that a stable frequency display may be read from the LEDs. Notice that an extra 1s display has been added in Fig. 12-35 to remind you that a 0 must be added to the right of the three active displays for a readout in hertz.

To make the displays look better, *display blanking* occurs during the count time of the reset-count-display sequence. The displays light normally with a stable readout during the display time. The display blanking waveform is a 0.1 s negative pulse generated by a 7404 inverter off the count/display waveform line. It causes the three displays to blank out for 0.1 s during the count time. The blanking does cause the displays to flicker. This problem could be cured by the use of latches to hold data on the inputs of the decoders.

For the most part commercial frequency counters operate just like the one in Fig. 12-35. Commercial counters usually have more displays and read out in kilohertz and megahertz. The experimental frequency counter needs an input signal of about 3 to 8 V to make it operate. Commercial counters usually have an amplifier circuit before the first wave-shaping circuit to amplify weaker signals to the proper level. Overvoltage protection is also provided with a zener diode. To get rid of the blinking of the display, commercial counters usually use a slightly different method of storing and displaying the contents of the counters. We used the powerline frequency of 60 Hz as our known frequency. Commercial frequency counters usually use an accurate high-frequency crystal oscillator to generate their known frequency.

JOB TIP

It is wise to use polite and ethical behavior when quitting a job.

Reset-count-display sequence

Some of the important specifications of commercial frequency counters are the *frequency range, input sensitivity, input impedance, input protection, accuracy, gate intervals,* and *display time.*

Frequency counter specifications

TEST

Supply the missing word(s) or number in each statement.

80. Refer to Fig. 12-35. FF1, FF2, FF3, and the NAND gate form a(n) ________ counter.
81. Refer to Fig. 12-35. The count time for this frequency counter is ________ s, while the display time is ________ s.
82. Refer to Fig. 12-35. The sampling time of the frequency counter is also called the ________ (count, display) time.
83. Refer to Fig. 12-35. The five-input AND gate generates a counter clear or ________ pulse. This is a ________ (negative, positive) pulse.
84. Refer to Fig. 12-35. The display blanking pulse is generated during the ________ (count, display) time. This is a ________ (negative, positive) pulse.
85. Refer to Fig. 12-35. The 7414 IC is called a hex ________ ________ inverter. The 7414 inverters are used for ________ shaping in this circuit.
86. Refer to Fig. 12-35. Each pulse from the unknown frequency that reaches the counter increases the frequency reading by ________ (1, 10, 100) Hz.
87. Refer to Fig. 12-35. The frequency range of this experimental counter is from a low of ________ Hz to a high of ________ Hz.
88. Refer to Fig. 12-35. The gate interval for the experimental frequency counter is

Sampling time (count time)

Display blanking

0.1 s while the display time is ________ seconds.

89. Refer to Fig. 12-35. The known frequency entering the experimental counter is ________ Hz.

Time-base clock

90. Refer to Fig. 12-35. The ________ (7447 decoder ICs, 74192 counter ICs) function as temporary memory devices to hold the highest count during the display time.

12-16 LCD Timer with Alarm

Most microwave ovens and kitchen stoves feature at least one timer with an alarm. Older appliances used mechanical timers, but modern microwaves and ranges feature electronic timers using digital circuitry. The concept of a timer system is sketched in Fig. 12-36(*a*). In this system, the keypad is the input and both the digital display and alarm buzzer are the output devices. The processing and storage of data occur within the digital circuits block in Fig. 12-36(*a*).

A somewhat more detailed block diagram of a digital timer is shown in Fig. 12-36(*b*). The digital circuits block has been subdivided into four blocks. They are the *time-base clock,* the *self-stopping down counter,* the *latch/decoder/driver,* and the *magnitude comparator.* The *input controls* block presets the time held in the down counter. The time base is an astable multivibrator which generates a known frequency. In this case, the signal is a 1-Hz square wave. The accuracy of the entire timer depends on the accuracy of the time-base clock. Activating the *start* input control causes the down counter to decrement. Each lower number is latched and decoded by the latch/decoder/driver. This block also drives the display.

The illustrations in Fig. 12-36 might be preliminary sketches used by a designer in visual-

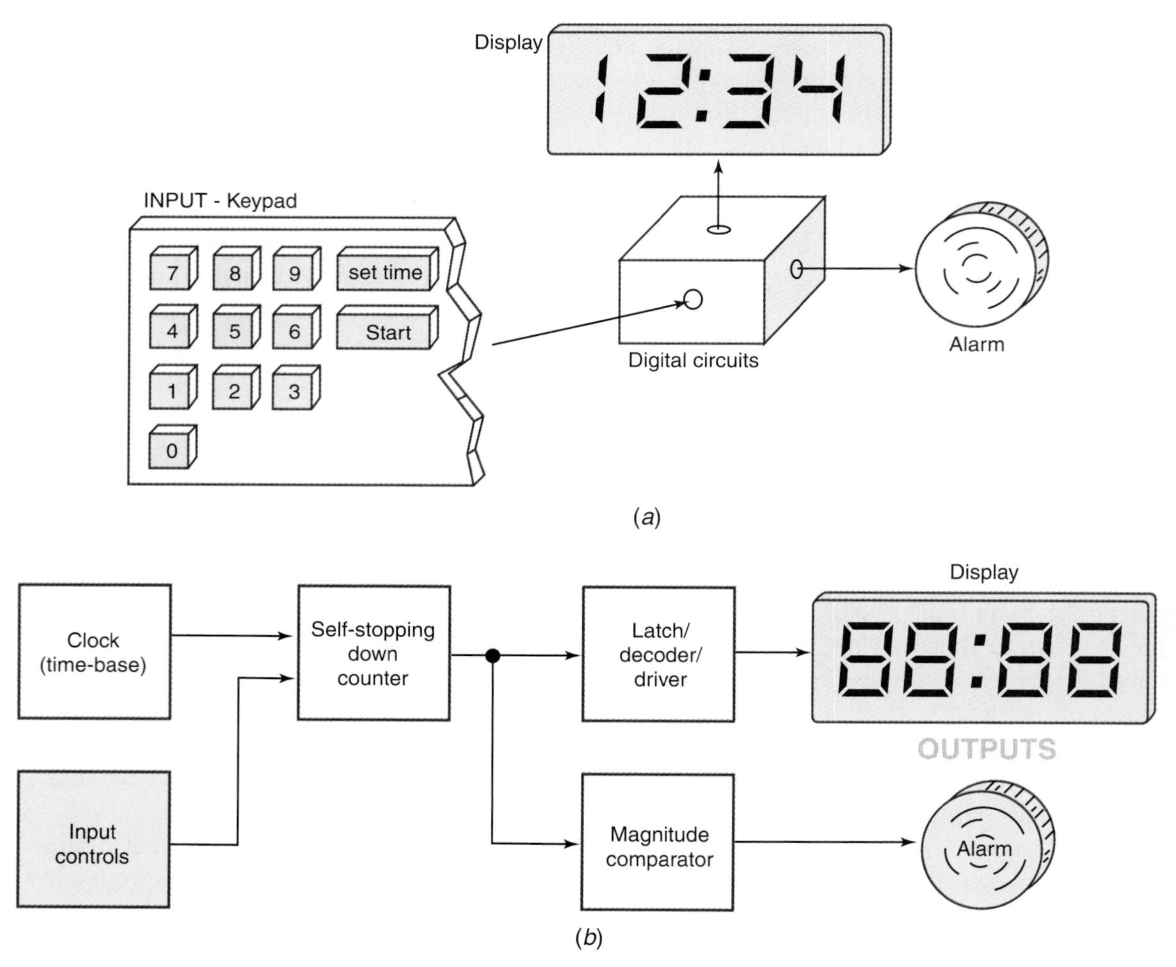

Experimental LCD timer circuit

Fig. 12-36 Digital timer system. (*a*) Concept sketch of timer with alarm. (*b*) Simple block diagram of timer with alarm.

izing a timer system. The designer might next decide on what type of input, output, and processing technologies to use to implement the system.

A somewhat more detailed block diagram of a digital electronic timer system is drawn in Fig. 12-37 on page 336. The designer decided to use a two-digit LCD along with low-power CMOS ICs. This system was designed with your lab trainers in mind, so the inputs are logic switches to simplify the input section. The designer decided on seconds as the time interval. Notice that each block roughly corresponds to an MSI digital IC or input/output device. A wiring diagram could then be developed from the detailed block diagram in Fig. 12-37.

The block diagram in Fig. 12-37 represents an experimental LCD timer with alarm that you might construct in the laboratory. The timer is operated as follows:

1. Set the load/start control to 0 (load mode).
2. Load the 1s counter by setting a BCD number using the top four switches.
3. Load the 10s counter by setting a BCD number using the bottom four switches.
4. A two-digit number should now be displayed on the LCD.
5. Move the load-start control to 1 (start down count mode).

The timer will start counting downward in seconds. The LCD shows the time remaining before the alarm sounds. When both counters reach zero, the LCD reads 00 and the alarm will sound. The final step is to disconnect the power to the circuit to turn off the alarm.

A wiring diagram for the experimental LCD timer circuit is detailed in Fig. 12-38 on pages 338 and 339. Notice that each IC is placed in the same relative position on the wiring diagram as in the block diagram in Fig. 12-37.

Detailed operation of the LCD timer circuit in Figs. 12-37 and 12-38 follows.

Time Base

Time-base clock

The *time-base clock* is a 555 timer IC wired as a free-running MV. It is designed to generate a 256-Hz square wave. The time-base clock in this experimental timer is not very accurate or stable. It can be calibrated by adjusting the value of resistor R_1. The nominal value of R_1 should be about 20 kΩ.

Divide-by-256-counter block

The second part of the time base is the *divide-by-256-counter* block. The function of this block is to output a 1-Hz signal. The divide-by-256-counter block is actually two 4-bit counters wired together. Figure 12-39 on page 340 shows the two 4-bit units wired as divide-by-16 counters. Note that the $\overline{CP}$ inputs are clock inputs and only the Q_D outputs are used. The first divide-by-16 counter divides the frequency from 256 to 16 Hz (256/16 = 16 Hz). The second counter divides the frequency down to the required 1-Hz output (16/16 = 1 Hz).

Self-Stopping Down Counters

Down counters

Experimental timer circuit

The two 74HC192 decade counters are the 75HCXXX series equivalent to the 74192 TTL IC detailed in Chap. 8, Section 8-7. When the load inputs to the 74HC192 counters are activated by a LOW, data at the data inputs (*A, B, C, D*) is immediately transferred into the counter's flip-flops. It then appears at the outputs of the counter (Q_A, Q_B, Q_C, Q_D). The data loaded should be in BCD (binary-coded decimal) form. When the load/start control goes HIGH, the 1-Hz signal activates the count down input of the 1s counter. The count decreases by 1 on each L-to-H transition of the clock pulse. The *borrow out* output of the 1s down counter goes from L-to-H when the 1s counter goes from 0 to 9. This decrements the 10s counter. The down counters are actually wired as a self-stopping down counter because of the *counter stop line* fed back to the CLR input of both 74HC192 counters. When this line goes HIGH, both counters stop at 0000.

8-Bit Magnitude Comparator

8-bit magnitude comparator

The 74HC85 4-bit comparators are shown cascaded in Fig. 12-38 to form an *8-bit-magnitude comparator.* Their purpose in this circuit is to detect when the outputs of the counters reach $0000\ 0000_{BCD}$. When both counters reach zero, the output of the 8-bit magnitude comparator ($A = B_{out}$) goes HIGH. This serves two purposes. First, it stops both 74HC192 counters at 0000. Second, the HIGH at the output of the comparator turns transistor Q_1 on. This allows current to flow up through the transistor, sounding the buzzer. The diode across the buzzer suppresses transient voltages that may be generated by the buzzer.

Experimental LCD timer

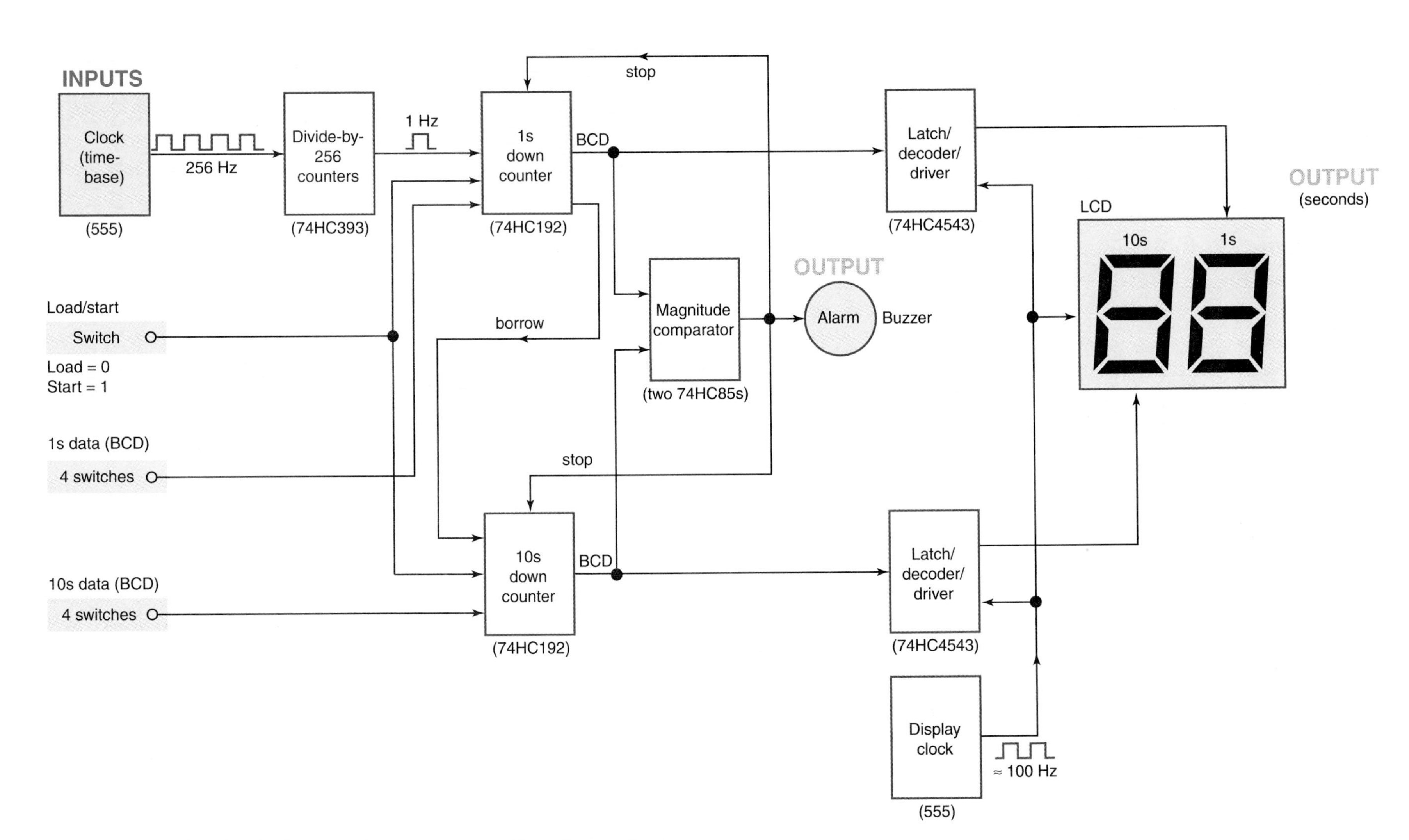

Fig. 12-37 Detailed block diagram of an experimental LCD timer with alarm.

Decoder/Driver

Decoder/driver

The two 74HC4543 ICs used in the timer circuit serve three purposes. The functions of the 74HC4543 IC are summarized in Fig. 12-40 on page 340. The latch disable (*LD*) input is permanently tied HIGH in the timer circuit (Fig. 12-38), which disables the latches. The BCD data flows through the latch to the BCD-to-seven-segment decoder. The decoder translates the BCD input to seven-segment code. Finally, the driver circuitry in the 74HC4543 chip energizes the correct segments on the LCD.

The *display clock* shown at the lower right in Fig. 12-38 generates a 100-Hz square wave. This is sent to the common (backplane) connection on the LCD and the Ph inputs of the 74HC4543 ICs. The LCD driver in the 74HC4543 chip sends inverted or 180° out-of-phase signals to the LCD segments that are to be activated. Segments that are not activated receive an in-phase square-wave signal from the LCD driver section of the 74HC4543 IC.

TEST

Answer the following questions.

91. Refer to Fig. 12-38. The accuracy of the entire timer depends on the frequency generated by the ________ clock.
92. Refer to Fig. 12-38. The initial numbers to be loaded into the counters in the timer must be entered in ________ (BCD, binary, decimal) form.
93. Refer to Fig. 12-38. What two components are considered output devices in this timer circuit?
94. Refer to Fig. 12-38. When the count reaches zero, the output ($A = B_{out}$) of the 8-bit magnitude comparator goes ________ (HIGH, LOW). This causes the counter stop line to go ________ (HIGH, LOW), stopping the counters. This also turns ________ (off, on) the transistor causing it to conduct electricity and sound the buzzer.
95. Refer to Fig. 12-38. The driver section of the 74HC4543 IC sends an ________ (in-phase, out-of-phase) square-wave signal to the LCD segments that are to be activated.
96. Refer to Fig. 12-38. The display clock sends a 100-Hz square-wave signal to the ________ inputs of both 74HC4543 ICs and the ________ connection on the LCD.
97. The timer circuit in Fig. 12-38 is calibrated in (minutes, seconds, tenths of a second).

12-17 Electronic Games

Electronics has been a popular hobby for more than a half century. A favorite task for many electronic hobbyists, young and old, is to construct electronic games. Electronic games and toys are also very popular with students studying electronics in high schools, technical schools, and colleges.

Electronic games may be classified as simple self-contained, computer, arcade, or TV games. The simple self-contained type includes the games and toys most often constructed by students and hobbyists. Several simple electronic games using SSI and MSI digital ICs will be surveyed in this section.

Simple Dice Game

Digital dice game

A block diagram of a simple *digital dice game* is sketched in Fig. 12-41 on page 340. When the push button is pressed, a signal from the clock is sent to the counter. The counter is wired to have a counting sequence of 1, 2, 3, 4, 5, 6, 1, 2, 3, and so on. The binary output from the counter is translated to seven-segment code by the decoder block. The decoder block also contains a seven-segment LED display driver. The output device in this circuit is an LED display. When the push-button switch is opened, the counter *stops at a random number* from 1 through 6. This simulates the roll of a single die. The binary number stored in the counter is decoded and shown as a decimal number on the display. This circuit could be doubled to simulate the rolling of a pair of dice.

A wiring diagram for the digital dice game is detailed in Fig. 12-42 on page 341. Pressing the input switch causes the counter to sequence through the binary numbers 001, 010, 011, 100, 101, 110, 001, 010, 011, etc. When the switch is opened, the last binary count is stored in the flip-flops of the 74192 counter. It is decoded by the 7447 IC and lights the seven-segment LED display.

The 555 timer IC is wired as an astable MV in Fig. 12-42. It generates a 600-Hz square-wave signal.

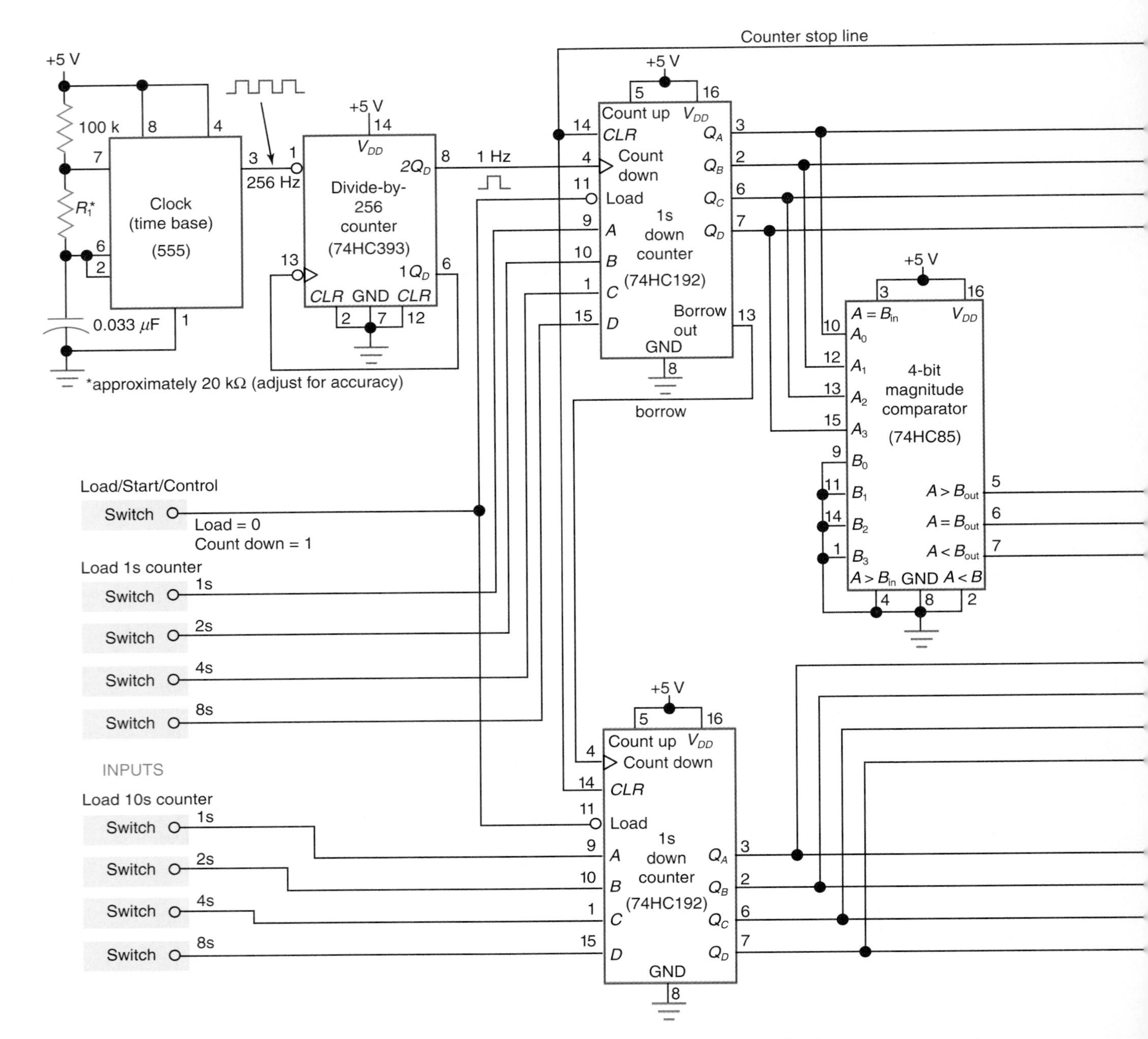

Fig. 12-38 Wiring diagram for an experimental LCD timer with alarm.

Experimental LCD timer with alarm

The 74192 IC is wired as a mod-6 (1 to 6) up counter. The three-input NAND gate is activated when the count reaches binary 111. The LOW signal from the NAND gate loads the next number in the count sequence, which is binary 001. It should be noted that the three outputs of the counter (Q_A, Q_B, Q_C) all go HIGH for only an extremely short time (less than a microsecond) while the counter is being loaded with 0001. Therefore, the temporary count of binary 111 never appears as a 7 on the LED display.

The 7447 BCD-to-seven-segment decoder chip translates the binary inputs (*A*, *B*, *C*) to seven-segment code. The 7447 IC drives the LED segments with active LOW outputs (*a* to *g*). The seven 150-Ω resistors serve to limit the current flow through the LEDs to a safe level. Note that the seven-segment LED display used in Fig. 12-42 is a common-anode type.

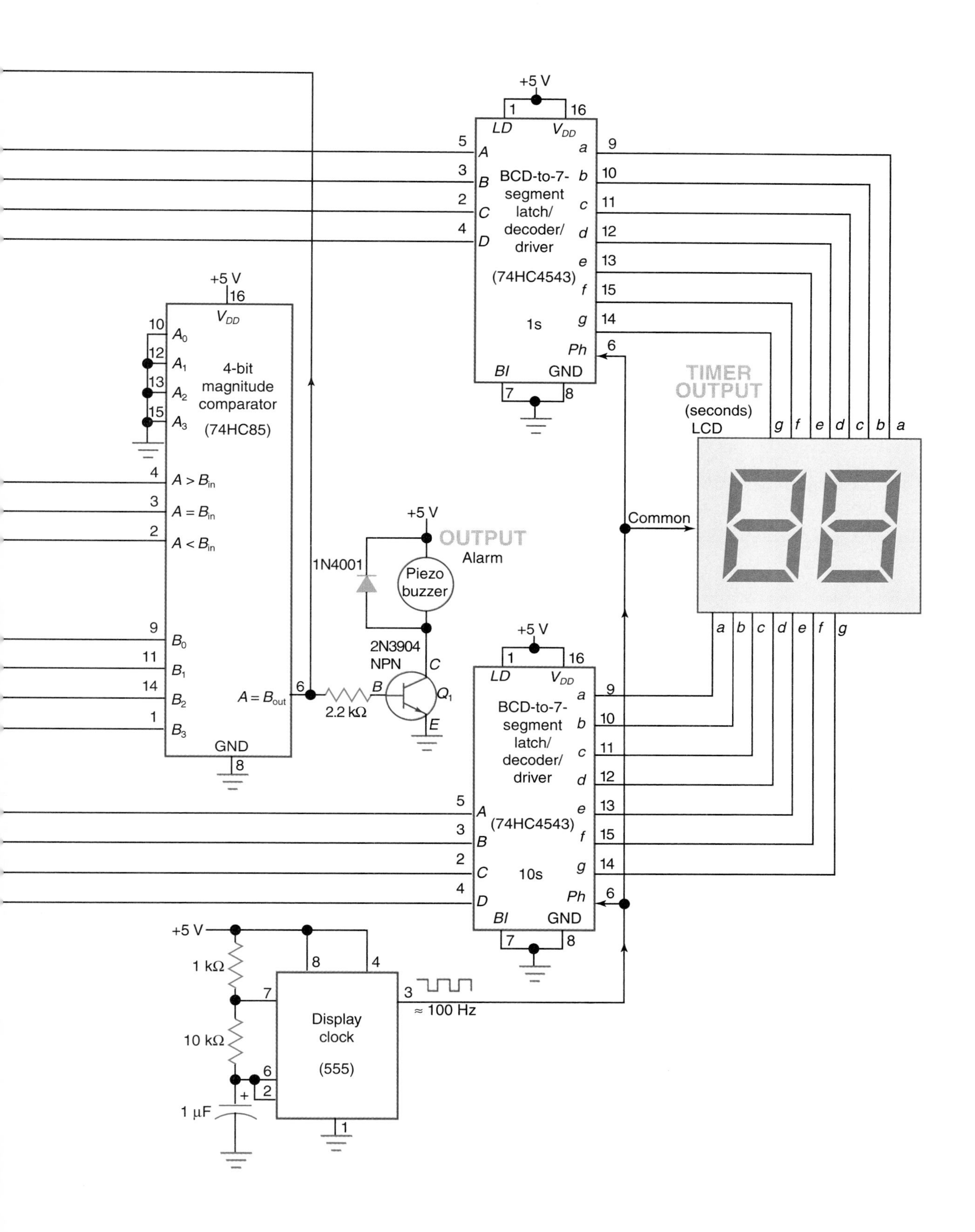
+5 V
BCD-to-7-segment latch/decoder/driver (74HC4543)
LD
VDD
1s
Ph
BI
GND
+5 V
VDD
4-bit magnitude comparator (74HC85)
A > Bin
A = Bin
A < Bin
A = Bout
GND
TIMER OUTPUT
(seconds)
LCD
Common
OUTPUT
Alarm
Piezo buzzer
1N4001
2N3904
NPN
Q1
2.2 kΩ
BCD-to-7-segment latch/decoder/driver
(74HC4543)
10s
1 kΩ
10 kΩ
1 μF
Display clock
(555)
≈ 100 Hz

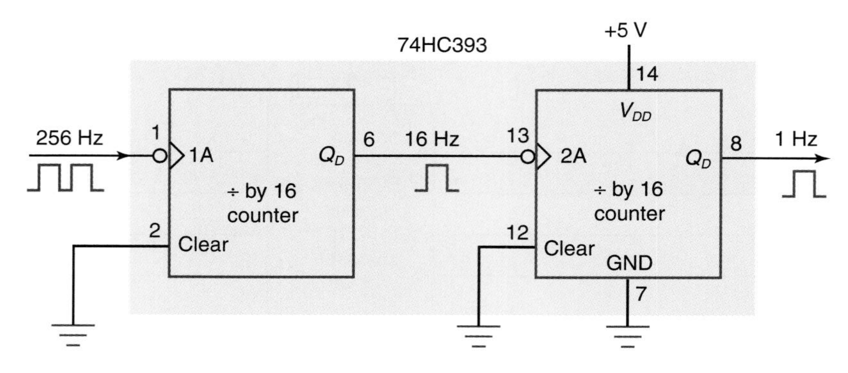

Fig. 12-39 Wiring a divide-by-256 block using two divide-by-16 counters.

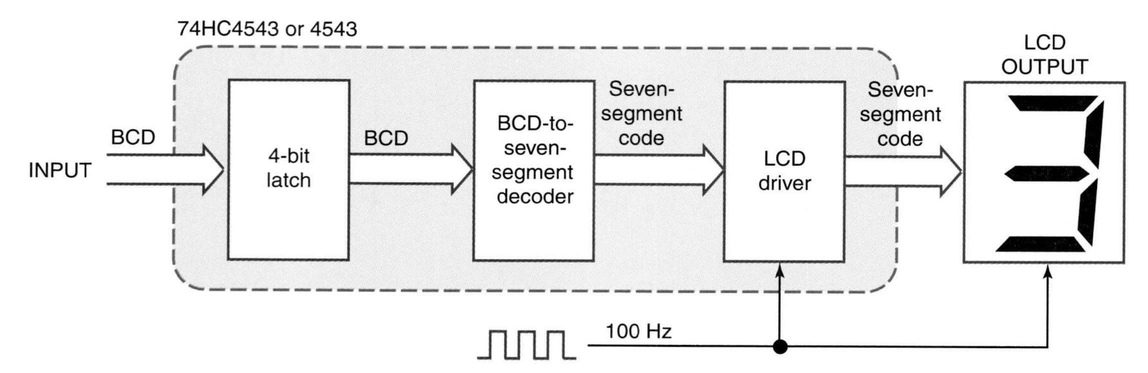

Fig. 12-40 Internal organization of the 74HC4543 IC including latch, decoder, and driver sections.

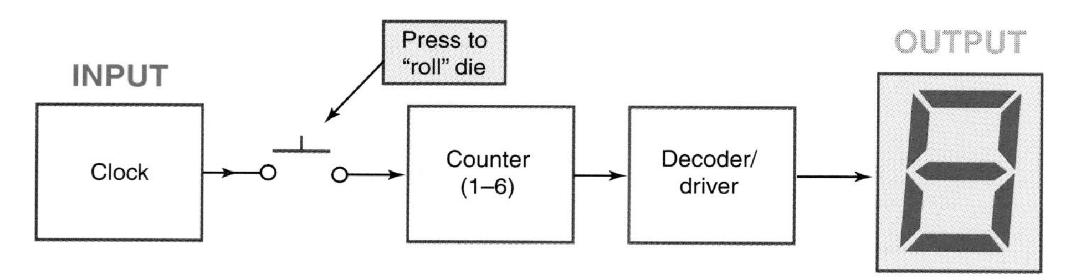

Fig. 12-41 Simple block diagram of a digital dice game.

Electronic dice simulation game

Another Dice Game

The digital dice game featured in Fig. 12-42 used TTL ICs with an unrealistic seven-segment display. A more realistic dice simulation is implemented in the circuit shown in Figs. 12-43 on page 342 and 12-44 on page 343.

A block diagram for a second digital dice game is sketched in Fig. 12-43(*a*). This unit uses individual LEDs for the output device.

Depressing the push button in Fig. 12-43(*a*) causes the clock to generate a square-wave signal. This signal causes the down counter to cycle through the count sequence 6, 5, 4, 3, 2, 1, 6, 5, 4, and so on. The logic block lights the proper LEDs to represent various decimal counts. The LED pattern for each possible decimal output is diagramed in Fig. 12-43(*b*).

The wiring diagram for the second digital dice game is detailed in Fig. 12-44. The circuit features the use of 4000 series CMOS ICs and a 12-V dc power supply. The push-button switch on the left is the input device, while the LEDs (*D*1 to *D*7) on the right form the output. Physically, the LEDs are arranged as shown at the lower right in Fig. 12-44.

When the "roll dice" switch is closed, the two NAND Schmitt trigger gates at the left in Fig. 12-44 produce a 100-Hz square-wave signal. The two NAND gates and associated re-

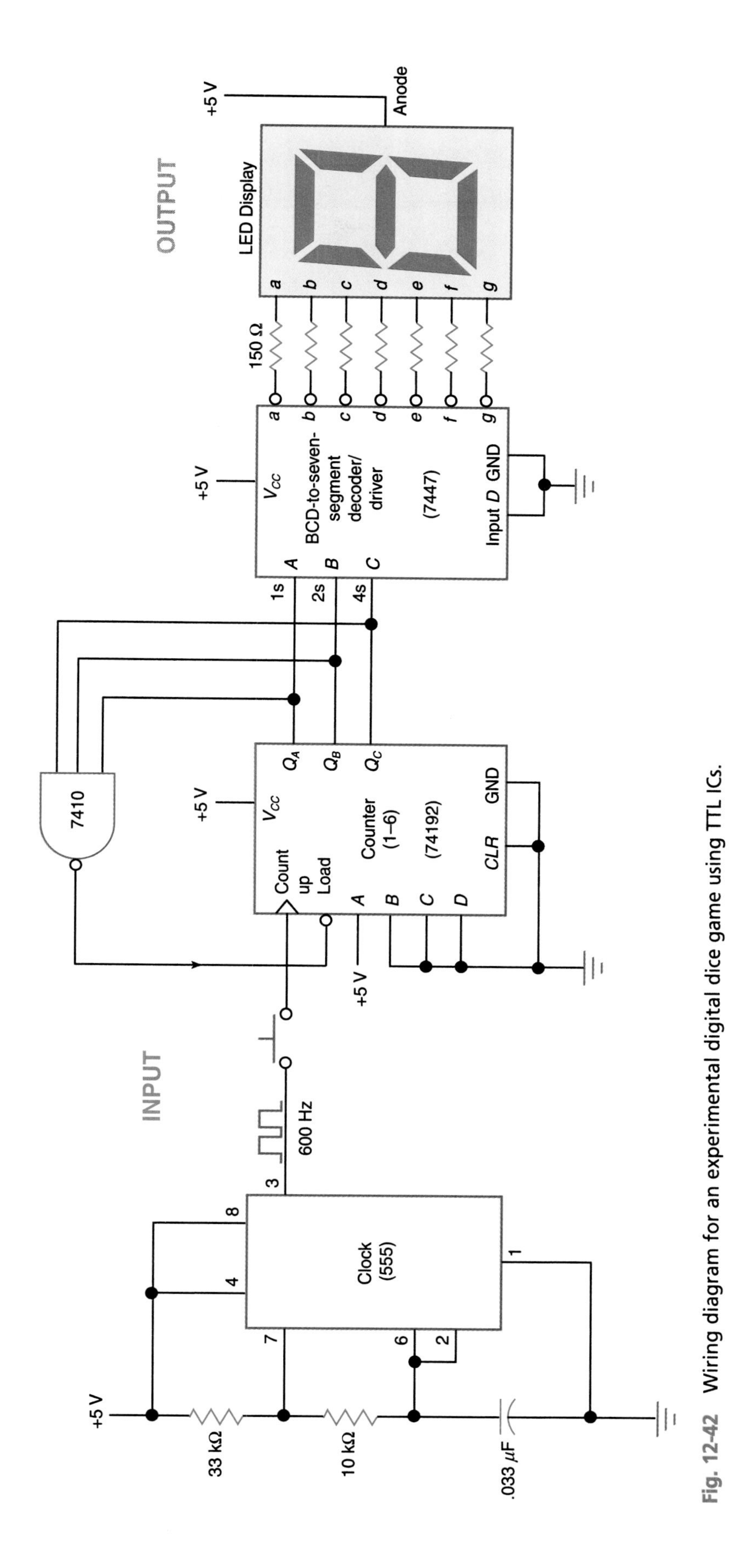

Fig. 12-42 Wiring diagram for an experimental digital dice game using TTL ICs.

Digital dice game circuit

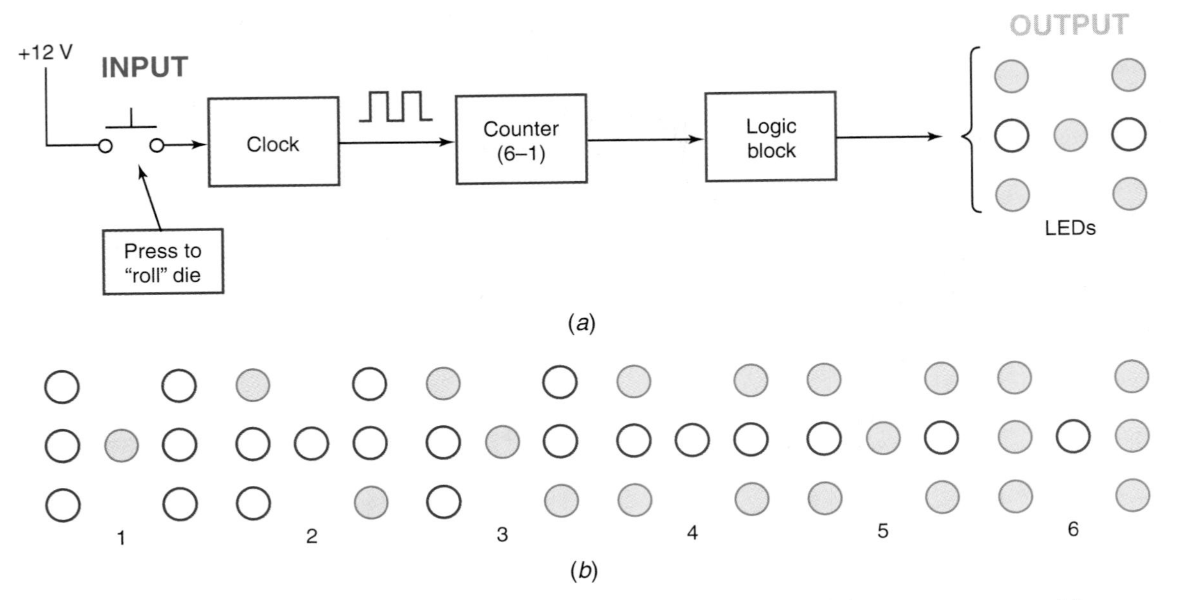

Fig. 12-43 Electronic dice simulation game. (*a*) Simple block diagram. (*b*) LED patterns used for representing dice rolls 1 through 6.

4029 presettable binary/decade up/down counter

sistors and capacitors are wired to form an astable MV. The 100-Hz signal is fed into the clock input of the *4029 presettable binary/decade up/down counter.* In this circuit, the 4029 IC is wired as a down counter whose outputs produce binary 110, 101, 100, 011, 010, 001, 110, 101, 100, etc.

Consider the situation when the down counter in Fig. 12-44 reaches binary 001. On the next LOW-to-HIGH transition of the clock pulse, the *carry out* output (pin 7) of the 4029 IC drops LOW. This signal is fed back and turns on transistor Q_1 This causes the *preset enable* input of the 4029 counter to go HIGH. When the preset enable input goes HIGH, the data at inputs *J*4, *J*3, *J*2, and *J*1 (the "jam inputs") is asynchronously loaded into the counter's flip-flops. In this example, binary 0110 is loaded on the preset pulse. Once the flip-flops have been loaded, the carry out pin goes back HIGH and transistor Q_1 turns off.

JOB TIP

On a new job, seek the help and advice of the positive leaders in your group.

Logic and output sections of dice game

The final sections of the dice game at the right in Fig. 12-44 contain many components. The table in Fig. 12-45 on page 344 will help explain the complexities of the *logic* and *output* sections of this dice game.

The input section of the table in Fig. 12-45 indicates the logic levels present at the output of the 4029 counter. The top line of the table shows a binary 110 (HHL) stored in the 4s, 2s, and 1s flip-flops of the counter. The middle column of the table lists only the components that are activated to light the proper LEDs.

Consider the first line of the table in Fig. 12-45. The output of the NAND gate goes LOW, which turns on the PNP transistor Q_2. The transistor conducts and all six LEDs (*D*2 to *D*7) on the right in Fig. 12-44 light. This would simulate a dice roll of 6.

Consider line 2 of the table in Fig. 12-45. The binary data equals 101 (HLH). The HIGH on the 1s line (pin 6) turns on transistor Q_5. Transistor Q_5 conducts and lights LED *D*1. The output of the NAND gate is HIGH, which causes both bilateral switches to be in the closed condition (low impedance from in/out to out/in). The bilateral switches pass the logic level from the 2s and 4s lines to the base of transistors Q_4 and Q_3. Transistor Q_3 is turned on by the HIGH and conducts. Light-emitting diodes *D*2, *D*3, *D*4, and *D*5 light. A decimal 5 is represented when these five LEDs (*D*1 to *D*5) are lit.

You may look over the remaining lines of the table in Fig. 12-45 to determine the operation of the logic and output sections of this CMOS digital dice game.

Quad bilateral switch

The 4016 IC used in Fig. 12-44 is described by the manufacturer as a *quad bilateral switch.* It is an electronically operated SPST switch. A HIGH at the control input of the 4016 bilateral switch causes it to be in the "closed" or "ON"

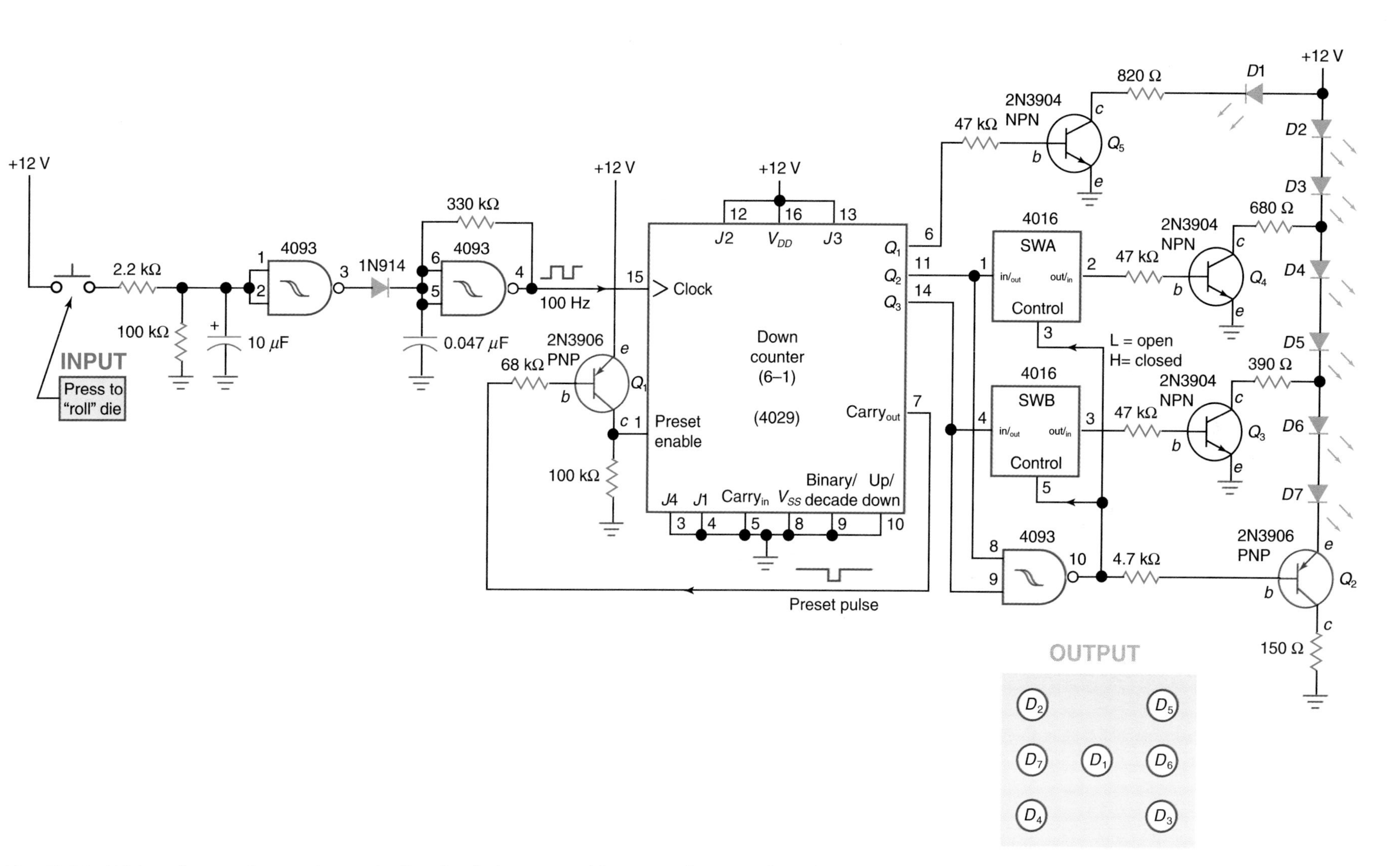

Fig. 12-44 Wiring diagram for an electronic dice simulation game. (*Courtesy of Graymark, Inc.*)

Electronic dice game circuit

INPUTS			ACTIVE COMPONENTS	OUTPUT	
4s (pin 14)	2s (pin 11)	1s (pin 6)		LEDs LIT	DECIMAL
H	H	L	NAND output LOW transistor Q_2 turned on	$D2$, $D3$, $D4$, $D5$, $D6$, $D7$	6
H	L	H	transistor Q_5 turned on bilateral switch SWB closed transistor Q_3 turned on	$D1$ $D2$, $D3$, $D4$, $D5$	5
H	L	L	bilateral switch SWB closed transistor Q_3 turned on	$D2$, $D3$, $D4$, $D5$	4
L	H	H	transistor Q_5 turned on bilateral switch SWA closed transistor Q_4 turned on	$D1$ $D2$, $D3$	3
L	H	L	bilateral switch SWA closed transistor Q_4 turned on	$D2$, $D3$	2
L	L	H	transistor Q_5 turned on	$D1$	1

Fig. 12-45 Explaining the logic and output sections of the electronic dice simulation game.

Logic and output sections of dice game

position. In the "closed" position the internal resistance from in/out to out/in terminals is quite low (400 Ω typical). A LOW at the control input of the bilateral switch causes it to be in the open or OFF position. The 4016 IC acts like an open switch when the control is LOW. Unlike a gate, a bilateral switch can pass data in either direction. It can pass either dc or ac signals. A bilateral switch is also referred to as a *transmission gate.*

Transmission gate

TEST

Answer the following questions.

98. Refer to Fig. 12-42. The 555 timer is wired as a(n) ________ multivibrator in this digital system.
99. Refer to Fig. 12-42. When the 74192 IC increases its count from 110 to 111, the output of the NAND gate goes ________ (HIGH, LOW) immediately loading ________ (binary) into the counter.
100. Refer to Fig. 12-42. List the possible digits that can appear on the seven-segment LED display when the input switch is released.
101. Refer to Fig. 12-44. Two gates packaged in the ________ (4016, 4029, 4093) IC are wired as a free-running MV in this digital dice game circuit.
102. Refer to Fig. 12-44. List the binary counting sequence of the 4029 IC in this circuit.
103. Refer to Fig. 12-44. When the output of the counter is binary 001, only $D1$ lights because only transistor ________ is turned on and conducts.
104. Refer to Fig. 12-44. When the output of the counter is binary 010, LED(s) ________ light because the bilateral switches are ________ (closed, open) and transistor ________ is turned on grounding the cathode of light-emitting diode of $D3$.
105. A(n) ________ is an IC device available in CMOS that acts much like a SPST switch that can conduct either ac or dc signals.

12-18 Programmable Logic Controllers (PLCs)

Programmable logic controller (PLC)

A *programmable logic controller (PLC)* is a specialized computer like device used to replace banks of electromagnetic relays in industrial process control. The PLC is also known as a *programmable controller (PC).* The title "PC" for programmable controller could be confused in common usage with "PC" used to mean personal computer. To avoid this confusion, we shall refer to the programmable controller as a programmable logic controller or PLC.

You can think of the programmable logic controller as a *heavy-duty computer system designed for machine control.* Like a general-purpose computer, the PLC is based on digital logic and can be field-programmed. The programming language is a bit different because the purpose of the PLC is to control machines. The PLC is used to time and sequence functions that might be required in assembly lines, robots, and chemical processing. It is designed to deal with the harsh conditions of the industrial environment; some of the physical environment problems could include vibration and shock, dirt and vapors, and temperature extremes. The PLC commonly has to interface with a wide variety of both input and output devices. Some input devices include limit and pressure switches, temperature and optical sensors, and analog-to-digital converters (ADCs). Some output devices include motors and solenoids, pneumatic valves, motors and cylinders, and digital-to-analog converters (DACs).

A simple block diagram of a programmable logic controller is sketched in Fig. 12-46. As a system, it looks much like classic computer architecture. What makes the PLC different from a general-purpose computer, however, is the type of inputs and outputs connected to the system. A PC system commonly has a keyboard or mouse for primary input while the PLC must interface with sensors, which detect the machine's action. The primary output from a PC is a monitor or printer, whereas the PLC must drive motors and solenoids. Notice from Fig. 12-46 that the programmer is shown as a separate module, which may or may not be connected to the processing unit. The programming device in Fig. 12-46 can be connected when an update is needed in the PLC and disconnected when the task is finished. Semiconductor memory devices within the pro-

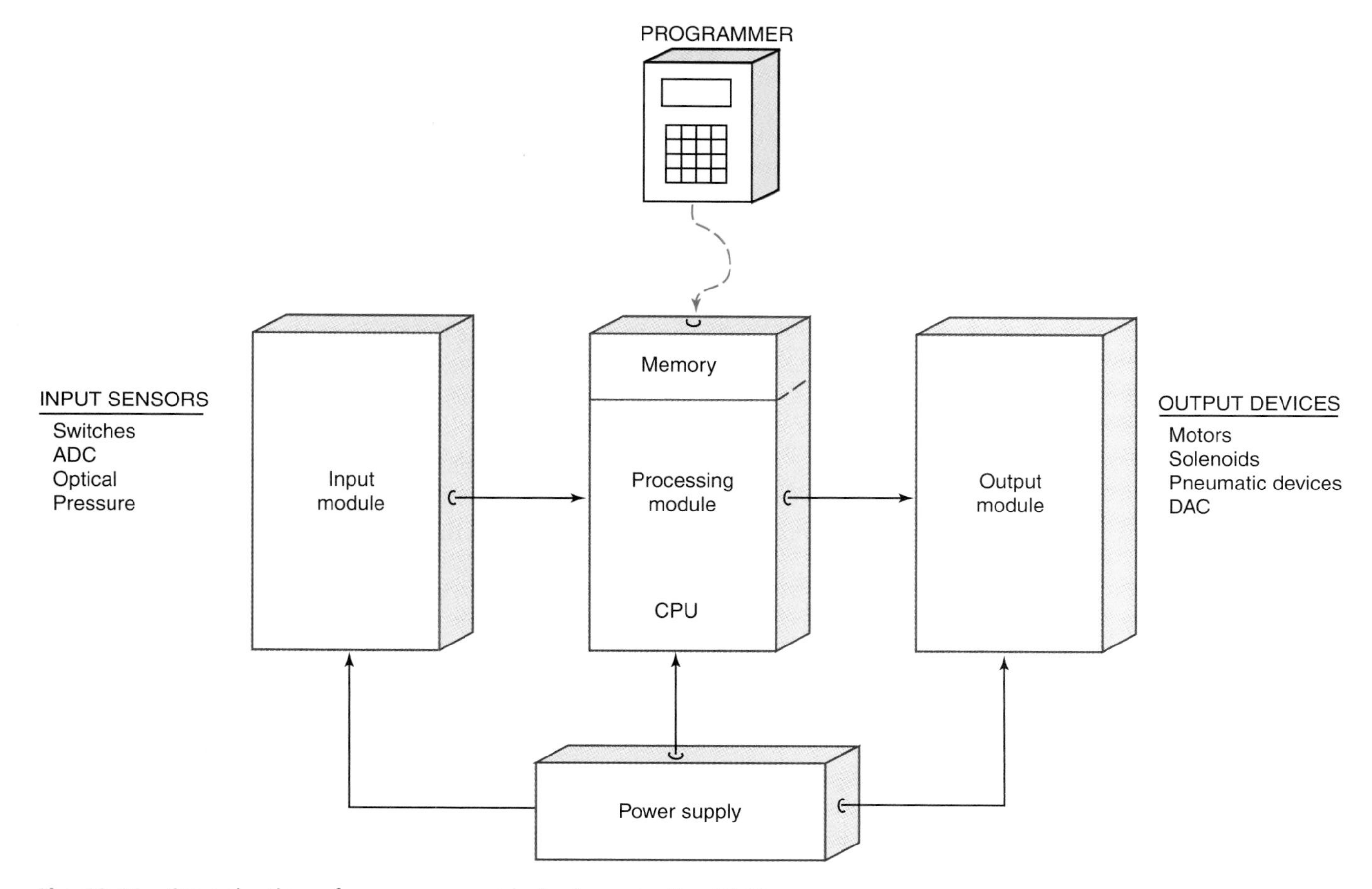

Fig. 12-46 Organization of a programmable logic controller (PLC).

Relay ladder diagram

Ladder logic diagram

cessing unit of the PLC hold the machine or process control program. In small PLCs, the input and output modules can be part of the device. In larger systems, the processing module, input module, output module, and power supplies are housed in separate heavy-duty, industrial-style enclosures. The programmer can be a dedicated terminal, general-purpose computer, or handheld programming device.

The *processing unit* of the PLC typically contains a CPU (a microprocessor) and semiconductor memory devices such as RAM and EEPROM or EPROM. The CPU communicates with memory and input/output (I/O) modules via the typical address, control and data buses. The input sensors and output devices are hardwired to the input and output modules. The architecture of the PLC and a PC look very much alike. Many PLCs have a simple machine-control language built permanently into their memory. The PLC programming language is simpler than the languages used to program general-purpose computers. Programmable logic controllers can be reprogrammed by the electricians and technicians that maintain the other industrial electrical-electronic devices in a factory or plant. The *instruction set* for a specific PLC may contain as few as 15 to as many as 100 instructions. Besides the normal arithmetic and logic functions associated with computer CPUs, specialized instructions are needed to sense and control output devices and to do the following tasks:

Examine an input bit for ON condition
Examine an input bit for OFF condition
Turn ON and latch an output
Turn OFF and latch an output
Turn ON for a certain time, then turn off.

Programmable logic controllers are closely associated with relay logic or hard-wired logic used prior to their introduction in the 1970s. A *relay ladder diagram* is a graphic method of describing how a circuit works. A *ladder logic diagram* is a graphic programming language developed from the relay ladder diagram and is useful in programming a PLC. Some examples of equivalent relay ladder diagrams, relay logic diagrams, and logic gate diagrams will be illustrated. These examples are from the excellent textbook, *Programmable Logic Controllers,* 2nd edition by Frank Petruzella, Glencoe/McGraw-Hill. Notice that each of the three types of diagrams have their own symbols and conventions. Each of the types of diagrams were developed by various manufacturers and users to suit their needs. For instance, the relay schematics were developed before digital logic as we know it became popular. The ladder logic diagrams were developed directly from the relay schematics used earlier.

Example 12-1 shows two input switches in series and the output device is a solenoid valve. The relay schematic is shown at the left, the ladder logic diagram in the center, and the familiar logic gate diagram on the right. We recognize that two switches in series is an AND situation as shown at the right with a Boolean expression of $AB = Y$. Note the symbols used in the relay schematic, ladder logic diagram, and gate logic are different but each represent the same task—the AND function.

Example 12-2 shows two input switches in parallel, and the output device is a solenoid valve. The relay schematic is shown at the left, the ladder logic diagram in the center, and the logic gate diagram on the right. We recognize that two switches in parallel is an OR situation as shown at the right with a Boolean expression of $A + B = Y$.

Example 12-3 shows two input switches in parallel with a normally open relay contact in series with both, and the output device is a

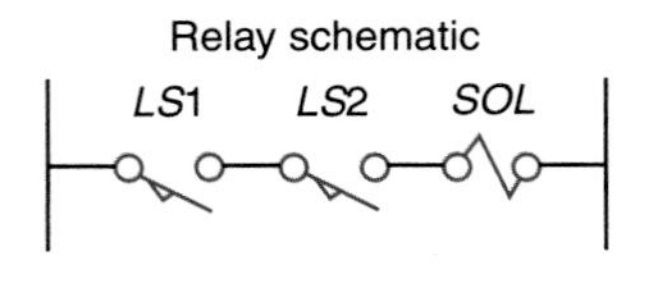

Ladder logic program

A B Y

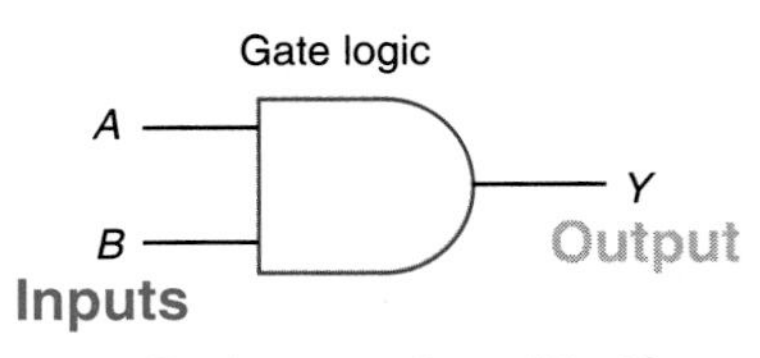

Boolean equation: $AB = Y$

Example 12-1 Two limit switches (LS) connected in series are used to control a solenoid (SOL). (*from Frank Petruzella, Programmable Logic Controllers, 2nd ed., New York: Glencoe/McGraw-Hill, 1998.*)

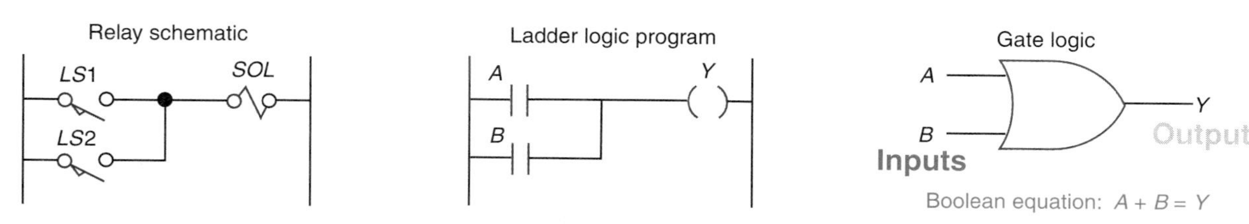

Example 12-2 Two limit switches (LS) connected in parallel are used to control a solenoid (SOL). *(from Frank Petruzella, Programmable Logic Controllers, 2nd ed., New York: Glencoe/McGraw-Hill, 1998.)*

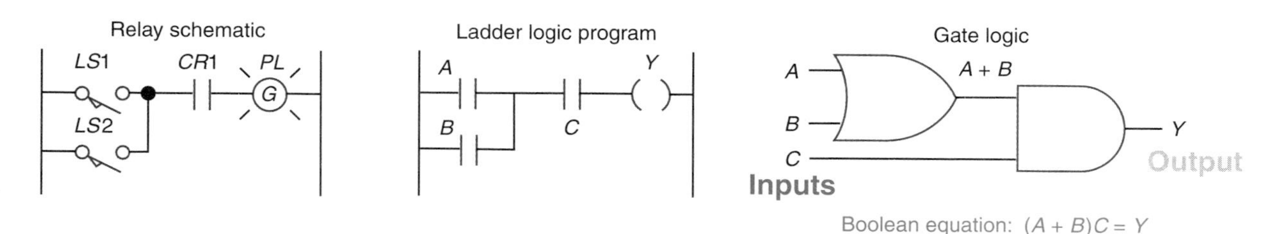

Example 12-3 Two limit switches (LS) connected in parallel are placed in series with a relay contact (CR) are used to control pilot lamp (PL). *(from Frank Petruzella, Programmable Logic Controllers, 2nd ed., New York: Glencoe/McGraw-Hill, 1998.)*

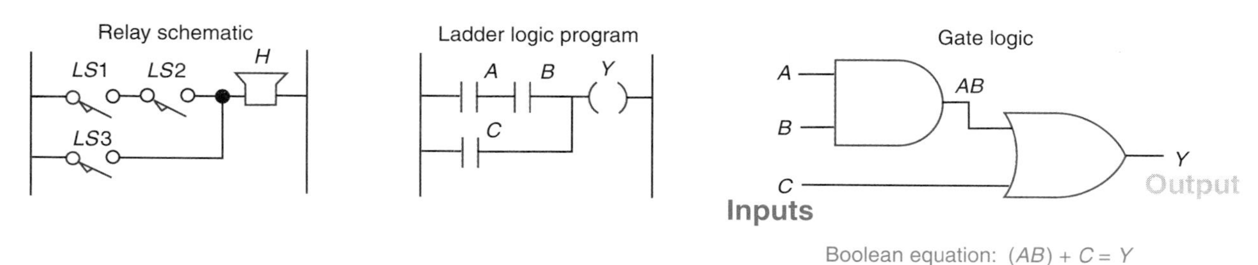

Example 12-4 Two limit switches (LS) connected in series with each other and in parallel with a third limit switch are used to control a warning horn (H). *(from Frank Petruzella, Programmable Logic Controllers, 2nd ed., New York: Glencoe/McGraw-Hill, 1998.)*

green pilot light. The relay schematic is shown at the left, the ladder logic diagram in the center, and the logic gate diagram on the right. We recognize that two switches in parallel (A and B) is an OR situation which feeds a series relay contact (AND situation). Both the Boolean expression $(A + B)C = Y$ and logic gate diagram are shown at the right.

Example 12-4 shows two input switches (A and B) in series with each other, and both are in parallel with a single switch (C). The output device in this example is a warning horn. The relay schematic is shown at the left, the ladder logic diagram in the center, and the familiar logic gate diagram on the right. We recognize that two switches (A and B) are in series, which is an AND situation while switch C is parallel with the two switches. Again, remember that the relay schematic, ladder logic diagram, and gate logic diagrams all represent the same logic function as described by the Boolean expression $(AB) + C = Y$.

In summary, a programmable logic controller (PLC) is a heavy-duty computer system used to replace older relay logic. PLCs are used in factories and plants to control machines, material handling, and chemical processing. PLCs are built to withstand the more harsh environment of a factory, warehouse, or processing plant. The language used to program a PLC has specialized instructions for evaluating inputs and generating outputs. Some PLC languages are based directly on relay ladder diagrams. Because the processing unit (CPU) of the PLC is a microprocessor, it can also perform arithmetic and logic functions as well as data handling and branch and subroutine calls typical of general-purpose computer languages. Some manufacturers of PLCs are Allen-Bradley Company, Cincinnati Milcron

Microcontroller

Company, Eaton Corporation (Cutler-Hammer products), Gould Inc., Honeywell, Inc., Square D Company, Texas Instruments, and Westinghouse Electric Company.

TEST

Answer the following questions.

106. The programmable controller (PC) is also commonly known as the ________ ________ ________ or PLC.
107. A programmable logic controller (PLC) is a heavy-duty computer system designed for ________ (general-purpose office use, machine control in factories).
108. Once programmed, inputs to a PLC would probably come from devices such as ________ (keyboard and mouse, limit switches, pressure switches, temperature and optical sensors).
109. Typically a PLC has a programming module connected to it ________ (always, occasionally during reprogramming).
110. Refer to Fig. 12-46. The power supply, processing, input, and output sections are referred to as modules because they are sometimes physically housed in separate enclosures in larger systems (T or F).
111. The programming language used with PLCs is commonly ________ (less, more) complex than general-purpose computer languages.
112. Given the relay schematic in Fig. 12-47, draw the ladder logic program that might be used with a PLC for this circuit.
113. Given the relay schematic in Fig. 12-47, draw a logic gate equivalent of this circuit using AND and OR symbols.
114. Given the relay schematic in Fig. 12-47, write the Boolean expression that describes the logic function of this circuit.

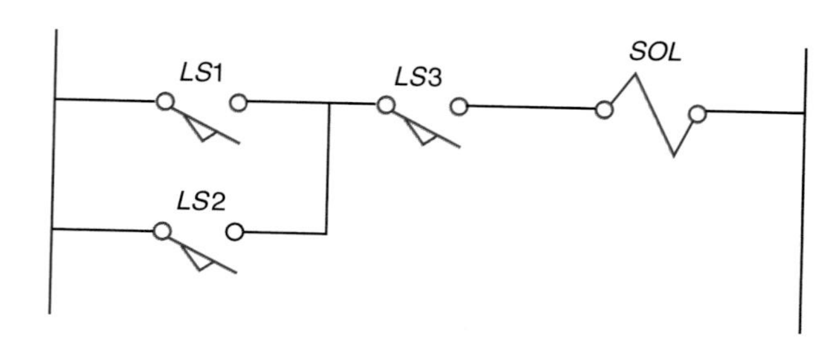

Fig. 12-47 Relay schematic diagram.

12-19 Microcontrollers

A *microcontroller* is considered to be a "computer on a chip." A single-package microcontroller contains a central processing unit (CPU), semiconductor memory (RAM for data memory and read-only memory for program memory), a clock generator, and input/output capabilities. The read-only memory used to store programs in a microcontroller can take the form of ROM, EPROM, EEPROM, or even flash EEPROM. Microcontrollers are low-cost, programmable, electronic devices that can be embedded in inexpensive appliances. Microcontrollers are popular in consumer products because of their extremely low cost: a simple microcontroller IC may cost only a few dollars. The features of microcontrollers vary widely; some are faster, some contain more memory, some have more input/output ports, and some have other characteristics that may be necessary for a specific application. The term "microcomputer" or "small computer" might be used to describe the microcontroller but is *not* common usage. The term "microcontroller" fits the jobs these small "computers on a chip" perform, which are *control functions.* Microcontrollers are not used as general-purpose computers.

Microcontrollers were developed shortly after their larger relatives, microprocessors. The same companies that developed microprocessor chips (for instance Intel and Motorola) also developed a line of microcontrollers. The first 8-bit microcontrollers appeared in the late 1970s, and some are still in use today. Microcontrollers sell in large volumes. Motorola, for example, recently announced that it had shipped its two billionth 68HC05 microcontroller. Microcontrollers are embedded in many everyday consumer products, such as cars, toys, TVs, VCRs, microwave ovens, and PC keyboards. For instance, a modern automobile may contain 10 microcontrollers whereas a high-tech home may contain many more than that.

Microcontrollers Compared to Microprocessors

When compared to a microprocessor-based system, a microcontroller has less semiconductor memory (RAM, ROM, EPROM, and/or

EEPROM), is lower in cost, and uses less printed circuit board space. Microcontrollers commonly address only a limited size memory. Microcontrollers usually have fewer commands in their instruction set than microprocessors. Microcontrollers typically are programmed to do several limited tasks efficiently and are usually not reprogrammed. Microcontroller programs are commonly held in read-only memory. Microcontroller-based systems rarely have complex input/output devices attached, such as keyboards, disk drives, printers, and monitors. Manufacturers support both their microcontroller and microprocessor product with software development tools and application notes (examples of typical applications).

Manufacturers of microcontrollers produce a wide variety of low-cost programmable devices. For instance, a recent *IC Master* (a manual that lists most of the world's ICs) used more than 60 pages to list the various microcontrollers available from manufacturers.

A Family of Microcontrollers

The chart in Fig. 12-48 illustrates a "family of icrocontrollers" from Microchip Technology, Inc. This family of devices is described by the manufacturer as the EPROM/ROM-based 8-bit CMOS Microcontroller Series. The PIC16C5X device is listed on the left side of the chart with columns showing some of the important characteristics of these low-cost microcontrollers. The operating frequencies of these units allow them to execute instructions very quickly. The *program memory size* is given in words (word size equals 12 bits for the 16C5X series) and is stored in either ROM or EPROM. The *data memory* or *RAM size* is very small ranging from 24 to 73 bytes. Because microcontrollers are *control devices,* they typically have many IC pins dedicated to either input or output (I/O pins). The number of I/O pins for the PIC16C5X microcontrollers range from 12 to 20. These I/O pins can be programmed to be either *inputs* or *outputs*. The PIC16C5X series of microcontroller ICs are CMOS devices and operate on low voltages. All of the ICs are available in a variety of packages including the traditional DIP (dual in-line package), SOIC (small-outline IC), and SSOP (shrink small-outline package). The SOIC and SSOP packages are small surface-mount packages. Remember that microcontrollers are embedded CPUs in everyday de-

PIC16C5X family of microcontrollers

SOIC package

SSOP package

	Clock	Memory			Peripherals		Features		
	Maximum Frequency of Operation (MHz)	Program Memory (x12 words) EPROM	Program Memory (x12 words) ROM	RAM Data Memory (bytes)	Timer Module(s)	I/O Pins	Voltage Range (Volts)	Number of Instructions	Packages
PIC16C52	4	384	—	25	TMR0	12	2.5–6.25	33	18-pin DIP, SOIC
PIC16C54	20	512	—	25	TMR0	12	2.5–6.25	33	18-pin DIP, SOIC; 20-pin SSOP
PIC16C54A	20	512	—	25	TMR0	12	2.0–6.25	33	18-pin DIP, SOIC; 20-pin SSOP
PIC16CR54A	20	—	512	25	TMR0	12	2.0–6.25	33	18-pin DIP, SOIC; 20-pin SSOP
PIC16C55	20	512	—	24	TMR0	20	2.5–6.25	33	28-pin DIP, SOIC; SSOP
PIC16C56	20	1K	—	25	TMR0	12	2.5–6.25	33	18-pin DIP, SOIC; 20-pin SSOP
PIC16C57	20	2K	—	72	TMR0	20	2.5–6.25	33	28-pin DIP, SOIC; SSOP
PIC16CR57B	20	—	2K	72	TMR0	20	2.5–6.25	33	28-pin DIP, SOIC; SSOP
PIC16C58A	20	2K	—	73	TMR0	12	2.0–6.25	33	18-pin DIP, SOIC; 20-pin SSOP
PIC16CR58A	20	—	2K	73	TMR0	12	2.5–6.25	33	18-pin DIP, SOIC; 20-pin SSOP

All PIC 16/17 Family devices have Power-On Reset, selectable Watchdog Timer, selectable code protect and high I/O current capability.

Fig. 12-48 General specifications for the PIC16C5X family of microcontrollers. (*Courtesy of Microchip Technology, Inc.*)

vices and the small package ICs are ideal for "hiding" them inside of products.

RISC

CISC

The PIC16C5X series of microcontrollers features RISC architecture using only 33 instructions in their instruction set. *RISC* means *reduced instruction set computing* as opposed to *CISC (complex instruction set computing).* RISC CPUs have fewer instructions but execute them faster. CISC CPUs have more instructions in their instruction set with some of these instructions executing complex tasks. The RISC architecture was developed to speed up the processors, but for complex operations many instructions are needed.

The PIC16C55 Microcontroller

As an example, the 28-pin DIP diagram for the PIC16C55 microcontroller is reproduced in Fig. 12-49(*a*). A description of the IC's pins is detailed in the chart in Fig. 12-49(*b*). Note especially from the pin diagram and pin-out descriptions the great number of I/O pins. They are organized into three ports (*A, B,* and *C*). Port A (4-bit port) consists of I/O pins RA0–RA3, while ports *B* and *C* are each 8-pin ports. Individual I/O pins can be programmed to be either an input or output.

Using a Microcontroller

A simple application of the PIC16C55 microcontroller is shown in the schematic diagram reproduced in Fig. 12-50(*a*) on page 352. The PIC16C55 has been programmed by Chaney Electronics, Inc. to display various light patterns on the nine-row by 10-column LED display board. A schematic of the 9 × 10 LED display is detailed in Fig. 12-50(*b*). To light the upper-left red LED on the display, the top row must go LOW (Y1 = 0) while the left column must go HIGH (X1 = 1). To light the entire horizontal row of yellow LEDs in the middle of the display in Fig. 12-50(*b*), all column inputs must be HIGH whereas only row 5 input must go LOW (Y5 = 0). You can see that the driver for this display must have 18 output pins which are available using the 16C55 microcontroller. Two I/O pins of the 16C55 IC are programmed to be inputs (RA_0 and RA_1) in this design. These pins (6 and 7) can be held LOW or HIGH based on the position of switches *S*2 and *S*3. The input conditions caused by switches *S*2 and *S*3 cause the microcontroller to execute one of four possible programs which produce four unique light displays. Switches *S*4 and *S*5 change the *RC* timing circuit, which is connected to the *CLKIN* input of the IC. Various positions of *S*4 and *S*5 change the frequency of the display (slow to fast). Switch *S*7 serves to reset the current program to its beginning. The four programs are stored as firmware in the microcontrollers read-only memory. The I/O pins have good drive capabilities (25 mA to sink, 20 mA to source), which allow them to drive the LEDs directly in this display.

To operate the light display in Fig. 12-50, close *S*1 and the 5-V voltage regulator IC (*Q*1) will drop the 9-V dc battery voltage to 5 V dc to power the circuit. Set the *S*2 and *S*3 programming switches to the appropriate program. Set the speed of the display with switches *S*4 and *S*5. Press the reset switch (*S*7) to start the selected program at the beginning of a routine. Closing switch *S*6 causes column 10 to light with the same pattern as column 9. Opening switch *S*6 turns off column 10 only.

This light display project using a microcontroller is available from Chaney Electronics, Inc. in kit form. The 16C55 microcontroller's read-only memory has been programmed by Chaney Electronics with the proper programs.

TEST

Answer the following questions.

115. A ________ (microcontroller, microprocessor) can be described as a "computer on a chip" because it contains a CPU, RAM, read-only memory, clock, and I/O pins within a single IC.
116. The microcontroller is most likely to appear in a ________ (CPU section of a PC, VCR).
117. The microcontroller is noted for its small size and extremely low cost (T or F).
118. Program memory in a microcontroller is held in a read-only memory device and is rarely programmed (T or F).
119. All microcontrollers from different manufacturers are alike in size, speed, packaging, instruction set, and function (T or F).
120. Microcontrollers can address ________ (very large, very small) amounts of RAM as compared to microprocessors.

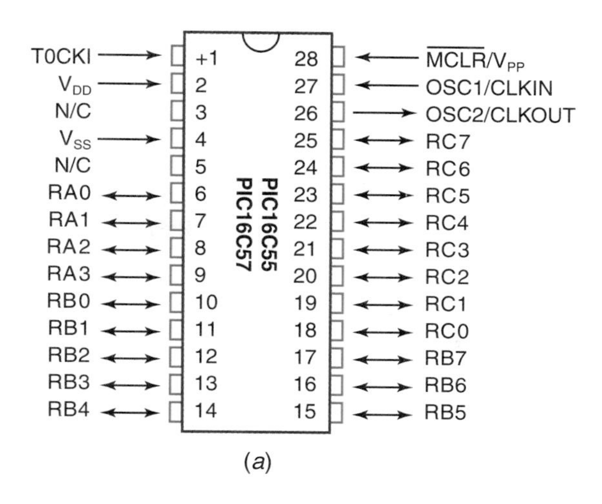

(*a*)

PIC16C55/C57 PINOUT DESCRIPTION

Name	DIP, SOIC No.	SSOP No.	I/O/P Type	Input Levels	Description
*RA*0	6	5	I/O	TTL	Bidirectional I/O port
*RA*1	7	6	I/O	TTL	
*RA*2	8	7	I/O	TTL	
*RA*3	9	8	I/O	TTL	
*RB*0	10	9	I/O	TTL	Bidirectional I/O port
*RB*1	11	10	I/O	TTL	
*RB*2	12	11	I/O	TTL	
*RB*3	13	12	I/O	TTL	
*RB*4	14	13	I/O	TTL	
*RB*5	15	15	I/O	TTL	
*RB*6	16	16	I/O	TTL	
*RB*7	17	17	I/O	TTL	
*RC*0	18	18	I/O	TTL	Bidirectional I/O port
*RC*1	19	19	I/O	TTL	
*RC*2	20	20	I/O	TTL	
*RC*3	21	21	I/O	TTL	
*RC*4	22	22	I/O	TTL	
*RC*5	23	23	I/O	TTL	
*RC*6	24	24	I/O	TTL	
*RC*7	25	25	I/O	TTL	
T0CKI	1	2	I	ST	Clock input to Timer0. Must be tied to V_{SS} or V_{DD}, if not in use, to reduce current consumption.
$\overline{MCLR}/V_{PP}$	28	28	I	ST	Master clear (reset) input/programming voltage input. This pin is an active low reset to the device. Voltage on $\overline{MCLR}/V_{PP}$ must not exceed V_{DD} to avoid unintended entering of programming mode.
*OSC*1/*CLKIN*	27	27	I	ST	Oscillator crystal input/external clock source input.
*OSC*2/*CLKOUT*	26	26	O	—	Oscillator crystal output. Connects to crystal or resonator in crystal oscillator mode. In *RC* mode, *OSC*2 pin outputs *CLKOUT* which has 1/4 the frequency of *OSC*1, and denotes the instruction cycle rate.
V_{DD}	2	3,4	P	—	Positive supply for logic and I/O pins.
V_{SS}	4	1,14	P	—	Ground reference for logic and I/O pins.
N/C	3,5	—	—	—	Unused, do not connect

Legend: I = input, O = output, I/O = input/output,
P = power, — = Not Used, TTL = TTL input,
ST = Schmitt trigger input

(*b*)

Fig. 12-49 PIC16C55 microcontroller IC. (*a*) Pin diagram (DIP or SOIC packages only). (*b*) Pinout description. (*Courtesy of Microchip Technology, Inc.*)

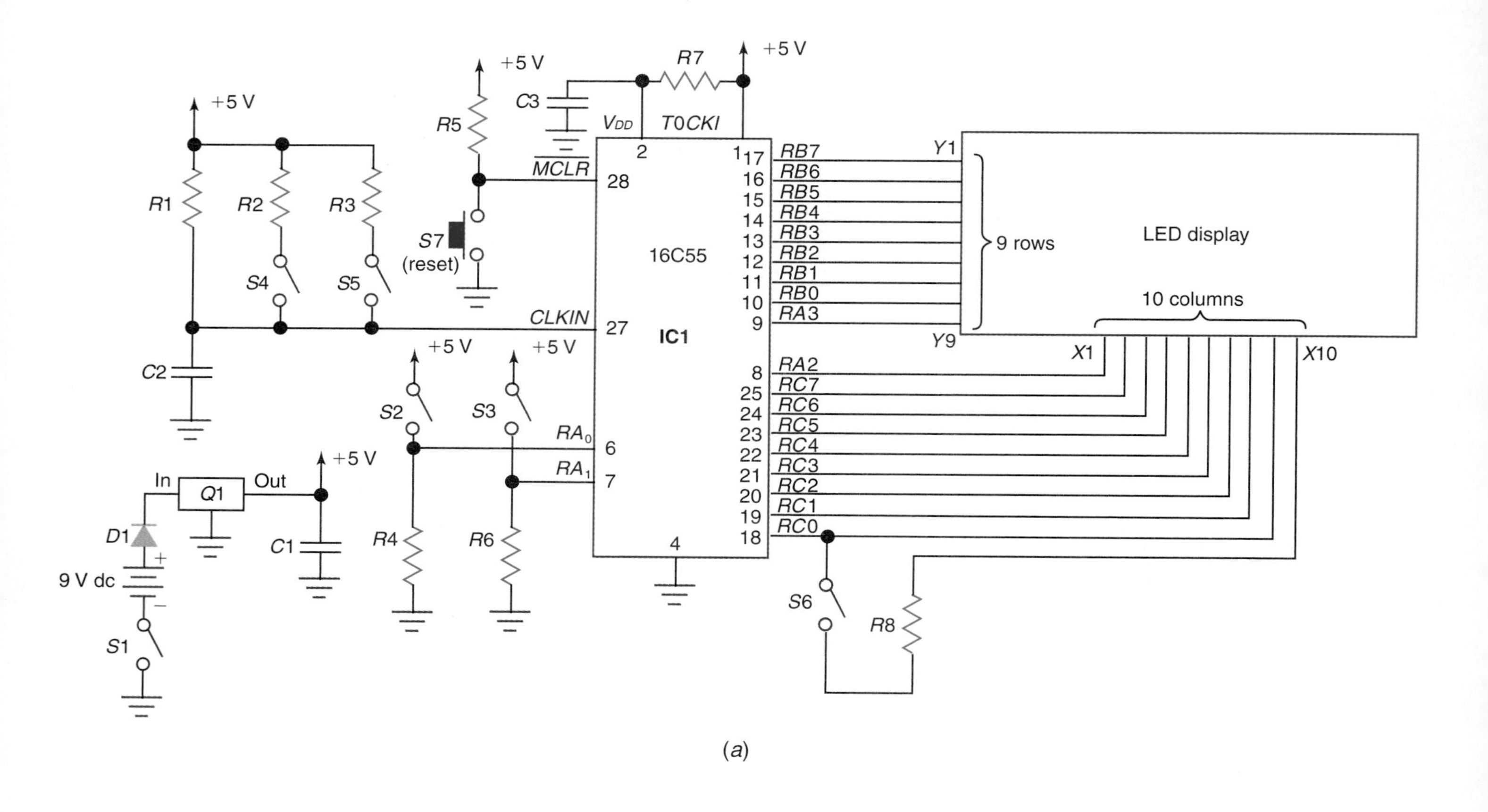

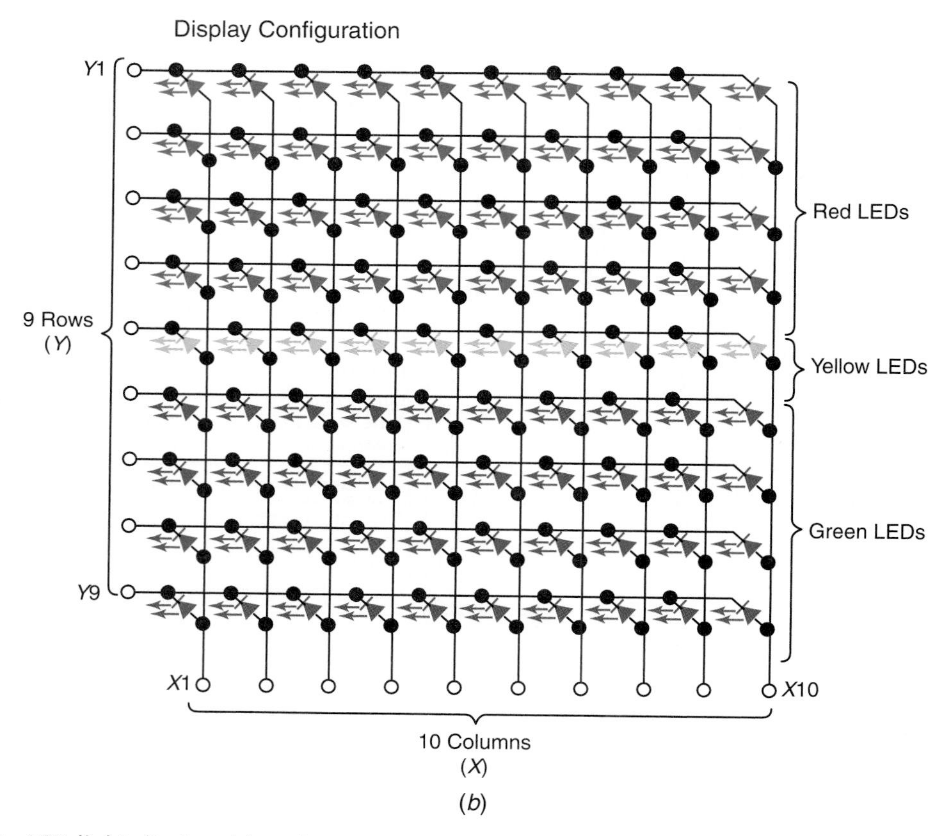

Fig. 12-50 LED light display driven by 16C55 microcontroller IC. (*a*) Wiring diagram of light display project. (*b*) Wiring diagram for nine-row by 10-column LED display. (*Courtesy of Chaney Electronics, Inc.*)

121. A _________ (microcontroller, microprocessor) is the device that is considered the CPU of a personal computer.
122. The PIC16C55 IC is a _________ (microcontroller, PLC) featuring an EPROM program memory that holds _________ words and a RAM data memory that will hold _________ bytes.
123. The PIC16C55 IC would probably cost _________ (less than 5, more than 50) dollars if purchased in small quantities.
124. The PIC16C55 IC features a RISC architecture using _________ (33, 72) instructions in its instruction set.
125. Refer to Fig. 12-50. The PIC16C55 microcontroller has a total of _________ (12, 20) I/O pins. In this light display project, _________ (2, 18) pins are programmed as inputs while _________ (2, 18) pins are programmed to be outputs for driving the LED display.
126. Refer to Fig. 12-50. The programs held in read-only memory in the 16C55 microcontroller are called _________ (firmware, hardware).
127. Refer to Fig. 12-50. To light just the nine LEDs in the left column, the *X*1 column must be _________ (HIGH, LOW) while all nine row inputs (*Y*1 through *Y*9) must be _________ (HIGH, LOW).
128. Refer to Fig. 12-50. The 16C55 microcontroller operates on _________ volts, which is set the voltage regulator IC (*Q*1).
129. Refer to Fig. 12-50. Switches *S*2 and *S*3 are connected to I/O pins that are programmed as _________ (inputs, outputs) and control which one of four programs are executed by the microcontroller.
130. Refer to Fig. 12-50. Activating switch *S*7 would cause the microcontroller to _________ (start at the beginning of the current program, turn off the unit).

Chapter 12 Review

Summary

1. An assembly of digital subsystems connected correctly forms a digital system.
2. Digital systems have six common elements: input, transmission, storage, processing, control, and output.
3. Manufacturers produce ICs that are classified as small-, medium-, large-, very large-, and ultra large-scale integrations.
4. A calculator is a complex digital system often based upon a single LSI IC. A typical modern calculator might use a CMOS IC, LCD, and be battery- and/or solar-powered.
5. The computer is one of the most complex digital systems. It is unique because of its adaptability, vast size, high speed, and stored program.
6. The microcomputer is slower and less expensive than its larger counterparts. The microcomputer is a microprocessor-based digital system.
7. The microcomputer makes extensive use of both ROM and RAM for internal storage. Floppy, optical, and hard disks are widely used for secondary bulk storage. Microcomputers support many peripheral input and output devices.
8. Microprocessing unit instructions are composed of the operation and operand parts. The MPU follows a fetch-decode-execute sequence when running a program.
9. Combinational logic gates can be used for microcomputer address decoding.
10. Three-state devices, such as buffers, must be used when several memories and microprocessors transfer information over a common data bus.
11. Multiplexers and demultiplexers can be used for data transmission. More complex UARTs may also be used for serial data transmission.
12. Data transmission can be either serial or parallel. Various interface ICs are available for sending and receiving parallel and serial data.
13. Errors occurring during data transmission can be detected using parity bits.
14. A digital clock and a digital frequency counter are two closely related digital systems. Both make extensive use of counters.
15. Many LSI digital clock chips are available. Most clock ICs need other components to produce a working digital clock.
16. Multiplexing is a commonly used method of driving seven-segment LED displays.
17. All digital systems are basically constructed from AND gates, OR gates, and inverters.
18. A frequency counter is an instrument that accurately counts input pulses in a given time interval and displays it in digital form. It constantly cycles through a reset-count-display sequence.
19. Block diagrams communicate the organization of a digital system. The most detailed block diagrams break the system down to the chip level.
20. Electronic games are popular construction projects. Many are simulations of older games like throwing dice.
21. A programmable logic controller (PLC) is a rugged computer system used in factories, warehouses, and chemical plants to control machines. PLCs are replacing hard-wired relay logic for machine control.
22. Ladder relay schematics, ladder relay logic diagrams, logic gate diagrams, and Boolean expressions can all be used to describe a control logic problem.
23. A microcontroller is a "computer on a chip" embedded in many everyday devices. Microcontrollers contain a CPU, a small RAM (data memory), read-only memory (program memory containing firmware), a clock, and I/O pins.
24. Microcontrollers are produced in huge quantities at very low prices.

Answer the following questions.

12-1. List the six elements found in most digital systems.

12-2. What do the following letters stand for when referring to ICs?

a. IC	e. VLSI	i. RAM
b. SSI	f. ULSI	j. SRAM
c. MSI	g. ROM	k. NVSRAM
d. LSI	h. PROM	l. WORM

12-3. The term "chip" usually is taken to mean a(n) ________ (IC, sliver of plastic) in digital electronics.

12-4. Inexpensive pocket calculators are usually based on an ________ (LSI, MSI).

12-5. The organization of the circuits within a calculator is called the ________ (architecture, dimensions) of the IC.

12-6. A calculator is *timed* by the ________ (internal clock, pressing of keys on the keyboard).

12-7. A simple calculator contains a rather large ________ (RAM, ROM).

12-8. ________ (Calculators, Computers, Both calculators and computers) contain a controller or control circuitry.

12-9. A ________ (computer, digital wristwatch) is usually based upon a single LSI IC.

12-10. The CPU of a computer contains what three sections?

12-11. The ________ (ALU, MUX) section of a computer performs calculations and logic functions.

12-12. The most complex digital system is a ________ (computer, digital multimeter).

12-13. An IC called a(n) ________ is the CPU of a microcomputer.

12-14. Refer to Fig. 12-5. The parts of a microcomputer system are connected by control lines, a(n) ________ bus, and a two-way ________ ________.

12-15. The input-store-print operation shown in Fig. 12-7 required three instructions, which use ________ bytes of program memory.

12-16. A gating circuit called a(n) ________ decoder is used to select one of many memory devices for sending or retrieving data via the data bus.

12-17. Classify these microcomputer peripheral devices as input, output, storage, or input/output units.

a. CRT monitor
b. Floppy disk drive
c. Keyboard
d. Mouse
e. Modem
f. Laser printer
g. Hard disk drive
h. Plotter

12-18. Microcomputer memory addresses are commonly listed in ________ (Gray code, hexadecimal).

12-19. What do the following letters stand for when referring to a microcomputer system?

a. CPU
b. PIA
c. PPI
d. UART

12-20. The baud rate is the number of ________ per second being transmitted serially through a data link.

12-21. The IEEE-488 standard is a common ________ (parallel, serial) interface standard for the data link between a computer and scientific instrumentation.

12-22. Draw the logic symbol and truth table for a three-state buffer.

12-23. A MUX/DEMUX system converts parallel input data to (asynchronous, serial) data for transmission.

12-24. A MUX/DEMUX system operates somewhat like two ________ (rotary, three-way) switches.

12-25. Errors in transmission can be detected by using a ________ (parity bit, 16-bit word).

12-26. An ________ (AND, XOR) gate can detect an odd number of 1s at its input.

12-27. Refer to Fig. 12-21. To perform subtraction, the add/subtract mode control must be ________ (HIGH, LOW).

12-28. Refer to Fig. 12-21. When activated, the ________ (clear, clock) input zeros both registers and the digital readout will display 0.

12-29. Refer to Fig. 12-21. Assume these inputs: clear = 1, load A = 1, load B = 1, add/subtract = 0. Pressing the number 3 on the keyboard and at the same time applying a positive pulse to the clock input will yield these results: display A = ________, display B = ________, and the digital display = ________.

12-30. Refer to Fig. 12-21. The _________ (7486, 74147) IC changes the data flowing from register B to the adder IC to its 1s complement form during subtraction.

12-31. Refer to Fig. 12-21. During subtraction, display A holds the _________ (minuend, subtrahend), display B contains the _________ (minuend, subtrahend), and the result or difference appears in decimal form on the seven-segment display.

12-32. Refer to Fig. 12-21. Pressing "8" on the keyboard activates input 8 on the 74147 IC with a _________ (HIGH, LOW), causing the outputs of the encoder to give the 4-bit 1s complement of decimal 8, which is _________.

12-33. Refer to Fig. 12-21. The _________ (7483 and 7447, 74194) ICs act as temporary memory devices in this adder/subtractor system.

12-34. Refer to Fig. 12-21. The _________ (7447, 7486) IC converts the binary output of the adder to decimal form in this adder/subtractor system.

12-35. A digital clock makes extensive use of _________ (counter, shift register) subsystems.

12-36. A known frequency is the main input to a digital _________ (clock, frequency counter) system.

12-37. Counters are used for counting upward and _________ (shifting data, storing data) in the digital clock system.

12-38. The National Semiconductor MM5314 clock chip _________ (directly drives, multiplexes) the output displays.

12-39. The multiplex oscillator's frequency in Fig. 12-26(*a*) is set by _________ (connecting an external capacitor and resistor to the correct IC pins; the factory and cannot be changed).

12-40. In practice, the segment driver block shown in Fig. 12-28(*b*) may consist of _________ (a VLSI chip, seven transistors with associated resistors).

12-41. The multiplexed displays in Fig. 12-30 are _________ (all turned on and off together to save power; turned on and off one at a time in rapid succession).

12-42. The *known frequency* entering the MM5314 clock chip in Fig. 12-30 is _________ Hz. This frequency comes from the _________ (oscillator, transformer).

12-43. Counters are used for counting upward and _________ (counting downward, dividing frequency) in a digital frequency counter.

12-44. The three J-K flip-flops (FF1, FF2, FF3) and the NAND gate in Fig. 12-35 function as a _________ (down counter, frequency divider).

12-45. The 7408 AND gate in Fig. 12-35 serves to _________ (clear, inhibit) the counters.

12-46. The frequency counter in Fig. 12-35 counts from a low of _________ Hz to a high of _________ Hz.

12-47. What IC(s) are being used as wave-shaping circuits in the frequency counter in Fig. 12-35?

12-48. Refer to Fig. 12-35. The unknown frequency is allowed to pass through the control gate for 0.1 s when the count/display waveform goes _________ (HIGH, LOW).

12-49. Refer to Fig. 12-35. The displays are blanked during the _________ portion of the count/display waveform.

12-50. Refer to Fig. 12-38. List two ICs that form the time-base clock section of the LCD timer.

12-51. Refer to Fig. 12-38. List the IC(s) that detect when the count of the timer reaches 00.

12-52. Refer to Fig. 12-38. When the count on the timer reaches 00, the output of the magnitude comparator goes _________ (HIGH, LOW). This turns on transistor Q_1, sounding the alarm, and activates the _________ _________ line.

12-53. Refer to Fig. 12-38. When the LCD reads 88, the signals on all of the lines from the 74HC4543 drivers to the displays are _________ (in phase, 180° out of phase) with the signal at the output of the display clock.

12-54. Refer to Fig. 12-38. The accuracy of the entire timer depends on the accuracy of the _________ clock.

12-55. Refer to Fig. 12-36(*b*). A commercial timer would probably use a(n) _________-controlled oscillator (astable MV) for the time-base clock to assure maximum accuracy.

12-56. Refer to Fig. 12-38. The 74HC4543 ICs have the _________ (decoder, driver, latch) section of the chip disabled in this circuit.

12-57. Refer to Fig. 12-38. The divide-by-256 block consists of _________ (two, four, 256) divide-by-16 counters.

12-58. Refer to Fig. 12-42. When the push button is _________ (closed, opened), the display will stop and indicate a random number from 1 to _________ (number) simulating the roll of a single die.

12-59. Refer to Fig. 12-42. This circuit uses _________ (CMOS, TTL) ICs.

12-60. Refer to Fig. 12-42. If a "1" shows on the seven-segment LED display, then outputs _________ (letters) of the 7447 IC are activated with a _________ (HIGH, LOW).

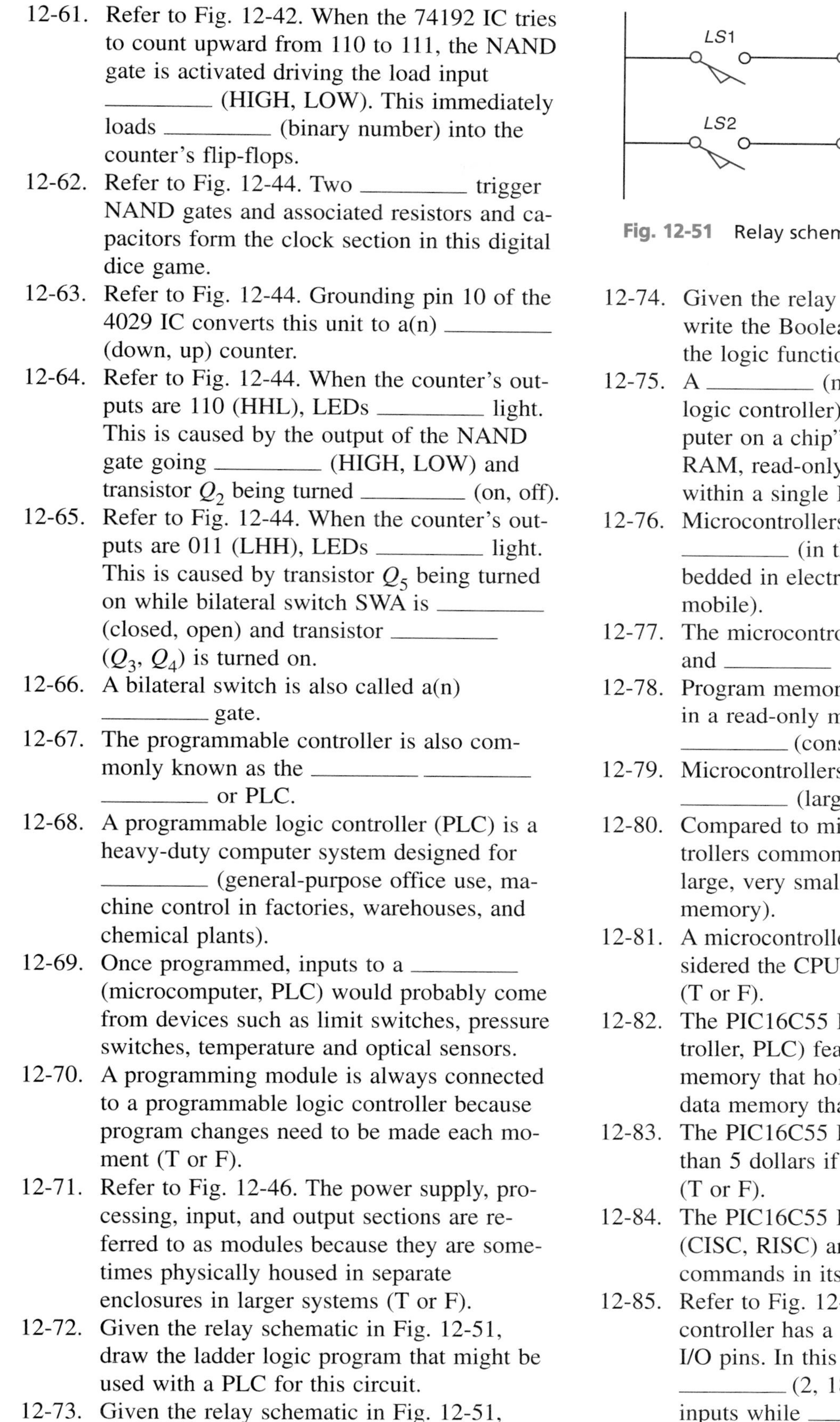

12-61. Refer to Fig. 12-42. When the 74192 IC tries to count upward from 110 to 111, the NAND gate is activated driving the load input ________ (HIGH, LOW). This immediately loads ________ (binary number) into the counter's flip-flops.

12-62. Refer to Fig. 12-44. Two ________ trigger NAND gates and associated resistors and capacitors form the clock section in this digital dice game.

12-63. Refer to Fig. 12-44. Grounding pin 10 of the 4029 IC converts this unit to a(n) ________ (down, up) counter.

12-64. Refer to Fig. 12-44. When the counter's outputs are 110 (HHL), LEDs ________ light. This is caused by the output of the NAND gate going ________ (HIGH, LOW) and transistor Q_2 being turned ________ (on, off).

12-65. Refer to Fig. 12-44. When the counter's outputs are 011 (LHH), LEDs ________ light. This is caused by transistor Q_5 being turned on while bilateral switch SWA is ________ (closed, open) and transistor ________ (Q_3, Q_4) is turned on.

12-66. A bilateral switch is also called a(n) ________ gate.

12-67. The programmable controller is also commonly known as the ________ ________ ________ or PLC.

12-68. A programmable logic controller (PLC) is a heavy-duty computer system designed for ________ (general-purpose office use, machine control in factories, warehouses, and chemical plants).

12-69. Once programmed, inputs to a ________ (microcomputer, PLC) would probably come from devices such as limit switches, pressure switches, temperature and optical sensors.

12-70. A programming module is always connected to a programmable logic controller because program changes need to be made each moment (T or F).

12-71. Refer to Fig. 12-46. The power supply, processing, input, and output sections are referred to as modules because they are sometimes physically housed in separate enclosures in larger systems (T or F).

12-72. Given the relay schematic in Fig. 12-51, draw the ladder logic program that might be used with a PLC for this circuit.

12-73. Given the relay schematic in Fig. 12-51, draw a logic gate equivalent of this circuit using AND and OR symbols.

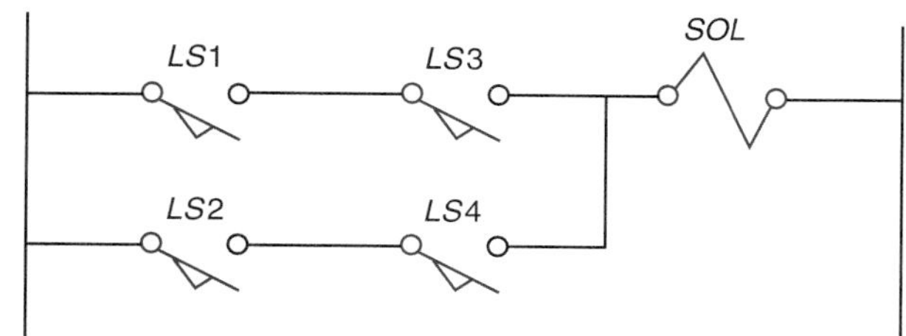

Fig. 12-51 Relay schematic diagram.

12-74. Given the relay schematic in Fig. 12-51, write the Boolean expression that describes the logic function of this circuit.

12-75. A ________ (microcontroller, programmable logic controller) can be described as a "computer on a chip" because it contains a CPU, RAM, read-only memory, clock, and I/O pins within a single IC.

12-76. Microcontrollers are most likely to appear ________ (in the CPU section of a PC, embedded in electronics devices in your automobile).

12-77. The microcontroller is noted for its small size and ________ (high, very low) cost.

12-78. Program memory in a microcontroller is held in a read-only memory device and is ________ (constantly, rarely) reprogrammed.

12-79. Microcontrollers are manufactured in ________ (large, small) volumes.

12-80. Compared to microprocessors, microcontrollers commonly address ________ (very large, very small) amounts of RAM (data memory).

12-81. A microcontroller is the device that is considered the CPU of a personal computer (T or F).

12-82. The PIC16C55 IC is a ________ (microcontroller, PLC) featuring an EPROM program memory that holds 512 words and a RAM data memory that will hold 24 bytes.

12-83. The PIC16C55 IC would probably cost less than 5 dollars if purchased in small quantities (T or F).

12-84. The PIC16C55 IC features a(n) ________ (CISC, RISC) architecture employing only 33 commands in its instruction set.

12-85. Refer to Fig. 12-50. The PIC16C55 microcontroller has a total of ________ (12, 20) I/O pins. In this light display project, ________ (2, 18) pins are programmed as inputs while ________ (2, 18) pins are programmed to be outputs for driving the LED display.

12-86. Refer to Fig. 12-50. The programs held in read-only memory in the 16C55 microcontroller may be called firmware (T or F).

12-87. Refer to Fig. 12-50. To light just the 10 red LEDs across the top of the display, the *Y*1 row input must be ________ (HIGH, LOW) while all 10 column inputs (*X*1 thru *X*10) must be ________ (HIGH, LOW).

12-88. Refer to Fig. 12-50. Switches *S*2 and *S*3 are connected to I/O pins that are programmed as ________ (inputs, outputs) and control which one of four programs are executed by the microcontroller.

12-89. Refer to Fig. 12-50. Activating switch *S*7 would cause the microcontroller to ________ (reset at the beginning of the current program, turn off the unit).

12-90. Refer to Fig. 12-50. The PIC16C55 microcontroller's I/O pins have enough drive capabilities to drive the LED display directly (T or F).

Critical Thinking Questions

12-1. List at least five common pieces of equipment that are considered digital systems.

12-2. List at least four devices you used or studied about that are considered digital subsystems.

12-3. Draw a block diagram of the organization of the five main sections of a computer. Show the flow of *program* information and *data* through the system.

12-4. The ________ (Hamming, Hollerith) code is an error-correcting code that uses several parity bits.

12-5. What part(s) of the adder/subtractor shown in Fig. 12-20 perform the following system functions:
a. input
b. output
c. storage
d. processing
e. control

12-6. How are the segment drivers shown in block form in Fig. 12-28 implemented in a working digital clock (see Fig. 12-30)?

12-7. The oscillator shown at the lower left in Fig. 12-28 is associated with what function within the digital clock?

12-8. Why was the experimental frequency counter shown in Fig. 12-35 included for study when it is not a practical piece of equipment?

12-9. What are some differences between the conceptual version of the digital timer in Fig. 12-36 and the working experimental timer in Fig. 12-38?

12-10. Why would the digital dice game shown in Fig. 12-44 probably be preferred over the simpler version in Fig. 12-42?

12-11. Refer to Fig. 12-44. When the preset pulse line goes ________ (HIGH, LOW), PNP transistor Q_1 turns on and the preset enable input to the 4029 counter is activated with a ________ (HIGH, LOW).

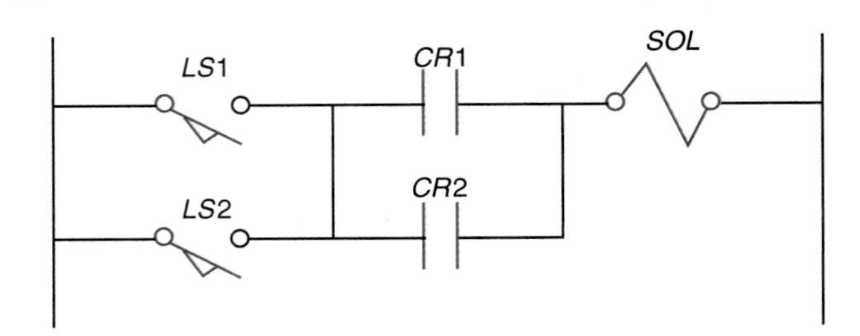

Fig. 12-52 Relay schematic diagram.

12-12. Refer to Fig. 12-44. When the counter's outputs are 100 (HLL), LEDs ________ light. The bilateral switches are closed because of the ________ (HIGH, LOW) on their control inputs. Only transistor ________ (Q_3, Q_4) is turned on grounding the cathode of LED D_5.

12-13. Why do PLCs simulate relay logic so closely?

12-14. Given the relay schematic in Fig. 12-52, draw the ladder logic program that might be used with a PLC for this circuit.

12-15. Given the relay schematic in Fig. 12-52, draw the logic gate diagram for this circuit (use AND and OR symbols) and write the Boolean expression that describes the logic of the circuit.

12-16. Describe how you might light all of the red LEDs (top four rows) on the display shown in Fig. 12-50(*b*).

12-17. At the option of your instructor, use circuit simulation software to (1) draw a four-row by four-column LED maxtrix (something like that shown in Fig. 12-50(*b*)), (2) use a word generator to program light patterns on the 4 × 4 LED display, (3) operate your 4 × 4 LED display, and (4) show instructor your light pattern generator.

Answers to Tests

1. control
2. input
3. 12 to 99
4. 10,000
5. 1960s
6. microcontroller IC
7. microprocessor
8. LSI
9. output
10. input
11. 1. power supply problems
    2. keyboard problems
12. peripheral
13. size, stored program, speed, multipurpose
14. program information and data
15. data
16. programs
17. address, control
18. modem
19. CRT monitor
20. mouse
21. operation, operand
22. 100, 101
23. MPU (microprocessor)
24. decode-execute
25. sequential
26. address decoder
27. three-state buffers or tristate buffers
28. 0, 8
29. in its high-impedance state
30. multiplexer, demultiplexer
31. 1111
32. PIA (peripheral-interface adapter)
33. UART (universal asynchronous receiver-transmitter)
34. baud
35. serial
36. even
37. XOR, XOR
38. 1
39. 0
40. XOR
41. encoder
42. 4-bit adder
43. LOW
44. 0
45. 0100, 0100, 8
46. subtraction
47. minuend, subtrahend
48. LOW, 1000
49. 74194
50. 7447
51. frequency divider, count accumulator
52. wave-shaping
53. wave shaping
54. control gate (start/stop control)
55. MOS
56. 60
57. prescale
58. 11 to 19
59. multiplexes
60. V_{SS}
61. has internal
62. resistor and capacitor
63. multiplexing
64. digit drivers
65. segment drivers
66. 12
67. *C*3 and *R*4
68. V_{DD}
69. 100
70. C_1
71. 60 Hz pulsating dc or ac
72. PROM
73. 1.0
74. LOW
75. Schmitt trigger inverters
76. counters
77. 0.1
78. wave-shaping
79. storing or accumulating
80. divide-by-6
81. 0.1, 0.9
82. count
83. reset, positive
84. count, negative
85. Schmitt trigger, wave
86. 10
87. 10, 9990
88. 0.9
89. 60
90. 74192 counter ICs
91. time-base
92. BCD
93. piezo buzzer, LCD (liquid-crystal display)
94. HIGH, HIGH, on
95. out-of-phase
96. Ph (phase), common (back place)
97. seconds
98. astable (free-running)
99. LOW, 0001
100. 1, 2, 3, 4, 5, 6
101. 4093
102. 110, 101, 100, 011, 010, 001
103. Q_5
104. *D*2 and *D*3, closed, Q_4
105. bilateral switch or transmission gate
106. programmable logic controller
107. machine control in factories
108. limit switches, pressure switches, and temperature and optic sensors
109. occasionally during reprogramming
110. T
111. less
112.

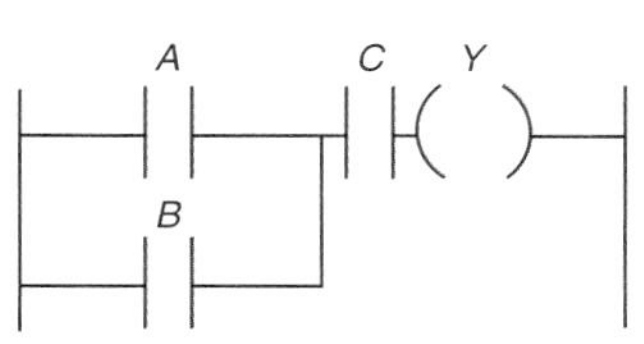

113.

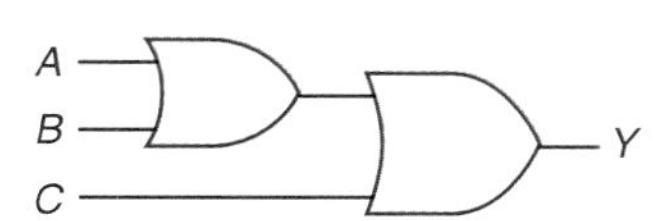

114. $(A + B)C = Y$
115. microcontroller
116. VCR
117. T
118. T
119. F
120. very small
121. microprocessor
122. microcontroller, 512, 24
123. less than 5
124. 33
125. 20, 2, 18
126. firmware
127. HIGH, LOW
128. +5
129. inputs
130. start at the beginning of the current program

Connecting with Analog Devices

Chapter Objectives

This chapter will help you to:

1. *Discuss* analog-to-digital and digital-to-analog conversion.
2. *Design* an op-amp circuit with a given gain.
3. *Analyze* the operation of several elementary D/A converter circuits.
4. *Answer* selected questions on a counter-ramp A/D converter with voltage comparator.
5. *Discuss* the operation of an elementary digital voltmeter (A/D converter).
6. *Identify* several other A/D converters including the ramp and successive-approximation types.
7. *List* several specifications associated with commercial A/D converters.
8. *Answer* selected questions about the commercial ADC0804 A/D converter IC.
9. *Analyze* several digital light meter systems that are based on the ADC0804 A/D converter IC.
10. *Answer* selected questions about a digital voltmeter system based on a commercial 3½-digit LCD single-chip A/D converter.

Interfacing

To this point in our study, most data entering or leaving a digital system has been digital information. Many digital systems, however, have *analog* inputs that vary *continuously* between two voltage levels. In this chapter we discuss the *interfacing* of analog devices to digital systems.

Most real-world information is analog. For instance, it was mentioned in Chap. 1 that time, speed, weight, pressure, light intensity, and position measurements are all *analog in nature.*

Analog-to-digital converter

A/D converter

D/A converter

Hybrid system

The digital system in Fig. 13-1 has an analog input. The voltage varies continuously from 0 to 3 V. The *encoder* is a special device that converts the analog signal to digital information. The encoder is called an *analog-to-digital converter* or *A/D converter.* The A/D converter, then, converts analog information to digital data.

The digital system diagramed in Fig. 13-1 also has a *decoder.* This decoder is a special type: it converts the digital information from the digital processing unit to an analog output. For instance, the analog output may be a continuous voltage change from 0 to 3 V. We call this decoder a *digital-to-analog converter* or *D/A converter.* The D/A converter, then, decodes digital information to analog form.

The entire system in Fig. 13-1 might be called a *hybrid system* because it contains both digital and analog devices. The encoders and

D/A converter

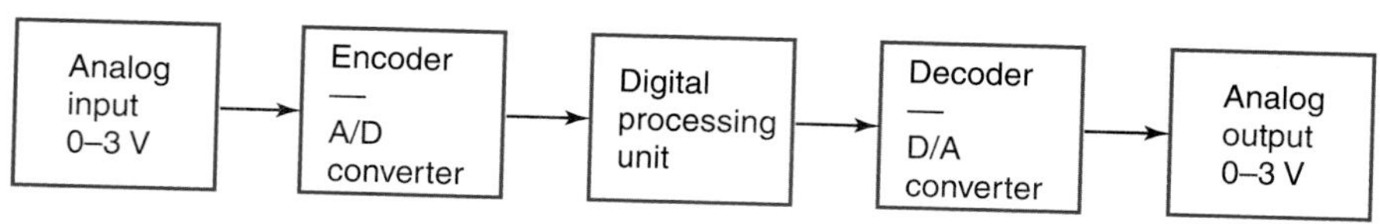

Fig. 13-1 A digital system with analog input and analog output.

decoders that convert from analog to digital and digital to analog are called *interface devices* by engineers and technicians. The word "interface" is generally used when referring to a device or circuit that converts from one mode of operation to another. In this case we are converting between analog and digital data.

Note that the input block in Fig. 13-1 refers to an analog voltage ranging from 0 to 3 V. This voltage could be produced by a transducer. A *transducer* is defined as a device that converts one form of energy to another. For instance, a photocell could be used as an input transducer to give a voltage proportional to light intensity. In this example, light energy is being converted into electrical energy by the photocell. Other transducers might include microphones, speakers, strain gauges, photoresistive cells, temperature sensors, potentiometers, and rev/min pickup coils.

13-1 D/A Conversion

Refer to the D/A converter in Fig. 13-1. Let us suppose we want to convert the binary from the processing unit to a 0- to 3-V output. As with any decoder, we must first set up a truth table of all the possible situations. Table 13-1 shows four inputs (*D, C, B, A*) into the D/A converter. The inputs are in binary form so the exact value of the inputs is not important. Each 1 is about +3 to +5 V. Each 0 is about 0 V. The outputs are shown as voltages in the rightmost column in Table 13-1. According to the table, if binary 0000 appears at the input of the D/A converter, the output is 0 V. If binary 0001 is the input, the output is 0.2 V. If binary 0010 appears at the input, then the output is 0.4 V. Notice that for each row you progress downward in Table 13-1, the analog output increases by 0.2 V.

JOB TIP

Use the company's evaluation process, networking, and good work to gain promotions.

Transducer

Table 13-1 Truth Table for D/A Converter

	Digital Inputs				Analog Output
	D	*C*	*B*	*A*	Volts
Row 1	0	0	0	0	0
Row 2	0	0	0	1	0.2
Row 3	0	0	1	0	0.4
Row 4	0	0	1	1	0.6
Row 5	0	1	0	0	0.8
Row 6	0	1	0	1	1.0
Row 7	0	1	1	0	1.2
Row 8	0	1	1	1	1.4
Row 9	1	0	0	0	1.6
Row 10	1	0	0	1	1.8
Row 11	1	0	1	0	2.0
Row 12	1	0	1	1	2.2
Row 13	1	1	0	0	2.4
Row 14	1	1	0	1	2.6
Row 15	1	1	1	0	2.8
Row 16	1	1	1	1	3.0

A block diagram of a D/A converter is shown in Fig. 13-2 on page 362. The digital inputs (*D, C, B, A*) are at the left. The decoder consists of two sections: the *resistor network* and the *summing amplifier*. The output is shown as a voltage reading on the voltmeter at the right.

Resistor network

Summing amplifier (scaling amplifier)

The resistor network in Fig. 13-2 must take into account that a 1 at input *B* is worth twice as much as a 1 at input *A*. Also, a 1 at input *C* is worth four times as much as 1 at input *A*. Several arrangements of resistors are used to do this job. These circuits are called *resistive ladder networks*.

The summing amplifier in Fig. 13-2 takes the output voltage from the resistor network and amplifies it the proper amount to get the voltages shown in the rightmost column of Table 13-1. The summing amplifier typically uses an IC unit called an *operational amplifier*. An operational amplifier is often simply called an *op amp*. The summing amplifier is also called a *scaling amplifier*.

Operational amplifier (Op amp)

The special decoder called a D/A converter consists of two parts: a group of resistors forming a resistive ladder network and an op amp used as the summing amplifier.

TEST

Supply the missing word in each statement.

1. A special encoder that converts from analog to digital information is called a(n) ________.
2. A special decoder that converts from digital to analog information is called a(n) ________.
3. A D/A converter consists of a(n) ________ network and a(n) ________ amplifier.
4. The name "op amp" stands for ________.

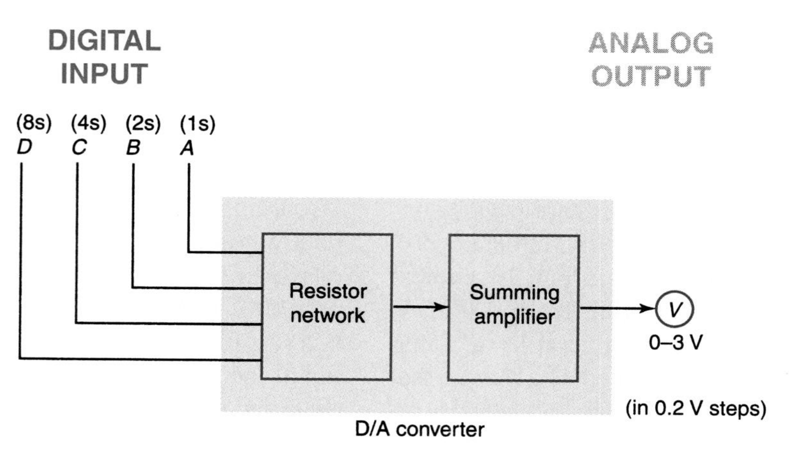

Fig. 13-2 Block diagram of a D/A converter.

Inverting input

Noninverting input

Op amp voltage gain

5. Refer to Fig. 13-2 and Table 13-1. If the binary input to the D/A converter is 0111_2, the analog output will be _________ volts.
6. Refer to Fig. 13-2 and Table 13-1. If the binary input to the D/A converter is 1111_2, the analog output will be _________ volts.
7. Refer to Fig. 13-2 and Table 13-1. If the binary input increases from 0001 to 0010, the analog output will increase by _________ volts.

13-2 Operational Amplifiers

The special amplifiers called *op amps* are characterized by high input impedance, low output impedance, and a variable voltage gain that can be set with external resistors. The symbol for an op amp is shown in Fig. 13-3(*a*). The op amp shown has two inputs. The top input is labeled an *inverting input.* The inverting input is shown by the minus sign (−) on the symbol. The other input is labeled a *noninverting input.* The noninverting input is shown by the plus sign (+) on the symbol. The output of the amplifier is also shown on the right of the symbol.

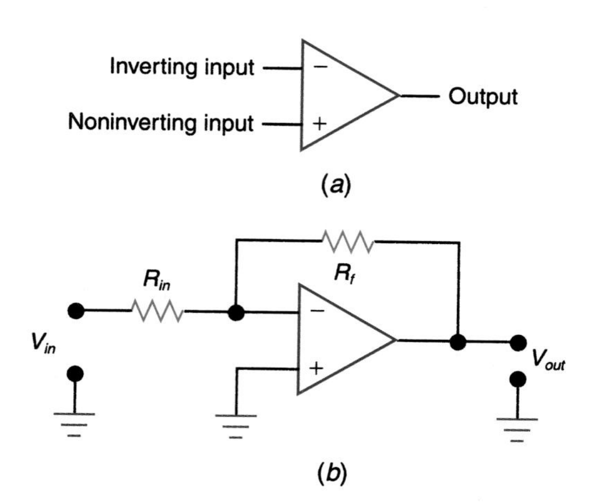

Fig. 13-3 Operational amplifier. (*a*) Symbol. (*b*) With input and feedback resistors for setting gain.

The operational amplifier is almost never used alone. Typically, the two resistors shown in Fig. 13-3(*b*) are added to the op amp to set the voltage gain of the amplifier. Resistor R_{in} is called the input resistor. Resistor R_f is called the feedback resistor. The *voltage gain* of this amplifier is found by using the simple formula

$$A_v \text{ (voltage gain)} = \frac{R_f}{R_{in}}$$

Suppose the values of the resistors connected to the op amp are $R_f = 10\ \text{k}\Omega$ and $R_{in} = 10\ \text{k}\Omega$. Using our voltage gain formula, we find that

$$A_v = \frac{R_f}{R_{in}} = \frac{10{,}000}{10{,}000} = 1$$

The gain of the amplifier is 1. In our example, if the input voltage at V_{in} in Fig. 13-3(*b*) is 5 V, the output voltage at V_o is 5 V. The inverting input is being used, and so if the input voltage is +5 V, then the output voltage is −5 V. The voltage gain of the op amp can also be calculated using the formula

$$A_v = \frac{V_{out}}{V_{in}}$$

The voltage gain for the circuit above is then

$$A_v = \frac{V_{out}}{V_{in}} = \frac{5}{5} = 1$$

The voltage gain is again found to be 1.

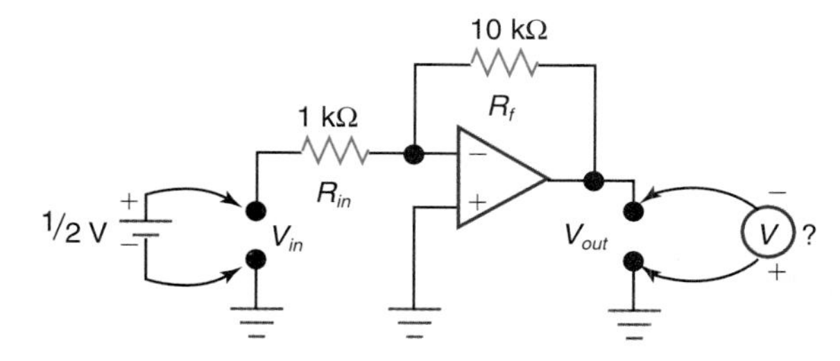

Fig. 13-4 Amplifier circuit using an op amp.

Suppose the *input and feedback resistors* are 1 kΩ and 10 kΩ, as shown in Fig. 13-4. What is the voltage gain for this circuit? The voltage gain is calculated as

$$A_v = \frac{R_f}{R_{in}} = \frac{10{,}000}{1000} = 10$$

The voltage gain is 10. If the input voltage is +0.5 V, then the voltage at the output is how many volts? If the gain is 10, then the input voltage of 0.5 V times 10 equals 5 V. The output voltage is −5 V, as measured on the voltmeter in Fig. 13-4.

You have seen how the voltage gain of an op amp can be changed by changing the ratio between the input and feedback resistors. You should know how to set the gain of an operational amplifier by using different values for R_{in} and R_f.

In summary, the op amp is part of a D/A converter; it is used as a *summing amplifier* in the converter. The gain of the op amp is easily set by the ratio of the input and feedback resistors.

Input and feedback resistors

TEST

Answer the following questions.

8. Refer to Fig. 13-3(*b*). The resistor labeled R_f in this op amp circuit is called the _________ resistor.
9. Refer to Fig. 13-3(*b*). The resistor labeled R_{in} in this op amp circuit is called the _________ resistor.
10. What is the voltage gain (A_v) of an op amp such as the one shown in Fig. 13-3(*b*) if R_{in} = 1 kΩ and R_f = 20 kΩ?
11. What is the output voltage (V_o) from the op amp in question 10 if the input voltage is +0.2 V?
12. What is the voltage gain (A_v) of an op amp such as the one shown in Fig. 13-3(*b*) if R_{in} = 5 kΩ and R_f = 20 kΩ?
13. What is the output voltage (V_o) from the op amp in question 12 if the input voltage is +1.0 V?

13-3 A Basic D/A Converter

A simple D/A converter is shown in Fig. 13-5. The D/A converter is made in two sections. The *resistor network* on the left is made up of resistors R_1, R_2, R_3, and R_4. The summing amplifier on the right consists of an op amp and

Summing amplifier

Resistor network

Fig. 13-5 A D/A converter circuit.

Basic D/A converter

a feedback resistor. The input (V_{in}) is 3 V applied to switches D, C, B, and A. The output voltage (V_o) is measured on a voltmeter. Notice that the op amp requires a dual power supply: a +10-V supply and a −10-V supply.

With all switches at GND (0 V), as shown in Fig. 13-5, the input voltage at point A is 0 V and the output voltage is 0 V. This corresponds to row 1, Table 13-1. Suppose we move switch A to the logical 1 position in Fig. 13-5. The input voltage of 3 V is applied to the op amp. We next calculate the gain of the amplifier. The gain is dependent upon the feedback resistor (R_f), (which is 10 kΩ), and the input resistor (R_{in}), which is the value of R_1, or 150 kΩ. Using the gain formula, we have

$$A_v = \frac{Rf}{R_{in}} = \frac{10{,}000}{150{,}000} = 0.066$$

To calculate the output voltage, we multiply the gain by the input voltage as shown here:

$$V_{out} = A_v \times V_{in} = 0.066 \times 3 = 0.2\ V$$

The output voltage is 0.2 V when the input is binary 0001. This satisfies the requirements of row 2, Table 13-1.

Let us now apply binary 0010 to the D/A converter in Fig. 13-5. Only switch B is moved to the logical 1 position, applying 3 V to the op amp. The gain is

$$A_v = \frac{Rf}{R_{in}} = \frac{10{,}000}{75{,}000} = 0.133$$

Multiplying the gain by the input voltage gives us 0.4 V. The 0.4 V is the output voltage. This satisfies row 3, Table 13-1.

Notice that for each binary count in Table 13-1 the output voltage of the D/A converter increases by 0.2 V. This increase occurs because of the increased voltage gain of the op amp as we switch in different resistors (R_1, R_2, R_3, R_4). If only resistor R_4 from Fig. 13-5 were connected by activating switch D, the gain would be

Ladder-type D/A converters

$$A_v = \frac{Rf}{R_{in}} = \frac{10{,}000}{18{,}700} = 0.535$$

R-2R ladder network

The gain multiplied by the input voltage of 3 V gives 1.6 V at the output of the op amp. This is what is required by row 9, Table 13-1.

When all switches are activated (at logical 1) in Fig. 13-5, the op amp puts out the full 3 V because the gain of the amplifier has increased to 1.

Any input voltage up to the limits of the operational amplifier power supply (±10 V) may be used. More binary places may be added by adding switches. If a 16s place value switch is added in Fig. 13-5, it needs a resistor with half the value of resistor R_4. Its value would then have to be 9350 Ω. The value of the feedback resistor would also be changed to 5 kΩ. The input would then be a 5-bit binary number; the output would still be an analog output varying from 0 to −3.1 V (in 0.1 V steps).

Trying to expand D/A converter in Fig. 13-5 to many bits results in an impractical range of resistor values and poor accuracy.

TEST

Answer the following questions.

14. Calculate the voltage gain of the op amp in Fig. 13-5 when only switch C (the 4s switch) is at logical 1.
15. Using the voltage gain from question 14, calculate the output voltage of the D/A converter in Fig. 13-5 when only switch C is at logical 1.
16. List two limitations of the basic D/A converter shown in Fig. 13-5 for large binary words.
17. Calculate the voltage gain of the op amp in Fig. 13-5 when both switches A and B are at logical 1 (Hint: use parallel resistance formula $R_T = (R_1 \times R_2)/(R_1 + R_2)$).
18. Using the voltage gain from question 17, calculate the output voltage from the D/A converter in Fig. 13-5 when both input switches A and B are at logical 1.

13-4 Ladder-type D/A Converters

Digital-to-analog converters consist of a resistor network and a summing amplifier. Figure 13-6 diagrams a type of resistor network that provides the proper weighting for the binary inputs. This resistor network is sometimes called the *R-2R ladder network.* The advantage of this arrangement of resistors is that only two values of resistors are used. Resistors R_1, R_2, R_3, R_4, and R_5 are 20 kΩ each. Resistors R_6, R_7, R_8, and R_f are each 10 kΩ. Notice that all the horizontal resistors on the "ladder" are ex-

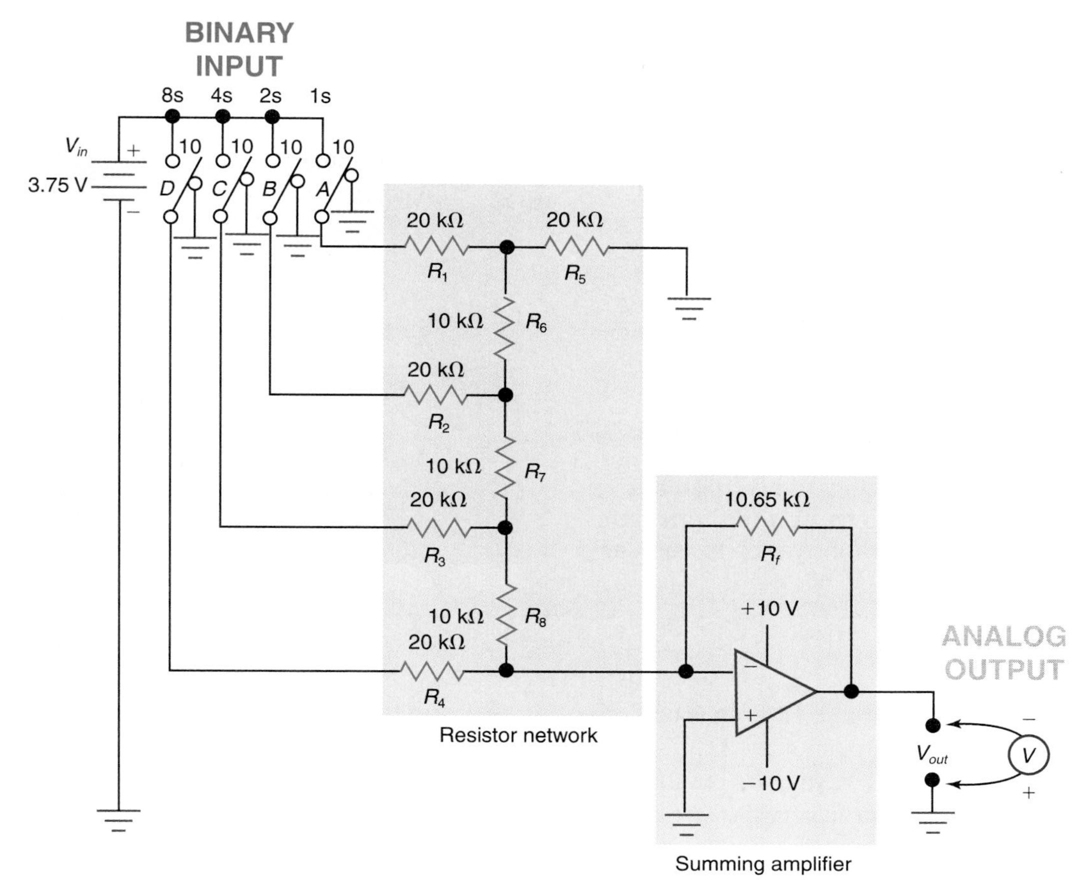

Fig. 13-6 A D/A converter circuit using an R-2R ladder resistor network.

D/A converter circuit

R-2R ladder resistor network

Table 13-2 Truth Table for D/A Converter

Binary Inputs				Analog Output
8s	4s	2s	1s	
D	*C*	*B*	*A*	Volts
0	0	0	0	0
0	0	0	1	0.25
0	0	1	0	0.50
0	0	1	1	0.75
0	1	0	0	1.00
0	1	0	1	1.25
0	1	1	0	1.50
0	1	1	1	1.75
1	0	0	0	2.00
1	0	0	1	2.25
1	0	1	0	2.50
1	0	1	1	2.75
1	1	0	0	3.00
1	1	0	1	3.25
1	1	1	0	3.50
1	1	1	1	3.75

actly twice the value of the vertical resistors, hence the title R-2R ladder network.

The summing amplifier in Fig. 13-6 is the same one used in the last section. Again notice the use of the dual power supply on the op amp.

The most valuable employees have good technical skills and positive attitudes.

The operation of this D/A converter is similar to the basic one in the last section. Table 13-2 details the operation of this D/A converter. Notice that we are using an input voltage of 3.75 V on this converter. Each binary count increases the analog output by 0.25 V, as shown in the rightmost column of Table 13-2. Remember that each 0 on the input side of the table means 0 V applied to that input. Each 1 on the input side of the table means 3.75 V applied to that input. The input voltage of 3.75 V is used because this is very close to the output of TTL counters and other ICs you may have used. The inputs (*D, C, B, A*) in Fig. 13-6, then, could be connected directly to the outputs of a TTL IC and

Did You Know?

Geostationary Earth Orbits
Business satellites in very high equator-level geostationary (GEO) earth orbits occupy a limited territory. There are already 150 of them up there, leaving room for only a few more. Pictured is a GEO satellite, courtesy of Intelsat. These satellites move at the same speed as the earth and therefore appear to "hover" in place. This is very useful to businesses such as cable TV operators. However, because these satellites are located so high above the earth, they also cause the slight time lag that can be an annoyance to cellular telephone users.

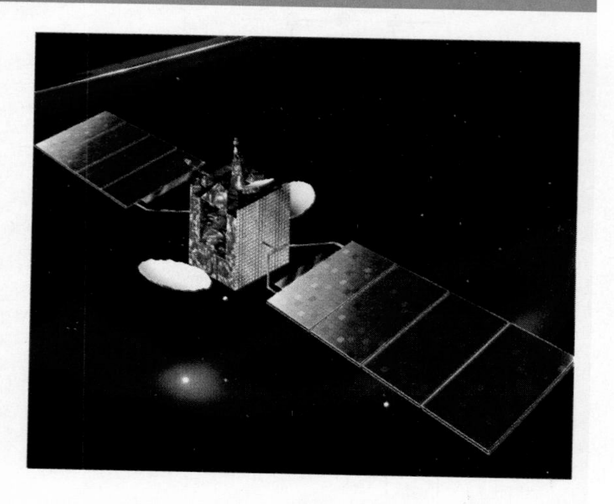

operate according to Table 13-2. In actual practice, however, the outputs of a TTL IC are not accurate enough; they have to be put through a level translator to get a very precise voltage output.

More binary places (16s, 32s, 64s, and so on) can be added to the D/A converter in Fig. 13-6. Follow the pattern of resistor values shown in this diagram when adding place values.

Two types of special decoders called digital-to-analog converters have been covered. The R-2R ladder-type D/A converter has some advantages over the more basic unit. The heart of the D/A converter consists of the resistor network and the summing amplifier.

TEST

Answer the following questions.

19. The digital-to-analog converter in Fig. 13-6 is a(n) _________ -type D/A converter.
20. Refer to Fig. 13-6. The gain of the op amp is greatest when all input switches are at logical _________ (0, 1).
21. Refer to Fig. 13-6 and Table 13-2. The gain of the op amp is *least* when switch _________ (*A*, *B*, *C*, *D*) is the only switch at logical 1.
22. Refer to Fig. 13-6 and Table 13-2. If the binary input to the D/A converter is 1011, the analog output voltage is _________ volts.
23. Refer to Fig. 13-6 and Table 13-2. If the binary input to the D/A converter increases from 0111 to 1000, the analog output voltage increases by _________ volts.

A/D converter

13-5 An A/D Converter

An analog-to-digital converter is a special type of encoder. A basic block diagram of an A/D converter is shown in Fig. 13-7. The input is a single variable voltage. The voltage in this case varies from 0 to 3 V. The output of the A/D converter is in binary. The A/D converter translates the analog voltage at the input into a 4-bit binary word. As with other encoders, it is well to define exactly the expected inputs and outputs. The truth table in Table 13-3 shows how the A/D converter should work. Row 1 shows 0 V being applied to the input of the A/D converter. The output is binary 0000. Row 2 shows a 0.2-V input. The output is binary 0001. Notice that each increase of 0.2 V increases the binary count by 1. Finally, row 16 shows that when the maximum 3 V is applied to the input, the output reads binary 1111. Notice that the truth table in Table 13-3 is just the reverse of the D/A converter truth table in Table 13-1; the inputs and outputs have just been reversed.

The truth table for the A/D converter looks quite simple. The electronic circuits that per-

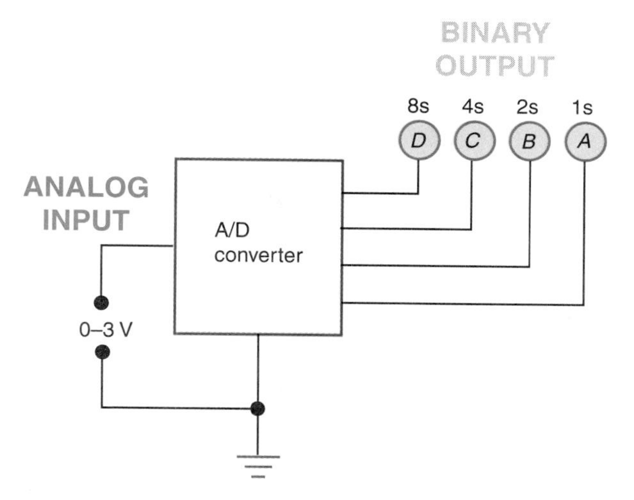

Fig. 13-7 Block diagram of an A/D converter.

form the task detailed in the truth table are somewhat more complicated. One type of A/D converter is diagramed in Fig. 13-8 on page 368. The A/D converter contains a *voltage comparator,* an AND gate, a binary counter, and a D/A converter. All the sections of the A/D converter except the comparator are familiar to you.

The analog voltage is applied at the left of Fig. 13-8. The comparator checks the voltage coming from the D/A converter. If the analog input voltage at *A* is *greater than* the voltage at input *B* of the comparator, the clock is allowed to *increase* the count of the 4-bit counter. The count on the counter increases until the feedback voltage from the D/A converter becomes greater than the analog input voltage. At this point the comparator stops the counter from advancing to a higher count. Suppose the input analog voltage is 2 V. According to Table 13-3, the binary counter increases to 1010 before it is stopped. The counter is reset to binary 0000, and the counter starts counting again.

Voltage comparator

TABLE 13-3 Truth Table for A/D Converter

	Analog Input	Binary Output			
		8s	4s	2s	1s
	Volts	*D*	*C*	*B*	*A*
Row 1	0	0	0	0	0
Row 2	0.2	0	0	0	1
Row 3	0.4	0	0	1	0
Row 4	0.6	0	0	1	1
Row 5	0.8	0	1	0	0
Row 6	1.0	0	1	0	1
Row 7	1.2	0	1	1	0
Row 8	1.4	0	1	1	1
Row 9	1.6	1	0	0	0
Row 10	1.8	1	0	0	1
Row 11	2.0	1	0	1	0
Row 12	2.2	1	0	1	1
Row 13	2.4	1	1	0	0
Row 14	2.6	1	1	0	1
Row 15	2.8	1	1	1	0
Row 16	3.0	1	1	1	1

Now for more detail on the A/D converter in Fig. 13-8. Let us assume that there is a logical 1 at point *X* at the output of the comparator. Also assume that the counter is at binary 0000. Assume, too, that 0.55 V is applied to the analog input. The 1 at point *X* enables the AND gate, and the first pulse from the clock appears at the CLK input of the counter. The counter advances its count to 0001. The 0001 is displayed on the lights in the upper right of Fig. 13-8. The 0001 is also fed back to the D/A converter.

According to Table 13-1, a binary 0001 produces 0.2 V at the output of the D/A converter. The 0.2 V is fed back to the *B* input of the comparator. The comparator checks its inputs. The *A* input is higher (0.55 V as opposed to 0.2 V), and so the comparator puts out a logical 1. The 1 enables the AND gate, which lets the next clock pulse through to the counter. The counter advances its count by 1. The count is now 0010. The 0010 is fed back to the D/A converter.

According to Table 13-1, a 0010 input produces a 0.4-V output. The 0.4 V is fed back to

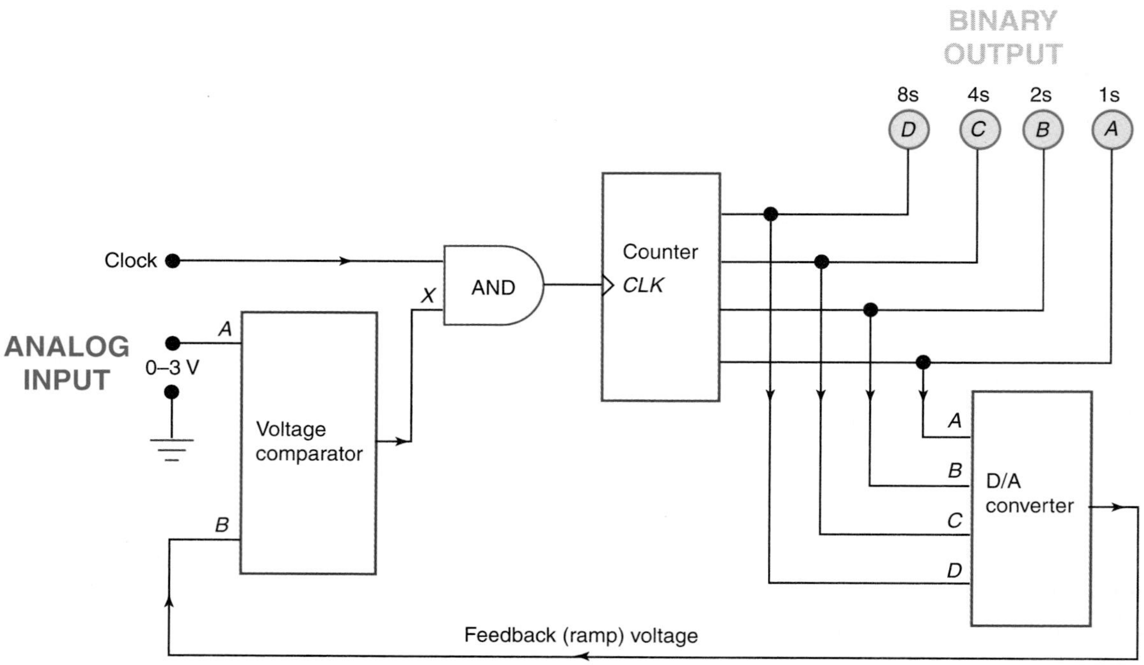

Fig. 13-8 Block diagram of a counter-ramp-type A/D converter.

the *B* input of the comparator. The comparator again checks the *B* input against the *A* input; the *A* input is still larger (0.55 V as opposed to 0.4 V). The comparator outputs a logical 1. The AND gate is enabled, letting the next clock pulse reach the counter. The counter increases its count to binary 0011. The 0011 is fed back to the D/A converter.

Counter-ramp A/D converter

According to Table 13-1, a 0011 input produces a 0.6 V output. The 0.6 V is fed back to the *B* input of the comparator. The comparator checks input *A* against input *B*; for the first time the *B* input is larger than the *A* input. The comparator puts out a logical 0. The logical 0 disables the AND gate. No more clock pulses can reach the counter. The counter stops at binary 0011. Binary 0011 must then equal 0.55 V. A look at row 4, Table 13-3, shows that 0.6 V gives the readout of binary 0011. Our A/D converter has worked according to the truth table.

Ramp

If the input analog voltage were 1.2 V, the binary output would be 0110, according to Table 13-3. The counter would have to count from binary 0000 to 0110 before being stopped by the comparator. If the input analog voltage were 2.8 V, the binary output would be 1110. The counter would have to count from binary 0000 to 1110 before being stopped by the comparator. Notice that it does take some time for the conversion of the analog voltage to a binary readout. However, in most cases the clock runs fast enough so that this time lag is not a problem.

You now should appreciate why we studied the D/A converter before the A/D converter. This *counter-ramp A/D converter* is fairly complex and needs a D/A converter to operate. The term "ramp" in the name for this converter refers to the gradually increasing voltage from the D/A converter that is fed back to the comparator. If you drew a graph of the voltage being fed to input *B* of the comparator, it would appear as a *ramp*.

TEST

Supply the missing word or words in each statement.

24. An A/D converter will translate a(n) _________ input voltage into a(n) _________ output.
25. Refer to Table 13-3. If the analog input voltage is 1 V, the binary output will be _________.
26. Refer to Fig. 13-8. When the voltage at point *B* is less than *A*, the output of the comparator at point *X* is _________ (HIGH, LOW). This causes the clock pulses to _________ (be blocked by, pass through) the AND gate.

27. The unit diagramed in Fig. 13-8 is a(n) __________ -type A/D converter.
28. Refer to Fig. 13-8. The feedback voltage from the D/A converter to input B of the __________ (counter, comparator) would appear as a ramp or "stair-step shape" waveform if observed on an oscilloscope.
29. Refer to Fig. 13-8. The comparator compares the __________ (binary values, voltages) at inputs A and B.
30. Refer to Fig. 13-8. The __________ (AND, XOR) gate blocks clock pulses from getting to the counter when the output of the comparator goes __________ (HIGH, LOW).

13-6 Voltage Comparators

In the last section we used a *voltage comparator*. We found that a comparator compares two voltages and tells us which is the larger of the two. Figure 13-9 is a basic block diagram of a comparator. If the voltage at input A is larger than at input B, the comparator gives a logical 1 output. If the voltage at input B is larger than at input A, the output is a logical 0. This is written $A > B = 1$ and $B > A = 0$ in Fig. 13-9.

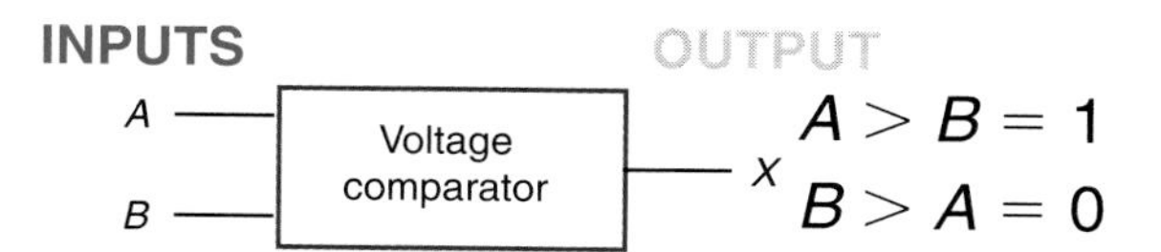

Fig. 13-9 Block diagram of a voltage comparator.

The heart of a comparator is an *op amp*. Figure 13-10(*a*) shows a comparator circuit. Notice that input A has 1.5 V applied and input B has 0 V applied. The output voltmeter reads about 3.5 V, or a logical 1.

Op amp

Voltage comparator

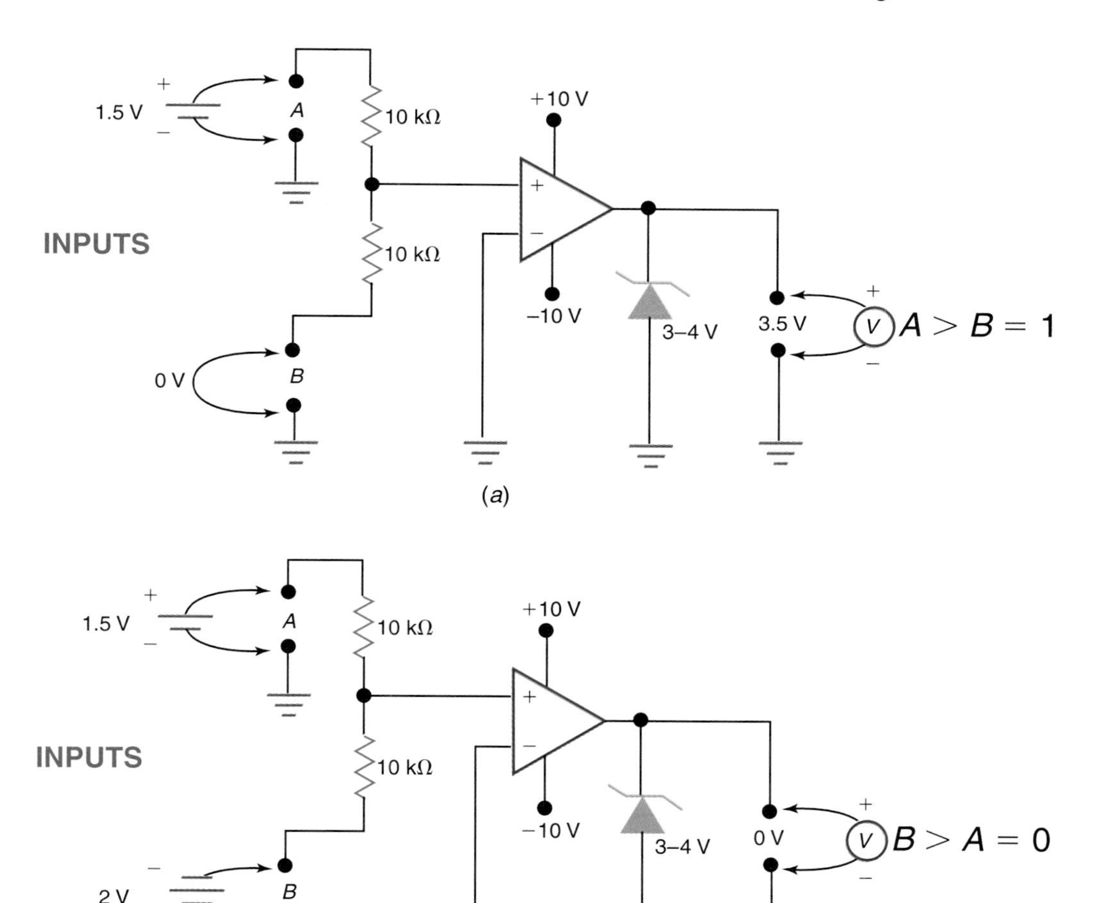

Fig. 13-10 Voltage comparator circuit. (*a*) With greater voltage at input A. (*b*) With greater voltage at input B.

Voltage comparator circuit

Figure 13-10(*b*) shows that the input *B* voltage has been increased to 2 V. Input *A* is still at 1.5 V. Input *B* is larger than input *A*. The output of the comparator circuit is about 0 V (actually the voltage is about −0.6 V), or a logical 0.

Zener diode

The comparator in the A/D converter in Fig. 13-8 works exactly like this unit. The *zener diode* in the comparator in Fig. 13-10 is there to clamp the output voltage at about +3.5 and 0 V. Without the zener diode the output voltages would be about +9 and −9 V. The +3.5 and 0 V are more compatible with TTL ICs.

TEST

Answer the following questions.

31. The comparator block shown in Fig. 13-8 compares two ________ (binary numbers, decimal numbers, dc voltages).
32. A voltage comparator circuit can be constructed using a(n) ________ IC, several resistors, and a zener diode.
33. Refer to Fig. 13-10 on page 369. When input *B* increases and becomes higher than input *A*, the output of the op amp will change from ________ (HIGH, LOW) to ________ (HIGH, LOW).

Elementary digital voltmeter

13-7 An Elementary Digital Voltmeter

One use for an A/D converter is in a *digital voltmeter.* You have already used all the subsystems needed to make an elementary digital voltmeter system. A block diagram of a simple digital voltmeter is shown in Fig. 13-11. The A/D converter converts the analog voltage to binary form. The binary is sent to the decoder, where it is converted to a seven-segment code. The seven-segment readout indicates the voltage in decimal numbers. With 7 V applied to the input of the A/D converter, the unit puts out binary 0111, as shown. The decoder activates lines *a* to *c* of the seven-segment display; segments *a* to *c* light on the display. The display reads as a decimal 7. Note that the A/D converter is also a decoder; it decodes from an analog input to a binary output.

A wiring diagram of an elementary digital voltmeter is shown in Fig. 13-12. Notice the voltage comparator, the AND gate, the counter, the decoder, the seven-segment display and the D/A converter. Several power supplies are needed to set up this circuit. A dual ±10-V supply (or two individual 10-V supplies) is used for the 741 op amps. A 5-V supply is used for the 7408, 7493, and 7447 TTL ICs and the seven-segment LED display. A 0- to 10-V variable dc power supply is also needed for the analog input voltage.

Let us assume a 2-V input to the analog input of the digital voltmeter in Fig. 13-12. Reset the counter to 0000. The comparator checks inputs *A* and *B*; *A* is larger ($A = 2$ V, $B = 0$ V). The comparator output is a logical 1. This 1 enables the AND gate. The pulse from the clock passes through the AND gate. The pulse causes the counter to advance one count. The count is now 0001. The 0001 is applied to the decoder. The decoder enables lines *b* and *c* of the seven-segment display; segments *b* and *c* light on the display, giving a decimal readout of 1. The 0001 is also applied to the D/A converter. About 3.2 V from the counter is applied through the 150-kΩ resistor to the input of the op amp. The voltage gain of the op amp is

$$A_v = \frac{R_f}{R_{in}} = \frac{47{,}000}{150{,}000} = 0.31$$

The gain is 0.31. The voltage gain times the input voltage equals the output voltage:

$$V_{out} = A_v \times V_{in} = 0.31 \times 3.2 = 1\ V$$

The output voltage of the D/A converter is

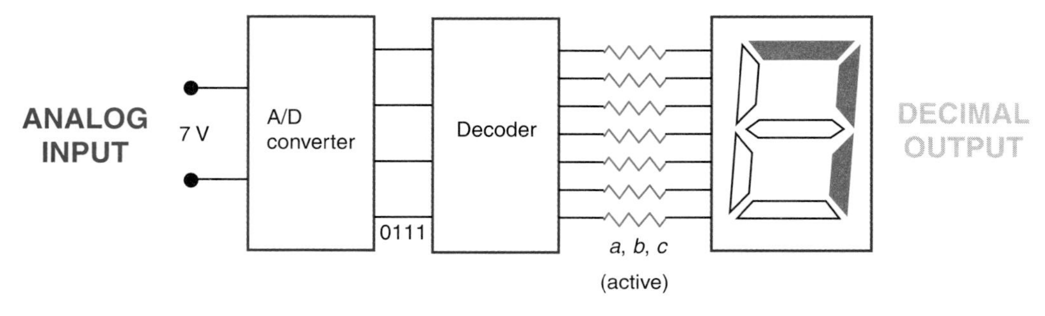

Fig. 13-11 Block diagram of an elementary digital voltmeter.

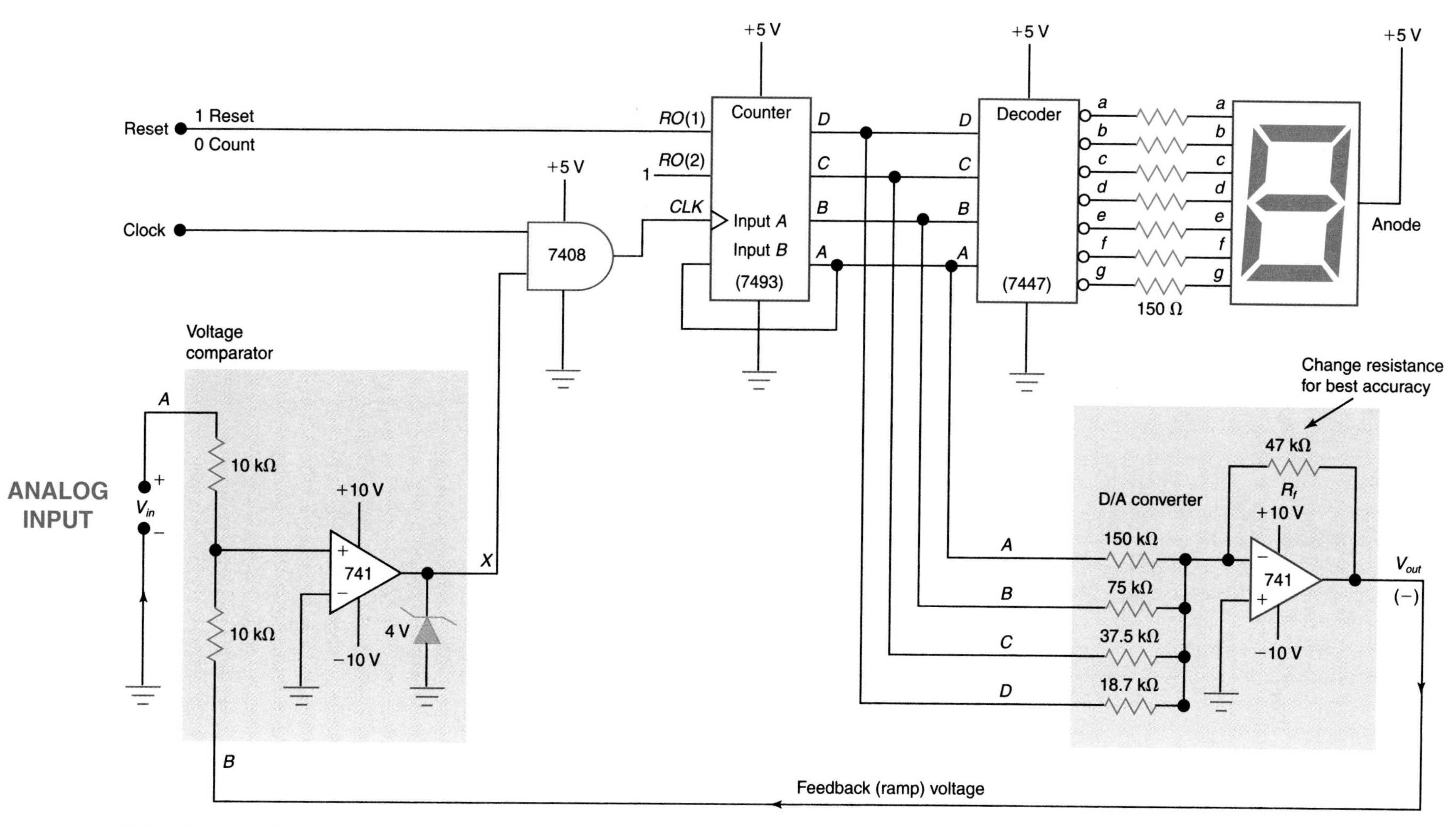

Fig. 13-12 Wiring diagram of an elementary digital voltmeter.

Elementary digital voltmeter

About Electronics

Microcontrollers are small, inexpensive "computers on a chip" containing a CPU, RAM, ROM, and input/outputs. The average American interacts with microcontrollers hundreds of times each day in appliances, computers, telephones, security systems, televisions, thermostats, radios, automobiles, "smart cards," and other products.

−1 V. The 1 V is fed back to the comparator.

Now, with 2 V still applied to the input, the comparator checks *A* against *B*; input *A* is larger. The comparator applies a logical 1 to the AND gate. The AND gate passes the second clock pulse to the counter. The counter advances to 0010. The 0010 is decoded and reads out as a decimal 2 on the seven-segment display. The 0010 also is applied to the D/A converter. The D/A converter puts out about 2 V, which is fed back to the *B* input of the comparator.

The display now reads 2. The 2 V is still applied to input *A* of the comparator. The comparator checks *A* against *B*; *B* is just slightly larger. Output *X* of the comparator goes to logical 0. The AND gate is disabled. No clock pulses reach the counter. The count has stopped at 2 on the display. This is the voltage applied at the analog input.

The digital voltmeter in Fig. 13-12 is an experimental circuit. The circuit is included because it demonstrates the fundamentals of how a digital voltmeter works. It shows how SSI and MSI ICs can be used to build more complex functions. It is a simple example of a *hybrid electronic system* containing both digital and analog devices.

Modern digital voltmeters and DMMs are based on LSI ICs. These specialized A/D converters are available from many manufacturers. *Large-scale-integrated digital voltmeter chips* include all the active devices on a single CMOS IC. Included are A/D converters, seven-segment decoders, display drivers, and a clock.

Ramp A/D converter

Hybrid electronic system

Ramp generator

Sawtooth waveform

LSI digital voltmeter IC

TEST

Answer the following questions.

34. One application for an A/D converter is in a(n) ________.
35. Refer to Fig. 13-12. The elementary digital voltmeter is considered a ________ (digital, hybrid) system because it contains both digital and analog ICs.
36. Refer to Fig. 13-12. With the counter *reset* to 0000, the feedback (ramp) voltage will be about ________ V.
37. Refer to Fig. 13-12. If the analog input voltage is 3.5 V and the counter is reset, how many clock pulses reach the 7493 IC before the counter stops?
38. Refer to Fig. 13-12. If the analog input voltage is 4.6 V, the display will read ________ V after the reset/count sequence.
39. Refer to Fig. 13-12. The op amp on the right is wired as a(n) ________ while the operational amplifier at the left functions as a voltage comparator.
40. Refer to Fig. 13-12. If the analog input voltage is 8.5 V, the display will read ________ V after the reset/count sequence.
41. Refer to Fig. 13-12. When input *B* of the voltage comparator becomes ________ (greater than, less than) the input voltage at *A*, the output goes LOW and the AND gate ________ (does not pass, passes) the clock pulses to the counter.

13-8 Other A/D Converters

In Sec. 13-5 we studied the counter-ramp A/D converter. Several other types of A/D converters are also used; in this section we shall discuss two other types of converters.

A *ramp A/D converter* is shown in Fig. 13-13. This A/D converter works very much like the counter-ramp A/D converter in Fig. 13-8. The *ramp-generator* at the left in Fig. 13-13 is the only new subsystem. The ramp generator produces a *sawtooth waveform*, which looks like a triangle-shaped wave in Fig. 13-14(*a*).

Suppose 3 V is applied to the analog voltage input of the A/D converter in Fig. 13-13. This situation is diagramed in Fig. 13-14(*a*). The ramp voltage starts to increase but is still lower than input *A* of the comparator. The comparator output is at a logical 1. This 1 enables the AND gate so that a clock pulse can pass through. In Fig. 13-14(*a*) the diagram shows three clock pulses getting through the AND gate before the ramp voltage gets larger than the input voltage. At point *Y* in Fig. 13-14(*a*)

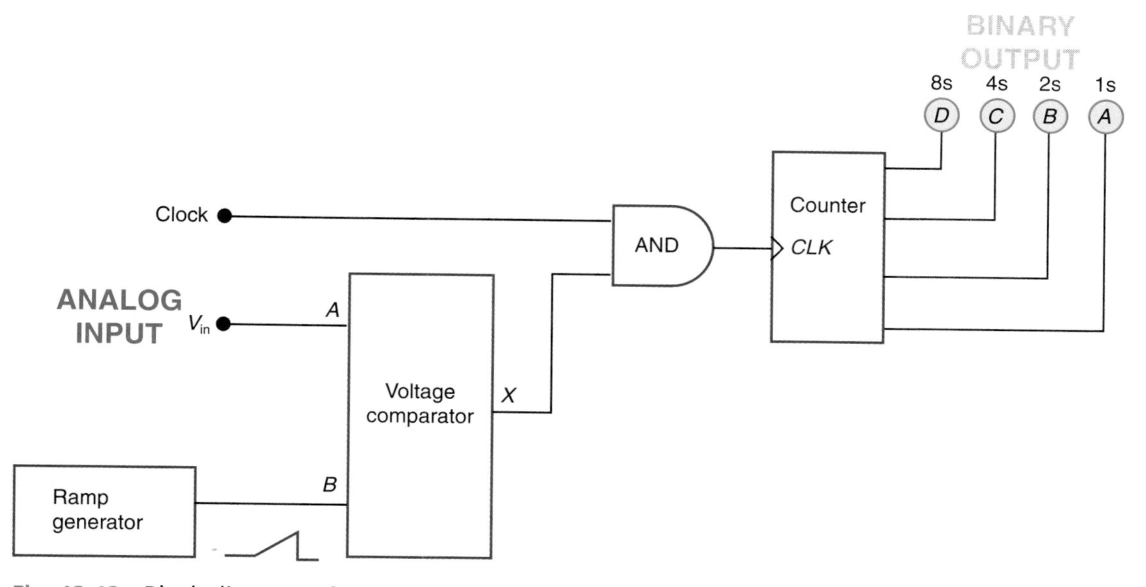

Fig. 13-13 Block diagram of a ramp-type A/D converter.

the comparator output goes to a logical 0. The AND gate is disabled. The counter stops counting at binary 0011. The binary 0011 means 3 V is applied at the input.

Figure 13-14(*b*) gives another example. The input voltage to the ramp-type A/D converter is 6 V in this situation. The ramp voltage begins to increase from left to right. The comparator output is at a logical 1 because input *A* is larger than the ramp generator voltage at in-

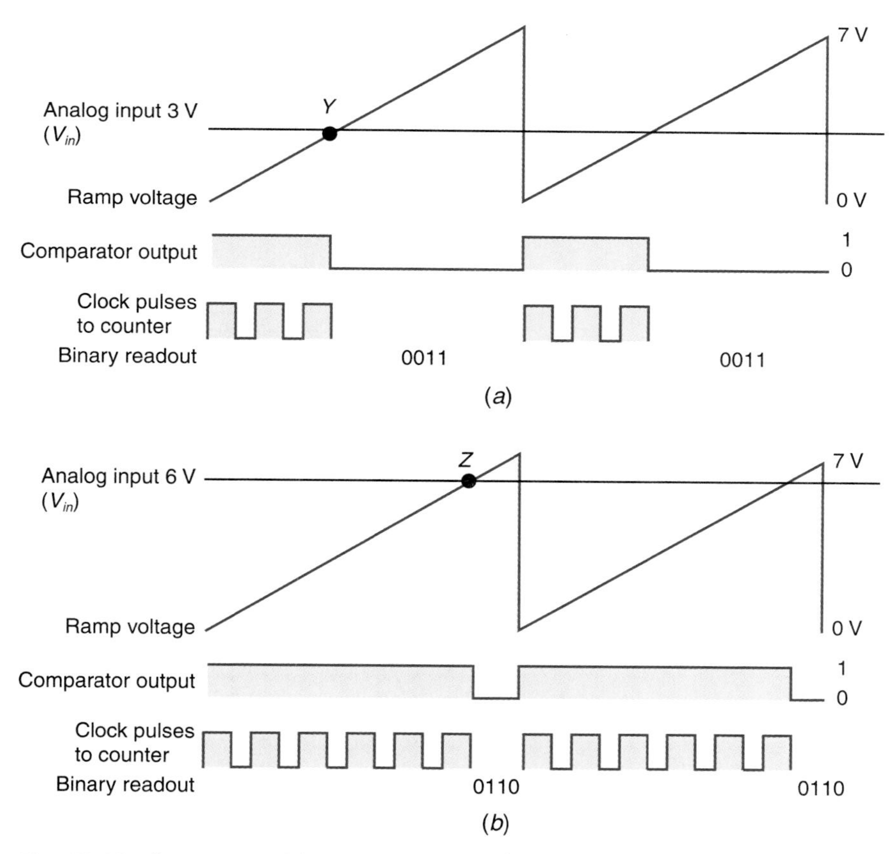

Fig. 13-14 Ramp-type A/D converter waveforms. (*a*) With 3 V applied. (*b*) With 6 V applied.

Ramp-type A/D converter

put *B*. The counter continues to advance. At point *Z* on the ramp voltage the ramp generator voltage is larger than V_{in}. At this point the comparator output goes to a logical 0. This 0 disables the AND gate. The clock pulses no longer reach the counter. The counter is stopped at binary 0110. The binary 0110 represents the 6-V analog input.

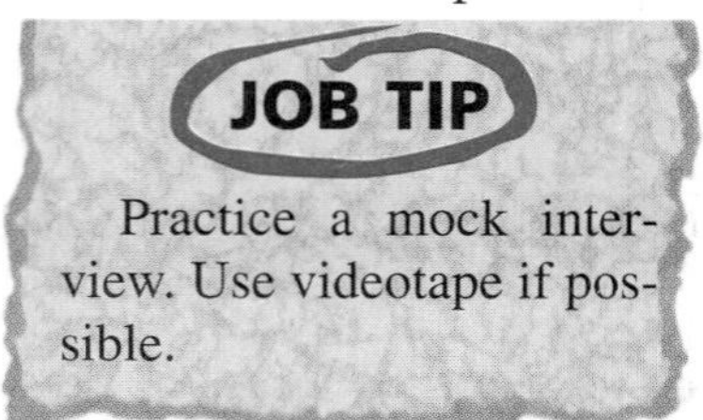

The difficulty with ramp-type A/D converters is the long time it takes to count up to higher voltages. For instance, if the binary output were eight binary places, the counter might have to count up to 255. To eliminate this slow conversion time we use a different type of A/D converter. A converter that cuts down on conversion time is a *successive-approximation A/D converter*.

Successive-approximation A/D converter

A block diagram of a successive-approximation A/D converter is shown in Fig. 13-15. The converter consists of a voltage comparator, a D/A converter, and a new logic block. The new logic block is called the successive-approximation logic section.

Suppose we apply 7 V to the analog input. The successive-approximation A/D converter first makes a "guess" at the analog input voltage. This guess is made by setting the MSB to 1. This is shown in block 1, Fig. 13-16 on page 375. This job is performed by the successive-approximation logic unit. The result (1000) is fed back to the comparator through the D/A converter. The comparator answers the question in block 2, Fig. 13-16: is 1000 high or low compared with the input voltage? In this case the answer is "high." The successive-approximation logic then performs the task in block 3. The 8s place is cleared to 0, and the 4s place is set to 1. The result (0100) is sent back to the comparator unit through the D/A converter. The comparator next answers the question in block 4: is 0100 high or low compared with the input voltage? The answer is "low." The successive-approximation logic then performs the task in block 5. The 2s place is set to 1. The result (0110) is sent back to the comparator. The comparator answers the question in block 6: is 0110 high or low compared with the input voltage? The answer is "low." The successive-approximation logic then performs the task in block 7. The 1s place is set to 1. The final result is binary 0111. This stands for the 7 V applied at the input of the A/D converter.

Notice in Fig. 13-16 that the items in the

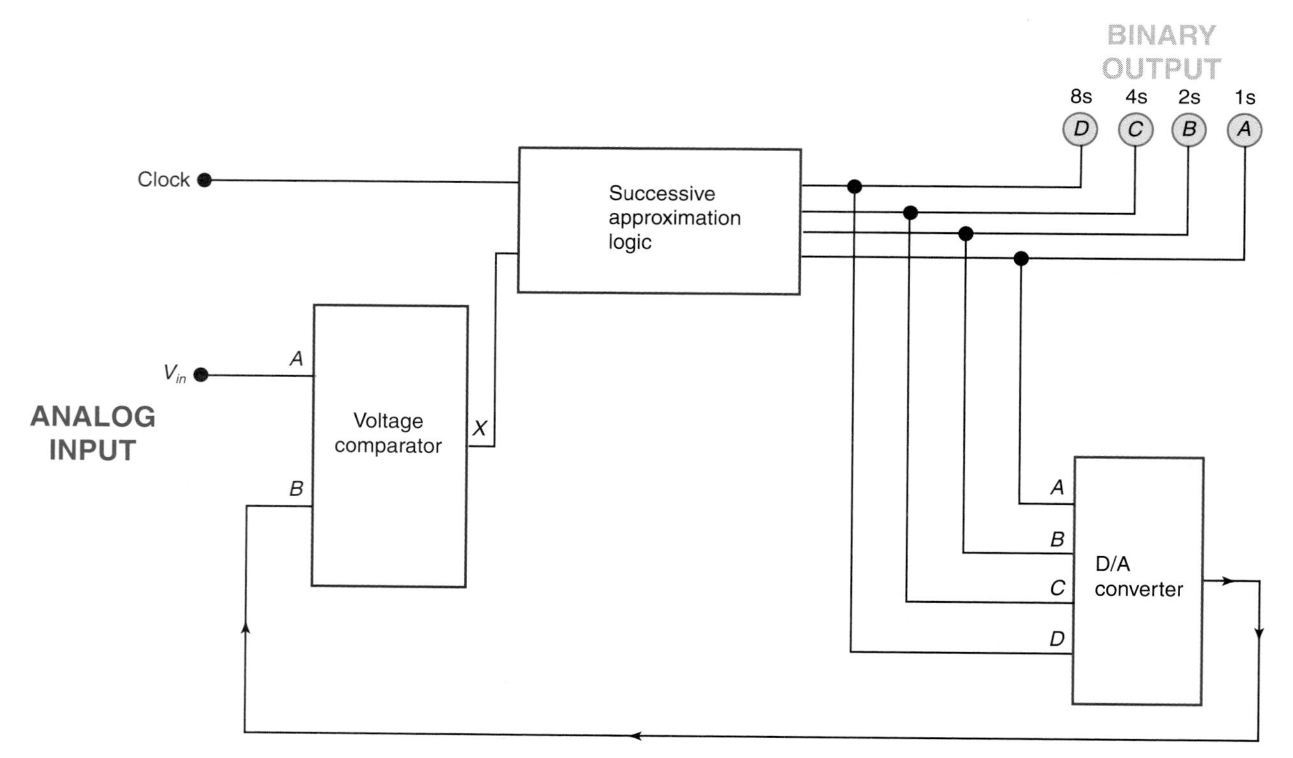

Fig. 13-15 Block diagram of a successive-approximation-type A/D converter.

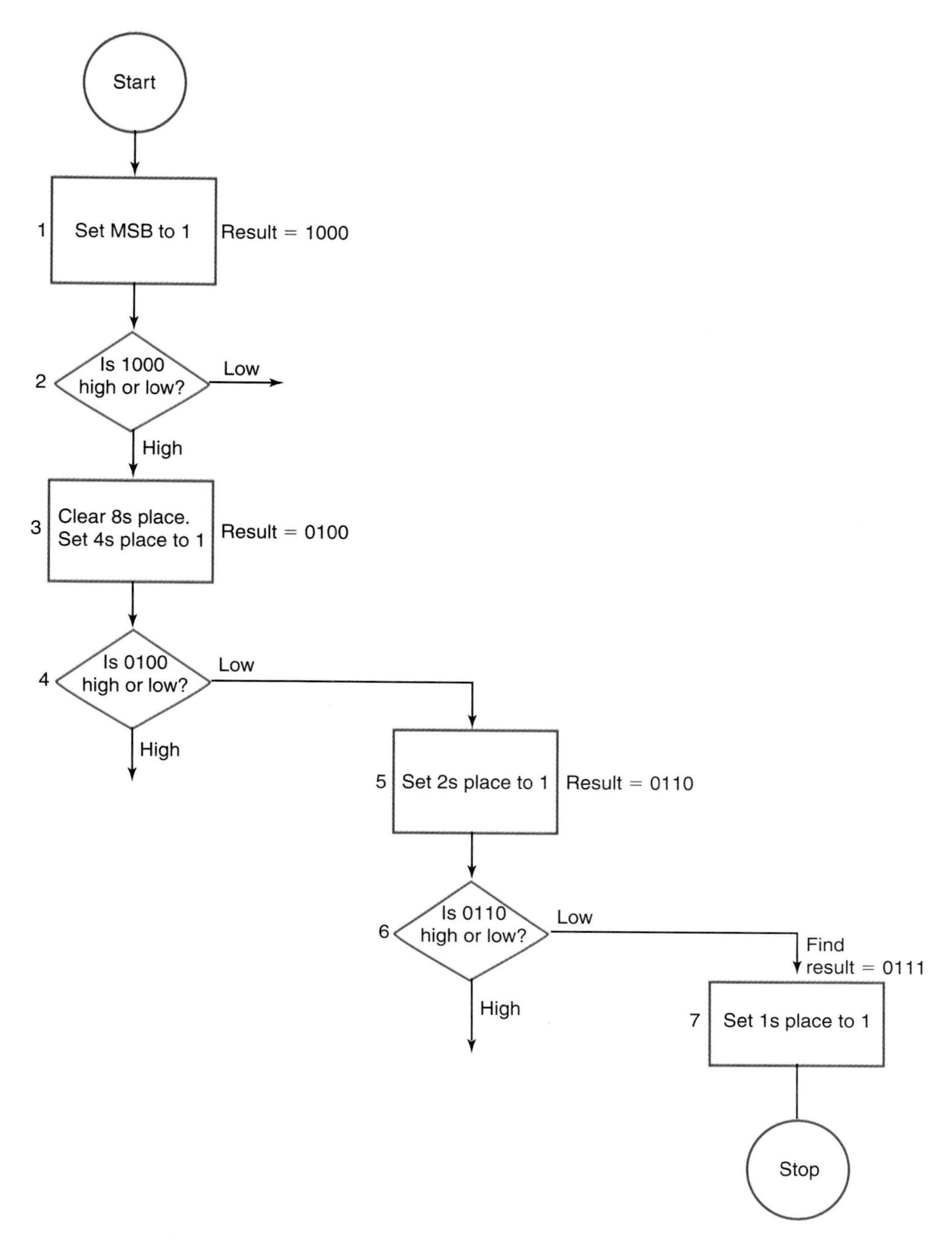

Fig. 13-16 Flowcharting the operation of the successive-approximation-type A/D converter.

Flowcharting

blocks are performed by the successive-approximation logic unit. The questions are answered by the comparator. Also notice that the task performed by the successive-approximation logic depends upon whether the answer to the previous question is "low" or "high" (see blocks 3 and 5).

The advantage of the successive-approximation A/D converter is that it takes fewer guesses to get the answer. The *digitizing* process is thus faster. The successive-approximation A/D converter is very widely used.

TEST

Answer the following questions.

42. List three types of A/D converter circuits.
43. The counter-ramp A/D converter uses a D/A converter to generate the ramp voltage fed back to the comparator, whereas

the ramp type uses a(n) ________ to do this job.

44. The successive-approximation A/D converter is ________ (faster, slower) than ramp-type units.
45. Refer to Fig. 13-13. The ramp generator produces a ________ (sawtooth, square) waveform.
46. Refer to Fig. 13-13. If the input voltage (V_{in}) is 2 V and the ramp voltage is 0 V, the output of the voltage comparator is ________ (HIGH, LOW) and the AND gate allows clock pulses to pass to the counter.
47. Refer to Fig. 13-15. When starting a new conversion, the ________ first sets the MSB to 1 and the voltage comparator checks to see if the input voltage (V_{in}) is higher than the feedback voltage at *B*.

A/D converter specifications

13-9 A/D Converter Specifications

Manufacturers produce a wide variety of A/D converters. One recent publication lists over 300 different A/D converters fabricated by many manufacturers.

Some of the more common specifications of A/D converters are detailed below.

Type of Output

Generally, A/D converters are classified as having either binary or decimal outputs. Analog-to-digital converters with decimal outputs are commonly used as digital voltmeters and used in digital panel meters and DMMs. Analog-to-digital converters with binary outputs have from 4 to 16 outputs. Analog-to-digital converters with binary outputs are common input devices to microprocessor-based systems. The latter are sometimes referred to as *μP-type A/D converters*.

μP-type A/D converters

Resolution

Resolution

Decimal-output A/D converters

The *resolution* of an A/D converter is given as the *number of bits* at the output for a binary-type unit. For decimal-output A/D converters (used in DMMs), the resolution is given as the *number of digits* in the readout (such as 3½ or 4½). Typical A/D converters with binary outputs have resolutions of 4, 6, 8, 10, 12, 14, and 16 bits. The errors that occur due to the use of discrete binary steps to represent a continuous analog voltage are called *quantizing errors.*

A 16-bit A/D converter has much finer resolution than a 4-bit unit because it divides the input or reference voltage into smaller discrete steps. For instance, each step in a 4-bit A/D converter would be one-fifteenth ($2^4 - 1 = 15$) of the input range. This would be a resolution of 6.7 percent ($1/15 \times 100 = 6.7$ percent). An 8-bit A/D converter has finer increments. An 8-bit unit has 255 ($2^8 - 1 = 255$) discrete steps. This provides a resolution of 0.39 percent ($1/255 \times 100 = 0.39$ percent). The 8-bit unit has better resolution than the 4-bit A/D converter. The resolution of a 16-bit converter is 0.0015 percent.

Accuracy

Accuracy

The resolution of an A/D converter can be thought of as the inherent "digital" error due to the discrete steps available at the output of the IC. Another source of error in an A/D converter might be an analog component, such as the comparator. Other errors might be introduced by the resistor network. The overall precision of an A/D converter is called the *accuracy* of the A/D converter IC.

The accuracies of typical A/D converter ICs with binary outputs range from ±½ LSB to ±2 LSB. Those with decimal outputs might range from 0.01 accuracy to 0.05 percent accuracy.

Conversion Time

Conversion time

The *conversion time* is another important specification of an A/D converter. It is the time it takes for the IC to convert the analog input voltage to binary (or decimal) data at the outputs. Typical conversion times range from 0.05 to 100,000 μs for A/D converter ICs with binary outputs. Conversion times for A/D converters with decimal outputs are somewhat longer and might typically be 200 to 400 ms.

Other Specifications

Four other common characteristics given for A/D converters are the power supply voltage, output logic levels, input voltage, and maximum power dissipation. Power supply voltages are commonly +5 V. However, some A/D converter ICs operate on voltages from +5 to +15 V. The output logic levels are either TTL, CMOS, or tristate. The input voltage range is commonly 5 V. Maximum power dissipation for an A/D converter IC might be in a range from about 15 to 3000 mW.

TEST

Supply the missing word or number in each statement.

48. An A/D converter with binary outputs is sometimes referred to as a _________ (meter, μP)-type unit.
49. The _________ of an A/D converter is given as the number of bits at the output of a binary-type unit.
50. An 8-bit A/D converter has greater resolution than a _________ (4, 12)-bit chip.
51. A typical conversion time for an A/D converter might be about _________ (110 μs, 1 s).
52. A typical A/D converter might have a maximum power dissipation of about _________ (850 mW, 10 μW).
53. The conversion time for meter-type A/D converters is _________ (longer, shorter) than for μP-type units.

13-10 An A/D Converter IC

A commercial A/D converter IC will be featured in this section. Figure 13-17(*a*) on page 378 shows the pin diagram for an *ADC0804 8-bit A/D converter IC*. The table in Fig. 13-17(*b*) lists the name and function of each pin on the ADC0804 IC.

The ADC0804 A/D converter was designed to interface directly with older 8080, 8085, or Z80 microprocessors. Some pin labels on the ADC0804 IC correspond to pins on popular microprocessors. For instance, the ADC0804 uses $\overline{RD}$, $\overline{WR}$, and $\overline{INTR}$ as pin labels which correspond to the *RD, WR,* and *INTR* pins on the 8085 microprocessor. The ADC0804 can also be interfaced with other older 8-bit microprocessors such as the 6800 and 6502. The $\overline{CS}$ control input to the ADC0804 A/D converter receives its signal (chip select) from the microprocessor address-decoding circuitry.

The ADC0804 is a CMOS 8-bit *successive-approximation A/D converter*. It has three-state outputs so that it can interface directly with a microprocessor-based system data bus. The ADC0804 has binary outputs and features a short conversion time of only 100 μs. Its inputs and outputs are both MOS- and TTL-compatible. It has an on-chip clock generator. The on-chip generator does need two external components (resistor and capacitor) to operate. The ADC0804 IC operates on a standard +5-V dc power supply and can encode input analog voltages ranging from 0 to 5 V.

The ADC0804 A/D converter IC can be tested using the circuit shown in Fig. 13-18 on page 379. The function of the circuit is to encode the difference in voltage between $V_{in}(+)$ and $V_{in}(-)$ compared to the reference voltage (5.12 V in this example) to a corresponding binary value. The resolution of the ADC0804 IC is 8 bits or 0.39 percent. This means that for each 0.02 V (5.1 V × 0.39 percent = 0.02 V) increase in voltage at the analog inputs the binary count increases by 1.

The "start switch" in Fig. 13-18 is first closed and then opened to start this free-running A/D converter. It is "free-running" because it continuously converts the analog input to digital outputs. The start switch should be left open once the A/D converter is operating. The $\overline{WR}$ input can be thought of as a clock input with the interrupt output ($\overline{INTR}$) pulsing the $\overline{WR}$ input at the end of each analog-to-digital conversion. A L-to-H transition of the signal at the $\overline{WR}$ input starts the A/D converter process. When the conversion is finished, the binary display is updated and the $\overline{INTR}$ output emits a negative pulse. The negative interrupt pulse is fed back to clock the $\overline{WR}$ input and it initiates another A/D conversion. The circuit in Fig. 13-18 will perform about 5,000 to 10,000 conversions per second. The conversion rate of the ADC0804 is high because it uses the successive-approximation technique in the conversion process.

ADC0804 8-bit A/D converter IC

The resistor (R_1) and capacitor (C_1) connected to the *CLK R* and *CLK IN* inputs to the ADC0804 IC in Fig. 13-18 cause the internal clock to operate. The data outputs (DB7-DB0) drive the LED binary displays. The data outputs are active HIGH three-state outputs.

What is the binary output in Fig. 13-18 if the analog input voltage is 1.0 V? Recall that each 0.02 V equals a single binary count. Dividing 1.0 V by 0.02 V equals 50 in decimal.

Successive-approximation A/D converter

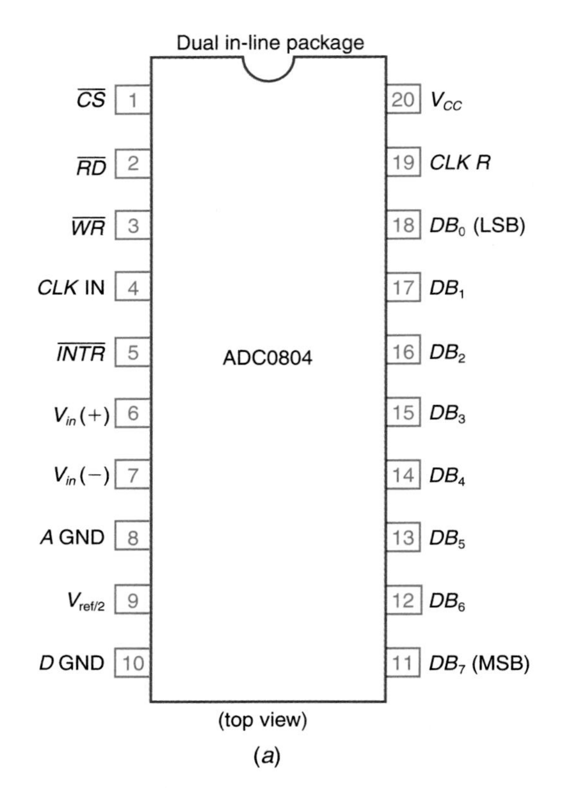

(a)

ADC0804 A/D converter IC

Pin No.	Symbol	Input/Output or Power	Description
1	$\overline{CS}$	Input	Chip select line from μP-control
2	$\overline{RD}$	Input	Read line from μP-control
3	$\overline{WR}$	Input	Write line from μP-control
4	*CLK* IN	Input	Clock
5	$\overline{INTR}$	Output	Interrupt line goes to μP interrupt input
6	V_{in} (+)	Input	Analog voltage (positive input)
7	V_{in} (−)	Input	Analog voltage (negative input)
8	*A* GND	Power	Analog ground
9	$V_{ref/2}$	Input	Alternate voltage reference (+)
10	*D* GND	Power	Digital ground
11	DB_7	Output	MSB data output
12	DB_6	Output	Data output
13	DB_5	Output	Data output
14	DB_4	Output	Data output
15	DB_3	Output	Data output
16	DB_2	Output	Data output
17	DB_1	Output	Data output
18	DB_0	Output	LSB data output
19	*CLK* R	Input	Connect external resistor for clock
20	V_{CC} (or ref)	Power	Positive of 5-V power supply and primary reference voltage

(b)

ADC0804 A/D converter IC

Fig. 13-17 ADC0804 A/D converter IC. (*a*) Pin diagram. (*b*) Pin labels and functions. (*Courtesy of National Semiconductor Corporation.*)

Converting decimal 50 to binary equals 00110010_2. The output indicators will show binary 00110010 (LLHHLLHL).

TEST

Supply the missing word in each statement.

54. The ADC0804 A/D converter is manufactured using ________ (CMOS, TTL) technology.
55. The ADC0804 IC is a ________ (meter, microprocessor)-type A/D converter.
56. The ADC0804 is an A/D converter with a resolution of ________.
57. The ADC0804 IC's inputs and outputs meet both MOS and ________ voltage-level specifications.
58. The conversion time for the ADC0804 IC is about ________ (100 μs, 400 ms).
59. Refer to Fig. 13-18. Components R_1 and C_1 are used by the ADC0804 IC's internal ________ (clock, comparator).
60. Refer to Fig. 13-18. If the analog input voltage is 2.0 V, the binary output is ________.
61. Refer to Fig. 13-18. An ________ (H-to-L, L-to-H) signal at the $\overline{WR}$ input to the ADC0804 IC starts a new A/D conversion.
62. Refer to Fig. 13-18. What output terminal of the ADC0804 IC produces a negative pulse immediately after each A/D conversion?

13-11 Digital Light Meter

The A/D converter is the electronic device used to encode analog voltages to digital form. These analog voltages are often generated by transducers. For instance, light intensity may be converted to a variable resistance using a photocell.

Digital light meter

A schematic diagram for a basic *digital light meter* is drawn in Fig. 13-19 on page 380. The ADC0804 IC is wired as a free-running A/D converter as in the last section. The push-button switch is pressed only once to start the A/D converter. The analog input voltage is being measured across resistor R_2. The photocell

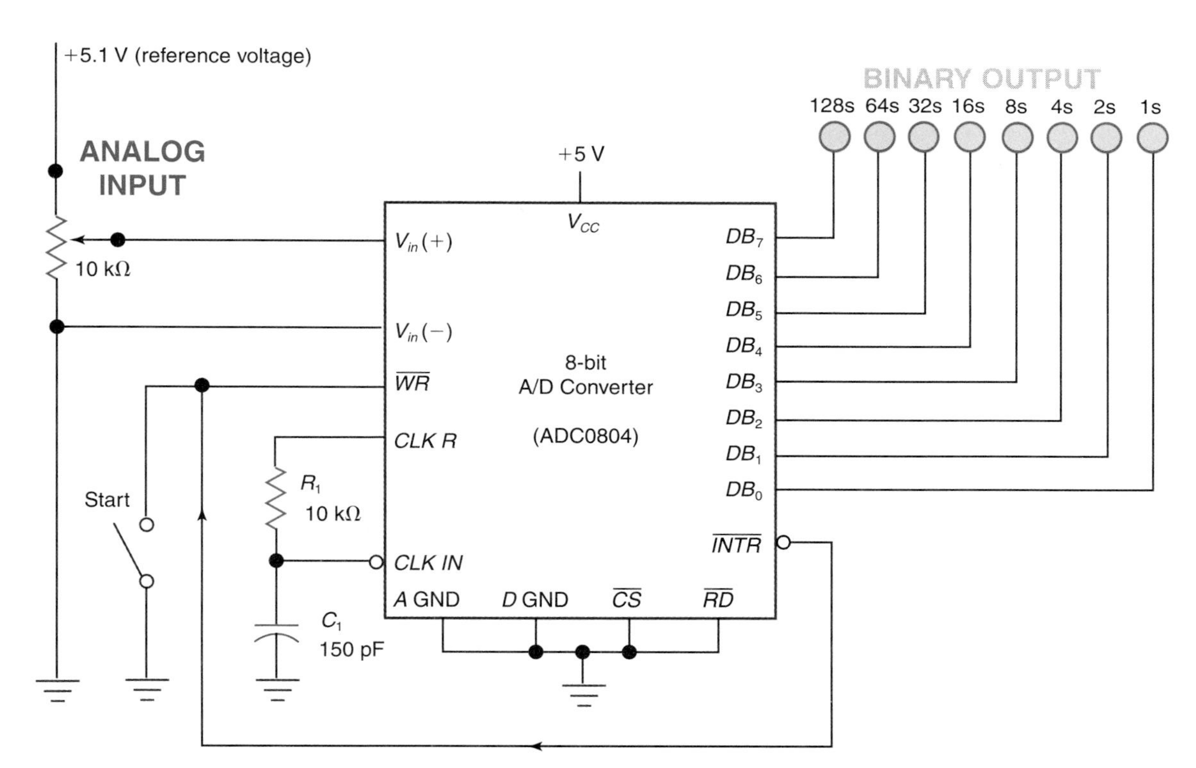

Fig. 13-18 Wiring diagram for a test circuit using the ADC0804 CMOS A/D converter IC.

Testing the ADC0804 A/D converter

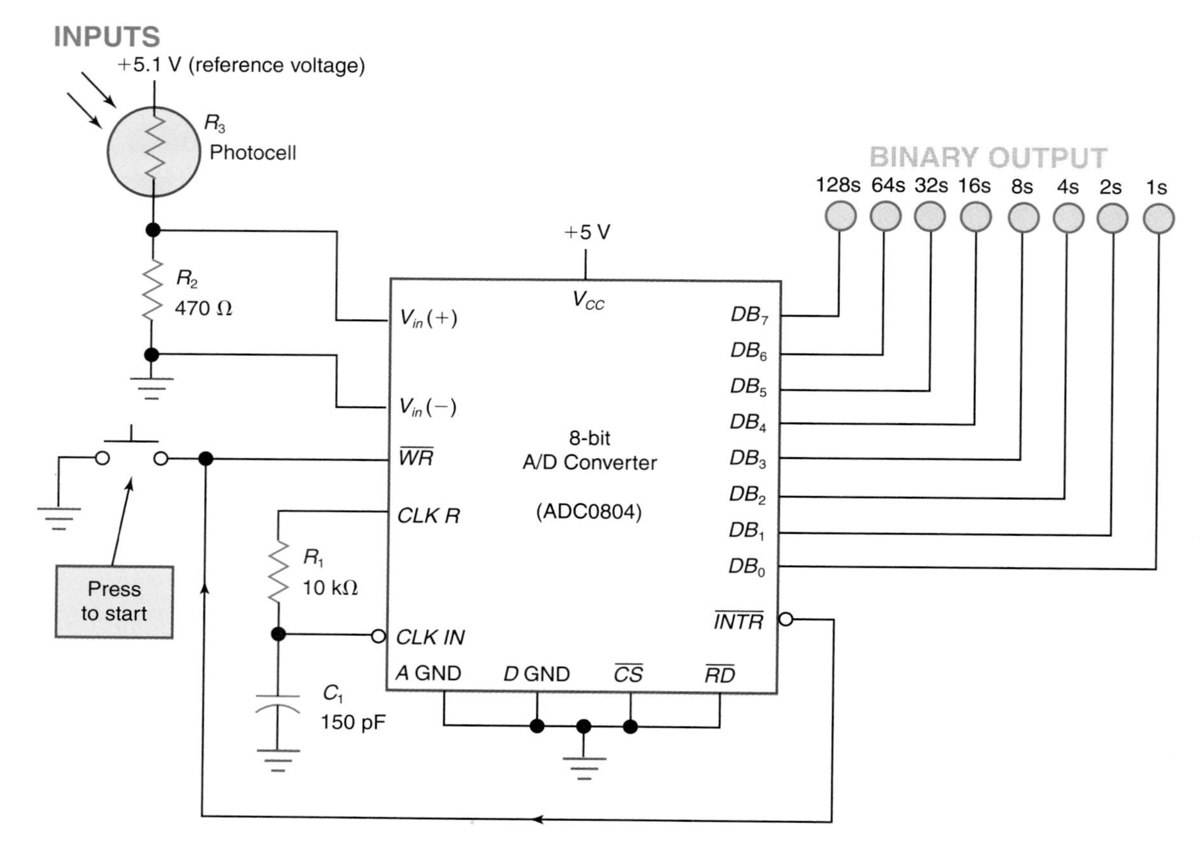

Fig. 13-19 Wiring diagram for a digital light meter using binary outputs.

Transducer

(R_3) is the light sensor or *transducer* in this circuit. As the light intensity increases, the resistance of the photocell (R_3) decreases. Decreasing the resistance of R_3 causes an increase in current through series resistances R_2 and R_3. The increased current through R_2 causes a proportional increase in the voltage drop across the resistor. The voltage drop across R_2 is the analog input voltage to the A/D converter. An increase in the analog input voltage causes an increase in the reading at the binary outputs.

Cadmium sulfide photocell

The *cadmium sulfide photocell* used in Fig. 13-19 is a variable resistor. As the intensity of the light striking the photocell increases, its resistance decreases. The photocell shown in Fig. 13-19 might have a maximum resistance of about 500 kΩ and a minimum of about 100 Ω. The cadmium sulfide photocell is most sensitive in the green-to-yellow portion of the light spectrum. The photocell is also referred to as a *photoresistor* or a *photoresistive cell.*

Photoresistor

Photoresistive cell

ADC0804 A/D converter

Other photocells may be used in Fig. 13-19. If the substitute photocell has different resistance specifications, you can change the value of resistor R_2 in the light meter circuit to scale the binary output as desired.

A second digital light meter circuit is drawn in Fig. 13-20. This light meter indicates the relative brightness of the light striking the photocell in decimal (0 to 9). The new light meter is similar to the circuit in Fig. 13-19. The new light meter has a clock added to the circuit. The clock consists of a 555 timer IC, two resistors, and a capacitor wired as an astable MV. The clock generates a TTL output with a frequency of about 1 Hz. This means the analog input voltage is only converted into digital form one time per second. The very low conversion rate keeps the output from "jittering" between two readings on the seven-segment LED display.

The 7447A IC decodes the four MSBs (DB7, DB6, DB5, DB4) from the output of the ADC0804 A/D converter. The 7447A IC also drives the segments on the seven-segment LED display. The seven 150-Ω resistors between the 7447A IC and seven-segment LED display limit the current through an "on" segment to a safe level.

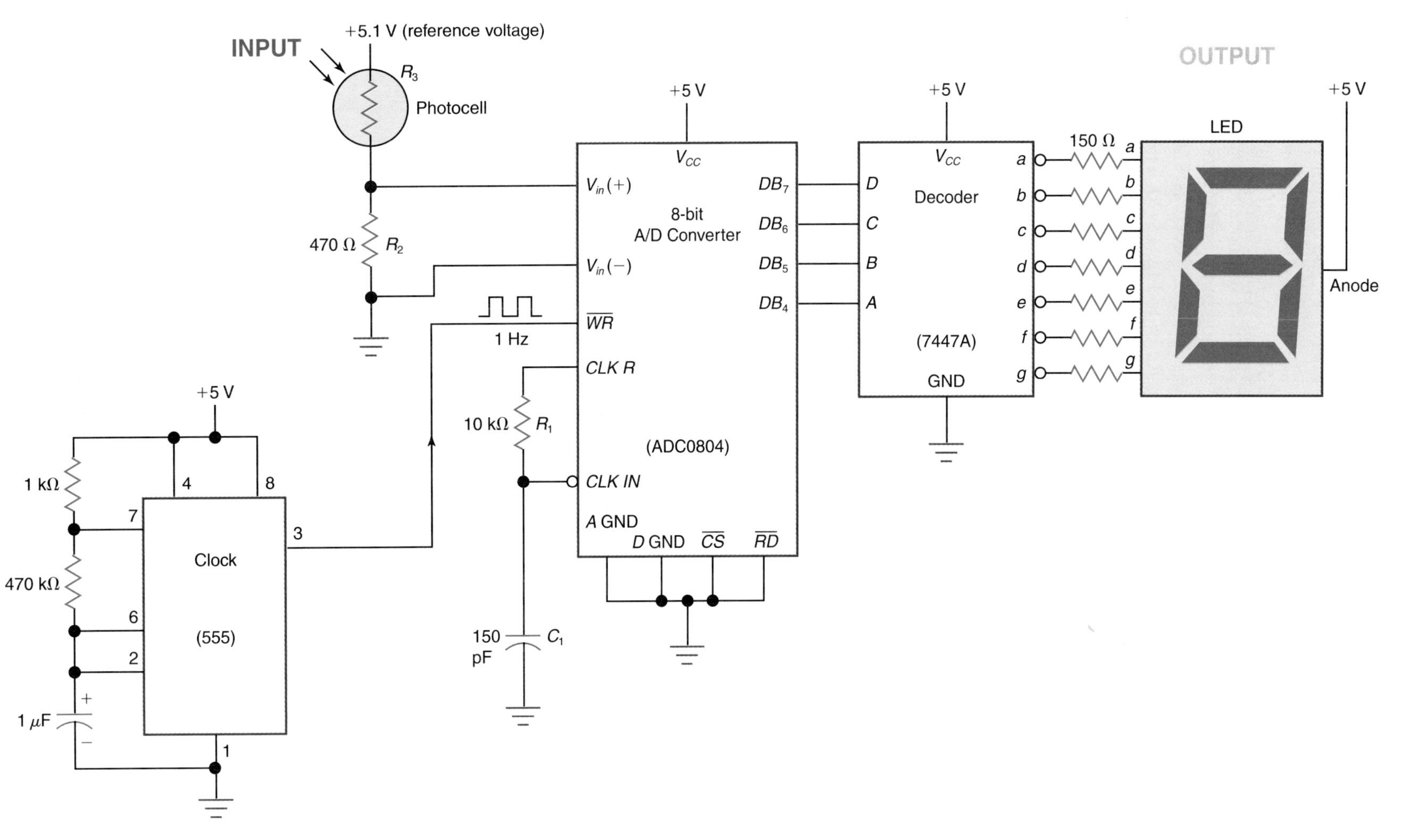

Fig. 13-20 Wiring diagram for a digital light meter circuit using a decimal display.

Digital light meter circuit

As in the previous circuit (Fig. 13-19), the output of the new light meter may have to be scaled so that low light reads 0 and high light intensity reads 9 on the seven-segment LED display. The value of resistor R_2 can be changed to scale the output. If R_2 is substituted with a lower-value resistor, the decimal output will read lower for the same light intensity. However, if the resistance value of R_2 is increased, the output will read higher.

TEST

Supply the missing word or number in each statement.

63. Refer to Fig. 13-19. As the light intensity striking the surface of the photocell increases, the binary value at the output of the light meter circuit ________ (decreases, increases).

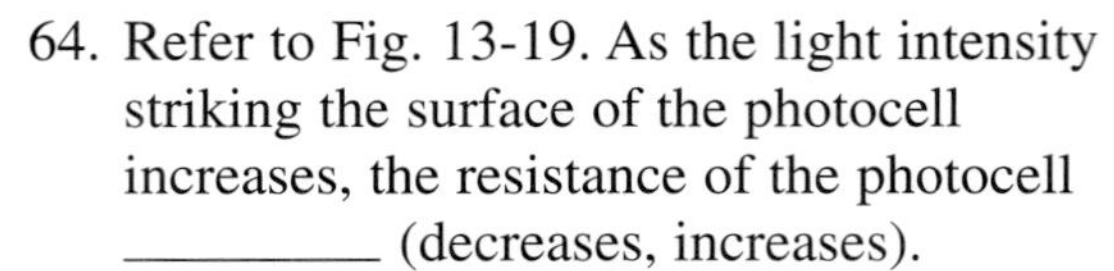

64. Refer to Fig. 13-19. As the light intensity striking the surface of the photocell increases, the resistance of the photocell ________ (decreases, increases).
65. Refer to Fig. 13-20. If current through series resistances R_2 and R_3 increases, the analog input voltage to the A/D converter ________ (decreases, increases).
66. Refer to Fig. 13-20. The conversion rate of the ADC0804 IC in this digital light meter circuit is about ________ (1, 400) A/D conversion(s) per second.
67. Refer to Fig. 13-20. Substituting R_2 with a resistor of a lower ohmic value would cause the output display to read ________ (higher, lower) for the same light intensity.
68. Refer to Fig. 13-20. The part labeled R_3 in the light meter circuit is a ________ (transducer, transformer) that converts light intensity into a variable resistance.

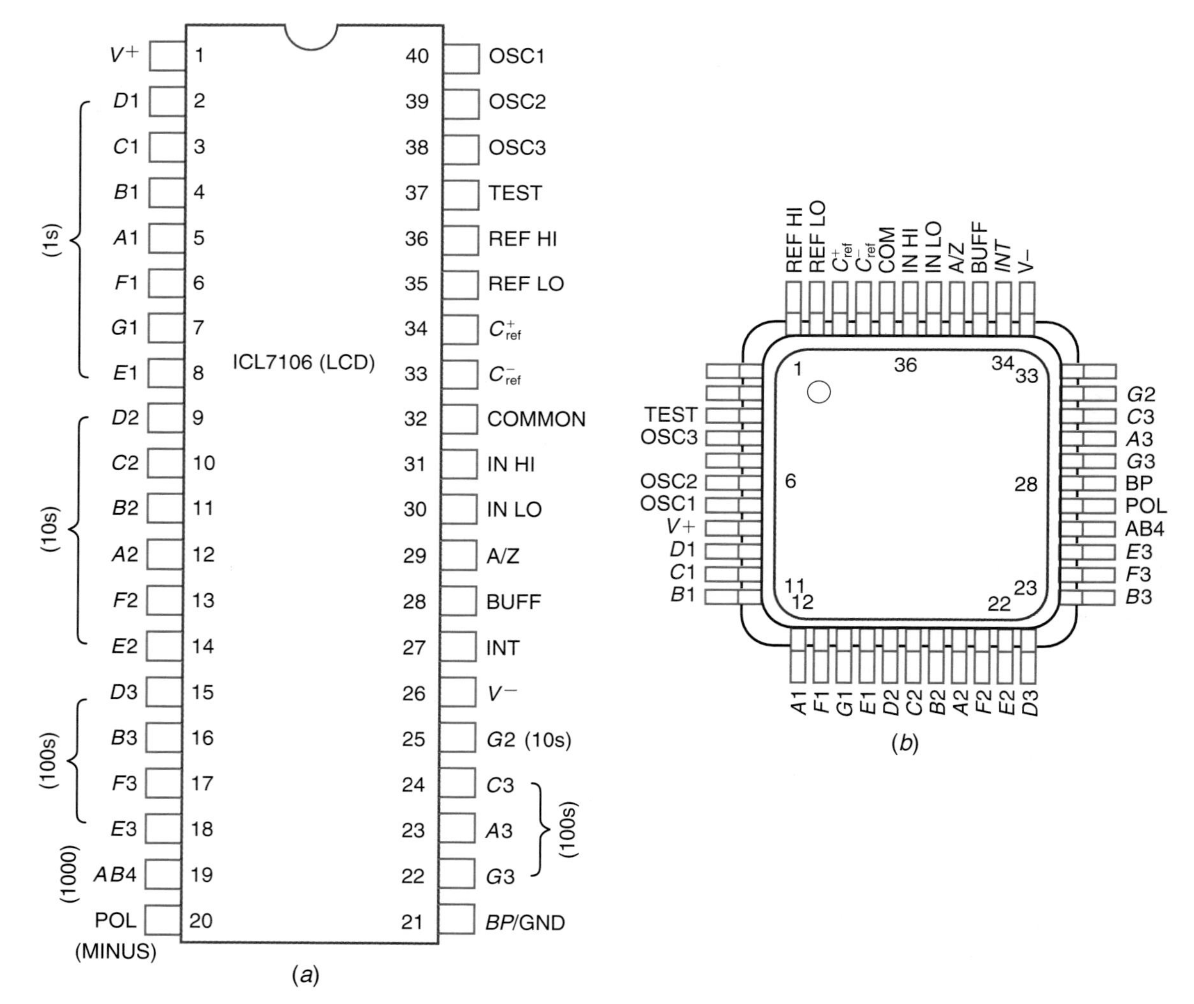

ICL7106 3½-digit A/D converter IC

Fig. 13-21 ICL7106 3½-digit LCD single-chip A/D converter IC. (*a*) Pin diagram—dual in-line package. (*b*) Pin diagram—surface-mount package. (*Courtesy of Harris Semiconductor.*)

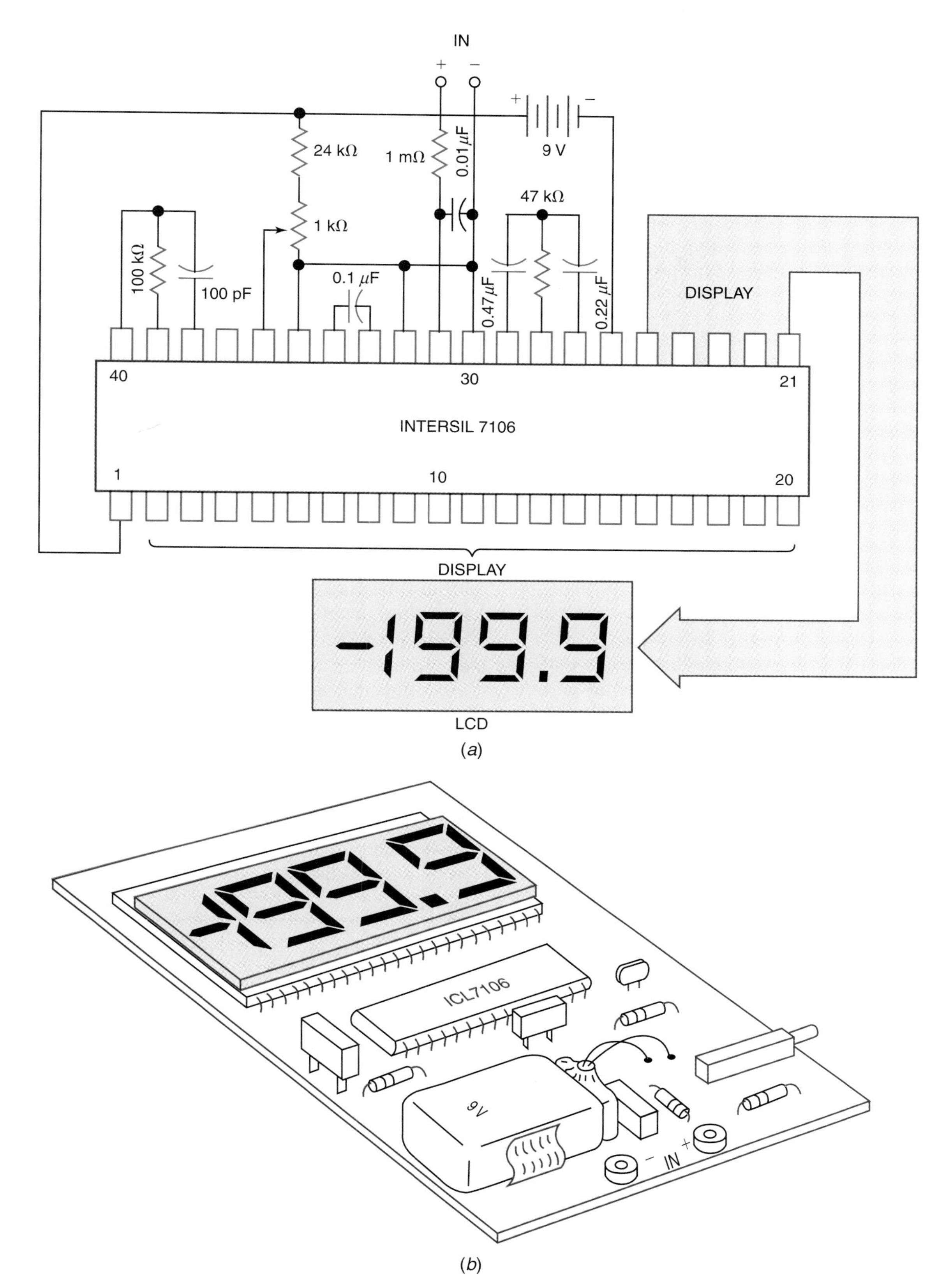

Fig. 13-22 A 3½-digit panel meter circuit. (*a*) Wiring diagram using the 7106 A/D converter IC. (*Courtesy of Harris Semiconductor.*) (*b*) Sketch of panel meter circuit mounted on a PC board.

A 3½-digital panel meter circuit

69. Refer to Fig. 13-20. The component labeled R_3 is a cadmium ________.

13-12 Digital Voltmeter

Display-type A/D converters

Digital voltmeters

Digital thermometers

Harris ICL7106 3½-digit LCD single-chip A/D converter

Surface-mount package

ICL7106 and ICL7107 3½-digit A/D converters

At least one manufacturer of chips groups A/D converters as being either microprocessor type or display type. The display-type A/D converters are used in constructing digital voltmeters, digital thermometers, and digital multimeters.

One display-type A/D converter will be featured in this section. Pin configurations for the *Harris Semiconductor ICL7106 3½-digit LCD single-chip A/D converter* are detailed in Fig. 13-21 on page 382. This complex CMOS IC is packaged in both a DIP and the newer *surface-mount package.* The top view of both IC package styles is shown in Fig. 13-21. Notice the location of pin 1 on the surface-mount package in Fig. 13-21(*b*). Pin 1 is immediately counterclockwise from the dot at the upper-left corner of the package. Note that the pin numbering is *not the same* on the DIP and the surface-mount package.

The ICL7106 A/D converter needs only 10 external passive components plus an LCD to make an accurate 3½-digit panel meter. A schematic diagram of such a panel meter using a 3½-digit LCD is drawn in Fig. 13-22(*a*) on page 383. A sketch of a completed panel meter is shown in Fig. 13-22(*b*). The circuit shown in Fig. 13-22(*a*) will measure voltage from 0 to 200.0 mV.

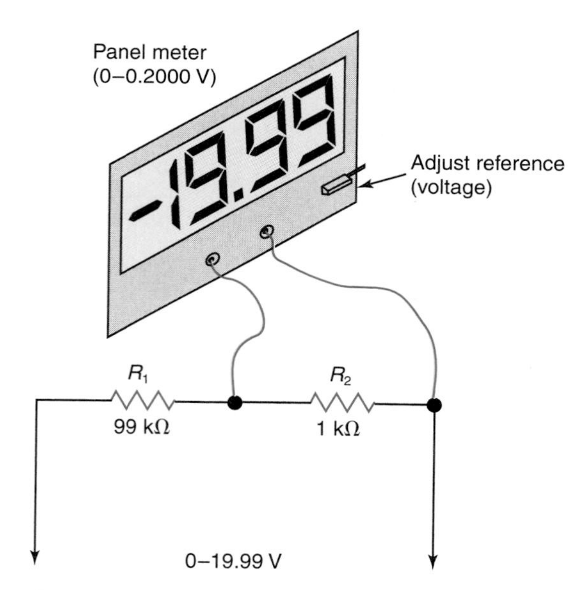

Fig. 13-23 Extending the range of the 0- to 0.2-V panel meter using external resistors.

The limited voltage range (0 to 0.2V) of the panel meter in Fig. 13-22 can be extended. The easiest method of extending the range of the voltmeter is illustrated in Fig. 13-23. The input voltage (0 to 19.99 V) is divided by 100 by the series resistors R_1 and R_2. The panel meter has a potentiometer used to adjust the reference voltage for accurate calibration even if resistors R_1 and R_2 are not extremely accurate. External components can be added to the panel meter in Fig. 13-22 for measuring current, ac voltage, and resistance.

The ICL7106 A/D converter is manufactured using CMOS technology. It typically consumes less than 10 mW of power and operates on a single 9-V battery. The ICL7106 features a built-in clock, voltage reference, decoders, and direct display drivers for 3½-digit seven-segment LCDs. The A/D converter is very accurate and has an auto zero feature and high input impedance.

Another version of the ICL7106 IC is the ICL7107 3½-digit LED single-chip A/D converter. It performs the same as the ICL7106 unit except it requires a +5V/−5 V power supply. The ICL7107 will directly drive a 3½-digit LED display. These chips can also be used to design digital thermometers.

TEST

Supply the missing word(s) in each statement.

70. Besides the dual in-line packaging of chips, a newer ________ package is becoming popular.
71. The ICL7106 IC is a ________ (display, microprocessor)-type A/D converter.
72. The ICL7106 IC contains internal decoding and drivers for a 3½-digit ________ (LED display, LCD).
73. The ICL7106 IC operates on a ________ (9 V, +5 V/−5 V) power supply.
74. Refer to Fig. 13-23. If the voltage range is to be extended to measure 2.000 V and if R_1 = 90 kΩ, then resistor R_2 would be ________ kΩ.

Chapter 13 Review

Summary

1. Special interface encoders and decoders are used between analog and digital devices. These are called D/A converters and A/D converters.
2. A D/A converter consists of a resistor network and a summing amplifier.
3. Operational amplifiers are used in D/A converters and comparators. Gain can be easily set with external resistors on the op amp.
4. Several different resistor networks are used for weighting the binary input to a D/A converter.
5. Common A/D converters are the counter-ramp, ramp-generator, and successive-approximation types.
6. A voltage comparator compares two voltages and determines which is larger. An operational amplifier is the heart of the comparator.
7. Common specifications used for A/D converters include such characteristics as type of output, resolution, accuracy, conversion time, power supply voltage, output logic levels, input voltage, and power dissipation.
8. The ADC0804 IC is a CMOS 8-bit A/D converter. It features fast conversion times, microprocessor compatibility, three-state outputs, TTL level inputs and outputs, and an on-chip clock.
9. A photocell can be used as a transducer to drive an A/D converter in a digital light meter circuit.
10. An A/D converter is at the heart of a digital voltmeter. Most commercial digital voltmeters and DMMs use complex meter-type A/D converter LSI ICs.

Chapter Review Questions

Answer the following questions.

13-1. An A/D converter is a special type of ________ (decoder, encoder).

13-2. A D/A converter is a(n) ________ (decoder, encoder).

13-3. The ________ (A/D, D/A) converter digitizes analog information.

13-4. The ________ (A/D, D/A) converter translates from binary to an analog voltage.

13-5. A D/A converter consists of a(n) ________ network and a summing ________.

13-6. The term "operational amplifier" is frequently shortened to ________.

13-7. The voltage gain of the operational amplifier in Fig. 13-3(*b*) is determined by dividing the value of ________ (R_f, R_{in}) by the value of ________ (R_f, R_{in}).

13-8. Draw a symbol for an operational amplifier. Label the inverting input with a minus sign and the noninverting input with a plus sign. Label the output. Label the +10-V and −10-V power supply connections.

13-9. Refer to Fig. 13-4. What is the gain (A_v) of the op amp in this diagram if R_{in} = 1 kΩ and R_f = 100 kΩ?

13-10. Refer to Fig. 13-4. With the input voltage at +½ V, the output voltage is ________ (+, −)5 V. This is because we are using the ________ (inverting, noninverting) input of the op amp.

13-11. Refer to Fig. 13-5. What is the voltage gain of the op amp in this circuit with only switch *A* at logical 1?

13-12. Refer to Fig. 13-5. What is the combined resistance of parallel resistors R_1 and R_2 if both switches *A* and *B* are at logical 1?

13-13. Refer to Fig. 13-5. What is the gain (A_v) of the op amp with switches *A* and *B* at logical 1? (Use the resistance value from question 13-12.)

13-14. Refer to Fig. 13-5. What is the output voltage when binary 0011 is applied to the D/A converter? (Use the A_v from question 13-13.)

13-15. The arrangement of resistors in Fig. 13-6 is called the ________ ladder network.

13-16. A high, or logical 1, from a TTL device is about ________ (0, 3.75, 8.5) V.

13-17. The _________ (A/D, D/A) converter is the more complicated electronic system.

13-18. Refer to Fig. 13-8. If point X is at a logical _________ (0, 1), the counter advances one count as a pulse comes from the clock.

13-19. Refer to Fig. 13-8. If input B of the comparator has a higher voltage than input A, the AND gate is _________ (disabled, enabled).

13-20. Refer to Fig. 13-9. If input A of the comparator equals 5 V and input B equals 2 V, then output X is a logical _________ (0, 1). Output X is about _________ (0, 4) V.

13-21. The primary component in a voltage comparator is a(n) _________ (counter, op amp).

13-22. Refer to Fig. 13-12. This digital voltmeter uses a _________ (counter-ramp, successive-approximation) A/D converter.

13-23. The _________ (ramp, successive-approximation) A/D converter is faster at digitizing information.

13-24. Devices such as microphones, speakers, strain gauges, photocells, temperature sensors, and potentiometers convert one form of energy to another and are generally called _________.

13-25. An A/D converter with binary outputs might be classified as a _________ (meter, microprocessor)-type unit.

13-26. Refer to Fig. 13-18. What is the resolution of the ADC0804 A/D converter?

13-27. A(n) _________ (8, 16)-bit A/D converter has a lower quantization error and is considered more "accurate."

13-28. Conversion times are somewhat longer for _________ (meter, microprocessor)-type A/D converters.

13-29. The ADC0804 (Fig. 13-17) has _________ (binary, decimal) outputs.

13-30. The A/D converter wired in Fig. 13-18 performs about _________ (3, 5,000 to 10,000) A/D conversions per second.

13-31. Refer to Fig. 13-18. If the analog input voltage is 3.0 V, the binary output is _________.

13-32. Refer to Fig. 13-20. Decreasing the light intensity striking R_3 causes the resistance of the photocell to _________ (decrease, increase).

13-33. Refer to Fig. 13-20. Decreasing the light intensity striking the photocell causes the decimal output to _________ (decrease, increase).

13-34. The ADC0804 IC in Fig. 13-20 performs about _________ (1, 10,000) analog-to-digital conversion(s) per second.

13-35. Refer to Fig. 13-20. If current through series resistances R_2 and R_3 decreases, the analog input voltage to the A/D converter _________ (decreases, increases).

13-36. The ICL7106 IC is a _________ (meter, microprocessor)-type A/D converter.

13-37. Refer to Fig. 13-22. This digital panel meter can measure from 0 to _________ dc volts.

13-38. The ICL7106 A/D converter IC features a built-in clock, voltage reference, decoders, and direct display drivers for 3½-digit seven-segment _________ (LCDs, LED displays).

Critical Thinking Questions

13-1. Calculate the gain of the op amp circuit in Fig. 13-4 if R_{in} = 1 kΩ and R_f = 5 kΩ. Using the calculated gain, what is the output voltage (V_{out}) if V_{in} = 0.5 V?

13-2. Refer to Fig. 13-5.
 a. What is the combined resistance of parallel resistors R_2 and R_3 if both switches B and C are at logical 1?
 b. Using the calculated resistance, what is the gain (A_v) of the op amp with switches B and C at a logical 1?
 c. What is the output voltage when binary 0110 is applied to the inputs of the D/A converter (use calculated A_v)?

13-3. Compare Tables 13-1 and 13-2. Explain the difference between the data in the two tables.

13-4. List the four sections of a counter-ramp A/D converter circuit.

13-5. List the fours sections of a ramp-type A/D converter circuit.

13-6. Refer to Fig. 13-23. The value of resistor R_1 would have to be increased to what resistance to extend the range of the panel meter to 200 V dc?

13-7. Compare the D/A converter resistor networks in Figs. 13-5 and 13-6. Why would the R-2R ladder resistor network in Fig. 13-6 be easier to expand from four to eight binary inputs?

13-8. Refer to Fig. 13-12. The inverting input of the 741 op amp in the D/A converter is being used, while in the voltage comparator op amp the _________ input is being employed.

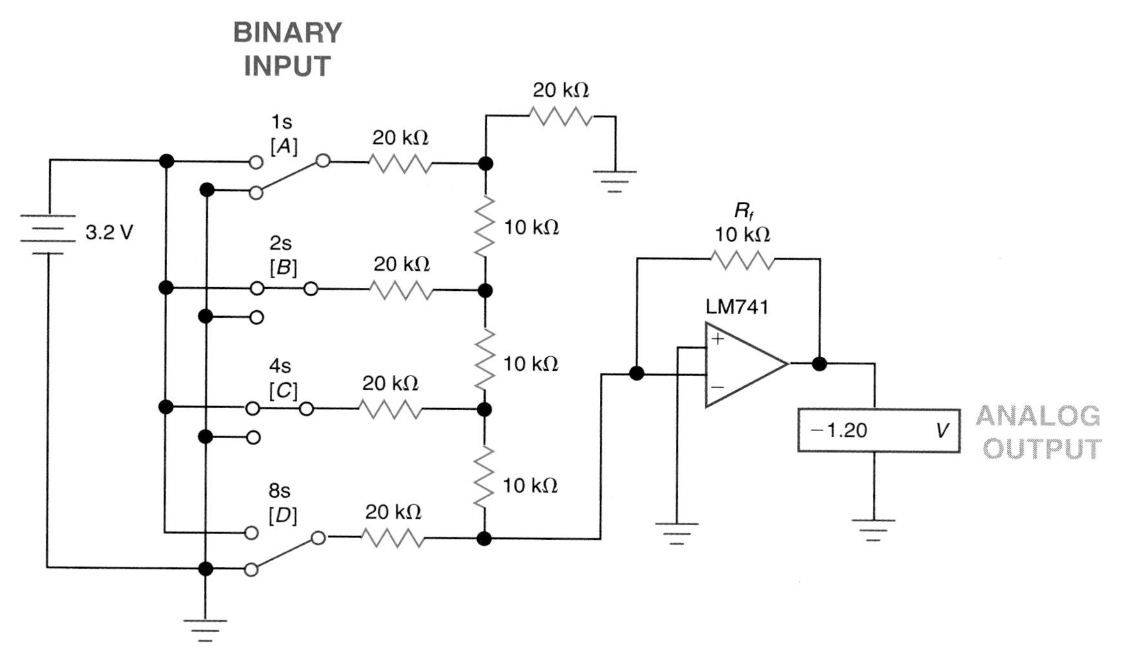

Fig. 13-24 Electronics Workbench® simulation circuit of a D/A converter circuit using R-2R resistor network and op amp (scaling amplifier).

13-9. Refer to Fig. 13-8. What would be the *resolution* of this A/D converter?

13-10. A digital voltmeter is one application of a(n) ________ (A/D, D/A) converter.

13-11. At the option of your instructor, use circuit simulation software to (1) draw the 4-bit D/A converter using the R-2R ladder resistor network and op amp detailed in Fig. 13-24, (2)

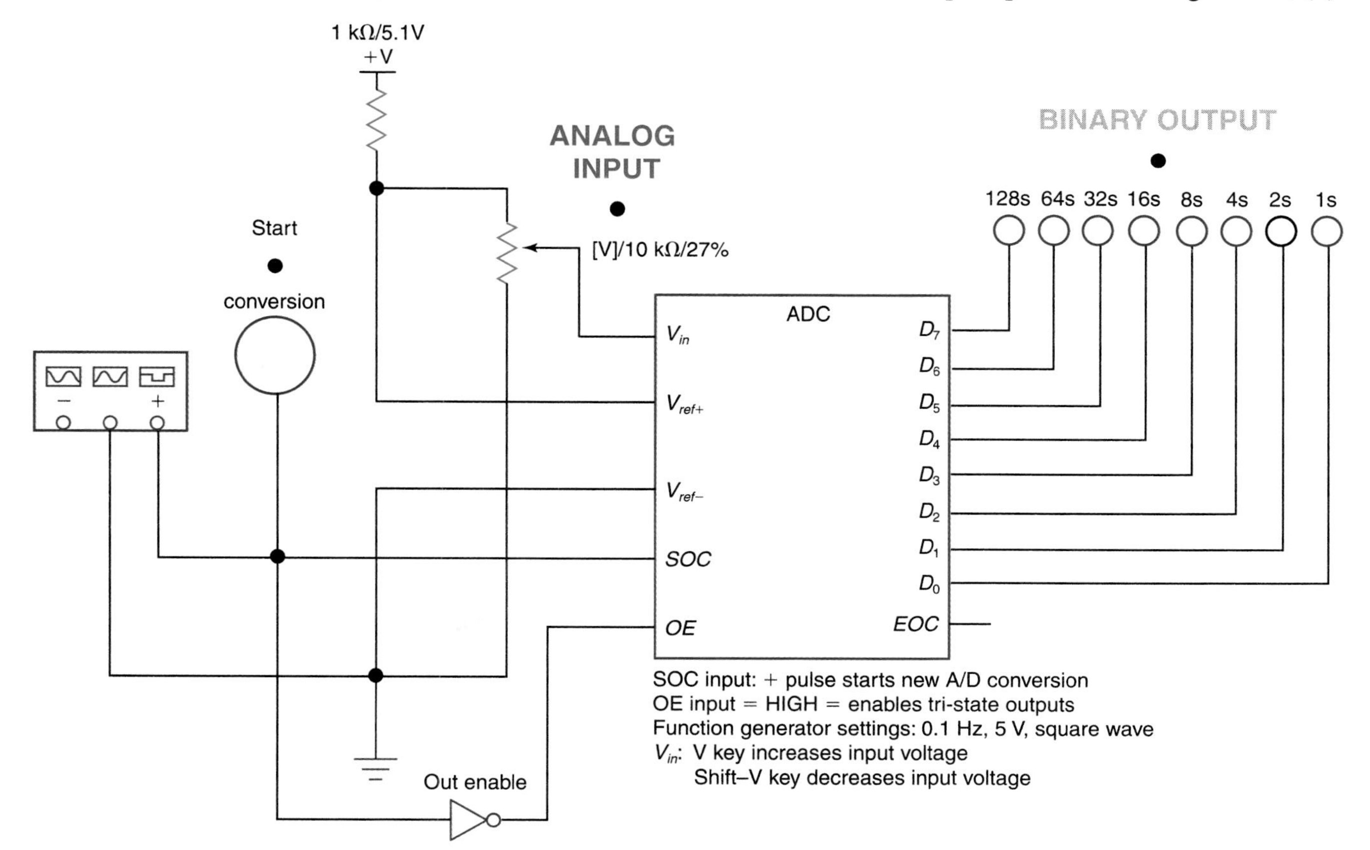

Fig. 13-25 Electronics Workbench® simulation circuit of an A/D converter circuit with an 8-bit binary readout.

operate the 4-bit D/A converter circuit, and (3) show instructor your working D/A converter.

13-12. At the option of your instructor, use circuit simulation software to (1) draw a 5-bit D/A converter using the R-2R ladder resistor network and op amp something like the unit in Fig. 13-24, (2) operate the 5-bit D/A converter circuit, and (3) show instructor your working 5-bit D/A converter.

13-13. At the option of your instructor, use circuit simulation software to (1) draw the generic 8-bit A/D converter circuit (with binary output) detailed in Fig. 13-25, (2) operate the 8-bit A/D converter circuit, and (3) show instructor your working A/D converter.

13-14. At the option of your instructor, use circuit simulation software to (1) add seven-segment displays to show hexadecimal readout for the generic 8-bit A/D converter circuit detailed in Fig. 13-25, (2) operate the 8-bit A/D converter circuit with hexadecimal displays, and (3) show instructor your working A/D converter with hexadecimal displays.

Answers to Tests

1. analog-to-digital converter (A/D converter)
2. digital-to-analog converter (D/A converter)
3. resistor, summing (scaling)
4. operational amplifier
5. 1.4
6. 3.0
7. 0.2
8. feedback
9. input
10. $A_v = 20$
11. $V_o = -4$ V
12. $A_v = 4$
13. $V_o = -4$ V
14. $A_v = 0.266$
15. $V_o = -0.8$ V
16. 1. low accuracy
    2. a large range of resistor values needed
17. $A_v = 0.2$
18. $V_o = -0.6$
19. ladder (R-2R ladder)
20. 1
21. *A*
22. 2.75
23. 0.25
24. analog, digital (binary)
25. 0101
26. HIGH, pass through
27. counter-ramp
28. comparator
29. voltages
30. AND, LOW
31. dc voltages
32. op amp
33. HIGH, LOW
34. digital voltmeter
35. hybrid
36. 0
37. four
38. 5
39. D/A converter
40. 9
41. greater than, does not pass
42. 1. counter-ramp
    2. ramp
    3. successive-approximation
43. ramp generator
44. faster
45. sawtooth
46. HIGH
47. successive-approximation logic
48. μP (microprocessor)
49. resolution
50. 4
51. 110 μs
52. 850 mW
53. longer
54. CMOS
55. microprocessor
56. 8 bits (0.39 percent)
57. TTL
58. 100 μs
59. clock
60. 01100100_2 (decimal 100)
61. L-to-H
62. $\overline{INTR}$
63. increases
64. decreases
65. increases
66. 1
67. lower
68. transducer
69. sulfide photocell
70. surface-mount
71. display
72. LCD (liquid-crystal display)
73. 9 V
74. 10 kΩ

Appendix A
Formulas and Conversions

Two's Complement Number Conversion Chart

2's Comp	Decimal	2's Comp	Decimal	2's Comp	Decimal	2's Comp	Decimal
11111111	−1	11011111	−33	10111111	−65	10011111	−97
11111110	−2	11011110	−34	10111110	−66	10011110	−98
11111101	−3	11011101	−35	10111101	−67	10011101	−99
11111100	−4	11011100	−36	10111100	−68	10011100	−100
11111011	−5	11011011	−37	10111011	−69	10011011	−101
11111010	−6	11011010	−38	10111010	−70	10011010	−102
11111001	−7	11011001	−39	10111001	−71	10011001	−103
11111000	−8	11011000	−40	10111000	−72	10011000	−104
11110111	−9	11010111	−41	10110111	−73	10010111	−105
11110110	−10	11010110	−42	10110110	−74	10010110	−106
11110101	−11	11010101	−43	10110101	−75	10010101	−107
11110100	−12	11010100	−44	10110100	−76	10010100	−108
11110011	−13	11010011	−45	10110011	−77	10010011	−109
11110010	−14	11010010	−46	10110010	−78	10010010	−110
11110001	−15	11010001	−47	10110001	−79	10010001	−111
11110000	−16	11010000	−48	10110000	−80	10010000	−112
11101111	−17	11001111	−49	10101111	−81	10001111	−113
11101110	−18	11001110	−50	10101110	−82	10001110	−114
11101101	−19	11001101	−51	10101101	−83	10001101	−115
11101100	−20	11001100	−52	10101100	−84	10001100	−116
11101011	−21	11001011	−53	10101011	−85	10001011	−117
11101010	−22	11001010	−54	10101010	−86	10001010	−118
11101001	−23	11001001	−55	10101001	−87	10001001	−119
11101000	−24	11001000	−56	10101000	−88	10001000	−120
11100111	−25	11000111	−57	10100111	−89	10000111	−121
11100110	−26	11000110	−58	10100110	−90	10000110	−122
11100101	−27	11000101	−59	10100101	−91	10000101	−123
11100100	−28	11000100	−60	10100100	−92	10000100	−124
11100011	−29	11000011	−61	10100011	−93	10000011	−125
11100010	−30	11000010	−62	10100010	−94	10000010	−126
11100001	−31	11000001	−63	10100001	−95	10000001	−127
11100000	−32	11000000	−64	10100000	−96	10000000	−128

Term	Definition	Symbol or Abbreviation
Access time	In memories, the time it takes to retrieve a piece of data from storage.	
Active HIGH input	Digital input which executes its function when a HIGH is present.	
Active LOW input	Digital input which executes its function when a LOW is present.	
A/D converter	Device for converting an analog voltage into a proportional digital quantity. Types include a microprocessor-compatible device with binary outputs or meter-type with decimal outputs.	ADC
Adder	Combinational logic circuit which generates sum and carry outputs from any set of binary inputs. Half-adders and full-adders are two fundamental adder circuits.	
Address	In computer systems, a number that represents a unique storage location.	
Alphanumeric	Consisting of numbers, letters and other characters. ASCII is a common alphanumeric code.	
Ampere	Base unit of current.	A
Analog	A branch of electronics dealing with infinitely varying quantities. Also referred to as linear electronics.	
Analog to digital	Conversion of an analog signal to a digital quantity such as binary.	A/D
AND gate	Basic combinational logic device where all inputs must be HIGH for the output to be HIGH.	
Anode	Positive section of a device such as a diode or LED.	
Arithmetic logic unit	Part of central processing unit of computer that processes data using arithmetic and logic operations.	ALU
American Standard Code for Information Interchange	One of the most widely used alphanumeric codes.	ASCII
Astable multivibrator	Device that oscillates between two stable states. Commonly called a free-running clock or multivibrator.	
Asynchronous	In digital circuits, meaning that operations are not executed in step with the clock.	
Base	Center section of a bipolar transistor used to control current flow from the emitter to collector.	
Baud	Unit of signal transmission speed in telecommunications equal to the number of discrete events per second.	Bd
BCD counter	A 4-bit counter that commonly counts from 0000 to 1001 and resets to 0000. A 4-bit binary counter would usually count from 0000 thru 1111 and then reset.	
Binary	Base 2 number system using numbers 0 or 1.	
Binary-coded decimal	A common code in which each decimal digit (0–9) is represented by a 4-bit group.	BCD

Term	Definition	Symbol or Abbreviation
Bistable multivibrator	A device having two stable states but it must be triggered to jump from one to the other. Also called a flip-flop.	
Bit	A single binary digit (0 or 1). Useful in representing on-off switching in digital circuits. An acronym for binary digit.	
Block diagram	A drawing using labeled blocks for functional sections of an electronic system.	
Boolean algebra	Mathematical system for representing logical statements. Very useful in digital electronics.	
Boolean expression	Mathematical representation of a logic function. Function could also be described using a truth table or logic circuit diagram.	$AB + C'D = Y$
Broadside loading	Parallel loading.	
Bubble	On logic symbol, it means an active LOW input or output.	
Buffer	Special solid state device used to increase the drive current at the output. Noninverting buffer has no logical function.	
Bus	In a computer system, parallel conductors used for communication between CPU, memories, and perpherial devices. Most systems have an address bus, data bus, and control bus.	
Byte	A 8-bit group that is commonly used to represent a number or code in computers and digital electronics.	
Cascading	Generally, the series connection of electronic devices with the output of the first feeding the input of the second. Term is used in both linear and digital electronics.	
Cathode	Negative section of a device such as a diode or LED.	K
Cathode-ray tube	Vacuum tube used in televisions, video monitors, and most oscilloscopes to display images.	CRT
CD-ROM	A read-only mass storage device based on the compact disk.	
Cell	In memories, a single storage element.	
Central processing unit	In computer system, the logic unit that performs logic arithmetic, control functions, and is the center of most data transfers.	CPU
Chip	An integrated circuit.	
Clock	Signal generated by an oscillator used to provide timing for a digital system such as a computer.	
Collector	The region of a bipolar transistor that receives the flow of current carriers.	C
Combinational logic	Use of logic gates to produce desired output immediately. No memory or latching characteristics.	
Complementary metal-oxide semiconductor	A popular technology for manufacturing ICs which features extremely low-power consumption. Uses opposite polarity field-effect transistors in its design.	CMOS
Current	Movement of charge in a specified direction. Base unit is ampere.	
Current sinking	Conventional current flow into LOW output of digital device. Current is "sinking" to ground.	
Current sourcing	Conventional current flow from HIGH output into load. Output is "sourcing" current.	

Term	Definition	Symbol or Abbreviation
Cylinder	On a hard disk drive, a series of identical tracks on various platters.	
D/A converter	Device for converting a digital quantity into a proportional analog voltage.	DAC
Data selector	Combination logic block that selects one-of-*X* data inputs and connects that information to the output. Also called a multiplexer.	
Decoder	A logic device that translates from binary code to decimal. Generally, it translates processed data into a digital system to another format such as alphanumeric.	
Decrement	To decrease the count by 1.	
Demultiplexer	Combination logic block that distributes data from single input to one-of-*X* outputs. Also called a distributor. Can change serial to parallel data.	DEMUX
D flip-flop	Flip-flop with at least set and resent modes of operation. Also called a data or delay flip-flop.	D, CLK → FF → Q, $\overline{Q}$
Digital	Branch of electronics dealing with discrete signal levels Signals are commonly HIGH or LOW and may be represented by binary numbers.	
Digital to analog	Conversion of a digital signal to its analog equivalent, such as a voltage.	D/A
Diode	Two-terminal semiconductor device. They usually allow current to flow in only one direction.	⊳|
Drive	Generally in computers, it refers to a mass storage device such as a floppy disk drive, hard disk drive, optical drive, or even a solid-state drive. Usually an electromagnetic or optical device which moves mass storage media under read/write heads.	
Dynamic RAM	Extremely common random-access (read/write) memory device whose memory cells need refreshing many times per second. A volatile memory.	DRAM
Dual-in-line package	Popular packaging method for ICs.	DIP
Edge triggering	In synchronous devices such as flip-flops, the exact time the device is activated such as on the rising (positive-edge) or falling edge (negative-edge) of the clock pulse.	
8421 BCD code	Four-bit BCD code with weighting of 8, 4, 2, and 1. See binary-coded decimal.	
Electrically erasable programmable read-only memory	A nonvolatile memory that can be programmed, electrically erased, and reprogrammed. Flash memories are a type of EEPROM.	EEPROM
Encoder	A logic device that translates from decimal to another code such as binary. Generally, it translates input information to a code useful to digital circuits.	
Emitter	The region of a bipolar transistor that sends the current carriers on to the collector.	E
Even parity	In data transmission, sending a parity bit that will make the number of 1s in a group even.	

Term	Definition	Symbol or Abbreviation
Extended Binary-Coded Decimal Interchange Code	An 8-bit alphanumeric code used mainly on mainframe computers	EBCDIC
Fan out	Output drive characteristic of logic device. The number of inputs of the logic family that can be driven by a single output.	
Field-effect transistor	Type of transistor where gate terminal controls the resistance of a semiconducting channel.	
Firmware	Computer programs and data held permanently in nonvolatile memory devices such as ROMs.	
Flash memory	A newer nonvolatile memory similar to the EEPROM. Its outstanding characteristics include very high density (small memory cell), low power, and nonvolatile but rewritable.	
Flip-flop	Basic sequential logic device having two stable states. Can serve as a memory device. Also called a bistable multivibrator.	
Floating input	An input not held HIGH or LOW which may "float" either HIGH, LOW or in between. Can cause problems.	
Frequency divider	A logic block that divides the input waveform's frequency by a certain number (such as divide-by-10). Counters commonly perform this function.	
Full-adder	Digital circuit with three inputs for carry in and two bits with sum and carry out outputs.	C_{in}, A, B — FA — E, C_o
Gain	A ratio of output to input. May be measured in terms of voltage, current, or power. Also known as amplification.	
Gate	Basic combinational logic device which performs a specific logic function (AND, OR, NOT, NAND, NOR).	
GND	Label for negative of power supply in TTL ICs and some CMOS ICs. Common ground.	
Half-adder	Digital circuit that will add two bits and output a sum and carry. Cannot handle a carry inputs.	A, B — HA — E, C_o
Hertz	The base unit of frequency. One cycle per second.	Hz
Hexadecimal	Base 16 number system using characters 0 thru 9, A, B, C, D, E, and F. Used to represent binary numbers 0000 through 1111.	Hex
Hysteresis	Unequal switching thresholds exhibited by some logic circuits making their outputs "snap action." Schmitt trigger logic devices exhibit this feature.	
Increment	To increase the count by 1.	
Instruction set	The complete set of commands responded to by a microprocessor, microcontroller, or PLC.	
Interfacing	The design of interconnections between circuits that shift the levels of voltage and current to make them compatible.	
Integrated circuit	Combination of many electronic components in a compact package that functions as an analog, digital, or hybrid circuit. Classified as to levels of circuit complexity (SSI, MSI, LSI, VLSI, or ULSI).	IC

Term	Definition	Symbol or Abbreviation
Inverter	Basic logic function where the output is always opposite the input. Also called the NOT function.	
J-K flip-flop	Flip-flop with at least set, reset, toggle and hold modes of operation. Very adaptable.	J, CLK, K, FF, Q, Q
Karnaugh map	A graphic method of reducing Boolean expressions to simpler forms.	K maps
Latch	Fundamental binary storage device. Also called a flip-flop.	
Large-scale integration	Used by some manufacturers to indicate the complexity of an integrated circuit. LSI usually means having a complexity of from 100 to 9999 gates.	LSI
Least significant bit	The bit position in a binary number with the least weight.	LSB
Light-emitting diode	Special PN junction that gives off light when current flows through it. Has lens to focus the light.	LED
Liquid-crystal display	Very low power display technology used in most battery operated devices. Nematic fluid in display changes reflectivity when energized changing display from silver to black characters.	LCD
Logic analyzer	An expensive test instrument that can sample and store many channels of digital information.	
Logic diagram	A schematic showing interconnection between logic devices like gates, flip-flops, etc.	
Logic family	A group of totally compatible digital ICs that can be interconnected with no interfacing problems. Common examples are the 7400 series TTL, 74HC00 series CMOS, and 4000 series CMOS.	
Logic function	The logical task needed to be performed. It might be represented by the name (such as AND), a logic symbol, a Boolean expression (such as $AB = Y$), and/or a truth table.	
Logic levels	In digital electronics, voltage ranges at which inputs to digital devices interpret signal as HIGH, LOW, or undefined. Voltage thresholds may be different for various logic families.	
Logic probe	Simple service tool which indicates logical 0s, logical 1s, or pulses in digital circuits.	
Logic subfamilies	Groups of related digital ICs that have similar characteristics but may vary in speed, power dissipation, and current drive capabilities. Examples might be 7400-, 74LS00-, 74F00-, 74ALS00-, and 74AS00-series TTL ICs. In some applications you may be able to substitute between subfamilies.	
Logic symbols	Two systems are used in the U.S. Traditional representations using the unique shaped logic gate symbols. The newer IEEE symbols using rectangle boxes.	&
Magnetic core memory	Older nonvolatile read/write memory system using ferrite cores as memory cells.	
Magnitude comparator	A combination logic block that compares two binary inputs A and B and activates one of three outputs ($A > B$, $A = B$, or $A < B$).	

Term	Definition	Symbol or Abbreviation
Maxterm Boolean expression	See product-of-sums.	
Medium-scale integration	Used by some manufacturers to indicate the complexity of an integrated circuit. MSI usually means having a complexity of from 12 to 99 gates.	MSI
Memory card	Packaging method for arrays of memory devices (such as flash memories). The cards are commonly about the size of a thick credit card with edge connectors. See PCMCIA.	
Metal oxide semiconductor	Technology used in the fabrication of integrated circuits using metal and an oxide (silicon dioxide) as an important part of the devices' structure.	
Microcontroller	An inexpensive IC which contains a microprocessor, limited RAM, ROM, and I/O. A small computer on a chip. They are usually embedded in a product.	
Microprocessor	An IC which forms the CPU of most microcomputers.	MPU
Minterm Boolean expression	See sum-of-products	
Minuend	The number the subtrahend is being subtracted from.	
Monostable multivibrator	Emits a single pulse when triggered. Also called a one-shot multivibrator.	
Most significant bit	The bit position in a binary number with the most weight.	MSB
Multiplex	In driving displays, to turn on/off one of several displays each for a short time in turn at a high enough frequency so they appear to be lit continuously. In general, transmitting several signals over common lines.	
Multiplexer	Combinational logic block selects one-of-*X* inputs and directs the information to a single output. Also called a data selector. Can change parallel to serial data.	MUX
Multivibrator circuits	Classified as bistable (flip-flops), monostable (one-shots), and astable (free-running clocks).	MV
NAND gate	Basic combinational logic device where all inputs must be HIGH for the output to be LOW. A not AND circuit.	[NAND gate symbol]
Nibble	One half a byte. A 4-bit binary word.	
Noise	In digital electronics, it is unwanted voltages induced in connecting wires and PC board traces that might affect input logic levels and therefore outputs in circuits.	
Noise immunity	A digital circuit's insensitivity to undesired voltages or noise. Also called noise margin in digital circuits.	
Nonvolatile memory	Memory which retains data even if the power is turned off.	
Nonvolatile RAM	Read/write memory that will hold its data even when the power is turned off.	
NOR gate	Basic combinational logic device where all inputs must be LOW for the output to be HIGH. A not OR circuit.	[NOR gate symbol]
NOT	Basic combinational logic device where the output is always the opposite from the input. Also called an inverter.	[inverter symbol]
Octal	Base 8 number system using characters 0 thru 7.	
Odd parity	In data transmission, sending a parity bit that will make the number of 1s in a group odd.	
Ohm	The base unit of resistance.	Ω

Term	Definition	Symbol or Abbreviation
1s complement	To for 1s complement from binary number invert each bit.	
Open collector	Digital circuit output which has no internal path to the positive of the power supply. Commonly used with an external pull-up resistor.	
Operational amplifier	An adaptable amplifier with inverting and non-inverting inputs featuring high input- and low output-impedance, and very high gain. Gain can be set by external components.	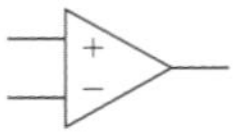 op amp
Optical disk drive	Very high capacity mass storage device which commonly stores data as surface pits. Reading is done by directing laser bean at pits/no pits and detecting the light bouncing from the reflective disk. Other optical recording methods are also used.	
Optoisolator	An interface device used to electrically isolate input from output by using a light beam to transfer data.	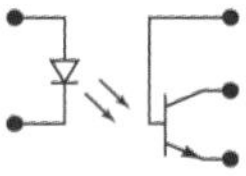
OR gate	Basic combinational logic device where the output goes HIGH when any or all inputs are HIGH.	
Oscillator	Electronic circuit that generates AC waveforms from a DC source.	
Oscilloscope	Test instrument that plots time against voltage drawing a graph or waveform on the screen. Oscilloscopes are available in either analog or digital models. Also called a scope.	
Parallel data	Transmission of data in groups at the same time over multiple lines.	
Parity	A system used to detect errors in binary data transmission.	
Parity bit	An extra bit sent with data bits to check for errors in transmission.	
PC	Commonly means personal computer but it may also be used to refer to a programmable controller or programmable logic controller.	
PCMCIA	Personal Computer Memory Card International Association sets standards for memory cards.	
Photo resistive cell	A photo sensitive resistor whose resistance decreases as the light striking the unit increases. A cadmium sulfide photo cell or photo resistor.	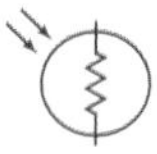 CdS
Plastic leaded chip carrier	A type of surface mount IC package with leads bent under the case.	PLCC
Platter	In a hard disk drive, a single hard disk. The drive may contain a stack of platters to increase storage capacity.	
Port	In computers, the circuits used to transfer data in and out of the system.	I/O
Product-of-sums	The form of a Boolean expression that looks like this: $(A + B)(C + D) = Y$. Implemented using an OR/AND logic diagram. Also called a maxterm Boolean expression.	
Program	List of instructions which tells computer what to do. May be written in a variety of computer languages.	

Term	Definition	Symbol or Abbreviation
Programmable logic controller	A specialized heavy-duty computer system used for process control in factories, chemical plants, and warehouses. Closely associated with traditional relay logic. Also called a programmable controller (PC).	PLC
Programmable read-only memory	Nonvolatile memory which is programmed once by the user or distributor.	PROM
Propagation delay	The time it takes the output of a digital device to change state after the input is activated. Usually measured in nanoseconds.	
Pull-up resistor	A resistor connected to the positive of the power supply to hold a point in the circuit HIGH when it is inactive.	
Radix	The base of a number.	
Random-access memory	Memory organization allowing for easy access to each bit, byte or word. RAM is commonly used to mean semiconductor read/write memory.	RAM
Relay	Electrical device which uses the force of an electromagnet to open/close contacts. Used for heavy-duty switching and isolation of circuits.	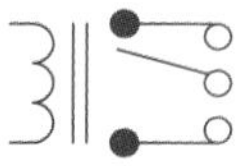
Resistance	Opposition to current flow. Measured in ohms.	R
Read	The process of sensing and retrieving data from a memory cell or cells.	
Read-only memory	Non-volatile memory which is not usually changed once it is programed. ROM commonly used to refer to mask-programmable read-only memory.	ROM
Reset condition	In a flip-flop, the normal output (Q) has been reset or cleared to 0.	
Rewritable optical disk	A very large capacity optical disk that can be rewritten to many times. Some versions are called PD rewritable optical disk or CD-E (compact disk erasable).	CD-E
Ripple counter	Simple binary counter where the changing state of the LSB flip-flop triggers the clock input of the next, etc. A time delay results from the rippling of the count from LSB to MSB.	
Ring counter	A recirculating shift register which is loaded with a pattern of 1s (such as a single 1) which continue to circulate around in the circle or repeated clock pulses.	
R-S flip-flop	Flip-flop with at least set, reset, and hold modes of operation. Fundamental latching (memory) circuit.	 S, R, FF, Q, $\overline{Q}$
Schmitt trigger	A circuit that exhibits hysteresis and is useful in signal conditioning in digital electronics.	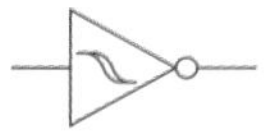
Semiconductor	Elements having four valence electrons and electrical properties between those of conductors and insulators.	
Sequential logic	A logic circuit whose logic states depend on asynchronous and synchronous inputs. Exhibit memory characteristics.	
Serial data	The transmission of data one bit at a time.	
Servo	General term for a motor whose either angular position or speed can be precisely controlled by a servo loop which uses feedback from the output back to the input for control.	

Term	Definition	Symbol or Abbreviation
Set condition	In a flip-flop, the normal output (Q) has been set to 1.	
Seven-segment display	Numeric display with seven segments. May be implemented with LED, LCD or VF technologies. A few letters can also be displayed for indicating hexadecimal numbers.	f e a d g b c
Shift register	A sequential logic block made up of flip-flops that allows parallel or serial loading and serial or parallel outputs as well as shifting bit by bit.	
Signal	The information transmitted within, to, and from electronics circuits.	
Silicon	A semiconductor element used in the manufacture of most solid-state devices such as diodes, transistors, and integrated circuits.	
Small-scale integration	Used by some manufacturers to indicate the complexity of an integrated circuit. SSI usually means having a complexity of less than 12 gates.	SSI
Solenoid	An actuator which converts electrical energy into linear motion. It is constructed as a hollow coil with a sliding iron core. In operation the spring-loaded iron core is "sucked into" the coil when current flows in the coil.	
Solid-state drive	A nonvolatile read/write memory device which would function like the hard disk drive in a computer system but consist of semiconductor memory (perhaps flash memory cards). Might be used to save power and weight in tiny portable systems.	
Source	Terminal of a field-effect transistor that sends current carriers to the drain.	S
Static RAM	Common random-access (read/write) memory device which stores data in a flip-flop like cell. Volatile memory.	SRAM
Stepper motor	A DC motor that jogs in short uniform angular movements in either direction given the proper digital signals. Common step angles might be 1.8°, 3.6°, 7.5°, and 15°. Two types are permanent-magnet and variable-reluctance stepper motors.	Stepper motor
Subtrahend	The number being subtracted from the minuend.	
Successive approximation	In D/A and A/D converters, a technique used to decrease conversion time.	
Sum-of-products	The form of a Boolean expression that looks like this: $AB + CD = Y$. Implemented using an AND/OR logic diagram. Also called a minterm Boolean expression.	
Surface-mount technology	A method of printed circuit fabrication in which the component leads are soldered on the component side of the board and the leads do not pass through holes in the PC board.	SMT
Synchronous	In digital circuits, meaning that operations are executed in step with the clock.	
Toggle	To change the opposite logic state. A pulse that changes logic circuits state to opposite condition. A mode of operation in a flip-flop where the output goes to the opposite state on each successive clock pulse.	

Term	Definition	Symbol or Abbreviation
Transistor	A solid-state amplifying or controlling device which commonly has three leads.	
Transistor-transistor logic	A type of digital IC fabricated using bipolar junction transistors.	TTL
2s complement	Notation commonly used to indicate sign and magnitude of a number using only 0s and 1s. To form 2s complement, take 1s complement of binary and add 1. Helpful with binary adders for binary subtraction.	
Trigger	A pulse that causes a logic device to be activated or change states.	
Truth table	Tabular listing of all inputs and resultant output conditions for a logic function or circuit.	*A B* | *Y* 0 0 | 0 0 1 | 0 1 0 | 0 1 1 | 1
2s complement subtraction	Method of subtraction using a 2s complement subtrahend added to the minuend. Used so adders can be used to perform subtraction.	
Ultra large-scale integration	Used by some manufacturers to indicate the complexity of an integrated circuit. ULSI usually means having a complexity of 100,000 or more gates.	ULSI
Universal shift register	Register with many features including serial-in/out, parallel-in/out, hold, and shift right or left.	
Vacuum fluorescent display	Low voltage triode vacuum tube display which commonly glows green (without filters).	VF
V_{CC}	Label for positive of power supply in TTL ICs and some CMO ICs (commonly +5V).	
V_{DD}	Label for positive of power supply in many but not all CMOS ICs (+3 to +18V).	
Very large-scale integration	Used by some manufacturers to indicate the complexity of an integrated circuit. VLSI usually means having a complexity of from 10,000 to 99,999 gates.	VLSI
Volatile memory	Memory which can store data only as long as power is applied.	
Volt	Base unit voltage.	V
Voltage	Electrical pressure.	V
Voltage comparator	An op amp circuit that compares a positive voltage input (*A*) with a negative voltage input (*B*) and indicates with a logic output which input is higher.	
Waveforms	A graphic representation of voltage verses time as might be viewed on an oscilloscope.	V time
Winchester drive	Historical name for a hard disk drive.	
Write	The process of recording data in a memory cell or cells.	
Write-once read-many	An optical CD recordable disk can be recorded on once using your PC and it then is permanent like a CD-ROM.	WORM
XNOR gate	Basic combinational logic device where an even number of HIGH inputs generates a HIGH output. A not XOR gate.	
XOR gate	Basic combinational logic device where an odd number of HIGH inputs generates a HIGH output.	

Index

M

N

O

P

R

S

T

U